J. Font Freide

Studies in Surface Science and Catalysis 147

NATURAL GAS CONVERSION VII

Studies in Surface Science and Catalysis

Advisory Editors: B. Delmon and J.T. Yates
Series Editor: G. Centi

Vol. 147

NATURAL GAS CONVERSION VII

Proceedings of the 7th Natural Gas Conversion Symposium,
June 6–10, 2004, Dalian, China

Edited by

Xinhe Bao and **Yide Xu**
State Key Laboratory of Catalysis,
Dalian Institute of Chemical Physics,
Dalian, China

2004

ELSEVIER

Amsterdam – Boston – Heidelberg – London – New York – Oxford – Paris – San Diego
San Francisco – Singapore – Sydney – Tokyo

ELSEVIER B.V.
Sara Burgerhartstraat 25
P.O. Box 211, 1000 AE
Amsterdam, The Netherlands

ELSEVIER Inc.
525 B Street, Suite 1900
San Diego, CA 92101-4495
USA

ELSEVIER Ltd
The Boulevard, Langford Lane
Kidlington, Oxford OX5 1GB
UK

ELSEVIER Ltd
84 Theobalds Road
London WC1X 8RR
UK

First edition 2004

Library of Congress Cataloging in Publication Data
A catalog record is available from the Library of Congress.

British Library Cataloguing in Publication Data
A catalogue record is available from the British Library.

ISBN: 0-444-51599-2
ISSN: 0167-2991 (Series)

♾The paper used in this publication meets the requirements of ANSI/NISO Z39.48-1992 (Permanence of Paper).
Printed in The Netherlands.

Preface

This volume contains peer-reviewed manuscripts presented at the 7th Natural Gas Conversion Symposium held in Dalian, China on June 6–10, 2004. This symposium continues the proud tradition of the previous meetings in Auckland, New Zealand (1987), Oslo, Norway (1990), Sydney, Australia (1993), Kruger National Park, South Africa (1995) Taormina, Italy (1998) and Alaska, USA (2001).

The Editors thank the International Advisory Board for choosing Dalian, China as the site for the 7th Natural Gas Conversion Symposium. The 7th Natural Gas Conversion Symposium has been stewarded by an Organizing Committee chaired by Xinhe Bao (Dalian Institute of Chemical Physics, DICP), Theo Fleisch (BP), Guoxing Xiong (DICP) and Yide Xu (DICP) and a Program Committee chaired by Xinhe Bao, and Yide Xu. Their essential individual contributions to the success of this symposium are acknowledged with thanks. The members of these various groups are separately listed below.

The Editors appreciate Professor Enrique Iglesia (University of California at Berkeley) for his valuable advice, timely assistance and friendly encouragement during the 7th NGCS in preparation.

The Editors thank the authors of these manuscripts for their contributions. These manuscripts have been divided into eight different topics, Industrial Processes, Economics, Technology Demonstration and Commercial Activities; Production of Hydrogen from Methane, Methanol, and Other Sources; Production of Synthesis Gas; Fischer-Tropsch Synthesis of Hydrocarbons; From Synthesis gas to Chemicals; From Natural Gas to Chemicals; Light Hydrocarbons: Production and Conversion; and Catalytic Combustion. In doing so, we have tried to help readers to find their specific interests within a set of topics that encompass those most important in the utilization of natural gas and the most recent scientific innovation and technological advances in these areas.

The Editors gratefully acknowledge the generous financial support given by the Sponsors, who are listed below.

The Editors acknowledge the talents and dedication of Mr. Patrick Fleisch, who carefully designed the symposium website and handled the details of the review, acceptance and submission processes for the manuscripts in this volume.

Editors

Xinhe Bao
Yide Xu

PROGRAM COMMITTEE

Xinhe Bao (Co-Chair), Dalian Institute of Chemical Physics (China)
Yide Xu (Co-Chair),Dalian Institute of Chemical Physics (China)

Carlos Apesteguia, INCAPE (Argentina)
C.T. Au, Hong Kong Baptist University (Hong Kong)
Manfred Baerns, Institute for Applied Chemistry, Berlin (Germany)
Alexis Bell, University of California at Berkeley (USA)
Wei Chu, Sichuan University (China)
Steven Chuang, University of Akron (USA)
Krijn de Jong, University of Utrecht (The Netherlands)
Rafael Espinoza, Conoco (USA)
Xianzhi Fu, Fuzhou University (China)
Mireya Goldwasser, Universidad Central de Venezuela (Venezuela)
Anders Holmen, Norwegian University of Science and Technology (Norway)
Philip Howard, BP (USA)
Graham Hutchings, University of Wales (UK)
Enrique Iglesia , University of California at Berkeley (USA)
Masaru Ichikawa, Hokkaido University (Japan)
Jay Labinger, California Institute of Technology (USA)
Jae Sung Lee, Postech (Korea)
Johannes Lercher, Technische Universitat Munchen (Germany)
Sheben Li, Lanzhou Institute of Chemical Physics (China)
Yongdan Li, Tianjin University (China)
Jianyi Lin, National University of Singapore (Singapore)
Changjun Liu, Tianjin University (China)
Ian Metcalfe, UMIST (UK)
Jens Norskov, Technical University of Denmark (Denmark)
Kiyoshi Otsuka, Tokyo Institute of Technology (Japan)
Sebastian C. Reyes, Exxon Research and Engineering (USA)
Julian Ross, University of Limerick (Ireland)
Jens Rostrup-Nielsen, Haldor Topsoe (Denmark)
Domenico Sanfilippo, Snamprogetti (Italy)
Jesus Santamaria, University of Zaragoza (Spain)
Jose Santiesteban, ExxonMobil Research and Engineering (USA)
Abdel Sayari, University of Ottawa (Canada)
Martin Schmal, COPPE/Federal University Rio de Janeiro (Brazil)
Lanny Schmidt, University of Minnesota (USA)
Shikong Shen, Beijing Petroleum University (China)
Stuart Soled, ExxonMobil (USA)
James J. Spivey, Louisiana State University (USA)
Jishan Suo, Lanzhou Institute of Chemical Physics (USA)
Samuel Tam, Nexant, Inc. (USA)
David Trimm, University of New South Wales (UK)
Theo Tsotsis, USC (USA)
Peter van Berge, SASOL (South Africa)
Constantinos Vayenas, University of Patras (Greece)
Stephen Wittrig, BP (USA)
Boqing Xu, Tsinghua University (China)

ORGANIZING COMMITTEE

Prof. Xinhe Bao, Dalian Institute of Chemical Physics (China)
Dr. Theo Fleisch, BP America (USA)
Prof. Guoxing Xiong, Dalian Institute of Chemical Physics (China)
Prof. Yide Xu, Dalian Institute of Chemical Physics (China)

Prof. Wenzhao Li, Dalian Institute of Chemical Physics (China)
Mr. Jeff Bigger, Syntroleum Corporation (USA)
Mr. Siliang Chen, Chinese Academy of Sciences (China)
Ms. Yan Chen, Dalian Institute of Chemical Physics (China)

INTERNATIONAL SCIENTIFIC ADVISORY BOARD

Carlos Apesteguia (Argentina)
Manfred Baerns (Germany)
Theo Fleisch (USA)
Anders Holmen (Norway)
Graham Hutchings (UK)
Enrique Iglesia (USA)
Ben Jager (South Africa)
Eiichi Kikuchi (Japan)
Wenzhao Li (China)
Jack Lunsford (USA)
Ian Maxwell (The Netherlands)
Claude Mirodatos (France)
Julian Ross (Ireland)
David Trimm (Australia)
Domenico Sanfilippo (Italy)
Lanny Schmidt (USA)
Jens Rostrup-Nielsen (Denmark)

FINANCIAL SUPPORT

Ruthenium Sponsors:
BP
ConocoPhillips
ExxonMobil
Shell

Cobalt Sponsors:
Johnson Matthey Catalysis

Sponsors:
Haldor Topsoe A/S
BASF

Ministry of Science and Technology of China
National Nature Science Foundation of China
PetroleumChina
Sinopec
Chinese Academy of Sciences
Sinopec, Dalian
Dalian Institute of Chemical Physics, Chinese Academy of Sciences

TABLE OF CONTENTS

Industrial Processes, Economics, Technology Demonstration and Commercial Activities

Production of Hydrogen from Methane, Methanol, and Other Sources

Production of Synthesis Gas

Fischer-Tropsch Synthesis of Hydrocarbons

From Synthesis Gas to Chemicals

Catalytic Combustion

From Natural Gas to Chemicals

Light Hydrocarbons: Production and Conversion

Industrial Processes, Economics, Technology Demonstration and Commercial Activities

Most Recent Developments in Ethylene and Propylene Production from Natural Gas Using the UOP/Hydro MTO Process

J. Q. Chen*[a], B. V. Vora[a], P. R. Pujadó[a], Á. Grønvold[b], T. Fuglerud[b], S. Kvisle[b]
[a]UOP LLC, P.O.Box 5017, Des Plaines, Illinois 60017-5017, USA
[b]Norsk Hydro ASA, P.O. Box 2560, N-3901 Porsgrunn, Norway
*Email: qianjun.chen@uop.com   Phone: 1-847-3913920

Condensing Methanol Synthesis and ATR - the Technology Choice for Large-Scale Methanol Production

Esben Lauge Sørensen* and Jens Perregaard
Haldor Topsøe A/S, Nymøllevej 55, DK-2800 Lyngby, Denmark
*Email: els@topsoe.dk Phone +45 4527 2000

Preferred Synthesis Gas Production Routes for GTL

Per K. Bakkerud*[a], Jorgen Nergaard Gol, Kim Aasberg-Petersen and Ib Dybkjaer
[a]Haldor Topsøe A/S, Nymøllevej 55, DK-2800 Kgs. Lyngby, Denmark
Email: pkb@topsoe.dk Phone: +45 4527 2000

Statoil`s Gas Conversion Technologies

Ola Olsvik* and Rolf Ødegård
Statoil Research and Technology, N-7005 Trondheim, Norway
* Email: ools@statoil.com Phone: +4797500398 Fax: +4773584345

The Shell Middle Distillate Synthesis Process: Technology, Products and Perspective

A. Hoek*[a] and L.B.J.M. Kersten[b]
[a]Shell Global Solutions International BV, Shell Research and Technology Centre Amsterdam, PO Box 38000, 1030 BN, Amsterdam, The Netherlands
[b]Shell International Gas Ltd, Shell Centre, York Road, London SE1 7NA, UK
* Email: arend.hoek@shell.com Phone: +31(0)206302815 Fax: +31(0)206303964

Large-Scale Gas Conversion through Oxygenates: Beyond GTL-FT

T.H. Fleisch* and R.A. Sills
BP E&P Technology, P.O. Box 3092, Houston, TX, 77253, U.S.A.
* Email: FleiscTH@bp.com Phone: 011 281 366-7311 Fax: 011 281 366-5355

Large Scale Production of High Value Hydrocarbons Using Fischer-Tropsch Technology

André P. Steynberg*[a], Wessel U. Nel[b] and Mieke A. Desmet[b]
[a] Sasol Technology R&D, 1 Klasie Havenga Road, P. O. Box 1, Sasolburg 1947, South Africa
[b] Sasol Technology, Baker Square West, 33 Baker Street, Rosebank, 2196,   South Africa
* Email: andre.steynberg@sasol.com Phone: +27 16 960 2590 Fax: +27 16 960 4015

Scale-up of Statoil's Fischer-Tropsch Process

Dag Schanke[a], Erling Rytter[a] and Fred O. Jaer[b]
[a]Statoil Research Centre, Arkitekt Ebbells vei 10, N-7005 Trondheim, Norway
[b]Statoil International E&P, N-4035 Stavanger, Norway

* Email: dsc@statoil.com Phone: (47) 73 58 45 24 Fax: (47) 73 58 43 45

Catalysis: Wrapping it in Concrete and Steel
Michael J. Gradassi
BP America, Inc., 501 Westlake Park Boulevard, Houston, Texas, 77079, USA
* Email: gradasmj@bp.com Phone: 281-366-5662 Fax: 281-366-3436

ITM Syngas Ceramic Membrane Technology for Synthesis Gas Production
Christopher M. Chen*[a], Douglas L. Bennett[a], Michael F. Carolan[a],
Edward P. Foster[a], William L. Schinski[b], Dale M. Taylor[c]
[a]Air Products and Chemicals, Inc., 7201 Hamilton Blvd, Allentown, PA, 18195 USA
[b]ChevronTexaco Energy Research and Technology Co, 100 Chevron Way, Richmond, CA, 94802 USA
[c]Ceramatec Inc., 2425 South 900 West, Salt Lake City, UT, 84119 USA
* Email: chencm@apci.com Phone: +1 (610) 481-3315 Fax: +1 (610) 706-6586

Gas Conversion – Its Place in the World
J.J.H.M. Font Freide*
BP Exploration & Production Technology, Chertsey Road, Sunbury-on-Thames, Middx TW16 7LN, United Kingdom
* Email: Fontfrei@bp.com Phone: <(44)-(0)1932-775601> Fax: <(44)-(0)1932-764120>

Application of FBD Processes for $C_3 - C_4$ Olefins Production from Light Paraffins
G. R. Kotelnikov*[a], S. M. Komarov[a], V. P. Bespalov[a], D. Sanfilippo[b], I. Miracca[b]
[a] OAO NII Yarsintez, prospekt Oktyabrya 88, 150040 Yaroslavl, Russia
[b] Snamprogetti S.p.A., Viale De Gasperi 16, 20097 – San Donato Milanese, Italy
* Email: yarsintez@yaroslavl.ru Phone: 085-255-0878 Fax: 085-255-0878

Production of Hydrogen from Methane, Methanol, and Other Sources

CO_x-Free Hydrogen and Carbon Nanofibers Production by Decomposition of Methane on Fe, Co and Ni Metal Catalysts
Jiuling Chen, Xingzheng Zhou, Lei Cao, Yongdan Li*
Department of Catalysis Science and Technology, School of Chemical Engineering, Tianjin University, Tianjin 300072, China
* Email: ydli@tju.edu.cn Tel.: +86-22-27405613

Hydrogen Production from Non-Stationary Catalytic Cracking of Methane: a Mechanistic Study Using *in operando* Infrared Spectroscopy.
E. Odier[a], Y. Schuurman[a], K. Barral[b], C. Mirodatos[a]*
[a]Institut de Recherches sur la Catalyse, 2. Avenue Albert Einstein 69626 Villeurbanne Cedex, France
[b]Air Liquide, Centre de Recherche Claude Delorme, BP 126, Les Loges en Josas, 78353 Jouy en Josas Cedex, France
* Email: mirodatos@catalyse.cnrs.fr Phone: [33] 472445366 Fax: [33] 472445399

Carbide- and Nitride-Based Fuel Processing Catalysts
J. J. Patt, S.K. Bej and L.T. Thompson*
Department of Chemical Engineering, Ann Arbor, MI 48109-2136, USA

* Email: ltt@umich.edu

One-Step-Hydrogen: a New Direct Route by Water Splitting to Hydrogen with Intrinsic CO_2 Sequestration

Domenico Sanfilippo*[a], Ivano Miracca[a], Ugo Cornaro[b], Franco Mizia[b], Alberto Malandrino[b], Valerio Piccoli[b], Stefano Rossini[b]
[a]Snamprogetti SpA, Viale De Gasperi 16 - 20097 S. Donato Milanese – Italy
[b]EniTecnologie SpA, Via Maritano 26 - 20097 S. Donato Milanese - Italy
*Email: domenico.sanfilippo@snamprogetti.eni.it Phone: +3902 52053558 Fax: +3902 52053575

Iron Oxide as a Mediator for the Storage and Supply of Hydrogen from Methane
S. Takenaka, C. Yamada, V. T. D. Son and K. Otsuka*
Department of Applied Chemistry, Graduate School of Science and Engineering, Tokyo Institute of Technology, 2-12-1 Ookayama, Meguro-ku, Tokyo, 152-8552, Japan
* Email: kotsuka@o.cc.titech.ac.jp Phone: 81-3-5734-2143 Fax: 81-3-5734-2879

Hydrogen from Stepwise Reforming of Methane: a Process Analysis
Lei Cao, Jiuling Chen, Shaowen Liu and Yongdan Li*
Department of Catalysis Science & Technology, School of Chemical Engineering, Tianjin University, Tianjin 300072, China
* Email: ydli@tju.edu.cn Phone: +86-22-27405613 Fax: +86-22-27405243

NO_x-Catalyzed Gas-phase Activation of Methane: the Formation of Hydrogen
Chao-Xian Xiao, Zhen Yan, Yuan Kou*
College of Chemistry and Molecular Engineering, Peking University, Beijing 100871, China
* Email: yuankou@pku.edu.cn Phone: +86 (10) 62757792 Fax: +86 (10) 62751708

Hydrogen Production by Ethanol Steam Reforming. Study of Cobalt-Doped Ce-Zr Catalysts.
J.C. Vargas, E. Vanhaecke, A.C. Roger*, A. Kiennemann
Laboratoire des Matériaux, Surfaces et Procédés pour la Catalyse, UMR 7515, ECPM - ULP, 25 rue Becquerel, 67087 Strasbourg Cedex 2, France
* Email: rogerac@ecpm.u-strasbg.fr Phone: 0390242769 Fax: 0390242768

Production of Synthesis Gas

Steam Reforming of Hydrocarbons. A Historical Perspective
Jens Rostrup-Nielsen
Haldor Topsøe A/S, Nymøllevej 55, DK-2800 Lyngby, Denmark
* Email: jrn@topsoe.dk Phone: +45-45 27 20 00 Fax: +45-45-272999

Development of Synthesis Gas Production Catalyst and Process
Fuyuki Yagi*[a], Shuhei Wakamatsu[a], Ryuichirou Kajiyama[a], Yoshifumi Suehiro[b], Mitsunori Shimura[a]
[a]R&D Center, Chiyoda Corporation, Kawasaki, Kanagawa, 210-0855, Japan
[b]Technology Research Center, Japan National Oil Corporation, Mihama, Chiba, 261-0025, Japan
* Email: fyagi@ykh.chiyoda.co.jp Phone: [81] 44-328-6265 Fax: [81] 44-328-6272

Novel Catalysts for Synthesis Gas Generation from Natural Gas
Shizhong Zhao[a]
[a]Süd-Chemie Inc., 1600 West Hill Street, Louisville, Kentucky 40210, USA
Email: jzhao@sud-chemieinc.com Phone: 1(502)634-6834 Fax: 1(502)634-7730

Monolith Composite Catalysts based on Ceramometals for Partial Oxidation of Hydrocarbons to Synthesis Gas
S. Pavlova*, S. Tikhov, V. Sadykov, Y. Dyatlova, O. Snegurenko, V. Rogov, Z. Vostrikov, I. Zolotarskii, V. Kuzmin, S. Tsybulya
Boreskov Institute of Catalysis, SB RAS, Pr. Lavrentieva, 5, 630090, Novosibirsk, Russia
*Email: pavlova@catalysis.nsk.su Phone: +7(3832)342672 Fax: +7(3832)343056

The Relative Importance of CO_2 Reforming during the Catalytic Partial Oxidation of CH_4-CO_2 Mixtures
N.R. Burke* and D.L. Trimm
CSIRO Petroleum, Ian Wark Laboratory, Bayview Ave, Clayton, Victoria, 3168, Australia
*Email: nick.burke@csiro.au Phone: +61 (0)3 92596845 Fax: +61 (0)3 92596900

Comparison of Microkinetics and Langmuir-Hinshelwood Models of the Partial Oxidation of Methane to Synthesis Gas
Xueliang Zhao, Fanxing Li and Dezheng Wang*
Department of Chemical Engineering, Tsinghua University, Beijing
100084 China (email: wangdz@flotu.org ; fax: ++86 10 62772051)

Methane Selective Oxidation into Syngas by the Lattice Oxygen in Ceria-Based Solid Electrolytes Promoted by Pt
V. Sadykov*[a], V. Lunin[b], T. Kuznetsova[a], G. Alikina[a], A. Lukashevich[a], Yu. Potapova[a,f], V. Muzykantov[a], S. Veniaminov[a], V. Rogov[a], V. Kriventsov[a], D. Kochubei[a], E. Moroz[a], D. Zuzin[a], V. Zaikovskii[a], V. Kolomiichuk[a], E. Paukshtis[a], E. Burgina[a], V. Zyryanov[c], S. Neophytides[d], E. Kemnitz[e]
[a]Boreskov Institute of Catalysis, Lavrentieva, 5, Novosibirsk, 630090, Russia
[b]Chemical Department of Moscow State University, Moscow, Russia
[c]Institute of Solid State Chemistry SB RAS Novosibirsk, Russia
[d]Institute of Chemical Engineering, Patras, Greece
[e]Institute for Chemistry, Humboldt- University, Berlin, Germany
[f]Samsung Advanced Institute of Technology, Suwon, Republic of Korea
*Email: sadykov@catalysis.nsk.sul Phone: (73832)343763 Fax: (73832)343056

Reforming of Higher Hydrocarbons
R. K. Kaila* and A. O. I. Krause
Helsinki University of Technology, Laboratory of Industrial Chemistry, P.O. Box 6100, FIN-02015 HUT, Finland, * kaila@polte.hut.fi

Study of the Mechanism of the Autothermal Reforming of Methane on Supported Pt Catalysts
M.M.V.M. Souza[a], S. Sabino[b], F.B. Passos[b], L.V. Mattos[c], F.B. Noronha*[c], M. Schmal[a]
[a]NUCAT/PEQ/COPPE-Universidade Federal do Rio de Janeiro, Brazil.
schmal@peq.coppe.ufrj.br.
[b]Universidade Federal Fluminense ,Brazil, fbpassos@vm.uff.br.

[c] Instituto Nacional de Tecnologia (INT), Brazil, fabiobel@int.gov.br.

Fischer-Tropsch Synthesis of Hydrocarbons

[b]Department of Chemistry, Moscow State University, 119992 Moscow, Russia
*Email: andrei.khodakov@ec-lille.fr Phone: 33 3 20 33 54 37 Fax: 33 3 20 43 65 61

Fischer-Tropsch Synthesis on Cobalt Catalysts Supported on Different Aluminas
Dag Schanke[a], Sigrid Eri[a], Erling Rytter[a], Christian Aaserud[c], Anne-Mette Hilmen[b,d], Odd Asbjørn Lindvåg[b], Edvard Bergene[b] and Anders Holmen[c*]
[a]Statoil R&D, Research Centre, Postuttak, N-7005 Trondheim, Norway.
[b]SINTEF Applied Chemistry, N-7465 Trondheim, Norway. [d]Present address: Shell Technology Norway, Drammensveien 147A, N-0277 Oslo, Norway.
[c]Dept Chemical Engineering, Norwegian University of Science and Technology (NTNU), N-7491 Trondheim, Norway.
*Email: holmen@chemeng.ntnu.no Phone: +47 73594151 Fax: +47 73595041

Effect of Water on the Catalytic Properties of Supported Cobalt Fischer-Tropsch Catalysts
J. Li*[a], B.H. Davis[b]
[a]College of Chemistry and Life Science, South Central University for Nationalities, No.5 Minyuan Road, Wuhan 430074, China
[b]Center for Applied Energy Research, University of Kentucky, 2540 Research Park Drive, Lexington, KY 40511-8410, USA
*Email: lij@scuec.edu.cn Phone: +86 (27) 87523016 Fax: +86 (27) 87532752

CeO_2-Promotion of Co/SiO_2 Fischer-Tropsch Synthesis Catalyst
H.-B. Shi, Q. Li, X.-P. Dai, C.-C. Yu, S.-K. Shen*
The Key Laboratory of Catalysis CNPC, University of Petroleum Beijing, Beijing 102249 China
* Email: skshen@bjpeu.edu.cn Tel: 010-89733784, Fax: 010-69744849,

Structure-controlled La-Co-Fe Perovskite Precursors for Higher C_2-C_4 Olefins Selectivity in Fischer-Tropsch Synthesis
L. Bedel [a], A.C. Roger [a], J.L. Rehspringer [b] and A. Kiennemann [a*]
[a] Laboratoire des Matériaux, Surfaces et Procédés pour la Catalyse, UMR 7515, ECPM, 25 rue Becquerel, 67087 Strasbourg cedex 2, France
[b] Groupe des Matériaux Inorganiques, UMR 7504, IPCMS, 23 rue du Loess, 67037 Strasbourg cedex, France
*Email: kiennemann@chimie.u-strasbg.fr Phone: +33 3 90 24 27 66 Fax: +33 3 90 24 27 68

Superior Catalytic Activity of Nano-sized Metallic Cobalt inside Faujasite Zeolite for Fischer-Tropsch Synthesis
Qinghu Tang, Ye Wang*, Ping Wang, Qinghong Zhang and Huilin Wan
State Key Laboratory for Physical Chemistry of Solid Surfaces, Department of Chemistry, Xiamen University, Xiamen 361005, China
* Email: yewang@jingxian.xmu.edu.cn

Fischer-Tropsch Synthesis: Effect of Water on Activity and Selectivity for a Cobalt Catalyst
Tapan K. Das, Whitney Conner, Gary Jacobs, Jinlin Li, Karuna Chaudhari and Burtron H. Davis*
Center for Applied Energy Research, University of Kentucky, 2540 Research Park Drive, Lexington, KY, 40511
*Email: davis@caer.uky.edu Phone: 859-257-0251 Fax: 859-257-0302

Larissa B. Shapovalova*, Gauhar D. Zakumbaeva, Aliya A. Zhurtbaeva.
The Institute of Organic Catalysis & Electrochemistry of the Ministry of Education and
Science of the Republic of Kazakstan; 142, Kunaev str., Almaty, 480100, Kazakstan;
*Email: orgcat@nursat.kz   fax: (007) 3272 915722

From Synthesis gas to Chemicals

Direct Dimethyl Ether (DME) Synthesis from Natural Gas
Takashi Ogawa, Norio Inoue, Tutomu Shikada, Osamu Inokoshi, Yotaro Ohno*
DME Development Co., Ltd, Shoro-koku Shiranuka-ch, Hokkaido, 088-0563 Japan
*Email: ohno@dme-development.com Phone: +81 (3) 35026671 Fax: +81 (3) 35026675

Integrated Synthesis of Dimethylether via CO_2 Hydrogenation
F. Arena *[a,b], L. Spadaro [a], O. Di Blasi [a], G. Bonura [a] and F. Frusteri [a]
[a] Istituto CNR-ITAE, Salita S. Lucia 39, I-98126 S. Lucia (Messina), Italy
[b] Dipartimento di Chimica Industriale e Ingegneria dei Materiali, Università degli Studi di
Messina, Salita Sperone 31, I-98166 S. Agata (Messina), Italy
*Email: Francesco.Arena@unime.it Phone: +39 090 6765606 Fax: +39 090391518

A Catalyst with High Activity and Selectivity for the Synthesis of C_{2+}-OH: ADM Catalyst
Modified by Nickel
D.-B. Li, C. Yang, H.-R. Zhang, W.-H. Li, Y.-H. Sun*, B. Zhong
State Key Laboratory of Coal Conversion, Institute of Coal Chemistry, Chinese Academy of
Sciences, Taiyuan, 030001 P.R.China
* E-mail: yhsun@sxicc.ac.cn Fax: +86-351-4041153

On the Catalytic Mechanism of $Cu/ZnO/Al_2O_3$ for CO_x Hydrogenation
Qi Sun
Süd-Chemie Inc. P. O. Box 32370, Louisville, KY 40232, USA,
Email: qsun@sud-chemieinc.com

Solid Superacids for Halide-free Carbonylation of Dimethyl Ether to Methyl Acetate
G.G. Volkova*, L.M. Plyasova, L.N. Shkuratova, A.A. Budneva, E.A. Paukshtis, M.N.
Timofeeva, V.A. Likholobov
Boreskov Institute of Catalysis, Pr. Akademika Lavrentieva 5, Novosibirsk, 630090, Russia
*Email: g.g.volkova@catalysis.nsk.su

A New Low-temperature Methanol Synthesis Process from Low-Grade Syngas
P. Reubroycharoen[a], T. Yamagami[a], Y. Yoneyama[a], M. Ito[b], T. Vitidsant[c],
N. Tsubaki[a*]
[a]Dept. of Material System & Life Science, School of Engineering, Toyama University,
Gofuku, Toyama, 930-8555, Japan
[b]Dept. of Applied Chemistry, The University of Tokyo, Tokyo, 113-8656, Japan
[c]Dept. of Chemical Technology, Chulalongkorn University, 10330, Thailand
*Email: tsubaki@eng.toyama-u.ac.jp

Enabling Offshore Production of Methanol by Use of an Isopotential Reactor
F. Daly* and L. Tonkovich
Velocys, Inc., 7950 Corporate Boulevard, Plain City, OH 43064, United States

*Email: daly@velocys.com Phone: 614/733-3300 Fax: 914/733-330

The Promotion Effects of Mn, Li and Fe on the Selectivity for the Synthesis of C_2 Oxygenates from Syngas on Rhodium Based Catalysts

H. -M. Yin, Y. -J. Ding*, H. -Y. Luo, W. –M. Chen, L.-W. Lin
Natural Gas Utilization and Applied Catalysis Laboratory, Dalian Institute of Chemical Physics, Chinese Academy of Sciences, 457[#] Zhongshan Rd., Dalian, 116023, China
*Email: dyj@dicp.ac.cn

Selective Synthesis of LPG from Synthesis Gas

Kenji Asami*, Qianwen Zhang, Xiaohong Li, Sachio Asaoka, and Kaoru Fujimoto
Faculty of Environmental Engineering, The University of Kitakyushu, Wakamatsu-ku, Kitakyushu 808-0135, Japan
*Email: asami@env.kitakyu-u.ac.jp

Catalytic Alcohols Synthesis from CO and H_2O on TiO_2 Catalysts Calcined at Various Temperatures

Dechun Ji, Shenglin Liu, Qingxia Wang and Longya Xu[*]
State Key Laboratory of Catalysis, Dalian Institute of Chemical Physics, Chinese Academy of Sciences, P.O. Box 110, Dalian 116023, P. R. China
*Email: lyxu@dicp.ac.cn

Synthesis of Gasoline Additives from Methanol and Olefins over Sulfated Silica

Shaobin Wang[a] and James A. Guin[b]
[a]Department of Chemical Engineering, Curtin University of Technology, GPO Box U1987, Perth, WA 6845, Australia, Email: wangshao@vesta.curtin.edu.au
[b]Department of Chemical Engineering, Auburn University, Auburn, AL 36849, USA

Synthesis, Characterization, and MTO Performance of MeAPSO-34 Molecular Sieves

Lei Xu, Zhongmin Liu*, Aiping Du,Yingxu Wei, Zhengang Sun
Dalian Institute of Chemical Physics, Chinese Academy of Sciences,
457 Zhongshan Road, P. O. Box 110, Dalian, 116023, China
*Email: zml@dicp.ac.cn

Bifunctional Catalysts for the Production of Motor Fuels and Aromatic Hydrocarbons from Synthesis Gas

V. M. Mysov*, K. G. Ione
Scientific Engineering Centre "Zeosit" SB RAS, Pr. Ak. Lavrentieva, 5, Novosibirsk 630090, Russia
* Email: mysov@batman.sm.nsc.ru Tel/Fax: (3832) 356251

Catalytic Combustion

Low NO_x Emission Combustion Catalyst Using Ceramic Foam Supports

J. T. Richardson*, M. Shafiei, T. Cantu, D. Machiraju, and S. Telleen
Department of Chemical Engineering, University of Houston
Houston, Texas, U.S.A. 77204-4792
*Email: jtr@uh.edu Phone: +011 (713) 7437473 Fax: +011 (713) 7434323

Preparation of LaCoO$_3$ with High Surface Area for Catalytic Combustion by
Spray-Freezing/Freeze-Drying Method

Seong Ho Lee[a], Hak-Ju Kim[a], Jun Woo Nam[a], Heon Jung[b], Sung Kyu Kang[b], and
Kwan-Young Lee[a,*]
[a]Department of Chemical and Biological Engineering, Korea University,
1, 5-ka, Anam-dong, Sungbuk-ku, Seoul 136-701, Korea
[b]Korea Institute of Energy Research, 71-2, Jang-dong, Yusung-gu, Taejon 305-343, Korea
*Email: kylee@prosys.korea.ac.kr Phone: 82-2-3290-3299 Fax: [82] 29266102

Effect of Support Structure on Methane Combustion over PdO/ZrO$_2$

L.-F. Yang*, Y.-X. Hu, D. Jin, X. -E He, C.-K. Shi, J.-X. Cai
Department of Chemistry, State Key Laboratory of Physical Chemistry for Solid Surface,
Xiamen University, Xiamen 361005, China
* Tel: +86-592-2185944; Fax: +86-592-2183047; E-mail: lfyang@xmu.edu.cn

The Catalytic Combustion of CH$_4$ over Pd-Zr/HZSM-5 Catalyst with Enhanced Stability and
Activity

Chun-Kai Shi, Le-Fu Yang, and Jun-Xiu Cai*
Department of Chemistry and State Key Laboratory for Physical Chemistry of Solid Surface,
Xiamen University, Xiamen 361005, P. R. China
*Email: jxcai@xmu.edu.cn Phone: +86 (592) 2185944 Fax: +86 (592) 2183047

Reverse Microemulsion Synthesis of Mn-Substituted Barium Hexaaluminate (BaMnAl$_{11}$O$_{19-\alpha}$)
Catalysts for Catalytic Combustion of Methane

Jinguang Xu[a,b], Zhijian Tian[a]*, Yunpeng Xu[a], Zhusheng Xu[a], Liwu Lin[a]
[a]Dalian Institute of Chemical Physics, Chinese Academy of Sciences, 457 Zhongshan Road,
Dalian 116023, China, * tianz@dicp.ac.cn (Z. Tian)
[b]Institute of Applied Catalysis, Yantai University, Yantai 264005, China

Effect of Drying Method on the Morphology and Structure of High Surface Area
BaMnAl$_{11}$O$_{19-\alpha}$ Catalyst for High Temperature Methane Combustion

Zhijian Tian*, Jinguang Xu, Junwei Wang, Yunpeng Xu, Zhusheng Xu, Liwu Lin
Dalian Institute of Chemical Physics, The Chinese Academy of Sciences, Dalian 116023,
China, *tianz@dicp.ac.cn (Z. Tian)

Synthesis of Nano-sized BaAl$_{12}$O$_{19}$ via Nonionic Reverse Microemulsion Method: I. Effect of
the Microemulsion structure on the Paticle Morphology

Fei Teng, Zhijian Tian*, Jinguang Xu, Guoxing Xiong*, Liwu Lin
Dalian Institute of Chemical Physics, Chinese Academy of Sciences, Dalian 116023, China.
tianz@dicp.ac.cn   (Z. Tian), gxxiong@dicp.ac.cn   (G. Xiong)

From Natural Gas to Chemicals

Selective Oxidation of CH$_4$ to CH$_3$OH using the Catalytica (bpym)PtCl$_2$ Catalyst: a
Theoretical Study

X. Xu*[,a, b], G. Fu[a], W. A. Goddard III *[, b] and R. A. Periana[c]
[a]State Key Laboratory for Physical Chemistry of Solid Surfaces; Department of Chemistry,
Center for Theoretical Chemistry, Xiamen University, Xiamen, 361005, China
[b]Materials and Process Simulation Center (139-74), Division of Chemistry and Chemical

Engineering, California Institute of Technology, Pasadena, California 91125, USA
[c]Department of Chemistry, University of Southern California, Los Angeles, California 90089, USA
*Email: xinxu@xmu.edu.cn Phone: +86-592-2181600 Fax: +86-592-2183047

Micro-plasma technology - Direct Methane-to-Methanol in Extremely Confined Environment -

Tomohiro Nozaki*, Shigeru Kado, Akinori Hattori, Ken Okazaki, Nahoko Muto
Mechanical and Control Engineering, Tokyo Institute of Technology
2-12-1 O-okayama, Meguro, Tokyo, JAPAN 152-8551
*Email: tnozaki@mech.titech.ac.jp

Towards a Better Understanding of Transient Processes in Catalytic Oxidation Reactors
Renate Schwiedernoch[a], Steffen Tischer[a], Hans-Robert Volpp[b], Olaf Deutschmann[a]
[a]Institute for Chemical Technology, University of Karlsruhe, Engesserstr. 20, 76313 Karlsruhe, Germany, deutschmann@ict.uni-karlsruhe.de
[b]Institute of Physical Chemistry, Heidelberg University, INF 253, 69120 Heidelberg, Germany

Direct Oxidation of Methane to Oxygenates over Supported Catalysts
S.A.Tungatarova, G.A.Savelieva, A.S.Sass, and K.Dosumov*
D.V.Sokolsky Institute of Organic Catalysis and Electrochemistry. Republic of Kazakhstan. 142, D.Kunaev st., Almaty 480100.
*Email: orgcat@nursat.kz Phone: 7 3272 916473 Fax: 7 3272 915722

Synthesis of Methanesulfonic Acid and Acetic Acid by the Direct Sulfonation or Carboxylation of Methane
Sudip Mukhopadhyay, Mark Zerella, and Alexis T. Bell*
Department of Chemical Engineering, University of California-Berkeley, CA 4720-1462, U.S.A
*Email: bell@cchem.berkeley.edu

New Study of Methane to Vinyl Chloride Process
Christophe Lombard and Paul-Marie Marquaire*
Département de Chimie Physique des Réactions, UMR 7630 CNRS - INPL ENSIC, 1 rue Grandville, BP 451, 54001 Nancy Cedex, France.
*Email: Paul-Marie.Marquaire@ensic.inpl-nancy.fr

Structure-Activity Relationships in the Partial Oxidation of Methane to Formaldehyde on FeO_x/SiO_2 Catalysts
F. Arena*[a], G. Gatti [b], S. Coluccia [b], G. Martra [b] and A. Parmaliana [a]
[a] Dipartimento di Chimica Industriale e Ingegneria dei Materiali, Università di Messina, Salita Sperone 31, 98166 S. Agata (Messina), Italy
[b] Dipartimento di Chimica IFM, Università di Torino, Via P. Giuria 7, 10125 Torino, Italy
*Email: Francesco.Arena@unime.it Phone: +39 090 6765606 Fax: +39 090391518

Methanol from Oxidation of Methane over MoO_x/La-Co-oxide Catalysts
X. Zhang, D.-H. He*, Q.-J. Zhang, B.-Q. Xu, Q.-M. Zhu
Key Lab of Organic Optoelectronics & Molecular Engineering of Ministry of Education，Department of Chemistry, Tsinghua University, Beijing 100084, China

*Email: hedeh@mail.tsinghua.edu.cn Phone: 86-10-62772592

SbO$_x$/SiO$_2$ Catalysts for the Selective Oxidation of Methane to Formaldehyde Using Molecular Oxygen as Oxidant

Haidong Zhang, Pinliang Ying, Jing Zhang, Changhai Liang, Zhaochi Feng and Can Li*
State Key Laboratory of Catalysis, Dalian Institute of Chemical Physics, Chinese Academy of Sciences, P. O. Box 110, Dalian 116023, China
* Email: canli@dicp.ac.cn Fax: (+86) 411-4694447

Highly Stable Performance of Catalytic Methane Dehydro-condensation to Benzene and Naphthalene on Mo/HZSM-5 by Addition and a Periodic Switching Treatment of H$_2$

Ryuichiro Ohnishi, Ryoichi Kojima, Yuying Shu, Hongtao Ma, and Masaru Ichikawa[a*]
[a]Catalysis Research Center, Hokkaido University, Kita-Ku, N-11, W-10, Sapporo 060-0811, Japan.
*Email: michi@cat.hokudai.ac.jp Phone: [81] 11-706-2912 Fax: [81] 11-706-4957

Reactions of Propane and n-Butane on Re/ZSM Catalyst

F. Solymosi*, P. Tolmacsov and A. Széchenyi
Institute of Solid State and Radiochemistry, the University of Szeged and Reaction Kinetics Research Group of the Hungarian Academy of Sciences; P.O. Box 168, H-6701 Szeged, Hungary
*Email: fsolym@chem.u-szeged.hu Phone: [36] 62-420-678 Fax: [36] 62-424-997

Dehydro-aromatization of CH$_4$ over W-Mn(or Zn, Ga, Mo, Co)/HZSM-5(or MCM-22) Catalysts

L.-Q. Huang, Y.-Z. Yuan, H.-B. Zhang*, Z.-T. Xiong, J.-L. Zeng, G.-D. Lin
Department of Chemistry & State Key Laboratory of Physical Chemistry for the Solid Surfaces, Xiamen University, Xiamen 361005, China
*Email: hbzhang@xmu.edu.cn Phone: +86 592 2184565 Fax: +86 592 2183047

Detailed Mechanism of the Oxidative Coupling of Methane

Y. Simon* , F. Baronnet, G.M. Côme and P.M. Marquaire
Département de Chimie Physique des Réactions, UMR 7630 - CNRS, ENSIC-INPL, 1 rue Grandville - BP 451, 54001 NANCY Cedex (France).
* Email:Yves.Simon@ensic.inpl-nancy.fr Phone: 33 383 17 51 22 Fax: 33 383 37 81 20

High Performance Methane Conversion into Valuable Products with Spark Discharge at Room Temperature

S. Kado*[a], Y. Sekine[b], K. Urasaki[b], K. Okazaki[a] and T. Nozaki[a]
[a] Department of Mechanical and Control Engineering, Tokyo Institute of Technology, 2-12-1 O-okayama, Meguro-ku, Tokyo 152-8552, Japan
[b] Department of Applied Chemistry, School of Science and Engineering, Waseda University, Okubo, Shinjuku-ku, Tokyo 169-8555, Japan
*Email: k-shigeru@kit.hi-ho.ne.jp Phone: +81 (3) 57342179 Fax: +81 (3) 57342893

Combined Single-Pass Conversion of Methane via Oxidative Coupling and Dehydroaromatization: a Combination of La$_2$O$_3$/BaO and Mo/HZSM-5 Catalysts

Y.-G. Li, H.-M. Liu, W.-J. Shen, X.-H. Bao, Y.-D. Xu*
State Key Laboratory of Catalysis, Dalian Institute of Chemical Physics, Chinese Academy of

Sciences, 457 Zhongshan Road, Dalian 116023, China
*Email: xuyd@dicp.ac.cn

Formation of Acetaldehyde and Acetonitrile by CH_4-CO-NO Reaction over Silica Supported Rh and Ru Catalysts
Nobutaka Maeda, Toshihiro Miyao and Shuichi Naito*
Department of Applied Chemistry, Kanagawa University, 3-27-1, Rokkakubashi, Kanagawa-ku, Yokohama, 221-8686, Japan
*Email: naitos01@kanagawa-u.ac.jp Phone: 81-45-481-5661 Fax: 81-45-491-7915

Methane Dehydroaromatization over Alkali-treated MCM-22 Supported Mo Catalysts: Effects of Porosity
L.-L. Su, Y.-G. Li, W.-J. Shen, Y.-D. Xu*, X.-H. Bao*
State Key Laboratory of Catalysis, Dalian Institute of Chemical Physics, Chinese Academy of Sciences, 457 Zhongshan Road, Dalian 116023, China.
* Email: xuyd@dicp.ac.cn ; xhbao@dicp.ac.cn

Middle Pressure Methane Conversion into C_2 Hydrocarbons on Supported Pt-Co Catalysts
L. Borkó* and L. Guczi
Department of Surface Chemistry and Catalysis, Institute of Isotope and Surface Chemistry, CRC, Hungarian Academy of Sciences, P. O. Box 77, H-1525, Budapest, Hungary
*Email: borko@sunserv.kfki.hu Phone: +361 3922534 Fax: +361392-2703

The Critical Role of the Surface WO_4 Tetrahedron on the Performance of Na-W-Mn/SiO_2 Catalysts for the Oxidative Coupling of Methane
S.-F. Ji*[a, b, c], T.-C. Xiao[c], S.-B. Li[b], L.-J. Chou[b], B. Zhang[b], C.-Z. Xu[b], R.-L. Hou[b] and M.L.H. Green[c]
[a]The Key Laboratory of Science and Technology of Controllable Chemical Reactions, Ministry of Education, Beijing University of Chemical Technology, 15 Beisanhuan Dong Road, P.O.Box 35, Beijing 100029, China. Email: jisf@mail.buct.edu.cn
[b]State Key Laboratory for Oxo Synthesis and Selective Oxidation, Lanzhou Institute of Chemical Physics, Chinese Academy of Sciences, 342 Tianshui Road, Lanzhou 730000, China
[c]Wolfson Catalysis Centre, Inorganic Chemistry Laboratory, University of Oxford, South Parks Road, Oxford, OX1 3QR, U.K.

Hydrogenolysis of Methyl Formate to Methanol on Copper-Based Catalysts and TPR Characterization
Y. Long, W. Chu*, Y. H. Ye, X. W. Zhang, X. Y. Dai
Department of Chemical Engineering, Sichuan University, Chengdu 610065, China
Email: chuwei65@yahoo.com.cn Phone: +86 28 85400591

Role of SnO_2 over Sn-W-Mn/SiO_2 Catalysts for Oxidative Coupling of Methane Reaction at Elevated Pressure
Lingjun Chou, Yingchun Cai, Bing Zhang, Jian Yang, Jun Zhao, Jianzhong Niu, and Shuben Li*
State Key Laboratory for Oxo Synthesis and Selective Oxidation, Lanzhou Institute of Chemical Physics, Chinese Academy of Sciences, Lanzhou 730000, P.R.China
Email: lzcpcn@ns.lzb.ac.cn Phone: +86 931 8277147 Fax: +86 931 8277088

Infrared Study of CO And CO/H_2 Adsorption on Copper Catalysts Supported on Zirconia Polymorphs

Z.-Y. Ma, R.-Xu, C.-Yang, Q.-N. Dong, W.- Wei, Y.-H. Sun*
State Key Laboratory of Coal Conversion, Institute of Coal Chemistry, Chinese Academy of Sciences, Taiyuan, 030001 P.R.China
Email: yhsun@sxicc.ac.cn, yangch68@sxicc.ac.cn Phone: +86 351 4049612

Nonoxidative Aromatization of Methane over Mo/MCM-49 Catalyst

Qiubin Kan, Dongyang Wang, Ning Xu, Peng Wu, Xuwei Yang, Shujie Wu and Tonghao Wu*
Department of Chemistry, Jilin University, Changchun 130023, China
*Email: wth@mail.jlu.edu.cn Phone: 86 431 8499140 Fax: 86 431 8949334

Selective Oxidation of Methane to C_1-Products Using Hydrogen Peroxide

T.M. Nagiev*, L.M. Gasanova, E.M. Mamedov, I.T. Nagieva, Z.Y. Ramasanova, and A.A. Abbasov
Institute of Chemical Problems, National Academy of Sciences of Azerbaijan,
29 H. Javid av., 370143 Baku, the Republic of Azerbaijan
*Email: chem.@dcacs.ab.az Phone: (99412) 394691 Fax: (99412) 970593

Methane Activation on Alumina Supported Platinum, Palladium, Ruthenium and Rhodium Catalysts

Ruth L. Martins[a], Maria A. Baldanza[a], Mariana M.V.M. Souza[a] and Martin Schmal[*a,b]
[a]NUCAT/PEQ/COPPE, Universidade Federal do Rio de Janeiro, C.P. 68502, 21945-970, Rio de Janeiro, Brazil. *E-mail: schmal@peq.coppe.ufrj.br
[b]Escola de Quimica, Universidade Federal do Rio de Janeiro, C.P. 68542, 21940-900, Rio de Janeiro, Brazil.

Light Hydrocarbons: Production and Conversion

Improvement of the Selectivity to Propylene by the Use of Cyclic, Redox-Decoupling Conditions in Propane Oxidehydrogenation

N. Ballarini[a], F. Cavani[a], A. Cericola[a], C. Cortelli[a], M. Ferrari[a], F. Trifirò[a], R. Catani[b] and U. Cornaro[c]
[a]Dipartimento di Chimica Industriale e dei Materiali, Viale Risorgimento 4, 40136 Bologna, Italy. cavani@ms.fci.unibo.it
[b]Snamprogetti SpA, Viale De Gasperi 16, 20097 San Donato Milanese MI, Italy.
[c]EniTecnologie SpA, Via Maritano 26, 20097 San Donato Milanese MI, Italy.

Oxidative Activation of Light Alkanes on Dense Ionic Oxygen Conducting Membranes

M. Rebeilleau[a], A.C. van Veen[a], D. Farrusseng[a], J.L. Rousset[a], C. Mirodatos*[a] Z.P. Shao[b], G. Xiong[b].
[a]Institut de Recherches sur la Catalyse, 2. Av. A. Einstein 69626 Villeurbanne Cedex France
[b]State Key Laboratory of Catalysis, Dalian Institute of Chemical Physics, Chinese Academy of Sciences, P.O. Box 110,Dalian 116023, People's Republic of China
*Email: mirodatos@catalyse.cnrs.fr Phone: 33 472445366 Fax: 33 472445399

Effect of Support on Performance of Movteo Catalyst for Selective Oxidation of Propane to

Moscow 119991, Russia.
[b]Institute of Problems of Chemical Physics, Russian Academy of Sciences, Chernogolovka, Moscow Region., 142432, Russia.
*Email: arutyunov@center.chph.ras.ru Phone: +007(095) 9397287 Fax: +007(095) 9382156

Dehydro-aromatization of Methane under Non-Oxidative Conditions over MoMCM-22: Effect of Surface Aluminum Removal
A. C. C. RODRIGUES AND J. L. F. MONTEIRO*
NUCAT/PEQ/COPPE/UFRJ, P.O. Box 68502 – 21945-970 – Rio de Janeiro – Brazil
*Email: monteiro@peq.coppe.ufrj.br

The Role of Coke in the Deactivation of Mo/MCM-22 Catalyst for Methane Dehydroaromatization with CO_2
Jie Bai, Shenglin Liu, Sujuan Xie, Longya Xu*, and Liwu Lin
State Key Laboratory of Catalysis, Dalian Institute of Chemical Physics, Chinese Academy of Sciences, P.O. Box 110, Dalian 116023, P. R. China
*Email: lyxu@dicp.ac.cn Phone: +86-411-4693292

Production of Isobutene from n-Butane over Pd Modified SAPOs and MeAPOs
Yingxu Wei, Gongwei Wang, Zhongmin Liu*, Lei Xu and Peng Xie
Natural Gas Utilization & Applied Catalysis Laboratory, Dalian Institute of Chemical Physics, Chinese Academy of Sciences, P.O.Box 110, Dalian 116023, P. R.China
*Email: zml@dicp.ac.cn   Phone: +86 411 4685510

Studies in Surface Science and Catalysis, volume 147
X. Bao and Y. Xu (Editors)

Most recent developments in ethylene and propylene production from natural gas using the UOP/Hydro MTO process

J. Q. Chen[a], B. V. Vora[a], P. R. Pujadó[a], Å. Grønvold[b], T. Fuglerud[b], S. Kvisle[b]

[a]UOP LLC, P.O.Box 5017, Des Plaines, Illinois 60017-5017, USA

[b]Norsk Hydro ASA, P.O. Box 2560, N-3901 Porsgrunn, Norway

ABSTRACT

The UOP/Hydro MTO Process utilizes a SAPO-34-containing catalyst that provides up to 80% yield of ethylene and propylene at near-complete methanol conversion. The process has great flexibility to produce a product with a range of ethylene/propylene ratios, depending on the operating conditions used. This paper will discuss recent improvements related to the technology which have increased the carbon selectivity from methanol to ethylene-plus-propylene to about 85-90%.

1. INTRODUCTION

As worldwide natural gas reserves continue to increase, natural gas is expected to become an increasingly important raw material for the chemical industry in the 21st century. Conversion of natural gas to syn-gas and higher value added products such as methanol, ammonia and transportation fuels is currently being practiced commercially. The combination of methanol production using state-of-the-art mega-scale methanol technology and methanol-to-olefins (MTO) technology developed by UOP and Norsk Hydro provides an economically attractive route from natural gas to ethylene and propylene, two large-tonnage, high-value petrochemicals. The MTO technology has been extensively demonstrated at a large-scale facility owned by Norsk Hydro in Norway. The process converts methanol or dimethyl ether (DME) to ethylene and propylene at about 75-80% carbon selectivity[1]. A world-scale MTO unit is currently under design[2,3]. This paper will discuss recent improvements related to the MTO technology which have increased the carbon selectivity from methanol to ethylene-plus-propylene to about 85-90%.

2. MTO CATALYST

The MTO process utilizes a molecular sieve, SAPO-34, that contains silicon, aluminium, and phosphorous. The structure of SAPO-34 along with small organic molecules that are key to the MTO process are shown in Figure 1. The small pore size (about 4 Å) of SAPO-34 restricts the diffusion of branched hydrocarbons, and this leads to high selectivity to the desired small linear olefins.

ZSM-5 molecular sieve used in methanol-to-gasoline (MTG) processes produces much lower ethylene and propylene yields[4], primarily due to the larger pore openings of the MFI structure (about 5.5 Å) (Fig. 2).

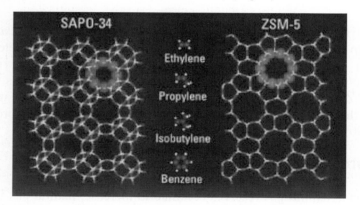

Fig. 1. Framework structure of SAPO-34 (CHA) and ZSM-5 (MFI) molecular sieves

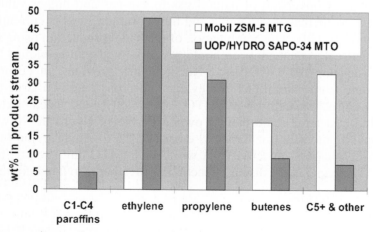

Fig. 2. Comparative maximum C_2= mode performance of SAPO-34 vs. ZSM-5

Another key feature of the SAPO-34 molecular sieve is its mild acidity which reduces paraffinic by-product formation through the hydride transfer reaction. This feature provides olefin purities of about 97% even without splitter columns for further purification, and makes it easy to upgrade the olefins to polymer grade.

3. PROCESS DESCRIPTION AND RECENT IMPROVEMENTS

The MTO process utilizes a circulating fluidized bed reactor that offers a number of advantages. The moving bed of catalyst allows the continuous movement of a portion of used catalyst to a separate regeneration vessel for the removal of coke deposits by burning with air. Thus, a constant catalyst activity and product composition can be maintained in the MTO reactor. A fluidized-bed reactor also allows for better heat recovery from the exothermic methanol-to-olefins reaction. Catalysts manufactured at the commercial scale have demonstrated the required selectivity, long term stability, and attrition resistance.

Conventional treating methods have been shown to be effective for removing by-products to the specification levels required for olefin polymerization processes.

Reactor operating conditions can be adjusted to the desired product requirements. Pressure is normally dictated by mechanical considerations, with lower methanol partial pressures resulting in higher overall yields. Thus, some yield advantage can be obtained by using a crude methanol feed that typically may contain around 20 wt% water. On the other hand, increased pressures may be favored for the production of higher propylene ratios. Temperature is an important control variable – higher temperatures are required to increase the proportion of ethylene in the product relative to propylene, but higher temperatures also result in a small reduction in the overall light olefin selectivity. Temperature severity effects are best appreciated in Figure 3 where operating temperatures range from 400 to about 550°C. Figure 3 also shows that the maximum (carbon-based) selectivity of ethylene plus propylene are just shy of 80% and occur at an ethylene-to-propylene ratio close to 1:1.

It may be appreciated though that, if butenes were included, the overall carbon selectivity would be in the order of 90%. This lends itself to a modified process scheme whereby the heavier olefins are fed to an ATOFINA/UOP Olefin Cracking Process unit (OCP) based on technology developed and demonstrated by ATOFINA and UOP[5]. In essence, C_4 to C_6+ olefins generated as byproducts from the MTO process unit can be fed to the OCP unit in which these heavier olefins are cracked to ethylene plus propylene, but with a

4

preponderance of propylene. The fractionation scheme in the MTO unit remains unchanged except that it has to be sized to accommodate the added circulation to and from the OCP unit; a small purge removes the heavier or more refractory components from the recycle loop. Figure 4 illustrates the carbon selectivity pattern from high-purity methanol (99.85 wt% methanol) to ethylene plus propylene with the incorporation of the OCP unit.

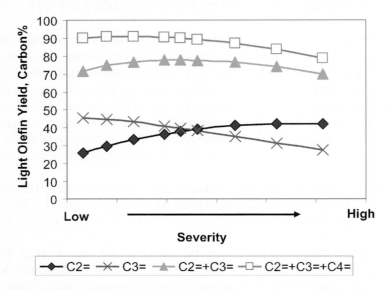

Fig. 3. Light olefin selectivities from an MTO unit with high-purity methanol feed

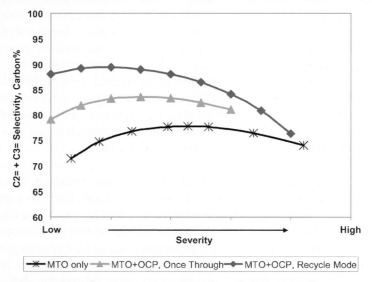

Fig. 4. Olefin selectivity vs. operating severity with and without OCP

The OCP unit makes use of a very robust catalyst that yields low levels of coke as byproduct from the olefin cracking reaction. Thus, although the catalyst is fully regenerable by using a standard carbon burn, it is not necessary to use a fluidized bed with catalyst transport as used in the MTO unit; either a fixed bed or a moving bed suffices in the OCP unit.

The incorporation of an MTO unit with an OCP unit on a once-through basis without recycle boosts the overall ethylene plus propylene selectivity to about 83% compared to 77-78% selectivity for an MTO unit by itself as in Figure 3 and 4 above. The addition of recycle around the OCP unit for maximum conversion of the heavy olefins leads to an overall yield of about 89% (Fig. 4). In the process, however, the points at which the maximum C_2=+C_3= yields are achieved shift to higher C_3=/C_2= ratios (on an overall MTO plus OCP basis) as corresponds to the preferential production of propylene from the OCP unit.

4. ECONOMICS

An MTO unit designed to produce 400,000 MTA of polymer-grade ethylene and 400,000 MTA of polymer-grade propylene will also produce about 175,000 MTA of C_4+ olefins which, although valuable in an integrated refining or petrochemical complex, may only warrant fuel value in some remote locations. If integrated with an OCP unit, the same MTO unit can produce about 426,000 MTA of polymer-grade ethylene and 500,000 MTA of polymer-grade propylene, with a 78% decrease in the amount of C_4+ byproduct to only 39,000 MTA. The ISBL cost of the OCP unit with recycle of unconverted olefins is about $50 million and, in addition, some extra capacity is required in the fractionation and recovery section of the MTO unit that amounts to another $10 million. Overall, inclusive of OSBL incremental allowances, license fees, catalysts, etc., inclusion of the OCP unit requires an incremental investment of about $70 million. Thus, the overall investment for the integrated MTO/OCP complex increases from about $425 million to about $495 million. Typical IRR rates are in the order of 20% or higher depending on the relative values assigned to the key products.

If additional propylene is desired, the production ratio of propylene to ethylene can be further increased by reducing the operating severity of the MTO unit and slightly increasing the size of the OCP unit. Thus, for example, with the same methanol feed rate as in the previous example, the MTO unit could be operated to produce about 340,000 MTA ethylene and 450,000 MTA propylene (propylene/ethylene ratio ~ 1.33) without OCP. The operation can be further shifted to produce about 370,000 MTA ethylene and 570,000 MTA propylene (propylene/ethylene ratio ~ 1.54) by incorporating an OCP unit to convert about 200,000 MTA of C_4+ olefins and reduce the C_4+ byproducts by almost 80%.

Note that, as seen in Figure 3, the amount of C_4+ olefins increases and the yield of $C_2=+C_3=$ decreases slightly as operating severity decreases (increased $C_3=/C_2=$ ratio) in an MTO unit; however, by incorporating an OCP unit the overall $C_2=+C_3=$ yield actually increases at the lower operating severity (higher $C_3=/C2=$ ratio) as shown in Figure 4.

5. CONCLUSIONS

The MTO process provides an economically attractive alternative for the production of light olefins, ethylene and propylene, and the corresponding polyolefins from methanol or DME that can be produced at low cost and in large quantities from natural gas or other hydrocarbon sources. The MTO process can be operated over a relatively broad range of $C_3=/C_2=$ ratios, but with slightly lower $C_2=+C_3=$ yields at the higher or lower $C_3=/C_2=$ ratios. The integration of an MTO unit with an OCP unit increases the carbon-based yield of ethylene plus propylene to almost 90% of the methanol feedstock. This essentially eliminates the need to market less desirable C_4+ byproducts and thereby provides a product slate focused entirely on ethylene and propylene with high propylene/ethylene ratios.

This paper has discussed some of the features of a typical process unit for the conversion of methanol and/or DME to light olefins. Though based on a relatively novel chemistry, the process configuration and recovery and purification techniques are akin to those currently practiced in industry, either for the production of olefins by fluidized catalytic cracking (FCC) of heavy hydrocarbons or by steam cracking of LPG or naphtha fractions

REFERENCES

[1] B.V. Vora, et al, Studies in Surface Science and Catalysis, 107, (1997) 87-98.
[2] a. Chemical Week, October 2, 2002
    b. European Chemical News, October 14-20, 2002
[3] London, CNI, 31-Mar-03
[4] Mobil data from 1992 "Eurogas Conference", Norway
[5] J.H. Gregor and W. Vermeiren, 5th EMEA Petrochemicals Technology Conference, June 25-26, 2003, Paris

Studies in Surface Science and Catalysis, volume 147
X. Bao and Y. Xu (Editors)
©2004 Elsevier B.V. All rights reserved.

Condensing methanol synthesis and ATR
- the technology choice for large-scale methanol production

Esben Lauge Sørensen and Jens Perregaard

Haldor Topsøe A/S, Nymøllevej 55, DK-2800 Lyngby, Denmark
Phone +45 4527 2000, e-mail: els@topsoe.dk

ABSTRACT

Large methanol plants are subject to the constraints imposed by size limitations on reactors and process equipment. To allow scale-up, the ability to process major amounts of synthesis gas (syngas) in comparatively small pieces of process equipment has become increasingly important. This paper demonstrates how this objective may be achieved by combining methanol synthesis technology consisting of Combined Methanol Synthesis and Condensation (CMSC) and syngas technology comprising an ATR (AutoThermal Reformer) operating at very low steam to carbon ratio.

1. SYNTHESIS GAS GENERATION

Over the last decade, the dominating trend in design of methanol production plants has been one of ever increasing capacities. This trend is naturally strongly related to the economics-of-scale, which results in lower production prices for large-scale plants than for small-scale plants. Since the construction of large methanol plants is particularly interesting where natural gas is available at low cost, this paper presents one cost efficient technology for addressing the "stranded gas" issue.

Feed gas for methanol synthesis units are normally generated by one of the three types of front-end processes: steam reforming (1), two-step processes where steam and oxygen-blown reforming are combined (2) and ATR (also called oxygen-blown reforming) (3). In the latter case, stoichiometric adjustment is normally achieved by removal of CO_2 (3a) or through hydrogen recovery and recycle (3b).

The merits and drawbacks of each of these processes are presented in detail by Dybkjær and Gøl, Ref. 1 and Dybjær and Christensen, Ref. 4, who conclude that the steam reforming based processes are only economic for small-scale plants. They also recommend the two-step process for the capacity range of 1500-7000 MTPD and the ATR process for capacities above 5000 MTPD.

Table 1
10,000 MTPY MeOH Plant
Key figures for selection of technology for synthesis gas generation

Syngas generation process	1	2	3 a	3 b.
Operation parameters				
S/C	2.5	1.9	0.6	0.6
Reforming outlet temperature, °C	925	1000	1050	1050
Front-end key figures				
1. NG flow to process (MTPD)	6800	5800	6400	5700
2. Oxygen flow to process (MTPD)	0	4600	6500	5800
3. Flow exit reforming section (MTPD)	25700	22700	17200	15200
4. Power; compression to 110 at g (MW)	82	64	60	62
Syngas properties				
5. Methane slip to methanol synthesis	4.87	1.13	1.95	1.67
7. CO/CO_2 ratio in MU gas	2.51	2.58	15.64	5.12
8. Converted gas dew point[*] @ 110 at g	191	198	211	202

The different types of processes have been compared in Table 1, which summarizes some of the most important process parameters that would result from producing 10,000 MTPD methanol by the investigated technologies. The same typical natural gas composition, carbon efficiency[+] and reforming pressure have been used for all the calculations. Reforming temperatures and S/C ratio have on the other hand been set to typical values for each of the different processes.

The data in Table 1 clearly support the conclusion in Ref. 1 that the ATR-based processes are favourable for very large plants. In such plants, the sheer flow rates of syngas through the reforming section and MU compressor will be dominating the overall economic function due to the large costs of these items. It may be argued that the high oxygen consumption is an inherent disadvantage of the ATR-based process, but on the other hand, the necessity of a traditional steam reformer completely vanishes. As the cost of a steam reformer varies more strongly with capacity than the costs of ATR and air separation units (ASU), the ATR-based process again turns out to be more attractive at high throughputs.

The front-end key figures in Table 1 do not clearly indicate whether syngas adjustment by hydrogen addition or by CO_2-removal is preferable. However,

---

[*] The converted dew point is defined as the equilibrium temperature at which condensation starts, meaning that the temperature at which the equilibrated methanol concentration curve indicated in Fig. 2 breaks.

[+] The carbon efficiency is in this paper defined as the ratio between the amount of methanol produced and the amount of CO and CO_2 entering the methanol synthesis.

the syngas key figures are definitely in favour of module adjustment by CO_2-removal if the methanol synthesis is of the CMSC type - partly because of the higher CO/CO_2 ratio, which reduces cost of catalyst, and partly because the process gas condenses more easily as the methanol production takes place. The latter property has in the table been quantified by means of the converted gas dew point.

It has therefore been concluded that a front-end design comprising ATR with CO_2-removal is optimal for the type of methanol synthesis investigated in this paper. The selected front-end process lay-out has been indicated in Fig. 1.

2. COMBINED METHANOL SYNTHESIS & CONDENSATION

Having identified the most appropriate syngas generation technology, we will now investigate how to achieve the optimal synthesis configuration.

Most conventional methanol synthesis units comprise a recycle loop. The explanation for this is that in order to maintain the reaction rate high, the methanol reactor is normally operated at a fairly high temperature ($>230°C$).

As can be seen in Fig. 2, this has a negative impact on the equilibrium concentration of methanol, meaning that good carbon efficiency can only be attained by recycle of unconverted gas.

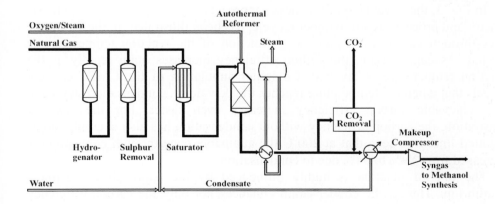

Fig. 1. Synthesis Gas Production by ATR and CO_2-removal

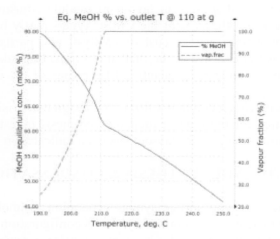

Fig. 2. Equilibrium content of methanol outlet methanol reactor

It has been demonstrated experimentally by Hansen et al, Ref. 2, that it is possible to overcome the thermodynamic limitation imposed by gas-phase equilibrium. It was further experimentally demonstrated that operation of the methanol reactor in two-phase conditions actually makes it possible to achieve a once-through carbon efficiency above 92%, thus opening the door for considering process lay-out without recycle of unconverted gas.

In Fig. 3, the advantage of this novel approach, e.g. using a CMSC synthesis reactor and process lay-out without recycle, has been illustrated. In the figure the once-through carbon efficiency[*] as a function of the reaction temperature is plotted. As can be seen, the carbon efficiency increases quite rapidly as the reaction temperature is lowered. Computer simulation of the CMSC-concept reveals that when the temperature reaches the dew point, condensation starts and the achievable carbon efficiency increases even faster. For the sake of illustration, the carbon efficiency without condensation has also been plotted as a dotted line. It is seen that at 200°C the attainable once-through conversion is increased from 89% to 94% due to condensation.

The rate of reaction is highly influenced by the approach to chemical equilibrium. Hence, the condensation enhances not only the carbon efficiency but also the rate of reaction.

---

[*] Assuming chemical equilibrium in the methanol reactor and a temperature of 43°C in the reactor effluent separator.

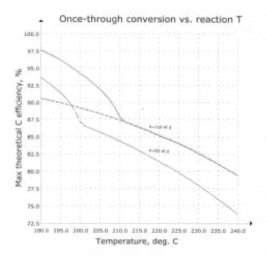

Fig. 3. Once-through conversion of syngas to methanol

None-the-less, fresh syngas is far from the dew point, and the initial conversion is therefore carried out at approximately same reaction temperatures as traditional processes. Upon formation of methanol, the dew point increases and the temperature can be decreased to take advantage of the condensation. In the CMSC process, the catalyst is therefore installed in two reactors operating at different temperature levels.

In order to assess the advantages of the proposed novel technology, we will in the following endeavour to compare it with state-of-the-art methanol technology currently in operation.

Table 2 summarises important process data for two traditional processes and the novel approach. In all the investigated cases, the process conditions have been set to obtain a carbon efficiency of 90%. In the CMSC process, the desired carbon efficiency is obtained by lowering the outlet temperature from the methanol reactor. In the conventional processes, the amount of recycle gas is adjusted to match the specified conversion. The process data have been prepared using the syngas produced by process 3a in Table 1 as feed to all the analysed synthesis configurations, and the reaction temperature has in each case been adjusted to obtain equal amounts of ethanol in the raw methanol.

Table 2
10,000 MTPY MeOH Plant
Key-figures for selection of synthesis configuration

Process description	Loop	Loop	CMSC
Synthesis pressure, at g	85	110	110
Converter flow rate, MTPD	22,200	18,300	11,900
Temperature outlet methanol reactor	255	236	244/201
Water in raw methanol, wt %	1.5	1.3	0.5
Power of MU compressor, MW	47	62	63
Power of recycle compressor, MW	2.2	1.0	n.a.

As shown in Table 2, the CMSC process features smaller syngas flow rates everywhere in the synthesis unit, elimination of all recycle equipment and a much improved product quality. For many applications, the amount of water in the raw product is so small that distillation of the product is rendered superfluous. It is noteworthy; that the mass flow rate through the methanol converter is merely 50% of the flowrate obtained using traditional designs. As the size of the methanol converter is often governing the plant capacity, the maximum achievable single-line plant capacity is doubled by the proposed technology.

3. CONCLUSION

It has been demonstrated that in order to obtain maximum single-line capacity, the proposed CMSC technology is the best choice. It has also been shown that the syngas, which is fed to the CMSC, must possess a high CO/CO_2 ratio to obtain the most economic lay-out. It is therefore very fortunate that the ATR technology, which is the best choice for construction of large-scale syngas production, also is the technology that produces the best syngas for the CMSC process. The water content in the raw methanol that results when combining ATR and CMSC is merely 0.5 wt%. Hence, combining ATR and CMSC may in many cases completely eliminate the requirements for a distillation unit.

REFERENCES

[1] I. Dybkjær, J.N. Gøl, Alternate Use of Natural Gas; Austr. GTL forum 2002
[2] J. Bøgild Hansen, J. Joensen, High Conversion of Synthesis Gas into Oxygenates, pp 457-467, Proceeding of the NGCS, Oslo, August 12-17, 1990.
[3] US Patent 5262 443, Nov. 16, 1993.
[4] I. Dybkjær, T.S. Christensen, Large-scale Conversion of Natural Gas to Liquid Fuels, NGCS, Alaska, 2001.

Studies in Surface Science and Catalysis, volume 147
X. Bao and Y. Xu (Editors)
©2004 Elsevier B.V. All rights reserved.

Preferred synthesis gas production routes for GTL

Per K. Bakkerud[*], Jorgen Nergaard Gol, Kim Aasberg-Petersen and
Ib Dybkjaer

Haldor Topsøe A/S, Nymøllevej 55, DK-2800 Kgs. Lyngby, Denmark
Phone: +45 4527 2000, e-mail: pkb@topsoe.dk

1. INTRODUCTION

Gas-to-liquid (GTL) is the conversion of natural gas to liquid fuels, mainly diesel, but the aspects described in this paper are also valid for methanol, another GTL product. GTL is attracting attention as a promising area for future development in the energy sector. Proven technology is available, and several large projects are in advanced stages of development for diesel production.

The preparation of synthesis gas is the most capital intensive part of a GTL complex. It may account for in the order of 60% of the total (including Air Separation Unit) investment in a GTL complex. Hence, there is a considerable incentive to optimise syngas production technologies for the purpose of cost reduction.

Processes for the production of synthesis gas (syngas) for GTL plants are based on steam reforming, partial oxidation or combinations hereof. The most attractive and economical technology today is oxygen-blown AutoThermal Reforming (ATR), which has been commercialised by Haldor Topsøe A/S at a steam to carbon (S/C) ratio of 0.6. A European plant has been in commercial operation for more than 2 years. Other plants are close to start-up or in advanced stages of engineering and/or construction in Qatar and Nigeria.

This paper focuses on potential, new technologies for synthesis gas production at large GTL plants:
- Next generation of ATR
- ATR in combination with gas-heated reforming
- Catalytic Partial Oxidation (CPO)
- Ceramic Membrane Reforming (CMR)

2. TECHNOLOGY STATUS

A typical process concept for the production of synthesis gas based on ATR is shown in Fig. 1. The key steps in the process scheme are desulphurisation, adiabatic prereforming, ATR and heat recovery.

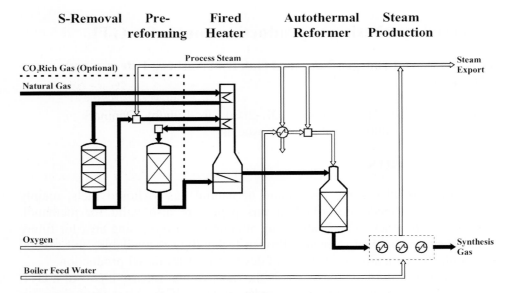

Fig. 1. Typical Process Flow Diagram for Synthesis Gas Production

The adiabatic prereformer converts the higher hydrocarbons by steam reforming into a mixture of methane, steam, hydrogen and carbon oxides.

$$C_nH_m + n\ H_2O \rightarrow nCO + (n + m/2)\ H_2 \tag{1}$$

$$CO + 3\ H_2 \rightleftharpoons H_2O + CH_4 \tag{2}$$

$$CO + H_2O \rightleftharpoons CO_2 + H_2 \tag{3}$$

The removal of higher hydrocarbons allows a higher preheat temperature to the ATR, thus saving oxygen. The prereformer is loaded with a highly active nickel catalyst with a ceramic support such as Mg Al$_2$ O$_4$.

The key part of the synthesis gas unit in a GTL plant is the autothermal reformer. The ATR reactor has a compact design consisting of burner, combustion chamber and catalyst bed placed in a refractory-lined vessel. The prereformed natural gas reacts with oxygen and steam in a substoichiometric mixture. The burner provides proper mixing of the feed streams in a turbulent diffusion flame. All the oxygen is consumed to extinction.

In the catalyst bed, the gas is equilibrated with respect to the methane steam reforming (reverse of 2) and shift reactions (3). The catalyst also ensures that soot precursors (e.g. ethylene and acetylene) known to be formed in the combustion chamber are destroyed. The produced synthesis gas is completely free of soot and oxygen.

The desired syngas composition for FT diesel production is often characterised by an H_2/CO ratio of about 2.0. This cannot be produced by the ATR-based process alone except at very low S/C-ratio and by adjusting the pre-heat temperatures and the ATR exit temperature. Instead CO_2 or other carbon-rich gases may be recycled to adjust the H_2/CO ratio to the desired value.

3. ATR AT LOWER S/C-RATIO

The ATR technology has considerable potential for further optimization, especially by reduction of the S/C-ratio. Lower S/C-ratio improves the syngas composition and reduces the CO_2 recycle. The result is reduced investment per barrel of product and the possibility for higher single-line capacity.

Reducing the S/C-ratio reduces the margin to carbon formation in the prereformer and to soot formation in the ATR. Operation for an extended period of time at S/C < 0.40 has been demonstrated in Topsøe's process demonstration unit in Houston, Texas.

In Table 1, comparative indexed flows for various key parameters are shown to illustrate the impact on single-line capacity by reducing S/C-ratio from 0,6 to 0,4.

Mechanical design optimization of the ATR-reactor and other main equipment has been carried out in parallel with development towards lower S/C-ratio. The optimised design in combination with the reduction of the S/C-ratio will result in a major increase in the single-line capacity by more than 25% within the next few years.

4. COMBINATION OF ATR AND PROCESS GAS-HEATED REFORMING

One of the main cost elements in the front-end of a GTL complex is the Air Separation Unit (ASU). The ASU has a high capital cost and the physical size limits the plant capacity. Reduction of the oxygen consumption will therefore have several effects.

Table 1
Index flow as function of S/C-ratio

S/C	0,6	0,4
NG Feed + Fuel	100	101
O_2	100	98
CO_2 recycle	100	56
H_2 +CO syngas	100	100
Exit ATR	100	93

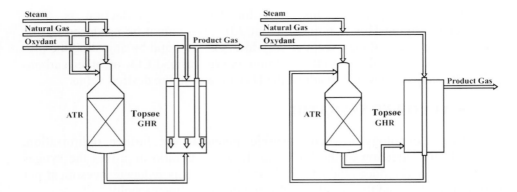

Fig. 2. Gas heated reforming in combination with ATR

High inlet temperatures and low outlet temperatures in the ATR will reduce oxygen requirement. Low outlet temperature will increase the H_2/CO-ratio with the result that a higher recycle of CO_2 is required. Although this isolated may increase the process carbon efficiency, the drawback is a high content of CO_2 and CH_4 in the syngas, both of which are inerts in the low temperature FT-synthesis

By Process Gas Heated Reforming (PGR), part of the heat in the ATR effluent is used for steam reforming and feed preheat in a heat exchange type of reactor. The high level of steam reforming (as compared to adiabatic prereforming alone) and the lower oxygen consumption increases by itself the H_2/CO-ratio. Similar as above, more CO_2 must be recycled to obtain the target H_2/CO-ratio. This increases the overall carbon efficiency of the plant.

There are two principally different layouts for incorporating gas heated reforming in combination with ATR; a parallel arrangement and a series arrangement, as is shown in Fig. 2.

In the parallel arrangement, the two reformers are fed independently, giving freedom to optimise the S/C-ratio individually. However, the PGR must operate at a higher temperature than in the series arrangement to obtain a low CH_4-leakage. In the series arrangement, all gas passes through the PGR unit and then the ATR. This will mean that the steam reforming catalyst may set the lower limit for the S/C-ratio. The main challenge in the series and the parallel arrangement is the risk of metal-dusting corrosion. Mechanical design, including choice of materials, is critical. A PGR designed by Topsøe, operating in a parallel arrangement with ATR at relative low S/C-ratio, has been in operation at Sasol's facilities in Secunda, South Africa since early 2003.

The results from a comparative study presented in Table 2 shows that due to better understanding of the operating parameters and due to new technology, a significant higher single train capacity is to be expected. The cost reduction

Table 2
Results from a comparative study

Case	Current	ATR	ATR with parallel PGR	ATRwith series PGR
S/C-ratio	0.6	0.4	0.4 / 1.1	0.4 / 0.55
Equivalent Capacity, bbl/d index	100	165	165	165
O_2 cons., ton per bbl produced index	100	92	82	81
Overall LHV efficiency index	100	105	108	109
SGU investment per bbl/d index	100	69	81	76
ASU investment per bbl/d index	100	83	76	74
SGU + ASU investment per bbl/d index	100	76	79	75

compared to current state of the art technology is substantial. It is noteworthy that the enhanced ATR technology still is competitive with the new reforming technologies. Comparing the overall LHV efficiencies reveals that the PGR technology is especially attractive if the price of energy is high.

5. CATALYTIC PARTIAL OXIDATION (CPO) AND CERAMIC MEMBRANE REFORMING (CMR)

The process concept in a flow sheet of a GTL-complex based on CPO or ATR is similar. The difference is that in CPO no burner is used and all the chemical reactions take place in a catalyst bed or monolith reactor.

It has been claimed that syngas can be produced in a CPO directly by the following exothermic reaction:

$$CH_4 + \frac{1}{2} O_2 \rightarrow CO + 2H_2 \tag{4}$$

This reaction would result directly in the wanted stoichiometry of the product gas and could be performed at low temperature. However, even though fundamental studies indicate that reaction (4) may take place at high temperature, it will in practice always be accompanied by steam reforming and shift reaction, as well as combustion of the reactant and product gases.

Table 3
Relative natural gas and oxygen consumption for CPO and ATR based front-ends for GTL-plants at S/C = 0.30.

Reactor	Feed temperature, reactor inlet	Oxygen consumption	Natural gas consumption
ATR	650	100	100
CPO	200	118	107

The product gas composition can therefore be predicted from thermodynamics and will, assuming the same operating parameters, not be different from ATR.

It is not possible to preheat the feed to the CPO to the same level as in an ATR due to the presence of a highly flammable mixture upstream the catalyst. The auto ignition temperature of this mixture will typically be about 250°C, depending on gas properties. This is a severe drawback in the potential of CPO for GTL plants. Table 3 illustrates the increased natural gas and oxygen consumption caused by the constraint in inlet temperature. Considering that the ASU-investment is a large fraction of the GTL-complex cost, CPO does not appear economical.

With CMR, the ASU is avoided. Air is introduced at one side of a membrane through which oxygen with 100% selectivity is transported to the other side, where it reacts with hydrocarbons to produce syngas.

The oxygen content on the syngas side is extremely low.

The mechanical integrity and the stability of the membrane are severe challenges in the development of CMR. It may render CMR non-economical, if air must be compressed to process pressure. With ambient air pressure, the technological principle appears promising in small and possibly medium scale, but the feasibility remains to be proven. It is difficult to see CMR competitive for large-scale plants, because the membrane area scales directly with capacity, as opposed to an ASU, which has a better economy of scale.

6. CONCLUSION

This paper demonstrates that the ATR technology holds promise for significant further improvement, both as a stand-alone technology and in combination with PGR in series arrangement. These technologies will be the dominant, at least for the next 5-10 year horizon. Of the new more radical technologies emerging, CMR shows promising results. However, with substantial issues still to be solved, it is not considered as a real competitor to ATR, or combinations of ATR/PGR, within the next 10 years.

REFERENCES

[1] Fundamentals of Gas-to-Liquids, Synthesis Gas Technology, Petroleum Economist, 2003.
[2] Recent developments in autothermal reforming and pre-reforming for synthesis gas production in GTL applications, Fuel Processing Technology (Elsevier), 2003, Kim Aasberg-Petersen et al.
[3] T. Nørgaard, M.F. Jensen, Unpublished results

Studies in Surface Science and Catalysis, volume 147
X. Bao and Y. Xu (Editors)

Statoil`s gas conversion technologies

Ola Olsvik and Rolf Ødegård

Statoil Research and Technology, N-7005 Trondheim, Norway

ABSTRACT

Gas conversion technologies play an important role with respect to bring gas to the market as both fuel and/or petrochemicals. Statoil has during the last 15 years gained wide experience by developing and/or demonstrating different important gas conversion technologies, e.g. methanol, GTL, and MTP.

1. INTRODUCTION

The global reserves of natural gas are estimated at some 6,000 trillion cubic feet (160 trillion cubic meters) of which 3500 trillion cubic feet can be classified as stranded gas (i.e. gas without direct access to a market) [1].

There are 5 main categories for use of natural gas: 1) Re-injection into oil reservoirs to enhance liquid production, 2) Pipe line transport to gas grid marked, 3) Direct use, e.g. for iron ore reduction, fermentation etc., 4) Local production of heat and power, 5) Conversion to other energy carriers or chemical bulk products   (e.g. LNG, ammonia, GTL or methanol).

Norway and Statoil have been at the forefront of development and use of technologies for gas-conversion and gas refining, and to this day, Statoil has developed and / or taken into use a number of gas-handling technologies.

The present paper will focus on Statoil`s technology developments and competence in the field of gas conversion; synthesis gas production, methanol production, gas to liquids (GTL), methanol to propylene (MTP), and cost-effective and environmental natural gas to energy carriers including concepts for CO_2 removal and injection for enhanced oil recovery (EOR).

2. GAS CONVERSION TECHNOLOGIES

2.1 Methanol

In the early 90`s Statoil had to find a non-flaring solution for an associated gas field (Heidrun) outside middle of Norway. Different gas solutions were

Fig. 1. The Statoil Methanol Plant

considered and in 1997 Statoil commissioned and started up a world scale
methanol plant (Tjeldbergodden). The synthesis gas production is based on
combined reforming technology and the methanol production is based on a
boiling water methanol reactor system [2]. With respect to increase the
production and efficiencies new technology elements have during the recent

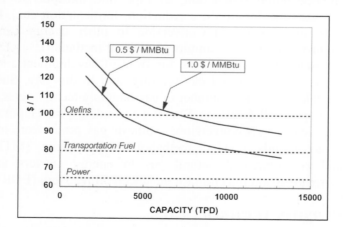

Fig. 2. Delivered methanol cost as a function of capacity for different natural gas prices[*].
Dashed lines show approximate threshold cost for methanol to enter new markets
(Equivalent cost for methanol to transportation fuels and power are based on a crude
price of 16 $/bbl).

[*]Assumptions: Present world scale methanol plant cost (2500 MTD) = 300 Million US $,
other capacities scaled by factor 0.7, fixed annual operating cost = 5% of total investment cost,
transport cost = 15 $/T (assuming large product carriers), and annual capital cost = 17% of
total investment cost (equivalent to ca. 12% IRR)

years been developed and implemented. The plant shows very high on stream factor (from 96 to 98,6 % over the last three years) and productivity and ranks as the most energy efficient (> 70 %) methanol production facility in the world. It should be kept in mind that a high-energy efficient methanol plant is a "stiff system" to operate and therefore it is more challenging to obtain a high on-stream factor than for less energy efficient facilities.

New business opportunities will materialize depending on the methanol production cost. Methanol is shown to be a potential fuel for both fuel cells and gas turbines both with respect to production cost (see Fig. 2) and technical use, and may become a feedstock in the production of olefins. The methanol to propylene route will be discussed in Chapter 2.3.

2.2 Gas To Liquids (GTL)

The need for converting large amounts of gas into liquefied energy carriers and the limited possibility in licensing Fischer-Tropsch (FT) technology Statoil commenced in 1986 the development of own proprietary FT technology for production of diesel and naphtha from gas [3]. This work resulted in advanced Statoil proprietary and patented Fischer-Tropsch slurry reactor and catalyst technology. Over the last 10-15 years the technology is further improved and tested in laboratory and pilot scale reactors. A semi-commercial Fischer-Tropsch demonstration unit is installed inside the PetroSA plant in Mossel Bay (based on Statoil proprietary FT technology). The plant will be started up late in 2003 [4].

Fig.3 shows that GTL is a strongly integrated process. Maximum yield for the total GTL process is achieved by combining the global stoichiometric

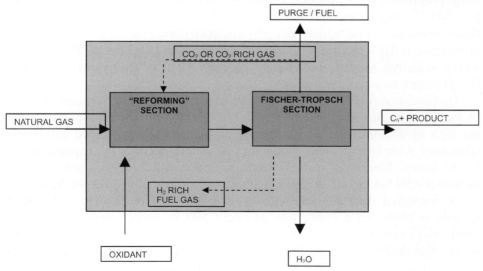

Fig. 3. A schematic sketch of a GTL process

Table 1
Characteristics of different synthesis gas technologies

Technology	T exit (oC)	O_2/C	S.N.	H_2/CO	Min. steam/carbon	CO_2/CO
POX	1250	0.65-0.8	1.6	<2.0	0-0.2	very low
ATR	1000	0.55-0-6	1.8	ca.2.2	0.6	low
CSR	880	0	2.6-3.0	>3.0	1.5-2.5	high
GHR+ATR	1000	ca.0.4	ca.2.0	>2.0	<1.5	medium
CSR+ATR	1000	ca.0.4	ca.2.0	>2.0	1.5	medium

number[†] (S.N.) and local ($H2$/CO) requirements. The selectivity to C_5+ products increases with decreasing H_2/CO-ratios. This makes the integration of the synthesis gas section with the FT- synthesis different than for example in methanol production. Some characteristics of different synthesis gas technologies are shown in Table 1. The H_2/CO ratio from the reforming section is strongly influenced by technolgy and operational conditions.

Previous work has shown that Autothermal Reforming (ATR) is a good choice for natural gas based, large scale, production of synthesis gas for the production of naphtha and diesel by the Fischer-Tropsch Synthesis [3]. New upcoming GTL projects have shown that the specific investments may be the order of 25.000 $/bbl/d [1].

2.3 Methanol To Propylene (MTP®)

Lurgi, Borealis and Statoil co-operate with the aim of demonstrating the Lurgi proprietary MTP® process converting Methanol To Propylene from natural gas. MTP® offers potential de-coupling of conventional steam cracker feed fluctuations, de-coupling from the ethylene/propylene cracker yields, and a way to increase today's relative limited methanol market [5]. The MTP® process is highly selective towards propylene and high yield of propylene is obtained (Lurgi proprietary zeolite catalyst).

Qualification (demonstration) and optimisation of the MTP process at Statoil Methanol Plant Site started early 2002. The results from the pilot plant confirm the high methanol conversion rates and propylene yields obtained earlier in a laboratory scale reactor. The obtained cycle length exceeds the expectations of 5-600 hours, based on results from lab scale experiments (conversion of methanol > 94 %). So far, an estimated propylene yield of 68% is obtained.

A simplified flow diagram of a MTP® plant is given in Fig. 4. The mass balance is based on a production of about 500 ktons per year of propylene. Crude methanol (1.667 million tons per year) is partly converted to DME and water (dimethylether) in an adiabatically operated pre-reactor. The pre-reactor is

---

[†]S.N. = Stoichiometric number = (H_2 - CO_2) / (CO+CO_2)

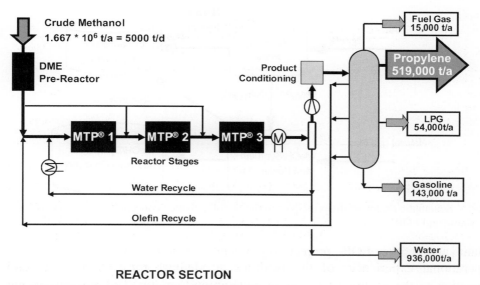

REACTOR SECTION

Fig. 4. Simplified Flow diagram of the MTP®-process.

loaded with a high-activity high selectivity commercial catalyst that gives almost an equilibrium mixture in the outlet of the reactor. Part of the methanol/DME/water mixture is fed into the first MTP®-reactor. Additional steam is added for further dilution of the mixture. More than 99% of the methanol and DME is converted to hydrocarbons, with propylene as the major product.

The remaining part of the methanol/DME mixture from the pre-reactor are sent to the second and third MTP®-reactor, where the mixture is converted in a similar way. Economical considerations show that this gas to propylene route is promising with a low to moderate natural gas price [5].

3. USE OF BYPRODUCTS TO ENHANCE OIL RECOVERY

Process schemes with the aim of integrate different gas refining plants with respect to decrease investment costs, increase energy efficiencies, and decrease CO_2 emissions are developed. Fig.5 shows a Statoil proprietary and patented scheme [6] for production of different gas based products, including heat and power, combined with enhanced oil recovery (EOR) by injection of N_2 (optionally CO_2, steam and heat).

The synthesis gas production step is cost intensive, mainly because of the cost of air separation (when an oxygen based synthesis gas concept is chosen e.g. ATR). Co-production of N_2 and O_2 will marginally increase the total investments costs, and this will improve the economics of the gas conversion

24

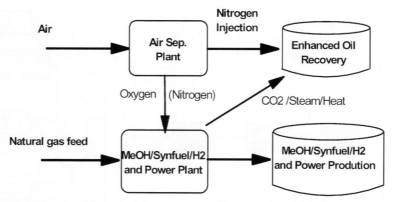

Fig. 5. Schematic sketch of the principles in the Statoil patented concept "Integrated Gas Refining and EOR"

plant and/or the EOR project significantly. Statoil has gained years of operational experiences of the facilities at Sleipner for CO_2 removal and injection of the resulting CO_2 in an underground aquifer. The present concept also includes the possibility to inject the CO_2 together with N_2 and steam/heat.

4. CONCLUSION

Gas handling solutions, both LNG and Chemical Gas Conversion, have now become a necessity for oil and gas companies. Gas conversion technologies play an important role with respect to bring gas to the market as both fuel and/or petrochemicals. Statoil has during the last 15 years gained wide experience by developing and/or demonstrating different important gas conversion technologies, e.g. methanol, GTL, and MTP.

REFERENCES

[1] R. Ødegård, and O. Olsvik, "Statoils Gas Management Technologies", presented at 12th International Oil, Gas and Petrochemical Congress in Teheran (Iran) 24 Feb. (2003).
[2] O. Olsvik, and R. Hansen:"The Statoil Mid Norway Methanol Plant. Operational Experience and Future Plans", Oil and Gas Journal, pp 46-52, 7 th Feb. (2000).
[3] O. Olsvik, and D. Schanke: "Statoil-PetroSA-RIPI GTL Technology", presented at the GTL workshop at 12th International Congress Oil, Gas and Petrochemical, Tehran (Iran) 22 Feb. (2003).
[4] D. Schanke, E. Rytter and F.O. Jaer, "Scale-up of Statoil's Fischer-Tropsch process" will be presented during the 7thNGCS, Dalian, China, June 2004.
[5] M. Rothaemel, J. Wagner, S.F. Jensen, and O. Olsvik: "Demonstrating The New Methanol to Propylene (MTP) Process", Presented at AIChE Spring National Annual Meeting, New Orleans (USA), March 30 (2003).
[6] O. Olsvik, E. Rytter, J. Sogge, R. Kvale, S. Haugen, and J. Grøntvedt, patent WO 03/0118958.

Studies in Surface Science and Catalysis, volume 147
X. Bao and Y. Xu (Editors)
©2004 Elsevier B.V. All rights reserved.   25

The Shell Middle Distillate Synthesis process: technology, products and perspective

A. Hoek[a] and L.B.J.M. Kersten[b]

[a]Shell Global Solutions International BV, Shell Research and Technology Centre Amsterdam, PO Box 38000, 1030 BN, Amsterdam, The Netherlands

[b]Shell International Gas Ltd, Shell Centre, York Road, London SE1 7NA, UK

1. INTRODUCTION

The Shell Middle Distillate Synthesis Process (SMDS), Shell's proprietary GTL process, is the result of a sustained R&D effort since the early 1970's. The Bintulu (Sarawak, Malaysia) plant is the first commercial SMDS facility. Since its startup in 1993 a wealth of operational and technological know-how was generated which, in combination with ongoing R&D, has led to continuous improvement. A 2nd generation commercial project will benefit from state-of-the-art technology as well as from the economy of scale.

2. TECHNOLOGY

Our technology is well known and indicated in the scheme below. The feed streams are treated natural gas (NG) and oxygen generated in an air separation unit (ASU). Synthesis gas is generated in a NG version of the Shell Gasification Process (SGP) which is a non-catalytic partial oxidation process operated with a flame. A smaller NG stream is converted in a Steam Methane Reformer (SMR)

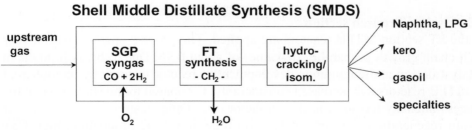

Fig. 1. SMDS process, schematic

Fig. 2. The Bintulu (Malaysia) SMDS GTL plant.

to generate hydrogen for downstream processing and to adapt the H2/CO ratio of the synthesis gas to the desired value. The synthesis gas is converted into paraffins in a proprietary Fischer-Tropsch (FT) process involving multi-tubular fixed bed (MTFB) reactors. The FT product is hydrocracked / hydro-isomerised in a dedicated process yielding naphtha, middle distillates and waxy raffinate (WR). Upon dewaxing the WR yields high quality lube base oils. Alternatively the FT product or fractions thereof can be hydrogenated to yield fully paraffinic specialties such as refined waxes, solvents and detergent feedstocks.

2.1. Synthesis gas production

The SGP process, due to its simplicity, enabled easy upscaling: unit capacity in current designs has increased considerably relative to the Bintulu plant. Other features are: low CO2 production, low methane slip, long burner lifetime, high availability and world-wide experience.

2.2. Fischer-Tropsch section

In the FT section MTFB reactors are applied. The process operates with a very high chain growth probability (α-value) of about 0.96. Due to breakthroughs in catalyst design and continuous development in reactor technology the potential of MTFB reactors has increased considerably. Compared to the Bintulu plant the process intensity has increased by a factor of 2. Other features are: once a year in-situ regeneration, 5 years catalyst life time, low CO2 production, high C5+ yield, deep syngas conversion.

Table 1
Long term FT catalyst performance potential

Parameter	MTFB	Slurry
C5+ (liquid) selectivity, %w/C1+	90-92	88-90
CO2 selectivity, %mol of CO converted	1	2-3
C5+ efficiency, %w	89-91	85-88
Catalyst consumption (19,000 bpd), m3/y	50	100-250

Table 2
Long term FT reactor design potential

Design Parameter	MTFB	Slurry
Reactor capacity, bpd	19,000	19,000
ID pressure shell, m	8	8-10
Total reactor height, m	21	40
Total reactor weight, tonnes	1400-1700	1800-2000
Cooling tubes internal arrangement	shell&tube HX	immersed tubes
Cooling system	steam drum boiling water	steam drum boiling water
Chemical heat production, MW	~300-350	~300-350

An in-house study based on current capabilities and existing R&D leads has indicated that MTFB technology equal potential when compared to slurry technology. It is concluded that MTFB technology will have higher C5+ (liquid) selectivity, lower CO2 production and lower catalyst consumption than slurry technology. This is also reflected in tables 1 and 2.

2.3. Downstream processing

In the downstream processing both the process conditions and the specially designed catalysts are optimised for the FT product being the feedstock and the required yields and properties of the products. The hydrocracking / hydro-isomerisation stage operates at low pressure and generates middle distillates with very good cold flow properties as well as WR with low wax content. Other features are: very low gas make, long catalyst lifetime, product flexibility.

3. PRODUCTS

In this section the focus will be on the products from the hydrocracking / hydro-isomerisation step and automotive gasoil more in particular. The naphtha consists mainly of isoparaffins but has a relatively low octane number. Rather than usage in the gasoline pool application as ethylene cracker feedstock seems (and is in practice) a better option: SMDS naphtha gives considerably higher ethylene and

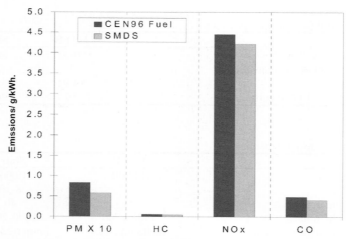

Fig. 3. Passenger car emissions for SMDS and crude oil derived diesel fuel.

propylene yield than the usual crude oil derived naphtha ethylene cracker feedstock. The properties of SMDS GTL ultra-clean transport fuel (diesel) make it an excellent product. The absence of sulphur and nitrogen compounds as well as of aromatics has a strong positive effect on the emissions as illustrated in Fig. 3. The emissions of particulate matter (PM), hydrocarbons (HC), NOx and CO are all lower for the SMDS GTL product. See Ref. 1. for more details. The benefit can be larger when the engines are especially tuned for the SMDS GTL product.

4. PERSPECTIVE

The perspective for a 2nd generation SMDS GTL plant can be viewed from different angles: project, product and relevance for society.

4.1. Projects

Several potential locations for a 2nd generation SMDS GTL plant are being considered with increasing focus on the Middle East. The expected capacity is in the order of 65,000-75,000 bpd or multiples thereof. As far as confidentiality allows the then actual project status will be reviewed during NGCS7.

4.2. Products

Until recently SMDS GTL ultra-clean transport fuel was applied as blending component allowing refiners to meet the tightening specifications. Only just the benefits of the product are becoming familiar to a wider audience. Examples are:

Fig. 4. German Prime Minister Schröder fueling up VW Golf with SMDS GTL ultra-clean fuel.

- In Thailand, Shell Pura Diesel (top grade) contains SMDS GTL material by guarantee.
- Several trials with neat SMDS GTL ultra-clean transport fuel are ongoing:
  o Berlin, Germany: 25 Volkswagen Golf cars (Fig. 4),
  o London, UK: London Bus #507 Victoria-Waterloo.

These developments can be seen as the start of a trend towards ultra-clean transport fuels which can first help to reduce emissions from internal combustion engines; in the next phase they will allow application of fuel cell based technology in cars. In this way GTL ultra-clean transport fuels can act as a bridge towards the foreseen hydrogen based economy.

4.3. Relevance for society

Apart from reduced emissions at the end user location and the economic benefit for the owner of the natural gas the efficiency or CO_2 emission of the total well to wheel cycle can be considered: in an independent 3rd party study the total well to wheel CO_2 emissions for the generation of SMDS GTL ultra-clean transport fuel and for that of crude oil derived diesel fuel was compared. The conclusion is that the SMDS GTL route leads to lower CO_2 emissions.

SMDS GTL can also contribute to a future sustainable energy cycle which does not generate CO_2 f.i. by using biomass which can be gasified to synthesis gas (Biomass to Liquid, BTL), see Ref. 2,3.

Shell continues to improve the efficiency of the SMDS technology: the current overall carbon efficiency of the process is in the order of 80%; the long term target is to increase this value by 5-10 percent points. This is to be achieved by technological advancement in all involved process steps.

REFERENCES

[1] R.H. Crark, I.G. Virrels, C. Maillard, M. Schmidt in Proceedings 3rd Int. Colloquium "Fuels", Tech. Akad. Esslingen, Ostfildern, Germany, Jan. 17-18, 2001

[2] SDE Project "The BIG-Fit concept", project no P1998-934, O-3450/2002-10164/KUG

[3] J. Wind, "Options for future transport fuels", 2nd Int. Workshop on Oil Depletion, Paris, France, May 26-27, 2003

Studies in Surface Science and Catalysis, volume 147
X. Bao and Y. Xu (Editors)

Large-scale gas conversion through oxygenates: beyond GTL-FT

T.H. Fleisch[*] **and R.A. Sills**

BP E&P Technology, P.O. Box 3092, Houston, TX, 77253, U.S.A.

ABSTRACT

The monetization of the world's large, remote gas reserves is the main driver for the advancement of GTL technologies where GTL has become synonymous for Fischer Tropsch (FT) technologies. Looking to other gas derived liquids beyond GTL-FT, the conversion of natural gas into oxygenates, the methanol anchored product family, is a chemical business today which is poised to transition into a high volume fuel and chemical business. At the 6th Natural Gas Conversion Symposium in 2001, the promise of oxygenates centering around methanol and DME for power generation, domestic home cooking/heating, transportation, as a hydrogen carrier and as an economical chemical intermediate, was presented [1]. The focus of this paper is the significant progress since 2001 in the global efforts to commercialize gas-to-liquids technologies for producing oxygenates primarily methanol and DME. Another oxygenate, DMM (dimethoxymethane) and its homologs, poly-oxomethylenes, have potential as gasoline and diesel fuel additive, respectively.

1. ADVANCEMENTS IN OXYGENATE TECHNOLOGIES

1.1. Methanol

Methanol is an important chemical business today with a global capacity of about 35 MMtpa. Global demand is expected to remain constant until 2007 because of phase-outs of MTBE, predominantly in North America. Despite this flat

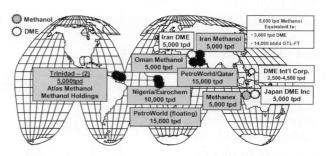

Fig. 1. MeOH/DME Plants under Construction or Consideration

demand, numerous plants are being built or are being evaluated. Most importantly, a step change has occurred away from the traditional 2500tpd world-scale plants to much larger plant sizes.

Specifically, a "mega-methanol" plant, called Atlas, twice the size of previous world scale plants at 5000tpd or about 1.7 MMtpy capacity is being built in Trinidad by BP/Methanex using Lurgi technology. This plant has set a new benchmark with numerous followers advocating even larger plants. The plant is scheduled for start-up in the first quarter of 2004.   Another doubling of capacity to more than 10,000tpd (4MMtpy) is expected within 5 years. Fig. 1 shows some of the proposed plants. It is quite apparent that all of these new large plants are built at locations with low-cost gas allowing lower production costs.

This trend of rapidly increasing plant sizes with access to low cost feedstock leads to rationalization of higher cost producers but also enables and requires the development of new, large markets. Fig. 2 shows schematically the relationship between the delivered fuel price of both methanol and DME in $/MMBTU as function of the better known delivered price in $/ton.   It takes 1.4 tons of methanol to produce 1 ton of DME in the well known dehydration reaction explaining the price correlations between MeOH and DME in Fig. 1. In the past few decades up to today, MeOH fetched an average delivered price of about $150/ton, which corresponds to about $7/MMBTU. With fuel prices typically below $4 to 5/MMBTU, MeOH could not compete in the fuel market. However, economy of scale and low cost feedstocks will drive delivered MeOH and DME costs well below $100/ton and $140/ton, respectively, making them competitive in the fuel market. The PetroWorld projects in Qatar and West Africa (Fig. 1.) with up to 15,000tpd capacity target the power generation market with their "crude methanol". From a market perspective, great strides have been made in the understanding of methanol as a turbine fuel for power generation. Commercial guarantees are now available from major turbine manufacturers for methanol and DME. Both exhibit record efficiencies and very low emissions [2].

Another potentially very large, new market for methanol could be MTO (methanol-to-olefins) technology. Again, the technology originally developed by Mobil (but never commercialized) with

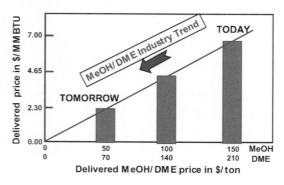

Fig. 2. Transition: Chemical to Fuel/Chemical Business

modified approaches by UOP/NorskHydro [3] and Lurgi's MTP (methanol to propylene) technology [4] basically awaits low cost methanol/DME to be commercially viable. The project led by Eurochem in Nigeria (Fig. 1) aims at the production of polypropylene and polyethylene from MeOH/DME using the UOP/NorskHydro technology at the 10,000tpd plant size. This plant size constitutes about 10% of the world methanol market but provides only enough feedstock for one world scale ethylene/propylene plant. Thus, both power generation and MTO can transition methanol to a large combined fuel and chemical business.

1.2. DME

Methanol has long been considered the centerpiece of a "Gas Refinery" or "Natural Gas Park" with a multitude of derivatives. DME, currently one of the least important derivatives, could emerge as the highest volume product because of it's multi-purpose applications. The potential of DME is best understood when one recognizes that DME is basically a synthetic LPG, although with some interesting twists such as high cetane compression ignition combustibility and easy reformability into hydrogen. Since the work by Amoco and Haldor Topsoe in the early nineties [1, 5, 6, 7], DME has become the center of studies and interest. Over the last three years, a global recognition of the potential of this multi-purpose fuel and chemical feedstock can be seen manifested by numerous events such as the formation of the International DME Association (IDA) in 2001 to promote public awareness and use of DME and the formation of the Japan DME Forum (JDF). The JDF coordinates multiple programs which are part of a large Japanese National DME Initiative, costing over $200 million in 2002-2005, to develop and commercialize DME manufacturing technology, shipping/distribution and marketing for multiple end-use applications such as diesel alternative, synthetic LPG and power generation. Japan is clearly leading the global DME effort at this time. The IDA initiated a website, www.aboutdme.org, to serve as a global DME information center. More recently, in 2002, the Korea DME Forum (KDF) was formed to advance the understanding and use of DME in Korea.

DME and methanol production costs are very similar because the final conversion of methanol into DME is a very high yield, very low cost step. Also, a plant has the option of co-producing both products.

Because of synergies in co-production coming mainly from higher single pass conversions. A DME plant can be actually slighter lower cost than the corresponding MeOH plant. Many technology providers offer closely integrated MeOH/DME plants. The DME Development Co., Ltd is demonstrating a slurry DME process at a large 100ton/d scale [8, 9], which should be operational early 2004, and ready for commercialization by late 2005. DME can be made from many feedstocks, including biomass with exciting efforts in Sweden (bio-DME) and coal in China, using available gasification technologies.

New DME developments in the past three years include:

- LPG Alternative for Domestic Use: A theoretical study conducted by BP and Arthur D. Little [10] showed that blends leading to gas-phase concentrations up to 10% DME in LPG are likely to be acceptable for residential and commercial applications, with no adjustments to combustion equipment. Higher DME blends will likely be acceptable with some adjustments. This study used Weaver Indices, which represent major combustion phenomena. An experimental study by Snamprogetti and ENI [11] concluded that mixtures of 15-20% DME in LPG were feasible without modification to infrastructure.

- Vehicle and Engine Studies. Numerous vehicle/engine research and development and demonstrations activities are being conducted and have been reported at IDA Meeting in 2002 [12] by McCandless and Boehman in the U.S.A.; Landalv and Sorenson in Europe; Goto, Kajitani, Oguma and Sato in Japan; Daeyup Lee in Korea; and Cipolat in South Africa.

- Production Technology Developments and Plant Designs and Economics have been reported by NKK, Lurgi, Haldor Topsoe and Toyo [12]. Developments to use biomass as feedstock in Sweden are progressing [12].

DME can be economic today but further cost reductions are important. Numerous studies have shown that at scales above 5000tpd DME and 7500tpd MeOH commercially attractive projects may be achievable.

1.3. DMM and "poly-DMM"

The completely sootless combustion of DME in diesel engines can be traced back to the absence of carbon-carbon bonds. The use of neat DME leads to excellent engine performance and ultra-low emissions but requires new infrastructure because of the LPG like properties of DME [5]. Unfortunately, although DME is completely soluble with diesel, it cannot be used as an additive to diesel for lowering exhaust emissions because of the high vapor pressure of DME. DME/diesel blends are not stable with DME slowly evaporating at room temperature. We, therefore, explored the viability of higher molecular weight analogs of DME, poly-oxymethylenes, of the general formula CH3—(OCH2)$_x$—OCH3 as such additives. These products can be manufactured from methanol, DME and formaldehyde at temperatures below 100°C in a catalytic distillation reactor over an acidic catalyst [13, 14]. We found that a product mixture with x ranging from 3 to 8, called DMM$_{3-8}$, gave us the right boiling range to blend with diesel without the need for any engine modification (Table 1). This blend also has exceptional high cetane of 76 and a good flash point of

150°F. DMM and DMM$_2$ have too low a boiling point, too low a FP and low cetane and are therefore not suited for blending. Higher analogs with x larger than 8 tend to crystallize at low temperatures and are not suited for blending either.

Property	DMM	DMM2	DMM3-8
BP (°F)	107	221	306-599
Mol.W.	76	106	136-286
Density (g/cc)	0.860	0.973	1.064
O$_2$ (Wt%)	42	45	47-50
FP15% (°F)	0	<75	150
Cetane No	28	41	76
Octane No	77	48	39

Table 1. Properties of DMM, DMM2 and DMM3-8

We have tested blends of DMM$_{3-8}$ with diesel and found dramatic reductions in soot (particulate matter) emissions with some additional reductions of NOx and unburned hydrocarbons. Independent studies at Southwest Research labs also showed that DMM containing diesel blends had lower emissions than ultralow sulfur diesel, FT diesel and biodiesel. A 15% DMM blend brought about a 50% reduction in particulate emissions. More recently, Snamprogetti has confirmed the production and use of "Oxo-diesel" (their terminology for poly-oxomethylene) as a very promising diesel additive [15].

1.4. Hydrogen

Use of DME as a source for producing H$_2$-rich fuel for power generation was studied by Bhattacharyya and others at BP [16, 17]. Catalytic steam reforming of DME to produce hydrogen (CH_3-O-CH_3 + 3H_2O ----> 2CO_2 + 6H_2) was accomplished at temperatures below 350°C in very high yields. Clay supported Ni/Cu/Zn catalysts were used. CO_2 can be separated from H$_2$ and sequestered, allowing for "CO_2 free" H$_2$ production. Production of H$_2$ from methanol is well understood and commercially practiced. Thus, both MeOH and DME hold promise as feedstocks for easy H$_2$ production ("liquid H$_2$").

2. GLOBAL COORDINATED ACTIVITIES

Creation of a new global business requires global collaboration. An annual World Methanol Conference serves the today's methanol business community. To promote DME technology development for both production and utilization of oxygenates a number of global organizations have been established over the last 3 years. As already mentioned, the IDA, JDF and KDF have been formed.

The single largest technical gas conversion effort has been initiated by BP through collaboration with the Universities of California at Berkeley and Caltech ("Methane Conversion Cooperative, MC2"), and Dalian and Tsinghua ("Clean Energy Facing the Future, CEFTF"). The 10 year, $30MM programs involve nearly 100 people focusing on advancing gas conversion technologies, especially in the oxygenate field with R&D in syngas, methanol, DME, DMM, etc.

3. CONCLUSIONS

Gas conversion is rapidly becoming a major gas utilization tool offering a third alternative to bringing gas to the market place after the well established pipeline and LNG options. Among the large scale, potential gas conversion technologies, GTL-FT is by far the most popular technology simply because it converts gas into conventional (though cleaner), easy to market liquid fuels such as naphtha, diesel and lubestocks. This paper tries to show that oxygenates have great potential as advanced clean fuels and chemical feedstocks of the 21st century. Methanol is the enabling product in a "Natural Gas Park" both as a potential large volume fuel and as feedstock for numerous other fuels and chemicals. DME is the most promising methanol derivative, due to its potential as a clean, multi-purpose fuel and chemical feedstock. Other potentially significant derivatives include DMM and poly-DMM. MeOH and DME as feedstocks for olefins could change the face of the chemical industry. MeOH and DME also have potential as "liquid hydrogen" because of their easy conversion into gaseous H$_2$. Significant global collaboration efforts are providing a foundation for creating a new global oxygenate business.

REFERENCES

[1] T.H. Fleisch, et al, Stud. Surf. Sci. Catal., 136 (2001) 423.
[2] A. Basu and J. Wainwright, Proceedings Petrotech 2001, New Dehli, January 2001.
[3] B.V. Vora et al, Stud. Surf. Sci. Catal., 136 (2001) 537.
[4] J. Haid and U. Koss, Stud. Surf. Sci. Catal., 136 (2001) 399.
[5] T.H. Fleisch, et al, SAE Paper No. 950061, Detroit, Michigan, March 1995.
[6] T.H. Fleisch, et al, Stud. Surf. Sci. Catal., 107 (1995) 117.
[7] T.H. Fleisch, et al, US Patent 5906664 (1999).
[8] Y. Ohno, Energy Total Engineering, Vol 20 (1997) 45.
[9] T. Ogawa et al., Stud. Surf Sci. Catal, NGCS VI (2004), this journal.
[10] D. Griffiths, 4th IDA Meeting, Copenhagen, April, 2002.
[11] M. Marchionna, et.al. , 4th IDA Meeting, Copenhagen, April, 2002.
[12] 4th IDA Meeting, Copenhagen, April 2002 and 5th IDA Meeting, Rome, December 2002.
[13] G. Hagen, and M.J. Spangler, US Patents No. 6160174, 6160186, 6166266 (2000).
[14] D.S. Moulton, and D.W. Naegeli, US Patent No. 5746785 (1998).
[15] M. Marchionna et al., Stud. Surf. Sci. Catal., 136 (2001) 489.
[16] A. Bhattacharyya and A. Basu, U.S. Patent 5, 498, 370, March 12, (1996).
[17] A. Bhattacharyya and A. Basu, U.S. Patent 5, 740667, April 21, (1998).

Studies in Surface Science and Catalysis, volume 147
X. Bao and Y. Xu (Editors)

Large scale production of high value hydrocarbons using Fischer-Tropsch technology

André P. Steynberg[a], Wessel U. Nel[b] and Mieke A. Desmet[b]

[a]Sasol Technology R&D, 1 Klasie Havenga Road, P. O. Box 1, Sasolburg 1947, South Africa

[b]Sasol Technology, Baker Square West, 33 Baker Street, Rosebank, 2196, South Africa

1. INTRODUCTION

Most of the recent interest for future GTL plants has been for the use of supported cobalt catalyst in slurry phase reactors. There are good reasons for this approach. Work is in progress to construct a plant at Ras Laffan, Qatar. This concept uses the Sasol Slurry Phase Distillate™ (Sasol SPD™) process to produce mainly diesel fuel with by-product naphtha.

Although the Sasol SPD™ process is ideally suited for the production of diesel, Sasol believes that it may offer an opportunity to produce chemicals in a synergistic manner. The concept of levering Fischer-Tropsch technology to produce chemicals is not new. In Sasolburg the product from the FT reactors is upgraded to high quality synthetic waxes and specialty paraffin products. In Secunda, in addition to fuel, several chemical products are co-produced finding application in the olefins, surfactants, solvents and polymer markets.

There is a common misconception that future large scale GTL plants will not be able to target higher value hydrocarbon products due to limitations imposed by the size of the markets. For example, Shell stated in 1995 that their future SMDS projects will be based on transportation fuels only [1]. However, for at least three important products namely ethylene, propylene and lubricant base oils (also known as waxy raffinate), the markets are large enough to sustain large scale co-production of these products.

Sasol believes that two approaches may be considered for the large scale production of higher value hydrocarbons via Fischer-Tropsch technology. Firstly, there is the option to modify the product upgrading schemes for the primary products from supported cobalt catalyst to produce olefins and lubricant base oils in addition to diesel fuel. Secondly, Sasol believes that its proprietary iron based Fischer-Tropsch processes can be used to further enhance the production of higher value hydrocarbon products with fuels as significant by-products. In this paper some insight is provided as to how the different Fischer-Tropsch technologies can be used to produce mainstream bulk chemical products.

2. OPTIONS FOR PRODUCTION OF HIGH VALUE HYDROCARBONS

In reviews of the low temperature Fischer-Tropsch (LTFT) technology [2] and high temperature Fischer-Tropsch (HTFT) technology [3] the opportunities to produce high value hydrocarbons in large quantities has already been mentioned. Considering first the Sasol SPD™ process, the following opportunities exist:

- o Convert the naphtha to lower olefins by steam cracking.
- o Convert the olefins and paraffins to alkylates.
- o Recover high value co-monomers (such as C6 and C8) from the product slate
- o Recover linear paraffins for sale as solvents or for conversion to olefins which can be converted into a variety of final products (such as detergent alkylates or detergent alcohols).
- o Produce lubricant oil base stocks from the primary wax product.

The Sasol SPD™ Process makes use of a supported cobalt catalyst in a slurry phase FT reactor. Another approach is to use a precipitated iron catalyst (also in a slurry phase reactor). The same product options are available for the LTFT iron catalyst. In general, Fe based LTFT technology would offer a higher concentration of olefins in the product than Co based LTFT. This approach has the disadvantage that it is more costly to achieve high reactant conversions. Another disadvantage, for natural gas applications, is that carbon dioxide is produced with the LTFT iron catalyst although this feature may be advantageous for low H_2/CO syngas derived from coal.

Another well proven Fischer-Tropsch process that can be considered is the HTFT process. Traditionally this process has been applied for the maximum production of gasoline but more and more use is being made of the highly olefinic products to produce higher value chemicals. The HTFT products contain more branched hydrocarbons than the highly linear products produced by LTFT synthesis. For a new application, the HTFT process can be reconfigured for maximum olefin production and using this approach the process can be simplified to decrease capital cost. Thus higher value products (mainly propylene and ethylene) can be produced using less capital.

2.1 Lower olefins from naphtha

Even for very large GTL plants, there is not sufficient naphtha produced to justify the construction of a world scale naphtha cracker. As a result it is the operators of existing naphtha crackers that will benefit from the high yields to lower olefins using Sasol SPD™ naphtha. Tests have been done to quantify the increased olefin yield [4]. The Sasol SPD™ naphtha used in these tests was a fully hydrogenated product. Its performance was compared with typical yields reported for a light petroleum naphtha.

The ethylene yield attainable with the Sasol SPD™ naphtha is about 8% higher than that of the light petroleum naphtha at relatively low severity. If the comparison were based on the total olefins yield, the SPD™ naphtha would yield about 8% more olefins at the same severity, and up to 12% more when compared with the highest yields observed during the pilot plant tests.

Using industry correlation techniques to estimate the best yields that could be achieved at optimum commercial conditions, it was predicted that for operation aimed at maximum ethylene production, the yield to ethylene would increase by about 10% relative to a light petroleum naphtha.

2.2 Detergent alkylates and paraffin processing

The olefin content and quality of the Sasol SPD process straight run product cut is greater than that obtained with typical paraffin de-hydrogenation processes used to convert paraffins to olefins that are alkylated to linear alkylbenzene (LAB). This makes the production of LAB an obvious candidate for higher value products.

Two options can be considered for the production of LAB. Firstly only the straight run olefins could be alkylated and the remaining stream could be re-combined with the residual hydrocarbon condensate. The second option is to include a process that de-hydrogenates the product paraffins (after the first pass olefin conversion) into olefins for recycle to the alkylation unit. This approach will increase the amount of LAB product and may result in improved economy of scale. The material balances presented in this paper assume that the first processing option is used.

Iron based LTFT product can also be considered for the production of LAB. In this case advantage can be taken from the higher olefin content of the feed from the iron based process.

2.3 Lubricant base oils

It has been reported that a substantial part of the primary products from a GTL process may be converted to lubricant base oils [5]. Although the world lubricant base oil market is not growing rapidly, it is large, estimated at some 40 million tons per annum [1], and a substantial fraction of Fischer-Tropsch base oils will be able to substitute product in the premium market. This is then also an obvious candidate for higher value products.

2.4 Light olefins

For the production of light olefins, the HTFT process offers many advantages. Using HTFT technology (with optimized catalysts) for the production of light olefins the amount of olefin product can be further enhanced by cracking (via fluidized catalytic cracking i.e. FCC) unwanted longer chain molecules to higher value shorter chain molecules. It is believed that an overall mass selectivity of 15-30% towards propylene is viable via such a two step approach. Overall selectivity towards ethylene can be as high as 10-20%. An illustrative mass balance, as shown in Figure 1, shows an overall selectivity of about 19% towards propylene. This assumption is based on current FT catalyst performance and the results from first round cracking tests. Similarly it assumed that 10% of the hydrocarbons can be recovered as ethylene and 13% as various butenes. About 23% will be diesel boiling range material while about 13% occurs in the gasoline range (excluding C4 components that may eventually end up in the gasoline). About 12% of the product is water soluble oxygenated hydrocarbons. The remainder of the product from the converted synthesis gas consists of ethane, LPG and fuel gas (that may be partly recycled to the syngas generator). The current proposal differs from previous publications, for example Ref. 3, where the possibility of using FCC-type technology to enhance lower olefin yields was not considered. It is anticipated that the modified HTFT technology could in future provide the most cost effective route to short chain olefin production. Given that there is a fast growing market for propylene it may be worthwhile to enhance the propylene production using a metathesis process to convert butylene and ethylene to propylene.

2.5 Combined LTFT and HTFT

Besides the higher olefin content already mentioned, the LTFT promoted iron catalyst also has a higher wax yield than cobalt catalysts. As a result of these two differences, the LTFT iron catalyst may offer advantages for the production of various chemicals, i.e. the production of lubricant base oils, and as mentioned earlier, production of olefins and olefin derived products such as detergent alkylates and alcohols. A key disadvantage associated with the LTFT iron catalyst use is that it is expensive to reach high syngas conversions. However, for conversions up to about 50% the capital cost expressed per unit product is similar to that for cobalt catalysts.

An attractive option to consider is a two stage process that uses LTFT iron catalyst for the first stage with HTFT technology as a second stage. This combination will allow for the production of lubricant base oils, olefin derived products such as detergent alkylates and alcohols, light olefins and oxygenated hydrocarbons (such as ethanol and methanol). All these products will be produced with a competitive advantage when compared to conventional processes. The lubricant base oils, detergent alkylates, co-monomers and other

Sasol SPD™ process

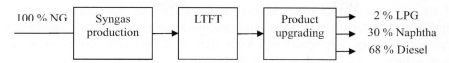

100 % NG → Syngas production → LTFT → Product upgrading →
- 2 % LPG
- 30 % Naphtha
- 68 % Diesel

Total product value: 100

Sasol SPD™ with chemicals co-production

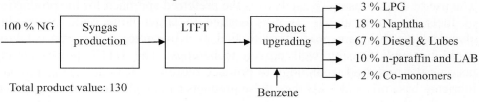

100 % NG → Syngas production → LTFT → Product upgrading →
- 3 % LPG
- 18 % Naphtha
- 67 % Diesel & Lubes
- 10 % n-paraffin and LAB
- 2 % Co-monomers

↑ Benzene

Total product value: 130

Sasol HTFT with light olefins co-production

100 % NG → Syngas production → HTFT → FCC and Product upgrading →
- 13 % Gasoline
- 23 % Diesel
- 10 % Ethylene
- 19 % Propylene
- 13 % Butylene
- 12 % Oxygenates
- 3 % Ethane
- 4 % LPG
- 4% Other

Total product value: 150

Sasol integrated LTFT and HTFT with chemicals co-production

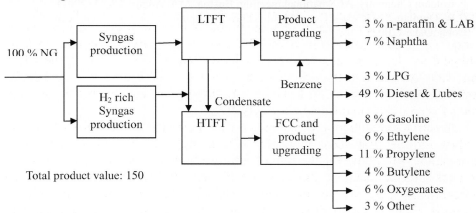

100 % NG → Syngas production / H₂ rich Syngas production → LTFT / HTFT → Product upgrading / FCC and product upgrading →
- 3 % n-paraffin & LAB
- 7 % Naphtha
- 3 % LPG
- 49 % Diesel & Lubes
- 8 % Gasoline
- 6 % Ethylene
- 11 % Propylene
- 4 % Butylene
- 6 % Oxygenates
- 3 % Other

Condensate

↑ Benzene

Total product value: 150

Fig. 1. Possible mass balances for Fischer-Tropsch process options

olefin derived products from the first stage (that is the iron based LTFT plant) are the high value products. The second stage provides a lower capital cost method to consume the residual reactants; consumes carbon dioxide that is generated in the upstream processes; and provides a way to produce additional commodity chemicals. The result is a high yield of total hydrocarbon products per unit of syngas and a high average product price.

3. CONCLUSION

The use of cobalt based FT catalysts is the preferred approach for the production of fuels from natural gas. If it is desired to target large scale commodity chemicals as the main products with fuel as a by-product then iron catalysts could be applied using well proven technology. The HTFT process might become the preferred technology to produce olefins. If it is desired to produce lubricant base oils and LAB then these products can be produced as co-products from the Sasol SPD™ process or by using LTFT iron catalyst in plants integrated with other processes.

REFERENCES

[1] P.J.A. Tijm, J.M. Marriott, H. Hasenack, M.M.G. Senden and Th. van Herwijnen, "The Markets for Shell Middle Distillate Synthesis Products", paper presented at Alternate Energy '95, Vancouver, Canada, May 2-4 (1995) 228.
[2] R.L. Espinoza, A.P. Steynberg, B. Jager and A.C. Vosloo, Applied Catalysis A: General 186 (1999) 13.
[3] A.P. Steynberg, R.L. Espinoza, B. Jager and A.C. Vosloo, Applied Catalysis A: General 186 (1999) 41.
[4] L.P. Dancuart, J.F. Mayer, M.T. Tallman and J. Adams, "Performance of the Sasol SPD Naphtha as Steam Cracking Feedstock", General Papers symposium, Division of Petroleum Chemistry, ACS national meeting, Boston, MA, August 18-22 (2002).
[5] C.W. Quinlan, "AGC-21 – ExxonMobil's Gas-To-Liquids Technology", Gastech 2002, Doha, Qatar (2002).

Studies in Surface Science and Catalysis, volume 147
X. Bao and Y. Xu (Editors)

Scale-up of Statoil's Fischer-Tropsch process

Dag Schanke[a], Erling Rytter[a] and Fred O. Jaer[b]

[a]Statoil Research Centre, Arkitekt Ebbells vei 10, N-7005 Trondheim, Norway

[b]Statoil International E&P, N-4035 Stavanger, Norway

1. BACKGROUND

Statoil has been involved in Fischer-Tropsch (F-T) based GTL technology development since the mid 1980s [1]. Significant untapped natural gas resources on the Norwegian Continental Shelf initiated a search for alternative ways of utilising natural gas. Statoil's current interest in GTL is focused on international gas utilisation where large reserves of natural gas are often found too far from the markets to be reached by pipeline.

The present paper will describe the major elements of Statoil's Fischer-Tropsch technology and the current efforts within scale-up and commercialisation.

2. BASIC CATALYST AND REACTOR TECHNOLOGY

In order to maximise distillate production from subsequent hydrocracking of the F-T wax, a low temperature, cobalt catalyst based technology has been selected. It is generally accepted that a slurry bubble column reactor offers the best performance in terms of economy of scale, throughput and yield [2], but also presents several technical challenges both regarding catalyst properties and critical details of the reactor design [3].

2.1. Catalyst development

In order to achieve the desired reactor productivities, a highly active cobalt catalyst is needed. In addition, the catalyst must be adapted to suit the requirements of the slurry reactor. Statoil has discovered that addition of small quantities of rhenium or certain other metals to alumina-supported cobalt catalysts will increase the activity significantly [4]. The effect of Re is illustrated in Fig. 1, showing that the effect of Re is most pronounced at small loadings and then levels out at higher amounts of Re. TPR-studies have shown that the increased activity resulting from addition of Re is to a large extent caused by an enhancement of the reduction of alumina-supported Co [5]. Some examples of the performance of Co-Re/Al$_2$O$_3$ catalysts are shown in Table 1. Note the improved selectivity in slurry reactors compared to a fixed-bed type of reactor, even at the higher reaction temperature.

Table 1.
Examples of catalytic performance of Co-Re/Al$_2$O$_3$ at 20 bar total pressure. (The results are not directly comparable due to different catalyst preparations, run length etc.)

	T (°C)	H$_2$/CO (feed)	GHSV (h^{-1})	CO conv. (%)	Reaction rate (g$_{HC}$/g$_{cat}$/h)	Selectivity % C$_5$+
Fixed bed	210	2.1	9200	45	0.82	84
Slurry CSTR	220	2.2	3000	70	0.41	89
Slurry Bubble Column Reactor	220	1.6	13000[*]	26	0.59	88

[*] 27% N$_2$ + H$_2$O in feed gas

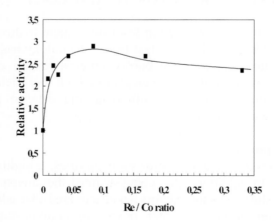

Fig. 1. Effect of Re on F-T activity of Co/Al$_2$O$_3$ [4]

2.2. Reactor concept

The selection of a slurry F-T reactor is based on three main advantages:

- Heat removal capability. Overall heat transfer coefficients for slurry reactors operating in the churn-turbulent flow regime are high enough to give several times lower heat transfer area than in a comparable tubular fixed-bed reactor. This allows higher per pass conversions and significantly lower reactor construction costs [2].

- Ability to use high activity, small particle size catalyst. The use of very active, small particle size catalysts is very important from a diffusion/reaction point of view. For reasonably active catalysts, it has been shown that reactant diffusion limitation in liquid-filled catalyst pores will lead to reduced C_5+ selectivities at particle sizes above ca. 200 microns or even less [6,7].

- Relative simple mechanical construction. The capacity limit for a slurry reactor is mainly set by the maximum economic vessel fabrication size. Reactor diameters above 10 m are technically feasible and the current design capacity is in the region of 15.000-18.000 bpd. Reactor capacities between 20.000 and 30.000 bbl/d should be possible in the future.

The advantages of slurry reactors for F-T synthesis have been known for many years [2,3]. The main challenge for a successful realisation of a slurry F-T reactor (Fig. 2) is to find a solution to the separation of product wax or catalyst from the slurry, whichever approach is followed. This is well illustrated by the operational difficulties encountered in the four F-T campaigns carried out in the U.S. DOE Process Demonstration Unit in LaPorte [8,9]. Statoil has developed a proprietary filtration process where high separation rates can be maintained at a low differential pressure without plugging or clogging of the filters and with negligible amounts of solids in the filtrate [10,11].

A prerequisite for successful operation of the wax separation system is the physical properties of the F-T catalyst. The nature of the slurry environment requires a very attrition and abrasion resistant catalyst, on par with or even beyond what is known from e.g. FCC catalysts. Other requirements include proper design of gas distribution devices and other reactor internals.

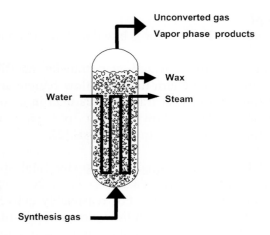

Fig. 2. Slurry reactor concept (schematical)

3. SCALE-UP OF CATALYST AND REACTOR TECHNOLOGY

3.1. Catalyst scale-up

Scaling up production of the supported cobalt F-T catalyst while satisfying the strict requirements for use in a slurry reactor is a non-trivial task. Starting from laboratory recipes for preparation of catalyst at typically <1 kg scale, a stepwise scale-up to a 1 ton/day order of magnitude production has been carried out in cooperation with a commercial catalyst supplier (Johnson Matthey Catalysts). The catalyst will be used in the semi-commercial demonstration plant (Ch. 3.2.).

Several challenges have been encountered underway and important issues include raw materials selection, development and production of a suitable alumina support as well as selection and adaptation of conditions and equipment for adding catalytic ingredients to the support and subsequent treatment steps (drying, calcination, reduction). Simple extrapolation of laboratory-scale preparative methods resulted in catalysts that were inferior to the initial laboratory-scale samples. These differences were resolved by collaborative work with the catalyst supplier (Johnson Matthey Catalysts) - investigating the fundamental mechanism involved in discrete process stages and linking these to specific features known to affect performance and new methods of manufacture.

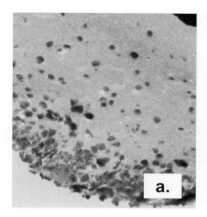

Fig 3. TEM micrographs showing Co distribution on alumina for varying manufacturing conditions. (Same magnification). The catalyst to the right (b) showed superior properties in terms of activity and selectivity.

The performance of the catalyst depends on the distribution of Co on the alumina carrier, as well as the Co particle size and distribution, degree of reduction, and the pore structure of the alumina. The relationship between these parameters is complex, e.g. a high dispersion catalyst is not necessarily a good catalyst in terms of activity, selectivity and stability. Examples of catalysts of varying quality are shown in Fig. 3.

3.2. Semi-commercial Demonstration Plant

Statoil and PetroSA are building a semi-commercial scale Fischer-Tropsch demonstration plant in Mossel Bay (South Africa), scheduled to come on stream by late 2003. The plant will use synthesis gas from the world's largest GTL plant in operation and will produce up to 1000 bbl of Fischer-Tropsch liquids per day. (Fig.4.) Iran's RIPI (Research Institute of the Petroleum Industries) is also participating in the project.

A combination of small pilot reactors, several purpose-built cold-flow columns of different size, simulation models and measurement of physical properties of wax/catalyst slurries have provided the basis for the design of the semi-commercial demonstration plant. The size of the demonstration plant is very advantageous from a reactor scale-up point of view due to several factors:

- Limited further hydrodynamic scale-up effects
- Mechanical design very similar to full scale
- All operational procedures will be the same as in a full scale reactor

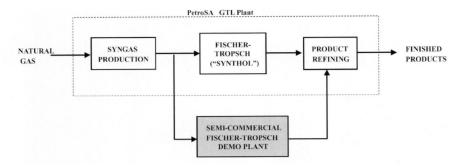

Fig. 4. Simplified block flow diagram showing integration of Statoil-PetroSA semi-commercial Fischer-Tropsch demonstration plant with existing PetroSA GTL plant in Mossel Bay.

In addition to proving the core F-T reactor and catalyst technologies, the plant will also serve to demonstrate and gain experience with all surrounding process and utility systems, such as product cooling/recovery systems, catalyst handling and process control. Locating the demonstration plant in an operating GTL environment also contributes to reducing risks associated with use of real plant synthesis gas and other process interfaces.

REFERENCES

[1] E. Rytter, Å. Solbakken and P.T. Roterud , AIChE Spring National Meeting (1990)
[2] J.M. Fox et al , Final report DOE Project No. DE-AC22-89PC89867 (June 1990)
[3] R.L. Espinoza, A.P. Steynberg, B. Jager and A.C. Vosloo, Appl. Catal. A: General 186 (1999) 13
[4] S. Eri, J.G. Goodwin Jr., G. Marcelin and T. Riis, U.S. Pat. 4 880 763 (1988)
[5] A.M. Hilmen, D. Schanke and A. Holmen, Catal. Lett. 38 (1996) 143
[6] D. Schanke , P. Lian, S. Eri, E. Rytter, B.H. Sannæs and K.J. Kinnari, Paper 73 presented at the 6th Natural Gas Conversion Symposium (2001)
[7] E. Iglesia, S.C. Reyes, R.J. Madon and S.L. Soled, Adv. Catal. 39 (1993) 221
[8] B. Bhatt, Topical Report (Final) DE-AC22-91PC90018 (Sept. 1995)
[9] B. Bhatt, Topical Report (Final) DE-FC22-95PC93052 (June 1999)
[10] E. Rytter, P. Lian, T. Myrstad, P.T. Roterud and Å. Solbakken, Eur. Pat. EP 627 959 (1993)
[11] E. Rytter, P. Lian, T. Myrstad and P.T. Roterud, UK Pat. GB 2 289 856 (1994)

Studies in Surface Science and Catalysis, volume 147
X. Bao and Y. Xu (Editors)

Catalysis: wrapping it in concrete and steel

Michael J. Gradassi

BP America, Inc., 501 Westlake Park Boulevard, Houston, Texas, 77079, USA

ABSTRCT

A simple spreadsheet tool is developed and used to generate an economic rationale to guide research towards a commercially attractive catalyst development.

1. INTRODUCTION

Successful development and application of a commercial catalyst requires a balance of several performance characteristics including its conversion and selectivity for the desired product. At development outset, it may not be clear what catalyst performance level is economically attractive.

Computer-based chemical process simulators can be quite useful to determine the attractiveness of a given catalyst's conversion and selectivity performance. However, these types of sophisticated tools may not be readily available or may be too complex to implement, especially at very early stages of catalyst development and experimentation.

This paper discusses the research guidance that can be obtained using a simple spreadsheet-based process model, together with a cash flow spreadsheet, to determine the economic benefit catalyst conversion and selectivity can create for an example Fischer-Tropsch process. The aim is to provide an experimenter with a perspective of the economic consequences that result from catalyst performance and an economic rationale to guide catalyst development research for a commercially attractive gas conversion process.

2. TECHNICAL APPROACH

It is assumed a Fischer-Tropsch catalyst is to be developed that converts syngas to liquid products. It is further assumed that it has selectivity to both hydrocarbons and CO_2, and that the hydrocarbon distribution follows the Anderson-Schultz-Flory model. Lastly, it is assumed that syngas is generated by known technology.

Table 1
Generic syngas and Fischer-Tropsch catalyst performance information

Syngas	Fischer-Tropsch (FT)
$xCH_4 + yO_2 + zH_2O \longrightarrow$ $aCO + bH_2 + cH_2O + dCO_2 + \Delta H_{Rxn}$	$xCO + yH_2 \longrightarrow aH(-CH_2-)_nH + bH_2O + cCO_2 + H_{Rxn}$ $C_n = (1-\alpha) * \alpha_{(n-1)}$

The spreadsheet tool is intended to help the catalyst researcher address performance issues that include what level of α might be targeted, what level of selectivity to hydrocarbons versus CO_2 might be targeted, and what conversion level per pass might be targeted. The tool will not address promoters, catalyst preparation, or any other laboratory preparation aspect. Information needed to develop the spreadsheet tool includes, stoichiometry, laboratory data and/or literature information, basic engineering, and a cash flow model.

3. ESTABLISHING THE BASE CASE PROCESS

The process Base Case is comprised of three basic steps: syngas generation, Fischer-Tropsch conversion, and product workup, as illustrated in Fig. 1.

In this example, syngas is assumed to be generated by autothermal reforming, according to the stoichiometry listed in Table 1. Other syngas technologies, such as steam methane reforming, could be assumed and likely would provide the same research guidance result.

In the next processing step, generated CO_2 and water are removed before the syngas is fed to the Fischer-Tropsch reactor, wherein a major portion of it is converted. The reactor effluent is cooled, dewatered, and stabilized to produce a finished C_{5+} hydrocarbon product. Subsequent hydrocracking and distillation to finished products is not considered, because it is not necessary for the research guidance task. The C_{4-} hydrocarbons, including methane and un-reacted syngas, are recycled to the syngas generation step. Alternatively, the C_2-C_4, hydrocarbons could have been assumed to be recovered and prepared as separately saleable products, subject to economic incentive to do so.

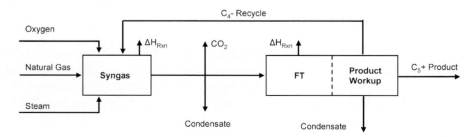

Fig. 1. Fischer-Tropsch process schematic

Table 2
Process model parameter performance assumptions

Parameter	Base Case	Sensitivity Values
α	0.93	0.90; 0.86; 0.80
CO Conversion Per Pass	70%	60%; 80%
CO_2 Selectivity	2%	0%; 4%; 8%

The process model Base Case assumptions listed in Table 2 are based on literature values from Ref. 1 and 2. The spreadsheet process model was built to determine liquid product yield sensitivities to changes in α, CO conversion per pass through the FT reactor, and selectivity of converted CO to hydrocarbons and CO_2. This latter parameter will be shown to be especially important, because CO converted to CO_2 diverts carbon away from revenue-generating product yield.

4. BASE CASE AND SENSITIVITY RESULTS

The spreadsheet process model results for the Base Case are shown in Fig. 2. The natural gas fresh feed rate was set to provide a process output of about 30,000 barrels per day of C_{5+} liquid, which has been shown to be the commercial threshold rate in Ref. 3. Because a sophisticated convergence technique was not used, the Base Case mass balance shows an inherent error of 0.5%, which is not material.

Process yield as a function of the sensitivity parameters listed in Table 2 is shown in Fig. 3. What it shows is that α can have a greater affect on yield than conversion per pass at a given catalyst CO_2 selectivity, e.g., a 0.01 increment in α can have as much, or more, of an effect on yield than a 10% increment in conversion per pass. Even when conversion per pass is low, recycle preserves the un-reacted syngas, making it available for liquid product production via successive passes through the FT reactor. Lastly, an increase in CO_2 selectivity by even 2% can destroy between 1000 and 1500 barrels per day of revenue-generating liquid product. This yield loss effect is equivalent to a 0.05 decrement in α versus the Base Case.

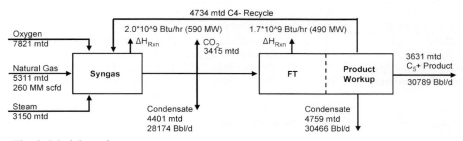

Fig. 2. Model results

52

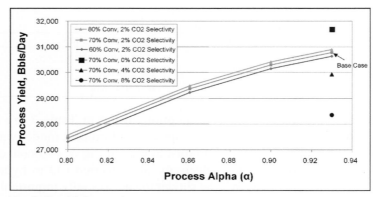

Fig. 3. Sensitivity results

So, on what parameter should the catalyst developer focus? Is it better to work to improve catalyst α, or to reduce catalyst CO_2 selectivity? What is the economic benefit? To address these questions, catalyst performance needs to be coupled to FT process value creation. This can be accomplished by establishing an FT process costing basis, followed by a cash flow analysis.

5. ESTABLISHING A COSTING BASIS

To determine the effect catalyst performance has on creating economic value, a process costing basis was established using FT plant capital cost allocation information from Ref. 3, together with the process model parameters that were chosen to determine their effect on total process plant capital cost, Table 3. Lastly, the resultant cost and material balance information was utilized to determine process economic value creation, as measured by NPV10, for the Base Case and each sensitivity case using methodology previously described in Ref. 4.

Table 3
Costing Basis

Process Element	% Capital	Costing Element	Rationale
Syngas	30	Oxygen Rate	Proportional to reformer duty
FT Synthesis	15	FT Reactor Feed Rate	Proportional to reactor capacity
Product Work-Up	10	C_{5+} Product Rate	Proportional to reactor capacity
Other Process Units	10	C_{5+} Product Rate	Proportional to distillation capacity
Utilities	15	FT ΔH_{Rxn}	Measure of overall heat duty
Offsites	20	(Held Constant)	Minor function of process capacity

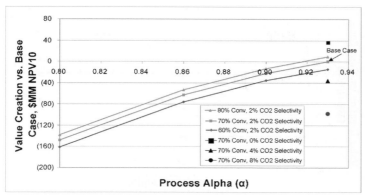

Fig. 4. Process value creation

6. CATALYST PERFORMANCE AND VALUE CREATION

The value creation results for each sensitivity case versus the Base Case are shown in Fig. 4. The overall shape of each curve is similar to that shown earlier for process yield but with the added advantage of showing what monetary value can be either created or destroyed by a developed catalyst system operating at various performance levels.

The results show that a 10% increment in conversion per pass has about the same effect on value creation as a 0.01 or 0.02 increment in process α at constant CO_2 selectivity and is worth about $10MM to $15MM in value creation. This result indicates the importance of each increment of catalyst α. The results also show that even a doubling of CO_2 selectivity from 2% to 4% can destroy as much value as a 0.05 decrement in α from the Base Case, about equivalent to a $40MM value destruction. A further doubling of CO_2 selectivity from 4% to 8% results in an additional $60MM value destruction for a total of $100MM versus the Base Case. Thus, for the example catalyst performance assumed, the catalyst developer would focus on maximizing α, while minimizing catalyst CO_2 selectivity.

7. TARGETING CATALYST PERFORMANCE

The tools illustrated herein can be used to identify catalyst performance targets that meet value creation objectives. To illustrate, assume a catalyst developer has developed a catalyst with an $\alpha = 0.86$ and CO_2 selectivity of about 6%. Locating this performance as shown in Fig. 5, the catalyst would lead to the creation of about $155MM less value than the Base Case. Further assume that the catalyst developer wishes to improve performance to increase value creation by at least $100MM. The illustration shows that this can be achieved by any combination

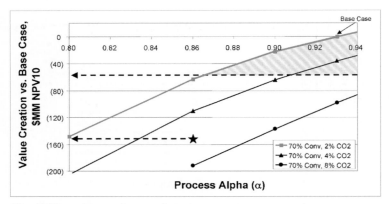

Fig. 5. Targeting catalyst performance

of α and CO_2 selectivity that lie in or above the crosshatched area.

8. SUMMARY

A spreadsheet-based FT process model was established assuming arbitrary FT catalyst performance. The model was designed to determine liquid product yield as a function of FT catalyst performance parameters. Process costing parameters were established and used to estimate process capital cost for combination sets of catalyst performance parameters. For each set, the FT process liquid product yield and corresponding estimated process capital cost were used to determine value creation as measured by NPV10 calculated using a cash flow model. It was shown how this combination of spreadsheet tools could be used to generate an economic rationale to guide research toward the development of a commercially attractive catalyst.

REFERENCES

[1] R. Zennaro, G. Pederzani, S. Morselli, S. Cheng, C. Bartholomew, Stud. Surf. Sci. Catal. 136 (2001) 513
[2] H. Ardus, G. Choi, A. Heath, S. Tam, Stud. Surf. Sci. Catal. 136 (2001) 519
[3] B. Ghaemmaghami, Oil and Gas Journal, Vol. 99, Iss. 11 (2001)
[4] M. Gradassi, Stud. Surf. Sci. Catal. 136 (2001) 429

Studies in Surface Science and Catalysis, volume 147
X. Bao and Y. Xu (Editors)

ITM Syngas ceramic membrane technology for synthesis gas production

Christopher M. Chen[*a], Douglas L. Bennett[a], Michael F. Carolan[a], Edward P. Foster[a], William L. Schinski[b], Dale M. Taylor[c]

[a]Air Products and Chemicals, Inc., 7201 Hamilton Blvd, Allentown, PA, 18195 USA

[b]ChevronTexaco Energy Research and Technology Co, 100 Chevron Way, Richmond, CA, 94802 USA

[c]Ceramatec Inc., 2425 South 900 West, Salt Lake City, UT, 84119 USA

1. INTRODUCTION

The ITM Syngas Team, led by Air Products and including ChevronTexaco, Ceramatec, and other partners, in collaboration with the U.S. Department of Energy, is developing Ion Transport Membrane (ITM) technology for the production of synthesis gas. ITM ceramic membranes are fabricated from non-porous, multi-component metallic oxides and operate at high temperatures with exceptionally high oxygen flux and selectivity. A conceptualization of the ITM Syngas technology is shown in Fig. 1. Oxygen from low-pressure air permeates, as oxygen ions, through the ceramic membrane and is consumed through chemical reactions, thus creating a chemical driving force that pulls oxygen ions across the membrane at high rates. The oxygen reacts with natural gas in a partial oxidation process to produce a hydrogen and carbon monoxide mixture (synthesis gas).

2. PROCESS AND APPLICATIONS

The ITM Syngas process is a breakthrough technology that combines air separation and high-temperature synthesis gas generation processes into a single ceramic membrane reactor, with significant savings in the capital cost of synthesis gas production. Because synthesis gas is a feedstock for a range of different processes, ITM Syngas represents a technology platform that has numerous applications, such as gas-to-liquids; hydrogen; clean fuels, including liquid transportation fuels; and chemicals (Fig. 1).

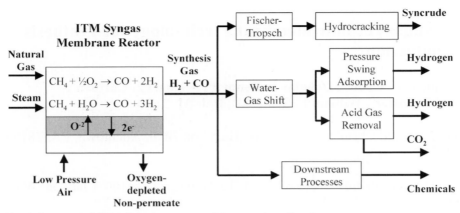

Fig. 1 Conceptual ITM Syngas process with several applications.

Process designs and cost estimates were developed for single-train Fischer-Tropsch Gas-to-Liquids plants producing 6,000 to 18,000 barrels/day. The capital cost of the synthesis gas process was projected to be 30% lower for ITM Syngas compared to conventional technology employing oxygen-blown autothermal reforming.

Applications to co-produce hydrogen and carbon dioxide, such as use of the hydrogen as a decarbonized fuel, were also evaluated. ITM Syngas processes producing 150 to 760 million std.cu.ft./day of hydrogen were evaluated, and resulted in a similar 30% reduction in synthesis gas process capital cost compared to current technology. The ITM Syngas technology was also evaluated for applications using less than 1 million std.cu.ft./day of hydrogen and where a CO_2 co-product was not desired. For this application pressure swing adsorption (PSA) was chosen to separate hydrogen and the cost of hydrogen was projected to be 20 to 40% less than delivered liquid hydrogen, which varies in price with delivery distance.

3. MEMBRANE FABRICATION

Since the initiation of work to develop planar ITM Syngas membranes, the active area of membrane articles has increased approximately 140 fold. The first planar articles were small disks that featured dense layer membrane thicknesses of 100-250 μm.. Next, rectangular "PDU wafers" were fabricated that had over 30 times the active area of the disks. These hermetic double-sided "wafers" had an active ITM membrane on both primary surfaces of the planar body and a common internal air flow system. The PDU wafers included all the critical engineering features found in commercial-scale wafers, and therefore provided an excellent sub-scale surrogate for the development of commercial-size wafer fabrication processes.

Fig. 2. Commercial-scale wafer showing two active regions (darker area) on either side of the central air manifold.

Fig. 3. Photograph of ceramic-sealed 3-wafer built from sub-scale "PDU wafers."

Development work began in 2001 to fabricate commercial-scale wafers. An example of a commercial-scale "SEP wafer" is shown in Fig. 2. A double-sided planar ITM Syngas wafer was designed that will utilize commercially practiced fabrication methods. The design features a symmetric planar structure composed of four separate active regions – two on each primary planar surface and flanking the central air manifold as shown in Fig.2 as the dark regions. Air flows inside the wafer from end to end in a series of long parallel channels built into the substructure of the wafer. Channel dimensions were optimized to minimize flow resistance and maximize membrane support. A porous layer was included on the exterior of the active ITM wafer regions (dark regions) to help support the thin ITM membrane and control reaction rates.

An ITM module can be created by stacking wafers vertically. An example ITM module is shown in Fig. 3, which shows a 3-wafer stack built from sub-scale "PDU wafers." The wafers were joined and sealed to ceramic "spacers" that create an open channel for the flow of natural gas and synthesis gas between wafers (Fig. 3). All ceramic parts were joined and sealed using a ceramic sealant suitable for operating temperatures up to 1100°C.

The basic fabrication methods used to construct ITM Syngas wafers are 1) tape casting, 2) laser cutting, 3) lamination, and 4) co-sintering. Although these processes are each practiced commercially, the sub-surface open channels found in these wafers required the development of new lamination techniques to

Fig. 4. SEM photo of channel cross-sections showing excellent layer-layer bonding.

58

prevent collapse of the ribs separating adjacent channels while assuring adequate contact pressure to prevent de-bonding during co-sintering. Figure 4 provides a photo of the interior region of a wafer, and shows excellent layer-layer bonding.

4. MATERIALS SELECTION

The ITM Syngas process places severe demands on the membrane material. The membrane must simultaneously meet the criteria of being thermodynamically stable in the high-pressure, reducing, natural gas feed and intermediate synthesis gas; being thermodynamically stable in the low-pressure, oxidizing air feed; having sufficient mixed electronic and oxygen ion conductivity to achieve economically attractive oxygen fluxes; and having the requisite mechanical properties (strength, creep, and expansion) to meet lifetime and reliability criteria. Through the careful tailoring of the membrane composition, a patented family of materials [1] developed by the ITM Syngas team meets these criteria.

The economics of the syngas production look most attractive when a high-pressure natural gas feed and a low-pressure air feed are used to produce a high-pressure synthesis gas product to match downstream process pressures, thus avoiding the expenses associated with compressing the air. A large hydrostatic pressure difference will exist across the membrane from the synthesis gas side of the membrane to the air side of the membrane. The requirement of a long service life for the membrane will call for the membrane material to have a sufficiently low creep rate at process temperatures to avoid excessive creep deformation or creep rupture of the membrane.

The family of materials which has been carefully designed to meet this condition are perovskites with the formula $(La_{1-x}Ca_x)_yFeO_{3-\delta}$ where $0<x<0.5$; $1.0<y$ and δ makes the compound charge neutral. The A/B cation ratio, y, has a

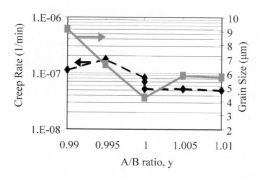

Fig. 5. Creep rate as a function of A/B cation ratio, y

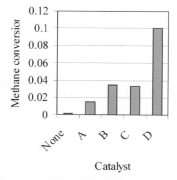

Fig. 6. Methane conversion at 900°C for a series of candidate catalysts.

large effect on the observed creep rates. Dense 50 x 4 x 2 mm bars of $(La_{1-x}Ca_x)_yFeO_{3-\delta}$ were prepared where y=0.98, 0.99, 1.00, 1.01 and 1.02 and x was maintained constant. Fig. 5 shows the 4 point bending creep rate [2] as a function of the A/B ratio, y, along with the grain size [3] of the samples. The creep rate decreases by factor of 2 as the A/B ratio increases from less than 1.0 to > 1.0. This is even more impressive when one considers that the creep rate can increase significantly with decreasing grain size [4]. The samples (A/B ratio of y ≥ 1) with the lowest creep rates, had a much smaller grain size than the samples (y < 1.0) with the fastest creep rates. Therefore, lower creep rates and longer membrane life can be obtained by using A/B ratios ≥ 1.0.

5. CATALYST DEVELOPMENT

The partially reformed ITM Syngas feed gas contains CH_4, H_2O, H_2 and CO. These species undergo oxidation and reforming reactions within the ITM Syngas membrane reactor to produce a mixture of H_2, CO, H_2O, CO_2 and CH_4. The endothermic methane reforming reactions can be promoted with appropriate catalyst and help to balance heat released by the exothermic oxidation reactions. To minimize mass and heat transfer resistances, it is preferred to place catalyst close to the active ITM membrane. Therefore, catalysts are under development that are able to catalyze the reforming reactions within the pores of the ITM membrane porous layer.

The reforming kinetics of candidate catalysts are initially screened in a non-fluxing mode: ITM membranes incorporating a catalyst are exposed to a steam-methane mix only, without an air feed, so only steam methane reforming can occur. Promising catalyst systems identified in the non-fluxing experiment are subsequently tested in an oxygen fluxing mode. Both platforms used disk shaped ITM membranes, 2 cm in diameter, consisting of a 200-500 μm thick porous layer sintered to a 100-250 μm thick dense layer. A 400-500 μm thick slotted layer with about 70% open area was sintered to the dense layer on the side opposite the dense layer to provide mechanical support. Candidate metal catalysts were impregnated into the porous layer using solutions of metal salts or organometallic compounds. The metal solutions were applied to the porous layer, the solvent allowed to evaporate, followed by calcining in air to decompose the salt or compound, leaving a metal or metal oxide.

A series of noble metal catalysts were screened initially in the non-fluxing experiment using a steam to methane ratio of 2.0 and a total flow of 1200 sccm. Fig. 6 shows the fraction of methane converted at 900°C for 4 candidate catalysts as compared to an uncatalyzed membrane. The addition of a catalyst significantly increases the methane conversion. Methane conversion with the highest activity catalyst, catalyst D, is approximately 2 orders of magnitude greater than that observed with an uncatalyzed membrane. Similar trends in catalyst activity were observed when the catalysts were tested in oxygen fluxing mode experiments.

6. CONCLUSIONS

Significant progress has been made in materials selection, membrane design and fabrication, and catalyst development that are necessary steps toward commercializing the ITM Syngas technology. The potential economic benefits of the technology continue to be very promising.

ACKNOWLEDGEMENT AND DISCLAIMER

This paper was written with support of the U.S. Department of Energy (DOE) under Cooperative Agreement DE-FC26-97FT96052. The Government reserves for itself and others acting on its behalf a royalty-free, nonexclusive, irrevocable, worldwide license for Governmental purposes to publish, distribute, translate, duplicate, exhibit and perform this copyrighted paper.

Neither Air Products and Chemicals, Inc. nor any of its contractors or subcontractors nor the DOE, nor any person acting on behalf of either:

1. Makes any warranty or representation, express or implied, with respect to the accuracy, completeness, or usefulness of the information contained in this report, or that the use of any information, apparatus, method, or process disclosed in this report (Technology) may not infringe privately owned rights; or

2. Assumes any liabilities with respect to the use of, or for damages resulting from the use of such Technology.

Reference herein to any specific commercial products, process, or service by trade name, trademark, manufacturer, or otherwise, does not necessarily constitute or imply its endorsement, recommendation, or favoring by the DOE. The views and opinions of authors expressed herein do not necessarily state or reflect those of the DOE.

REFERENCES

[1] P.N. Dyer, M.F. Carolan, D. Butt, R.H.E. VanDoorn, and R.A. Cutler, Mixed Conducting Membranes for Syngas Production, US Patent No. 6492290 B1 (2002).
[2] D. Cramer and D. Richerson, Mechanical Testing Methodology for Ceramic Design and Reliability, Marcel Decker, Inc., New York, 1998.
[3] American Society of Testing and Materials, ASTM C1211-92.
[4] A. Hynes and R. Doremus, Critical Reviews in Solid State and Materials Sciences, 21(2), 129-187 (1996).

Studies in Surface Science and Catalysis, volume 147
X. Bao and Y. Xu (Editors)

Gas conversion – its place in the world

J.J.H.M. Font Freide[*]

BP Exploration & Production Technology, Chertsey Road, Sunbury-on-Thames, Middx TW16 7LN, United Kingdom

ABSTRACT

Gas conversion to liquid transportation fuels has been a major technology effort over the past 20 years. The market is large and fungible but the hurdles to over come are many. Challenges to overcome are process as well as capital hurdles. Furthermore it needs to obtain its place in the options game, where pipelines, liquefied natural gas (LNG), power generation (gas-by-wire) and Compressed Natural Gas all have their role to play. The world scene has changed, with the imminent start-ups of two more world-scale plants, doubling the number of plants and tripling the production.

This paper will present an overview of BP's gas conversion activities and strategy, including a historical comparison with other options of bringing the gas to market, such as pipelines, LNG, methanol. This will show that gas conversion to transportation fuels has a role to play and will identify any further challenges, which may need to be faced.

1 INTRODUCTION

Gas conversion technologies need to obtain their place in the options game, where pipelines, liquefied natural gas (LNG), power generation (gas-by-wire) and compressed natural gas (CNG) all have their role to play. The world gas scene is constantly changing and gas to liquids need to adapt to this. All options for bringing gas to market will need to be considered to locate gas conversion in the the market place. Lessons-drawn from two of the major industries will be applied to gas conversion and its future mapped out.

2 OPTIONS OF BRINGING GAS TO THE MARKET PLACE

2.1 The market place

---

[*] Corresponding Author: **Email:** Fontfrei@bp.com **Phone:** 44 1932 775601 **Fax:** 44 1932 764120

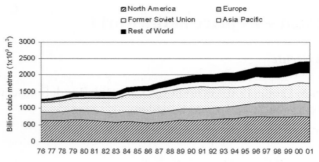

Fig. 1. World growth in gas consumption

Of the world's primary energy mix oil is still the dominant resource (40%; half of which goes into the transportation fuel market). Natural gas and coal each have roughly an equal share (25%), with the balance taken up by nuclear and hydroelectric.

Gas consumption has increased dramatically over the last 25 years; doubling from 1250 billion cubic meter (bcm) in 1972 to over 2500 bcm in 2002 (see Fig. 1). The bulk of the gas supplied to the market place is via pipelines (about 95%) for power generation, industrial or domestic use. Its transportation cost over the same period has been dramatically reduced (from $2 to about $ 0.4 per GJ per km). Another intriguing aspect of the gas market is that about 80% of the resource stays within the boundaries of the country where it has been found. This gives gas a more "local" aspect and it is definitely not a commodity such as the transportation fuels, diesel or gasoline. The remainder of the gas, about 5%, is supplied via liquefaction (LNG).

2.2 Physical conversion: liquefied natural gas

In the late fifties the first cargos of liquefied natural gas were shipped into the UK to test the concept. The first commercial project was Sonatrach's Camel

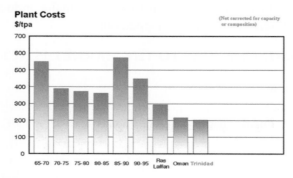

Fig. 2. LNG plant cost since 1965

in Algeria (1964): 0.6 million tonne per annum (mtpa) using 3 trains; followed by Phillip's Kenai plant in Alaska (1969): 1.5 mtpa, 1 train (1 bcm is 0.73 mtpa). Since then many more projects have seen the light and the cost of LNG production has come down due to economies of scale (see Fig. 2).

The LNG industry is currently characterized by a limited number of technology suppliers (Phillips/Conoco-APCI/IFP/Linde), with large-scale projects concentrating in a small number of locations, where it becomes beneficial to carry out expansions. Installed capacity is about 131 mtpa (180 bcm) with market demand following closely. It is a mature industry with an established, but small market (power generation).

2.3 Chemical conversion: methanol

Chemical conversion of natural gas into carbon monoxide and hydrogen (syngas), followed by syngas conversion into oxygenates, such as methanol, is also a well-established industry. This could be considered as an additional route of bringing gas to the market place, in this case in liquid form. Like LNG, over the past 30 years the methanol production capacity has dramatically increased and economies of scale are enabling larger plants to be built (see Fig. 3).

Its industry is characterized by many technology suppliers with numerous, but small projects, dotted all over the world, not necessarily connected to a vast gas reserve. Its end product is currently used mainly as a building block or "raw material" for the chemical industry. Installed capacity converts about 25 mtpa (35 bcm) of natural gas. Like LNG it is a mature industry, with an established market, but even smaller than LNG in the overall world (although large in chemical terms).

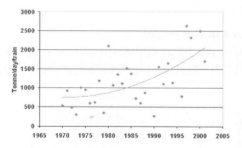

Fig. 3. World scale methanol production

Efficiency, %	Gas To Liquids	Methanol	LNG
Thermal	61	67	85
Carbon	77	83	85

Table 1. Efficiencies of gas to liquids, methanol and LNG

2.4 Technology Efficiencies

Besides product market, market size and plant size, one could also look at the efficiencies involved in the three technologies. From a generic point of view they are quite comparable (see Table 1).

The increase in efficiency, moving from gas to liquids, to methanol, to LNG technology is merely a reflection of the reduction in complexity of the relevant technology.

3 GAS CONVERSION: ITS ROLE AND PLACE IN THE WORLD

3.1 Gas conversion: transportation fuels

For a long time gas conversion into liquid transportation fuels has been seen as the Holy Grail for tackling remote or stranded gas, bringing it to the market place in liquid form and hence helping the world with its energy balance. Hence many patents and articles have been published over the years (see Figure 4), but not many plants have been built.

Its basic process is similar to methanol; only the syngas conversion is now carried out by using a Fischer-Tropsch catalyst to produce wax, which can be upgraded to premium transportation fuels. Currently there are two plants in operation (PetroSA's Mossel Bay and Shell's Bintulu; together about 40,000 barrels per day, bpd) and one is under construction (Sasol/QGPC's 35,000 bpd at Ras Laffan), with the Nigerian plant to follow. There are few, independent technology suppliers and many resource owners, who claim and monopolize their own technologies. In total there are about 55 suggested plans or projects, with a total capacity (if all were executed) of about 2 million bpd or 120 mtpa of gas consumed (166 bcm).

3.2 Gas conversion's place in the world

It has been an expensive toy for resource owners to develop gas conversion to tackle stranded gas and convert it into transportation fuels, serving a vast and

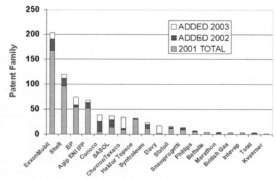

Fig. 4. Gas conversion patents filed of a number of companies

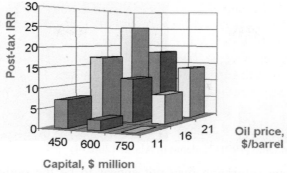

Fig. 5. Economics of a 30,000 bpd gas conversion plant

extensive market. There has been no real technology supply drive by technology suppliers, like in the methanol market, or real market pull, where gas was needed to fill a niche need, as in the case of LNG. The technology has remained costly (see Fig. 5) and is fairly complex.

Market entry is a key and vital aspect of gas to liquids, if it wants to play its role in bringing natural gas to the market place. This is currently restricted since there are only a few independent technology suppliers and the technology are expensive and only tested in a few plants. It is important to open up the market place, so that gas conversion can play its role alongside LNG and methanol to fulfill market needs.

4 BP AND GAS CONVERSION

In the light of the above review, BP's presence in the area is clear: the gas conversion reaction via Fischer-Tropsch offers the possibility of manufacturing clean transportation fuels from a huge and largely untapped resource. As such the manufacture of syngas is by far the most capital-intensive part of the plant and so a competitive technology needs to combine cost effective syngas generation, such as the Compact Reformer, with efficient syngas conversion.

4. 1 Syngas production

Many syngas producing options have been assessed and tested in our laboratories: to name but a few, catalytic partial oxidation, gas heated/convective reforming and partial combustion. To eliminate the use of pure oxygen a breakthrough was developed based on the compact reforming principle: it had a small footprint and was as efficient as any steam methane reforming. This was tested in Warrensville, Ohio, developed further through modeling and finally demonstrated at commercial scale in Nikiski, Alaska (see Fig. 6).

Fig. 6. Pilot plants in Ohio (1991-7), demo in Alaska (2002-4)

Although steam reforming may not appear to be the optimum choice for syngas conversion, its excess hydrogen produced can easily be used to fire the reformer and hence supply the energy needed to carry out the endothermic steam reforming reaction. On the other hand oxygen and combustion based routes may produce a syngas mixture which is short in hydrogen and hence additional hydrogen may need to be sourced. As long as the right amount of hydrogen to carbon monoxide ratio is supplied the Fischer-Tropsch syngas conversion will carry out its synthesis.

4.2 Syngas conversion

Since the early eighties BP has been working on its FT catalyst, either for fixed bed application or for a slurry bubble column. During the process the catalyst has been scaled up from the milligram, laboratory scale, to the tonnage scale needed for commercial production, as in Nikiski.

4.3 BP's approach

It was recognized that the only way to get the technology to the market place is to accept partners, who are operating in that market place. Hence BP has partnered with Davy Process Technology to develop and license the compact reformer and the FT conversion technologies. In addition to that we have teamed up with world-scale catalyst manufacturers, who add their expertise and capabilities for scaling-up and manufacturing the desired catalyst.

Fig. 7. Micro reactors in Sunbury (1985-03), pilots in Hull (1985-93, 2002-4)

Studies in Surface Science and Catalysis, volume 147
X. Bao and Y. Xu (Editors)

Application of FBD processes for $C_3 - C_4$ olefins production from light paraffins

G. R. Kotelnikov[a], S. M. Komarov[a], V. P. Bespalov[a], D. Sanfilippo[b], I. Miracca[b]

[a]OAO NII Yarsintez, prospekt Oktyabrya 88, 150040 Yaroslavl, Russia

[b]Snamprogetti S.p.A., Viale De Gasperi 16, 20097 – San Donato Milanese, Italy

ABSTRACT

New environmentally sound and economically effective process of light olefins production by paraffins dehydrogenation in a fluidized catalyst bed is reported. The mathematic models allowing to develop the optimum reactor design of a large unit capacity for both propylene, isobutylene and isoamylenes productions have been generated. The models adequacy has been fully confirmed while examining a large commercial olefin production units operation.

1. INTRODUCTION

In a joint effort of co-operation Yarsintez and Snamprogetti have developed the FBD (Fluidized Bed Dehydrogenation) technologies to obtain propylene (FBD-3) and isobutylene (FBD-4). The processes have been tested both for individual paraffins propane and isobutane and for $C_3 - C_4$ mixtures of various ratios. The paraffin feedstock is evaporated and heated to reaction temperature in a gas-gas exchanger. Then it flows through the reactor fluid bed countercurrently to the catalyst. In the reactor the feedstock is partially dehydrogenated, leaving the system through the gas-gas heat exchanger. A continuous transfer of catalyst takes place between the reactor and the regenerator. In the regenerator air is fed to burn the coke formed as by-product and some additional fuel, as required to satisfy the overall thermal balance. The effluent is cooled, filtered and sent to the stack. This paper deals with the conceptual design of the FBD reaction system, based on reaction chemistry and reactor hydrodynamics.

2. THE CHEMISTRY

The main reaction may be schematized as:

Paraffin $\Leftrightarrow$ Mono-olefin + H_2

Side reactions, mainly based on thermal cracking and condensation, may also take place, and a complex scheme of reactions leading to different olefinic and aromatic by-products, ultimately resulting in coke is established.

The extent of side reactions increase with temperature rise, resulting in reduced selectivity to the main product. The reactions are highly endothermic (~30 Kcal/mol) and the achievable conversion is limited by thermodynamics. The temperature required an equilibrium conversion of about 50% is higher than 600°C and it is increased by molecular weight reduction (Fig.1).

Catalysts tailored for each hydrocarbon were developed and reaction conditions optimized to maximize the yield for each paraffin.

It is clear from the stoichiometry of the reaction that high pressure adversely affects the reaction: as a consequence, it must be performed in the gas phase at the lowest pressure compatible with the process economics.

High endothermicity results in a high adiabatic temperature decrease. Heat supply to the reaction system is mandatory and characterizes the industrial processes presently commercialized.

In spite of operating at high temperatures, thermal reactions of dehydrogenation are slow. Furthermore, side reactions like isomerization, thermal cracking and coking, which are thermodynamically and kinetically favored over the main one, can occur in the required operating conditions. For this reason in the industrial practice the use of a catalyst is necessary in order to keep a suitable conversion per pass while obtaining high selectivity toward the desired olefin. The unavoidable coke formation on the catalyst face results in progressive reduction of catalytic activity requiring the periodic regeneration of the catalyst.

The FBD processes use a chromia/alumina catalyst [1, 2, 3]. The catalyst is produced in a form suitable to be used in a fluidized bed system.

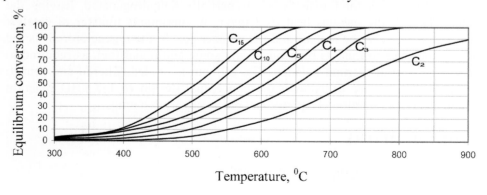

Fig. 1. Hydrocarbons equilibrium conversions.

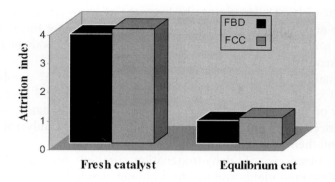

Fig. 2. Comparison of FBD and FCC catalyst properties.

It has a microspheroidal shape and it is very similar to a FCC catalyst as concerns its physical properties (e.g. attrition resistance) (Fig.2).

3. CONCEPTUAL DESIGN

A fluidized bed system is the most effective way to achieve steady process conditions supplying the reaction heat to the catalyst. Fluidized solids can easily be transferred in huge quantity in a closed loop, offering the opportunity to carry out reaction and regeneration in two separate vessels with pneumatic conveying of solids between them: a completely continuous and steady process can be thus obtained. The heat produced in the regenerator raises the temperature of the catalyst, which is immediately transferred back to the reactor. Heat is exchanged between the hot catalyst particles and the surrounding gas in the efficient way allowed by fluidized beds with a high heat transfer coefficient: indeed, the whole surface area of the catalyst forms the heat transfer surface.

The heart of the process is the reactor – regenerator system [4] (Fig. 3).

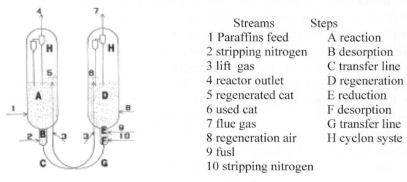

Streams	Steps
1 Paraffins feed	A reaction
2 stripping nitrogen	B desorption
3 lift gas	C transfer line
4 reactor outlet	D regeneration
5 regenerated cat	E reduction
6 used cat	F desorption
7 flue gas	G transfer line
8 regeneration air	H cyclon syste
9 fusl	
10 stripping nitrogen	

Fig. 3. Paraffin dehydrogenation cycle.

The catalyst circulates continuously from the reactor vessel bottom to the regenerator top and vice-versa by means of pneumatic transfer lines, originating in each vessel a countercurrent movement of gas and solid both in the reactor and in the regenerator.

In the regenerator the catalyst restores its initial activity by combusting the low amount of coke deposited on the surface: additional fuel is catalytically burned directly on the catalyst to satisfy the overall thermal balance.

The heat developed in the regenerator is stored by the catalyst itself as the sensible heat and then released to the reaction environment.

Being the reactions of dehydrogenation limited by equilibrium, the achievement of optimal yields and reactor volumes would be favored by a plug-flow movement of the gas phase, while in a fluidized bed the features of an ideal CSTR (Continuous Stirred Tank Reactor) are approached [5, 6].

The ideal dehydrogenation reactor should be a plug-flow unit, isothermal or with increasing gas temperature along the catalyst bed.

Specific nature of FBD dehydrogenation reactors is the significant change of gas flow volume along the catalyst bed.

It is well known that vertical or horizontal baffles with suitable openings can be inserted in a fluid bed to limit the free movement of solids, and therefore the phenomena of back mixing. The catalyst bed is split into a series of stages, each one is comparable to a CSTR.

Using a sufficient number of baffles, the fluidized bed reactor approaches very closely the behavior of a plug-flow reactor from the viewpoint of the gas phase. In this way the bed is no longer isotherm, but an axial increasing temperature profile is established with the highest temperature at the gas outlet, where hot catalyst arrives from the regenerator.

The two aforesaid conditions to optimize the conversion of the paraffins and the selectivity to the olefin have been met and it must be stressed that no other type of the reactor can meet both of them simultaneously.

The flowrate of circulating catalyst is determined by the large amount of heat required by the reaction rather than by catalyst de-activation: therefore the temperature profile created by the insertion of baffles allows a substantial reduction of this flowrate in comparison with isothermal conditions [7].

Another advantage of the FBD reactor is the low pressure drop in comparison with fixed bed reactor and the chance to substitute part of the catalyst while the plant is running. In this way the problem of catalyst aging with variation of yields is eliminated: periodical substitutions are made, also allowing for the fines lost through the cyclones, so that the average age of the catalyst remains constant in time.

The authors have already discussed their design against the existing processes in some publications [6, 8].

4. OPTIMIZATION THROUGH HYDRODYNAMICS

The main theoretical assumption is that in each section between two baffles may be schematized by a bi-phasic approach: a denser phase with most of the catalyst in the lower part and a leaner phase in the upper part. Cold model testing on "mock-ups" allowed obtaining the relationship between catalyst concentration and the catalyst circulation flowrate [8]:

$$\rho' = (\rho - D)/2 + \sqrt{((\rho - D)/2)^2 + D\rho_{sf}} \tag{1}$$

Where $D = \pm G_{circ.} \left(\rho_{sf} - \rho \right) / \left[F\rho_{sf} \left(W - W_{sf} \right) \right]$ (2)

The sign "plus" refers to concurrent and the sign "minus" refers to countercurrent flow of catalyst and gas.

The optimization of paraffins dehydrogenation reactors was performed by experimental selection of a Process Demonstration Unit (PDU), maintaining constant axial catalyst concentration through the reactor (i.e. constant mean concentration in each zone between two baffles), thus achieving a steadier profile of temperature change and hydrocarbon conversion along the reactor height. In this case the design parameters of each of the baffles were set close to flooding, maintaining a reasonable operating range, thus minimizing back mixing phenomena. As a result relatively uniform axial temperature variation (Fig. 4) and catalyst concentration (Fig. 5) have been obtained.

Respectively, the hydrocarbon conversion and selectivity were increased. In this case the selectivity raised by 5 – 8 % (Fig. 6).

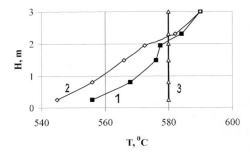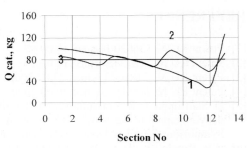

Fig. 4. Temperature distribution along the fluidized bed height
1 - identical design baffles (same open area); 2 - optimized baffles;
3 - no baffles (theoretical)
Fig. 5. Non-uniformity of catalyst distribution through reactor sections:
1 - identical design baffles (same open area); 2 - optimized baffles;
3 - theoretical target

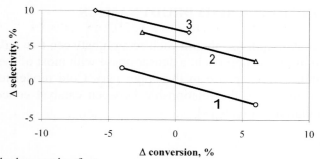

Fig. 6. Dehydrogenation factors:
1 - identical baffles; 2 - optimized baffles; 3 - bench scale isothermal reactor.

This optimization allowed to achieve in the PDU paraffin conversions close to equilibrium with >90% selectivity to the desired olefin.

Nomenclature

ρ - density at steady fluidization, kg/m^3
ρ_{sf}- density at starting point of fluidization, kg/m^3
ρ'- fluidized bed concentration at circulation, kg/m^3
$G_{circ.}$ – catalyst circulation, kg/hr.
F – apparatus free cross section area, m^2
W – gas velocity in open cross section of apparatus, m/sec.
W_{sf} - gas velocity in open cross section of apparatus at the starting point of fluidization.
H – Axial reactor height
Q_{cat} – Catalyst inventory in each section of the reactor
T – Reaction temperature

REFERENCES

[1] Patent RU № 1366200 issued 15.09.87
[2] Patent RU № 2127242 issued 10.03.99
[3] Patent RU № 2188073 issued 27.07.2000
[4] Patent RU № 2156161 issued 20.09.2000
[5] D. Sanfilippo, G.R. Kotelnikov, et al. Paraffins activation through fluidized bed dehydrogenation: the answer to light olefins demand increase. 5[th]Natural Gas Conversion Symposium. Giardini Naxos, Italy, September 1998
[6] G.R. Kotelnikov, D. Sanfilippo, et.al. New propylene production process. 6th Natural Gas Conversion Symposium. Girdwood Alaska, USA. June, 2001.
[7] D. Sanfilippo, G.R. Kotelnikov, et al. Make propylene and MTBE or MTBE-substitutes via FBD fluidized bed dehydrogenation technology. 17th World Petroleum Congress. Rio de Janeiro, Brazil. September 2002.
[8] G.R. Kotelnikov, S.M. Komarov, et al. Propylene production process by propane dehydrogenation in chromia – alumina fluidized bed. Neftekhimiya, 2001, v.41, No 6, p. 458 – 463. Russia.

Studies in Surface Science and Catalysis, volume 147
X. Bao and Y. Xu (Editors)

73

COx-free hydrogen and carbon nanofibers production by decomposition of methane on Fe, Co and Ni metal catalysts

Jiuling Chen, Xingzheng Zhou, Lei Cao, Yongdan Li[*]

Department of Catalysis Science and Technology, School of Chemical Engineering, Tianjin University, Tianjin 300072, China

ABSTRACT

Decomposition of methane was performed on Fe(Co,Ni)/Al_2O_3 catalysts to obtain COx-free hydrogen and carbon nanofibers. The stepwise heating experiment shows that the activity order for Fe/Al_2O_3, Co/Al_2O_3 and Ni/Al_2O_3 is a reverse of that of their melting points and is also a reverse of that of the initial temperatures for the deactivation. Doping of Cu improves the stability and activity of Ni/Al_2O_3 effectively at high temperatures, but the effect is not observable on Co/Al_2O_3. The obtained result is discussed in this paper.

1. INTRODUCTION

Fuel cells are an exciting technology for converting fuel directly into electricity. Hydrogen is regarded to be the most promising fuel. While for proton exchange membrane fuel cells, CO in the fuel is a strong poison when its concentration is above 10 ppm. For alkaline fuel cells, ppm level of CO_2 is a poison [1]. Most of hydrogen from current techniques, such as steam reforming and partial oxidation of methane, yield large amounts of COx (CO, CO_2) as by-products. Removal of them to ppm range is a complex and expensive process and requires bulky equipments [2].

Decomposition of methane has focused recently on the transitional metal catalysts, especially Fe, Co and Ni, because they produce only hydrogen and solid carbon and thereby eliminate the necessity for separation of COx from hydrogen. Owing to the rapid deactivation by carbon deposition, the catalysts have to be regenerated with oxygen or steam frequently [2,3]. In principle, it is difficult to avoid ppm-level COx in hydrogen products [3]. Carbon nanofibers or nanotubes can be produced from the reaction by optimizing the catalyst and reaction condition, and the formation of COx is prevented because of a

---

[*] Corresponding author: Tel.: +86-22-27405613, Email: ydli@tju.edu.cn

prolonged running time[4]. The reaction is endothermic, and a high temperature is favorable to the conversion. However, the deactivation of catalyst becomes fast with the increase of the reaction temperature.

Results in the published literature show that pure metals, Fe, Co and Ni, have a weak catalytic activity, however, the doping of a little promoter, such as Al_2O_3 and SiO_2, improves it effectively [2,4,5]. A catalyst derived from Feitknecht compound (FC) has been designed and applied in our lab. FCs were found to have brucite-like layers containing octahedrally coordinated bivalent and trivalent cations, as well as interlayer anions and water molecules [6]. The cations are supposed to have a uniform distribution. Due to the difficulty of solid phase diffusion of metal ions, the precursor results in a well mixed phase of active component and promoters. A maximum interaction is produced between components after careful calcination and reduction. In principle, metal catalyst particles of several nanometers with high loading content are obtained [6]. In this work, Fe, Co, Ni, Cu and Al are employed as bivalent and trivalent cations, respectively, and the promoters used are Al_2O_3 and Cu.

2. EXPERIMENTAL

2.1. Catalyst samples

Catalyst precursors, well-crystallized FCs, were prepared by coprecipitation from a mixed aqueous solution of nitrates with sodium carbonate. The precipitates were then washed with water, dried in air at 393 K for 5 h and calcined at 723 K for 10 h. Details were given elsewhere [4].

2.2. Methane decomposition

All reactions were performed under atmospheric pressure in a horizontal tubular reactor [4]. 100 mg of catalyst particles of 200-260 meshes was first reduced for 2 h at 973 K with a flow of H_2/N_2 of 1/3 (vol.) and a flow rate of 150 ml/min (STP). Then the temperature in the reactor was adjusted to a prescribed value in nitrogen and the feed was switched to the reaction gas, composed of 37 vol.% methane and 63 vol.% nitrogen. The total flow rate is 68 ml/min (STP). The composition of outlet gas was measured with a GC.

2.2.1. Reactions with stepwise heating

These reactions were first started at 723 K, and after 1 h, the temperature was increased by 10 K/min for 50 K, and the reaction was performed at each temperature for 1 h, respectively. The highest obtainable temperature on this experimental apparatus was 1123 K.

2.2.2. Reactions at constant temperature

These reactions were carried out at 1023 K on different catalyst samples.

When the conversion is lowered to near zero, the reactions were stopped. The reactor was cooled naturally to the ambient temperature and the carbon formed was unloaded from the reactor and characterized.

2.3 Charaterization

The morphology of obtained carbon with the catalyst particles was observed with a JEOL JEM-100 CXII transmission electron microscope (TEM).

3. RESULTS

Fig. 1 and 2 show the conversion of methane with temperature on Fe(Co or Ni)/Al$_2$O$_3$ in stepwise heating mode. A peak can be found for each one of the conversion curves on 2Co-1Al(3Co-1Al) and 2Ni-1Al(3Ni-1Al), ("xA-yB" means that the atomic ratio of A and B is x/y in the catalyst), and the temperature for them is 1073 K and 923 K, respectively. However, for Fe/Al$_2$O$_3$ catalysts, no peak can be seen and the conversion is increased with temperature. In the temperature range lower than the peaks of Ni/Al$_2$O$_3$, the activity of the three catalysts follows the order Ni/Al$_2$O$_3$ > Co/Al$_2$O$_3$ > Fe/Al$_2$O$_3$.

To observe the effect of promotion, Cu is doped to Co/Al$_2$O$_3$ and Ni/Al$_2$O$_3$. It can be seen from Fig. 3 that the peak of the conversion curve on 3Ni-1Al shifts from 923 K to 1023 K after doping of Cu and the conversion of methane at the peak is increased from 50 % to 70 %, however, the conversion is decreased in the low temperature range. For Co/Al$_2$O$_3$, the conversion of methane is decreased in the measured range and the peak of conversion curve for 7Co-3Cu-5Al appears at the maximum temperature measured.

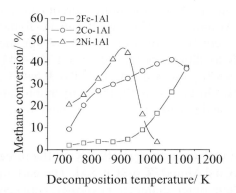

Fig. 1. Conversion of methane on three different catalysts, 2Fe-1Al, 2Co-1Al and 2Ni-1Al with step-wise heating.

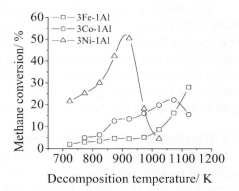

Fig. 2. Conversion of methane on three different catalysts, 3Fe-1Al, 3Co-1Al and 3Ni-1Al with step-wise heating.

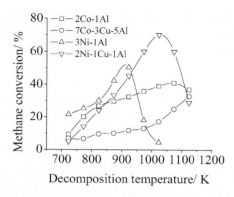

Fig. 3. Conversion of methane on Co/Al$_2$O$_3$ and Ni/Al$_2$O$_3$ catalysts with and without doping of Cu with step-wise heating.

Fig. 4. Conversion of methane on Co/Al$_2$O$_3$ and Ni/Al$_2$O$_3$ catalysts with and without doping of Cu at 1023 K.

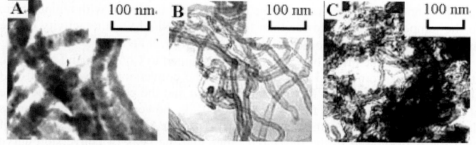

Fig. 5. The morphology of carbon formed at 1023 K on three different catalyst samples, 2Ni-1Cu-1Al, 2Co-1Al and 7Co-3Cu-5Al.

Fig. 4 shows the conversion of methane at 1023 K on four different catalyst samples, 2Co-1Al, 7Co-3Cu-5Al, 3Ni-1Al and 2Ni-1Cu-1Al. It can be found that except 7Co-3Cu-5Al, the initial conversion is all around 70%. After the reaction, the conversion decreases sharply to about 3 % in an hour on 3Ni-1Al, and it decreases to about 30 % and 23 % swiftly on 2Co-1Al and 7Co-3Cu-5Al, respectively, and then slowly to near 7 %. On 2Ni-1Cu-Al, the conversion can almost keep constant, 70%, for 12h, and then decreases to zero in the next 4-5h.

3.2 Morphology of carbon formed

Fig. 5 A-C shows the morphology of carbon formed at 1023 K on the three different catalyst samples, 2Ni-1Cu-Al, 2Co-1Al and 7Co-3Cu-5Al. The carbon formed on 2Ni-1Cu-Al is fibrous and their outer diameter is mostly between 70-100 nm, and the radial size of the catalyst particles falls also in the range [7]. However, on 2Co-1Al, the formed carbon is thin, hollow and fibrous and their outer diameter is between 10-30 nm. On 7Co-3Cu-5Al, a few nanofibers were observed, whose morphology is similar to that formed on 2Co-1Al, and many large lumps were also found with sizes larger than one hundred nanometers. And no carbon nanofibers were found to grow from the latter. These may be formed as the result of sintering of small catalyst particles.

4. DISCUSSION

4.1 Activity of Fe, Co and Ni

From the results given in Fig. 1 and 2, it is interesting to note that in the low temperature range, the activity order of Fe/Al_2O_3, Co/Al_2O_3 and Ni/Al_2O_3 is just a reverse of that of their melting point temperatures, which are 1809 K, 1768 K and 1726 K for Fe, Co and Ni, respectively [8], and is also a reverse of that of the initial temperatures for their deactivation. As metal particles of the size of several nanometers, Fe, Co and Ni being mixed with Al_2O_3 in this work, should have lower melting points than their corresponding bulk sized particles, due to the fairly high surface energy of the small particles [9]. It may also be possible for them to be quasi-liquid state to some extent even at the temperatures much lower than their bulk melting points [9]. The order of the initial formation temperatures of quasi-liquid state is likely to be Ni < Co < Fe under the same conditions. According to the bulk diffusion mechanism, the diffusion process of carbon in the bulk of metal particles is the controlling step and the growth rate of carbon nanofibers should be much higher on the metal particles of quasi-liquid state than that on those of solid state [7]. However, it has been found that too high temperatures can easily make quasi-liquid Ni catalyst particles splitting into smaller ones and because encapsulated by the growing carbon layers, which will result in the deactivation of the Ni particles [10]. This is likely the reason for the activity decrease of Co/Al_2O_3 and Ni/Al_2O_3 after 1023 K and 923 K, respectively.

4.2 Effect of Cu doping

The results in Fig. 3 and 4 show that the doping of Cu to Ni/Al_2O_3 can shift the temperature for its deactivation to a higher range and improve its stability effectively at 1023 K. However, for Co/Al_2O_3, the effect goes to the opposite direction. The melting point of Cu is 1357 K, much lower than Fe, Co and Ni and it can form a perfect alloy phase with Ni by any proportion [8]. Cu has a weak ability to adsorb and decompose methane and its enrichment on Ni-Cu alloy surface can effectively decrease the formation rate of carbon on the surface of Ni particles, which is helpful to prevent the catalyst particles from encapsulating by the carbon layers due to their too fast growing on the surface and slow diffusion rate in the bulk of metal particles [7,10,12]. The formation of quasi-liquid state of $Ni-Cu/Al_2O_3$ particles at around 973 K and the amalgamation of small particles of several nanometers into big ones of several tens nanometers have been observed [11]. The diameter of carbon nanofibers in

Fig. 5A, indicates that the same changes happen, which increase the stability of the catalyst because large Ni particles were thought to have a better matching than small particles for the formation rate and the diffusion rate of carbon [13].

The crystalline structure of Co is different from that of Cu below 695 K and Co-Cu alloy phase cannot be formed like Ni-Cu alloy in any proportion below 1385 K [8]. In the low temperature range, the existence of Cu may decrease the activity of Co, and, in the high temperature range, Cu may lead to the sintering of Co particles and to their deactivation, as seen in Fig. 5C.

5. CONCLUSION

FC precursors were used to obtain Fe/Al_2O_3, Co/Al_2O_3 and Ni/Al_2O_3 catalysts of the same structure to decompose methane. In the stepwise heating experiments, Ni/Al_2O_3 is the most active, then Co/Al_2O_3, and finally Fe/Al_2O_3. The order is just a reverse of that of their melting point temperatures and is also a reverse of that of the initial temperatures for their deactivation. The reason may be partly attributed to a lower temperature for the formation of quasi-liquid state on Ni particle than those on Co and Fe. The doping of Cu improves the stability and activity of Ni/Al_2O_3 effectively at high temperatures because the stability of the catalyst is enhanced by the enrichment of Cu on Ni-Cu alloy surface and the amalgamation of small particles into large catalyst particles of a certain diameter. While for Co/Al_2O_3, the doping of Cu induces a serious sintering of Co particles and makes them deactivate at high temperatures.

ACKNOWLEDGEMENTS

The financial supports from NSF of China under contract number 20006012 and the 973 program of the Chinese Ministry of Science and Technology under contract number G199902240 are gratefully acknowledged.

REFERENCES

[1] C. Song, Catal. Today, 77 (2002) 49.
[2] T. Zhang and M. Amirids, Appl. Catal. A, 167 (1998) 161.
[3] T.V. Choudhary, D.W. Goodman, Catal. Today, 77 (2002) 65.
[4] Y.D. Li., J.L. Chen, L. Chang and Y.N. Qin, Energy Fuels, 14 (2000) 1188.
[5] M.A. Ermakova, D.Y. Ermakova and G.G. Kuvshinov, Appl. Catal. A, 201 (2000) 61.
[6] F. Cavani, F. Trifiro and A. Vaccari, Catal. Today, 11 (1991) 173.
[7] D.L. Trimm, Catal. Rev. Sci. Eng., 16 (1977) 155.
[8] T.B. Massalski (eds.), Binary Alloy Phase Diagrams, Asm Intl, Portland, 1990.
[9] F. Benissad, P. Gadelle, M. Coulon and L. Bonnetain, Carbon, 26 (1988) 425.
[10] J.L. Chen, X.M. Li, Y.D. Li and Y.N. Qin, Chem. Lett., 32 (2003) 424.
[11] J.L. Chen, Y.D. Li, Y.M. Ma, Y.N. Qin and L. Chang, Carbon, 39 (2001) 1467.
[12] K.C. Khulbe and R.S. Mann, Catal. Rev. Sci. Eng., 24 (1982) 311.
[13] L.B. Avdeeva, O.V. Goncharova, D.I. Kochubey, V.I. Zaikovskii, L.M. Plyasova, B.N. Novgorodov, and Sh.K. Shaikhutdinov, Appl. Catal. A, 141 (1996) 117.

Studies in Surface Science and Catalysis, volume 147
X. Bao and Y. Xu (Editors)

Hydrogen production from non-stationary catalytic cracking of methane: a mechanistic study using *in operando* infrared spectroscopy

E. Odier[a]**, Y. Schuurman**[a]**, K. Barral**[b]**, C. Mirodatos**[*][a]

[a]Institut de Recherches sur la Catalyse, 2. Avenue Albert Einstein
69626 Villeurbanne Cedex, France

[b]Air Liquide, Centre de Recherche Claude Delorme, BP 126,
Les Loges en Josas, 78353 Jouy en Josas Cedex, France

ABSTRACT

The cyclic two-step process of methane cracking was investigated over Pt/CeO_2 catalyst at 400°C by *in operando* DRIFT spectroscopy. The production of H_2 by direct decomposition of CH_4 on the noble metal is improved by the capacity of the cerium oxide to store carbonaceous surface species. The 2D-IR correlation analysis shows that formates adspecies are formed from carbonyls diffusing from Pt sites and reacting with hydroxyls on partially reduced Ce sites. The comparison with other catalysts indicates that formates are the most suitable adspecies for an efficient reversible C storage during the H_2 production cycle.

1. INTRODUCTION

The short term industrial application of PEM fuel cells, either for on board or stationary application, requires new processes for producing hydrogen free of carbon monoxide. The conventional reforming of any hydrocarbon (steam reforming, partial oxidation, autothermal reforming) produces in addition to hydrogen CO as a major co-product. The latter has to be converted downstream into CO_2 through the complex and energy/space demanding steps of high and low temperature water gas shift (HT and LT WGS) and selective oxidation of CO in hydrogen rich mixture (SelOx) before reaching the low CO concentration that can be tolerated by the fuel cell (below 10 ppm). An alternative to produce directly CO free hydrogen is the non-stationary catalytic cracking of hydrocarbon followed by catalyst regeneration via a cyclic two-step process as proposed in [1-3] for the case of natural gas conversion :

$$CH_4 \rightarrow C_{deposited\ on\ catalyst} + 2H_2 \qquad \text{step I}$$

$$C_{deposited\ on\ catalyst} + O_2 \rightarrow CO_2 \qquad \text{step II}$$

A high capacity of C storage associated with fast dynamics appears as a prerequisite for such a process. At variance with Ni based systems which may store carbon as carbide and graphite at high temperature (700-800°C), a supported noble metal catalyst, which operates at lower temperature, requires C storage assistance from the support via spillover between the metallic phase and the support. For this case, a low temperature process is expected to be more efficient since the rate of adspecies diffusion on the support has to be comparable with the rate of carbon formation on metal during the cracking step. In order to investigate this low temperature domain and understand the related mechanism, various catalysts based on noble metals supported on oxides were screened under non stationary conditions. Optimized hydrogen yields were about 8 and 35% at 400 and 600°C, respectively. This paper focuses on selected Pt/CeO_2 formulas investigated in details by means of coupled kinetic measurements and *in situ* diffuse reflectance infrared spectroscopy (DRIFT).

2. EXPERIMENTAL

Material. High surface area (200 m^2/g) ceria powder was impregnated with aqueous solution of $Pt(OH)_2(NH_3)_4$. The precursor was calcined at 500°C for 8h, reduced in a hydrogen flow at 300°C for 15h and then kept under inert gas (Ar) before reaction. A commercial $0,5\%Pt/\gamma Al_2O_3$ catalyst was used as reference.
Testing procedure. The reaction was carried out at 400°C under atmospheric pressure in an *in situ* DRIFT cell as described in [3,4]. Experiments consisted in flowing alternatively 50% CH_4 in inert gas (during 1 min), then flushing pure inert gas and finally adding 20% O_2 in inert gas (as long as necessary to eliminate all the reversible C stored), all steps with a total flow rate of 20 ml/min (STP). The gas concentration at the reactor outlet was continuously monitored by on line mass spectrometry and DRIFT spectra continuously recorded during the transient operations with 4 cm^{-1} resolution and 0,67 s sampling interval. Absorbance was calculated from the spectra corrected from the background spectrum at the same temperature.

3. 2D IR CORRELATION ANALYSIS: BACKGROUND

The spectral intensity variation of absorbance $y(\nu,t)$ observed during a transient period T (corresponding to a chemical perturbation like methane switch on or off) can be analyzed with synchronous and asynchronous correlation intensities Φ and Ψ given by

$$\Phi(v_1,v_2)+i\Psi(v_1,v_2)= \frac{1}{\pi T} \int_0^\infty Y_1(\omega) \cdot Y_2^*(\omega)d\omega \tag{1}$$

where $Y_1(\omega)$ is the time-domain Fourier transform of $y(v_1,t)$, and $Y_2^*(\omega)$ is the conjugate of the Fourier transform of $y(v_2,t)$ [5]. The simple real time IR spectra recorded under CH_4 are complex because they consist of overlapping bands. They can be simplified by spreading peaks along the second spectral dimension. Transient 2D IR spectra analysis provides a clearer view of time-dependant phenomena by selectively highlighting the similarity (synchronous spectrum) or difference (asynchronous spectrum) of the time dependant behavior of IR intensity changes observed at different wavenumbers v.

4. RESULTS AND DISCUSSION

4.1. In operando DRIFT spectra primary analysis

Three series of IR spectra have been recorded under reaction with the following samples: 0,5%Pt/CeO$_2$ (A), CeO$_2$ (B), 0,5%Pt/Al$_2$O$_3$ (C). Under CH_4 flow, three main kinds of species are observed as listed and assigned in Table 1 [6]: gaseous species (CH_4, CO_2), adspecies on platinum (carbonyl –CO), and adspecies on cerium oxide (hydroxyls –OH, formates –CO$_2$H, carbonates –CO$_3$). Gaseous carbon dioxide is present at the beginning of the methane pulse (step I) for samples A and C because of the previous sample regeneration by oxidation (step II): the oxidized Pt surface leads to total oxidation of methane until Pt is reduced.

Table
Main IR bands observed during the CH_4 reaction step over A, B, C samples at 400°C

	Gas phase	
gaseous CH$_4$	3200-2800, 1360-1250 cm^{-1}	
gaseous CO$_2$	2390-2280 cm^{-1}	
	Solid phase (Pt)	
carbonyls on Pt	2061 cm^{-1}(C)	Pt
	2042, 1968 cm^{-1}(A)	Pt–C≡O
		Pt
	Solid phase (cerium oxide)	
type I hydroxyl groups	3720 cm^{-1} (A,B)	Ce OH
type II hydroxyl groups	3640 cm^{-1} (A,B)	Ce OH
unidentate carbonate	1066, 1456 cm^{-1} (A,B)	
bidentate carbonate	1028, 1560 cm^{-1} (B)	
bridged carbonate	1015, 1398 cm^{-1} (A)	
formate	1378, 1580 cm^{-1} (A)	

C-storage. The mechanism of CH_4 decomposition on Pt sites followed by C storage on oxide by spillover has already been described in [3, 4]. The three 1 samples are compared here to determine precisely which kinds of adspecies participate effectively to C storage. The Pt free sample B (CeO_2) shows that cerium oxide can activate methane and oxidize it: the C-H bond is broken to create carbonates (see Fig. 1). But for this case methane conversion is very low and this oxidation which could be initiated at 200°C [7] corresponds to the interaction of methane with surface lattice oxygen anions [8] created by the previous regeneration step. These carbonates are essentially irreversible (only a minor part can be outgassed as CO_2 during the regeneration step) and therefore cannot be considered as efficient C storage agents. They are also present in sample A (0, 5%Pt/CeO_2).

With sample C (0,5%Pt/Al_2O_3), the decomposition of CH_4 on Pt is rapidly stopped after methane admission because the non reducible alumina support is unable to store carbonaceous adspecies (conversion<1% CH_4). Only Pt carbonyls are formed, resulting from CH_4 partial oxidation by oxidized Pt sites. On sample A (0,5%Pt/CeO_2), Pt carbonyls are also detected but with different spectral characteristics: a shift of the maximum absorption (from 2061 (C) to 2042 (A) cm^{-1}) followed by a shoulder around 1968 cm^{-1} indicates that the nature and structure of carbonyls have changed (more reduced Pt acceptors and extension from single to multibonded carbonyls). The tight interaction between cerium oxide and Pt particles is likely to promote multiple surface dynamics. The carbonyl species, which in turn can either desorb into the gas phase or be

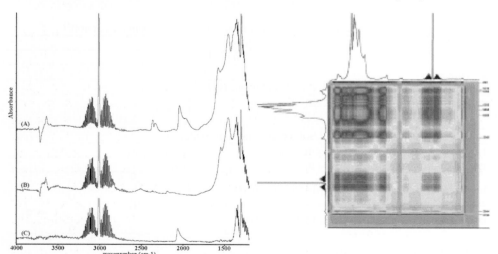

Fig. 1. DRIFT spectra at 400°C under CH_4 flow for samples A (0,5%Pt/CeO_2), B (CeO_2), C (0,5%Pt/Al_2O_3).

Fig. 2. 2D display of covariance obtained with 120 spectra recorded during 2 min on sample A under CH_4 flow at 400°C. The spectrum on axes corresponds to the arithmetic mean spectrum of the series.

stored on ceria surface by spillover [9], can only be formed thanks to oxygen provided by ceria, reacting with CH_x fragments coming from methane decomposition. This CO storage may lead to various oxidized carbonaceous adspecies developing at Pt particles vicinity during the cracking step, which explains the increased conversion (up to 15% at 400°C) without CO release into the gas phase (<2000 ppm). In addition to the carbonate species as mentioned above, the major difference for the A catalyst is the presence of formates which chemical formula corresponds to the storage of CO+OH on a reduced Ce surface site (Ce^{3+}). The presence of reduced ceria is confirmed during the regeneration step: at the beginning of the pulse, O_2 is firstly totally consumed before CO_2 and H_2O appear in the effluents and grow simultaneously up to a maximum. Then they are replaced by O_2 now in excess in the fixed bed. The oxygen mass balance calculated over the regeneration step proves that part of injected O_2 is used to the re-oxidation of the ceria ($Ce^{3+} \rightarrow Ce^{4+}$).

H-storage. OH groups are detected essentially in ceria containing samples (A and B). During the cracking step (Fig. 1) two types of hydroxyl groups can be distinguished : at high frequency (around 3720 cm^{-1}) form (type I) which is consumed after methane admission (negative peak) while a second one of lower frequency (around 3600 cm^{-1}) (type II) is accumulating. Another observation is that the IR bands of hydroxyls are larger in Pt containing sample A than for Pt free sample B. This may come from the fact that the Pt/ceria interaction improves the reducibility of the support by the fast decomposition of CH_4, H_2 or H_2O on Pt that provides reactive H atoms able to form OH groups on ceria [8].

Role of OH groups for C storage. Among the tested systems, only the Pt/CeO_2 catalyst displays both a significant C storage capacity and the ability to accumulate formates at the CeO_2 surface. It can therefore be concluded that the formation of formates contributes directly to an efficient C storage, though it requires hydrogen atoms, provided by CH_4 decomposition, which in turn decreases the H_2 yield. The other forms of carbonaceous adspecies participating to C storage - carbonates - are more oxidized but more stable, therefore less suitable for fast and reversible storage. The lower degree of oxidation of formates can be related to the redox properties of the ceria surface. When contacted by the reducing atmosphere during the cracking step (CH_4+H_2 mixing), ceria surface is partially reduced by providing oxygen atoms to platinum for oxidizing the carbon atoms arising from methane cracking. In the mean time, the oxygen migration from ceria bulk to surface, necessary to maintain a fully oxidized state, is not fast enough to locally avoid H storage as OH groups arising from the reaction of hydrogen (gas phase and/or Pt-H) with surface oxygen. It can therefore be concluded that formates formation requires partially reduced ceria containing oxygen vacancies and hydroxyl groups.

4.2. 2D correlation IR spectroscopy

The series of IR spectra recorded after introducing methane was analyzed through the 2D correlation method above described (Fig. 2) The synchronous correlation analysis revealed a strong correlation between the following IR bands transients : formates, hydroxyls (type I) and carbonyls on Pt. It comes therefore that OH groups of type I are required for formate formation. This statement, in agreement with Li *et al.* [10], supports the following mechanism:

where the active sites are high wavenumber OH groups of type I linked to Ce^{3+} ions (i.e. Ce^{4+} with oxygen vacancies), CO being provided by Pt carbonyls. Weaker correlations were also revealed between carbonyls and carbonates and between different forms of carbonates.

5. CONCLUSION

In operando IR spectroscopy and 2D analysis made it possible to prove that during methane cracking over Pt/CeO_2 catalyst, formate species are formed via the direct reaction of CO carbonyls with specific OH groups of ceria surface vibrating at 3720 cm^{-1}. The reactivity of the latter, associated with a high rate of surface diffusion [10], originates the catalyst capacity for storing carbon quantitatively and reversibly. A way for improving this storage capacity is therefore to increase the concentration of these reactive OH groups. This is possible by modifying acido-basicity and redox properties of the support, that involves in particular ceria doping with altervalent ions [to be published].

REFERENCES

[1] T. Zhang, M.D. Amiridis, Appl. Catal. A 167 (1998) 161
[2] T.V. Choudary, D.W. Goodman, Catal. Lett. 59 (1999) 93
[3] E. Odier, Y. Schuurman, H. Zanthoff, C. Millet, C. Mirodatos, Stud. Surf. Sci. Catal. 133 (2001) 327-332.
[4] E.Odier, C. Marquez-Alvarez, Y. Schuurman, H.W. Zanthoff, C. Mirodatos, Stud. Surf. Sci. Catal. 136 (2001) 483-488.
[5] I. Noda, Appl. Spectrosc. 47 (1993) 1329.
[6] A. Holmgren, B. Andersson, D. Duprez, Appl. Catal. B 22 (1999) 215.
[7] C. Bozo, N. Guilhaume, E. Garbowski, M. Primet, Catal. Today 59 (2000) 33.
[8] C. Li, Q. Xin, J. Phys. Chem. 96 (1992) 7714.
[9] T. Jin, T. Okuhara, G. J. Mains, J.M. White, J. Phys. Chem. 91 (1987) 3310.
[10] C. Li, Y. Sakata, T. Arai, K. Domen, K. Maruya, T. Onishi, J. Chem. Soc., Faraday Trans. 1, 85 (1989) 1451.
[11] D. Martin, D. Duprez, J. Phys. Chem. 101 (1997) 4428.

Studies in Surface Science and Catalysis, volume 147
X. Bao and Y. Xu (Editors)
85

Carbide- and nitride-based fuel processing catalysts

J. J. Patt, S.K. Bej and L.T. Thompson*

Department of Chemical Engineering, Ann Arbor, MI   48109-2136, USA

1. INTRODUCTION

Proton Exchange Membrane (PEM) fuel cells have emerged as leading candidates to provide primary and auxiliary power for stationary, automotive and portable electronic device applications. A key challenge to their use is the availability of sufficiently small and cost-effective H_2 sources. The principal functions accomplished by the fuel processor are the production of syngas and CO removal. Syngas, a mixture principally of H_2 and CO, is most often produced using steam reforming and/or partial oxidation reactions. The amount of CO must typically be reduced because the noble metals in fuel cell electrocatalysts are very susceptible to poisoning by as little as 10-100 ppm CO [1]. Presently CO removal is accomplished via water gas shift (WGS) and preferential oxidation reactions.

In this paper, we describe the the activities and selectivities of carbide and nitride based catalysts for methanol reforming (MSR) and WGS reactions, and characteristics of their active sites. Transition metal carbides and nitrides can be synthesized in high surface area form and they have catalytic properties similar to those of Pt-group metals [2, 3]. In addition, carbides are resistant to poisoning by sulfur compounds [4, 5].

2. EXPERIMENTAL

A series of carbide and nitride based materials were synthesized and evaluated as MSR and WGS catalysts. These catalysts were typically prepared in high surface area form using temperature programmed reaction methods. Additional details regarding synthesis of these materials are given elsewhere [6-8]. The BET surface areas for the Mo_2C and Mo_2N catalysts were 60 and 135 m^2/g, respectively.

The reaction rate measurements were carried out in a 4 mm ID quartz U-tube flow reactor using 25-130 mg of catalyst diluted with 30-35 mg of inert

---

* Corresponding author: Email: ltt@umich.edu

silica. The catalyst bed was held between quartz glass wool plugs. The temperature of the reactor was maintained using a furnace with digital temperature control. For MSR, the combined methanol-water mixture was fed using a HPLC pump and then vaporized. Experiments were conducted using methanol to water molar ratios of 1:1 to 1:3. Nitrogen (diluent) flow rates were adjusted using mass flow controllers. The methanol molar concentration in the feed was 11.6%. For WGS, the reactant contained 38.6% H_2 (99.99%), 15.9% N_2 (99.995%), 5.7% CO (99.5%), 6.3% CO_2 (99.8%), and 30.0% deionized H_2O. The gases were delivered by mass flow controllers and H_2O was added by bubbling the dry gas mixture through a heated saturator vessel. The gas lines were heated to 120 °C between the saturator and the catalyst bed to prevent condensation. Analysis of the reactor effluent was performed using an on-line gas chromatograph (HP 5890) with a thermal conductivity detector and Carboxen 1000 column. Rates were measured in the temperature range of 180 to 240 °C and atmospheric pressure.

The carbide and nitride catalysts were either pretreated in H_2 or a mixture of 15% CH_4 in H_2 for 4 h at 400-590 °C. The Cu-Zn-Al catalyst was pretreated at 200 °C in a mixture of 2% H_2 in N_2 for 4 h. Selected materials were characterized using thermal desorption spectroscopy, x-ray diffraction, and/or x-ray photoelectron spectroscopy.

3. RESULTS AND DISCUSSION

3.1. Methanol Steam Reforming

Methanol conversion rates and selectivities towards CO_2 for selected catalysts at 230 °C and a methanol to water molar ratio of 1:1 are shown in Figure 1. The methanol conversion was limited to ~25 mol%. The methanol

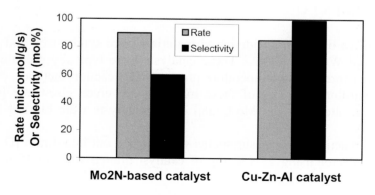

Fig. 1. Methanol conversion rate and CO_2 selectivity for Mo_2N based and Cu-Zn-Al catalysts at 230 °C and a methanol: H_2O molar ratio of 1:1.

conversion rate for the Mo_2N based catalyst was comparable to that for the Cu-Zn-Al catalyst, however, the selectivity was lower.

Since the selectivity toward CO_2 for the Mo_2N based catalyst was low, the methanol to water molar ratio was varied in order to understand its role on CO_2 selectivity. The results are shown in Figure 2. Again, the conversions were limited to ~ 25 mol%.

The higher concentration of water increased the rate of methanol conversion and CO_2 selectivity significantly. A comparison of the results shown in Figures 1 and 2 suggests that the surface of the Mo_2N based catalyst had a lower affinity for water compared to that for the Cu-Zn-Al catalyst. Therefore, a relatively higher concentration of water was necessary to achieve high selectivities.

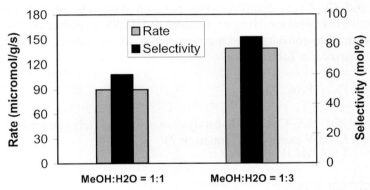

Fig. 2. Effect of methanol to water molar ratio on methanol conversion rate and CO_2 selectivity for Mo_2N based catalyst at 230 °C.

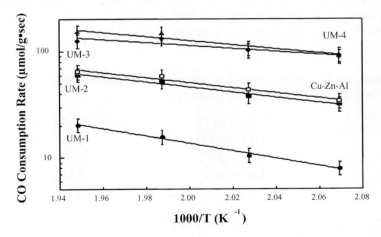

Fig. 3. Shift rates for selected carbide and Cu-Zn-Al catalysts with 38.6% H_2, 15.9% N_2, 5.7% CO, 6.3% CO_2 and 30.0% H_2O.

Table 1
Kinetic parameters for the power law equation

Catalyst	T (°C)	k (mol/g/s)	a	b	c	d
Mo_2C	240	32 ± 7	0.55 ± 0.05	0.19 ± 0.07	-0.02 ± 0.06	-0.01 ± 0.07
Cu-Zn-Al	240	278 ± 9	0.94 ± 0.05	0.72 ± 0.12	-0.38 ± 0.10	-0.42 ± 0.06

3.2. Water Gas Shift

The carbides and nitrides were also very active for the WGS reaction. In fact, several of the formulations substantially outperformed a commercial Cu-Zn-Al WGS catalyst (Figure 3).

Detailed studies of the WGS kinetics were performed at 240°C for the Mo_2C and Cu-Zn-Al catalysts. The reaction data was fit to power law and Langmuir-Hinshelwood-Hougen-Watson (LHHW) expressions by non-linear regression. Rate constants and reaction orders for the power law fit (equation 1) are summarized in Table 1.

$$r = k \, P_{CO}^{a} \, P_{H_2O}^{b} \, P_{CO_2}^{c} \, P_{H_2}^{d} \qquad [1]$$

Data for the Cu-Zn-Al catalyst was adequately described using the following LHHW expression (equation 2):

$$r = \frac{kP_{CO}P_{H_2O}}{(1 + K_{CO}P_{CO} + K_{H_2O}P_{H_2O} + K_{CO_2}P_{CO_2} + K_{H_2}P_{H_2})^2} \qquad [2]$$

The resulting parameters are listed in Table 2. This equation is representative of a dual-site process that is surface-reaction-limited and is influenced by adsorption of all four species.

Table 2
Kinetic parameters for LHHW expressions. The adsorption constants (K_i) are in units of atm^{-1} and the reaction rate (r) is expressed in $\mu mol/g/s$.

Constants	Catalyst	
	Mo_2C	Cu-Zn-Al
k	2157 ± 661	11900 ± 4260
K_{CO}	5.17 ± 1.22	1.54 ± 0.52
K_{H_2O}	4.49 ± 1.14	1.10 ± 0.63
K_{CO_2}	-	4.19 ± 1.70
K_{H_2}	-	1.78 ± 0.45

Rates for the Mo_2C catalyst were half-order with respect to CO and 0.2-order for H_2O. There was no dependence on partial pressures of the product species.

This is advantageous for running the catalyst at higher conversions where there is more CO_2 and H_2.

We developed the following LHHW expression (equation 3) to describe data for the Mo_2C catalyst (see Table 2):

$$r = \frac{kP_{CO}P_{H_2O}}{(1+K_{CO}P_{CO}+K_{H_2O}P_{H_2O})^2} \qquad [3]$$

Here also the surface reaction is rate-limiting. The rate expression suggests a dual-site mechanism where the adsorption of CO and H_2O are dominant, and the products do not compete for active sites. This expression is consistent with several mechanisms, including a regenerative mechanism where active sites are responsible for H_2O dissociation and CO oxidation.

In situ x-ray photoelectron spectroscopic results suggested that active sites on the carbide surfaces were linked to carbidic carbon (Figure 4). In particular, the WGS rates increased proportionally with the amount of carbidic carbon.

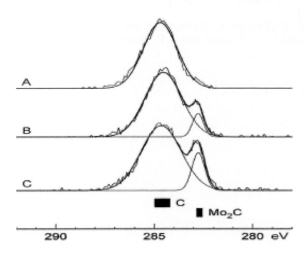

Fig. 4. X-ray photoelectron spectra (C 1s) for Mo_2C following A) passivation in 1% O_2 in He, B) pretreatment in H_2/Ar at 500 °C, and C) pretreatment in $CH_4/H_2/Ar$ at 500 °C.

4. CONCLUSION

Carbide and nitride-based catalysts have been demonstrated to be highly active for MSR and WGS, reactions that are important in the conversion of hydrocarbons and alcohols into syngas. The WGS activities have been tentatively linked to sites containing carbidic carbon, and the mechanism appears to be different from those for Cu-Zn-Al and noble metal/CeO_2 catalysts.

ACKNOWLEDGEMENTS

Financial support from the Department of Energy and Osram Sylvania was used to carry out work described in this paper.

REFERENCES

[1] K. Kordesch, and G. Simander, Fuel Cells and Their Application, VCH Publishers, NY (1996).
[2] R.B. Levy and M. Boudart, Science, 181 (1973) 547.
[3] S.T. Oyama, Catal. Today, 15 (1992) 179.
[4] B. Dhandapani, T.S. Clair and S.T. Oyama, Appl. Catal. A: General, 168 (1998) 219.
[5] S.T. Oyama, C.C. Yu and J. Ramanathan, J. Catal., 184 (1999) 535.
[6] M.K. Neylon, H.-H. Kwon, S. Choi, K.E. Curry and L.T. Thompson, Appl. Catal. A: General, 183 (1999) 253.
[7] J. Patt, D.J. Moon, C. Phillips and L. Thompson, Catal. Letters, 65 (2000) 193.
[8] M.K. Neylon, S.K. Bej, C.A. Bennett and L.T. Thompson, Appl. Catal. A: General, 232 (2002) 13.

Studies in Surface Science and Catalysis, volume 147
X. Bao and Y. Xu (Editors)
©2004 Elsevier B.V. All rights reserved.

One-Step-Hydrogen: a new direct route by water splitting to hydrogen with intrinsic CO_2 sequestration

Domenico Sanfilippo[a], Ivano Miracca[a], Ugo Cornaro[b], Franco Mizia[b], Alberto Malandrino[b], Valerio Piccoli[b], Stefano Rossini[b]

[a]Snamprogetti SpA, Viale De Gasperi 16 - 20097 S. Donato Milanese – Italy
[b]EniTecnologie SpA, Via Maritano 26 - 20097 S. Donato Milanese - Italy

1. INTRODUCTION

The enormous amount of energy available to mankind in the last centuries from fossil fuels has been a key pulling factor for the technological progress which has deeply influenced the life standards. In the last decades higher attention has been paid to the pollutants derived from an incomplete combustion and/or a fuel of poor quality and to the GHG (CO_2 accumulation) that are perturbing the ecosystem. There is consciousness, growing faster and faster, that the uncontrolled use of energy is undermining the "sustainable development approach" to mankind's future. On the emotional wave, a reaction is taking place targeting the so-called "H_2 economy". In its extreme (r) evolution, it leads to the end of the fossil fuels (coal, oil and natural gas) era as primary energy sources.

The authentic hydrogen economy should see the hydrogen produced by a process from renewable raw material (water!) and also renewable sources of energy. However it is predicted that hydrogen will be produced from fossil fuels as raw material and also as energy source, for decades from today. It turns out that CO_2 management is still an issue to be tackled with more and more efficient technologies.

2. CURRENT HYDROGEN PRODUCTION TECHNOLOGIES

The most commonly used technology to produce Hydrogen is the Steam Reforming of natural gas Eq. (1.a) followed by one or two (high temperature and low temperature) water gas shift steps Eq. (1.b)

$$CH_4 + [O_2/H_2O] \rightarrow H_2 + CO \qquad (1.a)$$

$$CO + H_2O \rightarrow H_2 + CO_2 \qquad (1.b)$$

Hydrogen is then recovered via a Pressure Swing Adsorption (PSA) while the exhausted stream is sent to the recovery of its thermal value. Other syngas preparation - like AutoThermal Reformer, Compact Reformer, and Partial Oxidation - can be used depending on the size of the plant.

In Fig. 1 the hydrogen production flow sheet, based on Steam Reforming, is reported. In addition a typical desulphurization (HDS) and prereforming units have been added. Three main sections can be identified: syngas preparation (Reaction Unit), hydrogen maximization (Shift Units) and hydrogen and CO_2 separation (Adsortpion and PSA).

About 90% of the carbon dioxide can be recovered upstream of the PSA separation via an amine based absorption. The cost of removal is commonly estimated to be in the 40-70 ($/ton) range [2].

3. ONE STEP HYDROGEN PROCESS

The One Step Hydrogen process is aimed not only to produce hydrogen from water splitting but also to tackle the carbon management issue targeting a zero carbon emission at lower costs than allowed by the Best Available Technologies (BAT).

As well known water possesses an oxidizing potential which can be utilized in a simple water splitting application where it reacts with selected metals (Al, Si, Zn, Mg) [3] to give pure hydrogen. So combining in a cycle the water oxidative potential and the reverse action by a reducing agent like hydrocarbons, it allows the hydrogen production and as well as the intrinsic carbon dioxide sequestration. This new process is under development in our company.

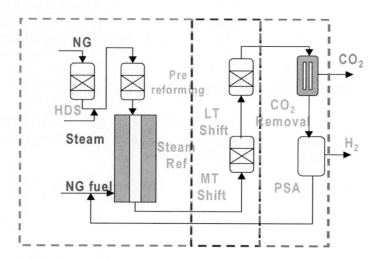

Fig. 1. "Carbon-free" hydrogen production via Steam Reforming and CO_2 adsorption

In the first step Eq. (2.a) a suitable oxide takes up the oxygen from water splitting producing hydrogen. The solid acts as an oxygen storage medium. Such "lattice" oxygen is in turn released through one or more elemental steps whose sum is equation Eq. (2.b). The process fits very well within a circulating fluid bed reactor which allows the movement of the solid from one reactive environment to the other. In terms of overall material balance, equations 2 show the chemical reactions in a quite generalized and simplified manner.

$$4\ MeO_x + 4\ H_2O \rightarrow 4\ H_2 + 4\ MeO_{x+1} \tag{2.a}$$

$$4\ MeO_{x+1} + CH_4 \rightarrow CO_2 + 2\ H_2O + 4\ MeO_x \tag{2.b}$$

$$CH_4 + 2\ H_2O \rightarrow 4\ H_2 + CO_2 \quad H = 39.5\ kcal/mol \tag{2}$$

The overall reaction Eq. (2) is endothermic and an energy supply is needed to close the energy balance. Heat can be supplied burning a small amount of hydrogen according to Eq. (2.c) or more generally via any other suitable exothermal reaction.

$$x\ (H_2 + 0.5\ O_2 \rightarrow H_2O) \tag{2.c}$$

The overall reaction becomes (3) whose thermal regime is dictated by the value of x.

$$CH_4 + (2\text{-}x)\ H_2O + 0.5x\ O_2 \rightarrow (4\text{-}x)\ H_2 + CO_2 \tag{3}$$

If x is approximately 0.68 the overall reaction can be thermoneutral. It is endothermic for lower x values while it becomes exothermic for higher x values.

Fig. 2 shows the plot of the relative stoichiometry and thermal balance as a function of x.

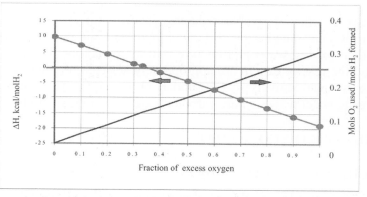

Fig. 2. Reaction heat and O_2 to H_2 ratio as function of (0.5x) from Eq. (2.c)

The decoupling of the overall reaction (3) into three separated reactions (2.a-c) allows the process to overcome the thermodynamics constraints of the overall equilibrium. The major breakthrough is however the exit of the two main products – hydrogen and carbon dioxide – from two different (see Eq 2.3) vessels, so that the carbon dioxide, once water is condensed, is pure and ready to be buried.

That is why the process has been called "One Step Hydrogen" having removed two of the three sections (including the syngas production step) in which a CO_2-free hydrogen production can be divided (see Fig. 1).

4. THE MATERIAL

The material is the core of the technology: it is required to exploit many different and highly demanding functions. The basic function is the oxidability by steam and the reducibility by natural gas (RedOx couple between H_2/H_2O and CO_2/NG) within a not extremely extended temperature range because it has to be attained during the cycling time.

The material should have mechanical properties and particle size suitable to resist to the severe stress of the circulation loop. All these characteristic should be maintained through cycle repetition, as well as its chemically stability.

We deeply investigated many chemical compounds able to cycle the oxidation states in a reasonably close range of conditions. It turned out that the best candidate was the iron oxide system. In fact water is able to add oxygen by displacing the RedOx couples Fe/FeO and FeO/Fe_3O_4 as per Eq. 4.a and 4.b directly forming hydrogen.

$$3\ Fe + 4\ H_2O \rightarrow Fe_3O_4 + 4\ H_2 \tag{4.a}$$

$$3\ FeO + H_2O \rightarrow Fe_3O_4 + H_2 \tag{4.b}$$

Natural gas or other derivable reducing species can remove oxygen by displacing the RedOx couples Fe_2O_3/FeO; Fe_3O_4/FeO and FeO/Fe forming a stream of carbon oxides, water and unconverted hydrocarbons. If the operating conditions and the reactor design are appropriately selected, the hydrocarbon is converted almost quantitatively to CO_2 and H_2O as per Eq. (4.c) through Eq. (4.f). Condensing water leaves a pure carbon dioxide stream ready to be stored.

$$4\ Fe_2O_3 + CH_4 \rightarrow 8\ FeO + CO_2 + 2\ H_2O \tag{4.c}$$

$$4\ Fe_2O_3 + 3\ CH_4 \rightarrow 8\ Fe + 3\ CO_2 + 6\ H_2O \tag{4.d}$$

$$4 \ Fe_3O_4 + CH_4 \rightarrow 12 \ FeO + CO_2 + 2 \ H_2O \tag{4.e}$$

$$Fe_3O_4 + CH_4 \rightarrow 3 \ Fe + \ CO_2 + 2 \ H_2O \tag{4.f}$$

The screening of many oxides led to the identification of iron as the best solid system to play with. However, several other levers has been activated to improve the actual material composition, as promoters, binder/carrier, etc. [4].

5. THE PROCESS

The logical choice to run continuously the three reactions Eq. (2.a÷c) is a circulating fluidized bed configuration as shown in Fig. 3 where the simplified sketch of the unit (left side) is paralleled to the evolution of the material and products formation (right side), thus making available pure hydrogen and pure carbon dioxide at battery limits.

The full reaction target can be accomplished through an optimized design, specifically of the internals, coherent with the reaction rates and the fluid dynamics constraints. The residence time in each process step has to be carefully assigned and maintained. The amount of oxygen exchanged per unit of mass is of paramount importance; it dictates the hydrogen production but it governs the more critical heat release during oxidation. If it is exceeds an optimum value,

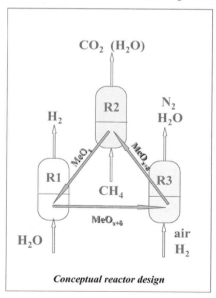

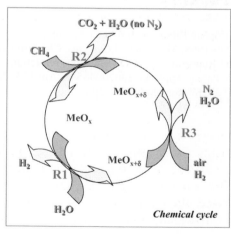

Fig. 3. Simplified sketch of the reaction vessels (left) in parallel with the reaction chemistry (right)

Fig. 4. One Step Hydrogen: simplified process scheme

the adiabatic temperature raise can be excessive while an optimal profile through the cycle has to be carefully maintained to reach the highest efficiency. The calculated efficiency η of the order of 75-80% when η is defined as $\eta = Q_{H2}*\Delta H_{H2} / Q_{CH4}*\Delta H_{CH4}$ being Q is the molar flow and ΔH is the Low Heating Value and it is comparable to the efficiency of a Steam Reforming depending of the valorization of the steam export.

The simplification of the production loop offered the One_Step Hydrogen Process (Fig. 4) is clearly evident when it is compared with the Best Available Technologies integration (BAT), shown in Fig. 1.

6. CONCLUSIONS

The technology under development allows the production of "carbon-free" hydrogen without the syngas intermediate and the amine adsorption: the technical effort is sustained by economical perspectives which look promising. The development is in progress through a scale-up route, governed by performing milestones and economical criteria to be met to proceed to the next gate. Actually intensive studies of the fluid-dynamic through mock-up models are in progress. Although the technology requires an extensive comparison with the other different routes to energy production/hydrogen use to be fully validated, preliminary economics seem to be promising.

REFERENCES

[1] European Environment Agency, Annual European Community:Greenhouse Gas Inventory 1990-1998, May 2000; J.G.Ayres, Chem.Ind., Nov. 4 (1996), 827
[2] SRI Report n.237, "CO_2 emissions reduction", November 2000
[3] US Patent Appl. A1 20020048548, April 25, 2002
[4] EP 34187 (September 19, 2001)

Studies in Surface Science and Catalysis, volume 147
X. Bao and Y. Xu (Editors)
97

Iron oxide as a mediator for the storage and supply of hydrogen from methane

S. Takenaka, C. Yamada, V. T. D. Son and K. Otsuka

Department of Applied Chemistry, Graduate School of Science and Engineering, Tokyo Institute of Technology, 2-12-1 Ookayama, Meguro-ku, Tokyo, 152-8552, Japan

ABSTRACT

Modified iron oxides (Cu-Cr-FeOx and Ni-Cr-FeOx) produced pure H_2 repeatedly through the reduction of Fe_3O_4 with CH_4 into Fe metal and the subsequent oxidation of the metal with H_2O.

1. INTRODUCTION

H_2 is expected to become an important energy carrier for sustainable energy consumption with a reduced impact on the environment. H_2 should be used as a fuel for H_2-O_2 fuel cells due to high conversion efficiency of chemical energy of H_2 into electricity without emission of any pollutant gases. However, one of the major obstacles to utilize H_2 as an energy carrier is the lack of safe, efficient and low cost storage systems suitable for various stationary and mobile applications.

With regard to this problem, we proposed a simple, safe and environmentally benign technology for the storage and supply of H_2 to fuel cells [1]. Our technology was based on a redox reaction of magnetite in Eq. (1).

$$Fe_3O_4 + 4H_2 \leftrightarrow 3Fe + 4H_2O \tag{1}$$

Fe_3O_4 was reduced with H_2 into Fe metal. The recovery of H_2 was done by the reverse reaction, i.e. the reoxidation of Fe metal with H_2O. By means of this method, 1 mol of Fe could store and regenerate ca. 1.3 mol of H_2, which corresponds to 4.8 wt% of Fe metal. The principle of our technology was analogous to the old steam iron process for the production of hydrogen rich fuel gas at high temperatures>1073 K and pressures>70 bars by using a gas mixture of CO, CO_2 and H_2 from coal and biomass for the reduction of iron ores [2].

Our method described above would stand on the conditions that we can get cheap and pure H_2 in order to reduce Fe_3O_4 into Fe metal. In contrast, if magnetite can be reduced directly with CH_4 (Eq. (2)) and the H_2 produced in Eq. (3) does not contain CO or CO_2, our technology could be more attractive from the standpoint of the storage and supply of pure H_2 to polymer electrolyte fuel cells (PEFC) by using natural gas as the primary energy source.

$$Fe_3O_4 + CH_4 \rightarrow 3Fe + CO_2 + 2H_2O \qquad (2)$$

$$3Fe + 4H_2O \rightarrow Fe_3O_4 + 4H_2 \qquad (3)$$

In the present study, we demonstrate the possibility of a new technology for the storage and supply of H_2 based on CH_4, i.e. Fe_3O_4 is reduced with CH_4 into Fe metal and subsequently Fe metal is oxidized with H_2O to form H_2 at temperatures lower than 873 K.

2. EXPERIMENTAL

Iron oxides alone and with foreign metal additives (M) were prepared by co-precipitation of their hydroxides from the aqueous solutions of Fe and M salts by hydrolysis of urea. The amounts of added foreign metals were adjusted to be 5 mol% of total metal cations. Hereafter, iron oxide samples without and with a foreign metal (M) were denoted as FeOx and M-FeOx, respectively. Reduction of the iron oxide samples with CH_4 and the subsequent oxidation of the reduced samples with H_2O were carried out by using a gas-flow system. For the reduction, CH_4 (101.3 kPa and 30 ml min^{-1}) was introduced into a tubular reactor packed with the iron oxide (0.2 g) at 473 K and the temperature at the reactor was raised to 1023 K by a rate of 3 K min^{-1}. The temperature was kept at 1023 K until the formation of CO and CO_2 could not be observed. After CH_4 was purged out with Ar, H_2O vapor (18.5 kPa) balanced with Ar (total flow rate = 70 ml min^{-1}) was contacted with the reduced iron oxide sample at 473 K and the temperature was raised linearly with time to 823 K by a rate of 4 K min^{-1}. The oxidation with H_2O was continued until the H_2 formation could not be observed. Unless otherwise noted, the reduction and the subsequent oxidation of the iron oxide samples were carried out repeatedly under the same conditions. Iron species for all the samples used in the present study were present as Fe_2O_3 phase mainly before the reaction. At the first redox cycle, Fe_2O_3 in the samples was reduced into Fe metal and subsequently the metal was oxidized into Fe_3O_4. Since the second cycle, the redox was performed between Fe_3O_4 and Fe metal. During the reactions, a part of the effluent gas was sampled out and analyzed by G.C. Detection limit of CO and CO_2 was ca. 100 ppm.

3. RESULTS AND DISCUSSION

Previously, we have reported that in the repeated redox reactions of Eq. (1), iron oxides modified with Al, V, Cr, Ga and Mo species could store and produce H_2 repeatedly, while pure iron oxides were deactivated quickly during the redox [3, 4]. The modification with these additives prevented the sintering of iron species during the repeated redox. Addition of Rh and Ir into iron oxides enhanced both the reduction with H_2 and the reoxidation with H_2O at lower temperatures. However, because of a high price of these metals, we should look for a cheaper alternative. In this work, we have looked for a favorable combination of non-precious metal additives to iron oxide for the repeated reduction with CH_4 and reoxidation of the reduced iron oxide with H_2O. Cu-Cr-FeOx and Ni-Cr-FeOx showed excellent performance for the redox. We demonstrate the redox performance of these samples in comparison with other samples based on Cr-FeOx.

Fig. 1 shows the results for the reduction with CH_4 and oxidation with H_2O for the Cr-FeOx modified with Ni, Cu, Rh and Ir. As shown in Fig. 1 (a), the formation of CO and CO_2 was observed due to the reduction of iron oxide samples with CH_4. Since the formation of H_2 was relatively slow compared to CO and CO_2, the hydrogen atoms in the reacted CH_4 might mostly be used for the reduction of the iron oxide samples. However, the quantitative analysis of H_2O by G.C. in this work was not satisfactory. Therefore, the formation rate of COx $(= CO + 2xCO_2)$ was used as a reduction index of the iron oxide samples in this work. The rates were plotted against temperature in Fig. 1 (a). Reduction of Cr-FeOx without any other additives proceeded slowly in the temperature range of 700 ~ 1023 K. Addition of Rh, Ir, Ni or Cu into Cr-FeOx dramatically improved the reduction with CH_4, irrespective of the kind of added metals. For the reduction of Cr-FeOx modified with Rh, Ir, Ni and Cu, the maximum formation rates of COx were observed at around 770, 840, 910 and 980 K,

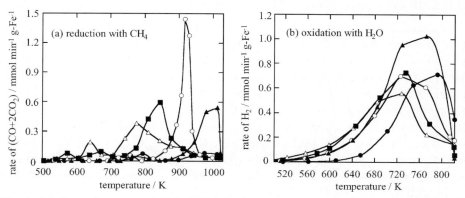

Fig. 1. (a) Formation rates of $(CO+2CO_2)$ in the reduction with methane for the Cr-FeOx modified with Ni, Cu, Rh and Ir. (b) Formation rate of H_2 in the subsequent oxidation with water vapor. The results for the first cycle. ●, Cr-FeOx; ○, Ni-Cr-FeOx; ▲, Cu-Cr-Fe Ox; △, Rh-Cr-FeOx and ■, Ir-Cr-FeOx.

respectively. During the reduction for these samples, the formation of H_2 (not shown in Fig. 1 (a)) was accelerated after CO and CO_2 were formed. This result suggests that H_2 was formed through the CH_4 decomposition catalyzed by the reduced iron oxides [5]. Therefore, the accumulation of carbons after the reduction of the samples could not be avoided in greater or lesser extent. The formation rates of H_2 at 1023 K due to CH_4 decomposition were higher in the order of Ir-Cr-FeOx > Rh-Cr-FeOx > Ni-Cr-FeOx > Cu-Cr-FeOx.

Fig. 1 (b) shows the results for the oxidation with H_2O of the iron oxide samples reduced with CH_4. The modification of the samples with Ni, Cu, Rh and Ir enhanced the oxidation with H_2O. The Cr-FeOx modified with Ni, Cu, Rh and Ir formed H_2 at lower temperatures than that for Cr-FeOx. The oxidation rates with H_2O were similar to each other among the four samples. During the oxidation with H_2O, a little formation of CO and CO_2 was also observed, although the results were not shown in Fig. 1 (b). These products must be formed by gasification with H_2O of the carbons deposited from CH_4. The total amounts of H_2, CO and CO_2 formed during the oxidation with H_2O were roughly estimated by integrating the formation rates of these products against the reaction time. The amounts of H_2, CO and CO_2 per 1 mol of Fe were 1.7, 0.05 and 0.18 mol for Ni-Cr-FeOx, 1.8, 0.01 and 0.11 mol for Cu-Cr-FeOx, 1.4, <0.01 and 0.04 mol for Ir-Cr-FeOx and 1.3, <0.01 and 0.01 mol for Rh-Cr-FeOx. The amounts of H_2 formed for Ni-Cr-FeOx and Cu-Cr-FeOx were higher than those for the other samples, probably because the H_2 produced by the gasification of deposited carbons with H_2O must have been counted.

Fig. 2 shows the results of the repeated redox cycles for Cu-Cr-FeOx sample. From the first to the 7th cycle, the redox reactions were repeated under the same conditions as those shown in Fig. 1. The results for the first, the third and the 7th cycle are shown in Fig. 2. It should be noted that the peak for the

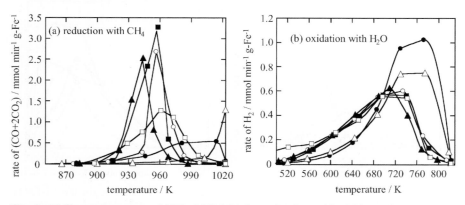

Fig. 2. (a) Formation rates of (CO+2CO$_2$) in the reduction with methane for Cu-Cr- FeOx. (b) Formation rate of H_2 in the subsequent oxidation with water vapor. ●, the 1st cycle; ○,the 3rd cycle; ▲, the 7th cycle; △, the 8th cycle; ■, the 10th cycle; □, the 11th cycle.

formation rate of COx during the reduction with CH₄ was shifted to lower temperatures with the repeated cycles, that is, the reduction of Cu-Cr-FeOx was accelerated with the repeated cycles. In addition, the kinetic curves of H_2 formation in the oxidation with H_2O did not change appreciably with the repeated cycles (2 ~ 7 cycles), except for the first one. The total amount of H_2 formed by the redox of Cu-Cr-FeOx at each cycle was kept almost constant (ca. 1.3 mol H_2 per 1 mol Fe). The high formation rate of H_2 at 720 ~ 780 K for the first oxidation may be due to the contribution of the gasification of deposited carbons with H_2O. Nevertheless, it should be noted that the formation of CO and CO_2 was not observed at all from the third to the 7th oxidation.

After the 7th oxidation, the deposited carbons on the sample were oxidized with O_2 at 823 K in order to evaluate the accumulated amount of carbons. The oxidation of deposited carbons produced CO_2 in addition to a trace of CO. The total amount of COx formed was estimated to be 15.6 mol per 1 mol of Fe.

After the oxidation of carbons described above, further three redox cycles of Cu-Cr-FeOx were repeated, which corresponds to the 8th to the 10th cycle in Fig. 2. The reduction rate of Cu-Cr-FeOx for the 8th cycle was very slow, but the reduction was improved for the 9th and 10th cycles. This observation was similar to that from the first to the third cycle, indicating that Cu-Cr-FeOx is a quite stable redox mediator for the storage and supply of H_2 from CH₄.

After the 10th oxidation, the Cu-Cr-FeOx was heated from 473 K to 1023 K under a stream of Ar instead of CH₄. CO and CO_2 were formed as shown in Fig. 2 (a) (the 11th cycle), although the maximum formation rate of them was lower compared to that under the CH₄ stream (the 3rd, 7th and 10th cycle). For the 11th oxidation, a similar kinetic curve of H_2 formation to those for the other cycles was observed. This result indicated that the deposited carbon itself could work as a reductant for Cu-Cr-FeOx.

Fig. 3 shows the results of the repeated redox for Ni-Cr-FeOx. Four redox

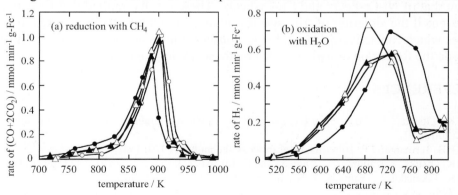

Fig. 3. (a) Formation rates of (CO+2CO₂) in the reduction with methane for Ni-Cr- FeOx. (b) Formation rate of H_2 in the subsequent oxidation with water vapor. ●, the 1st cycle; ○, the 2nd cycle; ▲, the 3rd cycle; △, the 4th cycle.

cycles were performed under the same conditions as those in Fig. 1. As shown in Fig. 3 (a), kinetic curves of the formation rates of COx for Ni-Cr-FeOx did not change appreciably during the first to the fourth cycles. It should be noted that Ni-Cr-FeOx could be reduced at lower temperatures than Cu-Cr-FeOx. However, the formation rate of H_2 at 1023 K due to CH_4 decomposition in the reduction of Ni-Cr-FeOx was approximately twice as high as that of Cu-Cr-FeOx. As for the oxidation with H_2O (Fig. 3 (b)), H_2 was formed at temperatures higher than 520 K. The formation rate of H_2 attained to the maximum at around 700 K. H_2 was formed continuously at temperatures higher than 770 K, where the continuous formation of CO and CO_2 was observed. The H_2 formation at temperatures > 770 K may be ascribed to the gasification of deposited carbons with H_2O. Thus, the total amounts of CO and CO_2 formed in the oxidation with H_2O for Ni-Cr-FeOx were higher than those for Cu-Cr-FeOx. The total amounts of H_2, CO and CO_2 per 1 mol of Fe were estimated to be 1.7, 0.05 and 0.18 mol for the 1st oxidation, 1.5, 0.02 and 0.08 mol for the 2nd run, 1.4, 0.02 and 0.09 mol for the 3rd run, and 1.3, 0.01 and 0.02 for the 4th run. The formation of CO and CO_2 during the oxidation was accelerated at temperatures higher than 760 K. However, at temperatures lower than 760 K, the oxidation of reduced Ni-Cr-FeOx with H_2O can produce pure H_2 without a trace of COx. In fact, the redox of the reduction with CH_4 at 873 K and the subsequent oxidation with H_2O at 673 K for Ni-Cr-FeOx produced pure H_2 repeatedly after the second cycle.

From the results described above, we conclude that Cu-Cr-FeOx and Ni-Cr-FeOx can store and supply pure H_2 repeatedly through the reduction with CH_4 and the subsequent oxidation with H_2O. Ni-FeOx and Cu-FeOx deactivated quickly in the first redox cycle, although they were reduced more easily with CH_4 than FeOx and Cr-FeOx. Cr-FeOx produced H_2 repeatedly through the oxidation with H_2O, although the redox of Cr-FeOx needed higher temperatures than that of Ni-Cr-FeOx and Cu-Cr-FeOx. These results implied that Cr cations prevented the sintering of iron species, while Cu and Ni species worked as the sites which can activate CH_4 and H_2O. Further studies on the conditions that can evade the formation of carbons from methane are in progress.

REFERENCES

[1] K. Otsuka, A. Mito, S. Takenaka and I. Yamanaka, Inter. J. Hydrogen Energy, 26 (2001)191.

[2] P.B. Terman and R. Biljetina, Coal Process Technol. 5 (1979) 114.

[3] K. Otsuka, T. Kaburagi, C. Yamada and S. Takenaka, J. Power Sources, 122 (2003) 111.

[4] S. Takenaka, C. Yamada, T. Kaburagi and K. Otsuka, Stud. Surf. Sci. Catal. 143 (2002) 795.

[5] R.T.K. Baker, J.J. Chludzinski, Jr and C.R.F. Lund, Carbon, 25 (1987) 295.

Studies in Surface Science and Catalysis, volume 147
X. Bao and Y. Xu (Editors)
103

Hydrogen from stepwise reforming of methane: a process analysis

Lei Cao, Jiuling Chen, Shaowen Liu and Yongdan Li*

Department of Catalysis Science & Technology, School of Chemical Engineering, Tianjin University, Tianjin 300072, China

ABSTRACT

Stepwise reforming of methane is compare with three traditional methane reforming routes, steam reforming, partial oxidation and autothermal reforming, with respect to their thermal efficiency and CO_2 co-production. It shows that the thermal efficiency of stepwise reforming is about 94%, with steam reforming at 91.3% and autothermal reforming at 93.9%, respectively.

1. INTRODUCTION

Apart from large scale industrial applications, hydrogen is the most promising fuel for fuel cells and other on-board applications. For low temperature fuel cells such as proton exchange membrane fuel cells (PEMFC), high purity H_2, in which CO and sulfur impurities within 10 ppm level, is demanded [1]. New routes for COx-free production of H_2 are in hot debate in literature.

Methane is a preferred source for H_2 production due to its high H/C ratio and its huge reservation [2]. The conventional processes for methane conversion to H_2 include steam reforming (SR), partial oxidation (PO) and autothermal reforming (ATR). These processes operate at high-temperature (>800℃) and produce CO, which is converted to CO_2 by water-gas shift in consequent steps [3, 4]. CO_2 absorption and trace amount of CO removal is highly energy consuming and bulky, adding the H_2 production cost by about 30% [5].

Direct decomposition of methane (DDM) produces H_2 and solid carbon, thus making it easier to get CO-free H_2. It was shown that, at temperatures above 823 K, Ni based catalysts exhibited stable performance for a few hours [6, 7] with CO concentration at around 50 ppm [8]. A concept of stepwise reforming (SWR) with CO_2 was proposed by Zhao et al. [9], and was developed by authors [10-11] to COx-free H_2 production. In this process CH_4 is first catalytically decomposed to H_2 and carbon, and in the second step the surface carbon is gasified by steam

---

* Corresponding author. Tel.: +86-22-27405613; E-mail address: ydli@tju.edu.cn

with catalyst regeneration. The operation for SWR varied from single fixed bed reactor [8] to two parallel reactors in cyclic manner [12] and to a micro-vibrational fluidized bed reactor [13].

A few papers compared the process efficiencies between SR and DDM [5, 14-15], and showed that with carbon black revenue, the economy of DDM is comparable to that of SR with CO_2 sequestration. In this work, a route of SWR of methane is proposed and compared with the conventional SR and ATR processes by thermal and mass balances employing a commercial software HSC-Chemistry 4.0. Simplified schemes for each process are given and the reforming efficiency is used as a criterion for comparison.

2. COMMERCIAL CH₄ REFORMING

The technology for H_2 production from natural gas is mature and practiced on a large scale for production of methanol, ammonia and H_2 for petrochemical and other applications [16-18]. Normally three steps are practiced for high purity H_2 production, as shown in Fig. 1: (1) syngas production by reforming; (2) CO conversion to CO_2 by water-gas shift and (3) a gas purification step for CO_2 sequestration and trace amount of CO removal. Traditionally, SR, PO and ATR processes are employed to produce synthesis gas. SR and PO processes are commercialized for a few decades, and ATR is still in plant scale test.

2.1 Steam reforming

$$CH_4 + H_2O \Rightarrow CO + 3H_2 \qquad\qquad \Delta H^0_{298} = 206.2 \text{kJ/mol}$$

In this process, steam reacts with CH_4 in the presence of a nickel catalyst to produce H_2, CO and CO_2 (Fig. 1a). To ensure a minimum concentration of CH_4 in the product, the process normally employs a H_2O/CH_4 ratio of 3 to 5 at a process temperature of around 815°C and pressures up to 3.5 MPa [5]. This reformer is well suited for stationary operation and can produce a relatively high concentration of H_2 (> 70% on a dry base).

2.2 Partial oxidation

$$CH_4 + \frac{1}{2}O_2 \Rightarrow CO + 2H_2 \qquad\qquad \Delta H^0_{298} = -36 \text{ kJ/mol}$$

Partial oxidation reformers react the natural gas with air to produce syngas. The initial oxidation reaction results in heat generation and a high temperature. The heat thus generated is used to preheat the reactant temperature to above 1000°C [3]. This process can be conducted with or without a catalyst.

2.3 Auto-thermal reforming

$$CH_4 + O_2 + H_2O \Rightarrow CO + CO_2 + H_2 \qquad\qquad \Delta H_{r,298} \approx 0$$

Auto-thermal reforming combines the heat effect of both SR and PO by inputting air and steam together (Fig. 1b). The reaction pathways and the relative extents of SR and PO reactions are controlled by a catalyst.

3. STEPWISE REFORMING

$$CH_4 \Rightarrow C + 2H_2 \qquad\qquad \Delta H^0_{298} = 74.8 \text{ kJ/mol}$$

$$C + O_2 + H_2O \Rightarrow CO + CO_2 + H_2 \qquad\qquad \Delta H_{r,298} < 0$$

The origin of the stepwise reforming comes from the DDM process. In DDM [19], carbon materials deposit on catalyst particles. However, in this mode the catalyst is consumable; unloading and reloading the catalyst lowers the economy. In the majority of related publications [12, 20-23] carbon produced was burned off in order to regenerate the catalyst. The combination of methane decomposition and catalyst regeneration is called SWR in literature [10].

By varying the carbon eliminating gas, a new route for SWR can be proposed with the following steps: (1) methane is first decomposed in one reactor with a catalyst; (2) part of the carbon is removed by a mixture of air and steam to produce additional H_2 and (3) the regenerated catalyst is switched back to decomposition and used as a heat carrier. The flow sheet is depicted in Fig. 1c.

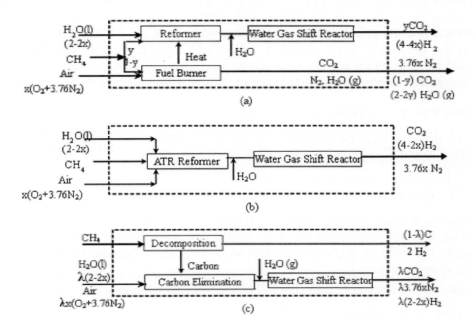

Fig. 1. Schematic diagrams of the idealized methane reforming processes, showing feeds and products from (a) SR, (b) ATR and (c) SWR.

Table 1
Calculated x and y parameters for each process at thermal neutral point

	SR	ATR	SWR (λ)
CH_4 input, /mol	1	1	1
CH_4 reformed, $y(\lambda)$/mol	0.76	1.0	0.5
CH_4 combusted, $1-y(\lambda)$/mol	0.24	0	0.5
Air to CH_4 ratio, x	0.48	0.44	0.285

4. ANALYSIS BASE AND ASSUMPTIONS

For process analysis, a dotted frame is defined for mass and thermal balances [3]. By adding extra amount of CH_4 as fuel for heat supply in SR and SWR, the processes within the dotted frame could be considered as thermal neutral.

The reactant and products' thermal condition is set as standard (298K and 1 atm) and all reactions are assumed to proceed with stoichiometry. In practice, a H_2O/CH_4 ratio of 3~3.5 is used for SR, but in this paper the extra steam not reacted with CH_4 is not included. Also the energy needed to heat the reactant gas, especially the water, and the heat recovery from the effluent is ignored.

A parameter y represents the amount of CH_4 used for reforming; x denotes the air/CH_4 ratio in the dotted frame. The mass flow in each process is set as 1 mole CH_4 input and the value of x and y/λ was derived by thermodynamic calculation and material balancing. In SWR process (Fig. 2c), CH_4 is first decomposed to carbon and H_2, then λ mole carbon is removed from the first reactor and is converted in another reactor to produce heat and additional H_2. So we can say here that λ mole of CH_4 is "reformed". By balancing the heat, the thermal neutral point for each process can be determined by the value of y/λ and x. The specific value of x and y/λ for each process is shown in Table 1.

5. PROCESS EFFICIENCIES

In order to set a base for process comparison, a thermal efficiency for CH_4 reforming is defined as follows [3]:

$$\text{Eff. } \% = \frac{\text{lower heating value of hydrogen produced}}{\text{lower heating value of methane used}} \times 100$$

Here the amount of CH_4 used is referred to the total amount consumed including the part combusted for heat supply. By mass balance, the overall reforming reactions to reach thermal neutral in these three processes could be written as one equation:

$$CH_4 + x(O_2 + 3.76N_2) + (2-2x)H_2O(l) = \lambda C + (1-\lambda)CO_2 + (4-2x)H_2 + 3.76xN_2 \quad \Delta H_{r,298} \approx 0$$

In Table 2, the efficiency of the SWR is compared to that of the SR and ATR. The ATR process is found to be more efficient (93.9%) than SR process (91.3%), and the SWR process the lowest (73%). But when the by-product carbon

Table 2
Comparison of hydrogen yield and reforming efficiencies for SR, ATR and SWR processes

	SR	ATR	SWR
Fuel: Methane	Methane	Methane	Methane
LHV of fuel, kJ/mol	802.531	802.531	802.531
LHV of H_2, kJ/mol	241.582	241.582	241.582
Reformer feeds (mol)			
Methane (total)	1.0	1.0	1.0
Methane for reforming, y/λ	0.76	1.0	0.5[a]
Methane for combustion, $1-y/\lambda$	0.24		0.5[b]
Air	2.285	2.09	0.6783
Water	1.521	1.115	0.715
Reformer products (mol)			
Carbon materials	—	—	0.50
Hydrogen	3.04	3.11	$2^{[c]}+0.715^{[d]}$
Carbon dioxides	0.76	1.00	0.5
Nitrogen	1.8	1.66	0.5358
Steam	0.48		
Reformer products (gas) , %			
Hydrogen	50	53.9	$100^{[c]}+40.84^{[d]}$
Carbon dioxides	12.5	17.3	28.56
Nitrogen	29.6	28.8	30.60
Steam	7.9		
Total	100	100	100
LHV of H_2 produced (kJ)	734.41	751.32	587.04
LHV of by-product (kJ)	17.2		196.75
Reforming efficiency, %	91.7	93.9	73
Efficiency with by-product considered	91.7	93.9	94
CO_2 co-produced, mol/molH_2	0.329	0.322	0.184

[a] In SWR process the amount of CH_4 reformed is referred to the carbon gasified; [b] Here it is referred to the amount of carbon as by-product; [c] Referred to hydrogen by methane decomposition; [d] referred to the hydrogen by carbon reforming

materials are taken into consideration, the efficiency of SWR is as high as 94%. The SR process produces the highest concentration of H_2 (Vol 80%); the ATR process is the lowest with about 53.9% H_2; the SWR process is in the middle with pure and 40.84% H_2. For CO_2 emission, the SR process is the highest with 0.329 mol CO_2 per mol H_2 followed by ATR with 0.32 mol CO_2 per mol H_2; the SWR process is the lowest with 0.18 mol CO_2 per mol H_2. It should be noted that for SWR, the H_2 is separated into two parts: one part for pure H_2 without CO, and the other part is a mixture of H_2 and CO_2 with a ratio of 1.43. These two grades of H_2 can find their applications in different areas.

More hints for operating feasibility is given, the gas flux of reforming 1 mole of CH_4 for each process is listed in Table 3. The SR has the largest gas flux,

Table 3
Comparison of gas flux of each process for reforming 1 mole of methane

	Gas Input /mole	Gas Output /mole	Gas Flux
SR	4.806	6.08	Large
ATR	4.205	5.77	Middle
SWR	2.3933	3.751	Small

thus indicates that the equipment cost should be high; the SWR has the lowest gas flux, at a level of about one half of the SR. Note: for SR the extra steam has not taken into account here. High gas flux lowers the thermal efficiency due to the heat exchange between the cold incoming reactants and the hot effluent. Furthermore, high flux of reactant and effluent means larger reactor size and investment.

6. CONCLUDING REMARKS

(1) As a novel route for CH_4 reforming to H_2, the SWR process is the most thermal efficient, followed by ATR and SR with by product revenue.
(2) H_2 free of CO and high value-added carbon material is co-produced by SWR; the operation is more flexibly and suitably used for small scale applications due to its lower mass intensity.
(3) The SWR process produces the lowest amount of CO_2 per mol of H_2 and has the smallest gas flux per mole CH_4 reformed compared to SR and ATR processes, thus makes it more environmentally benign and energy stint.

ACKNOWLEDGEMENTS

The financial supports from NSF of China (20006012) and a 973 project G199902240 are gratefully acknowledged.

REFERENCES

[1] C.S. Song, Catal. Today 77 (2002) 17
[2] R.A. Hefner III, Int. J. Hydrogen Energy, 27 (2002) 1
[3] S. Ahmed and M. Krumpelt, Int. J. Hydrogen Energy 26 (2001) 291
[4] A. Heinzel, B. Vogel and P. Hubner, J. Power Sources 105 (2002) 202
[5] M. Steinberg and H.C. Cheng, Int. J. Hydrogen Energy 1989, 14, 797
[6] T. Zhang and M. D. Amiridis, Appl. Catal. A 167 (1998) 161
[7] J.N. Snoeck, G.F. Froment, and M. Fowles, J. Catal. 169 (1997) 250
[8] T.V. Choudhary, C. Sivadinarayana, C. C. Chusuei, et al J. Catal. 199 (2001) 9
[9] J.B. Zhao, Y.D. Li and L. Chang, Tianranqi Huagong, 22(4) (1997) 48
[10] T.V. Choudhary and D.W. Goodman, J. Catal. 192 (2000) 316
[11] T.V. Choudhary and D.W. Goodman, Catal. Today, 77 (2002) 65
[12] V.R. Choudhary, S. Banerjee, and A.M. Rajput, J. Catal. 198 (2001) 136
[13] M.A. Ermakova, D.Yu. Ermakov, G.G. Kuvshinov, et al. J. Catal. 187 (1999) 77
[14] N.Z. Muradov, Energy & Fuels, 12 (1998) 41
[15] B. Gaudernack and S. Lynum, Int. J. Hydrogen Energy, 23 (1998) 1087
[16] J.N. Armor, Appl. Catal. A: 176 (1999) 159
[17] L.D. Andrews, J. Power Sources, 61 (1996) 113
[18] R. Ramachandran and R.K. Menon, Int. J. Hydrogen Energy, 23 (1998) 593
[19] Y.D. Li, J.L. Chen, Y.N. Qin and L. Chang, Energy & Fuels, 14 (2000) 1188
[20] M. Matsukata, T. matsushita, and K. Ueyama, Energy & Fuels, 9 (1995) 822
[21] D. Duprez and K. Fadili, Ind. Eng. Chem. Res., 36 (1997) 3180
[22] R. Aiello, J.E. Fiscus, H.C. Loye, et al. Appl. Catal. A: 192 (2000) 227
[23] K. Otsuka, S. kobayashi and S. Takenaka, Appl. Catal. A: 190 (2000) 261

Studies in Surface Science and Catalysis, volume 147
X. Bao and Y. Xu (Editors)

109

NOx-catalyzed gas-phase activation of methane: the formation of hydrogen

Chao-Xian Xiao, Zhen Yan, Yuan Kou[*]

College of Chemistry and Molecular Engineering, Peking University, Beijing 100871, China

ABSTRACT

NO_x-catalyzed oxidation of methane without any solid catalyst has been investigated, hydrogen selectivity of 27% has been obtained with an overall methane conversion of 34% and a free O_2 concentration of 1.7% at 700 °C.

1. INTRODUCTION

Methane is always used as the important resource of hydrogen, however, the partial oxidation of methane to syngas always makes coke formation at the presence of solid catalyst, therefore causes deactivation of the expensive catalyst. Gas-phase oxidation of methane without any solid catalyst can exceed this difficulty when NO_x is employed as catalyst especially.

Gas-phase oxidation of methane without the use of any solid catalyst opens up new possibilities for the direct conversion of methane to valuable chemicals, although results are still limited [1,2] as far as production of oxygenates is concerned when the catalyst employed is NO_x [3-9]. In this paper we show that the formation of hydrogen in significant amounts has always been ignored in recent reaction and mechanistic studies.

2. EXPERIMENTAL

The experiments were carried out under atmospheric pressure in a conventional gas flow system with an empty quartz tube reactor (inner diameter of 5 mm). The reactor was kept in a tubular furnace (20cm in length). The gases were controlled by mass flow controllers and premixed before entering the reactor with a N_2 balance. The temperatures indicated in this paper were the temperatures at the reaction zone of the reactor measured with a thermocouple. The amounts of H_2, O_2, CH_4, and CO in the products were analyzed by on-line gas chromatographs (GC122 and 102G, Shanghai General Analytical Instrument

Factory). The amounts of NO, NO_2, CH_4, CO, CO_2, HCHO, CH_3OH and C_2H_4 were determined by *in situ* or on-line FT-IR using a Bruker Vector 22 spectrometer with a gas cell of 10cm length. The concentrations were determined automatically using the OPUS Quantitative Method with appropriate calibration curves. CO and H_2 selectivities (S_{CO} and S_{H2} respectively) were calculated as follows:

$$S_{CO} = \frac{C_{CO(final)}}{C_{CH_4(initial)} - C_{CH_4(final)}} \times 100\%$$

$$S_{H_2} = \frac{C_{H_2(final)}}{2(C_{CH_4(initial)} - C_{CH_4(final)})} \times 100\%$$

3. RESULTS AND DISCUSSION

Varying the CH_4/O_2 ratio shows that lower CH_4/O_2 ratio, i.e., higher O_2 concentration in the feed gas leads to higher yield and lower selectivity of H_2 (see Fig. 1). When the O_2 concentration is 4% (CH_4/O_2 ratio = 5), 52% of the oxygen is converted, the free O_2 concentration after reaction is calculated to be 1.8% and the yield of hydrogen is only 3% at 700 °C. When the O_2 concentration reaches 10% (CH_4/O_2 =2), the H_2 yield is about 10% at 700-800 °C. In this case, the O_2 conversion is 83% and the free O_2 concentration is calculated to be 1.7%, a little different from the former. As the O_2 concentration increases from 4% to 10% (i.e., CH_4/O_2 ratio decreases from 5 to 2), the selectivity to H_2 goes down slightly, from 31% to 27%. The increase in H_2 yield is a consequence of the improved conversion of CH_4, which is less than 10% at CH_4/O_2 =5 but higher than 30% at CH_4/O_2 = 2 [1]. At CH_4/O_2 = 1, 98.5% of the

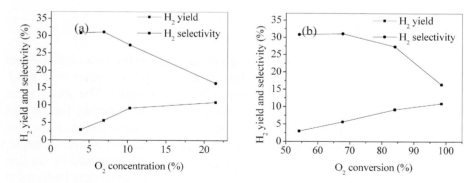

Fig. 1. Effects of O_2 (a) concentration and (b) conversion on H_2 yields and selectivities (CH_4 20%, NO 0.06%, flow rate 40 ml/min, reaction temperature 700 °C)

oxygen in the feed gas is converted, however the H_2 yield does not improve significantly, and especially the selectivity to H_2 is reduced to 16%, with the free O_2 concentration remaining in the reaction system calculated to be about 0.3%.

This means that low concentrations of free oxygen, for example, 1-2%, have little adverse effect on the formation of H_2. It is of particular importance to add here that, when free O_2 concentration is 1.7-1.9%, the H_2 concentration, which is carefully checked against a standard curve, is 3.9-4.5%, giving a H_2/O_2 ratio of about 2.

Studies with varying flow rates between 20 and 200 ml/min (residence times of 3.7s-0.4s) show that stable H_2 selectivity of around 27% can be obtained at 700-800 °C for residence times in the range of 0.6-3.7 s (see Fig. 2).

It is certain that NO_x plays a key role in catalytic conversion of CH_4. It has been found that the H_2 selectivity, increases from 2% to 13% with simultaneous increase of CH_4 conversion when the NO concentration changes from 0.02% to 0.16% under a higher flow rate of 200ml/min (employed in order to obtain reliable accuracy in the measurement of NO concentration). It can be seen from Fig. 3 (a) that the selectivity to CO goes up faster than that to H_2. It is likely that thermal decomposition of CH_4 may occur when the CH_4/NO ratio is greater than 500, giving a high H_2/CO ratio of about 4 (line I in Fig. 3 (b)). As the CH_4/NO ratio goes down, the partial oxidation of methane becomes dominant (line 2 in Fig. 3 (b)), however, complete oxidation of methane to CO_2 as a competitive side-reaction occurs simultaneously (line III in Fig. 3 (b)). Aromatics are also observed in the products (toluene, etc. [2]). This means that, by varying the reaction conditions, gas-phase catalytic oxidation of methane can in principle be used to produce a variety of chemicals. This is likely to be most valuable in those reactions where the presence of a solid catalyst presents separation problems at the end of the reaction. Furthermore, since a product derived from

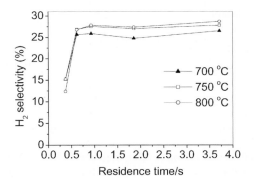

Fig. 2. Effect of residence time on H_2 selectivity at different reaction temperatures (CH_4 20%, O_2 10%, NO 0.06%)

the activation of methane is always believed to be commensurate with the surface structure of a catalyst, gas-phase catalytic oxidation of methane may provide an excellent reference point for comparing what is truly occuring on the surface.

Previous reports in the literature do not give any rationalization for the formation of hydrogen via NO_x-catalyzed gas-phase oxidation of methane. Based on a further consideration of the kinetic model proposed by Bromly [10], it can be seen that several elementary reactions may be responsible for the production of H_2 [6, 10]:

$$CH_4 + H \rightarrow CH_3 + H_2 \tag{1}$$

$$CH_3O + H \rightarrow CH_2O + H_2 \tag{2}$$

$$HNO + H \rightarrow H_2 + NO \tag{3}$$

$$HONO + H \rightarrow H_2 + NO_2 \tag{4}$$

$$CH_3NO_2 + H \rightarrow CH_2NO_2 + H_2 \tag{5}$$

In these reactions, HNO, HONO, and CH3NO2 are believed to be important intermediates [1-3, 8].

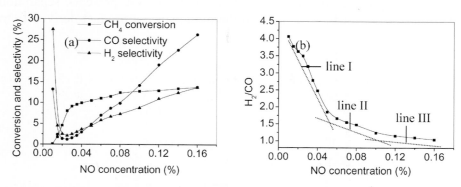

Fig. 3. Effect of NO concentration on (a) CH_4 conversion, CO and H_2 selectivities, (b) H_2/CO ratio (CH_4 12%, O_2 4%, flow rate 200 ml/min, reaction temperature 700 °C)

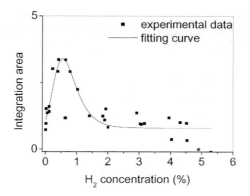

Fig. 4. Correlation between the area of $\upsilon_{C=O}$ IR peak of HCHO and observed H_2 concentration at different temperatures and flow rates (CH_4 20%, O_2 10%, NO 0.06%)

After carefully checking the H_2 concentration, the reaction results obtained at different temperatures and flow rates indicate that, as shown in Fig. 4, the increase of H_2 concentration parallels with the increase of the $\upsilon_{C=O}$ peak area in the IR spectrum of HCHO when the H_2 concentration is lower than 0.5%, suggesting that the reaction (4) is most likely to be the elementary step responsible for the formation of H_2. On the other hand, low yields of HCHO are always correlated with higher H_2 concentrations, suggesting that the decomposition of HCHO is also a possible route to produce H_2: [4, 8]

$$HCHO + NO_2 \rightarrow CHO + HONO \tag{6}$$

$$CH_2O + H \rightarrow CHO + H_2 \tag{7}$$

$$CHO + H \rightarrow CO + H_2 \tag{8}$$

In addition, gas-phase oxidation of methane to form different chemicals by a catalyst other than NO_x opens up an intriguing interest for further study.

REFERENCES

[1] Z. Yan, C.-X. Xiao and Y. Kou, Catal. Lett., 85, 1-2 (2003), 129.
[2] Z. Yan, C.-X. Xiao and Y. Kou, Catal. Lett., 85 (2003), 135.
[3] K. Tabata, Y. Takemoto, and E. Suzuki, Catal. Rev., 44 (2002), 1.
[4] T.E. Layng, and R. Soukup, Ind. Eng. Chem., 20 (1928), 1052.
[5] B.H. Mcconkey, and P.R. Wilkinson, I&EC Proc. Des. & Dev., 6 (1967), 436.
[6] K. Tabata, Y.H. Teng, and Y. Yamaguchi, J. Phys. Chem. A, 104 (2000), 2648.
[7] L. Han, S. Tsubota, and M. Haruta, Chem. Lett., (1995), 931.
[8] K. Otsuka, R. Takahashi, and I. Yamanaka, J. Catal., 185 (1999), 182.

[9]   Y. Teng, Y. Yamaguchi, and T. Takemoto, PCCP, 2 (2000), 3429.
[10] J. H. Bromly, F. J. Barnes, S. Muris, and B. S. Haynes, Combust. Sci. Tech., 115 (1996), 259.

Studies in Surface Science and Catalysis, volume 147
X. Bao and Y. Xu (Editors)
115

Hydrogen production by ethanol steam reforming. Study of cobalt-doped Ce-Zr catalysts.

J.C. Vargas, E. Vanhaecke, A.C. Roger[*], A. Kiennemann

Laboratoire des Matériaux, Surfaces et Procédés pour la Catalyse, UMR 7515, ECPM - ULP, 25 rue Becquerel, 67087 Strasbourg Cedex 2, France

ABSTRACT

Insertion of cobalt into Ce-Zr fluorite oxides is studied to provide active and selective catalysts for ethanol steam reforming. The controlled reduction of cobalt generates nanometric cobalt particles in strong interaction with the host oxide it comes from. High selectivity to hydrogen can be achieved compared to Ce-Zr fluorite oxides.

1. INTRODUCTION

In the last decade considerable efforts have been expended to develop the conversion of the abundant primary sources of energy into secondary ones that are less polluting, e.g. hydrogen, which may become an important fuel in the future for use as an energy carrier for electric vehicles and electric power plants. There exist several routes for hydrogen production.

An important source of hydrogen for stationary fuel cell applications is natural gas, while for transportation; methanol and gasoline are being considered [1]. In recent years, the production of hydrogen from these fossil-fuel-dependent sources has increased [2].

Renewable energy technologies could significantly contribute to solve the problems of energy supply, environmental protection and regional development. The production of hydrogen from non fossil-fuel-dependent sources is being developed. The attractively of bioethanol, produced by biomass fermentation, as hydrogen source is high (high hydrogen content, non-toxicity, safety storage). The development of active, selective and stable catalysts for ethanol steam reforming is a key point.

Various catalysts have been reported in the literature to be efficient in ethanol steam reforming: noble metals based catalysts on various supports, transition metals (Co, Ni, Cu and Zn) based catalysts, oxide catalysts with a wide range of redox and acid-base properties [3][4]. However, only recent works [5][6] are dedicated to the search for the active phase, the determination of which is necessary to obtain optimum catalyst design.

The aim of this work is to combine the properties of metal (cobalt) with that of the oxide by studying of cerium-zircon oxide as host for cobalt. The active catalyst is generated by controlled reduction of the precursor mixed oxide.

2. EXPERIMENTAL

2.1. Preparation of catalysts

A series of $Ce_2(Zr_{(1-x)}Co_x)_2O_{8-\delta}$ was synthesised by a sol-gel like method based on the thermal decomposition of mixed propionates. The value of x was varied between 0 and 0.625. The starting materials for Ce, Zr and Co were cerium (III) acetate hydrate, zirconium (IV) acetylacetonate and cobalt (II) acetate hydrate, which all led exclusively to propionate precursors in propionic acid. The starting salts were dissolved in boiling propionic acid in a concentration of 0.12 $mol.L^{-1}$ in cation. Boiling solutions were mixed and the solvent was evaporated until a resin was obtained. The resin was heated at 2 $K.min^{-1}$ under air until 773 K and maintained 6 h.

2.2. Catalytic test

Catalytic studies of ethanol steam reforming were carried out at atmospheric pressure in a U-shaped quartz reactor. Catalyst was reduced *in situ* under a 3 $mL.min^{-1}$ flow of pure hydrogen, heated at 2 $K.min^{-1}$ until 713 K and maintained 12 h. After reduction, inert gases (N_2:Ar=1:4) were admitted at 2.1 $L.h^{-1}$. When no more H_2 was detected, the reacting gases (6:1 molar H_2O:EtOH solutions at 0.9 $L.h^{-1}$ in gas) were admitted on the catalyst. The catalytic reaction was carried out using a temperature from 713 to 913 K, on 0.16 g of catalyst (GHSV = 26000 h^{-1}). The catalyst was maintained 1 hour at 713 K, 1 hour at 813 K, and 1 hour at 913 K (slope : 2 $K.min^{-1}$). Gas products were analyzed by on-line μ-GC.

2.3. Catalyst characterisation

The mixed oxides were characterized by XRD (room temperature, Siemens D5000 diffractometer using the Cu Kα radiation), SEM (Jeol JSM-6700E field emission microscope) using EDXS, and elemental analysis. Specific surface area was determined by isotherm of adsorption of nitrogen using BET method (Coulter SA 3100 apparatus). The reducibility of the precursor mixed oxides was studied by thermo-programmed reduction (TPR) on 50 mg of catalyst heated from room temperature to 1173 K at 15 $K.min^{-1}$ under 10% H_2/He with a total gas flow of 50 $mL.min^{-1}$.

3. RESULTS AND DISCUSSION

Catalyst series was studied before and after catalytic test. In Fig. 1 are presented the diffractograms of the six fresh catalysts.

The method of catalyst preparation allows the crystallization of the mixed oxides in the fluorite structure as low as 500 °C. By XRD, only one phase is detected (for x < 0.45), corresponding to the main reflections of cubic fluorite CeO_2 ((111), (200), (220) and (311) planes), with a smaller lattice parameter, in accordance to the variation of the ionic radii of $Ce_2(Zr_{(1-x)}Co_x)_2O_{8-\delta}$ with respect to CeO_2. Among the series the cubic lattice parameter slightly increases with the cobalt content. For the more cobalt-loaded catalysts (x=0.625 and x=0.45) an additional phase of cobalt oxide Co_3O_4 is detected. Until x=0.375 $(Ce_2Zr_{1.25}Co_{0.75}O_{8-\delta})$ cobalt cations are integrated in the fluorite lattice without any phase rejection.

The catalysts have been studied by SEM and TEM. The powders are made of grains of nanometric size (~5nm) and appear highly homogeneous.

The specific surface areas of mixed oxides are high, around 100 $m^2.g^{-1}$.

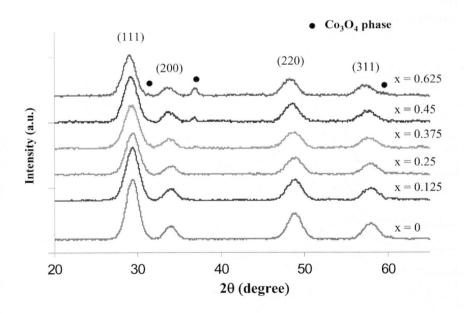

Fig. 1. Diffractograms of the fresh catalysts at various cobalt loading (x).

The $Ce_2(Zr_{1-x}Co_x)_2O_{8-\delta}$ mixed oxides being prepared to be then reduced to generate metal cobalt particles, particular attention has been paid to their reducibility. First their hydrogen consumption versus temperature was followed by TPR. The TPR profiles are presented in Fig. 2.

The cobalt free $Ce_2Zr_2O_8$ (x=0) reduces at 793 K. The H_2 consumption corresponds to the reduction of 58 % of Ce^{4+} into Ce^{3+}. Assuming that insertion of cobalt instead of zirconium in the mixed oxide (x≠0) does not affect the reduction rate of cerium [7], the two reduction steps at 593 and 933 K are attributed to cobalt reduction from the mixed oxide. These two peaks increase with the cobalt content. The reduction of cobalt in the mixed oxide at 593 K helps the cerium reduction which then occurs at lower temperature than in the cobalt free oxide (693 K instead of 793 K). The particular TPR profile of x=0.625 is due to the reduction of the additional Co_3O_4 phase, non integrated in the fluorite structure. For x=0.45, despite the Co_3O_4 detected by XRD, the reduction behaviour is similar to those of the pure solid solutions. In that case, the Co_3O_4 grains are thought to be in strong interaction with the Ce-Zr-Co oxide, which is consistent with the very small particle size.

After reductive pre-treatment at 713 K no cobalt metal phase is evidenced by XRD. However, for x=0.25 magnetic measurements under hydrogen evidence the formation of metallic Co° nanometric particles [7].

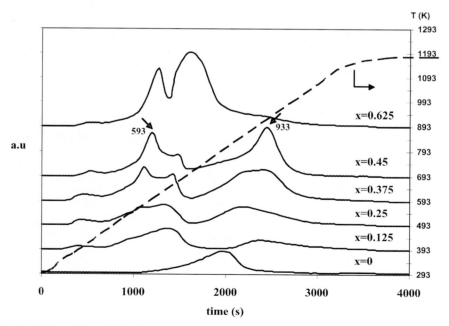

Fig. 2. TPR profiles of the fresh catalysts at various cobalt loading (x).

After in situ reduction at 713 K the ethanol steam reforming was carried out. The catalytic results are presented in Table 2.

Ethanol conversion was always 100%, except for the cobalt free catalyst ($Ce_2Zr_2O_8$ fluorite, x=0) at 713 K. This catalyst presents a non negligible activity (87 % of EtOH conversion at 713 K), however the selectivity into H_2 is much lower than cobalt containing fluorites. The undesirable products as ethylene are favored. The cobalt containing catalysts clearly favor the hydrogen production (see Fig. 3). As soon as cobalt is introduced in the Ce-Zr oxide (x=0.125) the H_2 production is increased from 44 % to 67 % of the gas phase. That corresponds to an increase of the hydrogen yield from 0.06 g_{H2} h^{-1} g_{cat}^{-1} to 0.18 g_{H2} h^{-1} g_{cat}^{-1} at 713 K (see Fig. 3). The increase of cobalt content in the fluorite series leads to a maximum hydrogen yield at 813 K for x=0.45 (the most Co-loaded system in which almost all Co is integrated in the fluorite oxide). Additional cobalt does not enhance the hydrogen production (x=0.625 compared to x=0.45). This is in accordance with a beneficial effect of a strong metal/oxide interaction claimed by Llorca et al. [5,6], and generated by the extraction of Co° from the host fluorite precursor oxide.

Table 2
Catalytic behaviour of $Ce_2(Zr_{1-x}Co_x)_2O_{8-\delta}$

Fresh Catalyst	T (K)	selectivity (molar %)							
		H_2	CO_2	CO	CH_4	C_2H_4	C_3H_6	C_2H_4O	C_2H_6O
$Ce_2Zr_{0.75}Co_{1.25}O_{8-\delta}$	713	70.3	24.5	2.6	2.6	-	-	-	-
x=0.625	813	73.4	19.8	4.9	1.9	-	-	-	-
	913	72.4	18.0	9.3	0.4	-	-	-	-
$Ce_2Zr_{1.1}Co_{0.9}O_{8-\delta}$	713	71.6	22.9	4.6	0.9	-	-	-	-
x=0.45	813	70.7	22.4	5.0	1.9	0.1	-	-	-
	913	70.7	20.2	8.2	0.9	-	-	-	-
$Ce_2Zr_{1.25}Co_{0.75}O_{8-\delta}$	713	70.5	19.4	5.8	0.7	0.6	-	-	3.1
x=0.375	813	70.9	21.6	5.1	2.2	-	-	-	-
	913	71.5	20.0	7.1	1.3	-	-	-	-
$Ce_2Zr_{1.5}Co_{0.5}O_{8-\delta}$	713	69.5	18.9	4.7	0.7	1.0	0.1	-	5.3
x=0.25	813	69.6	22.0	5.6	2.8	0.1	-	-	-
	913	68.2	21.5	8.9	1.4	-	-	-	-
$Ce_2Zr_{1.75}Co_{0.25}O_{8-\delta}$	713	67.6	17.8	2.9	0.7	2.9	0.2	0.1	7.9
x=0.125	813	63.7	19.3	5.2	4.6	5.7	0.1	0.2	1.1
	913	70.3	16.6	11.3	1.5	0.1	-	0.1	-
$Ce_2Zr_2O_8$	713	43.9	12.7	0.6	0.8	33.1	2.7	3.0	3.3
x=0	813	31.6	19.1	1.9	16.1	30.4	0.5	0.4	-
	913	45.1	32.3	3.6	9.5	9.5	0.1	-	-

120

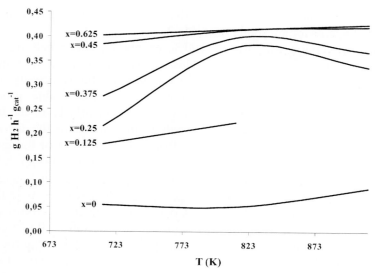

Fig. 3. Hydrogen yield as a function of reaction temperature

4. CONCLUSION

Efficiency of fluorite type defined structure as precursor of cobalt-based catalyst active in ethanol steam reforming was demonstrated. The controlled reduction of cobalt in the mixed fluorite at 713 K generates small Co° particles in strong interaction with the host oxide it comes from. The maximum hydrogen production corresponds to the maximum cobalt insertion in the fluorite Ce-Zr lattice. Since not inserted in the host lattice, additional cobalt has no beneficial effect on the H_2 yield.

ACKNOWLEDGEMENTS

Financial supports of ECOS-Nord and COLCIENCIAS are recognized.

REFERENCES

[1] A.N. Fatsikostas, D.I. Kondarides, X.E. Verykios, Chem. Commun., (2001) 851.
[2] J.N. Armor, Appl. Catal. A, 176 (1999) 159.
[3] J. Llorca, P.R. de la Piscina, J. Sales, N. Homs, Chem. Commun., (2001) 641.
[4] F. Aupêtre, C. Descorme, D. Duprez, Catal. Commun., 3 (2002) 263.
[5] J. Llorca, J.A. Dalmon, P.R. de la Piscina, N. Homs, Appl. Catal. A, 243 (2003) 261.
[6] J. Llorca, J.A. Dalmon, P.R. de la Piscina, J. Sales, N. Homs, Appl. Catal. B, 43 (2003) 365.
[7] E. Vanhaecke, J.C. Vargas, A.C. Roger, A. Kiennemann, Proc. 1st European Hydrogen Energy Conference, Grenoble, France (2003).

Studies in Surface Science and Catalysis, volume 147
X. Bao and Y. Xu (Editors)

Steam reforming of hydrocarbons.
A historical perspective

Jens Rostrup-Nielsen

Haldor Topsøe A/S, Nymøllevej 55, DK-2800 Lyngby, Denmark

ABSTRACT

After the industrial break-through of steam reforming, main problems were linked to sulphur poisoning, carbon formation and attempts to minimize the reformer size. The understanding of the catalysis followed the industrial developments with most studies carried out by industrial groups. Today, the hydrogen economy has created great interest in academia and industry to explore the limits of the reaction.

1. INDUSTRIAL DEVELOPMENTS

The steam reforming reaction was introduced to industry in 1930 more than 70 years ago [1, 2, 3, 4], although the reaction had not been subject to many scientific studies [1, 3]. The reforming process was adopted mainly in the US where natural gas was easily available as feedstock whereas reformers in Europe were initially reduced to operate on propane and LPG.

The industrial break-through of the steam reforming technology occurred in 1962 when ICI succeeded in starting two tubular reformers operating at pressure (15 bar) and using naphtha as feedstock [5]. The first reformer designed by Topsoe was started in 1956 and the first naphtha reformer in 1965 followed by a hydrogen plant in 1966 operating at 42 bar [1, 4]. The introduction of high pressure reforming meant that the energy consumption of the ammonia synthesis could be decreased significantly [6] which was further decreased to less than 28 GJ/t by integration of the steam turbine cycle and process and application of better catalysts. As a result, a number of natural gas based ammonia plants were built in the US primarily by MW Kellogg in the mid sixties [7].

At the same time in Europe, steam reforming of naphtha became the key technology in the build-up of the UK towns gas industry replacing low pressure gasification processes. Apart from the tubular reforming technology provided by ICI [5] and Topsøe [8] low temperature steam reforming of naphtha in an adiabatic reactor was used to produce methane-rich gas [1, 3]. This process was pioneered by British Gas [9] originally using methanol as fuel. The steam reforming of naphtha was used also for ammonia plants in countries like India [10] and for hydrogen in refineries in particular in Japan [11], where natural gas was not available. The steam reforming process also became the technology for syngas for methanol and other petrochemical products, oxo-alcohols, acetic acid, etc. [1, 12]. During the energy crisis in the 1970'ties, low temperature steam reforming followed by methanation was used to convert naphtha into substitute natural gas [4]. During the 1970'ties and 80'ties, it was predicted that the steam reforming of natural gas would form the basis for a new industry for direct reduction of iron-ore [13, 14] which however did not fully develop. Steam reforming was also applied to supply hydrogen for a series of prototype fuel cell plants started in the same period [15].

Today, the steam reforming process is well-established for the production of hydrogen and syngas for large scale commodity chemicals [16]. The need for hydrogen is increasing in refineries due to demands for low sulphur, low aromatics and hydrocracking of heavy feedstocks [17]. The steam reforming of natural gas and liquid fuels (logistic fuels) may play a key role for the introduction of a hydrogen economy. Hydrogen for fuel cells will be an important element [18], but steam reforming is a so a key reaction in the anode chamber of high temperature fuel cell [19]. In spite of efforts to produce hydrogen by schemes involving solar energy, bio-fuels etc., steam reforming of natural gas remains the cheapest route to hydrogen in the near term [17].

The steam reforming reaction may be applied as a mean for chemical recuperation [20] as pioneered at the ADAM/EVA system [20] for nuclear energy and now considered for solar energy. Chemical recuperation may also be applied with high temperature fuel cells (SOFC) coupled to gas turbine schemes [16].

The manufacture of syngas in the first step in the conversion of remote natural gas into liquids for instance via the FT synthesis. However, for large scale syngas plants, oxygen-blown autothermal reforming becomes more feasible [16]. Also for very small scale operation, oxidative processes might become more economic because of simplicity and compactness [18].

2. TOWARDS THE LIMITS

Steam reforming involves the risk of *carbon* formation in particular at operation with low steam to carbon ratios or at conditions aiming at low H_2/CO ratios. The potential for carbon formation could be assessed by the principle of equilibrated gas [12]. The mechanism of carbon formation via whisker carbon was studied by primarily electron microscopy [1, 21] and TGA [22, 23].

The thermodynamic limits were surpassed by using noble metals [22, 24] or by using a partly sulphur passivated nickel catalyst [25] as applied in the

SPARG process [26]. Both solutions were introduced to industry during the 1980'ties at conditions at which thermodynamic would predict carbon formation.

When steam reforming of naphtha was introduced in the mid 60'ties, *sulphur* poisoning was a problem [27,28]. The typical event was that the inlet layer was sulphur poisoned which resulted in non-converted naphtha passing to the hot part of the reforming tube where carbon would be formed by pyrolysis and the development of hot tubes [12].

Around 1970, precise knowledge was obtained on the chemisorption of sulphur on nickel – not the least through LEED-studies [29]. It was possible to estimate [12] that the true sulphur limits ca. 10 ppb and not 0.2 ppm (the analytic limit in the 60'ties). For normal naphthas, the sulphur limit could easily be achieved over new catalysts and absorption masses [27].

At first, alkali *promoted catalysts* were used for the tubular steam reforming of naphtha [5, 8], but it was possible to eliminate the potassium promotion and the problems associated with alkali by using active magnesia [1, 30]. At the same time, it was possible to achieve much higher activity because alkali was shown to drastically reduce the activity of the nickel surface [1]. It meant that the alkali-free catalyst could convert heavy naphtha with high contents of aromatics and hence less reactivity [10]. It was shown that if the feedstock was properly desulphurised, the catalyst could also steam reform kerosene and gas oil with the formation of only C_1-components [1, 10, 31]. The sulphur problem was solved with the introduction of a *prereformer* operating as a low temperature adiabatic reformer [16, 31]. The prereformer converts the higher hydrocarbons before entering the preheater to the tubular reformer. Without the higher hydrocarbons, it became possible to preheat to high temperatures without the risk of pyrolysis. This principle was introduced also to natural gas based reformers during the 1980'ties [16, 31]. The preformer was also applied in schemes for high temperature fuel cells [18].

The steam reformer is highly efficient, but the tubular reformer is very costful. Therefore there has been a strong effort to *minimize the size* of the tubular reformer. The steam reforming process may appear straightforward because the product gas is determined by thermodynamic equilibrium. It is however a complex coupling of heat transfer and catalysis and computer models are required to estimate tube wall temperatures and the risk for carbon formation. These models are not too sensitive to reactions kinetics as the gas is close to equilibrium in most of the tube [12].

The first kinetic studies were strongly influenced by diffusion restrictions [2, 12] and it was not until the Temkin group in the 1960'ties [32] studied the reaction on nickel foil that more precise information was achieved. Still empirical kinetics [33] were sufficient to develop highly sophisticated reactor models for the tubular reformer [34, 35].

With proper desulphurisation and the use of a prereformer, the catalyst activity is rarely a limiting factor [16]. The availability of better materials for reformer tubes and a better understanding of the stress problems allowed an increase of the heat transfer and hence a reduction of the number of tubes for a given service. With a prereformer, it is possible to increase the inlet temperature to the tubular reformer and thus reduce the size of reformer. Today reformers are designed for operation at average heat fluxes [17,36] almost two times higher than what was the industrial practice 20 years ago.

It is possible to increase the amount of heat transferred to the process gas in the reformer from about 50% to about 80% of the supplied heat when using a convective heat exchange reformer [16,36]. This resulted in a more compact piece of equipment, and it is possible to design for no-steam export. This type of *convective reformers* were used for fuel cell demonstration plants during the 1970'ies and 80'ties [37]. Today, it is applied for minor hydrogen plants [17, 36]. More compact designs can be developed for small scale reformers. It is a challenge to take advantage of the huge surplus of activity by using catalysed hardware and non-tubular designs [38] as explored in micro-channel reformers [17, 39, 40]. Another route may be membrane reforming at low temperature [17, 41].

Steam reforming of hydrocarbons in compact units may become the vehicle for introduction of a *hydrogen economy* as the preferred choice for decentralized units at gas stations or for fuel cells.

3. TOWARDS COLLABORATIVE EFFORTS

The development of the steam reforming process has not been based on initial understanding of the catalysis. The progress was driven mainly by an "inductive approach" with feedback from industrial operation and pilot tests [4, 35]. The initial vague ideas on the mechanism [1, 42, 43] were formulated by industrial groups around 1970 whereas only few academic scientists were involved in studies of the reaction [32, 44, 45, 46]. Research on nickel catalysts was more related to hydrogenolysis [47, 48] and later methanation [49, 50] became of interest in SNG from coal in the late 1970'ties. In the 1990'ties, the discussion about the greenhouse effect caused a series of studies on CO_2-reforming [24, 50] although CO_2-reforming may hardly solve the greenhouse problem and that stoichiometric reforming is rarely feasible [16]. Lately, the prospects of a hydrogen economy have caused a number of projects on compact small scale steam reformers [18, 39, 40].

Early steam reforming studies [1] at well-defined conditions aimed at describing parameters influencing the nickel surface area and the specific activity. This was later supplemented by ultra-high vacuum studies of methane activation on well-defined surfaces [51, 52, 53]. Recent studies [16, 54] have led to a more consistent understanding of the mechanism at the atomic level. DFT-calculations, adsorption studies [54] and in-situ high resolution electron microscopy [55] have shown that step sites play a key role in methane activation as well as for nucleation of carbon. This had led to a more rational approach to catalyst promotion [16]. It has revived old ideas of B5 sites [1, 56]. The development since the late 1980'ties has been characterised by significant contributions from academia [33, 50, 57, 58, 59, 60, 61, 62] and a balanced collaboration between academic and industrial research teams [16, 53, 54, 63, 64, 65, 66, 67].

It was the fundamental understanding of carbon formation and sulphur poisoning in terms of ensemble control [25] which formed the basis for the introduction of the SPARG process in the 1980'ties [26]. The present knowledge at the atomic scale [16, 67] may lead to more advanced reforming systems and it has contributed to the general knowledge of metal catalysis.

REFERENCES

[1] J.R. Rostrup-Nielsen "Steam Reforming Catalysts", Danish Technical Press, Copenhagen 1975.
[2] J.P. van Hook, Catal. Rev.-Sci. Eng. 21 (1981) 1.
[3] M. Appl, H. Gössling, Chem. Z. 96 (3) (1972) 135.
[4] J.R. Rostrup-Nielsen, Catal. Today, 71 (2002) 243.
[5] G.W. Brigder, Chem. Eng. Prog. 53 (1972) 39.
[6] I. Dybkjær in "Ammonia" (A. Nielsen, ed), Springer, Berlin 1995, p. 199.
[7] H.C. Mayo and J.A. Finneran, Oil Gas J., 66 (12) (1968) 78.
[8] H. Topsøe, Inst. Gas Eng. J, 6 (1966) 401.
[9] H.S. Davies, K.J. Humphries, D. Hebden, D.A. Percy, Inst. Gas Eng. J., 7 (1967) 708.
[10] J.R. Rostrup-Nielsen, Ammonia Plant Saf. 15 (1973) 82.
[11] J.R. Rostrup-Nielsen, Chem. Eng. Prog., 73 (1977) 87.
[12] J.R. Rostrup-Nielsen "Catalysis, Science and Technology" 5, Springer, Berlin 1984, 1.
[13] H.J. Töpfer, Gas Wasserfach 117 (1976) 412.
[14] H. Topsøe, J.R. Rostrup-Nielsen, Scan. J. Metallurgy, 8 (1979) 168.
[15] E.A. Gilles, Chem. Eng. Prog. 76 (10) (1980) 88.
[16] J.R. Rostrup-Nielsen, J. Sehested, J.K. Nørskov, Adv. Catal., 47 (2002) 65.
[17] J.R. Rostrup-Nielsen, T. Rostrup-Nielsen, Cattech, 6 (2002) 150.
[18] J.R. Rostrup-Nielsen, K. Aasberg-Petersen, in "Fuel Cell Handbook", vol. 3, Wiley, New York 2003 p.159.
[19] B. Paetsch, P.S. Patel, H.C. Maru, B.S. Baker, Abst. Fuel Cell Seminar 1986, 143.
[20] H. Fedders, R. Harth, B. Höhlein, Nucl. Eng. Des. 34 (1975) 119.
[21] R.T.K. Baker, M. Barber, P. Harris, F. Feates, R. Waite, J. Catal., 26 (1972) 51.
[22] L.S. Lobo, D.L. Trimm, J.L. Figueiredo, Proc. 5th ICC, Palm Beach 1972, p. 1125.
[23] J.R. Rostrup-Nielsen, D.L. Trimm, J. Catal., 48 (1977) 155.
[24] J.R. Rostrup-Nielsen, J.-H. Bak-Hansen, J. Catal., 144 (1993) 38.
[25] J.R. Rostrup-Nielsen, J. Catal., 85 (1984) 31.
[26] N.R. Udengaard, J.-H. Bak Hansen, D.C. Hanson, J.A. Stal, Oil Gas J., 90 (1992)62.
[27] J.A. Lacey, S.K. Mukherjee, Inst. Gas Eng. J., 7 (1967) 472.
[28] S. Morita and T. Inoue, Int. Chem. Eng., 5 (1965) 180.

126

[29] M. Perdereau, C.R. Acad. Sci. Ser. C., 267 (1968) 1107.
[30] J.R. Rostrup-Nielsen, P.B. Tøttrup in "Symposium on Science of Catalysis and its Application in Industry", FPDIL, Sindri 1979, p. 380.
[31] J.R. Rostrup-Nielsen, I. Dybkjær, T.S. Christiansen, Stud. Surf. Sci. Catal.113 (1998). 81
[32] I.M. Bodrov, L.O. Apel'baum, M. Temkin, Kinet. Catal., 5 (1964) 614.
[33] J. Xu, G.F. Froment, AIChE J. 35 (1989) 88.
[34] A.M. Groote, G.F. Froment, Rev. Chem. Eng., 11 (2) (1995) 145.
[35] J.R.Rostrup-Nielsen, L.J. Christiansen, J.-H. Bak Hansen, Appl. Catal. 43 (1988) 287.
[36] I. Dybkjær; S.E.L. Winter Madsen, Int. J. Hydrocarbon Eng., 3 (19) 556 (1997/98).
[37] H. Stahl, J.R. Rostrup-Nielsen, N.R. Udengaard, Abst. Fuel Cell Seminar, 1985, 83.
[38] R. Charlesworth, A. Gough, C. Ramshaw, Proc. 4th UK Heat Transf. Conf., IMechE (1995) p. 85.
[39] C. Cremers, U. Stimming, J. Find, J.A. Lercher, K. Haas-Santo, O. Görke, K. Schubert, Abst. Fuel Cell Seminar, 2002, p. 643.
[40] P. Irving, T. Moeller, Q. Ming, A. Lee, Abst. Fuel Cell Seminar 2003, p. 652.
[41] E. Kikuchi, Catal. Today 25 (1995) 333.
[42] D.A. Dowden, C.R. Schnell, G.T. Walker, 4th ICC, Moscow 1968, prep. 62.
[43] S.P.S. Andrew, Ind. Eng. Chem. Prod. Res. Dev. 8 (1969) 321.
[44] S.A. Balashova, T.A. Slovokhotova, A.A. Balandin, Kinet. Katal. 7 (1966) 303.
[45] J.R. Ross, M.C.F. Steel, J.Chem.Soc. Faraday Trans., 1, 69 (1973) 10.
[46] J. Barcicki, A. Denis, W. Grzegorizyk, D. Nazimek, T. Borowiecki, React. Kinet. Catal. Lett., 5 (1976) 471.
[47] J.H. Sinfelt, Catal. Rev. 3 (1969) 175.
[48] G.A. Martin, J.Catal. 60 (1979) 345.
[49] C.H. Bartholomew, R.J. Farrauto, J. Catal. 45 (1976) 41.
[50] J.G. McCarthy, H. Wise, J. Catal. 57 (1979) 406.
[51] M.C.J. Bradford, M.A. Vannice, Catal. Rev.-Sci. Eng. 41 (1999) 1.
[52] T.P. Beebe Jr., D.W. Goodman, B.D.Kay, J.T. Yates Jr., J. Chem. Phys., 87 (1987) 2305.
[53] J.H. Larsen, I. Chorkendorff, Surf. Sci. Rev. 35 (1999) 163.
[54] H.S. Bengaard, I. Alstrup, I. Chorkendorff, S. Ullmann, J.R. Rostrup-Nielsen, J.K. Nørskov, J. Catal., 187 (1999) 238.
[55] S. Helveg, C. López-Cartes, J. Sehested, P.L. Hansen, B.S. Clausen, J.R. Rostrup-Nielsen, F. Abild-Pedersen, J.K. Nørskov, Nature (in print).
[56] R. van Hardeveld, A. van Montfort, Surf. Sci. 17 (1969) 90.
[57] A.J.H.M. Kock, P.K. de Bokx, E. Boellaard, W. Klop, J.W. Geus, J. Catal. 96(1985) 468.
[58] Z. Zhang, X.E. Verykios, Catal. Lett. 38 (1996) 175.
[59] J.H. Bitter, K. Seshan, J.A. Lercher, J.Catal. 171 (1997) 279.
[60] D.L. Trimm, Catal.Today 37 (1997) 233.
[61] C. Li, Y. Fu, G. Bian, T. Hu, Y. Xie, J. Zhan, J. Nat. Gas. Chem. 12 (2003) 167.
[62] J. Wei, E. Iglesia. This volume.
[63] I. Alstrup, M.T. Taveres, C.A. Bernardo, O. Sørensen, J.R. Rostrup-Nielsen, Matr. Corr. 49 (1998) 367
[64] J. Sehested, A. Carlsson, T.V.W. Janssens, P.L. Hansen, A.K. Datye, J. Catal. 197 (2001) 200.
[65] M.V. Twigg, J.T. Richardson, Trans. Chem. E. 180A (2002) 183.
[66] J.W. Snoeck, G.F. Froment, M. Fowles, Ind. Eng. Chem. Res. 41 (2002) 4253.
[67] F. Besenbacher, I. Chorkendorff, B.S. Clausen, B. Hammer, A.M. Moelenbroek, J.K. Nørskov, Science 279 (1998) 191.

Studies in Surface Science and Catalysis, volume 147
X. Bao and Y. Xu (Editors)

Development of synthesis gas production catalyst and process

Fuyuki Yagi[*a], **Shuhei Wakamatsu**[a], **Ryuichirou Kajiyama**[a],
Yoshifumi Suehiro[b], **Mitsunori Shimura**[a]

[a]R&D Center, Chiyoda Corporation, Kawasaki, Kanagawa, 210-0855, Japan

[b]Technology Research Center, Japan National Oil Corporation, Mihama, Chiba, 261-0025, Japan

ABSTRACT

The CO_2 reforming catalyst and process have been developed. The commercial size CO_2 reforming catalyst has been demonstrated its activity and stability in the GTL pilot plant tests for more than 1,100 hours, and the first GTL oil from natural gas in Japan was produced in November 2002. In these tests, synthesis gas was generated by CO_2 and H_2O reforming with low CO_2/Carbon and H_2O/Carbon ratios in feed gas composition. The proprietary catalyst which has high resistance against carbon deposition could allow the stable operation under CO_2 and H_2O reforming conditions. The CO_2 reforming simulator could well represent the results of the pilot plant tests. The proposed process showed superior economics on the feasibility studies of applications to GTL processes.

1.INTRODUCTION

Recently, Gas To Liquid (GTL) processes are greatly paid attention as effective technologies for utilization of natural gas fields on middle or small scales which are not economical for LNG production[1-4]. In these processes, the capital cost of synthesis gas production section is considered to account for around 60% of the total capital cost[3, 4]. Therefore the development of more compact syngas production processes with high energy efficiency has been expected.

New CO_2 reforming catalyst and process have been developed on GTL national project by Japan National Oil Corporation(JNOC) since 1999. The process can directly generate the stoichiometrically balanced synthesis gases which have excess of neither H_2 nor CO as suitable feed gases for Fisher-Tropsch (FT) synthesis, methanol and dimethyl ether and so forth. In this process, natural gas is reformed in a tubular reformer with CO_2 and H_2O as

Fig. 1. Overview of JNOC GTL pilot plant (7BPD)

reforming agents so an expensive oxygen plant isn't needed.

The proprietary catalyst with high resistance against carbon deposition can produce synthesis gases by lower CO_2/Carbon and H_2O/Carbon molar ratios in the feed gas composition with low energy consumption. The pilot plant tests of this process were successfully carried out at Yufutsu gas field in Japan last year.

2. EXPERIMENTAL

Accelerated catalysts life tests (Dry reforming) to develop the commercial catalyst in the bench scales were employed. The commercial size ring catalysts of 16mm outer diameter, 8mm inner diameter and 16mm length were prepared by loading of novel metals on metal oxide supports. An average side crushing strengthof the catalysts was 500N/piece. These catalysts were crushed into 2-4mm size, and 50cc of them were loaded to the reactor. The reaction tests were carried out with feed CO_2/CH_4 ratio of 1.0, the reaction conditions at 850℃, 2.0MPaG and GHSV=6000 1/HR. Under the dry reforming conditions, the product synthesis gass have high thermodynamic potential for carbon formation. Nielsen et. al. defined the carbon activity as the indication of carbon formation possibility[5-7]. If this value exceeds to 1.0, carbon will deposit in the gas-phase. Under dry reforming condition, the carbon activity at the bottom of the catalyst bed was 2.6.

For the pilot plant operations, about 120L of commercial size catalysts were charged in mono-tube reformer which has 12.0m length and 11.0cm inner diameter, and were reduced at around 800℃ under atmospheric pressure before starting the reaction tests. Synthesis gas with $H_2/CO=2$ was generated directly by CO_2 and H_2O reforming of natural gas. Feed H_2O/CO_2/Carbon ratio was 1.13/0.40/1, and reaction conditions were at 890℃, 1.9MPaG and GHSV=3000

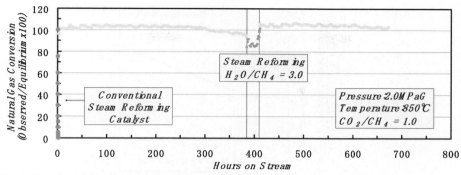

Fig. 2. Results of Accelerated Catalyst Life Tests

1/HR. The total content of higher hydrocarbons in natural gas, for example ethane and propane, was around 13% and total sulfur content was around 1ppm.

3. RESULTS AND DISCUSSION

The typical results of the dry reforming are shown in Fig 2. A conventional steam reforming catalyst deactivated immediately under this condition.

The CO_2 reforming catalyst could maintain a stable activity for about 300 hours. These results supported the robustness of thecatalyst against carbon deposition. After about 380 hours on stream, the catalyst regeneration test was carried out by the steam reforming, and it was confirmed that the catalyst could be regenerated. This result indicates that the morphology of carbon deposited on catalyst surface seems to be a material like an atomic carbon that is easily converted into carbon monoxide by steam. And we think the steam plays a significant role to remove the carbon precursor from the catalyst surface.

Fig.3 shows the relation between the feed gas volume and its composition to obtain the constant volume of synthesis gas having H_2/CO molar ratio of 2.0. The volumes of synthesis gases were estimated by the equilibrium calculation of steam reforming and water-gas shift reactions at 850°C and 2.1MPa. At the point of nearby feed molar ratio of $CH_4 : CO_2 : H_2O=1.0 : 0.4 : 1.0$, the feed gas volume is minimized and it is predicted that preferable operations with lower energy consumption can be accomplished. But in the case of CO_2 and H_2O reforming with conventional reforming catalysts, the selection of these operating conditions is prohibited by higher potential for carbon formation, and particular technologies for decoking will be required [5)-9)].

The results of the pilot plant operation are shown in Fig. 4. In this figure, the conversions of natural gas are presented by comparison with the percentage of the equilibrium values at the same reaction conditions.

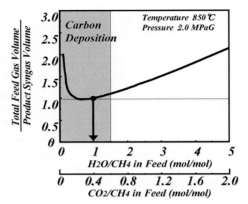

Fig. 3. Optimal CO₂/CH₄ and H₂O/CH₄
Molar Ratio in Feed Gas to Produce Syngas(H₂/CO=2.0)for FT Synthesis

The stable operation for more than 1,100 hours has been attained under the target reaction condition. During the operation, little change of the temperature and pressure drop of the catalyst bed was observed. After the tests, the catalyst was discharged from the reformer and the carbon content of thecatalysts was determined. The amounts of the carbon deposition on the catalysts at the all positions in the reformer were less than 0.1wt%. From the results, the carbon-free operation has been demonstrated.

The simulator to predict the temperature profile and gas compositions in the catalyst bed have also been developed for the design of the tubular reformer (Fig. 5). The heat–input of the reformer tube from the burners is calculated by computational fluid dynamics, and the endothermic heat-flux is calculated by the simulator. The modeling of reactions for reforming and carbon formation on the catalyst surface is conducted with consideration from the laboratory experimental results and information on literatures[10]. Catalysts life under optional operation conditions can be estimated because the carbon formation rate is taken into the consideration. Representative elementary reactions are

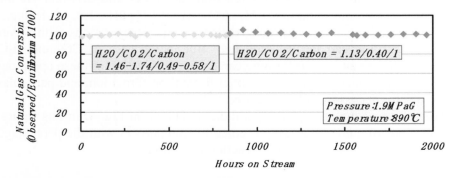

Fig. 4. Results of Pilot Plant Tests for CO₂ Reforming

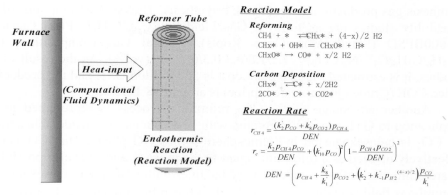

Fig. 5. CO₂ Reforming Simulator

shown in Fig 5. The reaction rate of CH_4 and the carbon formation reaction rate are expressed by the rate constant on elementary reactions and the partial gas pressures.

Typical simulation results of the pilot plant tests are shown in Fig. 6. This figure stands for the typical axial temperature profiles of the catalyst bed.

The plots represent the observed value, and corresponding curves present the calculation results. As shown in this figure, calculation results fitted with observed value well. The gas composition of thereformer outlet also agreed with the observed values. Through the analysis of the pilot plant test results by the simulator, it was confirmed that it could be effectively used for the CO_2 reformer design.

The feasibility studies of the new GTL process for the utilization of the middle and small size of natural gas fields in the Southeast Asia have been done in the GTL national project program. Almost 85 % number of natural gas fields in the world is middle and small sizes. Especially in the Southeast Asia, a lot of gas fields of those sizes containing a certain amount of CO_2 are found. In those natural gas fields, the CO_2 reforming process will be successfully applied to the

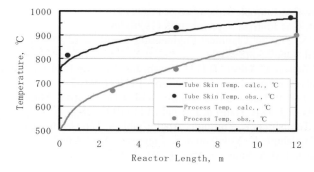

Fig. 6. Estimation of Temperature Profile in Reformer

synthesis gas production as feed for FT synthesis. The basic assumptions of the feasibility studies are as follows, Gas Field Size ; 1.75TCF, Plant Capacity; 15000BPSD (Stand Alone/Grass Roots), Natural Gas Composition; CH_4 /C_2H_6/C_3H_8/C_4^+ /N_2/CO_2 = 77.5/1.0/0.3/0.3/0.8/20.1(vol%). As the result of the studies, it is estimated that the plant cost is 560 million US$, and the product oil price COE (Crude Oil Equivalent price) is around 18 US$/bbl.

Another advantage of the CO_2 reforming process is represented in the application to GTL process integrated with LNG plants. A considerable amount of CO_2 is beingexhausted in atmosphere form LNG plants. That CO_2 can be effectively utilized as a reforming agent in the process and, it is also possible to share utility and offsite supplementary facilities with LNG plant. For example, a natural gas field containing 3.5 vol% of CO_2 can be utilized to GTL process of 15000BPSD size. With combination of LNG and GTL plants, more than 1.0 US$/bbl of cost reduction is expected on COE comparing with the stand alone case.

4. CONCLUSION

The activity and stability of the CO_2 reforming catalyst has been substantiated through the pilot-scale operation. We started further pilot plant tests from July 2003 With the successful completion of those tests, we would like to make the catalyst and process promising candidates for utilization of the middle or small scale of natural gas fields.

ACKNOWLEDGEMENTS

Grateful acknowledgements for the financial support are made to Japan National Oil Corporation.

REFERENCES

[1] M.J. Corke, Oil & Gas Journal, September 21 (1998) 71.
[2] M.J. Corke, Oil & Gas Journal, September 28 (1998) 96.
[3] P.J. Lakhapate, V.K. Prabhu, Chemical Engineering World, 35 (2000) 77.
[4] Oil & Gas Journal. June 15 (1998) 34.
[5] I. Alstrup, B.S. Clausen, C. Olsen, R.H.H. Smith, J.R. Rostrup-Nielsen, Studies in Surface Science and Catalysis (Natural Gas Conversion V) 119 (1992) 5.
[6] N.R. Udengaard, J.-H.B. Hansen, D.C. Hanson, Oil & Gas Journal, March 9 (1992) 62.
[7] J.R. Rostrup-Nielsen, J.-H.B. Hansen, J. Catal., 144 (1993) 38.
[8] O. Yamazaki, K. Tomishige, K. Fujimoto, Appl. Catal. A, 136 (1996) 49.
[9] Y-G. Chen, K. Tomishige, K. Yokoyama, K. Fujimoto, Appl. Catal. A, 165 (1997) 335
[10] M.C.J. Bradford and M.A. Vannice, Appl. Catal. A, 142 (1996) 97.

Studies in Surface Science and Catalysis, volume 147
X. Bao and Y. Xu (Editors)

133

Production of synthesis gas from natural gas using ZrO$_2$-supported platinum

Mariana M.V.M. Souza[a] and Martin Schmal[*a, b]

[a]NUCAT/PEQ/COPPE, Universidade Federal do Rio de Janeiro, C.P. 68502, 21945-970, Rio de Janeiro, Brazil. E-mail: mariana@peq.coppe.ufrj.br

[b]Escola de Química, Universidade Federal do Rio de Janeiro, C.P. 68542, 21940-900, Rio de Janeiro, Brazil.

ABSTRACT

Reforming of methane with steam, partial oxidation of methane and a combination of both reactions were carried out using Pt/Al$_2$O$_3$, Pt/ZrO$_2$ and Pt/10%ZrO$_2$/Al$_2$O$_3$ catalysts, in temperature range between 350-900°C. For steam reforming, the H$_2$/CO product ratio is strongly influenced by the water-gas shift reaction, with the Pt10Zr sample showing the highest activity and stability. When oxygen is present in the feed, the total combustion is the main reaction at low temperatures. The catalyst with 10% of ZrO$_2$ also presented the best performance for partial oxidation and oxy-steam reforming.

1. INTRODUCTION

Renewed attention in both academic and industrial research has recently been focused on alternative routes for conversion of natural gas to synthesis gas, a mixture of H$_2$ e CO, which can be used to produce chemical products with high added value. The steam reforming of methane (SRM) is the most important industrial process for the production of synthesis gas. SRM is a very energy-intensive process because of the highly endothermic property of the reaction; the high H$_2$/CO ratio (3:1) is useful for processes requiring a H$_2$-rich feed such as ammonia synthesis and petroleum refining process [1, 2].

Methane partial oxidation (POM) is an advantageous route for synthesis gas production for both economical and technical reasons: it makes the process less energy and capital cost intensive because of its exothermic nature and the lower H$_2$/CO ratio (about 2) is more favorable with respect to downstream processes such as methanol synthesis and Fischer-Tropsch synthesis of higher hydrocarbons [3-6].

On the other hand, oxy-steam reforming, which integrates SRM and POM simultaneously, has many advantages such as: low-energy requirements due to the opposite contribution of the exothermic methane oxidation and endothermic steam reforming; low specific consumption; variable H_2/CO ratio regulated by varying the feed composition [7, 8].

We have previously reported work on CO_2 reforming of methane showing that Pt/ZrO_2 and $Pt/ZrO_2/Al_2O_3$ are effective formulations for this reaction [9]. This paper gives results on the testing of these novel zirconia-supported catalysts for the steam reforming and partial oxidation of methane, as well as a combination of this both reactions.

2. EXPERIMENTAL

2.1. Catalyst Preparation

Al_2O_3 and ZrO_2 supports were prepared by calcination of γ-alumina (Engelhard) and zirconium hydroxide (MEL Chemicals) at $550°C$ for 2h under flowing air. $10\%ZrO_2/Al_2O_3$ was prepared by impregnation over alumina powder with a nitric acid solution of zirconium hydroxide; the mixture was stirred for 2h at $90°C$, dried and calcined at the same conditions as the bare supports [9]. The catalysts were prepared by incipient wetness technique, using an aqueous solution of H_2PtCl_6 (Aldrich), followed by drying at $120°C$ for 16 h and calcination in air at $550°C$ for 2 h. The platinum content was around 1wt%, which was measured by atomic absorption spectrometry. Prepared catalysts will be referred to as PtAl for Pt/Al_2O_3, PtZr for Pt/ZrO_2 and Pt10Zr for $Pt/10\%ZrO_2/Al_2O_3$.

2.2. Catalyst testing

Catalyst testing was performed in a flow tubular quartz reactor loaded with 20mg of catalyst, under atmospheric pressure. The total feed flow rate was held constant at 200 cm^3/min (WHSV= 600,000 cm^3/h.gcat), with flowing He. The activity tests were performed at different temperatures, ranging from 350 to $900°C$ in steps of $50°C$ that were kept for 30 min at each temperature. The loss in catalyst activity at $800°C$ was monitored up to 70 h on stream. The reaction products were analyzed by on-line gas chromatograph (CHROMPACK CP9001), equipped with a Hayesep D column and a thermal conductivity detector.

3. RESULTS AND DISCUSSION

3.1. Steam Reforming

The comparison of catalyst activities for steam reforming under stoichiometric feed ($CH_4:H_2O$=1:1) was carried out as a function of temperature

(400-800°C) and the results are displayed in Figure 1A and B, in terms of CH_4 conversion and H_2/CO product ratio. The experimental CH_4 conversion increased almost linearly with increasing reaction temperature from 400 to 650°C for PtAl and Pt10Zr catalysts, with PtAl showing the highest conversions at high temperatures. The PtZr catalyst presented low conversions up to 600°C; simultaneously this catalyst exhibited the highest H_2/CO ratio in this temperature range. The observed high H_2/CO ratio (>8.0) suggests that water-gas shift (WGS) reaction occurs to a great extent with SRM, as already observed in the literature [7, 10]. At the same time, the decrease in H_2/CO ratio with increasing temperature is consistent with the fact that the WGS reaction is thermodynamically unfavorable at higher temperatures.

The stability of the catalysts for steam reforming of methane at 800°C is displayed in Figure 2. The PtAl showed an oscillatory behavior around 72% of conversion, while the PtZr deactivated at a rate of 0.25%/h, probably due to deposition of inactive carbon. The Pt10Zr sample presented the highest activity and stability, giving a constant conversion of about 81% throughout the duration of the experiment (40h). These results are comparable to the reported values for Pt/ZrO$_2$ of Hegarty et al. [10].

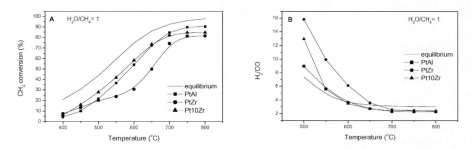

Fig. 1. CH_4 conversion (A) and H_2/CO product ratio (B) of Pt catalysts for steam reforming of methane as a function of temperature.

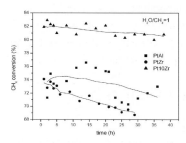

Fig. 2. CH_4 conversion as a function of time on stream at 800°C for steam reforming of methane.

3.2. Partial Oxidation

The effect of temperature on the CH_4 conversion and H_2/CO product ratio for partial oxidation of methane, with O_2/CH_4 feed ratio of 0.5, is displayed in Figure 3A and B. The activities of supported Pt catalysts were influenced by the nature of the support, with a better activation of PtZr at low temperatures. At temperatures higher than 550°C, PtAl and Pt10Zr exhibited activities greater than PtZr. As temperature rises from 500 to 800°C, the H_2/CO ratio decreases from 2.5-3 to 1.5. At low temperatures the high H_2/CO ratio is associated with the reforming of methane with steam produced by total combustion, which suggests that the steam reforming is faster than reforming with CO_2. At higher temperatures, the H_2/CO ratio is mainly related to the partial oxidation of methane, with simultaneous RWGS reaction.

The catalyst stabilities for partial oxidation of methane at 800°C are shown in Figure 4. The catalysts exhibited similar initial activities and the deactivation rates of PtAl and PtZr were very close: 0.21%/h and 0.18%/h, respectively. The deactivation is related to the deposition of inactive carbon over the active surface and the amount of coke on these catalysts, as quantified by thermogravimetric analysis carried out in an oxygen-containing atmosphere, was about 0.6-0.8 mg coke/h.gcat. On the other hand, the Pt10Zr remained stable up to 54h of reaction, with a deactivation rate of only 0.03%/h.

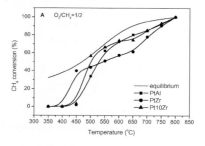 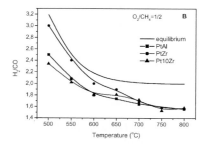

Fig. 3. CH_4 conversion (A) and H_2/CO product ratio (B) of Pt catalysts for partial oxidation of methane as a function of temperature.

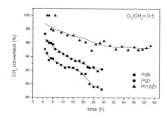

Fig. 4. CH_4 conversion as a function of time on stream at 800°C for partial oxidation of methane.

3.3. Oxy-steam reforming

The comparison of catalyst activities in terms of CH_4 conversion is presented in Figure 5 for oxy-steam reforming of methane. The PtZr showed the highest activity at 400°C, with 20% of CH_4 conversion, maintaining this conversion up to 550°C. The Pt10Zr was the most active catalyst over the mid-temperature range (450-600°C) while at higher temperatures PtZr exhibited slightly better activity than PtAl and Pt10Zr.

The composition profiles for PtZr catalyst are shown in Figure 6. The O_2 conversion is 100% over the whole temperature range. The production of steam and CO_2 decreases rapidly with the temperature increase due to the exothermic nature of total combustion. The lower production of H_2 and CO up to 600°C is also related to the occurrence of total combustion. At 800°C, the H_2/CO product ratio is 1.4, showing the influence of the water-gas shift reaction that occurs simultaneously with partial oxidation. These results suggest that the reaction sequence is probably initiated by methane combustion, followed by reforming of the remaining methane with CO_2 and H_2O, which is in agreement with reported studies of oxy-steam reforming over supported noble metal catalysts in the literature [11, 12].

The effect of time on stream at 800°C for oxy-steam reforming is displayed in Figure 7. PtZr showed the highest initial activity, but deactivated at a rate of 0.86%/h during 28 h on stream. Also in this case the Pt10Zr exhibited the best performance, with CH_4 conversion decreasing from 82 to 68% in the first 10h, maintaining this conversion for over 60h.

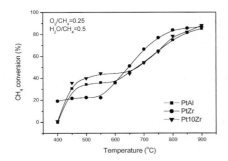

Fig. 5. CH_4 conversion as a function of time on stream at 800°C for oxy-steam reforming of methane.

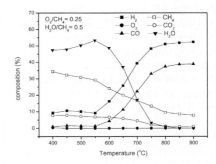

Fig. 6. Composition profiles for oxy-steam reforming of methane as a function of temperature over the PtZr catalyst.

138

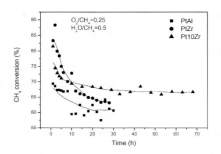

Fig. 7. CH$_4$ conversion as a function of time on stream at 800°C for oxy-steam reforming of methane.

4. CONCLUSIONS

The Pt/10%ZrO$_2$/Al$_2$O$_3$ material proved to be an effective catalyst for both steam reforming and partial oxidation of methane, besides CO$_2$ reforming. The higher stability of this catalyst is closely related to its coking resistance, which has been attributed to Pt-Zr^{n+} interactions at metal-support interface [9, 13]. The interfacial sites on Pt-ZrO$_x$ are active for CO adsorption and CO$_2$ dissociation, providing active species of oxygen that may react with carbon formed by CH$_4$ decomposition on the metal particle, suppressing carbon accumulation. Moreover, zirconia is a well-know oxygen supplier, and its oxygen mobility is fast, which helps to keep the metal surface free of carbon.

REFERENCES

[1] S.C. Tsang, J.B. Claridge and M.L.H. Green, Catal. Today 23 (1995) 3.
[2] M.A. Peña, J.P. Gómez and J.L.G Fierro, Appl. Catal. A 144 (1996) 7.
[3] D.A. Hickman and L.D. Schmidt, Science 259 (jan.1993) 343.
[4] D.A. Hickman, E.A. Haupfear and L.D. Schmidt, Catal. Lett. 17 (1993) 223.
[5] D.A. Hickman and L.D. Schmidt, J. Catal. 138 (1992) 267.
[6] S. Liu, G. Xiong, H. Dong and W. Yang, Appl. Catal. A 202 (2000) 141.
[7] Z.-W. Liu, K.-W. Jun, H.-S. Roh and S.-E. Park, J. Power Sources 111 (2002) 283.
[8] S. Ayabe, H. Omoto, T. Utaka, R. Kikuchi, K. Sasaki, Y. Teraoka and K. Eguchi, Appl. Catal. A 241 (2003) 261.
[9] M.M.V.M. Souza, D.A.G. Aranda and M. Schmal, J. Catal. 204 (2001) 498.
[10] M.E.S. Hegarty, A.M. O'Connor and J.R.H. Ross, Catal. Today 42 (1998) 225.
[11] D. Dissanayaki, M.P. Rosynek and J.H. Lunsford, J. Phys. Chem. 97 (1993) 3644.
[12] E.P.J. Mallens, J.H.B. Hoebink and G.B. Marin, Catal. Lett. 33 (1995) 291.
[13] M.M.V.M. Souza, D.A.G. Aranda and M. Schmal, Ind. Eng. Chem. Res. 41 (2002) 4681.

Studies in Surface Science and Catalysis, volume 147
X. Bao and Y. Xu (Editors)

Microkinetic model assisted catalyst design for steam methane reforming

De Chen[*][a], Erlend Bjørgum[a], Rune Lødeng[b], Kjersti Omdahl Christensen[a], and Anders Holmen[a]

[a]Department of Chemical Engineering, Norwegian University of Science and Technology (NTNU), N-7491 Trondheim, Norway.

[b]SINTEF Applied Chemistry, N-7465 Trondheim, Norway.

ABSTRACT

Steam reforming catalysts have been designed by means of a microkinetic model. A lower binding energy of C-M results in a lower conversion, but much lower potential of carbon formation. A preferred binding energy window of C-M of 160-169 kcal/mol and H-M of 64-67 kcal/mol has been identified, which has been used as a guideline for a selection of promotors and dopands for Ni catalysts.

1. INTRODUCTION

The chemistry of steam methane reforming (SMR) and carbon formation has attracted renewed attention due to the increasing demand for synthesis gas for GTL processes as well as for hydrogen production in different scales. A key problem for improved development is to avoid carbon formation and several approaches are followed to develop catalysts with high carbon formation resistance. In general, catalyst development has so far been mainly based on trial-and-error procedures, or heuristic methods [1]. More recently, deterministic methods including microkinetic modeling have started to evolve in catalyst design. The present description presents our initial work on deterministic design of catalysts for steam methane reforming.

A microkinetic model for methane reforming has previously been established, based on detailed experimental results from the Tapered Element Oscillating Microbalance (TEOM) reactor and literature [2]. It was found that a phenomenological model, namely Bond Order Conservation (BOC) [3], can predict rather reliable activation energies of elementary reaction steps. Most importantly, it relates kinetic parameters directly to the properties of the active surface. In addition, during the past years there has been a large progress in DFT calculations for detailed understanding of elementary reaction steps in methane reforming [4]. These progresses can make it possible to (semi) quantitatively design catalysts for steam methane reforming.

The present work deals with analyzing the influence of C-, H-, and O-Ni binding energies on the activity of SMR reaction and carbon formation. An assessment of the model simulation will be given based on the experimental data provided by TEOM and by literature. The design of new Ni catalyst modified with promoters and alloys is discussed.

2. CATALYST DESIGN FOR SMR

A microkinetic model consisting of 13 surface elementary reactions from methane leading to synthesis gas has previously been developed [2]. The mechanism of filamentous carbon formation on Ni catalysts, which is a dominating type of carbon formed on Ni during SMR, has also been implemented in the model [2]. It includes formation of surface carbon via surface elementary reactions, segregation, diffusion of carbon through the Ni particle, and precipitation at the support side of the particle. At steady state, the coking rate equals the rate of carbon diffusion through the Ni particles. It is directly proportional to the driving force (ΔC) for the diffusion, namely $\Delta C = C_{Ni,f} - C_{sat}$, where $C_{Ni,f}$ is the solubility of surface carbon in the Ni particle at the front side, C_{sat} is the solubility of carbon filaments at the support side. At the zero driving force, the condition is related to a coking threshold. The $C_{Ni,f}$ is in equilibrium with the surface atomic carbon, and can be described by a Langmuir type equation [2]. The site coverage of surface carbon depends on a kinetic balance of all the elementary reaction steps, which can be calculated by means of the microkinetic model [2]. For a given catalyst, C_{sat} is constant at each temperature. A low carbon site coverage results in a low $C_{Ni,f}$, thus a small diffusion driving force, yielding a low potential for carbon formation.

The activation energies of all the elementary reaction steps in the microkinetic model have been calculated by means of the BOC method. The detailed formulations for calculating adsorption heat and activation energy by means of BOC have been reviewed by Shustorovich [3]. An example of the relation between the activation energy ($\Delta E^{*}_{AB,g}$) of a dissociation step $AB_g \rightarrow A_s + B_s$ and the binding energies is presented in Equation (1).

$$\Delta E^{*}_{AB,g} = \frac{1}{2}\left(D_{AB} + \frac{Q_A Q_B}{Q_A + Q_B} - Q_{AB} - Q_A - Q_B \right) \tag{1}$$

Where Q_A, Q_B and Q_{AB} are the binding energy of A, B and AB on the metal surface, respectively, and D_{AB} is the dissociation energy in the gas phase. The binding energies of the intermediates can be estimated by the binding energy of C, H, and O on the metal surface taking into account configuration of intermediates. The typical binding energies of C-, H- and O-Ni are 171, 63 and 115 kcal/mol, respectively, which are used in the modelling on Ni catalysts [2,3]. The BOC based microkinetic model can then easily be used to evaluate the effects of binding energy on the conversion and surface carbon coverage, thus the potential of carbon formation.

Effects of the carbon binding energy on SMR are studied at the hydrogen binding energy of 63 kcal/mol, while the effects of hydrogen binding energy are studied at the carbon binding energy of 171 kcal/mol. The simulation indicates that the binding energy of O-M does not affect the reaction, although some oxygen containing intermediates involved in the reaction are bonded to the metal surface via oxygen atoms. It indicates that those intermediates are not involved in rate determining steps. However, the effects of both carbon and hydrogen bond energy on the steam reforming and carbon formation are rather significant. Fig. 1 shows a sensitive window for C-M and H-M binding energy, respectively. Within a window with carbon binding energy less than 171 kcal/mol, a decrease in carbon binding energy decreases methane conversion, but it reduces the carbon site coverage more significantly. As discussed above, a low site coverage means a low potential for carbon formation. When the carbon binding energy is larger than 171 kcal/mol, changes in carbon binding energy is not so sensitive to the conversion, but the surface coverage is very high. An increase in hydrogen bond energy up to 63 kcal/mol results in a relatively small increase in conversion and a small decrease in carbon site coverage. Further increase in hydrogen bond energy reduces significantly both activity and carbon site coverage. It indicates that too high hydrogen binding energy causes an inhibition of strongly adsorbed hydrogen for the steam reforming and the carbon formation.

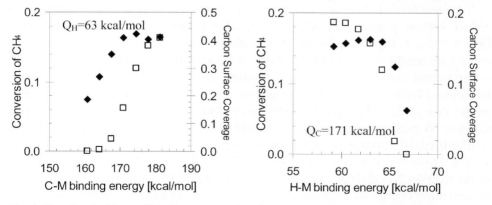

Fig. 1. Simulated effects of binding energy of carbon (Q_C) and hydrogen (Q_H) on the methane conversion (♦) and the carbon site coverage(□). Reaction conditions: 630 °C, P_{tot}=20 atm, S/C=1, P_{CH4}=3.8 atm , W/F0=0.067 g,h/mol.

A principle goal of the catalyst design for steam methane reforming is to achieve high activity for synthesis gas production and low activity for carbon formation. Fig. 1 clearly indicates that both carbon and hydrogen bond energies are important parameters for the catalyst design. However, it seems to be difficult to achieve high SMR activity and low carbon formation activity simultaneously. A balance has to be made between lower activity and high carbon formation resistance. The detailed optimization of the catalyst design depends on the criteria of the process, which is beyond the scope of the present work. Generally speaking, the carbon binding energy window of 160-169 kcal/mol and hydrogen binding energy window of 64-67 kcal/mol seems to be desirable for steam methane reforming, where the activity is still relatively high, but the potential of carbon formation is much lower.

There are many ways to formulate catalysts in order to achieve the specified windows, such as selecting different metals, adding, surface doping, etc. There have been a lot of efforts on studying different metals as catalysts for steam reforming. Reported binding energies [3] of selected metals are collected in Table 1 and 2. It is very interesting to note that Ni is the most active catalyst for steam methane reforming due to both suitable carbon and hydrogen binding energy, as shown in Fig. 1. This then explains well why Ni catalysts have long been used for steam reforming catalyst. However, Ni has a relatively high carbon potential, as shown in Fig. 1. Pd or Co might be the alternatives for high carbon formation resistance. It should be pointed out that a precise design of different metals as reforming catalysts needs knowledge of carbon solubility and diffusivity in metal crystals, and such knowledge is unfortunately very limited so far. Anyhow, Ni is still the most attractive reforming catalyst due to its lower cost and high activity. The catalyst design for the modification of Ni catalysts is of particular interest.

Promoters: Potassium has long been recognized as a good promoter to reduce carbon potential of reforming catalysts. Our experimental results also clearly show in Fig. 2 that promoter K reduces significantly both coking rate and coking threshold of Ni catalyst during SMR at 630 °C. A model fitting to the experimental data gives a carbon binding energy of 168.3 kcal/mol and a hydrogen binding energy of 58 kcal/mol for the K promoted Ni catalyst. The model can predict the experimental conversions and the coking rate very well for both Ni catalysts. For example, the experimental methane conversion on the K modified Ni catalyst is 14.7% at S/C of 1.3, while the model prediction is 15.2%. However, the TOF on promoted Ni catalyst was found to be six times lower than on Ni catalysts. These are mainly caused by an increase of 11 kJ/mol in activation energy of methane dissociation. It is rather close to the result from a DFT calculation [4], where it was found that potassium increases the energy barrier for methane dissociation on Ni surface by ca. 20 kJ/mol at θ_K=0.125ML.

In addition, both DFT calculations and experiments indicate that K preferably adsorbs on the steps, and that the difference in the carbon binding energy between Ni step sites (211) and (111) is 27 kJ/mol [4]. It is therefore expected that the presence of K reduces both carbon and hydrogen binding energies, The decrease in carbon binding energy reduces the conversion and reduces the carbon formation more significantly, while the decrease in hydrogen binding energy increases slightly the carbon formation. The fitted carbon binding energy of 168.4 kcal/mol is located well inside the desirable window. This example is a good assessment of the model assisted catalyst design.

Dopands: It has long been known that doping with a small amount of a second metal can improve the carbon resistance of Ni catalysts [1]. The BOC theory also provides a framework for calculation of atomic adsorption heats on bimetallic surfaces [3]. The calculated carbon binding energies on Ni/M bimetallic surfaces are presented in Table 1. The hydrogen binding energy can also be calculated in a same way, but the data are not listed here. The binding energy of a bimetallic surface varies between the two extremes of pure metal 1 and metal 2, depending on the number of metal atoms of the second type [3]. The data in the table can be used as a database for selection of dopands based on the binding energy window as discussed above. Ruban et al. [5] have calculated surface segregation energies for different alloys. Possible formation of surface alloys or bulk alloys is also listed in Table 1. This is also an important parameter for catalyst design. All the grey marked values in the Table 1 are located in the binding energy window, indicating that Au, Ag, Pt, Pd and Co can be used as dopands. Ni_2Pd might be expected to have similar behavior as K modified Ni catalysts as shown in Fig. 2. Co might also be a good dopand due to the low cost. However, it is difficult to draw a conclusion due to formation of Ni/Co bulk

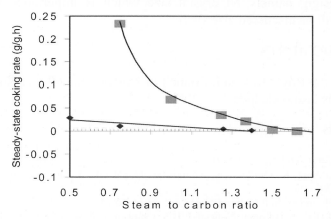

Fig. 2. Coking rate at 630 °C, P_{tot}=20 atm, P_{CH4}=3.8 atm , W/F_0=0.067 g,h/mol on Ni catalysts: (■) 11 (wt%) $Ni/CaAl_2O_4$, (♦): 14 (wt%) $Ni/1.5(wt\%)$ $K/CaAl_2O_4$. The experiments were performed in TEOM by step wisely decrease in steam to carbon ratio.

Table 1
BOC atomic heats of adsorption, Q_0 (kcal/mol) of carbon on Ni/M bimetallic surfaces. (Grey marked area is the favorable window indicated by the microkinetic model)

M	Composition of three-fold hollow site				Surface alloy	Bulk alloy
	Ni_3	Ni_2M	NiM_2	M_3		
Cu	171.3	157.4	140.0	120.2	+	
Au	171.3	161.8	151.4	140.3	+	
Au	171.3	163.1	154.4	145.3	+	
Pt	171.3	164.5	157.5	150.3	+	
Pd	171.3	167.5	163.8	160.3	+	
Co	171.3	168.0	164.9	162.0		+
Fe	171.3	181.3	190.9	200.0		+
Mo	171.3	175.0	178.9	183.0		+

alloy, which can change the carbon solubility and diffusion. The analysis pointed out that Au is efficient to lower the C-M bond energy, which is in good agreement with literature observations [4]. However, Au might not be a good choice for high temperature applications, since Au will diffuse into the bulk phase [6].

3. CONCLUSIONS

Microkinetic model assisted catalyst design seems to be a powerful tool for catalyst development. Model simulations indicate that the use of Ni alloys and alkali promoters reduce the potential of carbon formation, but it also reduces the activity for methane reforming. Co, Pd, Pt, Ag and Au are the candidates as dopands for Ni catalysts to form surface alloys. Another important parameter for the catalyst design, namely Ni crystal size which is importance for C_{sat} in Equation (1), will be discussed elsewhere.

ACKNOWLEDGMENTS

The support by the Royal Research Council of Norway, Statoil and Norsk Hydro ASA is gratefully acknowledged.

REFERENCES

[1] D.L. Trimm, Catal. Today, 49 (1999) 3.
[2] D. Chen, R. Lødeng, K. Omdahl, A. Anundskås, O. Olsvik, A. Holmen. Stud. Surf. Sci. Catal., 139 (2001) 93.
[3] E. Shustorovich and H. Sellers, Surf. Sci. Rep., 31 (1998) 1.
[4] H.S. Bengaard, J.K. Nørskov, J. Sehested, B.S. Clausen, L.P. Nielsen, A.M. Molenbroek, J.R. Rostrup-Nielsen, J. Catal., 209 (2002) 365.
[5] A.V. Ruban, H.L. Skriver, J.K. Nørskov, Phys. Rev B, 59 (1999) 15990.
[6] P.M. Holmblad, J.H. Larsen, and I. Chorkendorff, J. Chem. Phys., 104 (18) 1996, 7289.

Studies in Surface Science and Catalysis, volume 147
X. Bao and Y. Xu (Editors)
145

Influence of calcination temperature on the primary products of methane partial oxidation to synthesis gas over Rh/Al$_2$O$_3$ catalyst

W. Z. Weng*, C. R. Luo, J. M. Li, H. Q. Lin and H. L. Wan*

State Key Laboratory for Physical Chemistry of Solid Surfaces, Department of Chemistry, Institute of Physical Chemistry, Xiamen University, Xiamen, 361005, China

ABSTRACT

The reaction pathway of methane partial oxidation to synthesis gas over Rh/Al$_2$O$_3$ calcined at 600 and 900 °C was studied by using the *in situ* time-resolved FTIR spectroscopy to follow the primary reaction products over the catalysts. The results of TPR and XPS characterizations suggest that the difference in the reaction mechanism of methane partial oxidation to synthesis gas over the two catalysts can be related to the difference in the reducibility of the supported Rh species, which may affect the concentration of O^{2-} species on the surface of the catalysts under the reaction conditions.

1. INTRODUCTION

The reaction pathway of partial oxidation of methane (POM) to synthesis gas over supported transition metal catalysts has been extensively studied in recent years. The factors determining the formation of primary products are a subject of great interest. In this work, a comparative study using *in situ* time-resolved FTIR (*in situ* TR-FTIR) spectroscopy to follow the primary product of POM reaction over the Rh/Al$_2$O$_3$ catalysts calcined at 600 and 900 °C, respectively, was carried out. The results of TPR and XPS characterizations on the reducibility and oxidation state of the Rh species over the catalysts are also presented. It is expected that these experiments will provide us with some useful information to understand the factors determining the mechanism of POM reaction over supported metal catalysts.

Table 1
BET surface area and metal dispersion of the catalysts

Catalyst	BET Surface Area (m^2/g)	CO/Rh
1wt% Rh/Al$_2$O$_3$-600	208.1	1.18
1wt% Rh/Al$_2$O$_3$-900	125.7	0.06

2. EXPERIMENTAL

Rh/Al$_2$O$_3$ catalysts were prepared by the method of incipient wetness impregnation, using RhCl$_3$•nH$_2$O as the precursor compound for the metal, When the solvent was evaporated, the solid material was dried at 110 °C and calcined at 600 and 900 °C, respectively, in air for 4 h. The BET surface area (Table 1) of the catalysts was measured by nitrogen adsorption, using a Micromeritics Tristar 3000 system. The sample was degassed at 300 °C for at least 4 h under vacuum before the measurement. The dispersion of the metal (Table 1) was determined by CO chemisorption at 35 °C by assuming a 1/1 stoichiometry. Experiments were performed on a Micromeritics ASAP 2010 system. Before in contact with CO, the catalyst sample (~0.4 g) was reduced under a flow of H$_2$ (25 ml/min) at 600 °C for 60 min followed by evacuation at 360 °C for 90 min and cooling to 35 °C under vacuum.

The *in situ* TR-FTIR experiments were performed on a Perkin Elmer Spectrum 2000 FTIR spectrometer equipped with a MCT detector and a home built high temperature *in situ* IR cell with quartz lining and BaF$_2$ windows. The catalyst sample in the IR cell was reduced for 45 min with H$_2$ (99.999%) at 600 °C before it was switched to a flow of simulated POM feed (CH$_4$/O$_2$/Ar = 2/1/45, volume ratio) at 500 °C. The IR spectra of the surface and the gas phase species formed during the reaction were continuously recorded at a resolution of 16 cm^{-1} with an average of 2 scans. The reference spectrum in the IR experiments was that of the catalyst prior to the admission of CH$_4$/O$_2$/Ar mixture. The catalytic performance of the catalysts for the POM to synthesis gas was studied using CH$_4$/O$_2$/Ar = 2/1/45 mixture (volume ratio) as a reaction feed. The reaction was carried out in a fixed bed quartz tube reactor (5 mm inside diameter) at atmospheric pressure. The catalyst (15 mg) was reduced at 600 °C for 40 min before it was switched to the reaction feed. The data were collected after 30 min on stream. The TPR experiment was performed on a GC-TPR set-up. The rate of hydrogen consumption was monitored by a TCD. Prior to the TPR experiment, each calcined catalyst sample was prereduced for 30 min under flowing H$_2$/Ar (5/95) mixture at 600 °C followed by reoxidizing for 30 min under a flow of O$_2$/Ar (5/95) mixture at 500 °C. XPS experiments were performed on a Phi Quantum 2000 system equipped with Al$K\alpha$ source. Rh 3d binding energies were determined using the Al 2p binding energy at 73.9 eV as a reference.

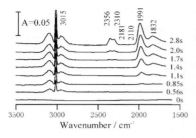 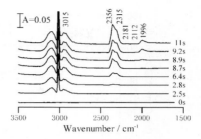

Fig. 1. *In situ* TR-FTIR spectra of the reaction of CH₄/O₂/Ar (2/1/45) mixture over H₂-reduced 1wt% Rh/Al₂O₃-600 at 500 °C.

Fig. 2. *In situ* TR-FTIR spectra of the reaction of CH₄/O₂/Ar (2/1/45) mixture over H₂-reduced 1wt% Rh/Al₂O₃-900 at 500 °C.

3. RESULTS AND DISCUSSION

3.1. *In situ* time-resolved FTIR (TR-FTIR) characterization on the POM reaction over the catalysts

The Rh/Al_2O_3 catalysts calcined at 600 and 900 °C were denoted as Rh/Al_2O_3-600 and Rh/Al_2O_3-900, respectively. The *in situ* TR-FTIR spectroscopic investigation on the reaction of $CH_4/O_2/Ar$ mixture over the H_2-reduced 1wt%Rh/Al_2O_3 catalyst was performed at 500 °C. The corresponding IR spectra were shown in Fig. 1 and 2. It was found that, when a H_2-reduced 1wt%Rh/Al_2O_3-600 sample was switched to a flow of $CH_4/O_2/Ar$ mixture at 500 °C (Fig. 1), the IR bands of linear- (1991 cm⁻¹) and bridge- (1832 cm⁻¹) bonded CO species on the metallic Rh sites were observed at 0.85 s. The IR bands of CO_2 (2310, 2356 cm⁻¹) could only be detected after 1.4 s on stream. This observation indicates that CO is the primary product of POM reaction over the H_2-reduced 1wt%Rh/Al_2O_3-600 catalyst. In contrast to the case over 1wt% Rh/SiO_2-600 catalyst, the reaction of $CH_4/O_2/Ar$ mixture over a H_2-reduced 1wt%Rh/Al_2O_3-900 catalyst at 500 °C led to the formation of CO_2 (2315, 2356 cm⁻¹; 2.8 s) as the primary product (Fig. 2). The IR band of adsorbed CO (~1996 cm⁻¹) and those of gaseous CO (2112, 2181 cm⁻¹) were detected at 8.9 and 11 s, respectively. This result clearly indicates that CO_2 is the primary product of the POM reaction over H_2-reduced 1wt%Rh/Al_2O_3-900 catalyst.

3.2. Effect of space velocity on the catalytic performance of the catalysts

Effect of gas hourly space velocity (GHSV) on the catalytic performance of the POM to synthesis gas at 600 °C over H_2-reduced 1wt% Rh/Al_2O_3-600 and 1wt% Rh/Al_2O_3-900 catalysts is shown in Fig. 3. The higher conversion value compared to the equilibrium data, particularly for the Rh/Al_2O_3-600 catalyst, may be resulted from the fact that the actual temperature of the catalyst zone is higher than 600 °C. For the 1wt% Rh/Al_2O_3-600 catalyst, as the GHSV increased from 1×10^5 to 6×10^5 h⁻¹, the conversion of methane decreased, while

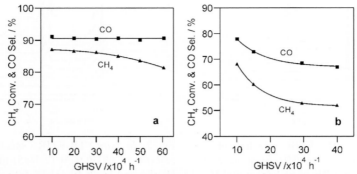

Fig. 3. Effect of GHSV on the catalytic performance of POM to synthesis gas over a. 1wt% Rh/Al$_2$O$_3$-600 and b. 1wt% Rh/Al$_2$O$_3$-900 catalysts.

the selectivity to CO was almost unchanged. For the 1wt% Rh/Al$_2$O$_3$-900 catalyst, however, both the conversion of methane and the selectivity of CO decreased with increasing GHSV. These results are consistent with the mechanisms proposed for the POM reaction over the catalysts based on the *in situ* TR-FTIR studies.

3.3. TPR and XPS characterizations of the catalysts

Considering the data of metal dispersion as well as the results of TR-FTIR studies and effect of GHSV on the POM reaction over the Rh/SiO$_2$ [1] and Rh/Al$_2$O$_3$ catalysts, it can be concluded that the difference in the primary reaction product of POM to synthesis gas over Rh/Al$_2$O$_3$-600 and Rh/Al$_2$O$_3$-900 catalysts can not be simply related to the difference in surface area of Rh. The possible reason for the difference in the POM performance over the two catalysts may have resulted from the difference in the metal-support interaction, which affects the reducibility of supported Rh species. Fig. 4 displays TPR profiles of Rh/Al$_2$O$_3$ catalysts calcined at 600 and 900 °C. For each TPR profile, two reduction peaks can be distinguished. The peaks at low reduction temperatures (108~137 °C) come from a RhO$_x$ species insignificantly affected by the support, while the peaks with at reduction temperatures (196~256 °C)

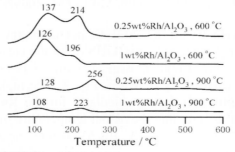

Fig. 4. TPR profiles of Rh/Al$_2$O$_3$ catalysts.

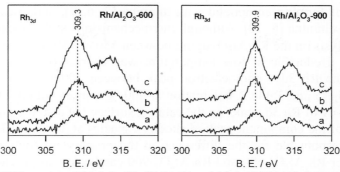

Fig. 5. Rh 3d XPS spectra of (a) 0.5wt%, (b) 1wt% and (c) 2wt% Rh/Al$_2$O$_3$ catalysts.

Table 2
Quantification of Rh 3d$_{5/2}$ XPS signal for 2wt% Rh/Al$_2$O$_3$

	2wt% Rh/Al$_2$O$_3$-600		2wt% Rh/Al$_2$O$_3$-900	
Rh 3d$_{5/2}$ B. E. (eV)	308.0	309.8	308.0	309.8
FWHM	3.1	2.8	2.4	2.4
Percentage (%)	54.0	46.0	18.6	81.4

may be assigned to the RhO$_x$ species intimately interacting with the Al$_2$O$_3$ surface [2]. Obviously, calcinations at 900 °C favors the formation of RhO$_x$ species at high reduction temperature. The lower intensities of the TPR peaks for Rh/Al$_2$O$_3$-900 as compared to Rh/Al$_2$O$_3$-600 suggests that some of the Rh species in the former case can not be reduced by H$_2$ at temperatures below 600 °C.

Fig. 5 displays the XPS spectra of the Rh$_{3d}$ region for the alumina supported Rh catalysts after calcined in air at 600 and 900 °C. The XPS spectra for 2wt% Rh/Al$_2$O$_3$ can be deconvoluted into two peaks. The binding energies, full-width-at-half-maxima (FWHM), and percentages of the signal intensity for each component peak are given in Table 2. According to the literature, the component peak with binding energy at 308.0 eV can be assigned to hexagonal Rh$_2$O$_3$ [3-5], while that at 309.8 eV may be attributed to Rh^{4+} present in RhO$_2$ or non-stoichiometric Rh$_2$O$_3$, where O/Rh ratio is in excess of 1.5 in the outermost layer of Rh$_2$O$_3$ [5].

The effect of thermal aging on the reducibility of the Rh/Al$_2$O$_3$ catalyst has been extensively studied. It was found that heating a Rh/Al$_2$O$_3$ sample at temperature above 600 °C in an oxidizing environment caused not only the oxidation of supported Rh particles, but also the formation of an oxidized form of Rh which was difficult to reduce to metal by H$_2$ at the temperature below 600 °C [2, 6, 7]. A number of explanations have been proposed for the nature of the high temperature oxidizing interaction including the diffusion of Rh into the bulk of Al$_2$O$_3$ [7, 8], formation of Rh interacted strongly with Al$_2$O$_3$ and even to the extent of forming compound with alumina [2, 6, 9], and encapsulation of rhodium by alumina [8, 10]. Although the mechanism is still unclear, the results of XPS analysis on the Rh/Al_2O_3 aging between 550 to 950 °C indicated that the Rh species irreducible at low temperature was characterized by the binding energy of the Rh $3d_{5/2}$ photoelectron peak between 309.3~309.9 eV [11]. As shown in Fig. 5 and Table 2, calcination at 900 °C causes not only the decrease in the intensity of Rh 3d peaks on the catalyst surface but also the increase in the proportion of the component peak of rhodium species at 309.8 eV. These factors should be responsible for the difference in the primary products of the POM reaction over Rh/Al_2O_3-600 and Rh/Al_2O_3-900 catalysts. Our previous study on the POM reaction over Rh/SiO_2 and Ru/SiO_2 catalysts suggested that significant difference in the reaction mechanisms on the two catalysts can be well explained by the difference in the surface concentration of O^{2-} species over the catalysts under the reaction conditions mainly due to the difference in the affinity of the metal for oxygen [12-14]. Based on the results of TPR and XPS characterizations shown above, the mechanistic difference for the POM reaction over Rh/Al_2O_3-600 and Rh/Al_2O_3-900 catalysts may be resulted from the formation of Rh species of different reducibility, which affects the concentration of O^{2-} species on the surface of the catalysts under the reaction conditions of methane partial oxidation to synthesis gas.

ACKNOWLEDGEMENTS

This project is supported by MSTC (grant No. G1999022408) and NSFC (grant No. 20021002).

REFERENCES

[1] W.Z. Weng, C.R. Luo, J.J. Huang, Y.Y. Liao and H.L. Wan, Top. Catal., 22 (2003) 87.
[2] C.P. Hwang, C.T. Yeh and Q. Zhu, Catal. Today, 51 (1999) 93.
[3] T.L. Barr, J. Phys. Chem., 82 (1978) 1801.
[4] M. Peuckert, Surf. Sci., 141 (1984) 500.
[5] Z. Weng-Sieh, R. Gronsky and A. T. Bell, J. Catal., 170 (1997) 62.
[6] H.C. Yao, S. Japar and M. Shelef, J. Catal., 50 (1977) 407.
[7] C. Wong and R.W. McCabe, J. Catal., 119 (1989) 47.
[8] J.G. Chen, M.L. Colaianni, P.J. Chen, J.T. Yates Jr and G.B. Fisher, J. Phys. Chem., 94 (1990) 5059.
[9] D. Duprez, G. Delahay, H. Abderrahim and J. Grimblot, J. Chim. Phys., 83 (1986) 465.
[10] R.W. McCabe, R.K. Usmen, K, Ober, H.S. Gandhi, J. Catal., 151 (1995) 385.
[11] S. Suhonen, M. Valden, M. Hietikko, R. Laitinen, A. Savimäki and M. Härkönen, Appl. Catal. A. Gen., 218 (2001) 151.
[12] W.Z. Weng, M.S. Chen, Q.G. Yan, T.H. Wu, Z.S. Chao, Y.Y. Liao, and H.L. Wan, Catal. Today, 63 (2000) 317.
[13] W.Z. Weng, Q.G. Yan, C.R. Luo, Y.Y. Liao and H.L. Wan, Catal. Lett., 74 (2001) 37.
[14] W.Z. Weng, M.S. Chen and H.L. Wan, Chem. Rec. 2 (2002) 102.

Studies in Surface Science and Catalysis, volume 147
X. Bao and Y. Xu (Editors)
151

Dynamic behaviour of Ni-containing catalysts during partial oxidation of methane to synthesis gas

Yu. P. Tulenin, M. Yu. Sinev[*], V. V. Savkin and V. N. Korchak

Semenov Institute of Chemical Physics, Russian Academy of Sciences
4 Kosygin Street, Moscow 119991, Russia; E-mail: sinev@center.chph.ras.ru

ABSTRACT

A dynamic behaviour (oscillations and temperature waves) is observed during a partial oxidation of methane over metal Ni-containing catalysts. The analysis of kinetic features suggests that oxygen and carbon stored in the catalyst play an important role in the appearance of such phenomena.

1. INTRODUCTION

The analysis of dynamic behaviour, including transition processes and self-oscillations, gives valuable information about the intrinsic mechanisms of catalytic reactions [1,2]. Recently we have observed a synergistic behaviour and reaction rate oscillations during methane oxidation in a combined catalytic bed containing oxide catalyst (mixed Mg-Nd oxide) and metal wire (Ni-Cr alloy) [3-5]. Such dynamic phenomena are accompanied by a shift in selectivity from methane coupling products (C_2-hydrocarbons) formed over oxide catalyst alone to synthesis gas ($CO+H_2$) produced in a binary oxide-metal system. Later similar phenomena were observed [6,7] over single Ni metal catalysts. Although the shape of oscillation curves in both cases are very similar, it is important to notice that in the conditions where oscillations were observed in mixed oxide-metal system (relatively low temperatures and flow rates), no reaction over a low surface area metal component taken alone could be detected [3,4]. This strongly suggested that methane activation takes place over the surface of oxide component and some intermediates (likely free methyl radicals) further react over the surface of metal forming final products. As we demonstrated, namely the process localised on the metal surface is responsible for the complex dynamic behaviour of the system [5]. Since at higher temperatures similar dynamic phenomena can be observed over Ni-containing catalyst alone [6,7], it is very likely that in these conditions methane can be activated over low surface area metals, and following reaction steps proceed over the same surface in the same way as in mixed oxide-metal systems. These observations can shed light upon the mechanism of methane partial oxidation to synthesis gas, which is still under discussion. Two alternative mechanisms of CO and H_2 formation during methane oxidation with molecular oxygen are extensively discussed in literature (see, for instance, a review in [8]):

/i/ "direct" - CO and H_2 are formed as primary products from CH_4 and O_2;

/ii/ "indirect" or "sequential" - methane is first oxidised to CO_2 and water; then it undergoes two endothermic reforming reactions with these primary products.

Justified notions about the reaction pathways may be derived from the "conversion-selectivity" relationships, from the analysis of temperature profiles along the reactor axis (as it is done, for example in [9]), from experiments with isotope-labelled reactants [10-12]. If the process proceeds in oscillatory regime, additional information can be obtained by the analysis of reaction kinetics on a short-time scale defined by a characteristic time of oscillations. The latter approach is used in this paper, where we present new data obtained over low-surface area Ni-containing catalysts (metal Ni foam and Ni-Cr alloys) active in partial oxidation of methane to synthesis gas and displaying a complex dynamic behaviour in a wide range of reaction conditions. A possible mechanism of the appearance of oscillatory behaviour will be also discussed.

2. EXPERIMENTAL

Catalysts. Two types of metal Ni-containing catalysts were used in this work: /a/ bulky metal nickel foam (0.16 g, ~ 5 mm length) and /b/ Ni-Cr alloys. In the latter case the reaction proceeded on a chromel-alumel thermocouple (TC) tip which simultaneously played a role of a temperature-sensitive element.

Reactor and analysis of gas mixtures. Experiments were carried out in a quartz flow reactor (3.5 mm i.d.). Concentrations of components of the reaction mixture were measured by an on-line mass-spectrometry (MS). Gas probe leaked into the MS-ionisation chamber (RIBER QMM17) through a quartz diaphragm placed right after the catalyst layer and metal capillary with differential pumping. The sampling system provided with a high temporal resolution (better than 0.1 s.). The initial gas mixture contained (in vol. %) 75 CH_4, 20 O_2 and 5 Ar (added as an internal standard for a MS-analysis).

Temperature. A heating system described in [4] was utilised. It provides with a uniform temperature profile along the catalyst bed and a sharp temperature drop before and after the catalyst. This enables to diminish the probability of side processes in pre- and post-catalytic void volumes. When reaction was carried out over a TC tip, the latter was simultaneously used to permanently record a reaction temperature. In the case of bulky Ni-foam catalyst, temperature was measured and recorded using a TC attached to the front layer of the catalyst.

3. RESULTS

Reaction over a chromel-alumel TC. At relatively low flow rates of gas mixture (< 20 ml/min.) methane oxidation starts at furnace temperatures above 700°C; total oxidation products (CO_2 and water) are predominant. At increasing flow rate CO and H_2 appear in the reaction mixture and the process transfers into an oscillatory regime, in which periods of relatively low activity alternate with sharp splashes of methane conversion and product formation.

Period of oscillations sharply decreases at increasing gas flow rate and furnace temperature. Fig. 1 demonstrates a typical example of temperature and gas concentration variations: the maximum methane consumption corresponds to the maximum rate of temperature growth. O_2-conversion curve has a two-peak shape: the first one coincides with that on methane conversion curve and the second (more intense) - with the maximum temperature. The total amount of products rises along with temperature, but further details differ from one product to another. CO and H_2 show similar behaviour: their highest content is achieved at the highest rate of temperature growth. The second (lower) maximum nearly coincides with the maximum temperature. Two peaks in CO_2 formation curve have nearly the same altitudes. A single flat maximum on a water formation curve is shifted to later times and corresponds to shoulders on O_2-consumption and temperature curves. It is important to notice that total oxidation products (CO_2 and water) keep forming during the period of low activity.

The data of Fig. 1 enables to calculate the instantaneous balance, i.e. the amount of each element (Δn), e.g. carbon and oxygen, stored in the catalyst (or released from it) within a unit time of reaction:

$$\Delta n = W * \{v * C - \Sigma(v_i * C_i)\} / 22400 \qquad (1)$$

where W - flow rate, ml s.$^{-1}$; v - number of atoms of certain element in a reactant molecule; C - mole fraction of this reactant in the feed gas; v_i - number of atoms of this element in a product i molecule; C_i - mole fraction of this product in the reaction mixture.

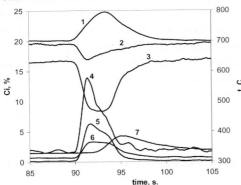

Fig. 1. Typical shape of temperature and concentration oscillations over a chromel-alumel thermocouple: ambient furnace temperature 750°C, initial mixture CH_4 : O_2 : Ar = 75 : 20 : 5, flow rate 40 ml/min.; 1 - temperature; 2 - CH_4 (x 0.3); 3 - O_2; 4 - H_2; 5 - CO; 6 - CO_2; 7 - H_2O

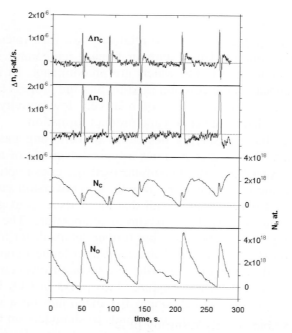

Fig. 2. Amounts of carbon and oxygen stored in a chromel-alumel thermocouple in a course of oscillations in the conditions of Fig. 1.

The amount of a certain element stored in the catalyst (N_i) can be calculated by summing Δn values for a given period. The results of such calculations show (see Fig. 2) that oxygen storage in the catalyst exceeds by far the storage of carbon and that the maximum N_O value corresponds to the minimum in carbon content in the sample. The maximum amount of oxygen stored in the catalyst within one cycle is 6.0×10^{-6} g-at or 3.6×10^{18} atoms. This number exceeds by far the number of atoms on the surface of the thermocouple tip and corresponds to a substantial degree of bulk metal oxidation (about 10 %).

Reaction over a Ni-foam catalyst. General features of the reaction over a Ni-foam are qualitatively very similar to the above described. However, because of much bigger size of this sample, the amounts of stored carbon and especially oxygen are much greater. During one period of oscillations, 4.1×10^{-5} g-at., or 2.5×10^{19} atoms of oxygen are stored and eliminated by the catalyst, which corresponds to 1.5% of Ni atoms in the catalyst bulk.

4. DISCUSSION

Mechanisms of self-oscillations in heterogeneous oxidative catalysis are widely discussed in literature (see, for instance, [1, 2]). A big difference in the catalytic performance of the solid in reduced and oxidised states is considered as a possible reason for the appearance of such behaviour. The data obtained in this work indicate that it is more correct to discuss the state of the system in terms of carbon and oxygen content in the catalyst (N_C and N_O respectively). The comparison of the data given in Figs. 1 and 2 indicates that the highest rate of methane consumption corresponds to the highest oxygen content in the catalyst, i.e. the oxidised state of the solid is more active in methane transformation. The latter process, in its turn, leads to a rapid reduction of the solid and to the drop in its capability of activating methane.

On the other hand, the N_C value keeps increasing in the period of decreasing oxidation degree. As indicated by the data of Fig. 2, there is no direct correlation between the N_C and N_O values, but both them approach their minima by the end of the period of low activity. So, we may conclude that every splash of activity starts when carbon and oxygen contents in the catalyst are minimal. The comparison of N_C and N_O evolutions with product formation during the period of low activity suggests that reaction between stored oxygen and deposits formed from hydrocarbon is one possible process that cleans the catalyst surface and restores its activity. After a phase of these relatively slow reactions of deposits, oxidation of the catalyst takes place and the cycle repeats.

Another possible activity restoring process is a relaxation of carbon stored in the catalyst. As it was recently demonstrated [13-14], methane interaction with reduced Ni and Co surfaces results in the formation of "endothermic" carbon, likely in metal carbide or amorphous form. It undergoes an exothermic transformation into a graphite-like form within the period of few minutes. If such processes take place also in the course of catalytic methane oxidation, they may cause a periodic poisoning (by blocking an active metal surface with amorphous carbon or metal carbide) and restoration of activity (due to a relaxation of such deposits into compact graphite crystallites or filaments).

Despite a possible role of carbon deposits in the temporary poisoning, it is important to notice that their amount stored/released in a single cycle of oscillations is relatively small. This suggests a difference with a dry reforming of methane, which can proceed via decomposition of methane, formation of carbon deposits and their further interaction with gaseous CO_2 [13]. This fact could be an evidence for a "direct" mechanism of CO and H_2 formation in the conditions of oxidative conversion of methane. This is in a good agreement with a mentioned above simultaneous (synchronic) formation of CO and H_2. Earlier the conclusion about a higher probability of "direct" mechanism was derived from the results of isotope-labelling experiments [11, 12].

Special remarks should be made on the dynamic regime over a Ni-foam catalyst. It was observed that methane oxidation is often associated with temperature waves moving along the catalyst/reactor axis. One may assume that, when the reaction starts in the front layer of the catalyst and proceeds with a

156

high rate, total consumption of oxygen is taking place. If so, the most of the catalyst stays in a CH_4-rich environment, and the formation of large amounts of carbonaceous deposits is highly probable. As soon as the reaction transfers into a period of low activity, O_2 concentration in next layers becomes higher causing a burn of deposits, restoration of activity and, as a result, to an appearance of a hot reaction zone moving along the catalyst bed. The processes that proceed in catalyst layers remaining before the hot zone, are similar to those taking place during the period of low activity over a TC tip and discussed above.

It is very likely that temperature oscillations measured by a TC placed in the front layer of the catalyst just reflect in some complex manner this wave process which involves the whole bulk of catalyst. Moreover, since a TC itself can act as a catalyst, it can distort the behaviour of the second catalyst. This is why a more attention has to be paid to the selection of an appropriate technique for temperature measurements. In particular, in the future work we plan to use a dynamic photometry specially developed for this application.

REFERENCES

[1] V.I. Bykov, Modelling of Critical Phenomena in Chemical Kinetics, Nauka, Moscow, 1988.
[2] M.M. Slin'ko and N.I. Eaeger, Oscillating Heterogeneous Catalytic Systems. Stud. Surf. Sci. Catal., v.86, Elsevier, 1994.
[3] Yu.P. Tulenin, M.Yu. Sinev and V.N. Korchak, 11th Int. Congress on Catalysis, June 30 - July 5, 1996, Baltimore, ML, USA, Programme and Book of Abstracts, P-275.
[4] Yu.P. Tulenin, M.Yu. Sinev, V.V. Savkin and V.N. Korchak, Stud. Surf. Sci. Catal., 110 (1997) 757.
[5] Yu.P. Tulenin, M.Yu. Sinev, V.V. Savkin, V.N. Korchak and Y.B. Yan, Kinetika i Kataliz (Russ. Kinetics and Catalysis), 40 (1999) 405.
[6] Y.H. Hu and E. Ruckenstein, Ind. Eng. Chem. Res., 37 (1998) 2333.
[7] X.L. Zhang, D.M.P. Mingos and D.O. Hayward, Catal. Lett., 72 (2001) 147.
[8] O.V. Krylov and V.S. Arutyunov, Oxidative Transformations of Methane (Russ.). Nauka Moscow, 1998.
[9] A.M. O'Connor and J.R.H. Ross, Catal. Today, 46 (1998) 203.
[10] A.Guerrero-Ruiz, P. Ferreira-Aparicio, M.B. Bachiller-Baeza and I. Rodriguez-Ramos, Catal. Today, 46 (1998) 99.
[11] Yu.P. Tulenin, V.V. Savkin, M.Yu. Sinev, V.N. Korchak and O.V. Krylov, 4th World Congress on Oxidation Catalysis. September 16-21, 2001, Berlin/Potsdam - Germany, Book of Extended Abstracts, v. II, p. 297.
[12] Yu.P. Tulenin, V.V. Savkin, M.Yu. Sinev and V.N. Korchak, Kinetika i Kataliz (Russ. Kinetics and Catalysis), 43 (2002) 912.
[13] V.Yu. Bychkov, O.V. Krylov and V.N. Korchak, Kinetika i Kataliz (Russ. Kinetics and Catalysis), 43 (2002) 94.
[14] V.Yu. Bychkov, Yu.P. Tulenin, O.V. Krylov and V.N. Korchak, Kinetika i Kataliz (Russ. Kinetics and Catalysis), 43 (2002) 775.

Studies in Surface Science and Catalysis, volume 147
X. Bao and Y. Xu (Editors)

157

Synthesis gas production by partial oxidation of methane on Pt/Al$_2$O$_3$, Pt/Ce-ZrO$_2$ and Pt/Ce-ZrO$_2$/Al$_2$O$_3$ catalysts

P. P. Silva[a], F. de A. Silva[a], A. G. Lobo[b], H. P. de Souza[b,c], F. B. Passos[c], C. E. Hori[a], L. V. Mattos[b], F. B. Noronha[b]

[a]FEQUI – UFU Av. João Naves de Ávila, 2160 - Bloco 1K - CEP 38400-902 Uberlândia.

[b]Instituto Nacional de Tecnologia (INT) - Av. Venezuela, 82 - Centro - CEP: 20081-310, Rio de Janeiro/RJ/Brazil, e-mail: fabiobel@int.gov.br.

[c]Universidade Federal Fluminense, Rua Passos da Pátria 156, Niterói-24210-240

ABSTRACT

The partial oxidation of methane was studied on Pt/Al$_2$O$_3$, Pt/Ce-ZrO$_2$ and Pt/Ce-ZrO$_2$/Al$_2$O$_3$ catalysts. The oxygen transfer capacity of the samples and the degree of coverage of the alumina surface were evaluated by oxygen storage capacity (OSC) and infrared spectroscopy of adsorbed CO$_2$, respectively. Pt/Ce$_{0.75}$Zr$_{0.25}$O$_2$ catalyst was more active and stable than Pt/Al$_2$O$_3$ and Pt/Ce$_{0.75}$Zr$_{0.25}$O$_2$/Al$_2$O$_3$ catalysts on partial oxidation of methane. These results were explained by the high oxygen storage/release capacity of Pt/Ce$_{0.75}$Zr$_{0.25}$O$_2$ catalyst, which allowed a removal of carbonaceous deposits from the active sites. Pt/Ce$_{0.75}$Zr$_{0.25}$O$_2$/Al$_2$O$_3$ catalyst also presented a high oxygen exchange capacity. However, this sample deactivated during the reaction. The low degree of coverage of alumina surface obtained for this catalyst avoided the occurrence of the redox mechanism of carbon removal.

1. INTRODUCTION

Currently, the gas-to-liquids technology (GTL) is an alternative to make the use of natural gas located at remote regions economically viable. This technology is based on the conversion of natural gas to a synthesis gas prior to the liquid production through the Fischer-Tropsch Synthesis [1]. Partial oxidation of methane is a technology that fulfills the requirements to a syngas with H$_2$/CO =

2, the ratio necessary for a GTL plant. Mattos et al. [2] studied the partial oxidation of methane over $Pt/CeZrO_2$ catalysts. The results showed that the oxygen transfer capacity of the support and the metal dispersion were fundamental to improve the activity and the stability of the catalyst in this reaction. Since ceria-zirconia materials usually do not provide a very high surface area, we decided to deposit ceria and ceria-zirconia on alumina. The aim of this work was to evaluate the performance of Pt/Al_2O_3, $Pt/CeZrO_2$ and $Pt/CeZrO_2/Al_2O_3$ catalysts in the reaction of partial oxidation of methane.

2. EXPERIMENTAL

The Al_2O_3 support was prepared by calcination at 1173 K for 6 hours. The $Ce_{0.75}Zr_{0.25}O_2$ support was obtained by a co-precipitation method [3] with a Ce/Zr atomic ratio of 3:1. For $Ce_{0.75}Zr_{0.25}O_2/Al_2O_3$ support (13.7 wt % of Ce-ZrO_2), the Ce-ZrO_2 oxide was co-precipitated on Al_2O_3, using the method described above. The supports were calcined at 1073 K for 1 h in a muffle oven. Then, the catalysts were prepared by incipient wetness impregnation of the supports with an aqueous solution of H_2PtCl_6 (Aldrich) and were dried at 393 K. The samples were calcined under air (50 cm^3/min) at 673 K for 2 h. All samples contained 1.5 %wt of platinum. X-ray diffraction measurements were done using a RIGAKU diffractometer with a CuKα radiation. Oxygen storage capacity (OSC) measurements were carried out in a micro-reactor connected to a quadrupole mass spectrometer (Balzers, Omnistar). Infrared spectroscopy of adsorbed CO_2 analyses were performed in a Magna 560-Nicolet apparatus.

Partial Oxidation of methane was performed in a quartz reactor at atmospheric pressure. Prior to reaction, the catalyst (20 mg of the catalyst diluted with 36 mg of SiC) was reduced under H_2 at 773 K for 1h and then heated to 1073K under N_2. The reaction was carried out at 1073 K and WHSV = 260 h^{-1} using a total flow rate of 100 cm^3/min and a $CH_4:O_2$ ratio of 2:1. The exit gases were analysed using a gas chromatograph (Agilent 6890) equipped with a thermal conductivity detector and a Carboxen 1010 column (Chrompack).

3. RESULTS AND DISCUSSION

3.1. X-ray diffraction (XRD)

The XRD data obtained for Al_2O_3, $Ce_{0.75}Zr_{0.25}O_2$ and $Ce_{0.75}Zr_{0.25}O_2/Al_2O_3$ supports are presented in Fig. 1. The Al_2O_3 support presented the characteristics lines of γ-alumina. For $Ce_{0.75}Zr_{0.25}O_2$ support, it was observed the presence of lines at 2θ= 29.0 and 33.5°, which indicates that zirconia was incorporated into CeO_2 lattice and formed a solid solution with a cubic symmetry. Hori et al [3] studied the effect of ZrO_2 addition to CeO_2 on the phase composition of Pt/Ce-ZrO_2 catalysts, using XRD experiments. They observed that the addition of 25 % of ZrO_2 did not result in a separate zirconia phase, but the ceria lines shifted from $2\theta = 28.6°$ to $2\theta = 29.0°$ and from $2\theta = 33.1°$ to $2\theta = 33.5°$. According to these authors, this shift occurred due to a change in the lattice parameter of ceria and it is an indication of the formation of a $CeZrO_2$ solid solution. The diffraction pattern of $Ce_{0.75}Zr_{0.25}O_2/Al_2O_3$ support presented a line at about $2\theta = 28.8°$, indicating that a solid solution was formed on this material, although in this case, it is possible that not all the zirconia was incorporated into the ceria lattice.

3.2. Oxygen storage capacity (OSC)

The oxygen storage capacity (OSC) measurements are presented in Table 1. $Pt/Ce_{0.75}Zr_{0.25}O_2$ and $Pt/Ce_{0.75}Zr_{0.25}O_2/Al_2O_3$ catalysts showed a high oxygen consumption. Several studies reported that cerium oxide has a very high oxygen exchange capacity [4]. This capacity is associated to the ability of cerium to act as an oxygen buffer by storing/releasing O_2 due to the Ce^{4+}/Ce^{3+} redox couple [4]. The incorporation of ZrO_2 into the CeO_2 lattice promotes the CeO_2 redox properties. The presence of ZrO_2 strongly increases the oxygen vacancies of the support due to the high oxygen mobility of the solid solution formed, which was identified by our XRD data.

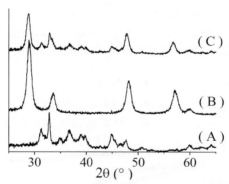

Fig.1. X-ray diffraction patterns of (A) Al_2O_3, (B) $Ce_{0.75}Zr_{0.25}O_2$ and (C) $Ce_{0.75}Zr_{0.25}O_2/Al_2O_3$ supports

Table 1
OSC (O_2/CeO_2 molar ratio) and CO_2-IR

Samples	OSC	OD/g $_{Al2O3}$
Pt/Al_2O_3	0	24.56
$Pt/Ce_{0.75}Zr_{0.25}O_2$	1.70	-
$Pt/Ce_{0.75}Zr_{0.25}O_2/Al_2O_3$	3.50	20.06

3.3. Infrared spectroscopy of adsorbed CO_2

Table 1 also presents the values of the optical density of the band at 1235 cm^{-1} calculated through CO_2 infrared spectra. According to the literature [5], for $Pt/CeO_2/Al_2O_3$ catalysts, these values are a quantitative measurement of the alumina surface that is not covered by ceria. The results obtained revealed a low degree of coverage of the alumina surface by cerium-zirconium oxide. This could be an indication that during the preparation of the $Ce_{0.75}Zr_{0.25}O_2/Al_2O_3$ sample, probably islands of ceria-zirconia particles were formed rather than uniformly dispersed on the alumina surface.

3.4. Partial Oxidation of Methane

Fig. 2 shows the methane conversion versus time on stream (TOS) for all the samples. $Pt/Ce_{0.75}Zr_{0.25}O_2$ catalyst presented the highest initial conversion and practically did not loose its activity after 24 h. On the other hand, Pt/Al_2O_3 and $Pt/Ce_{0.75}Zr_{0.25}O_2/Al_2O_3$ catalysts exhibited similar activity and strong deactivation during the reaction. Furthermore, a significant change in the selectivity towards CO and CO_2 was observed for Pt/Al_2O_3 and $Pt/Ce_{0.75}Zr_{0.25}O_2/Al_2O_3$ catalysts during the reaction (Fig. 3). The production of CO_2 increased and the selectivity to CO decreased as the CH_4 conversion decreased. Thus the deactivation of these catalysts was followed by a decrease on CO selectivity. This effect was less significant for $Pt/Ce_{0.75}Zr_{0.25}O_2$ catalyst.

We have previously observed the same results for Pt/Al_2O_3, Pt/ZrO_2 and $Pt/Ce-ZrO_2$ catalysts [2]. Pt/Al_2O_3 and Pt/ZrO_2 catalysts strongly deactivated during the reaction, while $Pt/Ce-ZrO_2$ catalyst presented a good stability. Moreover, the CO_2 selectivity increased as methane conversion decreased.

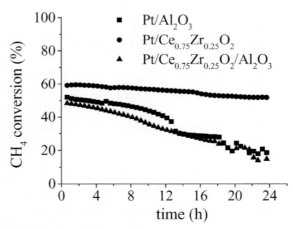

Fig. 2. Methane conversion versus time on stream.

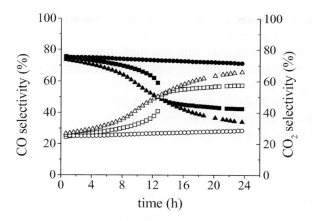

Fig. 3. CO selectivity for (■) Pt/Al$_2$O$_3$, (●) Pt/Ce$_{0.75}$Zr$_{0.25}$O$_2$ and (▲) Pt/Ce$_{0.75}$Zr$_{0.25}$O$_2$/Al$_2$O$_3$ catalysts and CO$_2$ selectivity for (□) Pt/Al$_2$O$_3$, (○) Pt/Ce$_{0.75}$Zr$_{0.25}$O$_2$ and (△) Pt/Ce$_{0.75}$Zr$_{0.25}$O$_2$/Al$_2$O$_3$ catalysts versus time on stream.

These results were explained by a two step mechanism of the partial oxidation of methane. In the first stage, combustion of methane occurs producing CO$_2$ and H$_2$O. In the second step, synthesis gas is produced via carbon dioxide and steam-reforming reaction of the unreacted methane. According to Stagg et al. [6], the support participates in the dissociative adsorption of CO$_2$ near the metal particles, transferring oxygen to the metal surface and accelerating the removal of carbon from the metal. Since the production of carbon and CO$_2$ increased on the Pt/Al$_2$O$_3$ and Pt/ZrO$_2$ catalysts during the reaction, the second step of the mechanism was inhibited. O' Connor et al [7] also reported that the reaction pathway for the partial oxidation of methane appeared to be methane combustion followed by reforming of the remainder of the methane by the resultant CO$_2$ and steam.

Then, for Pt/Al$_2$O$_3$ and Pt/Ce$_{0.75}$Zr$_{0.25}$O$_2$/Al$_2$O$_3$ catalysts, the increase of carbon deposits around or near the particle metal affects the CO$_2$ dissociation and inhibits methane conversion, and consequently the CO$_2$ reforming step of the partial oxidation of methane reaction. On the Pt/Ce$_{0.75}$Zr$_{0.25}$O$_2$ catalyst, the selectivity towards CO practically did not change during reaction due to the redox mechanism of carbon removal promoted by the support. This result could be assigned to high oxygen exchange capacity of Pt/Ce$_{0.75}$Zr$_{0.25}$O$_2$ catalyst, as revealed by OSC measurements. The higher amount of oxygen in the proximity of metal particles promotes the removal of carbon deposits from the metallic surface [2]. Although Pt/Ce$_{0.75}$Zr$_{0.25}$O$_2$/Al$_2$O$_3$ catalyst showed a higher oxygen exchange capacity, this sample deactivated during the reaction. Since the redox mechanism of carbon removal takes place at metal/ceria-zirconia interfacial perimeter, this result indicates that the low degree of coverage of alumina surface by $CeZrO_2$ caused a preferential deposit of the platinum on the Al_2O_3. For this catalyst there was not the formation of an interface between platinum metal particles and $CeZrO_2$, causing a suppression of the carbon removal mechanism. This hypothesis is confirmed by the similar deactivation behavior between $PtAl_2O_3$ and $Pt/Ce_{0.75}Zr_{0.25}O_2/Al_2O_3$ catalysts.

4. CONCLUSIONS

The results obtained showed that the $Pt/Ce_{0.75}Zr_{0.25}O_2$ catalyst presented the highest activity and stability on partial oxidation of methane. This result was attributed to the high oxygen exchange capacity of the support, which keeps the metal surface free of carbon. The strong deactivation observed for $Pt/Ce_{0.75}Zr_{0.25}O_2/Al_2O_3$ catalyst was assigned to the low degree of coverage of alumina surface, which avoids the occurrence of the redox mechanism for carbon removal.

ACKNOWLEDGEMENTS

The authors wish to acknowledge the financial support of the CNPq/CTPETRO (462530/00-0) and CAPES and CNPq scholarships.

REFERENCES

[1] V.K. Venkataraman, H.D. Guthrie, R.A. Avellanet, and D.J. Driscoll, Stud.Surf.Sci.Catal., 119 (1998) 913.
[2] L.V. Mattos, E.R. de Oliveira, P.D. Resende, F.B. Noronha and F.B. Passos, Catal. Today, 77 (2002) 245.
[3] C.E. Hori, H. Permana, K.Y. Simon Ng, A. Brenner, K. More, K.M. Rahmoeller and D. Belton, Appl. Catal. Envir. B, 16 (1998) 105.
[4] M.H. Yao, R.J. Baird, F.W. Kunz and T.E. Hoost, J. Catal. 166 (1997) 67.
[5] R. Fréty, P.J. Lévy, V. Perrichon, V. Pitchon, M. Primet, E. Rogemond, N. Essayem, M. Chevrier, C. Gauthier and F. Mathis, Third Congress on Catalysis Polution Control (CAPOT 3), 2 (1999) 265.
[6] S.M. Stagg-Williams, F.B. Noronha, G. Fendley, D.E. Resasco, J.Catal., 194 (2000) 240.
[7] A.M. O'Connor and J.R.H. Ross, Catal. Today 46 (1998) 203.

Studies in Surface Science and Catalysis, volume 147
X. Bao and Y. Xu (Editors)

An investigation of methane partial oxidation kinetics over Rh-supported catalysts

Ivan Tavazzi[*] , Alessandra Beretta, Gianpiero Groppi, Pio Forzatti

Dipartimento di Chimica, Materiali e Ingegneria Chimica, Politecnico di Milano, Piazza Leonardo da Vinci 32, 20133 Milano, Italy

ABSTRACT

The kinetics of CH_4 partial oxidation was investigated in a structured reactor at high space velocity under well-controlled operating conditions. High temperatures favored the production of CO and H_2. At increasing space velocity, CO and H_2 selectivities showed a moderate and a strong decrease respectively. Increase of O_2/CH_4 feed ratio caused a shift to higher temperatures of methane conversion and syngas production. These effects could be accounted for by a simplified kinetic scheme, which includes methane deep oxidation, steam and dry reforming, WGS and the consecutive oxidations of CO and H_2.

1. INTRODUCTION

A number of novel applications of catalytic partial oxidation are presently studied. NG partial oxidation represents a potential solution for the stationary small scale production of H_2. It has also been proposed for staged combustors (wherein a fuel rich catalytic combustion and a fuel lean combustion are realized in sequence) for gas turbines with advanced performances. Partial oxidation of hydrocarbon fuels is applied to the on-board syngas production for SOFC-based auxiliary power units. Such applications exploit the exothermic character of the process and the possibility to run it in small reaction volumes with reduced thermal inertia. The pioneering work by Schmidt and coworkers on methane partial oxidation first demonstrated the outstanding performances of Rh-based catalysts in autothermal short contact time reactors, consisting of insulated foam monoliths; yields to CO and H_2 as high as 90% were observed at few ms contact time at about 1000°C [1]. Many authors have confirmed the potentialities of Rh catalysts for methane partial oxidation and addressed with different approaches the study of the process mechanism and kinetics. A comprehensive review was provided by [2]. It appears, though, that under the severe operating conditions which characterize the autothermal applications, activity tests are scarcely informative on the real reaction kinetics whose effect can hardly be decoupled from heat and mass transfer limitations, as well as from thermodynamics.

In this work, experiments of methane partial oxidation were performed in an annular reactor, which was specifically designed for investigating the kinetics of very fast reactions [3]; it allows to control the extent of mass transfer limitations, to analyze the process far from the thermodynamic equilibrium constraints and to realize well-controlled temperature conditions with minimum axial gradients. A quantitative analysis of the observed effects was addressed by developing a simplified kinetic scheme, which reduces the complexity of the reacting system [2] into few main molecular stoichiometries.

2. EXPERIMENTAL

Methane partial oxidation was studied over a 0.5% w/w $Rh/\alpha-Al_2O_3$ catalyst [4]. Tests were realized in a structured annular reactor; it consists of an inner ceramic tube (o.d.=4.75 mm), coaxially inserted into an outer quartz tube (i.d.=9 mm), giving rise to an annular duct through which the gas flows. A short (few cm) and very thin (<20 μm) catalyst layer was washcoated on the ceramic tube [3]; the importance of intra-porous mass transfer limitations was thus reduced. Very high space velocities (up to 10^7 Nl/(kg$_{cat}$ h)) were easily realized by feeding high flow rates and the conversion of methane was kept well below the equilibrium value at all temperatures [5]. Almost isothermal conditions were observed during the experiments; under conditions of high exothermicity (main production of CO_2 and H_2O), temperature gaps along the catalyst were always below 20°C, but most of the tests were realized with temperature differences below 10°C and 5°C. This was partly due to the effective dispersion of the reaction heat by radiation from the thin catalytic layer. Also, diluted feed streams ($CH_4/O_2/N_2$ = 4/2/94 and $CH_4/O_2/N_2$ = 4/2.67/93.33) were used at this scope.

At the reference space velocity of $2,6*10^6$ Nl/(kg$_{cat}$ h), partial oxidation tests were carried out to investigate the effect of temperature in the range 400 ÷ 920 °C. At constant temperature, the effect of space velocity (GHSV) was explored, from 10^6 to $3.1 \ 10^7$ Nl/(kg$_{cat}$h), (i.e. from 0.12 to 4 ms contact time). Blank experiments verified that homogeneous reactions did not occur under these conditions.

4. RESULTS

4.1 Effect of temperature and CH_4/O_2 ratio

Figure 1 reports the observed effect of reaction temperature. With a stoichiometric feed (CH_4/O_2=2), appreciable conversion of reactants was observed at 450°C; at increasing temperature, O_2 conversion grew rapidly and

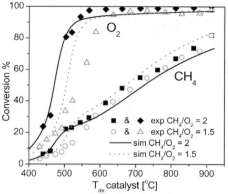

Fig. 1. CH$_4$ and O$_2$ conversion vs. T. CH$_4$/O$_2$ = 2 and 1.5. GHSV = 2.6*10^6 Nl/(kg*h)

was almost complete (>98%) at about 570°C. Above this temperature, CH$_4$ conversion further increased; this suggested the existence of secondary reactions of methane which did not involve O$_2$. Besides, at increasing methane conversion, temperature profiles along the catalyst agreed with the presence of an inlet exothermic zone, followed by an endothermic zone. Concerning the product distribution (Figure 2a,b), H$_2$O and CO$_2$ were mainly formed below 550-600°C, while CO and H$_2$ were observed in small traces. At increasing temperature, the concentration of the deep oxidation products passed through a maximum, then it decreased significantly above 600°C; a progressive production of CO and H$_2$ was observed instead, and these species became the most abundant products at the highest temperature investigated. The observed effects grew stronger the hypothesis of an indirect pathway to partial oxidation species, consisting of a highly active total oxidation of methane and slower endothermic reactions of steam and dry reforming. However, the results could not exclude the existence of direct routes of formation of CO and H$_2$, especially at the lower temperatures wherein O$_2$ was still present and consecutive reactions could be active.

A second set of experiments was performed with the decreased CH$_4$/O$_2$ ratio of 1.5; methane activation was significantly delayed, in line with literature reports on a negative reaction order with respect to O$_2$ of CH$_4$ oxidation under rich conditions [2]. On increasing the temperature, O$_2$ conversion reached an

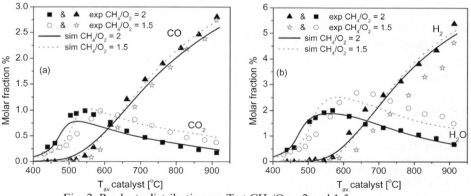

Fig. 2. Products distribution vs. T at CH$_4$/O$_2$ = 2 and 1.5.

asymptotic value (97%) above 650°C whereas CH_4 conversion monotonically increased showing similar values to those obtained with $CH_4/O_2=2$. Looking at the product distribution, an analogous delay in the formation of CO_2 and H_2O was observed, but at high temperature, a marked increase of their production was observed as a result of the excess amount of O_2. On the other hand, the concentration of CO and H_2 was always lower at the lower CH_4/O_2 feed ratio; it is believed that this could be due to inhibition of H_2O on the rate of steam reforming.

4.2 Effect of space velocity

The effect of GHSV in the range of ultra-high space velocity was investigated at two average catalyst temperature of 661°C and 761 °C. The measured reactant conversions, the selectivities of CO and H_2, the H_2/CO molar ratio are reported in Figure 3 (a-d).

At the lower values of space velocity and contact times of 2 ÷ 3 ms, conversion of O_2 was complete. However, as space velocity increased the conversion of O_2 significantly decreased; data showed that it was independent on temperature, indicating that the O_2 consumption rate was controlled by

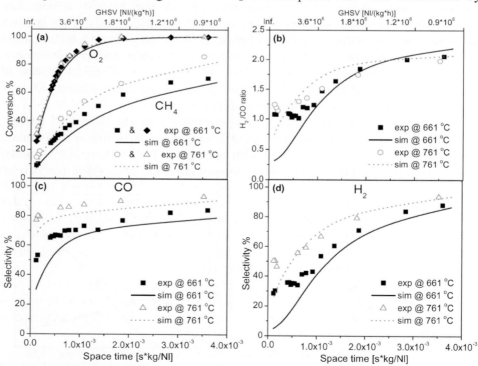

Fig. 3. Effect of GHSV at 661 and 761 °C. (a) reactants conversion, (b) H_2/CO ratio, (c) and (d) CO and H_2 selectivity. $CH_4/O_2/N_2 =4/2/94$.

external diffusion limitations, thus implying a low surface concentration. Also CH_4 conversion decreased progressively with shortening contact times and the effect of temperature was important. Concerning the product distribution, either the selectivity of H_2 and that of CO decreased with increasing GHSV. This effect was interpreted as an additional evidence of the existence of secondary routes to syngas. However, H_2 selectivity was almost halved, while CO selectivity decreased only moderately. As a result, a significant decrease of the H_2/CO ratio was observed; from the value of 2 obtained at 3-4 ms (the equilibrium value at high temperature), it dropped to 1 at sub-millisecond contact times. These trends indicated that the net production of H_2 and CO followed partly independent paths. The evidence was remarkable that, even at contact times in the order of 0.1 ms in the presence of residual O_2, the production of syngas was important, which could indicate either a role of methane reforming reactions at the very initial stage of reactant conversion or a possible role of direct reactions between O_2 and CH_4 to CO and H_2.

4.3 Theoretical analysis

A debate is still open in the literature about the kinetic scheme (direct or indirect) which is responsible for syngas formation at short contact times [2]. The experiments herein reported prove the importance of secondary reactions such as steam reforming. Still they do not exclude the possible presence of direct routes to syngas. Besides, a purely indirect scheme (methane combustion + methane reforming) could not account for the observed effects of temperature and space velocity on the product distribution.

A kinetic analysis of data was preliminarily addressed by assuming an indirect-consecutive scheme consisting of complete CH_4 combustion, secondary reactions (steam and dry reforming, reverse water gas shift), which are responsible for CO and H_2 production, and consecutive oxidations of CO and H_2 (Table 1). The rate equation and the parameters for CH_4 combustion were derived by data fitting in this work; analogous kinetic expressions were adopted for the consecutive combustions of CO and H_2, but much faster reaction rates were assumed as compared to CH_4 combustion, in line with the expected higher reactivity of CO and H_2. Concerning the intrinsic kinetics of secondary reactions, independent estimates had been obtained after dedicated experiments in a micro fixed bed reactor and have been included in this work [6].

A heterogeneous, isothermal, plug-flow model of the annular reactor was applied to simulate the partial oxidation tests; its governing equations have been previously reported [5] and consist of the mass balances for the reacting species in the bulk phase and at the gas-solid interface.

Model predictions are reported in Figures 1, 2 and 3 as solid and dotted lines. Despite of some inaccuracies, a good match was found between calculated and experimental trends at varying temperature, CH_4/O_2 ratio and contact time.

Surprisingly, though neglecting a direct route to syngas, the assumption of an indirect kinetic scheme could account for a significant production of syngas even at the lowest contact times, thus showing that secondary reactions over the Rh-catalyst are in principle fast enough to justify H_2 production at fractions of 1 ms. The role of consecutive oxidations of CO and H_2 was also found important to adequately describe the product distribution; without these additional routes the indirect scheme predicts a H_2/CO ratio which varies from the stoichiometry of steam reforming ($H_2/CO=3$) at very short contact time to the equilibrium value ($H_2/CO=2$), which is just the opposite of the experimental trend.

Further analysis is in progress; experiments and modeling are directed toward a better definition of the kinetic scheme and of the rate equations.

Table 1 - Consecutive kinetic scheme

I) $\quad CH_4 + O_2 \rightarrow CO_2 + 2H_2O$	1) $r_{TO} = k_{TO}\, P_{CH4}P_{O2}/(1+K_{ads}.P_{O2})^2$
	2) $r_{SR} = k_{SR}(P_{CH4}P_{H2O}-P_{CO}P^3_{H2}/K_{eq,SR})/$
$\left\{\begin{array}{l} CH_4 + H_2O \rightarrow CO + 3H_2 \\ CO + H_2O \rightarrow CO_2 + H_2 \end{array}\right.$	$\qquad\qquad (1+K_{AdsH2O}\,{}^*P_{H2O})$
II) $\left\{\begin{array}{l} CH_4 + CO_2 \rightarrow 2CO + 2H_2 \\ H_2 + \tfrac{1}{2}O_2 \rightarrow H_2O \\ CO + \tfrac{1}{2}O_2 \rightarrow CO_2 \end{array}\right.$	3) $r_{DR} = k_{DR}(P_{CH4}P_{CO2}-P^2_{CO}P^2_{H2}/K_{eq,DR})$
	4) $r_{RWGS} = k_{SH}(P_{CO2}P_{H2}-P_{CO}P_{H2O}/K_{eq,RWGS})$
	5) $r_{H2\ comb.} = k_{H2}P_{O2}P_{H2}/(1+K_{ads}.P_{O2})^2$
	6) $r_{O2\ comb.} = k_{CO}P_{CO}P_{O2}/(1+K_{ads}.P_{O2})^2$

$k_{TO} = 25.55\ \exp[-4015\ (1/T-1/773)]\ {}^{(*)}$ $\qquad k_{SR} = 4.5\ 10^{-1}\exp[-8257\ (1/T-1/873)]\ {}^{(**)}$

$k_{DR} = 1.57\ 10^{-3}\exp[-11180\ (1/T-1/673)]\ {}^{(**)}$ $k_{RWGS} = 2.12\ 10^{-3}\exp[-10080\ (1/T-1/673)]\ {}^{(**)}$

$k_{H2\ comb} = 10^4\ \exp[-2800(1/T - 1/773)]\ {}^{(*)}$ $\qquad k_{CO\ comb} = 200\ \exp[-3000(1/T - 1/773)]\ {}^{(*)}$

$K_{Ads.\ O2} = 550\ \exp[2500(1/T - 1/773)]\ {}^{(*)}$ $\qquad K_{Ads.\ H2O} = 10\ \exp[5410(1/T - 1/873)]\ {}^{(*)}$

r_j [mol/(s*g$_{cat}$)]; (*) This work; (**) [6]

Financial by MIUR – Rome is gratefully acknowledged.

REFERENCES

[1] K.L. Hohn, L.D. Schmidt, Appl. Catal. A 211 (2001) 53.
[2] P. Aghalayam, Y.K. Park, D.G. Vlachos, Catalysis, 15 (2000) 98.
[3] G. Groppi, W. Ibashi, E. Tronconi, P. Forzatti, Chem. Eng. J. 82 (2001) 57.
[4] L. Basini, K. Aasberg-Petersen, A. Guarinoni, M. Østberg, Cat. Today 64 (2001) 9.
[5] I. Tavazzi, A. Beretta, G. Groppi, P. Forzatti, Proceedings of the DGMK Conference "Innovation in the manufacture and use of Hydrogen", oct. 15-17, 2003, Dresden.
[6] L. Majocchi, PhD thesis, Politecnico di Milano (1999).

Studies in Surface Science and Catalysis, volume 147
X. Bao and Y. Xu (Editors)

Microchannel gas-to-liquids conversion – thinking big by thinking small

Terry Mazanec[a], Steve Perry[a], Lee Tonkovich[a], and Yong Wang[b]

[a]Velocys Inc., 7950 Corporate Blvd., Plain City, OH, 43064, USA

[b]Pacific Northwest National Laboratory, 902 Battelle Blvd, Richland, WA 99352, USA

1. INTRODUCTION

Oil resources in the developed world are declining rapidly. Abundant reserves of natural gas are available at remote sites and in offshore fields, but it is not cost effective to transport these to market. A cost competitive gas-to-liquids (GTL) process would unlock the world's natural gas supplies for use in motor vehicles. Global adoption of GTL technology would enable many more nations to supply liquid fuels, thereby increasing the number of sources of oil for international trade and reducing both costs and political uncertainties.

Upgrading natural gas to clean diesel fuel comprises (1) natural gas conversion to syngas, and (2) Fischer-Tropsch synthesis to produce waxy liquids. Productivity in each of these steps is limited by the heat and mass transport capabilities of the reactors. Commercialization is inhibited also by the immense scale on which GTL plants must be designed in order to take advantage of the economies of scale with conventional technologies. Small reactors that lend themselves to modularization have significant advantages in that they can be applied to large or small gas resources and can be situated on offshore platforms where space is limited.

Microprocessing technology (MPT) increases reactor productivity by greatly enhancing heat and mass transfer. An MPT reactor consists of a large number of tightly grouped microchannels, making it smaller than a conventional reactor. Heat generating microchannels are interlayered with cooling channels to provide short paths for heat transfer. The individual microchannels are less than 1 mm wide so that heat and mass transfer distances are small and rates are high. Close packing of small channels increases the surface area for heat transfer compared to conventional reactors, greatly increasing the amount of heat that can be transferred per unit volume of the reactor. Internal mixing provides additional flexibility in the design of MPT reactors.

An important advantage of MPT reactors is the very short residence time needed in the hot zone. This advantage is extremely important for reactions such as steam reforming since the reactions that produce coke are slower than reforming reactions. Shortening time in the hot zone favors reforming reactions over the coke-forming polymerization reactions. M. W. Kellogg used this concept to advantage in its "Millisecond" Crackers, first commercialized in 1985, in which the residence times are about 100 milliseconds. Bench scale MPT reactors developed by Velocys for steam reforming are operated with contact times of <10 milliseconds, leading to further productivity and cost advantages.

2. STEAM REFORMING

A steam reforming microchannel reactor consists of a steam reforming channel and an adjacent channel containing a heat source in close thermal contact. In some reactor designs they share a common wall. The heat source is the hot gas from a combustion stream. Heat exchange occurs through the wall separating these two zones as shown in Fig. 1. The steam reforming zone contains a catalyst to accelerate the conversion of methane and steam to synthesis gas.

The design of the microchannel reactor can be varied in numerous ways. The size and shape of the individual microchannels can be varied to affect parameters such as contact time, process intensity, volume of catalyst needed, heat flux, and pressure drop. Several options are available for the flow pattern.

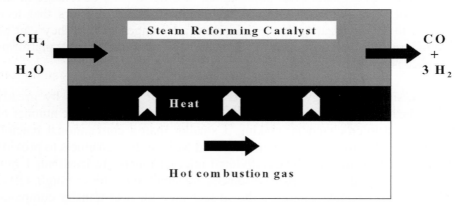

Fig. 1. Schematic of microchannel steam reformer reactor showing heat flow from the hot combustion stream to the process gas.

Fig. 1 shows a co-current flow pattern, but counter-current and cross-flow patterns are also being investigated. The coupling of the combustion process to the steam reforming process can be varied by modifying the position of the combustion reactions relative to the steam reforming zone and their rates. Positioning a combustion catalyst adjacent to the steam reforming zone can improve heat transfer and conversion in the combustion process, but this geometry limits separate control of the two processes. Optimization of the design must take into account the process performance features as well as the cost and ease of fabrication. Early recognition of the interaction of design and fabrication has permitted Velocys to identify and overcome several problems of early designs.

Equilibrium limits the methane conversion and CO selectivity achievable in steam reforming. The highest conversions and selectivities are possible at high temperature. Moreover the steam reforming reaction is endothermic, absorbing 49.3 kcal/mol at standard conditions as per Eq. (1).

$$CH_4 + H_2O \Leftrightarrow CO + 3 H_2 \qquad \Delta Ho = 49.3 \text{ kcal/mol} \qquad (1)$$

Thus, the central problem in steam reforming is to rapidly transport heat to the process stream to supply heat for the reaction. The heat transfer must be accomplished without producing carbon. To preserve the product against re-equilibration and Boudouard carbon formation, Eq. 2, it is desirable to cool the product as quickly as possible.

$$2 CO \Leftrightarrow CO_2 + C \qquad (2)$$

The Velocys® microchannel reactor has been developed to approach equilibrium at high temperature in a very short contact time. Conversion and selectivity results in a microchannel reactor are shown as a function of contact time in Fig. 2. At contact times below 10 ms the approach to equilibrium remains high - within about 10 C of equilibrium – for a 2:1 S:C ratio feed gas at 20 atm pressure. This demonstrates the rapid heat flux and kinetics achievable in a microchannel device. Average heat fluxes as high as 25 W/cm^2 in these tests; peak flux values above 50 W/cm^2 have been observed in other tests. On a volumetric basis, heat fluxes in a laboratory reactor at Velocys have been determined to be > 75 W/cm^3, almost one hundred times greater than conventional fired reformers (< 1 W/cm^3).

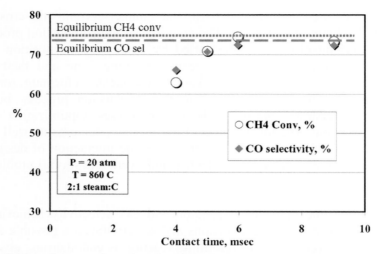

Fig. 2. SMR performance in a microchannel reactor at short contact time.

In Fig. 3 the results of a longer test are presented, showing longer term process and catalyst stability. In this test the reactor was intentionally operated at very rapid flow rates so that the process would not reach equilibrium. Under these conditions the stability of the catalyst can be assessed, since any decline in activity will be readily apparent. When operating under conditions where equilibrium is reached this is not as easily discerned since a small loss of catalyst activity will not necessarily result in a change in conversion or selectivity. The data show that the catalyst activity and selectivity are not subject to rapid deactivation, and that the process can be operated steadily in a microchannel reactor without plugging or fouling the system.

3. COMBUSTION

In order to balance the heat consumption of the steam reforming reaction, combustion must be integrated into the reactor. The performance of the combustor portion of the reactor must also meet commercial targets for fuel conversion, CO discharge, and NO_X production. The microchannel combustion system developed by Velocys achieves or exceeds these targets. Performance of the combustion portion of an integrated multichannel microchannel reactor containing both steam reforming and combustion functions is summarized in Fig. 4. The conversion of fuel is quantitative and the CO exhaust level is within acceptable standards. Furthermore, not shown in Fig. 4, are the NO_X levels, which remain between 5 and 10 ppm throughout the experimental demonstration.

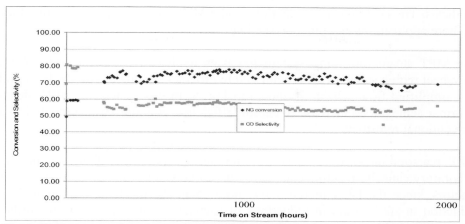

Fig. 3. Performance of a microchannel reactor during early tests for process and catalyst stability. Reaction conditions: temperature: 850 C, pressure: 20 atm, steam:carbon 3:1; equilibrium conversion 83%; equilibrium selectivity: 62%.

An important feature of microchannel reactor systems is their reduced size. Compared to a conventional steam reformer the Velocys® microchannel reformer is only about 1/25th as large. This translates into a smaller footprint required, which makes microchannel reactors particularly well suited to installation on off-shore oil platforms and in restricted spaces among existing equipment.

4. PROCESS EFFICIENCY

MPT technology has the potential to reduce costs in a variety of natural gas upgrading processes including steam reforming, FT synthesis and methanol

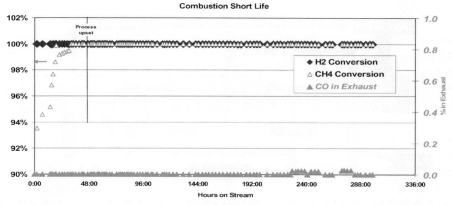

Fig. 4.Combustion performance of a microchannel reactor showing high fuel conversion and low CO production in the exhaust gas.

synthesis. The cost and size reductions make MPT technology ideally suited for offshore applications and plant expansions.

The MPT steam reforming system has been evaluated in an integrated GTL flowsheet with a standard FT reactor and hydrocracking process on the back end. The simplified flow scheme is shown in Fig. 5. In this system the MPT steam reformer achieved an increase in carbon efficiency of about 10% compared to a conventional reformer.

MPT reactors increase capacity by 'numbering up' rather than scaling up. Laboratory capacity reactors encompass the same reactor dynamics as do larger reactors, since the latter merely have more channels, not larger channels. The reduction in size results in reductions of catalyst inventory, leading to further cost savings. Mass manufacturing takes advantages of mass production to reduce costs and eliminate on-site expenditures.

5. CONCLUSION

Microchannel reactor systems show great promise in their application to gas-to-liquids applications. Both the steam reforming and the Fischer-Tropsch processes are amenable to improvement by use of a microchannel reactor. Velocys has built and operated microchannel steam reformers that show exceedingly high heat fluxes and process intensification. Their performance has been shown to be stable for several months of operation, demonstrating both device and process stability. The advantages of microchannel steam reformers are most obvious where space constraints are significant, but microchannel systems also have performance advantages compared to conventional systems that make them attractive for any GTL installation.

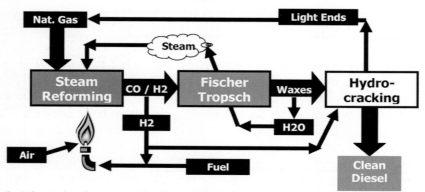

Fig. 5. Schematic of a gas to liquids process utilizing microchannel steam reformer and Fischer-Tropsch reactors.

Studies in Surface Science and Catalysis, volume 147
X. Bao and Y. Xu (Editors)
175

Combined steam reforming and dry reforming of methane using AC discharge

Huy Le, Trung Hoang, Richard Mallinson, Lance Lobban

School of Chemical Engineering and Materials Science,
100 E. Boyd, Room T335, Norman, OK, 73019, USA

1. INTRODUCTION

Today the utilization of natural gas for its environmental impact has received worldwide attention. Methane is the principal constituent of natural gas and carbon dioxide is a major green house gas that is frequently present in natural gas [1 – 4]. The use of these two components together could enhance their value as an environmentally friendly chemical and fuel feedstock.

Methane conversion to C_2 hydrocarbon and syngas, especially hydrogen, using non-equilibrium plasmas has recently attracted research interest [5 – 8]. The highly energetic electrons cause activation of molecules, creating free radicals. The radicals then react with other molecules or radicals. Due to the low bulk gas temperature, the chain reaction's life is short, inhibiting the products from achieving thermodynamic equilibrium product distributions.

This paper will discuss the combined dry and steam reforming of methane using the non-equilibrium plasma generated by an AC discharge.

2. EXPERIMENTAL

A schematic of the AC plasma tubular reactor with a point to point electrode configuration is shown in Fig. 1. The discharge was generated between two metal electrodes in a quartz tube by an AC power supply unit. The top electrode was a 1/8-inch stainless steel tube. The bottom electrode was a stainless steel rod. In the electrode configuration study, the bottom rod electrode was replaced by a 1/4-inch or 1/8-inch stainless steel tube. The gas gap (GG) between the two electrodes was 15 mm, unless otherwise specified. The reactor's inside diameter (ID) was 10 mm. Methane and carbon dioxide at just above atmospheric pressure flowed axially downward inside the reactor. Product gases only exited through the bottom tube electrode in the case of the hollow bottom electrode study (Figure 1). De-ionized liquid water at room temperature was pumped into the reactor via the top electrode by a Kd Scientific syringe pump model 100. The power system has

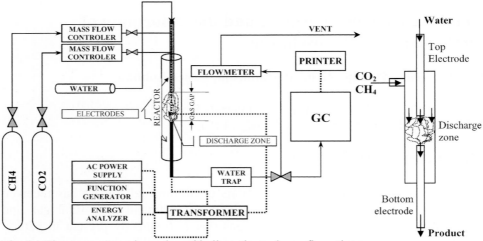

Fig. 1 – Plasma reactor schematic and hollow electrode configuration

been described previously [6,10].Temperatures measured by Infrared were less than 600 K [9, 10].

The product selectivities were calculated based on both methane and carbon dioxide. The energy consumption, *eV/molecule of C converted*, was based on the total input power and the total amount of CH_4 and CO_2 reacted. Likewise, the term "*eV/mol H_2 produced*" is the energy consumption to produce a molecule of hydrogen. These numbers are measured on the low voltage side and include system losses.

3. RESULTS

3.1 Gas Gap

The experimental conditions for the series of gas gap experiments were 200 sccm feed flow rate (varying residence time), 37.5 mole % CH_4, 37.5 mole % CO_2, and 25 mole % H_2O, 200 Hz, 10 mm ID reactor. The gas gap between the two electrodes was varied from 5.5mm to 15mm. The input power was also varied to maintain the discharge. As the gas gap increases, the feed conversions increased which is due to the increase in the residence time (from 0.13 seconds to 0.35 seconds) and the input power (from 8.5W to 24W). In methane conversion using a non-equilibrium plasma, the energy consumption usually increases with increasing feed conversion due to lower probability of the highly energetic electrons activating a previously un-reacted feed molecule at higher conversions. In this case, even though the conversions increase, the energy consumption remains relatively stable around 8 eV/molecule of C converted, which is 50% lower than previously reported [6]. It has been experimentally observed that as the gas gap increased, the visible discharge volume, resembling an oval shape, increased as well. The reason for the energy consumption not increasing is likely because increasing the gas gap decreases the amount of gas bypassing the discharge space.

The hydrogen selectivity increased from 58% to 70% with increasing gas gap. The acetylene selectivity shows a similar increase. However, the ethylene selectivities decreased slightly from 10% to 5%. The ethane selectivity was very low, less than 5%. Because the H balance, which is the amount of hydrogen in the reacted methane accounted for in the gas phase products, is less than 100%, at this CO_2 mole fraction, water production instead of consumption is inferred. The vapor product (in this case separated from the liquid with a trap at $40°C$) of the experiment with 15mm gas gap was also analyzed by Mass Spectrometry to detect methanol, ethanol, and any other heavy species in the product. There was no methanol (MW =32) or ethanol (MW=46) observed. This result shows that under these experimental conditions, organic oxygenates are not produced in the AC corona discharge.

3.2 Feed Composition Study

The feed of combined steam and dry reforming under AC discharge has been studied to find the best composition in terms of power consumption and methane conversion. Residence time was kept constant at 0.35s. Three series of experiments with different mole fractions of CO_2: 7.5%, 15% and 25% were conducted at 300Hz. In each series, the mole fractions of water ranged from 12.5% to 37.5%.

As shown in Fig. 2, the conversion of methane and carbon dioxide increases at higher mole fraction of water in the feed. The lowest power consumption is 9eV per molecule C converted and 7eV per molecule of hydrogen produced. The highest conversion 45% obtained at the maximum mole fraction of water 37.5% and CO_2 25% had slightly higher power consumption 10eV. The results suggest that adding more water and CO_2 into the feed improve the conversions. This is consistent with the results in steam reforming. In steam reforming, the consumed power could reach 10eV but only 19% conversion of methane was obtained [10]. The maximum conversion in steam reforming was 80%; however, the power consumption is also as high as 40eV [10]. Therefore, the presence of CO_2 in the feed has reduced the power consumed by the system. It can be explained that, besides the dry reforming, CO_2 has provided more O and then OH radicals in the system. Thus, more CH_x (x=1,2,3) radicals were produced because of the more hydrogen abstraction by the O species. As a result, higher conversion with lower power consumption was observed.

The distribution of products at different feed composition was also analyzed and is shown in Fig. 3. Adding more water increased H_2 and CO selectivity, while decreasing ethylene and ethane selectivities. It is noted that the selectivity of ethane is very low in most cases and there are no traces of it at high mole

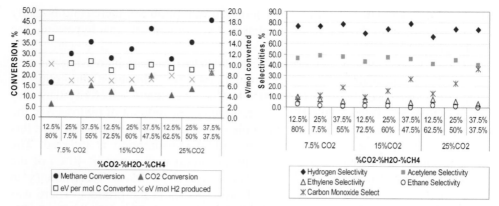

Fig. 2 – Conversion and power consumption

Fig. 3 – Product selectivities vs. feed compositions

fraction of water. These trends were due to increasing the conversion of methane and CO_2 when adding more water. Furthermore, from study of the pathways of plasma reactions [10], ethane is a primary product. At higher concentrations of water and CO_2, more hydrogen abstracting species were produced and thus more ethane and ethylene were converted. Hydrogen and acetylene selectivities were around 80% and 45% respectively. The more CO_2 present in the feed, the more CO was produced. At the highest conversion of methane of 45%, the highest CO selectivity of 40% was obtained and all C_2 selectivities decreased as expected.

3.3 Electrode study

To understand the bypassing that may occur during the reaction, hollow bottom electrodes have been studied. In this configuration, because gases exit through the bottom hollow electrode, they were better mixed than in the case of coaxial exit gases through the bottom of the reactor. In addition, all of the gas must pass through at least some part of the discharge zone. Therefore, in this configuration there is less bypassing than in the case of gas exiting through the bottom of the reactor. A series of experiments were conducted with three types of electrodes: the standard steel rod, a 1/4-inch steel tube and a 1/8-inch steel tube. The total flow rate was kept the same for each experiment at 200 sccm with 75 sccm CH_4, 50 sccm CO_2 and 75 sccm H_2O. Other parameters also kept constant: 10mm reactor ID, 15 mm gas gap, 25 W power input and 300Hz frequency.

By changing the bottom electrode from a solid rod to a hollow tube, conversions of methane and carbon dioxide each increase about 5%. The discharge shape and stability appear "better" than in the case of the solid rod electrode. Product distribution changed with an interesting trend. Acetylene selectivity decreases in changing from the bottom solid rod electrode to the hollow tube electrode. However, carbon monoxide selectivity acted in a different way. The smaller the hollow tube diameter, the more CO was produced. This perhaps can be explained by increased mixing of the product gas generated after entering the discharge zone, lower in the reactor below the discharge zone, and re-entering the discharge zone to exit through the bottom hollow electrode. In other words, the residence time in this situation was a little longer due to mixing. These results are consistent with the pathways study [10], because CO is produced in part from the oxidation of carbon formed from decomposition of C_2H_2. In addition, CO is also produced from CO_2 and this can explain higher conversion of carbon dioxide in the hollow electrode experiments.

In conclusion, the hollow bottom electrodes help discharge stability and increase the CO fraction in the product. This CO can also supply more H_2 product when the water gas shift reaction is taken into account. CO selectivity is about 60% in this case compared with 45% in the solid electrode's case. The energy consumption decreased from 10eV to 9eV per molecule carbon converted in this experimental condition. The lowest promising energy consumption is about 5eV per molecule of hydrogen produced if the water gas shift reaction is included.

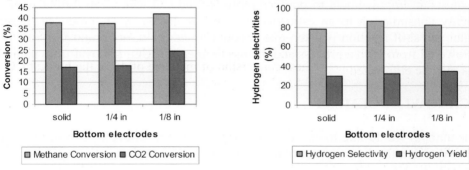

Fig. 6 – Methane conversion and Hydrogen selectivities vs. bottom electrodes

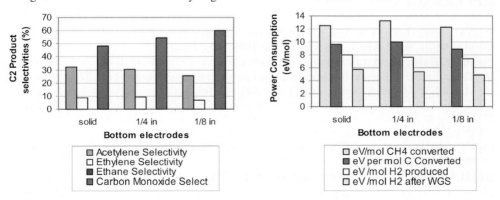

Fig. 7 – C_2 selectivities and Power consumption vs. bottom electrodes

4. SUMMARY

In the low temperature plasma combination of steam and dry reforming, H, OH, and O radicals help activate more methane molecules and increasing hydrogen abstraction from hydrocarbon species, prolonging the chain reactions. The O and OH radicals also help convert carbon to carbon oxides, preventing the hydrogenation of coke to produce methane and removing solid carbon from the discharge space. As a result, methane conversion in the combined steam and dry reforming of methane are much higher than that in the pure methane discharge or in the steam reforming discharge under similar experimental conditions. The results show a significant reduction in power consumption over either steam or dry reforming on their own. At the same consumed power, higher conversion is observed than in steam reforming alone, despite further dilution of methane. The lowest power consumption is 8eV per molecule of C converted, which is 50% lower than previously reported in combined steam reforming with partial oxidation of methane [6]. The study of the bottom electrode configuration suggests an improvement to produce more H_2 and CO which can further reduce energy consumption to an estimated 5eV per molecule of H_2 produced, if the water gas shift reaction is taken into account. In addition, net production of CO_2 is lowered from such a process and the requirement for CO_2 removal from natural gas may be eliminated.

ACKNOWLEDGEMENT

The authors would like to express their appreciation to ChevronTexaco, Inc. and PetroVietnam for their support of this research

REFERENCES

[1] C.J. Liu; G.H. Xu; T.M. Wang. Fuel Process Technol. 1999, 58, 119-134
[2] MA. Malik; S.S.J. Malik. Nat. Gas Chem. 1999, 8(2), 166-178
[3] M.C.J. Bradford; M.A. Vannice. Catal. Rev.-Sci. Eng. 1999, 41 (1), 1-42.
[4] Y. Li; G. Xu; C. Liu; B. Eliasson; B. Xue. Energy & Fuels 2001, 15, 299-302
[5] M. Tsui; T. Miyao; S. Naito. Catal. Lett.2000, 69, 195.
[6] K. Supat; S. Chavadej; L. Lobban; R. Mallinson. ACS Symposium Series, 2003, 852, 83.
[7] L. Yao; M. Okumoto; A. Nakayama; E. Suzuki. Energy & Fuels 2001, 15, 1295-1299.
[8] Y. Li; C. Liu; B. Eliasson; Y. Wang. Energy & Fuels 2002, 16, 864-870.
[9] J. Luque; M. Kraus and A. Wokaun; K. Haffner; U. Kogeischatz and B.J. Eliasson. Appl. Phys. 2003, 93(8), 4432
[10] H. Le, M.S. Thesis, University of Oklahoma, 2003.

Studies in Surface Science and Catalysis, volume 147
X. Bao and Y. Xu (Editors)
181

Self-stabilization of Ni catalysts during carbon dioxide reforming of methane

De Chen[*a], Rune Lødeng[b], and Anders Holmen[a]

[a]Department of Chemical Engineering, Norwegian University of Science and Technology (NTNU), N-7491 Trondheim, Norway.

[b]SINTEF Applied Chemistry, N-7465 Trondheim, Norway.

ABSTRACT

Methane dry reforming has been studied in a Tapered Element Oscillating Microbalance (TEOM) on an industrial Ni (11 wt%)/$CaAl_2O_4$ catalyst at 650 °C, total pressures of 1 bar and 5 bar and at a CO_2/CH_4 ratio of 1. The effects of partial deactivation and gasification on the stabilization of the catalyst have been addressed, and a microkinetic model was used to understand the mechanism of the self-stabilization.

1. INTRODUCTION

Synthesis gas with a CO/H_2 ratio of about 1 can be produced by catalytic reforming of methane with carbon dioxide (dry reforming). However, the large potential for carbon formation is a main drawback for the use of CO_2. Large efforts have been focused on the development of catalysts that incorporates a kinetic inhibition of carbon formation under conditions where deposition is thermodynamically favorable. Although noble metal catalysts have been reported to be less active for carbon formation, Ni catalyst is more attractive in industrial applications due to its availability and lower cost. Several methods have been proposed for reducing carbon formation on Ni catalysts, such as sulphur passivation, alkaline or alkaline earth metal as promoters and alloy catalysts [1]. The carbon deposition–gasification (CDG) treatment has also been shown to suppress carbon formation [2]. It has been shown that a simultaneous study of deactivation and carbon formation in TEOM gives much more detailed understanding of the mechanism of the catalyst deactivation and the carbon formation. The present work deals with a study of methane dry reforming on Ni in a TEOM reactor. The effect of CDG on the carbon formation and dry reforming has also been studied. Phenomenon of self-stabilization caused by partial deactivation and CDG are reported.

2. EXPERIMENTAL

Tapered Element Oscillating Microbalance (TEOM) reactor has proved to be a useful tool for studying carbon formation and deactivation simultaneously by coupling with on-line GC analysis [3]. The catalyst (typically 15 mg with a particle size of 0.43 – 0.60 mm) was installed with quartz particles as solid diluent. Dry reforming was performed in a TEOM reactor on an industrial Ni (11 wt%)/$CaAl_2O_4$ catalyst at 650 °C, total pressures of 1 bar and 5 bar and at a CO_2/CH_4 ratio of 1. The total surface area is 5.5 m^2/gcat and the Ni surface area is 0.33 m^2/gcat measured by H_2 chemisorption. The catalyst was reduced in a mixture of H_2 and Ar (20/20 ml/min) with a temperature program at a rate of 2 °C/min from room temperature to 670 °C and at this temperature for 5 hrs. The reaction was carried out in a mixture of $Ar/N_2/CH_4/CO_2$ (65/7/38/38 ml/min). Deposited carbon were removed by CO_2 or H_2 at the reaction temperature and pressure in a mixture of Ar/CO_2 or H_2 (110/38 ml/min). If the carbon was removed by CO_2, the catalyst was reduced in a mixture of Ar/H_2 (110/38 ml/min) before further reaction. The CDG cycle was repeated several times.

3. RESULTS AND DISCUSSION

The carbon formation and deactivation during methane dry reforming on the fresh Ni catalyst at a total pressure of 1 bar are shown in Fig. 1. It indicates a rapid decrease in activity corresponding to a very fast increase in initial carbon formation. After about 5 h on stream, the catalyst approaches a pseudo steady-state value, where the coking rate reaches a value close to zero. The maximum carbon content is about 30 wt% at the conditions of Fig. 1. The figure illustrates an interesting phenomenon, namely that catalyst can self-stabilize at a certain condition for both carbon formation and reaction, even on the industrial steam reforming catalyst used in the present work. Development of stable catalysts, especially with high coke resistance, for dry reforming of methane has attracted large interests, and many different catalysts and stability tests have often been reported at atmosphere pressure [1]. The tendency of conversion changes, as shown in Fig. 1 was also often reported previously [4], which can possibly be explained by the observations in the present work.

The deactivation functions (r/r_0) for the conversion of CO_2 and CH_4 and carbon formation are presented in Fig. 1 (B), where r and r_0 are the reaction rate at time t and zero, respectively. Although CO_2 reacts much faster than CH_4 due to the reverse water gas shift reaction, as shown in Fig. 1(A), the deactivation rates are identical for both conversions of CO_2 and CH_4. The fact that the fast initial deactivation of dry reforming corresponds well with the fast initial carbon formation indicates that carbon formation is the main cause for the deactivation. Catalyst regeneration, where most of the catalyst activity is recovered, as shown in

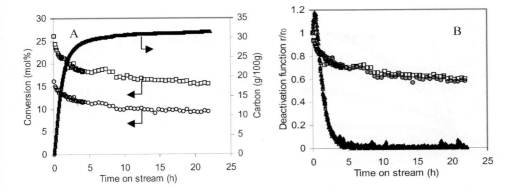

Fig. 1, (A) Conversion of CH_4 (○) and CO_2 (□) and carbon formation (△), (B) Corresponding deactivation function r/r_0 during dry reforming of methane on Ni catalyst at 650 °C, P_{tot} =1 bar, P_{CH4}=0.24 bar, CH_4/CO_2=1, W/F_0=0.15 g,h/mol

Fig. 2(B) and (D) also supports this conclusion. However, the carbon formed involves at least two types, namely filamentous and encapsulating carbon. In principle, filamentous carbon does not strongly deactivate the catalysts at relatively low carbon content, while encapsulating carbon does. Although TEM studies indicate that filamentous carbon is the dominating type of carbon, encapsulating carbon also seems to form on the Ni surface blocking the access of the reactants. Fig. 1(B) clearly indicates that the deactivation rate of carbon formation is much faster than the deactivation of dry reforming. Therefore, it can be concluded that the self-stabilization of the Ni catalyst is mainly induced by the faster deactivation of the carbon formation.

Both the initial carbon formation rate and deactivation rate determine the final carbon content. The operation conditions will affect the final carbon content significantly. An example is shown in Fig. 2, where the coking rate is much higher because of the high pressure (5 bar). The experiment was stopped after 13 min corresponding to 54 wt% carbon, due to a high pressure drop. Fig. 2 (A) clearly shows that the cycle of carbon deposition and gasification significantly reduced the potential of carbon formation, when the carbon was removed by CO_2. The gasification of carbon by CO_2 reduces also slightly the activity for dry reforming, as shown in Fig. 2B. When the carbon was removed by H_2, the conversions were almost not changed, but the carbon formation was increased. The reductions in carbon formation by carbon gasification using CO_2 were also observed at atmospheric pressure.

The general strategy for synthesis gas production is to operate the reforming reaction in a carbon free window. However, it is very difficult to achieve carbon free operation during dry reforming, due to the high potential of carbon formation. For example, it has been measured that the CO_2/CH_4 ratio must be as high as 2.6 in

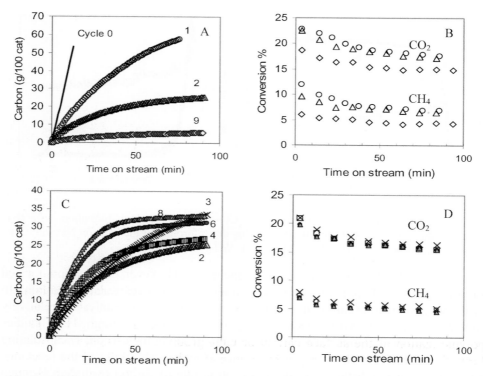

Fig. 2, Effect of CCDG on dry reforming of methane (B,D) and carbon formation (A,C) on Ni catalyst, at 650 °C, Ptot=5 bar, P_{CH4}=0.24 bar, CH_4/CO_2=1, WHSV=0.15 g/mol,h. CDG o:line, 1: ○, 2: △, 3:×, 4:□, 6: ●, 8: ▲, 9:◊. During CDG 1-3 and 9 (A,B), the carbon was gasified by CO2, while during CDG 4-8, the carbon was gasified by H2.

order to maintain carbon free operation on the Ni catalyst at the total pressure of 5 bar. Therefore, many techniques, such as using noble metal catalysts, sulphur passivating and using co-feed with steam, have been developed to achieve carbon free operation at mild conditions. The present work points out that the phenomena of self stabilization possibly can be used for maintaining proper operation of dry reforming, as long as the final carbon content is controlled at a relatively low amount. The final carbon content is the crucial parameter for proper operation of dry reforming. Too high carbon content will result in pore blocking of the catalysts, destroying the catalyst pellets mechanically, filling the void space in the reactor and thus causing pressure drop.

In order to understand the mechanism of the self-stabilization, a microkinetic model developed previously [5] is used. The model includes 13 surface elementary reactions leading to synthesis gas, as well as steps involved in the formation of filamentous carbon, such as segregation of carbon at the gas-Ni interface, carbon diffusion through the Ni particles, and precipitation at the support side. The diffusion of carbon through the Ni particles can be described with the following equation:

$$r = \frac{D_C}{d_{Ni}} a_{Ni} (C_{Ni.f} - C_{sat})$$

(1)

D_C is the effective diffusivity for the carbon diffusion through nickel, d_{Ni} is the effective diffusion path, a_{Ni} is the specific surface area of Ni, $C_{Ni,f}$ is the carbon concentration in the Ni particle near the gas-Ni interface, and C_{sat} is the saturation concentration of filamentous carbon in the Ni. The carbon potential depends mainly on the driving force of carbon diffusion, namely the concentration gradient ($C_{Ni,f}-C_{sat}$) in Eq. (1). The details of microkinetic simulation have been reported previously [5], and the simulated results are presented in Fig. 3 as an example.

The surface carbon coverage predicted by the model is rather high at dry reforming conditions, about 0.28 at the total pressure of 5 bar. Since it has been assumed that a carbon atom occupies two Ni atoms, it means that more than half of the active sites are covered by surface atomic carbon. Such high carbon site coverage results in a relatively fast formation of encapsulating carbon via polymerization of the surface carbon or CH_X group. The encapsulating carbon formation deactivates both dry reforming and carbon formation. However, the effects of sintering of Ni crystals can also not simply be excluded. So the deactivation during dry reforming is described by a reduction in specific surface area as shown in Fig. 3, caused by both encapsulating carbon and possible sintering. When the surface area decreased with time on stream, the simulated surface carbon coverage decreased accordingly, resulting in a decrease in

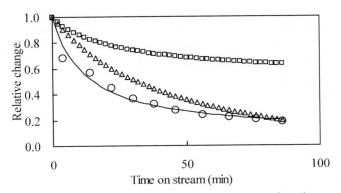

Fig. 3 Relative surface area (S/S_0) ($\circ$), deactivation function of methane conversion (r/r_0) ($\square$) and carbon formation (r_C/r_{C0}) ($\triangle$) with time on stream. Data are simulated by microkinetic modeling with C_{sat} of 93 mol/m^3. Conditions: 650 °C, P_{tot}=5 bar, P_{CH4}=0.24 bar, CH_4/CO_2=1, WHSV=0.15 g/mol,h. Catalysts after first CDG by CO_2

activity of methane dry reforming and carbon formation. Most importantly, carbon formation deactivates much faster than the main reaction, mainly caused by a faster lowering in the driving force for the carbon diffusion. At a certain degree of deactivation, the driving force closes zero, and no further carbon formation takes place. By this way, the deactivation stabilizes the catalyst and suppresses further carbon formation, as shown in Fig. 1 and 2. It should be pointed out that the rate of the deactivation of carbon formation depends very much on the value of the C_{sat} in the Eq. (1). The higher C_{sat}, the faster deactivation of carbon formation. Our previous work [6] indicated that the C_{sat} is a function of Ni crystal size, and a smaller crystal size results in a higher C_{sat}. Therefore, it might be possible to use small sized Ni crystals to achieve a low final carbon content.

The phenomenon of reduction in carbon formation by CDG process is consisted with previous report [2]. However, it is in contrast to the previous report [2] that a decrease in the initial conversion was observed in the present work. Microkinetic simulation indicates a decrease in the surface area of Ni, which is also in good agreement with the previous report [2]. Based on the results in Fig. 3, the reduced coking rate and conversion after CDG can then possibly be explained by the decrease in surface area of Ni. Some encapsulating carbon might remain on the surface after CDG with CO_2. The increase in carbon formation after CDG with hydrogen can possibly be explained by a slightly lower C_{sat}, caused by a sintering during CDG. Anyhow, the reason for the difference in the effect of gasification by CO_2 and H_2 needs to be further investigated.

4. CONCLUSIONS

It has been found that a Ni catalyst can be self-stabilized by means of partial deactivation or by the treatment of carbon formation and gasification. Microkinetic modeling indicated that the faster deactivation of carbon formation caused by the faster reduction in the driving force for carbon diffusion is the main reason for the self-stabilization. This phenomenon can possibly be used to maintain proper operation at mild conditions.

REFERENCES

[1] M.C.J. Bradford and M.A. Vannice. Catal. Rev.-Sci. Eng. 41 (1999) 1.
[2] M. Ito, T. Tagawa and S. Goto. Appl. Catal. A., 177 (1999) 15.
[3] D. Chen, A. Grønvold, H.P. Rebo, K. Moljord, A. Holmen. Appl. Catal. A.,137 (1996) L1.
[4] V.C.H. Kroll, H.M. Swaan, C. Mirodatos, J. Catal., 161(1996) 409.
[5] D. Chen, R. Lødeng, A. Anundskås, O. Olsvik, A. Holmen. Stud. Surf. Sci. Catal., 139 (2001) 93.
[6] D. Chen, K.O. Christensen, Z. Yu, B. Tøtdal, N. Latorre, A. Monzón, A. Holmen, submitted.

Studies in Surface Science and Catalysis, volume 147
X. Bao and Y. Xu (Editors)

Co/TiO$_2$ catalyst for high pressure dry reforming of methane and its modification by other metals

Katsutoshi Nagaoka[a], Kazuhiro Takanabe[b], Ken-ichi Aika[*][a, b]

[a]CREST, JST(Japan Science and Technology)

[b]Department of Environmental Chemistry and Engineering, Interdisciplinary Graduate School of Science & Engineering, Tokyo Institute of Technology, Yokohama, Japan.

ABSTRACT

Catalytic behaviour of 0.5wt%Co/TiO$_2$, prepared from TiO$_2$ containing only anatase phase, in CH$_4$/CO$_2$ reforming at 2 MPa was strongly affected by reduction temperature. Co/TiO$_2$ reduced at relatively lower temperatures (<1123 K) lost all of its activity at the beginning of the reaction, whereas the catalyst reduced at higher temperatures (1123 K) showed relatively stable activity with only a slight amount of deposited coke. However, the latter catalyst lost all of activities at a higher space velocity after the reaction for 20 h, mainly due to oxidation of metallic Co. The addition of trace amounts of noble metals (Pt or Ru) improved the catalytic stability drastically by retarding the oxidation. It was confirmed that the strong resistant nature of Co/TiO$_2$ to coke deposition was retained after the addition of the noble metals.

1.INTRODUCTION

CO$_2$ reforming of methane (CH$_4$ + CO$_2$ → 2CO + 2H$_2$) is attractive for generating synthesis gas with H$_2$/CO ratio of 1 or less, and associated with the utilization of natural gas fields containing CO$_2$ via gas to liquid (GTL) process. Methane is preserved at high pressures in natural gas fields and most of Fischer-Tropsch synthesis and methanol synthesis from synthesis gas (H$_2$/CO) are performed at high pressures in the GTL process. Hence, from an economical point of view, the CH$_4$/CO$_2$ reaction should also be carried out under high pressures (≥1 MPa) in such processes. The major drawback for the reaction is coke formation, which causes deactivation of catalysts and pressure drop due to plugging of reactor. Since it is well known that higher pressure favors coke deposition, the real effectiveness can only be discussed from the results under high pressures. Our group reported that only Ru/TiO$_2$ catalyst in the various supported Ru catalysts is resistant to coking [1] under both atmospheric pressure and 2.0 MP (industrial condition for GTL process).

Based on these results, titania supported Co catalysts were studied. Co was chosen due to its low price, compared to noble metals. After preliminary experiments [2], catalitic behavior of 0.5wt% Co/TiO$_2$ catalysts was investigated in this contribution. In addition, Co/TiO$_2$ was modified by the noble metals in order to improve the catalytic stability and the catalytic behavior of the modified Co/TiO$_2$ catalysts was reported.

2.EXPERIMENTAL

0.5wt%Co/TiO$_2$ was prepared by an incipient wetness impregnation with TiO$_2$ (Ishihra Sangyo, A -100, 100% Anatase form), which was calcined for 15 h at 773 K in flowing air (30ml min^{-1}), and an aqueous solution of Co(NO$_3$)$_2$·6H$_2$O (Wako Pure Chemical Industries, Ltd.) (0.005g Co per ml). The catalyst was denoted simply as Co/TiO$_2$. To prepare Co/TiO$_2$ with additives, aqueous solutions of H$_2$PtCl$_6$·6H$_2$O, or RuCl$_3$·nH$_2$O, were co-impregnated with Co(NO$_3$)$_2$·6H$_2$O. The Co loading was set to 0.5wt% for Co-Pt/TiO$_2$ (Pt/Co=0.01, or 0.05 in molar ratio) and Co-Ru/TiO$_2$ (Ru/Co=0.01 or 0.05 in molar ratio).

0.25-1.0 g of the catalysts were loaded into a tubular inconel reactor passivated by aluminum diffusion coating for activity tests. After the catalysts were reduced *in situ* in H$_2$ at 1123 K for 1 h under 0.1 MPa, the temperature was reduced in He to 1023 K. Subsequently, the pressure was increased to the reaction pressure (2 MPa). A CH$_4$/CO$_2$ mixture (CH$_4$/CO$_2$=1, total flow rate of 50 ml min^{-1}) was passed through a bypass and then introduced into the reactor (SV=3000–12000 ml g^{-1} h^{-1}). The reaction products were analyzed by gas chromatography with thermal conductivity detectors (TCDs).

The amounts of coke deposited on the catalysts were quantified by temperature-programmed oxidation (TPO) with O$_2$/He. For that purpose, the catalysts were transferred to a tubular quartz reactor. O$_2$/He (5/95 vol/vol with a total flow rate of 50 ml min^{-1}) was fed to the reactor at room temperature, and the catalyst was heated up to 1273 K with a heating rate of 10 K min^{-1}. The deposited coke was oxidized to CO and/or CO$_2$, which were subsequently converted to CH$_4$ by a methanizer, and finally detected by a flame ionization detector (FID).

The same apparatus as for TPO was used for TPR measurements and 0.4 g of the calcined catalyst were used. The catalyst was heated up to 1273 K with a heating rate of 10 K min^{-1} in H$_2$/Ar (5/95 vol/vol with a total flow rate of 30 ml min^{-1}) and the H$_2$ uptake was recorded by a TCD.

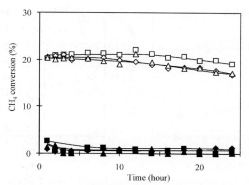

Fig. 1. CH_4 conversion *vs.* time on stream for Co/TiO$_2$ reduced at various temperatures (reaction condition: $CH_4/CO_2=1$; 1023 K; 2.0 MPa; SV=6000 ml g^{-1} h^{-1}). Reduction temperature: (▲) 973 K; (◆) 1023 K; (■) 1073 K; (△) 1123 K, (◇) 1173 K, (□) 1223 K.

3. RESULTS AND DISCUSSION

3.1. Catalytic activity of Co/TiO$_2$ reduced at different tempearatures

The catalytic behavior of Co/TiO$_2$ was strongly influenced by reduction temperature (see Fig. 1). The catalysts reduced below 1123 K showed fairly low activities. A large amount of deposited carbon was detected on the catalysts (≥2.9 wt%C). On the other hand, the Co/TiO$_2$ reduced at 1123 K or above kept relatively stable activities for 24 h. The catalysts showed strong resistance to coke formation (< 0.25 wt%C). However, it must be mentioned that slight deactivation was observed for the Co/TiO$_2$ reduced at 1123 K and was accelerated with increasing SV (see Fig. 2). Since the amount of deposited coke was negligible at the highest SV (0.01 wt%C), the cause of the slow deactivation

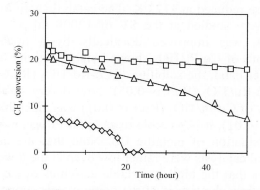

Fig. 2. CH_4 conversion at different SVs *vs.* time on stream for Co/TiO$_2$ reduced at 1123 K (reaction condition: $CH_4/CO_2=1$; 1023 K; 2.0 MPa). SV: (□) 3000 ml g^{-1} h^{-1}; (△) 6000 ml g^{-1} h^{-1}; (◇) 12000 ml g^{-1} h^{-1}.

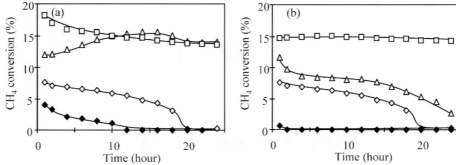

Fig. 3. CH$_4$ conversion *vs.* time on stream for Co-Pt/TiO$_2$ (a) and Co-Ru/TiO$_2$ (b) (reaction condition: CH$_4$/CO$_2$=1; 1023 K; 2.0 MPa; SV=12000 ml g^{-1} h^{-1} except for Pt/TiO$_2$ and Ru/TiO$_2$ of 6000 ml g^{-1} h^{-1}). Additive/Co: (◇) only Co; (△) 0.01; (□) 0.05; (◆) only noble metal (loading of noble metals is identical to that of Co-(Pt or Ru)/TiO$_2$ (additive/Co=0.05)).

could be oxidation of metallic Co and/or sintering of the Co particles. If the deactivation of Co/TiO$_2$ is due to the oxidation of the metallic Co, the catalyst can be regenerated by H$_2$ reduction. After Co/TiO$_2$ lost its activity in the CH$_4$/CO$_2$ reaction (SV=12000 ml g^{-1} h^{-1}) for 24 h (see Fig. 2), the catalyst was re-reduced at 1123 K and, then, initial activity was measured. By the regeneration, approximately 75% of the initial conversion was restored (not shown). This suggests that slow oxidation of Co/TiO$_2$ is mainly ascribed to oxidation of metallic Co.

3.2. Effect of addition of noble metals to Co/TiO$_2$

It is well known that the addition of the noble metals can promote the reducibility of the base metals and stabilize their degree of reduction during the catalytic process [3,4]. Thus, small amounts of Ru or Pt were added to Co/TiO$_2$ and the catalysts were reduced at 1123 K. The catalytic activity and stability of modified Co/TiO$_2$ were studied at the SV of 12000 ml g^{-1} h^{-1} (Fig. 3). The stability of Co/TiO$_2$ was improved drastically by a trace amount of Pt (Pt/Co=0.01) and both of Co-Pt/TiO$_2$ showed stable activity after the reaction for 24 h. On the other hand, the stability of Co-Ru/TiO$_2$ (Ru/Co=0.01) was not so high and only Co-Ru/TiO$_2$ (Ru/Co=0.05) showed very stable activity. Since the catalytic stability of Co-Pt/TiO$_2$ (Ru/Co=0.01) was much higher than that of Co-Ru/TiO$_2$ (Ru/Co=0.01), it is possible to say that Pt was more effective than Ru in preventing oxidation of metallic Co. Note that the activities of TiO$_2$ supported noble metal catalysts were rather low even at the lower SV of 6000 ml g^{-1} h^{-1}. This indicates that the high and stable activities of Co/TiO$_2$ with noble metals do not result from the nature of the noble metals supported on TiO$_2$ but also from the synergistic effects between noble metals and Co supported on TiO$_2$.

Table 1

The amounts of coke deposited over TiO_2 supported metal catalysts

metal	SV ($ml\,g^{-1}\,h^{-1}$)	Coke amount (wt%)
Co	12000	0.01
Co-Pt (Pt/Co=0.01)	12000	0.21
Co-Pt (Pt/Co=0.05)	12000	0.08
Pt	6000	0.18
Co-Ru (Ru/Co=0.01)	12000	0.25
Co-Ru (Ru/Co=0.05)	12000	n.d.
Ru	6000	0.04

After the activity tests, the amounts of deposited coke were determined and compiled in Table 1. The amounts of the coke were not significant for all Co/TiO_2 with noble metals, revealing that the strong resistance of Co/TiO_2 to coking was retained after the addition of noble metals.

In order to understand why deactivation caused by metallic Co was inhibited by the addition of noble metals, TPR was performed over the catalysts (see Fig. 4). Four peaks were observable for Co/TiO_2 (589, 660, 747, and 1030 K) and the ratio of O/Co calculated from the net amount of H_2 uptake for the first three peaks and all the peaks were approximately 1.4 and 3.6, respectively. Hence, H_2 uptakes observed below and above 873 K would be ascribed to reduction of Co species and TiO_2, respectively. This suggests that all Co in the catalyst is present in a metallic state after the H_2 reduction. No peak was

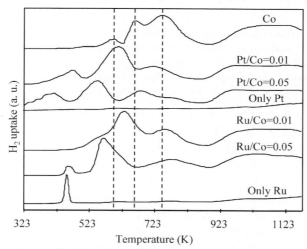

Fig. 4. TPR profiles of Co/TiO_2, $Co-Pt/TiO_2$, Pt/TiO_2 (loading of Pt is identical to that of $Co-Pt/TiO_2$ (Pt/Co=0.05)), $Co-Ru/TiO_2$, and Ru/TiO_2 (loading of Ru is identical to that of $Co-Ru/TiO_2$ (Ru/Co=0.05)).

observable for Pt/TiO$_2$, implying that Pt might already be reduced before the TPO measurement, whereas a single peak showing the formation of metallic Ru was viewable at 453 K for Ru/TiO$_2$. All Co/TiO$_2$ with noble metals presented three peaks for reduction of Co oxides. The addition of Pt shifted all peaks for Co oxides to lower temperatures. On the other hand, the third peaks were not affected by Ru addition and only the other two peaks were shifted to lower temperatures. It is summarized that the third peaks are possibly attributed to very small Co oxide clusters which has strong interaction with TiO$_2$. Therefore, our observations reveal that the peak for Co oxides having strong interaction with TiO$_2$ was shifted only by the addition of Pt. Note also that peak maxima affected by the addition of noble metals were shifted to lower temperature as the amounts of noble metals increased. On the other hand, the temperature of reduction for TiO$_2$ was also decreased by the addition of noble metals, but the decreased was independent of the sorts and amounts of noble metals. These results suggest that theaddition of noble metals results in the lowering of the reduction temperature of Co oxides and TiO$_2$, presumably due to hydrogen spillover from noble metal surface [5]. However, it must be mentioned that the reduction of Co oxides having strong interaction with TiO$_2$ was promoted only by the addition of Pt. Therefore, it can be mentioned that Pt promotes the reduction of Co oxides more efficiently than Ru, which results in well enhanced catalytic stability of Co/TiO$_2$, even if the amount of Pt was very small (Pt/Co=0.001). During the CH$_4$/CO$_2$ reaction, the hydrogen spillover would keep the metallic Co reduced and thus the catalysts show stable activities.

REFERENCES

[1] K. Nagaoka, M. Okamura, K. Aika, Catal. Commun. 2 (2001) 255.
[2] K. Nagaoka, K. Takanabe, K. Aika, Chem. Commun. (2002) 1006.
[3] N. Wagstaff, R. Prins, J. Catal. 59 (1979) 434.
[4] C. Raab, J.A. Lercher, J.G. Goodwin, J.Z. Shyu, J. Catal. 122 (1990) 406.
[5] G. Jacobs, T.K. Das, Y. Zhang, J. Li, G. Racoillet, B.H. Davis, Appl. Catal. A: General 233 (2002) 263.

Studies in Surface Science and Catalysis, volume 147
X. Bao and Y. Xu (Editors)

Carbon dioxide reforming of methane over supported molybdenum carbide catalysts

D. Treacy and J.R.H. Ross

Centre for Environmental Research, University of Limerick, Ireland.

ABSTRACT

Supported Mo_2C materials have been prepared, characterised and tested as catalysts for the CO_2 reforming of CH_4. The order of increasing reactivity was: Mo_2C-SiO_2<Mo_2C-ZrO_2<Mo_2C-Al_2O_3. Raman analysis indicated that the presence of easily reducible polymolybdate species on the surface of the precarbided material coincided with higher reactant conversions. Deactivation of the catalyst samples was shown to be due to C deposition; for example, the Mo_2C-Al_2O_3 sample deactivated completely after 4 h on stream. However, the amount of C deposited on the Mo_2C-ZrO_2 sample was insignificant and the catalyst remained relatively stable. Characterisation of the precarbided MoO_3-ZrO_2 material showed that low calcination temperatures. It was also found that the addition of Bi as promoter encouraged higher reactant conversions.

1. INTRODUCTION

The CO_2 reforming of CH_4 has received increased interest over the past decade due to the ability of the reaction to produce synthesis gas with low H_2/CO ratios (~1); higher ratios can then be obtained by combining the process with steam reforming or partial oxidation. The main problem associated with CO_2 reforming is that carbon deposition is significant with almost all the catalysts which have been tested for the reaction. Although it has been claimed that special formulations allow Ni-containing materials to be used, most such catalysts succumb to C deposition under more extreme operating conditions or after long periods of testing [1]. Noble metal catalysts, in particular Pt-ZrO_2, have been shown to have high activities for CO_2 reforming while resisting significant C deposition [2,3]. Noble metals, however, are expensive and hence their potential for use as CO_2 reforming catalysts is limited. Alternative catalyst formulations reported to have catalytic properties similar to the noble metals are the group VI transition metal carbides, WC and Mo_2C, and York *et al.* [4,5,6] have shown that unsupported Mo_2C and WC have activities as CO_2

reforming catalysts. However, they found that these catalysts deactivated after short periods under test at ambient pressures. We have shown that this deactivation is caused by the formation of MoO_2, this being catalytically inactive for the CO_2 reforming reaction; the MoO_2 is formed under CO_2 reforming conditions due to the thermodynamic unstability of the carbide phase under the reaction conditions prevalent at the entrance of the catalyst bed where there is a high partial pressure of CO_2 relative to that of CO [7]. In this paper, we report an investigation of the effect of support material on the activity and stability of Mo_2C catalysts. We also report the effects on the activities of such materials of various preparation conditions, e.g. calcination temperature, as well as the presence of promoters.

2. Experimental.

Catalyst Preparation. The supports used in this study were ZrO_2 (Norton), Al_2O_3 (Alkan) and SiO_2 (Alfa). Each supports were pretreated by calcination at 800°C for 15 h in a flow of 50 cm^3min^{-1} of air. An appropriate amount of ammonium heptamolybdate (Johnson Matthey) was then added and left to equilibrate. When appropriate, metal additives (Co, Ni or Bi) were added to the catalysts by co-impregnation by solutions containing the relevant metal nitrates and the ammonium heptamolybdate. Following impregnation, unless otherwise stated, the material was calcined at temperatures increasing to 600°C, using a ramp rate of 5 $Kmin^{-1}$, in a flow of 50 cm^3min^{-1} of air.

The desired quantity of the calcined material (0.33g) was placed between two pieces of quartz wool in a fixed bed quartz microreactor (6mm outer diameter, 4mm inner diameter and 540mm in length). The carbide was then formed by temperature-programmed reaction (reduction/carburisation) in the catalyst-testing rig following a procedure outlined by Lee et al. [8]. In outline, the temperature was ramped from room temperature to 750°C at a ramp rate of 1 $Kmin^{-1}$, the sample being maintained in flow of 20 %v/v CH_4/H_2.

Catalyst Testing. The CO_2 reforming of CH_4 was carried out in the same quartz plug flow reactor. The gas streams were analysed by a GC (Varian) equipped with Carboxen 1000 and Porapak Q columns in parallel and a TCD detector, using Ar as the carrier gas. Following preparation, each carbided material was brought to reaction temperature in a flow of N_2. The reactants were then introduced, the composition typically being $CO_2/CH_4/N_2 = 50/50/5$ $cm^{-1}min^{-1}$. The N_2 was included as an internal standard.

Characterisation by Raman Microprobe Spectroscopy. The 514.5 line of an Ar^+ ion laser was used for excitation. Measurements were made at room temperature.

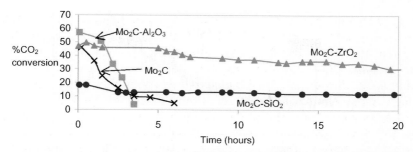

Fig. 1. CO_2 reforming of CH_4 using (1) Mo_2C, (2) 5wt% Mo_2C-Al_2O_3, (3) 5wt% Mo_2C-ZrO_2 and (4) 5wt% Mo_2C-SiO_2. [950°C, CO_2/CH_4/N_2: 50/50/5 ml/minute, 0.33g catalyst].

3. RESULTS AND DISCUSSION

3.1 Catalyst Activity

Fig. 1. shows the results of CO_2 reforming of CH_4 at 950°C and atmospheric pressure using the 5 wt% Mo_2C-Al_2O_3, 5 wt% Mo_2C-ZrO_2 and 5 wt% Mo_2C-SiO_2 catalysts. The performance of a sample of unsupported Mo_2C under similar reaction conditions is also shown for comparison. The Mo_2C-Al_2O_3 sample gave the highest initial activity, with a CO_2 conversion of 63%, while Mo_2C-ZrO_2 also gave a relatively high CO_2 conversion of 48%. Mo_2C-SiO_2 was a poor CO_2 reforming catalyst, giving a CO_2 conversion of 18%. However, the Mo_2C-Al_2O_3 deactivated almost completely after 4 h of operation whereas the ZrO_2- and SiO_2-supported materials remained comparatively stable over the period of the test; indeed, the Mo_2C-ZrO_2 sample was active even after 80 h at 950°C (not shown). The unsupported Mo_2C sample deactivated after 7 h. We have previously shown [7] that the deactivation of the unsupported carbide catalyst was due to the formation of the inactive MoO_2 phase under conditions where re-oxidation by CO_2 was thermodynamically favoured. Naito *et al.* [9] have argued that the deactivation of Mo_2C-supported catalysts is due not to MoO_2 formation but to C deposition on the catalyst surface. TG analysis carried out in our laboratory has shown that the level of C deposition on the Mo_2C- ZrO_2 material during CO_2 reforming is relatively insignificant in comparison with that on the Mo_2C-Al_2O_3 material [10]. These results are similar to those which we have reported previously [3] for Pt supported on Al_2O_3 and ZrO_2; coke was not deposited to any significant extent on the Pt-ZrO_2 sample while large amounts of coke were detected during CO_2 reforming on the Pt-Al_2O_3 sample. The reaction on the Pt- ZrO_2 appeared to take place at the interface between the Pt and the support and to involve an oxidation-reduction mechanism in which the zirconia takes part in the reaction [11]. There may be a similar explanation for the beneficial effect of the zirconia as a support for the Mo carbide phase.

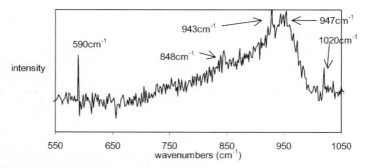

Fig. 2. Raman spectrum of 5wt% $Mo_2C-Al_2O_3$.

3.2 Raman Spectroscopy

Raman spectroscopy was used to examine whether there was any difference in the surface interactions between the support and the MoO_3 prior to the formation of the carbide phase. Fig. 2 shows the Raman spectrum for the 5wt%$MoO_3-Al_2O_3$ material prior to reduction/carburisation.

Although the spectrum has a poor signal to noise ratio, it is similar to that reported by Cheng and Schrader [12]. The bands at 590 and 943 cm^{-1} are due to the presence of the polymolybdate species, $Mo_7O_{24}^{6-}$, while the broad band centred at around 970cm^{-1} is due to the presence of a different polymolybdate species, $Mo_8O_{26}^{4-}$. The weak bands at 1020 and 848cm^{-1} also indicate the presence of crystalline MoO_3 [13]; these bands are significantly smaller than that due to the $Mo_8O_{26}^{4-}$ species. In agreement with the results of Cheng and Schrader, there was no evidence for the formation of any aluminium molybdate compounds for the sample used here (calcined at 600°C). Abello *et al.* [14] also showed, using XRD, TPR and NH_3-TPD, that $Al_2(MoO_4)_3$ formation was prevented when $MoO_3-Al_2O_3$ was calcined at 600°C. We therefore conclude that the predominant species on the surface of the alumina prior to carburisation are adsorbed polymolybdate species and that no amorphous $Al_2(MoO_4)_3$ is formed.

Fig. 3 shows the equivalent Raman spectrum for the 5wt% MoO_3-ZrO_2 catalyst prior to carburisation. The band at 953cm^{-1} indicates the presence of adsorbed polyoxomolybdate while the bands at 749, 942 and 1000cm^{-1} can be

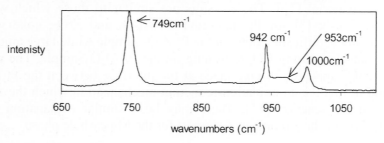

Fig. 3. Raman spectrum of 5wt% MoO_3-ZrO_2.

ascribed to the presence of bulk $Zr(MoO_4)_2$ species. There was no evidence of the presence of crystalline MoO_3 species [15]. A Raman spectrum of the precarbided MoO_3-SiO_2 catalyst (not shown here) shows that the predominant molybdenum species on the surface of the catalyst was a silicomolybdate species [10]. Polymolybdate surface species do not interact strongly with the support and are therefore highly polarized and easily reduced [16]. The strong interactions created when Mo reacts with the support material decreases the covalency of the Mo-O bonds and therefore stabilises the Mo species; this in turn decreases the reducibility of any such compounds [17]. The presence of the easily reducible polymolybdate species on Al_2O_3 and, to a lesser extent, on ZrO_2 enables a significant quantity of Mo_2C to be produced during the temperature programmed reduction/carburisation step of the preparation. This in turn leads to the formation of a larger number of active sites on the surface of the resultant catalyst. On the other hand, the Mo-Si material is difficult to reduce; the amount of Mo_2C produced during the temperature programmed reduction/carburisation step for this sample (and to a lesser extent with the zirconia-supported material) is therefore probably less than that with the most active Mo_2C-Al_2O_3 catalyst. TPO experiments [10] support this conclusion: with the Mo_2C-Al_2O_3 sample, 46% of the total possible amount of carbide was formed during the TPR preparation steps while, with Mo_2C-ZrO_2 sample, the amount of carbide formed was only 35%.

3.3 Optimisation of the Mo_2C-ZrO_2.sample and the effect of promoters

As reported above (Fig. 1), the Mo_2C-ZrO_2 sample exhibited a much higher stability than did unsupported Mo_2C or Mo_2C-Al_2O_3. Characterisation using techniques such as Raman, TPR and TG analysis to be reported elsewhere [10] showed that high concentrations of surface polymolybdate species could be achieved by using low calcination temperatures (400°C) by preventing the occurrence of strong Mo-Zr interactions. These Mo-Zr interactions (i.e. the formation of $Zr(MoO_4)_2$) inhibit the formation of the active carbide phase during the reduction/carburisation preparation step. For example, a 5wt% Mo_2C-ZrO_2 sample calcined at 400°C converted 11% more CO_2 than did the sample calcined at 600°C [10]. Other methods which encouraged the presence of polmolybdate species included the use of low Mo loadings and short calcination times [10].

We have also examined a series of Mo_2C-ZrO_2 catalysts doped with 1wt% metal oxide promoters (Ni, Bi, Co) to investigate their potential improvement to the activity of the catalyst. The results of this study are summarised in Table 1. The presence of Ni and Co did not significantly change the % CO_2 conversion (relative to the unprompted Mo_2C-ZrO_2 catalyst). However, it was found that the 1wt%Bi-5wt%Mo_2C-ZrO_2 catalyst gave a CO_2 conversion of 74.6%, this representing an increase of 16.6% relative to the unpromoted material. An explanation of this promotion is as yet unclear and further characterisation needs

Table 1

CO₂ reforming of CH₄ using (1) 5wt% Mo_2C-ZrO_2, (2) 5wt%Mo_2C-1wt%Ni-ZrO_2, (3) 5wt%Mo_2C-1wt%-Bi-ZrO_2 and (4) 5wt%Mo_2C-1wt%Co-ZrO_2, [950°C, 50/50/5 ml/minute $CO_2/CH_4/N_2$, 0.33g catalyst]. All catalysts were calcined at 400°C prior to carburisation. Results recorded after 3 hours of testing.

Mo phase	Promoter	Support	%CO₂ conversion
Mo_2C	-	ZrO_2	58
Mo_2C	Bi	ZrO_2	74.6
Mo_2C	Ni	ZrO_2	59.4
Mo_2C	Co	ZrO_2	55.6

to be done; however, it has been suggested by Lietti *et al.* [18] that Bi changes the redox properties of the surface Mo and an effect of this sort may have a role here.

4. CONCLUSIONS

Of all the catalysts tested Mo_2C-ZrO_2 was the most promising due to its stability and reasonably high activity. The activity of this catalyst can be improved by increasing the reducibility of the precarbided MoO_3 surface material. This could be achieved using a number of methods, including low calcination temperatures and the addition of small amounts of a Bi promoter to the catalyst.

REFERENCES

[1] K. Seshan, H.W. ten Barge, W. Hally, A.N.J. van Keulen and J.R.H. Ross, Stud. Surf. Sci. Catal., 81 (1994) 285.
[2] A.M. O'Connor and J.R.H. Ross, Catal. Today, 46 (1998) 203.
[3] J.R.H. Ross, A.N.J. van Keulen, M.E.S. Hegarty and K. Seshan, Catal. Today, 30 (1996) 193. 65.
[4] A.P.E. York, J.B. Claridge, C. Marquez-Alvarez, A.J. Brungs, S.C. Tang and M.L.H. Green, Stud. Surf. Sci. Catal., 110, (1997) 711.
[5] J.B. Claridge, A.P.E. York, A.J. Brungs, C. Marquez-Alvarez, J. Sloan, S.C. Tsang and M.L.H. Green., J. Catal. 180 (1998) 85.
[6] A.J. Brungs, A.P.E. York and M.L.H. Green, Catal. Lett., 57 (1999).
[7] D. Treacy, E. Xue, J.R. H. Ross, paper presented at 6th NGC Symposium, Alaska, June (2001).
[8] J.S. Lee, S.T. Oyama, M. Boudart, J. Catal. 103 (1987) 125.
[9] S. Naito, M. Tsuji, T. Miyao, Catal. Today, 77 (2002) 161.
[10] D. Treacy, J.R.H. Ross, *in preparation.*
[11] A.M. O'Connor, F.C. Meunier, J.R.H. Ross, Stud. Surf. Sci. Catal., 119 (1998) 819.
[12] C.P. Cheng, G.L. Schrader, J. Catal. 60 (1979) 276.
[13] D.S. Kim, K. Segawa, I.E. Wachs, J. Catal. 136 (1992) 539.
[14] M.C. Abello, M.F. Gomez, O. Ferretti, Appl. Catal. A: Gen., 207 (2001) 421
[15] B. Zhao, Z. Wang, H. Ma, Y. Tang, J. Mol Catal. A, 108 (1996) 167.
[16] M. Del Arco, S.R.G. Currazan, V. Rives, V.J. Llambias, P. Melet, J. Catal. 141(1993) 48.
[17] J.L Brito, J. Laine, J. Catal. 139 (1993) 540.
[18] L. Lietti, G. Ramis, G. Busca, F. Bregani, P. Forzatti, Catal. Today, 61 (2000) 187.

Studies in Surface Science and Catalysis, volume 147
X. Bao and Y. Xu (Editors)
©2004 Elsevier B.V. All rights reserved.

Methane oxidation to synthesis gas using lattice oxygen in La$_{1-x}$Sr$_x$FeO$_3$ perovskite oxides instead of molecular oxygen

R.-J. Li[a], C.-C. Yu[a], W.-J. Ji[b], S.-K. Shen[*a]

[a]The Key Laboratory of Catalysis CNPC, University of Petroleum, Beijing 102249, China

[b]Department of Chemistry, Nanjing University, Nanjing 210093, China

ABSTRACT

In this paper, partial oxidation of methane to synthesis gas using lattice oxygen instead of molecular oxygen was investigated. A perovskite La$_{1-x}$Sr$_x$FeO$_3$ (x=0, 0.1, 0.2, 0.5) oxide was used as an oxygen provider. The redox cycle experiments show that methane can be oxidized to CO and H$_2$ by the lattice oxygen of La$_{1-x}$Sr$_x$FeO$_3$ (x=0, 0.1, 0.2, 0.5) oxide, while the lost lattice oxygen can be supplemented by re-oxidation.

1. INTRODUCTION

Catalytic partial oxidation of methane (POM) to synthesis gas is a slightly exothermic, highly selective, and energy efficient process. It is an attractive alternative route to the conventional steam reforming process. However, oxygen supply by cryogenic distillation of air needs additional investment and operation expense. To avoid this problem, intensive efforts have been made to use membrane reactors for POM [1-5].

Recently, we have proposed a novel POM process to synthesis gas using lattice oxygen of perovskite oxides rather than molecular oxygen [6,7]. In this process a suitable perovskite oxide catalyst is circulated between two reactors. In one reactor, methane is oxidized to synthesis gas by lattice oxygen of the oxide, and in the other reactor, reduced oxide is re-oxidized by air. Since the reaction occurs between methane and lattice oxygen in the absence of molecular oxygen, this process has no risk of explosion and no requirement for pure

---

[*] Corresponding Author: Email: skshen@bjpeu.edu.cn Phone: +86 (10) 89733784 Fax: +86 (10) 69744849

oxygen supply. It can be performed in either a cyclic fluid-bed reactor or a switched fixed-bed reactor. Here we report the preparation, characterization and evaluation of $La_{1-x}Sr_xFeO_3$ (x=0, 0.1, 0.2, 0.5) perovskite oxides in POM. The characteristics of active oxygen species and redox process are discussed.

2. EXPERIMENTAL

2.1. Catalyst preparation

The perovskite oxide $La_{1-x}Sr_xFeO_3$ (x=0, 01, 02, 05) catalysts were prepared by the auto-combustion method [6,7]. The catalyst precursor was calcined at 900 deg C for 5 h. XRD showed that the $La_{1-x}Sr_xFeO_3$ oxides (x=0, 0.1, 0.2, 0.5) feature the typical crystalline structures of perovskite-type oxide [7].

2.2. Reactor setup and experimental procedure

The experiments were carried out in a fixed-bed micro-reactor, which consisted of a quartz tube (4.5 mm ID and 350 mm long) placed into a vertical electrically heated tube furnace. The effluent gas was continuously analyzed by an on-line AMETEK Q200MN quadrupole mass spectrometer.

The redox cycle experiments were carried out with the following steps: (1) 10mol% O_2/Ar (4 min); (2) Ar (3 min); (3) 10mol% CH_4/Ar (4-4.5 min); (4) Ar (3 min). During step (1) the reduced catalyst was regenerated, thereby became oxygen-enriched; in step (3) CH_4 reacted with the lattice oxygen supplied by the recovered catalyst, generating the products comprising CO_2, CO, H_2 and H_2O. To prevent both CH_4 and O_2 reacted directly in gas phase, the catalyst bed was purged by argon for 3 min in steps (2) and (4). The direct switching reactions from 11mol% O_2/Ar to 11mol% CH_4/He were performed at 900 deg C, atmospheric pressure and total gas flow rate of 23 Nml min^{-1}. No significant backmixing occurs in the reactor under the described reaction conditions.

3. RESULTS AND DISCUSSION

The redox cycle experiments over the $La_{1-x}Sr_xFeO_3$ catalysts (x=0, 0.1, 0.2, 0.5) were carried out. Fig. 1 shows the transient responses after switching from Ar to 10mol%CH_4/Ar over $LaFeO_3$ and $La_{0.8}Sr_{0.2}FeO_3$ oxides. The results indicated that most of the CH_4 was converted to CO_2 and H_2O in a few seconds initially, then the CO_2 intensity declined dramatically and the CH_4 intensity increased sharply. About 15-second reaction in the CH_4/Ar flow, CO and H_2 appeared, and their intensities increased steadily on stream, while the CH_4 intensity decreased gradually. After 60-second reaction in CH_4/Ar, the intensities of CH_4, H_2 and CO reached their steady-state levels respectively and the signal of CO_2 disappeared. However, after the $La_{0.8}Sr_{0.2}FeO_3$ and $LaFeO_3$ catalyst were pre-treated with CH_4/Ar at 900 deg C for 1 min, the main products were H_2 and CO and no CO_2

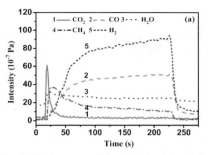 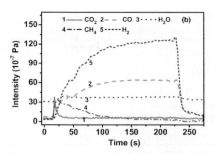

Fig. 1. Responses of the transient from Ar to 10mol%CH₄/Ar over (a) La$_{0.8}$Sr$_{0.2}$FeO$_3$ and (b) LaFeO$_3$ oxides at 900 deg C.

or H_2O was produced. The results indicated that in the absence of molecular oxygen, methane first reacted with the more reactive oxygen species to form CO_2 and H_2O, after these oxygen species were consumed, then the less reactive oxygen species reacted selectively with methane to from CO and H_2. The structure of the LaFeO$_3$ oxide changed only slightly after being treated with methane at 900 deg C for 1 min. The diffraction pattern of the LaFeO$_3$ (Fig.4 (b)) was essentially identical to that of the fresh catalyst.

With decreasing Sr^{2+} content, the amount of CO_2 formed declines considerably and most of the CH$_4$ is selectively oxidized to CO and H_2 (see Table 1). The LaFeO$_3$ sample has the highest selectivity to synthesis gas, it is the best catalyst among the tested La$_{1-x}$Sr$_x$FeO$_3$ series. For the perovskite oxides La$_{1-x}$Sr$_x$FeO$_3$ studied, by the substitution of trivalent La^{3+} with bivalent Sr^{2+} in A-sites, oxygen vacancies and/or a fraction of Fe^{4+} would be created to maintain overall charge neutrality for the oxide. This perhaps accounted for the lowest selectivity to CO and H_2 obtained over the La$_{0.5}$Sr$_{0.5}$FeO$_3$ sample.

After the catalyst reacted with methane, the reduced catalyst was re-oxidized by 10mol% O$_2$/Ar for 4 min at 900 deg C. In the first 90 seconds, no oxygen was detected in the effluent gas. The oxides restored oxygen very efficiently. During the re-oxidation step, the signal of CO_2 and/or CO was not detected over the tested La$_{1-x}$Sr$_x$FeO$_3$ (x=0, 0.1, 0.2, 0.5) oxides. The results implied that almost no carbon deposited on the catalysts during the methane reaction step under the reaction conditions.

The effect of reaction temperature on product distribution, such as CO, H_2, CO_2 and H_2O over the LaFeO$_3$ oxide are presented in Fig. 2. A very limited quantity of CO_2 was produced at the temperatures of 810 deg C, 860 deg C and 900 deg C respectively. It was found that at temperatures lower than 810 deg C, only a small fraction of methane was converted to the desired products. At temperatures higher than 860 deg C, the reaction in 11 mol%CH$_4$/Ar yielded a large amount of synthesis gas, with >98% selectivity to H_2 and CO and >1.94 H_2/CO ratio during the whole reaction period. The rate of oxygen supply during

Table1
The product distribution over $La_{1-x}Sr_xFeO_3$ oxides in 10mol% CH_4/Ar at 900 deg C

Catalyst	$LaFeO_3$	$La_{0.9}Sr_{0.1}FeO_3$	$La_{0.8}Sr_{0.2}FeO_3$	$La_{0.5}Sr_{0.5}FeO_3$
H_2 $(10^{-6}mol)$	326	236	253	120
CO $(10^{-6}mol)$	168	125	133	63.7
CO_2 $(10^{-6}mol)$	3.3	8.3	8.6	10.0
H_2O $(10^{-6}mol)$	11.9	24.0	23.9	23.6
CO selectivity(%)	98.1	93.8	93.9	86.4
H_2 selectivity (%)	96.5	90.7	91.4	83.6

Reaction time: 185 s

the CH_4 reaction step was strongly affected by the reaction temperature. The rate of oxygen migrating from bulk to surface increased with increasing temperature, and much more amount of CH_4 can be selectively oxidized to CO and H_2. This implies that bulk lattice oxygen diffuses to surface region generally requires rather high activation energy.

Fig. 3 shows the quantitative product distribution obtained from the reaction in 10mol% CH_4/Ar at 900 deg C on the $LaFeO_3$ catalyst. The first reaction cycle was conducted on the fresh catalyst; and the following cycles were carried out after re-oxidation of the catalyst. With increasing cycle numbers, the yield of products only slightly decreased.

It can be seen that the sample after redox cycles and re-oxidation still maintains the perovskite structure (see Fig. 4). The diffraction pattern of the reoxidized $LaFeO_3$ (Fig. 4 (b)) was very simliar to that of the fresh catalyst (Fig. 4 (a)). However, besides the perovskite structure, the La_2O_3 structure pattern (JCPDS 05-602) appears in the $LaFeO_3$ catalyst after reaction with CH_4 for 4 min at 900 deg C. With some lattice oxygen reacting with CH_4, part of the Fe^{4+} and/or Fe^{3+} in the $LaFeO_3$ catalyst is possibly reduced to lower valence iron, leading to partial disintegration of the perovskite structure. However, when it was re-oxidized at the reaction temperature in 10 mol% O_2/Ar for a few minutes, the catalyst could be restored to the perovskite-type structure.

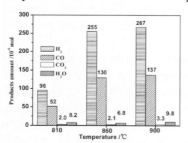

Fig. 2. Effect of reaction temperature on the product distribution over $LaFeO_3$ oxide.

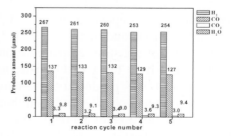

Fig. 3. Effect of reaction cycle number on the product distribution over $LaFeO_3$ oxide.

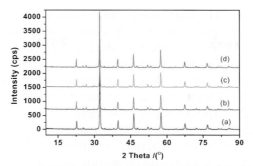

Fig. 4. XRD patterns of (a) fresh LaFeO₃; (b) LaFeO₃ after five redox cycles in 10 mol%CH₄/Ar-10%mol O₂/Ar and reoxidized in O₂/Ar atmosphere at 900 deg C; (c) LaFeO₃ reacted with 10 mol% CH₄/Ar for 4 min at 900 deg C; (d) LaFeO₃ reacted with 10 mol% CH₄/Ar for 1 min at 900 deg C.

Fig. 5 shows the product responses during the direct repeated switching reactions between O₂/Ar oxidation for 6 s and CH₄/He reduction for 30 s over the La₀.₈Sr₀.₂FeO₃ catalyst at 900 deg C. Compared to the redox cycle experiments with switching from inert argon to CH₄ (see Fig. 1(b)), there was no change in the product distribution of direct switching from O₂ to CH₄ for the first cycle. In the second redox cycle, only a small quantity of CO₂ was produced, and CO and H₂ were the dominant products. When the lattice oxygen restored in the re-oxidation step was not sufficiently restored, the oxidation level of the catalyst may not reach the initial state, and only the most reactive oxygen species can be removed. High selectivity to CO and H₂ was achieved in the third redox cycle. Calculations show that the CH₄ conversion is about 43%, the selectivities to H₂ and CO are over 93% with the H₂/CO ratio being ca. 1.90, nearly the stoichiometric ratio of the partial oxidation of methane.

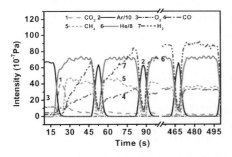

Fig. 5. The switching reactions between 11mol% O₂/Ar (6 s) and 11 mol% CH₄/He (11 s) over the La₀.₈Sr₀.₂FeO₃ catalyst (0.10 g) at 900 deg C.

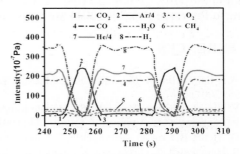

Fig. 6. The switching reactions between 11mol% O₂/Ar (11 s) and 11 mol% CH₄/He (25 s) over the pre-treated La₀.₈Sr₀.₂FeO₃ oxide (0.25 g) at 900 deg C.

To achieve high CH_4 conversion, the switching reactions were also performed under other conditions. Fig. 6 shows the results of the switching reactions between O_2/Ar and CH_4/He over the $La_{0.8}Sr_{0.2}FeO_3$ catalyst pre-treated in CH_4/He for 2 min at 900 deg C. The catalyst pre-treated with CH_4 provided mainly selective oxygen species. About 10-minute later, methane conversion reached ~90% and the selectivities to CO and H_2 maintained ~93%, with a H_2/CO ratio close to 2.0. The results of switching reaction indicate that methane can be highly selectively oxidized to CO and H_2 by the lattice oxygen of $La_{0.8}Sr_{0.2}FeO_3$ under appropriate reaction conditions and the consumed lattice oxygen can be supplemented by re-oxidation procedure.

4. CONCLUSION

In the present study, a highly selective approach to producing synthesis gas by partial oxidation of methane with lattice oxygen on $La_{1-x}Sr_xFeO_3$ oxides in cyclic reaction is suggested. In the absence of molecular oxygen methane first reacts with the more reactive but non-selective lattice oxygen to form CO_2 and H_2O. After these oxygen species were consumed, the less active but selective lattice oxygen reacts with methane to form CO and H_2. It was found that under an appropriate reaction condition, methane can be oxidized to CO and H_2 by the lattice oxygen of $La_{1-x}Sr_xFeO_3$ perovskite oxide with selectivity higher than 93% and the consumed lattice oxygen can be supplemented by re-oxidation procedure. It is proposed that partial oxidation of CH_4 to syngas by using lattice oxygen of oxides instead of molecular oxygen may be avilable

ACKNOWLEDGEMENTS

The supports by SINOPEC (No. X502015) and Natural Science Foundation of Jiangsu Province (BK2001201) are gratefully acknowledged.

REFERENCES

[1] A.F. Sammells, M. Schwartz, R.A. Mackay, T.F. Barton, and D.R. Peterson, Catal. Today, 56 (2001) 31
[2] Y. Hiei, T. Ishihara, Y. Takita. Solid State Ionics, 86-88(1996)1267.
[3] P. Tsiakaras, C.G. Vayenas, J. Catal., 144 (1993) 333.
[4] D. Tichit, B. Cog, H. Armendariz, F. Figureueras. Catal. Lett., 38(1996)109.
[5] S. Hamakawa, T. Hayakawa, K. Suzuki, K. Murata, K. Takehira, S. Yoshino, J. Nakamura, T. Uchijima. Solid State Ionics, 136-137(2000)761.
[6] R. Li, C. Yu, and S. Shen, Journal of Natural Gas Chemistry, 11 (2002)137
[7] R. Li, C. Yu, X. Dai, and S. Shen, Chin. J. Catal., 23(2002)549

Studies in Surface Science and Catalysis, volume 147
X. Bao and Y. Xu (Editors)

Partial oxidation of methane to synthesis gas in a dual catalyst bed system combining irreducible oxide and metallic catalysts

J.J. Zhu, M.S.M.M. Rahuman, J.G. van Ommen[*], L. Lefferts

Faculty of Science and Technology, University of Twente, P.O. Box 217, 7500 AE, Enschede, The Netherlands. Email: j.g.van ommen@utwente.nl

ABSTRACT

Operation of partial oxidation of methane to synthesis gas over yttrium-stabilized zirconia (YSZ) at very high temperatures (>900°C) slightly improves the selectivity to synthesis gas, which is caused by some activity of YSZ for steam and dry reforming of methane. $LaCoO_3$ perovskite is not active in methane reforming, but very active in deep oxidation. This perovskite can be reduced to a highly dispersed cobalt catalyst by CH_4 at high temperatures (>900°C), which is very active in methane reforming. The dual bed concept combines these two catalysts and demonstrates that synthesis gas with an equilibrium composition can be produced at high GHSV. The exposure of metal catalyst to O_2 in the second catalyst bed is avoided because oxygen is converted completely in the first catalyst bed. Therefore, deactivation of metal catalyst via evaporation of its volatile oxides can be prevented.

1. INTRODUCTION

As an alternative to steam reforming, the catalytic partial oxidation of methane (CPOM) to synthesis gas has attracted significant attention. Metallic catalysts, Rh, Pt, Ir, Co and Ni on different supports have been extensively researched. Two general mechanisms have been proposed for CPOM on those metallic catalysts. An indirect mechanism involves methane combustion, followed by steam and CO_2 reforming [1]. A direct partial oxidation mechanism on noble metal catalysts at high temperatures and very short contact time (millisecond) is claimed by Schmidt [2]. Under these extreme conditions the thermodynamic equilibrium can not be reached because of kinetic limitation [3]. All metallic catalysts are suffering from deactivation by evaporation and sintering at very high temperatures [4]. Moreover, these extremely high temperatures promote formation of volatile metal oxide in oxidizing atmospheres [5]. Metal loss in the form of volatile oxide, which results in deactivation of catalyst, is also a serious problem in the ammonia oxidation process operated at about the same conditions [6, 7]. Compared to metal catalysts, metal oxides show excellent stability at the expense of activity for partial oxidation of methane. Steghuis [8] and Stobbe [9] reported that methane is oxidized over YSZ mainly via reaction $CH_4+O_2 \rightarrow CO+H_2+H_2O$, though parallel deep oxidation occurs as well. It was confirmed that CO, CO_2, H_2 and H_2O are primary products of CPOM.

The objective of the present work is to determine if the performance of YSZ catalyst can be improved in term of selectivity by increasing the operation temperature. The role of consecutive reforming reactions is discussed as well. A dual catalyst bed concept for synthesis gas production is proposed and demonstrated in lab scale. Features of this system are discussed.

2. EXPERIMENTAL

2.1. Catalyst preparation and catalytic reaction
Preparation of catalysts, YSZ12C (12wt% Y_2O_3 in ZrO_2; S_g=4.5 m^2/g) and $LaCoO_3$ (S_g=1.8 m^2/g), and catalytic testing were described in our previous work [10].

2.2. Characterization
H_2-TPR experiments were carried out in a high temperature TPR setup. Before reduction, the catalysts were heated at 200°C in flowing helium for 1 h. Then, a mixture of 5% H_2 in Ar (25ml/min.) was passed through the sample and the temperature was raised from 200°C to 900°C with a rate of 10°C/min.

3. RESULTS

3.1. Oxidation of CH_4 in gas phase
Figure 1 shows the results of CH_4 oxidation in the empty rector. Methane oxidation can be neglected when reaction temperature is below 900°C, whereas oxidation of methane occurred significantly in the higher temperature region.

3.2. Methane oxidation on $LaCoO_3$
Conversion and yields as a function of reaction temperature in CPOM over $LaCoO_3$ are shown in Figure 2. As expected, $LaCoO_3$ perovskite is a very active catalyst for combustion at lower temperature. Oxygen was consumed completely at 600°C with 25% methane conversion. In a wide temperature window (<750°C), only deep oxidation products, CO_2 and H_2O were detected. CO and H_2 production started from 750°C, and their yields increased dramatically with increase of temperature. At about 950°C, complete CH_4 conversion and 100% selectivity to synthesis gas were obtained.

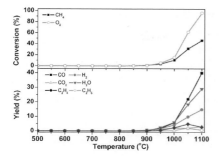 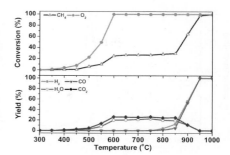

Fig.1 Oxidation of methane in gas phase. F=170 ml/min. CH₄:O₂:He=2:1:14

Fig.2 Catalytic performance of LaCoO₃ at different temperatures. 300 mg Catalyst. F=170 ml/min. CH₄:O₂:He=2:1:14

Pre-reduction of the catalyst at 1100°C in H_2 gave identical results. When the reaction temperature was decreased, the catalyst gave again deep oxidation although at slightly lower temperatures.

The reducibility of $LaCoO_3$ was measured by H_2-TPR (Figure 3). The reduction profile of $LaCoO_3$ consists of two peaks at approximately 700°C and 876°C respectively. The hydrogen consumption for the first reduction step (peak at 700°C) was approximately half of that for the second step (peak at 876°C). This implies that the first step is a one-electron reduction process ($Co^{3+} + e \rightarrow Co^{2+}$) whereas the second step is a two-electron reduction step process ($Co^{2+} + 2e \rightarrow Co^0$).

The XRD pattern showed a cubic perovskite structure of the fresh $LaCoO_3$ sample. In contrast, only diffraction lines of La_2O_3 are present in the XRD pattern of the sample after CPOM at 950°C for 3 h.

3.3. CPOM over YSZ

Figure 4 shows the conversion and yields as a function of temperature in CPOM over YSZ12C. Compared with $LaCoO_3$, a quite different performance was observed on YSZ catalyst. Besides four major oxidation products, CO, H_2, CO_2 and H_2O, trace of hydrocarbons were detected over the whole temperature window. After all oxygen was exhausted at around 950°C, yields of CO and H_2 increased, while those of CO_2 and H_2O decreased with increasing temperature, moreover methane conversion increased slightly. The effects of addition of steam and CO_2 to the feed are shown in figure 5. Obviously, the product distribution had been changed after addition of H_2O to the feed, moreover methane conversion was improved slightly. Increasing CO and H_2 yields indicate that steam reforming of methane ($CH_4+H_2O \rightarrow CO+3H_2$) occurred under this condition.

Compared with the effect of steam, addition of CO_2 to the feed influenced significantly not only product distribution but also methane conversion (Fig.5). Methane conversion, CO and H_2O yields were increased simultaneously, in

208

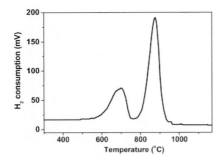

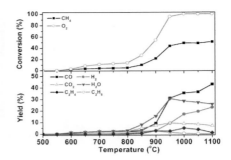

Fig. 3 H₂-TPR spectrum of LaCoO₃

Fig. 4. Catalytic performance of YSZ12C. 0.3g Catalyst. F=170 ml/min; CH₄:O₂:He=2:1:14.

contrast, CO_2 and H_2 remained unchanged. This can be the result of simultaneous CO_2 reforming of methane ($CH_4+CO_2\rightarrow CO+H_2$) as well as reverse water-gas shift reaction ($CO_2+H_2\rightarrow CO+H_2O$), which is thermodynamically favorable at high temperatures.

Separate experiments showed that the activity of the empty reactor for CO_2 and steam reforming of methane was one order of magnitude smaller than that of YSZ12C.

3.4. Dual catalyst bed system

Since water and CO_2 are produced over YSZ12C via $CH_4+O_2\rightarrow CO+H_2+H_2O$ and parallel combustion $CH_4+2O_2\rightarrow CO_2+2H_2O$, it is clear that to produce synthesis gas with an equilibrium composition, steam and CO_2 reforming of methane is indispensable. Sufficient activity for reforming can be introduced by using a second catalyst bed with an active reforming catalyst.

The dual catalyst bed concept is demonstrated by using 0.3g YSZ in the first reactor and 0.3g pre-reduced LaCoO₃ in the second reactor. CPOM was carried

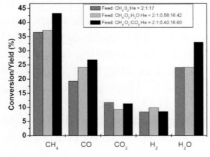

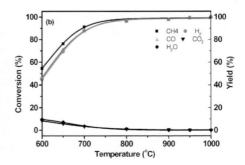

Fig. 5 Influence of CO_2 or H_2O addition to the feed on conversion and yields on YSZ12C at 950°C. Catalyst: 300mg; F=170 ml/min.

Fig. 6. Final conversion and yields as a function of temperature in the second reactor in dual bed system. Feed gas: CH₄/O₂=2:1; F=150 ml/min.

out in the dual catalyst system at GHSV of 2×10^4 h^{-1} (GHSV= (F_{CH4} + F_{O2})/ (V_{oxide}+ V_{metal})). After reaction in the first reactor at 1000°C, about 49.3% CH$_4$ and 100% O$_2$ were converted, and CO, H$_2$, CO$_2$, H$_2$O and hydrocarbons were produced with yields of 41.1 %, 27.3 %, 7.8 %, 21.7 % and 0.5 % respectively. The effluent of the first reactor was introduced to the second reactor directly. Figure 6 shows the influence of reaction temperature in the second reactor on total conversion and yields. Synthesis gas with equilibrium composition was produced at above 800°C in the second catalyst bed.

4. DISCUSSION

Steghuis [8] investigated catalytic performances of many oxides, such as ZrO$_2$, YSZ, TiO$_2$, La$_2$O$_3$ in CPOM. YSZ was reported as the most active and selective catalyst for CPOM. It's found that four major products, CO, H$_2$, CO$_2$ and H$_2$O are primary products, produced via direct partial oxidation of methane and parallel deep oxidation. They didn't observe reforming reactions occurring on YSZ in those experiments (500~900°C). However, after O$_2$ is exhausted completely above 900°C, the variation of product distribution implies that some reactions occur between products and/or between product and methane (Fig. 4). In Figure 5, it's clearly shown that methane reforming occurs to a certain extent on YSZ at high temperatures (>900°C). However, it's impossible to produce synthesis gas with an equilibrium composition using YSZ catalyst only, because its activity for reforming is far too low.

Comparison of methane conversion and product distribution between the empty reactor and YSZ12C (Figs 1 and 4) shows that, despite significant conversion in the empty reactor, the catalyst is mainly controlling the reactions because the product distribution is completely different.

Unlike on YSZ, oxidation of methane over pervoskite LaCoO$_3$ gave CO$_2$ and H$_2$O only. Similar catalytic performances of LaCoO$_3$ at different temperature were also observed by Slagtern[11] This indicates that LaCoO$_3$ is not active for steam and CO$_2$ reforming reactions. However, after prolonged exposure to mixture of methane, steam and carbon dioxide at higher temperature (>900°C), LaCoO$_3$ can be reduced to a dispersed cobalt catalyst (Co/La$_2$O$_3$), which is known to be active for reforming reactions. As reported by Lago[12],this process is reversible. The reduced catalyst is therefore well suited to be used as the second catalyst in the dual bed concept.

Dual bed system combines both advantages of good stability of YSZ oxide catalyst as well as high reforming activity of the metallic catalyst. Methane is selectively oxidized over a stable oxide catalyst (YSZ) in the first catalyst bed. This results in a lower temperature rise over the catalyst bed, because less heat is liberated compared to initial complete deep oxidation on metal catalysts. CO$_2$ and H$_2$O produced in the first catalyst bed react with unconverted CH$_4$ over cobalt metal catalyst in the second catalyst bed. Synthesis gas with an equilibrium composition is produced. Importantly, since oxygen is consumed completely in the first catalyst bed under certain conditions, exposure of metallic catalyst to O_2 in the second catalyst bed is absolutely prevented. Therefore, catalyst deactivation via sintering and evaporation in the form of volatile metal oxides is circumvented completely.

5. CONCLUSIONS

YSZ has some activity for methane reforming at high temperatures (>900°C), but it is not able to convert all H_2O and CO_2 with CH_4 to synthesis gas with an equilibrium composition. $LaCoO_3$ perovskite is not active in methane reforming, but very active in deep oxidation. This perovskite can be reduced to a highly dispersed cobalt catalyst by CH_4 at high temperatures (>900°C), which is very active in methane reforming. The dual bed concept combines these two catalysts to carry out CPOM. The present results demonstrate that synthesis gas with an equilibrium composition can be produced under high GHSV in this system. The direct partial oxidation on YSZ in the first catalyst bed results in a much flatter temperature profile. The exposure of metal catalyst to O_2 in the second catalyst bed is avoided because oxygen is converted completely in the first catalyst bed. Therefore, deactivation of metal catalyst via evaporation of its volatile oxides can be prevented.

ACKNOWLEDGEMENTS

This work was performed under the auspices of the Netherlands Institute for Catalysis Research (NIOK). The financial support from STW under project no. UCP-5037 is gratefully acknowledged.

REFERENCES

[1] A.T. Ashcroft, P.D.F. Vernon, and M.L.H. Green, Nature 344 (1990) 319.
[2] D.A. Hickman and L.D. Schmidt, Science 259 (1993) 343.
[3] A.P.E. York, T. Xiao, M.L.H. Green, Topics in Catal. 22 (2003) 345.
[4] S. Albertazzi, P. Arpentinier, F. Basile, D. Gallo, G. Fornasari, D. Gary, and A. Vaccari, Appl. Catal. A, 247 (2003) 1.
[5] H. Jehn, J. Less Common Metals, 100(1984) 321.
[6] M.M. Karavayev, A.P. Zasorin, N.F. Kleshchev, Catalytic oxidation of ammonia, Khimia, Moscow, 1983.
[7] T.H. Chilton, Chem. Eng. Progr. Monogr. Ser. 3(1960) 56.
[8] A.G. Steghuis, J.G. van Ommen, J.A. Lercher, Catal. Today, 46 (1998) 91.
[9] E.R. Stobbe, Ph.D thesis, University of Utrecht, 1999.
[10] J.J.Zhu, M.S.M.M. Rahuman, J.G. van Ommen, L.Lefferts, Appl. Catal. A (2003) in press
[11] A. Slagtern, U. Olsbye, Appl. Catal. A, 110 (1994) 99.
[12] R. Lago, G. Bini, M.A. Peria, and J.L.G. Fierro, J. Catal. 167(1997) 198.

Studies in Surface Science and Catalysis, volume 147
X. Bao and Y. Xu (Editors)
©2004 Elsevier B.V. All rights reserved.

New catalysts for the syngas production obtained by hydrotalcite type precursor containing silicate

F. Basile[* a], **S. Albertazzi**[a], **P. Arpentinier**[b], **P. Del Gallo**[b], **G. Fornasari**[a], **D. Gary**[b], **V. Rosetti**[a] and **A. Vaccari**[a]

[a]Dip. Chimica Industriale e dei Materiali, Università di Bologna, V.le Risorgimento 4, 40136 Bologna, Italy.

[b]Air Liquide, Centre de Recherche Claude Delorme, 1, chemin de la Porte des Loges, BP 126, Les Loges-en-Josas - 78354 Jouy-en-Josas Cedex, France.

[*] E-mail: basile@ms.fci.unibo.it; Phone: +39 051 2093663; Fax: +39 051 2093680.

1. INTRODUCTION

The direct transformation of CH_4 to liquid is difficult and nowadays not economic [1]. Therefore, the interest towards the indirect pathway involving the catalytic partial oxidation of CH_4 to syngas (CPO) is increasing [2]. Ni and/or Rh based catalysts (ex-HT) derived by hydrotalcite precursor have been claimed as effective catalysts in the CPO reaction [3-6]. In this work, new stable ex-HT catalysts are prepared by inserting silicate instead of CO_3^{2-} anions in the interlayer. The amount of silicates is not limited by the charge balance of the cations, since, an excess of silicates leads to the formation of polisilicate anions. The insertion of silicates affects the structure of the calcined samples, since, unlike CO_3^{2-}, the silicates remain in the structure and contribute to the formation of the final catalysts.

2. EXPERIMENTAL

Three catalysts have been obtained by HT/silicates precursors based on Ni and/or Rh as active phases (Tab. 1). The solution of the nitrates of the metals and the silicate solution were simultaneously dropped into a water solution at pH 10 under a N_2 atmosphere to avoid CO_3^{2-} present in the air. The precipitation was carried out at constant pH (9.9-10.1) and constant T (50-60°C). Then the precipitate (HT) was filtred, washed with water, dried overnight at 110°C and finally calcined at 900°C for 12 hours. The catalysts before and after reactionwere

Table 1
List of the studied catalysts (*determined by ICP).

	Composition (a.r.)	SA fresh (m^2g^{-1})	SA spent (m^2g^{-1})	wt% Rh (*)	wt% Ni (*)
Rh	$Rh_{0.5}/Mg_{71}/Al_{28.5}$	106	110	0.536	/
Ni	$Ni_8/Mg_{60}/Al_{32}$	112	104	n.d.	n.d.
Rh/Ni	$Rh_{0.15}/Ni_8/Mg_{60}/Al_{31.85}$	114	120	0.138	6.56

characterised by XRD and BET surface area analyses. ICP and TGA analyses were also carried out on some samples.

3. RESULTS AND DISCUSSION

All the catalysts have high specific surface area (Tab. 1), without significant changes after reaction. The TGA carried out on the Ni precursor (Fig. 1) shows mainly three endothermic steps corresponding to weight loss:

1st, 2nd) $\quad [Mg^{2+}_{0.60}Ni^{2+}_{0.08}Al^{3+}_{0.32}(OH^-)_2](SiO_3)^{2-}_x m(H_2O)$ $\Rightarrow$
$[Mg^{2+}_{0.60}Ni^{2+}_{0.08}Al^{3+}_{0.32}(OH^-)_2](SiO_3)^{2-}_x$

The first step corresponds to the loss of the water of crystallisation, while the second one concerns the loss of the water placed in the interlayer.

3rd) $\quad [Mg^{2+}_{0.60}Ni^{2+}_{0.08}Al^{3+}_{0.32}(OH^-)_2](SiO_3)^{2-}_x.$ $\Rightarrow$
$[Mg^{2+}_{0.60}Ni^{2+}_{0.08}Al^{3+}_{0.32}O^{2-}](SiO_3)^{2-}_x$

The 3rd is related to the formation of $(Mg/Al/Ni)SiO_4$. The weight decreases from 83% to 63% in the 3rd step and calculations confirm that the weight loss can be attributed to the oxide formation from a silicate HT.

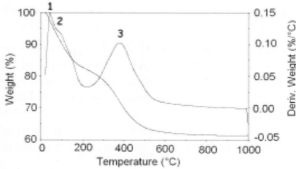

Fig. 1. TGA pattern of Ni sample.

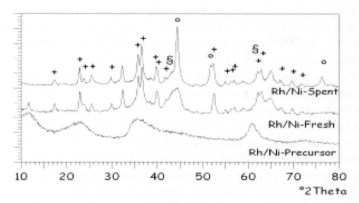

Fig. 2. XRD patterns of precursor, fresh and spent samples of Ni/Rh catalyst; ([+] = $(Mg/Al)_2SiO_4$, [o] = Ni^0, [§]= MgO).

All the catalysts mainly show the reflections of a $(MMgAl)SiO_4$ mixed silicate lattice with a forsterite type structure, together with a small amount of $(M/Mg/Ni)O$ type structure. These structures were retained even after reaction and only the spent catalysts with Ni and Ni/Rh show a significant change in the intensity of the mixed oxide lattice and new reflections attributable to the Ni^0. This observation suggests that a large part of the Ni was present in the mixed oxide phase and that it was reduced before reaction.

To discriminate among the catalytic activity of the investigated samples, the results of the tests far from the thermodynamic equilibrium ($CH_4/O_2/He = 2/1/20$ v/v, $T_{oven}= 500°C$, $\tau=65$ ms) are compared in Fig. 3.

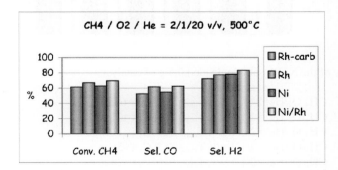

Fig. 3. Comparison of the catalytic performances of Rh samples (prepared with CO_3^{2-} or silicates), Ni and Ni/Rh samples (0.5g of catalyst, $CH_4/O_2/He = 2/1/20$v/v, $T_{oven}= 500°C$, $\tau=65$ms).

Ni/Rh catalyst ($Rh_{0.15}/Ni_8/Mg_{60}/Al_{31.85}$) showed the best values of methane conversion and selectivity in CO and H_2, confirming the Rh/Ni synergy [3]. All the catalysts have been also subjected to deactivation tests keeping the catalyst in hard reaction conditions and controlling the deactivation with tests at low temperature. After 6 days the methane conversion and synthesis gas selectivity were constant. Finally, the ex-HT silicate catalyst containing only Rh was compared with that obtained by ex-HT carbonate with the same Rh/Mg/Al atomic ratio (Rh-carb), showing that the catalyst containing silicate was more active and selective towards syngas than that prepared with CO_3^{2-}. Since the Ni/Rh sample shows the most interesting result, the details of the tests at different reaction conditions are reported in Fig. 4. High methane conversion and syngas selectivity were obtained even at low residence time (28 ms), due to the increase of temperature that favors the consecutive reforming reactions (usually hindered by short contact times). Finally, tests at low residence time have been carried out reducing the amount of the Ni/Rh catalyst to 0.1 g without any relevant change in the catalytic activity, also in the most discriminating tests. This test points out that, high yields in syngas are obtained even using a residence time of 5.6 ms.

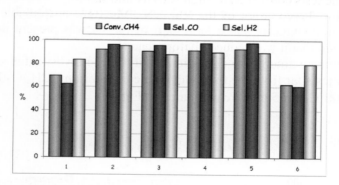

Fig. 4. Activity of the Ni/Rh sample, loaded 0.5 g of catalyst;

1) T_{oven}=500°C, $CH_4/O_2/He$=2/1/20v/v, T_{max}=592°C, τ=65ms
2) T_{oven}=750°C, $CH_4/O_2/He$ = 2/1/4 v/v, T_{max}=818°C, τ=65 ms;
3) T_{oven}=750°C, $CH_4/O_2/He$ = 2/1/1 v/v, T_{max}=827°C, τ=112ms;
4) T_{oven}=750°C, $CH_4/O_2/He$ = 2/1/1 v/v, T_{max}=883°C, τ=56 ms;
5) T_{oven}=750°C, $CH_4/O_2/He$ = 2/1/1 v/v, T_{max}=916°C, τ=28 ms;
6) T_{oven}=500°C, $CH_4/O_2/He$ = 2/1/20 v/v, T_{max}=630°C, τ=65 ms, tested after 18 days of time-on-stream in the pilot plant.

This is an important aspect, which will boost the possibilities of a real application in the industrial plant. On these bases the sample was used for additional tests in a pilot plant, using 160 g of catalyst at 7 atm and a residence time of 300 ms, for 18 days of time-on-stream with constant results. The spent catalyst obtained by the top of the reactor, therefore subjected to the most severe conditions, was tested in the laboratory plant giving results similar to the fresh sample, confirming the high activity and stability of this catalyst. Thanks to the composition flexibility of the precursor this new class of materials can be further improved or adapted to the different conditions of the processes based on CPO reaction.

4. CONCLUSIONS

Catalysts with innovative structure were obtained by HT precursor prepared using silicate as interlayer anion. The calcined catalysts showed two phases: a mixed oxide type structure and a new M-silicate structure, where the cations of the active metal are partially soluted. The Rh catalyst prepared by HT silicate was more active and selective than the one prepared by CO_3^{2-}. The $Rh_{0.15}/Ni_8/Mg_{60}/Al_{31.85}$ showed the most interesting result in terms of syngas yield and it was tested in a pilot plant. The tests confirmed its high activity and stability even after 18 days of time-on-stream. The amount of the $Rh_{0.15}/Ni_8/Mg_{60}/Al_{31.85}$ loaded catalyst was reduced down to 0.1 g without any change of activity, improving the possibilities of an application in the low residence time processes.

REFERENCES

[1] E.E. Wolf, Van Nostrand Reinhold (Ed.), Methane conversion by oxidative processes: fundamental and engineering aspects, Catalysis series (1992).
[2] P. Chaumette, P. Courty, Energy Prog. 27 (1987) 1.
[3] F. Basile, G. Fornasari, F. Trifirò and A. Vaccari, Catal. Today, 77 (2002) 215.
[4] F. Cavani, F. Trifirò, A. Vaccari, Catal. Today, 11 (1991) 173.
[5] F. Basile, L. Basini, G. Fornasari, M. Gazzano, F. Trifirò, A. Vaccari, Chem. Comm. (1996) 2435.
[6] V. Rives, Layered Double Hydroxides: present and future, (2001) ed. Nova Science Publisher, NY.
[7] V.R. Choudhary, B. Prabhakar, A.M. Rajput, J. Catal. 139 (1993) 326.

Studies in Surface Science and Catalysis, volume 147
X. Bao and Y. Xu (Editors)
©2004 Elsevier B.V. All rights reserved.   217

Novel catalysts for synthesis gas generation from natural gas

Shizhong Zhao

Süd-Chemie Inc., 1600 West Hill Street, Louisville, Kentucky 40210, USA

ABSTRACT

New calcium aluminates, $CaO \cdot xAl_2O_3$ or CA_x, catalyst shapes and formulation have been developed to fit various reactor designs and process requirements for the production of synthesis gas (syngas). The CA_x has been demonstrated to be a good catalyst support for both base and precious metals for syngas applications. Excellent mechanical and catalytic properties have been achieved. Catalysts Ni/CA_x and Rh/CA_x reach equilibrium conversion in steam reforming of CH_4 at temperatures as low as 460°C and 480°C respectively and gas hour space velocity (*GHSV*) 25,000 that is 10 times that of a typical commercial reformer. The Ni/CA_x catalysts show higher activity than the $Ni/\alpha-Al_2O_3$ catalysts in both steam reforming and CO_2 and steam mixed reforming of CH_4. The CA_x catalysts are more resistant to coke formation. The CA_x support is available for industrial applications.

1. INTRODUCTION

The need for lower *H₂/CO* ratio and cheaper syngas energizes the exploitation of syngas processes and catalysts. Commercial production of syngas from hydrocarbons has been dominated by steam reforming in large tubular reformers where large size catalyst particles are employed to reduce pressure drop since heat transfer is usually the rate limiting factor. The syngas generation section is very expensive due to the size of the reformer. It accounts for 60-70% of the capital and operation costs of a gas-to-liquid (GTL) plant [1]. Reducing the cost of syngas becoms a crucial factor for whether the GTL process is economically feasible. Reactor size reduction is a solution in this regard that can be achieved by improving reactor design and applying different reactions. The first route can be accomplished by enhancing heat transfer such as employing smaller reactor tubes. Regarding the second route, partial oxidization is a faster and less exothermic reaction compared to steam reforming. Therefore, the reactor can be operated at much higher space velocities [2]. Catalyst shapes with much higher geometric surface areas are required. Fischer-Tropsch synthesis that employs a Co and Fe-based catalyst requires syngas H_2/CO ratio of 2.1 and 1.7 respectively [1]. Steam reforming of natural gas provides syngas with $H_2/CO>3$. Partial oxdization and modified steam reforming processes such as autothermal reforming (by adding O_2) and mixed reforming (by adding CO_2) are more suitable. For instance, depending on the amount of CO_2 addition, the H_2/CO ratio of the syngas generated by steam and CO_2 mixed reforming changes from less than 1 to larger than 3, which makes mixed reforming a viable process when natual gas resources contain considerable amount of CO_2.

The basic requirements for a syngas generation catalyst are activity, coking resistance, thermal stability, and robustness of the catalyst particles. The lower the H_2/CO ratio, the higher the coking potential in syngas generation. Coking resistance is critical in insuring carbon-free operation. Mechanical properties of the pellets that are determined by the support largely are vital for catalyst performance. Engineering the support into different shapes with good mechanical and catalytic properties is as challenging as catalyst formulation development. Physical and catalytic properties of CA_x and performance of CA_x supported catalysts in steam and CO_2 mixed reforming and steam reforming of CH_4 are discussed.

2. EXPERIMENTAL

The CA_x supports were made by a Süd-Chemie proprietary compounding method to achieve desirable microstructures and geometric shapes. The catalysts were made by the impregnation method. Coking resistance was determined by steam reforming of n-hexane in a 1-inch ID fixed bed reactor with 18x20 mesh catalyst particles at inlet and outlet temperature at 540°C and 820°C respectively, and 24Mpa pressure. During the test, H_2O/C ratio was decreased following a schedule. The H_2O/C ratio right before the pressure-drop increase was reported as the coking formation index. Activity of steam reforming of CH_4 was determined in a 0.76-inch ID fixed bed reactor at $T = 400°C\text{-}540°C$, $P = 35\text{kPa}$, $H_2O/C = 3$, and $GHSV = 25,000$. Steam and CO_2 mixed reforming of CH_4 was carried out in a 1-inch ID reactor at 820°C, 100kPa, $H_2O/CH_4 = 0.76$, $CO_2/CH_4 = 0.82$, and $GHSV = 25,000/h$. Activity was calculated by $k' = GHSV \times \ln\dfrac{1}{1 - x_{CH_4}}$.

3. RESULTS AND DISCUSSION

3.1 Calcium Aluminate

The CA_x made by the Süd-Chemie proprietary method have a multiphase structure as showed in Fig. 1a. By varying the amount of Ca, one can form

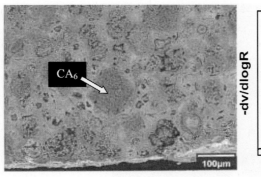 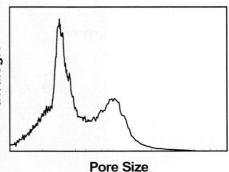

Pore Size

Fig. 1. Calcium aluminate: a. SEM image   b. Pore size distribution

various CA_x phases. The most common CA_x phases are monoaluminate (CA), bialuminate (CA_2), and hexaluminate (CA_6). As indicated in Table 1 the CA_6 phase has the highest melting point, crystal density, porosity, and compressive strength. Thermodynamically, it is the most stable phase in the CA_x family [3], therefore, the most desirable phase in most of catalytic applications where better porosity, thermal stability, and mechanical strength are beneficial. In Fig. 1a there are many porous islands that were identified by X-ray diffraction to be CA_6. The unique morphology resulted in bi-modal pore size distributions as showed in Fig. 1b. The small pores provide large surface area for metal dispersion while the large ones supply an access network for the reactants [4]. More balanced metal dispersion and mass-transfer resistance were achieved.

3.2 Various Shapes of Calcium Aluminate Support

Fig. 2 displays various shapes of CA_x based supports and catalysts. Some shapes in the picture have been successfully used in commercial syngas generation. Suitable catalyst shapes can be selected to meet reactor design and specific process requirements. The diameter of the catalysts range from 2mm to 8mm for the spheres, 8mm to 33mm for the rings, and 4mm and larger for the tablets. The rings have 1 to 10 holes and/or ribs for geometric surface area maximization. Excellent mechanical properties have been achieved for all the shapes. For instance, typical crush strengths of 8mm Raschig rings, 8mm ribbed

Table 1
Properties of various calcium aluminate phases [3]

Phase	Melting Point ($^{\circ}C$)	Crystal Density (g/cc)	Porosity (%)	Compressive Strength (Mpa)
CaO•Al$_2$O$_3$	1600	2.98	5.4	116
CaO•2Al$_2$O$_3$	1750	2.91	2.7	172
CaO•6Al$_2$O$_3$	1830	3.38	8.1	186

Fig. 2. Various shapes of calcium aluminate catalyst and support

rings, and 33mm 6-hole rings are 18kg, 16kg, and 140kg respectively. Small catalyst particles have attracted significant attention in recent years in applications such as (a) steam reforming where a small diameter tubular reactor is employed, (b) autothemal reforming and partial oxidation where large geometry surface area is required, (c) applications where less heat transfer in the catalyst bed by conduction are desired and (d) small reformer units (such as the ones providing syngas to fuel cells). The equivalent volume diameter and equivalent specific surface diameter of 6x6mm tablet, 5mm and 2mm sphere, 8x8x3 Raschig ring, and 8x8x2mm ribbed ring were calculated and compared in Table 2. The relative geometric surface area was calculated by estimating void fraction using $\varepsilon = 0.373 + 0.308\dfrac{D_p}{D_T}$ where D_p is the volume equivalent diameter of the particle

and D_T the diameter of the reactor tube. Relative pressure drop was estimated using the equation in M. Leva's [5] article. Compared to the 6x6mm tablet, the 8x8x2 mm ribbed ring causes only 32% of the pressure drop but provides 2% higher geometric surface area and the equivalent specific surface diameter is only 63%. The less complex Raschig ring retains 88% of the geometry surface and reduces pressure drop to about half. Small spheres offer larger geometric surface area at the cost of pressure drop. However, the small spheres are favorable in applications such as millisecond contacted time partial oxidization [2].

Table 2

Relative geometric surface area and pressure drop in a 2-inch tube

Catalyst Shape	Tablet	Sphere	Sphere	Ring	Ribbed
Catalyst Dimension (mm)	6x6	5	2	8x8x3	8x8x2
Equivalent Volume Diameter (mm)	6.9	5	2	8.7	8.0
Equivalent Specific SA Diameter (mm)	6.0	5	2	5.7	3.8
Relative Geometry Surface Area	1.0	1.2	3.2	0.88	1.02
Relative Pressure Drop	1.0	1.4	4.3	0.49	0.32

3.3 Catalyst Activity

Catalyst activities of Pt/CA_x, Rh/CA_x, Pd/CA_x (1.0wt%), Ir/CA_x (1.4wt%), Ni/CA_x (13.1wt%), and $Ni/\alpha-Al_2O_3$ (13.5 wt%) in steam reforming of CH_4 were plotted in Fig. 3. Equilibrium conversion has been reached on Ni/CA_x at *GHSV* 25,000 and as low as 460°C. The *GHSV* applied is about 10 times that of a typical commercial steam reformer while the temperature is several hundred degrees lower.

More data for higher *GHSV* and temperatures will be available in the future. At a much lower metal loading Rh/CA_x achieves equilibrium at a temperature slightly higher than on the Ni/CA_x catalyst (at about 480°C) indicating that CA_x is a good support for precious metals. Very limited information on CA_x supported precious metal catalysts is available in the literature. In addition, depending on the method by which the CA_x is made its catalytic properties can be quite different. The CA_x used in this study is commercially available for industrial applications. Activity of Ir/CA_x is lower than that of Rh/CA_x even at 40% higher metal loading. The activity of Pt/CA_x and Pd/CA_x is much lower than that of Ni/CA_x and Rh/CA_x. The activity of Pt/CA_x at 540°C and Pd/CA_x at 600°C is equivalent to that of Rh/CA_x at 420°C.

In steam reforming Ni/CA_x showed higher activity and coking resistance than $Ni/\alpha-Al_2O_3$. At 427°C Ni/CA_x is about 32% more active than $Ni/\alpha-Al_2O_3$. Part of the high activity of the Ni/CA_x catalysts can be attributed to larger pore volume and higher surface area. The pore volume and surface area of Ni/CA_x are 3 and 2 times those of $Ni//\alpha-Al_2O_3$ respectively. Nickel can be more finely dispersed on the CA_x support as confirmed by XRD, SEM, and H_2 chemisorptions [6]. Nickel oxide crystallite size on $\alpha-Al_2O_3$ was only slightly larger than that on CA_x. However, after NiO was reduced Ni crystallite size increased 130% on CA_x and 200% on $\alpha-Al_2O_3$. This is due to the strong interaction between Ni and $\alpha-Al_2O_3$ [6]. Carbon formation was detected at $H_2O/C = 5.5$ on $Ni/\alpha-Al_2O_3$ but not until $H_2O/C = 3.0$ on Ni/CA_x. Steam and CO_2 mixed reforming tests were carried out

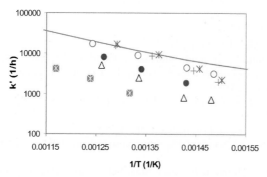

Fig. 3. Catalyst activity in steam reforming *: Ni, +: Ni/Al_2O_3; ○: Rh; •: Ir; Δ: Pt, ■: Pd

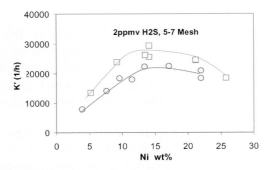

Fig. 4. Catalyst activity in mixed reforming □: CA_x, ○: Al_2O_3

on both Ni/CA_x and Ni/Al_2O_3 at various Ni loadings as illustrated in Fig. 4. Typically, the resulting H_2/CO ratio of the syngas was 1.6. No carbon deposition was found on the spent CA_x catalysts. Again, Ni/CA_x was more active than Ni/Al_2O_3 at similar Ni loading. Significantly higher active has been reported on one of the Ni/CA_x catalysts than a conventional catalyst in a commercial steam reforming application [7].

4. CONCLUSION

A variety of catalysts in various shapes have been developed based on CA_x for syngas generation processes. Depending on process conditions and reactor configrations, a suitable shape can be selected. Calcium aluminate has been applied as a catalyst support for both base metals and precious metals. Excellent catalytic activity has been demonstrated on both Ni/CA_x and Rh/CA_x in steam reforming of CH_4. Superior steam reforming activity, steam and CO_2 mixed reforming activity, carbon formation resistance of Ni/CA_x catalysts over Ni/Al_2O_3 catalysts have been realized. Syngas with a H_2/CO ratio as low as 1.6 has been generated using a Ni/CA_x catalyst without carbon formation.

REFERENCES

[1] M.E. Dry, Catalysis Today 71 (2002) 227.
[2] K.L. Hohn and L.D. Schmidt, Applied Catalysis A: General 211(2001) 53.
[3] R.E. Fisher (ed.), Advances in Refractories Technology (Ceramic Transactions 4), The American Ceramic Society, 1989.
[4] K. Shoji, R. Mogami, T. Numaguchi, T. Matsuhisa, H. Yanaru, Y. Nishioka, and Y. Izawa, US Petant 5,773,589.
[5] M. Leva, Chemical Engineering Progress 43(1947) 549.
[6] S. Zhao, 2003 AIChE Spring National Meeting, New Orleans USA, March 30-April 3, 2003.
[7] F. Watanabe, K. Shoji, T. Numaguchi, Stud. Surf. Sci. Catal. 145(2003) 501.

Studies in Surface Science and Catalysis, volume 147
X. Bao and Y. Xu (Editors)
©2004 Elsevier B.V. All rights reserved.

223

Monolith composite catalysts based on ceramometals for partial oxidation of hydrocarbons to synthesis gas

S. Pavlova[*], S. Tikhov, V. Sadykov, Y. Dyatlova, O. Snegurenko,
V. Rogov, Z. Vostrikov, I. Zolotarskii, V. Kuzmin, S. Tsybulya

Boreskov Institute of Catalysis, SB RAS, Pr. Lavrentieva, 5, 630090,
Novosibirsk, Russia

ABSTRACT

Microchannel ceramomet monoliths of a high thermal and mechanical stability have been synthesized via hydrothermal treatment of powdered mixtures containing aluminum and additives (oxides or/and Ni-Cr(Co) alloy) followed by calcination. The phase composition and textural properties of monoliths depend on the additive nature affecting the aluminum reactivity toward oxidation and the interaction of remaining Al^0 as well. Microchannel ceramomet monoliths and catalysts based on them ensure a high performance and stability in CPO of hydrocarbons at short contact times.

1. INTRODUCTION

The catalytic partial oxidation (CPO) of hydrocarbons (HC) at millisecond contact times is an attractive alternative to steam reforming for production of syngas in compact reactors [1]. Extremely short contact times determine the use of monolith catalysts having a low pressure drop. CPO can proceed indirectly via combustion of a part of HC to CO_2 and water followed by subsequent steam and dry reforming of remaining HC. These reactions would occur in two distinct zones of monolith: an inlet exothermic combustion zone where oxygen is present and a rear endothermic zone without oxygen. Monolithic catalysts based on traditional ceramic supports are destroyed as a result of a steep temperature gradient along a monolith. To prevent the temperature gradient, metallic high thermal conductivity monoliths of a foam or a honeycomb type can be used. However, preparation of metallic monolithic substrates is a rather demanding task,. Secondary supports and active components are usually loaded by using successive multistage procedures. Hence, the design of efficient and thermostable monolithic catalysts for CPO is an urgent problem.

Recently, composite catalysts based on Al_2O_3/Al porous ceramometals were shown to be promising systems for methane steam reforming [2]. For such systems, values of thermal conductivity coefficients (1-4 W/m·K) are intermediate between these of foam metals and conventional ceramic catalysts [3]. In this work, original approaches to design of efficient composite honeycomb monoliths based on ceramometals for CPO are considered.

2. EXPERIMENTAL

To synthesize composite ceramometal monoliths, the aluminum powder (a commerical PA-4 grade) as well as additives: powder of LaNiO$_x$ perovskite (prepared by a ceramic method or plasma thermolysis), CeO$_2$ (prepared by decomposition of nitrate) and Ni-Cr (20% of Ni) or Ni-Co-La (28 and 26% of Ni and Co, respectively, La - balance) alloys were used as raw materials. Two types of catalysts were prepared. The first type catalysts contain as an additive either mixed oxide (LaNiO$_x$) or alloy (Ni-Cr, Ni-Co). In the second one, CeO$_2$ as well as Ni-Cr alloy were used. The amount of an additive was 25-40 w.%. To form the transport microchannels along the monolith, easily burned organic fibers were inserted into the cermet matrix before hydrothermal treatment stage (HTT) by a proprietory procedure. The mixture of aluminium and additives loaded into the special die was subjected to HTT at 200°C followed by calcination in air at 900-1200°C. For some samples, to increase their catalytic performance, platinum and LaNiO$_3$ was introduced by the incipient wetness impregnation with corresponding solutions followed by drying and calcination in air at 900°C. As a reference sample, α-Al$_2$O$_3$ honeycomb monolith supported catalyst containing LaNiO$_x$ promoted with 0.5% Rh was used.

The phase composition was analyzed with URD-63 diffractometer using CuK$_\alpha$ radiation in the range 2θ = 10-70°. The specific surface area (SSA) was determined by the express BET method using Ar thermodesorption data. A pore volume and a mean pore size were estimated by the nitrogen adsorption isotherms measured at 77 K on a an ASAP-2400 installation.

The reducibility of the samples containing ceria and Ni-Cr alloy was characterized by H$_2$-TPR as in [4]. The testing of monolithic catalysts in CPO was carried out in the autothermal mode using a plug-flow quartz reactor and a feed containing 24% CH$_4$ in air. The contact time were was varied in the range of 0.14-0.07 sec. The inlet and outlet feed composition was analyzed by GC.

3. RESULTS AND DISCUSSION

3.1. The catalytic activity of catalysts based on LaNiO$_x$ or Ni-containing alloys.

Some characteristics of synthesized ceramometal monoliths are presented in Table 1. Mechanically strong monoliths (crushing strength 4-10 MPa/monolith) with diameters in the range of 20-50 mm having through microchannels are

Fig.1. Microchannel ceramometal monolith.

formed as a result of powdered component encapsulation into the Al_2O_3/Al matrix during HTT followed by calcination (Fig.1).

The catalytic data show a high activity of unpromoted monoliths containing $LaNiO_x$ or Ni-Cr(Co) alloys comparable with the activity of the conventional α-Al_2O_3 supported catalyst (Table 1). A lower starting temperature and a higher methane conversion and syngas selectivity are observed for the catalysts calcined at 900^0C containing lantanum nikelate prepared by the plasma thermolysis. Such an effective performance of these catalysts is due to the formation of highly dispersed metallic Ni due to perovskite reduction under the reaction conditions. However, after calcination at $1000-1100^0C$, the monolith activity decreases due to sintering, and even the sample promoted by $2\%Pt+CeZrO_x$ shows a poor activity (Table 1).

The catalysts based on Ni-Co and Ni-Cr alloys ensure a high level of methane conversion and syngas selectivity even after calcination at 1200^0C, but they have a high starting temperature that can be due to a low dispersion of

Table 1
Some characteristics of composite ceramometal honeycomb monoliths based on $LaNiO_x$ or Ni-containing alloys and their activity in POM.

Additive	Channels number per cm^2	Calcination temperature ^0C	Catalyst activity			
			Contact time, sec	Starting temperature ^0C	Xmeth., %	$CO+H_2$, Vol.%
LaNiRh[1]		900	0.14	400	83	44
$LaNiO_x$ (ceramic)	27	1000	0.08	530	88	46
$LaNiO_x$*	21	900	0.1	400	90	52
$LaNiO_x$*[,2]	21	1100	0.1	585	59	31
$LaNiO_x$*	53	900	0.13	465	94	54
			0.08		89	52
NiCo	50	1200	0.13	800	80	44
NiCo[2]	50	1200	0.14	635	87	44
NiCr	27	1200	0.11	700	80	45
NiCr[2]	27	1200	0.11	495	97	54

[1] - 0.5%Rh+8%$LaNiO_3$/α-Al_2O_3; * - perovskites were prepared by plasma thermolysis;
[2] - promoted by 2%Pt+$CeZrO_x$;

metallic Ni and Co. Addition of promoters improves the performance of these catalysts and decreases substantially the starting temperature. No cracking of monoliths or deterioration of their performance was observed after numerous start-ups of the reaction with the temperature excursions from ~ 400 to 1200^0C. The pressure drop across the monoliths was found to be comparable with that ensured by the foam-type substrates.

3.2. The catalysts based on Ni-Cr alloy and CeO_2.

The phase composition and textural properties of samples are presented in Table 2. In the X-ray diffraction patterns of the initial CeO_2 and Ni-Cr alloy, the reflections of CeO_2 correspond to cubic ceria oxide [JCPDS 43-1002], and those of initial Ni-Cr alloy correspond to metal Ni [JCPDS 04—0850]. The lines of Cr_2O_3 [JCPDS 38-1479] and NiO [JCPDS 44-1159] additionally appear after calcination of the alloy at 900°C. In the X-ray diffraction patterns of all monoliths, peaks corresponding to the 111, 200, 220 reflections of metal aluminum and the intense peaks corresponding to reflections typical for α-Al_2O_3 [JCPDS 05—712] are observed. The lines of Ni and NiO are observed in the patterns of samples containing Ni-Cr alloys. In accordance with earlier results [5], these data show that only a part of aluminum is oxidized, and mainly α-Al_2O_3 is formed in the course of HTT followed by calcination at 1100°C. Addition of CeO_2, Ni-Cr alloy or their mixture resulted in variation of the aluminum reactivity toward oxidation depending on CeO_2 concentration. CeO_2 decreases the metal aluminum reactivity and for the Ce_{40} sample the lowest intensity of α-Al_2O_3 reflections is observed. This phenomenon can be explained by hampering the process of aluminum oxidation by water due to interaction

Table 2

The characterictics of the ceramometal monoliths based on Ni-Cr alloy and CeO_2.

Sample	Initial chemical composition	Phase composition	SSA_{Ar} m²/g	SSA_{N2} m²/g	V_{pore} cm³/g	D_{pore} Å
CeO_2		CeO_2	83			
Ce_{40}	$CeO_2 - 40^1$	Al^0; α-Al_2O_3; γ-Al_2O_3 CeO_2	10	10	0.065	260
$Ce_{30}Ni_{10}$	$CeO_2 - 30$ Ni-Cr – 10	Al^0; α-Al_2O_3; γ-Al_2O_3 CeO_2; Ni^0; NiO; Al_3Ni	7.5	8	0.05	244
$Ce_{20}Ni_{20}$	$CeO_2 - 20$ Ni-Cr – 20	Al^0; α-Al_2O_3; γ-Al_2O_3; CeO_2; Ni^0, NiO; Al_3Ni	7	6	0.035	245
$Ce_{10}Ni_{30}$	$CeO_2 - 10$ Ni-Cr – 30	Al^0; α-Al_2O_3; γ-Al_2O_3; CeO_2; Ni^0; NiO; Al_3Ni; Al_3Ni_2; $AlCr_2$	4	3.5	0.022	252
Ni_{40}	Ni-Cr – 40	Al^0; α-Al_2O_3; γ-Al_2O_3; Ni^0; NiO; Al_3Ni; Al_3Ni_2; $AlCr_2$	1	1	0.004	156
Ni-Cr (900°)	Cr- 20 Ni -80	Ni^0; NiO; Cr_2O_3;	0.3	-	-	-

[1] - balance – Al^0 powder; component content – weight percents

of hydroxylated CeO_2 with the surface of the metallic aluminum particles and aluminum hydroxide layer generated by oxidation. In [5] the same effect was observed for TiO_2 as an additive. As a result, for different CeO_2 concentrations, at the same conditions of HTT, various amounts of aluminum hydroxide (α-Al_2O_3 after calcination) are formed and for the sample Ni_{40} without CeO_2, the metal aluminum oxidation is the most pronounced. In parallel with the different extents of the aluminum oxidation, different degrees of interaction of remaining metal aluminum with the Ni-Cr alloy are observed in the course of high temperature calcination. For $Ce_{30}Ni_{10}$ and $Ce_{20}Ni_{20}$, Al_3Ni alloy [JCPDS 02-0416] is only formed (Table 2). For $Ce_{10}Ni_{30}$ and Ni_{40} containing more than 20% Ni-Cr alloy, Al_3Ni, Al_3Ni_2 [JCPDS 14-0648] and $AlCr_2$ [JCPDS 29-0016] alloys are formed.

The specific surface area of samples, their pore volume and mean pore diameter increase with the CeO_2 content (Table 2). This effect can be due to a high initial dispersion of CeO_2 as well as due to stabilization of γ-alumina produced as a result of Al^0 oxidation by cerium [5].

TPR spectra of samples are presented in Fig. 2. In the spectra of Ce_{40} sample three peaks of hydrogen consumption at 390, 505 and 780°C are observed. Addition of Ni-Cr alloy lowers CeO_2 reduction temperature (samples $Ce_{30}Ni_{10}$, $Ce_{20}Ni_{20}$, $Ce_{10}Ni_{30}$). This can be due to Ni(NiO) ability to facilitate the reduction of CeO_2 as a result of the hydrogen spillover from Ni to oxide [6]. In the spectra of Ni_{40}, the low temperature intense peak at 460°C with a shoulder at 525°C can be assigned to the reduction of NiO [6] contained in the sample according to X-ray phase analysis (Table 2) whereas the peak at the 690°C can be assigned to the reduction of partially oxidized Ni-containing alloy. Some shifts of the low temperature peaks in the spectra of $Ce_{30}Ni_{10}$ and $Ce_{20}Ni_{20}$ are observed evidencing the appearance of highly reactive oxygen in these samples.

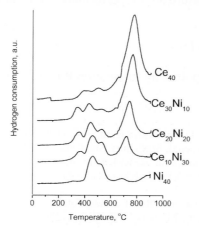

Fig. 2. TPR spectra of cermet monoliths based on CeO_2 and Ni-Cr-alloy.

228

Table 3

Parameters of POM in the autothermal mode over catalysts based on ceramometal monoliths containing CeO_2 and Ni-Cr alloy. 24% methane in air, contact time – 0.07 sec, inlet gas temperature - 430°C.

Catalyst*	T^1_{start} °C	T^2_{start} °C	$T_{monolith}$, °C		X_{CH4} %	$CO+H_2$, Vol.%
			inlet	Outlet		
C_{40}	490	450	1170	785	83	47
$C_{30}N_{10}$	470	400	1070	735	87	48
$C_{20}N_{20}$	460	420	1150	770	87	49
$C_{10}N_{30}$	470	420	1130	754	85	49
N_{40}	500	450	1154	780	83	48

- • - catalysts with supported $LaNiO_3$ promoted by 0.2%Pt.

The data on CPO activity of catalysts obtained by impregnation of cermet monoliths with La, Ni and Pt-containing solution is presented in Table 3. All catalysts show high methane conversions and syngas selectivities. However, the starting temperature depends on the composition of cermet monoliths: the samples containing both CeO_2 and NiCr alloy have low starting temperatures. In the second cycle of the reaction, the starting temperature is lowered due to prior reduction of supported lanthanum nickelate to nickel metal.

4. CONCLUSION

Microchannel ceramometal monoliths of a high thermal and mechanical stability have been synthesized via hydrothermal treatment of powdered mixtures containing aluminum and additives (oxides or/and Ni-Cr(Co) alloy) followed by calcination. The phase composition and textural properties of monoliths depend on the nature of additive affecting the aluminum reactivity toward oxidation and the interaction of remaining Al^0. Microchannel ceramometal monoliths and catalysts based on them ensure a high performance and stability in CPO at short contact times.

ACKNOWLEDGEMENTS

This work was in part supported by Integration Project 39 of SB RAS.

REFERENCES

[1] P.M Torniainen, X. Chu, L.D. Schmidt, J.Catal., 146 (1994) 1.
[2] S.F. Tikhov, V.A. Sadykov, et al, Stud. Surf. Sci. Catal. V. 136 (2001) 105.
[3] L.I. Kuznetsova, V.N. Ananin, et al, React.Kinet.Catal.Lett. V. 43 (1991) 545.
[4] A.V. Simakov, S.N. Pavlova, V.A. Sadykov et al, Chemistry for Sustainable Development, 11 (2003) 263.
[5] S.F. Tikhov, Yu.V. Potapova, et al, React. Kinet. Catall. Lett., V.77 (2002) 267.
[6] T. Takeguchi, S. Furukawa and M. Inoue, J. Catal. 202 (2001) 14.

Studies in Surface Science and Catalysis, volume 147
X. Bao and Y. Xu (Editors)

The relative importance of CO_2 reforming during the catalytic partial oxidation of CH_4-CO_2 mixtures

N.R. Burke[*] and D.L. Trimm

CSIRO Petroleum, Ian Wark Laboratory, Bayview Ave, Clayton, Victoria, 3168, Australia

ABSTRACT

Carbon dioxide (CO_2) reforming has been found to contribute significantly to the production of synthesis gas by the partial oxidation of mixtures of methane and CO_2 over supported rhodium catalysts. Under partial oxidation conditions, up to an additional 11% conversion of methane was obtained on the addition of CO_2 as compared to reaction without CO_2. The catalyst was found to restructure under operating conditions, but this had no effect on activity or selectivity.

1. INTRODUCTION

The production of synthesis gas (syngas) from natural gas is an important industrial reaction that has been extensively studied [1, 2]. Various processes based on steam reforming, dry reforming and partial oxidation have been developed [1].

The industrial development of dry reforming has been limited, despite the fact that some gas fields contain significant amounts of CO_2. However, a large scale development is under construction for methanol manufacture [3].

Both steam and dry reforming are endothermic reactions, and the combination with exothermic partial oxidation is an obvious route to lower energy production of syngas. Ashcroft et al [4] showed that supported precious metal systems could catalyse both reactions, and several alternative systems have since been explored [5-7]. Partial oxidation is known to be a very fast reaction, carried out at high temperatures (ca 1000°C) and short residence times (ca 10 ms) [8]. Dry and steam reforming, on the other hand are traditionally much slower reactions, involving temperatures of the order of 800°C and residence times of the order of 1s [1]. The question then arises as to the importance of reforming reactions under partial oxidation conditions.

Ruckenstein and Wang [6] report that the H_2:CO ratio obtained from combined partial oxidation and dry reforming depends on the O_2:CO_2 ratio in

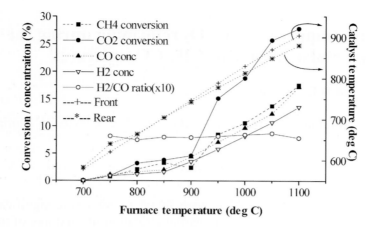

Fig. 1: CH$_4$ dry reforming (1:1 CH$_4$/CO$_2$ ratio)

the feed. This infers an equal probability of reaction and supports the suggestion of Rostrup Nielsen et al [1] that the products will reflect thermodynamic equilibria.

Mattos et al [9] on the other hand, find steam and dry reforming to be important only after O$_2$ depletion, albeit at lower temperatures and longer contact times.

Product spectra obtained from very fast reactions can reflect kinetics as much as thermodynamics and this may well account for the findings of the various authors. As a result, the present study was carried out to establish the importance of dry reforming during the very fast partial oxidation of mixtures of CH$_4$ and carbon dioxide.

2. EXPERIMENTAL

The experimental study focused on a 5% Rh catalyst supported on a γ-alumina (Norton) impregnated α-alumina blown foam monolith (Vesuvius, Australia) mounted in a flow reactor. The cylindrical catalyst was 10mm long by 17mm wide. The catalyst was prepared by coating the monolith with a 3% slurry of γ-alumina, calcining the resultant material at 600°C in air for 6 hours then impregnating the coated monolith with the nitrate salt of rhodium (Engelhard) and again calcining for 6 hours at 500°C.

The catalyst was held in a quartz-lined steel tubular reactor which could be heated electrically to ca 1300K as required. Gas flows were controlled by Brooks mass flow controllers and products were analysed using a Varian 3400 gas chromatograph equipped with a thermal conductivity detector.

Gas mixtures containing 1:1 mixtures of CH_4 and CO_2 and 20% N_2 internal standard (total gas flow of 5L/min) were used to investigate the effect of temperature on dry reforming over the Rh catalyst under conditions normally associated with catalytic partial oxidation (GHSV=270000h^{-1}).

A standard partial oxidation gas mixture of 20% N_2 with a $CH_4:O_2$ ratio of 1.8 was used as a benchmark reaction. Once initiated by the furnace, this reaction was autothermal. Stepwise substitutions of O_2 with either N_2 or CO_2 allowed determination of the impact of dry reforming on syngas production. The furnace was used to ensure catalyst temperatures were the same for N_2 and CO_2 addition testing allowing more accurate estimation of the effect of CO_2.

All conversion and gas concentration results had errors of less than 3%.

Catalysts were characterized prior to and after use by SEM, XRD and total surface area.

3. RESULTS AND DISCUSSION

Initial experiments were based on partial oxidation of CH_4 ($CH_4:O_2$ ratio of 1.8). Partial oxidation at a contact time of 15ms (GHSV=270000h^{-1}) gave CH_4 conversions of 83% and selectivity to syngas of 87%. These values were in accordance with those reported in the literature [8, 10].

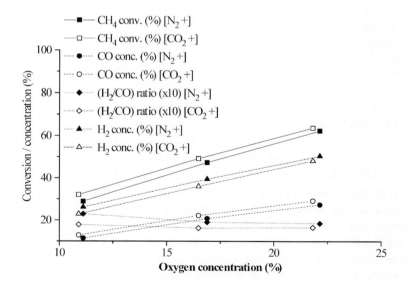

Fig. 2: Combined partial oxidation/dry reforming N_2 and CO_2 replacement experiments. CO_2 addition – open symbols; N_2 addition - closed symbols.

Attention was then focused on dry reforming of CH_4 over the catalyst, the results of which are summarised in Fig. (1). It was necessary to heat the reactor, since only endothermic dry reforming was under investigation. Even at the short contact times normally associated with partial oxidation, some CH_4 was converted to give a H_2:CO ratio of ca 1. At the highest temperatures, more CH_4 was converted than the value corresponding to syngas, indicating that some pyrolysis was occurring [11]. Under the most extreme conditions only ca 17% of CH_4 and ca 30% of CO_2 was consumed, and the temperature drop across the catalyst, caused by the endothermic reactions, was small (ca 25°C).

More CO was produced in the temperature range 900-1100°C (furnace temperature) than H_2, and the H_2:CO ratio fell slightly as temperature increased. The rate of CO_2 reforming increases with temperature, but this should not affect the H_2:CO ratio. However, the reverse water gas shift (WGS) reaction became more significant with increasing temperature, leading to more CO and less H_2.

The relative rate increase in CH_4 conversion with temperature was about the same as that reported by Yokota et al [12], but much smaller than that expected on extrapolation of other results [13].

Attention was then focused on combined partial oxidation and dry reforming (Fig. (2)). Reaction was initiated under partial oxidation conditions using CH_4:O_2 ratios of 1.8 and higher and replacing the removed O_2 with either N_2 or CO_2. The temperature inside the catalyst bed decreased as O_2 decreased, as did CH_4 conversion and syngas production. However, even at CH_4:O_2 ratios as high as 6 the reaction was self-sustaining. At higher CH_4:O_2 ratios the heat generated in the consumption of O_2, whether by combustion or partial oxidation, was insufficient to sustain reaction. Other workers [7] have reported a similar syngas/temperature effect. The equilibrium temperatures for equivalent O_2 contents were slightly higher for the N_2 addition experiments. To compensate for these differences temperature was controlled at the higher of the two equilibrium temperatures, allowing more accurate comparisons.

The system in which CO_2 replaced O_2 (open symbols in Fig. (2)) always gave higher CH_4 conversion and CO concentrations than the equivalent system containing extra N_2 (closed symbols in Fig. (2)). This shows that CO_2 reforming plays a significant part in syngas production under the conditions employed. The contribution of dry reforming can be estimated from the difference between the N_2 addition and CO_2 addition experiments (Table (1)). The relative significance of dry reforming is seen to increase as the O_2 concentration decreases. This is not unexpected but the result also reflects the fact that temperature decreases as O_2 decreases. Differences in the activation energies of the two reactions would contribute to the observed result. H_2 concentration in the absence of CO_2 was higher than when CO_2 was added as a reactant. This can be explained by the difference in the H_2/CO ratio of partial oxidation (ratio of 2) and dry reforming (ratio of 1). The H_2/CO ratios for N_2 and CO_2 addition are shown in Fig. (2). The CO_2 addition experiments have a lower ratio for all conditions tested.

One unexpected result was the difference between dry reforming in the absence (very small conversion) and presence (significantly higher conversion) of O_2. The most satisfactory explanation is based on local heating of the catalyst. Thermodynamic calculations showed that equilibrium conditions were not reached under the conditions of dry reforming alone [13]. When oxygen was added, the temperature of the catalyst rose, and dry reforming became more important. These results were consistent with reforming occurring in the catalyst bed after partial oxidation had depleted the methane, but the absence of accurate temperature measurements through the catalyst does not allow exact findings.

Fresh and used catalysts were examined using SEM, XRD and N_2 adsorption surface area analysis. XRD showed little difference between the fresh and used catalyst, but surface areas of the used catalysts were some 30% lower than those of the fresh catalysts. A dramatic difference between fresh and used catalysts can be seen from the SEM images (Fig. (3)). The fresh catalyst has a very even coating of Rh (brightest material), whereas the used catalyst contained Rh distributed in smaller, discrete particles. It was suspected that evaporation and recondensation of Rh resulted in this change in morphology. The melting point of Rh is 1964^0C and sintering and migration have been reported to take place at temperatures $\frac{1}{3}$ to $\frac{1}{2}$ of the melting temperature of some metals. Thus it is not unreasonable to expect temperature activated sintering of Rh under conditions of partial oxidation. Careful adjustment of the $CH_4:O_2$ ratio and the $CH_4:CO_2$ ratio for feeds of CH_4, CO_2 and O_2 could make use of the exothermicity of the partial oxidation reaction and the endothermicity of the reforming reaction and could minimise hot spots and prevent loss of activity due to sintering.

A small amount of carbon was formed on the walls of the reactor tube. None was seen on the catalyst itself, in agreement with previous work [1,12,13] The lack of carbon on the catalyst suggested that carbon was formed in the gas phase rather than on the catalyst.

4. CONCLUSION

The importance of dry reforming during partial oxidation of methane has been estimated. Improvements of up to 11% CH_4 conversion and 13% CO production were seen on addition of CO_2 to the system. Lower concentrations of H_2 on addition of CO_2 was evidence of dry reforming.

There was evidence of sintering on the catalyst. Sintering was probably the result of the exothermic partial oxidation and/or combustion reactions. Careful control of the $CH_4:O_2$ ratio in the feed gases can allow autothermal operation and minimize deactivation caused by hot spots.

Table 1.
Improvements in CH_4 conversion and syngas production in combined partial oxidation / dry reforming tests.S

Conditions				Relative improvements on CO_2 addn.		
O_2 (%)	CH_4 (%)	Added N_2 or CO_2 (%)*	T (°C)	CH_4 conv. (%)	H_2 (%)[+]	CO (%)
11.0	49.2	19.6	725	11	-13	13
16.5	49.2	14.6	725	4	-9	7
22.0	49.2	9.1	725	2	-4	7

*: 20% N_2 plus added N_2 or CO_2.
[+]: negative numbers indicate lower concentrations with CO_2 addition.

Fig. 3: a) Fresh Rh/Al_2O_3 catalyst; b) Used Rh/Al_2O_3 catalyst

REFERENCES

[1] J.R. Rostrup-Nielsen, J. Sehested and J.K. Nørskov, Adv. Catal., 47 (2002) 65.
[2] M.V. Twigg, (ed), Catalyst Handbook, Wolfe, London, 2002.
[3] H. Holm-Larsen, Studies in surface science and catalysis, 136 (2001) 441.
[4] A.T. Ashcroft, A.K. Cheetham, M.L. Green and P.D.F. Vernon, Nature, 352 (1991) 225.
[5] V.R. Choudhary, B.S. Uphade and A.A. Belhekar, J Catal, 163 (1996) 312.
[6] E. Ruckenstein and H.Y. Wang, Catal. Lett.,73 (2001) 99.
[7] A.M. O'Connor and J.R.H. Ross, Catal. Today, 46 (1998) 203.
[8] D.A. Hickman and L.D. Schmidt, J Catal, 138 (1992) 267.
[9] L.V. Mattos, E.R. de Olivera, P.D. Resende, F.B. Noronha and F.B. Passos, Catal. Today, 77 (2002) 245.
[10] L. Basini, K. Aasberg-Peterson, A. Guarinoni, M. Ostberg, Catal Today, 64 (2001) 9.
[11] A. Holmen, O. Olsvik and O.A. Rokstado, Fuel Processing Technology, 42 (1995) 249.
[12] S. Yokota, K. Okumura and M. Niwa, Catal. Lett., 84 (2002) 131.
[13] J.R. Rostrup-Nielsen and J-H. Bak Hansen, J. Catal., 144 (1993) 38.
[14] O. Deuschmann and L.D. Schmidt, AIChE journal, 44 (1998) 2465.

Studies in Surface Science and Catalysis, volume 147
X. Bao and Y. Xu (Editors)
235

Comparison of microkinetics and Langmuir-Hinshelwood models of the partial oxidation of methane to synthesis gas

Xueliang Zhao, Fanxing Li and Dezheng Wang[*]

Department of Chemical Engineering, Tsinghua University, Beijing 100084, China (email: wangdz@flotu.org; fax: ++86 10 62772051)

ABSTRACT

The reactions of methane and O_2, H_2O, CO_2 comprise a multiple reaction system. In modeling by a Langmuir-Hinshelwood model, the judicious choice of reactions is important. In modeling by microkinetics, many kinetic parameters are needed and one should use information from many independent sources.

1. INTRODUCTION

There is an increasing tendency to use elementary step kinetics to describe the reactions of methane and O_2, H_2O, CO_2, etc. because of increasing computing power. Although this offers opportunities to analyze catalysts in their working state, in contrast to Langmuir-Hinshelwood (LH) kinetics where the state of the catalyst is embedded in adsorption isotherms, LH kinetics will still be used, and may be in the majority in chemical reaction engineering, because LH kinetics is more familiar and tried, and commercial softwares usually need a special interface to handle the large reaction networks of microkinetics.

In this work, LH [1,2] and microkinetics [3] models of the partial oxidation of methane (POM) to synthesis gas are compared. The kinetics models used in the literature are analyzed to consider their capabilities to faithfully capture important kinetic features. LH models are based on some assumptions that are not well supported by present research results, and some LH models have used an adjustable parameter to make them agree with many of the reaction characteristics. It is useful to understand the utility and limits of the LH models which offer the advantage of ease of calculations, but may lead to errors under some conditions and can be difficult to implement in complex reaction systems.

2. EXPERIMENTAL

The kinetic models were simulated in a one-dimensional plug flow reactor (PFR) model. The microkinetics of POM are performed in monolithic reactors [3], but the simpler model is chosen to give a more direct comparison of the microkinetics and LH models since the interest here is the comparison of the intrinsic kinetic behavior. When possible, comparisons are made to other experimental results, from conventional reactors, a TAP reactor previously described [4], and surface science experiments [5].

2.1. Reactor equations

Methane oxidation is volume changing and its LH modeling is easier with flow rates as the variables of the reactor equations, which for a PFR are:

$$\frac{dF_i}{dV} = \sum_i r_i$$; F_i is the molar flow rate of gas species i, dV is a volume element

of the reactor and $\sum r_i$ sums over all reactions involving species i.

$$\frac{dT}{dV} = \frac{\sum_i r_i \Delta H_{rxn,i}}{C_i * Cp_i}$$; T is reactor temperature, $\Delta H_{rxn,i}$ is reaction heat i, C_i is the

concentration and $C_{p,i}$ is the heat capacity of gas species i. The reactions are:

$$CH_4 + 2\,O_2 \longrightarrow CO_2 + 2\,H_2O \qquad r_1 = \frac{k_1 x_{CH_4} x_{O_2}}{(1 + K^c_{CH_4} x_{CH_4} + K^c_{O_2} x_{O_2})^2} + \frac{k_1 x_{CH_4} x_{O_2}^{0.5}}{(1 + K^c_{CH_4} x_{CH_4} + K^c_{O_2} x_{O_2})}$$

$$CH_4 + H_2O \rightleftharpoons CO + 3\,H_2 \qquad r_2 = \frac{k_3 / p_{H_2}^{2.5}(p_{CH_4} p_{H_2O} - p_{H_2}^3 p_{CO} / K_{eq,3})}{(1 + K_{CO} p_{CO} + K_{H_2} p_{H_2} + K_{CH_4} p_{CH_4} + K_{H_2O} p_{H_2O} / p_{H_2})^2}$$

$$CH_4 + 2\,H_2O \rightleftharpoons CO_2 + 2\,H_2 \qquad r_3 = \frac{k_4 / p_{H_2}(p_{CO} p_{H_2O} - p_{H_2} p_{CO_2} / K_{eq,4})}{(1 + K_{CO} p_{CO} + K_{H_2} p_{H_2} + K_{CH_4} p_{CH_4} + K_{H_2O} p_{H_2O} / p_{H_2})^2}$$

$$CO + H_2O \rightleftharpoons CO_2 + H_2 \qquad r_4 = \frac{k_5 / p_{H_2}^{3.5}(p_{CH_4} p_{H_2O}^2 - p_{H_2}^4 p_{CO_2} / K_{eq,5})}{(1 + K_{CO} p_{CO} + K_{H_2} p_{H_2} + K_{CH_4} p_{CH_4} + K_{H_2O} p_{H_2O} / p_{H_2})^2}$$

The rate expressions most often used in the literature are those of Xu and Froment and Trimm and Lam [1,2]. These are also used here to facilitate a comparison with other workers. Most workers use efficiency factors too, but this is not done here because the aim is to look at chemical trends, i.e., to compare and assess whether there is accurate modeling of the intrinsic reaction chemistry by the different rate expressions rather than the accurate modeling of reactors.

Microkinetics differs from LH kinetics in the explicit accounting of gas and solid phases, in contrast to the LH pseudo-homogeneous rate expressions. This results in partial differential equations (PDEs) as the reactor equations. Some workers apply the steady state hypothesis to *all* surface species to simplify these to ordinary differential equations (ODEs). This requires that all these species are short-lived and in low concentrations, but there are kinetic regimes in POM where this is not so. The reactor equations for a PFR with microkinetics are:

$$\frac{\partial C_i^{surface}}{\partial t} = \sum_i r_i \; ; \; C_i^{surface} \text{ is the concentration of surface species i.}$$

$$\frac{\partial C_i^{gas}}{\partial t} + \frac{\partial (v \cdot C_i^{gas})}{\partial x} = \sum_i r_i \; ; \; v \text{ is gas velocity, which is not taken out of the}$$

differential operator because of a volume changing reaction. $\sum r_i$ sums over all sources/sinks of species i. C_i^{gas} is the concentration of gas i. Our $\sum r_i$ uses adsorption/desorption terms as the source/sink terms for gas species, which differs from Schmidt and coworkers [6,3] who used a mass transfer term.

$$\sum_i C_i c_{p,i} \left(\frac{\partial T}{\partial t} + \frac{\partial (v \cdot T)}{\partial x} \right) = \sum_i r_i \Delta H_{rxn,i} \text{ is the energy equation (adiabatic reactor).}$$

The elementary reactions and kinetic parameters are the same as in the work of Schwiedernoch et al. [3] except that methane desorption is assumed not to occur. These and the kinetic parameters of the LH model above are in the cited references.

The microkinetic parameters vary over orders of magnitude and an explicit numerical difference scheme was unstable. This indicates the need to use an algorithm designed for stiff equations, but these are well developed for ODEs. Thus, the PDEs were solved by the method of lines [7] with the spatial variable discretized to give a set of ODEs. The resulting stiff ODEs were solved by the semi-implicit extrapolation technique of Bader and Deuflhard [8] (B-D) using codes by Press et al. [9]. For the equations of the microkinetics model, the step size had to be $< 4 \times 10^{-6}$ s to ensure stability. The LH kinetics was not a stiff set, but when using explicit differencing schemes, too large a step size can give negative concentrations and the B-D implicit scheme gave better performance.

3. RESULTS AND DISCUSSION

For a general discussion, it is assumed that the reaction mechanism is the same over group 8-10 metals. The literature on methane oxidation has been reviewed by Pena et al. [10]. For this discussion, two reaction features are of importance because of the failure of the LH kinetics to show them: (1) under fuel lean conditions, the excess oxygen results in very high selectivity to H_2O and CO_2

[11,12], and (2) under POM conditions in PFRs, the temperature rises to a maximum near the reactor inlet before decreasing [13,14]. Concerning the first feature, the extrapolation of LH kinetics to beyond the narrow conditions under which they were developed, i.e., the use of conditions that include fuel lean conditions is in order to show the *reason(s)* why the kinetic expressions failed to capture important kinetic features, i.e, the extrapolations were *not* made to argue (trivially) that the LH expressions are incorrect in the extrapolated regions.

Fig. 1 shows simulated selectivity and temperatures as a function of %O_2 in the (O_2+CH_4) feed using LH kinetics used by Froment and coworkers [15], Mirodatos and coworkers [16], Avci et al. [17] and de Smet et al. [18]. Many research groups who used this kinetics indicated an important difficulty with LH kinetics, viz., developing reliable kinetics is very work-intensive and many workers choose to use published data. The O_2-rich feed region in Fig. 1 shows a failure of the LH kinetics. The selectivities shown are %H_2 in (H_2+H_2O) and %CO in (CO+CO_2). These show that in the O_2-rich feed region, the reaction selectivity is towards H_2 and CO rather than H_2O and CO_2. This is due to the

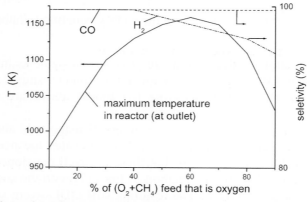

Fig. 1. Selectivity and maximum temperature (at reactor outlet) in the LH model.

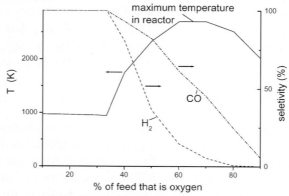

Fig. 2. Selectivity and maximum temperature (at reactor outlet) in the microkinetics model.

presence of steam reforming (r_2) in the reaction set of LH kinetics as a very fast reaction. Its presence reacts H_2O into H_2 and CO, and the presence of water gas shift (r_3) and large amounts of H_2 shifts CO_2 to CO. This selectivity is incorrect. In excess O_2, oxidation reactions dominate to give H_2O and CO_2. It might be argued that LH kinetics developed in a limited range should not to be used with excess oxygen, but there is a situation of excess oxygen on the catalyst even when the inlet gas is fuel rich which affects reactor temperature. To the *catalyst*, the much faster adsorption of O_2 makes it seem as if the gas phase is O_2-rich.

Fig. 1 also shows, incorrectly, that the reactor temperature does not get very high, and nor is its maximum near the inlet but it is at the outlet. These features of the simulation were also reported by Avci et al. [17]. Fig. 2 from the microkinetics simulation shows that the temperature can get very high with excess oxygen, in principle. In these simulations, an inlet temperature of 900 K and an adiabatic reactor were used. De Groote and Froment [15] forced simulation results into closer agreement with experiments by an adjustable parameter ("degree of reduction") that limits steam reforming and water gas shift rates near the reactor inlet. We do not use this parameter to flash out that the fundamental problem is a very fast steam reforming, which is endothermic, near the reactor inlet. Conventional and TAP reactor experiments [12,19,4] have shown a kinetics regime where the reaction order in oxygen is zero or equivalently that oxygen adsorption is much faster than methane adsorption, which implies adsorbed oxygen is in excess. When this is the situation near the reactor inlet, but the LH kinetics regardless includes a fast steam reforming, this LH modeling will not be able to give total oxidation of methane and its very large contribution of reaction heat, and the simulated temperature is not high.

We conclude that the choice of reactions in the LH model of a multiple reaction system is important. The problem can be solved by including CO and H_2 oxidation as even faster reactions. But this defeats the utility of the LH model which main advantage is a small number of kinetic parameters. It would also be better for the kinetic parameters for steam reforming to be determined *in situ* under POM conditions, but it is not clear how this can be done.

The microkinetics parameters of Schwiedernoch et al. [3] gave the correct trends of high CO_2 and H_2O selectivity and a high reactor temperature under O_2-rich conditions (Fig. 2). The details are not given here, but we comment that this is due to the subsequent gas-phase oxidation of CO and H_2, i.e., with these parameters, the main reaction pathway for POM is the direct partial oxidation pathway. This is not surprising as these parameters are descendent from Schmidt and coworkers whose reaction systems proceed mainly by the direct partial oxidation pathway [20]. The unsatisfactory aspect of this is that microkinetics parameters are claimed to be valid for wider ranges of conditions, and thus, they should show the oxidation-reforming pathway under kinetic regimes where experiments show its existence. This can be resolved by using information from many independent sources, which indicates that some of these parameters need revision: (1) the sticking probability of oxygen used is small but much evidence indicate that oxygen adsorption is very fast [12,19,21], (2) the sticking probability of methane used is too large where evidence exist that it is slow at lower temperatures and it is activated [5,22], and (3) CO oxidation is not very much slower than CO desorption [4,19].

4. CONCLUSION

The LH model of POM is incorrect when used for an O_2-excess catalyst. It gives temperatures that are too low, and needs to add CO and H_2 oxidation reactions. The microkinetics model needs revision to give the oxidation-reforming mechanism as the main pathway under some reaction conditions.

ACKNOWLEDGEMENTS

Work funded by the Natural Science Foundation of China (grant 20076045) and the Ministry of Science and Technology (G1999022408).

REFERENCES

[1] J. Xu and G.F. Froment, AIChE J. 35 (1989) 88.
[2] D.L. Trimm and C-W. Lam, Chem. Eng. Sci., 35 (1980) 1504.
[3] R.Schwiedernoch, S.Tischer, C.Correa, O.Deutschmann, Chem. Eng. Sci. 58 (2003) 633.
[4] DZ.Wang, ZL.Li, CR.Luo, WZ.Weng, HL.Wan, Chem. Eng. Sci. 58 (2003) 887.
[5] J.D. Beckerle, Q.Y. Yang, A.D. Johnson and S.T. Ceyer, J. Chem. Phys. 86 (1987) 7236.
[6] O. Deutschmann and L.D.Schmidt, AIChE, J. 44 (1998) 2465.
[7] W.E. Schiesser, The Numerical Method of Lines, Academic Press, San Diego, 1991.
[8] G. Bader and P. Deuflhard, Numerische Mathematik, 41 (1983) 373.
[9] W.H. Press, S.A. Teukolsky, W.T. Vetterling and B.P. Flannery, Numerical Recipes in C (2nd ed). Cambridge Univ. Press, Cambridge, 1992.
[10] M.A. Pena, J.P. Gomez and J.L.G. Fierro, Appl. Catal. A, 144 (1996) 7.
[11] M. Lyubovsky and L. Pferfferle, Catal. Today 47 (1999) 29.
[12] (a) D. Wang, F. Monnet and C. Mirodatos, Studies in Surface Sci. and Catal. 130 (2000) 3561. (b) D. Wang, O. Dewaele, A. Degroot and G.F. Froment, J. Catal. 159 (1996) 418.
[13] Pettre, M., Eichner, C.H. and Perrin, H., Trans. Far. Soc. 23 (1946) 257.
[14] D.Dissanayake, M.P.Rosynek, K.C.C.Kharas and J.H.Lunsford, J. Catal. 132 (1991) 117.
[15] A.M.Degroote and G.F. Froment, Appl. Catal. A, 138 (1996) 245.
[16] T.Ostrowski, A.Giroir-Fendler, C.Mirodatos and L.Mleczko, Catal. Today 40 (1998) 181.
[17] A.K. Avci, D.L. Trimm and Z.I. Önsan, Chem. Eng. Sci. 56 (2001) 641.
[18] C.R.H. de Smet, M.H.J.M. de Croon, R.J. Burger, G.B. Marin and J.C. Schouten, Chem. Eng. Sci. 56 (2001) 4849.
[19] M. Soick, D. Wolf and M. Baerns, Chem. Eng. Sci. 55 (2000) 2875.
[20] D.A. Hickman and L.D. Schmidt, A.I.Ch.E., J. 39 (1993) 1164.
[21] D.F. Padowitz and S.J. Sibener, Surf. Sci. 254 (1991) 125.
[22] D. Wang, O. Dewaele and G.F. Froment, J. Mol. Catal. A 136 (1998) 301.

Studies in Surface Science and Catalysis, volume 147
X. Bao and Y. Xu (Editors)
©2004 Elsevier B.V. All rights reserved.

Methane selective oxidation into syngas by the lattice oxygen in ceria-based solid electrolytes promoted by Pt

V. Sadykov[a], V. Lunin[b], T. Kuznetsova[a], G. Alikina[a], A. Lukashevich[a],
Yu. Potapova[a,f], V. Muzykantov[a], S. Veniaminov[a], V. Rogov[a], V. Kriventsov[a],
D. Kochubei[a], E. Moroz[a], D. Zuzin[a], V. Zaikovskii[a], V. Kolomiichuk[a],
E. Paukshtis[a], E. Burgina[a], V. Zyryanov[c], S. Neophytides[d], E. Kemnitz[e]

[a]Boreskov Institute of Catalysis, Lavrentieva, 5, Novosibirsk, 630090, Russia
[b]Chemical Department of Moscow State University, Moscow, Russia
[c]Institute of Solid State Chemistry SB RAS Novosibirsk, Russia
[d]Institute of Chemical Engineering, Patras, Greece
[e]Institute for Chemistry, Humboldt- University, Berlin, Germany
[f]Samsung Advanced Institute of Technology, Suwon, Republic of Korea

ABSTRACT

Nanocomposites based upon ceria doped by Zr, Zr+La or Sm with supported Pt nanoparticles efficiently convert methane into syngas by their lattice oxygen. In red-ox cycles with pure methane as reagent, the surface carbon build-up is observed, which is lower for Sm -doped samples possessing a higher surface and lattice oxygen mobility. When stoichiometric amounts of CO_2 or H_2O are present in the feed, the catalysts efficiently operate in methane steam and dry reforming at high space velocities without deactivation.

1. INTRODUCTION

Catalytic oxidation of methane into syngas by oxygen transferred through the membranes comprised of complex oxides with mixed ionic-electronic conductivity, or by the lattice oxygen of ceria-based mixed oxides reoxidized by air is now considered as a viable option in the natural gas conversion technologies. In both cases, the oxygen mobility and methane activation are important factors for those processes.

This paper presents results of work aimed at increasing the oxygen mobility in the lattice of ceria by incorporating dopant cations (Zr, La, Sm). To increase the rate of methane dissociation, Pt was supported. Since both in cyclic red-ox processes and methane selective oxidation in membrane reactors steam and dry reforming could occur concomitantly, effects of CO_2 and H_2O addition on the methane transformation are considered as well.

2. EXPERIMENTAL

Dispersed samples of ceria-based solid solutions were prepared by the complex polymerized method (PCM) and annealed at 500-700 °C [2]. Samples are denoted as CeZr - 50 mol.% CeO_2-50 mol.% ZrO_2, CeZrLa$_x$ - Ce$_{0.5-x/2}$Zr$_{0.5-x/2}$La$_x$, x=0.1-0.3, CeSm$_y$ - Ce$_{1-y}$Sm$_y$, y=0.05, 0.1, 0.15, 0.2, 0.4. Pt (1.4 wt.%) was supported by the incipient wetness impregnation from the water solutions of H_2PtCl_6 followed by drying and air calcination at 500 °C.

Samples were characterized by XRD, X-ray radial electronic distribution (RED) and EXAFS using a synchrotron radiation, TEM, FTIRS of lattice modes and adsorbed CO. Specific surface area estimated by BET routine varied in the range of 20-70 m^2/g. Oxygen mobility was estimated by using the isotope exchange both in dynamic and isothermal conditions [3]. Methane oxidation by the lattice oxygen (feed 1% CH_4 in He) and its steam and dry reforming (feeds 1% CH_4 + 1% H_2O or 1% CO_2 in He) were characterized in isothermal and TPR regimes up to 900 °C. Details of the experimental procedures are given in [1-3].

3. RESULTS AND DISCUSSION

3.1. Bulk structure

Ceria prepared by PCM is characterized by the lattice parameter **a**=5.411 Å reasonably agreeing with the well-know data [4]. In the wide range of Ce substitution (up to 50 mol.% Zr, up to 30 mol. % La or 40 mol.% Sm), ceria-based solutions maintain the fluorite structure. For CeZr sample the lattice parameter decreases up to 5.282 Å, while for CeSm$_{0.4}$ it increases up to 5.453 Å due to a smaller Zr^{4+} and larger Sm^{3+} radius, respectively [4]. Addition of a larger La cation to CeZr increases the lattice parameter of a cubic phase (**a**=5.331 Å for CeZrLa$_{0.1}$), but two solid solutions based upon the fluorite structure (**a**=5.309 Å and 5.440 Å for CeZrLa$_{0.2}$; **a**=5.344 Å and 5.479 Å for CeZrLa$_{0.3}$) probably differing by the La content are formed at 20-30 mol.% La. For all samples XRD particle sizes are in the range of 70-300 Å being far below mean TEM sizes (~ 1000 Å) of non-porous platelet-like particles. This suggests the domain structure of those platelets. Indeed, high-resolution TEM revealed particles to be comprised of domains with sizes ~ 50-100 Å coherently stacked in doped samples but having more disordered boundaries in pure ceria.

Broadening and splitting of IR bands observed even for pure ceria (Fig.1a) imply some distortion of the coordination polyhedra leading to appearance of differing Me-O distances. Indeed, EXAFS and RED data directly reveal such a distortion in the structure of ceria: two Ce-O distances - 2.24 Å (coordination number CN 3.0) and 2.39 Å (CN 5.7) as well as two Ce-Ce distances in the range of 3.84-3.86 Å (CN 8.0-8.5) are observed. For $CeSm_{0.2}$ the splitting of IR bands becomes less visible. Zr addition shifts the IR band maximum to higher frequencies due to the lattice contraction. Incorporation of more bulky La into Ce-Zr mixed oxide causes the opposite effect due to the lattice expansion.

In the IR hydroxyl region, a band of hydrogen bonded bulk hydroxyl (3550 cm^{-1}) is detected for ceria (Fig. 1b). The specific absorption in this range decreases for CeZr and further declines with La or Sm addition. This suggests that coordination polyhedra distortion in pure ceria could be due to the presence of residual hydroxyls in the lattice.

According to SAXS data [2], in pure CeO_2 extended defects with typical sizes ~ 20-40 Å assigned to microstrains generated by hydroxyls dominate. More extended defect regions (80-170 Å) are also detected which could be assigned to wide-angle grain boundaries. For CeZr only defects with sizes ~20-40 Å appear, and their intensity is higher than that in CeO_2 due to the lattice microstrains caused by incorporated small Zr^{4+} cations. La addition decreases the density of extended defects, probably via releasing the lattice strains due to generation of anion vacancies. For all CeSm samples only extended defects with sizes ~ 20-40 Å are observed, their integral intensity being comparable with that for ceria. Though Sm (+3) cation is bigger than Ce^{4+} cation, anion vacancies generated by this dopant [4] appear to relax the microstrains as well. Similarly, anion vacancies and Me^{3+} segregation at domain boundaries of CeO_2 [5] seems to be responsible for domains coherent stacking, thus explaining the absence of extended defects assigned to wide-angle grain boundaries.

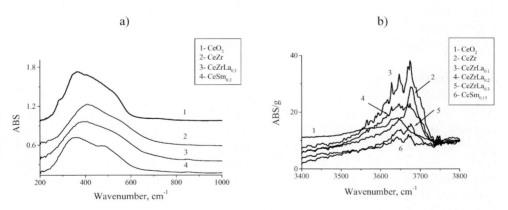

Fig. 1. IR spectra of the lattice modes (a), FTIR spectra of hydroxyl region (b) of ceria-based solid solutions.

3.2. Surface properties

The surface of doped ceria samples is characterized by increased density of the surface Lewis acid sites (detected as carbonyl bands at 2170-2180 cm^{-1} by FTIRS of adsorbed CO) [2] and Broensted acid centers (hydroxyl bands at 3600-3750 cm^{-1}) (Fig. 1b). This certainly correlates with the increased density of microstrains and anion vacancies for doped samples. The density of both Lewis and Broensted surface sites declines with the content of Me^{3+} dopant, thus indicating the surface rearrangement due to clustering of point defects [4,5]. For Pt supported catalysts both metallic Pt^0 (detected as linear carbonyls at ~2060 cm^{-1}) and oxidic forms ($Pt^{+\delta}$ -CO bands at ~2110 cm^{-1}) are observed thus indicating a strong interaction of Pt with support. This agrees with the absence of TEM detected Pt particles for oxidized samples. For Pt/CeSm system the share of Pt^0 species is higher than that for Pt/CeZr(La) samples correlating with a lower density of acid sites responsible for anchoring oxidic Pt species.

3.3. Isotope exchange

Mainly 3rd type of exchange (i.e. two oxygen atoms of oxides are involved in the process [3]) was observed for ceria-containing samples (Table 1). This is the evidence of a high mobility of the surface/bulk oxygen of those complex oxides. A higher rate of the oxygen exchange and a bigger amount of exchangeable oxygen in CeSm system as compared with CeZr-based samples are due to the surface and bulk oxygen vacancies generated by Sm doping [4]. No simple correlation of the oxygen mobility with the density of bulk extended defects or surface acid sites is observed. For $CeSm_y$ samples, X_e was found to go through the maximum at y=0.2, thus correlating with the density of isolated anion vacancies reaching a maximum at this Sm content [4]. Pt addition only slightly affects the rate of exchange, while X_e increases considerably implying Pt incorporation into the surface layer of oxide thus disordering it [1].

3.4. Reduction by methane

For doped ceria samples deep oxidation of CH_4 dominates up to ~700 °C while syngas is generated at higher temperatures. For Pt-supported samples reaction is strongly accelerated, low-temperature syngas selectivity being increased. Thus, for Pt-supported CeZr and $CeSm_{0.2}$ samples intense reduction starts at 400 °C with CO_2 and H_2O evolution due to the presence of weakly bound surface oxygen, and syngas (H_2:CO ratio ~2 [1]) appears as this oxygen is consumed (Fig. 2). The rate of CO evolution goes through the maximum for both samples. For Pt/CeZr sample CH_4 conversion starts to rise again at ~ 750 °C suggesting methane decomposition into H_2 and carbon. Indeed, subsequent reoxidation of samples by oxygen revealed CO_2 evolution equal to 2.25 and 0.7 mmol/g for Pt/CeZr and Pt/CeSm samples, respectively. Hence, carbons build-up is bigger for CeZr sample with a lower oxygen mobility and a higher dispersion of Pt species strongly interacting with this support. Repeated TPR experiments followed by samples reoxidation with O_2 at 500 °C revealed a good reproducibility of runs with all Ce^{4+} cations being reduced to Ce^{3+} state.

In reactions of methane steam and dry reforming (Fig. 3), activation of CH_4 appears not to be affected by the presence of mild oxidants in the feed, since its conversion starts at the same temperatures as in CH_4 TPR (Fig. 2). In addition, at high (~700-900 °C) temperatures, presence of CO_2 or H_2O in the feed ensures complete and stable transformation of methane into syngas even at high space velocities without samples coking (Fig. 4). This means that ceria-based Pt-promoted systems are also very efficient and stable catalysts in the reactions of methane steam and dry reforming apparently due to their ability to activate both methane and mild oxidants (water and carbon dioxide).

Table 1
Data on isothermal isotope oxygen exchange

Sample	T, °C	P/Pa	^aX_e	^{b}Rate/10^{15} O_2/m^2 s
CeZr	400	140	1.0	0.2
	500	140	1.0	1.5
Pt/CeZr	450	105	2.5	0.5
	500	110	7.0	2.4
$CeSm_{0.2}$	450	140	7.0	1.4
	600	140	12.0	14.5

^{a} - Amount of exchangeable oxygen, monolayers; ^{b} - rate of heteroexchange at the experiment pressure.

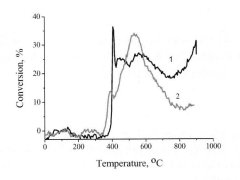
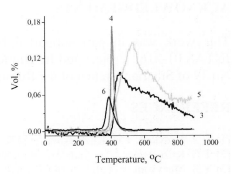

a)   b)

Fig. 2. CH_4 conversion (a) and concentration of CO (3,5) and CO_2 (4, 6) (b) during CH_4-TPR runs over Pt/CeZr (1, 3-4) and Pt/$CeSm_{0.2}$ (2, 5-6). GHSV 50,000/h, heating rate 10°/min.

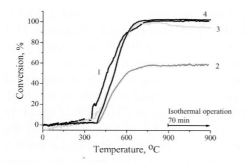 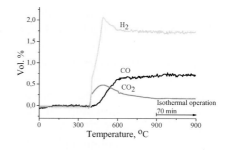

Fig. 3. Conversions of CH_4 (1, 3) and H_2O (2) or CO_2 (4) during CH_4-TPR run with addition of H_2O (1 %) or CO_2 (1 %) to the mixture of 1 % CH_4 in He over Pt/CeZr. GHSV 50,000/h, heating rate 10°/min.

Fig. 4. Concentrations of products during CH_4-TPR run with addition of 1 % H_2O to the mixture of 1 % CH_4 in He over Pt/CeZr. GHSV 50,000/h, heating rate 10°/min.

4. CONCLUSIONS

Nanocomposites comprised of ceria-based solid electrolytes and Pt nanoparticles are efficient catalysts of methane selective oxidation into syngas by their lattice oxygen. By tuning their composition and preparation procedure, mobility and reactivity of the lattice oxygen can be tuned via affecting the defect structure thus improving syngas yield. These systems also demonstrate a high and stable performance in the reactions of methane steam and dry reforming.

ACKNOWLEDGEMENTS

This work was in part supported by RFBR/INTAS Project No IR-97-402, INTAS 01-2162 Project and Integration Project No 8.17 of Presidium RAS and No 39 of Siberian Branch of the Russian Academy of Sciences.

REFERENCES

[1] V.A. Sadykov, T.G. Kuznetsova, et al React. Kinet. Catal. Lett. 76 (2002) 83.
[2] T.G. Kuznetsova, V.A. Sadykov, et al, Stud. Surf. Sci. Catal. 143 (2002) 659.
[3] V.S. Muzykantov et al. Kinetika I Kataliz 44 (2003) 349
[4] H. Inaba, H. Tagawa, Solid State Ionics 83 (1996) 1.
[5] G. Petot-Ervas, C. Petot et al, Mat. Res. Soc. Symp. Proc. 756 (2003) EE 4.4.

Studies in Surface Science and Catalysis, volume 147
X. Bao and Y. Xu (Editors)

Reforming of higher hydrocarbons

R. K. Kaila[*] and A. O. I. Krause

Helsinki University of Technology, Laboratory of Industrial Chemistry, P.O.
Box 6100, FIN-02015 HUT, Finland, [*]kaila@polte.hut.fi

ABSTRACT

Steam reforming (STR) and autothermal reforming (ATR) of n-heptane were
studied on a 15 wt-% Ni/Al_2O_3 catalyst. The optimal temperature for H_2
production is 700 °C, where both the conversion of n-heptane and the selectivity
to H_2 were high. At lower temperatures, thermal cracking of n-heptane occurred.

The catalyst was active and selective, but coke accumulation was observed.
Oxygen present in the ATR experiments reduced coke accumulation, which
slowed the deactivation. According to the thermodynamics, the optimal O_2/C
ratio was 0.34 mol/mol at 700 °C.

1. INTRODUCTION

In future, the demand for hydrogen will increase when new applications for fuel
cells are commercialised. Nowadays, hydrogen and synthesis gas ($H_2 + CO$) are
widely produced by steam reforming (STR) of methane in stationary systems.
Methane can easily be utilised in the vicinity of natural gas pipeline systems, but
it must be compressed when it is transported and stored in vehicles. Even then,
the volumetric hydrogen density remains low, though the H/C ratio of methane
is high.

Therefore, an easily deliverable and safely storable hydrogen source, such
as gasoline or diesel, would be preferred in mobile applications. And, the
hydrogen generation could be carried out anywhere using an on board reformer
[1]. STR of liquid hydrocarbons is a large scale commercial process, and it has
been practiced for the last 40 years in locations where natural gas is not
available [2]. Higher hydrocarbons have a high volumetric hydrogen density.
Moreover, the infrastructure exists already [3], which would ease the
changeover to fuel cells and a hydrogen based society.

Unfortunately, coke is formed in a large amount on the conventional nickel
catalyst when higher hydrocarbons are used in STR [4]. Furthermore,

commercial hydrocarbon fuels, particularly diesel, contain sulphur that causes poisoning of the catalyst [4]. Fischer-Tropsch gasoline, however, does not contain sulphur and would therefore be a potential feedstock for reforming. In addition, it has a low octane number and high H/C ratio due to its high linearity and low aromatic content [5].

In this work, STR and the thermodynamically more feasible autothermal reforming (ATR) of higher hydrocarbons were studied on a commercial nickel catalyst with n-heptane as the model compound for gasoline. ATR requires less energy than STR, since the endothermic STR is combined with the exothermic partial oxidation (POX) reaction.

2. EXPERIMENTAL

STR and ATR reactions were studied in a continuous system in a tubular fixed bed reactor. n-Heptane and water were vaporised and mixed with the gas flow before the reactor. The n-heptane concentration in the feed was 3 mol-%. Water was fed in excess to prevent coke formation on the catalyst.

Reforming reactions were studied on a commercial 15 wt-% Ni/Al_2O_3 catalyst (particle size 0.2–0.3 mm) that was prereduced for 1 hour with H_2/Ar flow (1 mol/mol, 200 cm^3/min (NTP)). Thermal cracking was studied under the same reaction conditions as STR and ATR but without catalyst.

Thermodynamics of STR, ATR and thermal cracking of n-heptane were calculated with HSC Chemistry version 3.02 [6].

2.1. Steam reforming

STR reactions were studied at 500–700 °C in a quartz glass reactor under atmospheric pressure. Argon was used as inert gas. The H_2O/C ratio was 3 mol/mol. The total flow rate of the feed was 100–130 cm^3/min (NTP), and the gas hourly space velocity (GHSV) was 1.6–2.1 x 10^5 1/h. Before use, the catalyst (0.05 g) was reduced at 900 °C.

2.2. Autothermal reforming

ATR experiments were performed at 650–725 °C in a stainless steel reactor. The pressure was increased to 4 barg to reduce coke accumulation. Nitrogen (in air) was used as inert gas. On the basis of thermodynamics, the O_2/C ratio was set to 0.25 and 0.34 mol/mol, and the H_2O/C ratio, respectively, to 3.37 and 2.92 mol/mol. The total flow rate of the feed was 200–300 cm^3/min (NTP), and the gas hourly space velocity (GHSV) was 2.4–4.7 x 10^5 1/h. Before use, the Ni/Al_2O_3 catalyst (0.05–0.1 g) was reduced at 700 °C.

2.3. Product analysis

The product gas was analysed by two online gas chromatographs (GC) (Hewlett-Packard, HP 5890). Hydrocarbons were separated in a DB-1 column and detected with a FID on one GC, and H_2, O_2, N_2, Ar, CO, CO_2, CH_4 and H_2O were separated in a packed column (activated carbon with 2% squalane) and detected with a TCD on the other GC. Since helium was used as carrier gas, detection of H_2 with TCD was only possible at higher concentration (>5%) and then with limited accuracy. The dry gas flow was measured with a gas meter, and the composition of the total product flow was determined from material balances based on the inert gas.

3. RESULTS AND DISCUSSION

3.1. Thermodynamics

The enthalpy of the endothermic STR reaction increases considerably when higher hydrocarbons are used as feed. This increases the energy demand of the system. ATR, which can be operated almost thermoneutrally, then becomes an interesting alternative to STR. Since air is fed to the autothermal reformer as well, exothermic POX Eq. (2) takes place in addition to STR Eq. (1) [7].

$$C_7H_{16} + 7\,H_2O \rightleftharpoons 7\,CO + 15\,H_2 \qquad \Delta H^{\circ}_{298} = 1106.78\ kJ/mol \qquad (1)$$

$$C_7H_{16} + 3.5\,O_2 \rightleftharpoons 7\,CO + 8\,H_2 \qquad \Delta H^{\circ}_{298} = -586.01\ kJ/mol \qquad (2)$$

Water gas shift (WGS) reaction Eq. (3) and methanation Eq. (4) proceed as side reactions and contribute to the product distribution.

$$CO + H_2O \rightleftharpoons CO_2 + H_2 \qquad \Delta H^{\circ}_{298} = -41.14\ kJ/mol \qquad (3)$$

$$CO + 3\,H_2 \rightleftharpoons CH_4 + H_2O \qquad \Delta H^{\circ}_{298} = -206.16\ kJ/mol \qquad (4)$$

According to thermodynamics the optimal reaction temperature for H_2 production by reforming of n-heptane is close to 700 °C. At lower temperatures methane and coke are formed, whereas at higher temperatures C_3–C_4 hydrocarbons are formed. Coke formation can be reduced by increasing the reaction pressure. Then, however, H_2 and CO formation decreases and more hydrocarbons are formed.

In addition, coke and hydrocarbon formation can be decreased by adding oxygen to the feed. Unfortunately, this decreases the H_2/CO ratio of the product stream due to POX. The O_2/C ratio of 0.34 mol/mol is the optimal feed ratio for a thermoneutral ATR reaction at 700 °C. When oxygen is fed in excess POX takes place more significantly, leading to higher amounts of CO_2 and H_2O. Despite the presence of oxygen, the metallic nickel is not oxidised at temperatures above 250 °C.

In pure POX, without steam in the feed, coke formation is increased and the amount of H_2 is significantly lower. Therefore, both steam and air are needed for the hydrogen production.

3.2. Thermal cracking

In the thermal experiments, n-heptane was cracked to lighter hydrocarbons. Ethene and methane were the main products in STR conditions, when only water, n-heptane and argon were fed to the reactor. The conversion of n-heptane increased with temperature, but did not reach the thermodynamic limit under the studied conditions. The distribution of the hydrocarbons formed was similar to that reported earlier [8]. At lower temperatures the selectivities to ethene and methane were higher. In the conditions of ATR, the main products of thermal reactions were CO and CO_2.

3.3. Reforming

In the reforming experiments, the main reaction products were H_2, CO, CO_2 and CH_4. As expected, the conversion of n-heptane increased with temperature and O_2/C feed ratio, and decreased with the space velocity. This can be seen for ATR from Fig. 1. The optimal temperature for H_2 production from n-heptane by STR and ATR was close to 700 °C, where the conversion of n-heptane reached its maximum value and the selectivity to H_2 was high.

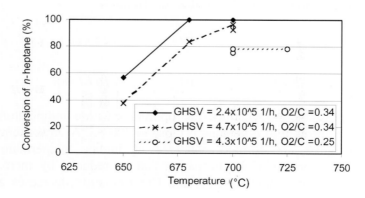

Fig. 1. n-Heptane conversion as a function of temperature in ATR.

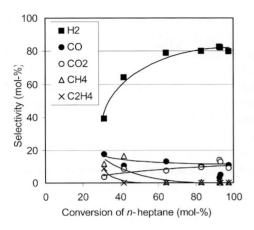

Fig. 2. The selectivity of STR products as a function of n-heptane conversion on 0.05 g Ni/Al$_2$O$_3$; H$_2$O/C = 3 mol/mol, V$_{feed}$ = 100–130 cm^3/min (NTP).

Fig. 3. The selectivity of ATR products as a function of n-heptane conversion on 0.05–0.1 g Ni/Al$_2$O$_3$; O$_2$/C = 0.25–0.34 mol/mol, H$_2$O/C = 3.37–2.92 mol/mol, V$_{feed}$ = 300 cm^3/min (NTP).

The selectivity to H$_2$ increased, whereas the selectivity to CH$_4$ decreased, when the conversion of n-heptane was improved (Figs. 2 and 3). This may suggest that, in reforming, n-heptane is first converted to methane, which then reacts further to synthesis gas. When air was fed to the reactor, the selectivity to H$_2$ decreased at high conversion levels from 80% (STR) to 53% (ATR) and more CO$_2$ was formed due to POX, which is in accordance with thermodynamics.

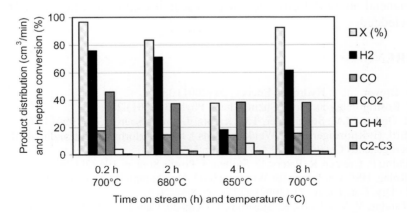

Fig. 4. ATR of n-heptane; p = 4 barg, GHSV = 4.7 x 10^5 1/h, O$_2$/C = 0.34 mol/mol and H$_2$O/C = 2.92 mol/mol.

The reaction temperature affected, not only the n-heptane conversion, but also the selectivity to H_2. At lower temperatures light hydrocarbons and coke were formed via thermal cracking (Fig. 4) since the reaction rate of reforming was slower. Although coke was formed on the Ni/Al_2O_3 catalyst, especially at lower temperatures, the catalyst deactivated only slightly during the ATR experiment (8 h). The ATR reaction was more stable as a function of time than STR, which lasted at its best only 4 hours. In STR, the reactor pressure started to rise significantly due to coke formation, which was verified by the measured carbon content of the used catalyst. Less coke was formed on the catalyst when air was fed to the reactor and the system was pressurised, as would be expected from the thermodynamics.

4. CONCLUSIONS

It is possible to produce hydrogen conventionally with high selectivity by STR of higher hydrocarbons. Unfortunately, coking of the STR catalyst was significant. Hence, modifications of the catalyst are needed in order to improve the catalyst stability.

ATR of n-heptane is a good alternative to STR, which is a highly endothermic reaction. ATR can be carried out almost adiabatically, which makes the on board hydrogen production feasible. Furthermore, the air feed reduces significantly the catalyst deactivation caused by coke formation.

ACKNOWLEGDEMENTS

The financial support from Technical Research Centre of Finland is gratefully acknowledged.

REFERENCES

[1] L.F. Brown, Int. J. Hydrogen Energy, 26 (2001) 381.
[2] J.R. Rostrup-Nielsen, Phys. Chem. Chem. Phys., 3 (2001) 283.
[3] J.M. Ogden, M.M. Steinbugler, T.G. Kreutz, J. Power Source, 79 (1999) 143.
[4] G. Ertl, H. Knözinger and J. Weitkamp (eds.), Steam reforming. In Handbook of Heterogeneous Catalysis, Vol 4, Weinheim 1997, 1824.
[5] C. Palm, P. Cremer, R. Peters, D. Stolten, J. Power Source, 106 (2002) 231.
[6] A. Roine, HSC Chemistry for Windows 3.02, Outokumpu Research Oy, 1997.
[7] M.E. Dry, Cat. Today, 71 (2002) 227.
[8] G. Taralas, V. Vassilatos, K. Sjöström, J. Delgado, Can. J. Chem. Eng., 69 (1991) 1413.

Studies in Surface Science and Catalysis, volume 147
X. Bao and Y. Xu (Editors)

Study of the mechanism of the autothermal reforming of methane on supported Pt catalysts

M.M.V.M. Souza[a], S. Sabino[b], F.B. Passos[b], L.V. Mattos[c], F.B. Noronha[c], M. Schmal[a]

[a]NUCAT/PEQ/COPPE-Universidade Federal do Rio de Janeiro, Brazil. schmal@peq.coppe.ufrj.br.

[b]Universidade Federal Fluminense, Brazil, fbpassos@vm.uff.br.

[c]Instituto Nacional de Tecnologia (INT), Brazil, fabiobel@int.gov.br.

ABSTRACT

Autothermal reforming of methane was investigated on Pt/CeO_2, Pt/ZrO_2 and Pt/Ce-ZrO_2 catalysts. Temperature programmed surface reaction experiments (TPSR) showed the reaction proceeds through an indirect mechanism. A carbon removal mechanism was observed for the catalysts that presented oxygen storage capacity, favoring the stability of the catalyst in the reaction.

1. INTRODUCTION

Steam methane reforming (SMR) and non-catalytic partial oxidation has been the commercial technology for producing syngas from natural gas. However, both technologies are not suited to GTL plants due to the required H_2/CO ratio for FT synthesis. Autothermal reforming (ATR) fulfills the requirements to a syngas with H_2/CO ratio of 2, the ratio necessary to GTL plant. Moreover, ATR has relative compactness, lower capital cost and greater potential for economy of scale [1]. ATR combines non catalytic partial oxidation with steam reforming in one reactor. The mechanism of the combined reactions (partial oxidation and steam reforming of methane) is not reported in the literature. In order to project good catalysts to the autothermal reforming, it is imperative to know the mechanism of this reaction. The aim of this work is to study the mechanism of the autothermal reforming on supported Pt catalysts.

2 EXPERIMENTAL

2.1. Catalyst preparation

The ZrO_2 support was prepared by calcination of zirconium hydroxide (MEL Chemicals) at 823 K, for 2h. The CeO_2 support was obtained by calcination of $(NH_4)_2Ce(NO_3)_6$ (ALDRICH) at 773 K, for 2 h. The Ce-ZrO_2 (18 mol % of CeO_2 and 82 mol % of ZrO_2) was supplied by MEL Chemicals and was calcined at 1073 K for 1 h. The catalysts were prepared by incipient wetness impregnation of the supports with an aqueous solution of H_2PtCl_6 (Aldrich) and were dried at 393 K. Then, the samples were calcined under air (50 cm^3/min) at 673 K, for 2 h.

2.2. Oxygen storage capacity (OSC)

Oxygen storage capacity (OSC) measurements were carried out in a micro-reactor coupled to a quadrupole mass spectrometer (Balzers, Omnistar). The samples were reduced under H_2 at 773K, for 1h, and heated to 1073K in flowing He. Then, the samples were cooled to 723K and a 5%O_2/He mixture was passed through the catalyst until the oxygen uptake was finished.

2.3. Temperature Programmed Surface Reaction (TPSR)

After the activation procedure, the sample (300 mg) was purged in He at 1073 K for 30 min, and cooled to room temperature. Then the sample was submitted to a flow of $CH_4/H_2O/O_2/He$ (4:2:1:25) at 200 cm^3/min while the temperature was raised up to 1073 K at heating rate of 20 K/min.

2.4. Reaction conditions

Reaction was performed in a quartz reactor at atmospheric pressure. The reaction was carried out at 1073 K and WHSV = 160 h^{-1} over all catalysts. A reactant mixture containing methane, oxygen and water (CH_4:O_2 ratio of 4:1 and CH_4:H_2O ratio of 2:1) diluted in helium was used at a flow rate of 200 cm^3/min. The exit gases were analyzed using a gas chromatograph (Chrompack CP 9001) equipped with a thermal conductivity detector and a Hayesep D column.

3. RESULTS AND DISCUSSION

3.1. Oxygen storage capacity (OSC)

Table 1 presents the oxygen storage capacity (OSC) measurements. Pt/CeO_2 and Pt/$Ce_{0.18}Zr_{0.82}O_2$ catalysts exhibited high and similar oxygen consumption whereas Pt/ZrO_2 catalysts practically did not present oxygen uptake. Cerium oxide has an important oxygen storage capacity [2]. This property is attributed to the ability of cerium to act as an oxygen buffer by storing/releasing O_2 due to the Ce^{4+}/Ce^{3+} redox couple [2]. The CeO_2 redox properties are

Table 1

Oxygen uptakes measured at 723K

Samples	O_2 uptake ($\mu mol/g_{catal}$)
Pt/CeO_2	194
Pt/ZrO_2	9
$Pt/Ce_{0.18}Zr_{0.82}O_2$	220

enhanced by the incorporation of ZrO_2 into CeO_2 lattice. The ZrO_2 addition increase the oxygen vacancies of the support due to the high oxygen mobility of the solid solution formed.

3.2. Temperature Programmed Surface Reaction (TPSR)

Figure 1 shows the TPSR curves obtained for $Pt/Ce_{0.18}Zr_{0.82}O_2$ catalyst. The TPSR profiles of all catalysts revealed consumption of methane and oxygen followed by CO_2 and H_2O formation around 720K. However, the production of H_2 and CO was only observed above 790K. These results are consistent to the indirect mechanism. According to this mechanism, in the first step, combustion of methane occurs producing CO_2 and H_2O. In the second one, synthesis gas is produced via carbon dioxide and steam-reforming reaction of the unreacted methane. The same mechanism was observed on the partial oxidation of methane on supported Pt catalysts [3].

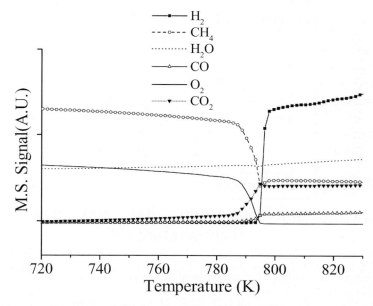

Fig. 1- Temperature Programmed Surface Reaction of Pt/Ce-ZrO_2 catalyst.

3.3 Authotermal Reforming of Methane

The product distribution on the methane autothermal reforming, as a function of reaction temperature, is shown in Figure 2 for the Pt/ZrO_2 catalyst. The O_2 conversion is 100% over the whole temperature range. The production of steam and CO_2 decreases rapidly as the temperature increases due to the exothermic nature of total combustion. The lower production of H_2 and CO up to 600°C is also related to the occurrence of total combustion. Figure 3 shows the methane conversion versus time on stream (TOS) for all the samples. Pt/ZrO_2 catalyst showed the highest initial conversion but deactivated during the reaction. On the other hand, Pt/CeO_2 and $Pt/Ce_{0.18}Zr_{0.82}O_2$ catalysts were quite stable.

Recently, it has been reported in the literature that the support has an important role on the mechanism of the CO_2 reforming [4-6] and the partial oxidation of methane [3,7]. A strong deactivation on the Pt/Al_2O_3 catalyst was observed on both reactions. The performance of Pt/ZrO_2 catalyst depended on the reaction investigated. Several authors have reported that Pt/ZrO_2 catalyst exhibit a good stability on the CO_2 reforming of methane [4-6]. The performance of the Pt/ZrO_2 catalyst has been associated with the low carbon

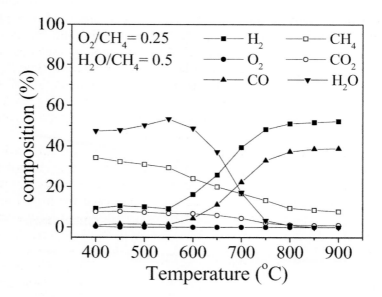

Fig. 2. Composition profiles for autothermal reforming of methane as a function of temperature over the Pt/ZrO_2 catalyst.

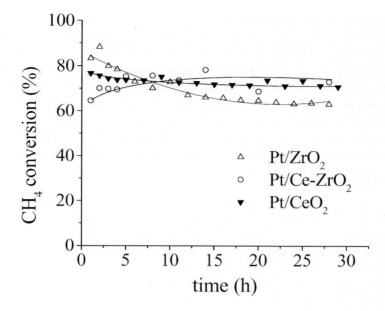

Fig. 3. Autothermal reforming of methane on platinum catalysts at 1073K.

deposit. A two independent path mechanism has been proposed to explain the stability of the Pt/ZrO_2 catalyst on the CO_2 reforming of methane. This mechanism involves the occurrence of CH_4 decomposition on metal particle and CO_2 dissociation on oxygen vacancies of the support at the same time [4,5]. The deactivation of the Pt/ZrO_2 catalyst was stronger on partial oxidation of methane as compared to that observed on CO_2 reforming of methane. As described previously [3], the second step of partial oxidation of methane involves the occurrence of both CO_2 and steam reforming of methane.

The strong deactivation of the Pt/ZrO_2 catalyst on partial oxidation of methane was attributed to disappearance of the oxygen vacancies of support caused by the reoxidation of the reduced zirconium oxide by water, which is produced during the first step of the mechanism proposed for partial oxidation [7]. Furthermore, many studies have shown that the addition of promoters, such as CeO_2 to ZrO_2 support promoted the activity and stability of the Pt/ZrO_2 catalysts on the CO_2 reforming of methane. The addition of ceria to zirconia greatly promoted the stability of the catalyst due to the higher oxygen exchange capacity of the support. The higher rate of oxygen transfer keeps the metal surface free of carbon. This promotional effect on the CO_2 reforming of methane affects the partial oxidation of methane, which comprehends two steps: combustion of methane and CO_2 and steam reforming of unreacted methane.

Therefore, the performance of the Pt/ZrO_2, Pt/CeO_2 and $Pt/Ce_{0.18}Zr_{0.82}O_2$ catalysts on the autothermal reforming of methane can be explained taking into account the two step mechanism proposed. Since water can reoxidize the reduced support, the deactivation of the Pt/ZrO_2 catalyst could be attributed to the disappearance of the oxygen vacancies of this support. This could inhibit the second step of the mechanism suggested. On the other hand, Pt/CeO_2 and $Pt/Ce_{0.18}Zr_{0.82}O_2$ catalysts are very stable. In this case, the oxidation of CeO_2 and $Ce_{0.18}Zr_{0.82}O_2$ supports did not affect the stability due to the higher amount of oxygen vacancies of these supports, as revealed by the OSC measurements.

4. CONCLUSIONS

Autothermal reforming proceeds through a two-step mechanism on platinum catalysts supported on CeO_2, ZrO_2 and $Ce-ZrO_2$: combustion of methane followed by CO_2 and steam reforming of unreacted methane. The OSC capacity showed to be an essential property in order to keep the catalysts stable, as the oxygen released/storage allowed a continuous removal of carbonaceous deposits from the metal surface.

ACKNOWLEDGEMENTS

The authors wish to acknowledge the financial support of the FNDCT/CTPETRO/PETROBRAS (65.04.1600.13; 21.01.0257.00) program.

REFERENCES

[1] D.J. Wilhelm, D.R. Simbeck, A.D. Karp and R.L. Dickenson, Fuel Proc. Technology 71 (2001) 139.
[2] M.H. Yao, R.J. Baird, F.W. Kunz and T.E. Hoost, J. Catal. 166 (1997) 67.
[3] L.V. Mattos, E.R. de Oliveira, P.D. Resende, F.B. Noronha and F.B. Passos, Catal. Today, 77 (2002) 245.
[4] S.M. Stagg-Williams, F.B. Noronha, E.C. Fendley, D.E. Resasco, Journal of Catalysis 194 (2000) 240.
[5] F.B. Noronha, E.G. Fendley, R.R. Soares, W.E. Alvarez, and D.E. Resasco, Chem. Eng. Journal 82 (2001) 21.
[6] M.M.V.M. Souza, D.A.G. Aranda and M. Schmal, J. Catal. 204 (2001) 498.
[7] L.V. Mattos, E.R. de Oliveira, D.E. Resasco, F.B. Passos, F.B. Noronha, Fuel Processing Technology 83 (2003) 147.

Studies in Surface Science and Catalysis, volume 147
X. Bao and Y. Xu (Editors)

Cobalt supported on carbon nanofibers – a promising novel Fischer-Tropsch catalyst

G. L. Bezemer, A. van Laak, A. J. van Dillen and K. P. de Jong

Department of Inorganic Chemistry and Catalysis,
Utrecht University, Sorbonnelaan 16, 3584 CA Utrecht, The Netherlands

ABSTRACT

The potential of carbon nanofibers supported cobalt catalysts for the Fischer-Tropsch reaction is shown. Using the wet impregnation method cobalt on carbon nanofiber catalysts were prepared with cobalt loadings varying from 5 to 12 wt%. The cobalt particle size of the catalysts varied with increasing loading from 3 to 13 nm. Cobalt particles were localized both at the external and at the internal surface of the fibers. The activity at 1 bar syngas varied with increasing loading from 0.71 to 1.71 10^{-5} $mol_{CO} \cdot g_{Co}^{-1} \cdot s^{-1}$. This might indicate that smaller particles are less active in the Fischer-Tropsch reaction, but may also be provoked by different fractions of cobalt present inside the fibers. Stable activity of 225 $g_{CH2} \cdot l_{cat}^{-1} \cdot h^{-1}$ for 400 h was obtained at pressures of 28-42 bar syngas. A C_{5+} selectivity of 86 wt% was found, which is remarkably high for an unpromoted catalyst.

1. INTRODUCTION

In the Fischer-Tropsch (FT) reaction synthesis gas (CO/H_2) is converted into hydrocarbons. By using synthesis gas produced from natural gas, coal or biomass, transportation fuels can be produced from feedstocks other than crude oil. Currently the FT process is commercially operated by several companies.

Cobalt is well known for its activity in the FT reaction. Almost all research on cobalt catalysts has been devoted to oxidic supports such as Al_2O_3, SiO_2 or TiO_2. A drawback of oxidic supports is that cobalt may react with the support to irreducible mixed compounds during either synthesis or catalysis [1]. Due to this reactivity FT synthesis using cobalt particles smaller than 4 nm has not been reported on oxidic supports [2,3].

To circumvent the problem of cobalt reacting with the support the use of carbon as support has been explored [4,5]. However the carbon materials used were not very pure and ill defined. Therefore it may be advantageous to explore graphitic carbon as a support material. Carbon nanofibers (CNF) is a very promising, novel graphitic support material with a potential use in many applications [6]. It consists of interwoven fibers of graphitic carbon from a high purity, a high mechanical strength and a high chemical inertness. The use of CNF for the FT reaction has been reported by van Steen *et al.* who studied copper and potassium promoted iron catalysts [7]. Here, we report on the interesting catalytic properties of Co/CNF prepared by wet impregnation. In a first attempt to study this system we varied the cobalt loading from 5 to 12 wt%.

2. EXPERIMENTAL

2.1 Catalyst preparation

Carbon nanofibers of the fishbone-type with a diameter of about 30 nm were grown out of synthesis gas using an earlier described method [8]. After growth impurities were removed by successively refluxing the sample in KOH [1 M] and in concentrated HNO_3. By treatment in HNO_3 the polarity of the surface is significantly increased due to the introduction of oxygen-containing surface groups. Due to the breaking up of the fibers resulting from oxidation of defect rich regions, the inner tubes are further opened [9,10]. Furthermore, the surface groups act as sites of interaction with the metal precursor.

Cobalt was deposited on thus activated CNF using a wet-impregnation method adapted from Reuel *et al.* [4]. In a flask the required amount of $Co(NO_3)_2.6H_2O$ dissolved in 10 ml toluene/ethanol (2:1 v/v) was added to 1.0 g of CNF. Under continuous stirring the solvent was evaporated at ambient temperature applying a dynamic vacuum. The catalyst precursors were dried at 60 °C in air and subsequently reduced in a pure hydrogen flow (200 ml/min) at 300 °C for two hours. The samples were carefully passivated at 150 °C using a CO_2/N_2 flow (1:5 v/v). Catalysts with cobalt loadings of 5, 8 and 11.9 wt% were prepared, further denoted as Co5, Co8 and Co12.

2.2 Catalyst Characterization

The reduced and passivated catalysts were characterised using SEM, TEM, XPS and XRD. SEM was performed using a Philips XL30 FEG apparatus. TEM images were obtained using a Philips Tecnai-20 FEG TEM operated at 200 kV. XPS data were obtained with a Vacuum Generators XPS system using-monochromatic Al K_α radiation. XRD measurements were performed with a Nonius PDS 120 powder diffractometer system using Co $K\alpha_1$ radiation (λ=1.78897 Å).

2.3 Catalytic testing

The FT measurements were carried out at 220 °C and at 1 bar CO/H_2 (1:2 v/v). A high GHSV (3500 h^{-1}) was used in combination with line tracing at 130 °C to prevent product condensation. Typically 50 mg of catalyst particles (0.5-1.0 mm) was diluted with 500 mg SiC (0.2 mm) to achieve isothermal plug-flow conditions. Hydrocarbon products (C_1-C_{16}) were analysed with an FID on a Varian 3800 GC with a fused silica CP-Sil 5CB column. Selectivities of the catalysts were compared at the same CO-conversion (2%), which was achieved by tailoring the GHSV. A linear relation between GHSV and CO-conversion was found.

The catalyst loaded with 11.9 wt% cobalt was also tested at a pressure of 28-42 bar, which is closer to conditions used in industry. After re-reduction the catalyst was tested at 214 °C using a GHSV of 1700 h^{-1} for 400 h.

3. RESULTS AND DISCUSSION

3.1 Characterization of the catalysts

The bulk density of the purified CNF was approximately 0.64 g/ml with a pore volume of 0.40 ml/g. The material obtained consisted of fibers of graphitic carbon with a diameter of about 30 nm and a specific surface area of 150 m^2/g. The fibers are interwoven and form skeins with meso- and macropores. The 3D structure of the support bodies with long interwoven fibers inside the CNF skein is clearly shown (Fig. 1).

The catalysts were further studied using XRD. With XRD characteristic graphite diffractions were found (30, 50, 64 and 95 ° 2θ) together with cobalt oxide reflections (43 and 73 ° 2θ) as can be seen in Fig. 2. By applying the Scherrer equation on the line width at 43 ° 2θ the cobalt particle size was calculated. The particle sizes, which are given in Table 1, increased from 3 to 13 nm with increasing cobalt loading.

Particle sizes were also measured using TEM. Representative images of Co5 and Co8 are shown in Fig. 3 and Fig. 4. Small particles are visible on the fibers

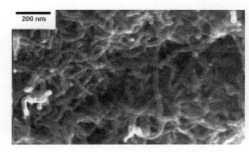

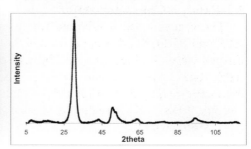

Fig. 1. SEM micrograph of purified CNF   Fig. 2. XRD spectrum of Co5

Table 1
Properties of catalysts studied

Name	Co loading Co/C (at/at)	TEM (nm)	XRD (nm)	XPS (Co/C)	Act. (CTY)	TOF	α	CH$_4$ (wt%)
Co5	0.0107	3-4	3	0.0016	0.71	1.3	0.61	46
Co8	0.0177	4-7	6	0.0275	1.60	5.9	0.62	41
Co12	0.0272	-	13	0.0344	1.71	13.6	0.62	39

XPS = Co/C atomic ratio   TOF = $10^{-3} \cdot s^{-1}$
CTY = $10^{-5} \, mol_{CO} \cdot g_{Co}^{-1} \cdot s^{-1}$

and with EDX it was confirmed that they consist of cobalt and cobalt oxide. Particle sizes found with TEM confirmed the XRD results (Table 1).

TEM images show that the fibers are not solid, but have an inner core with a diameter of approximately 8 nm. Fig. 3 indicates that a significant part of the cobalt particles is localized in the inner tube because a string of particles follows the course of the fiber core. Upon tilting the fiber along its axis during TEM analysis, these particles are not moved, which proves their location in the inner tube. This phenomenon was most pronounced with Co5. The deposition of cobalt and other metals inside carbon nanotubes using wet impregnation techniques has been reported earlier [11,12]. To our best knowledge, this is the first report on cobalt particles deposited in the central tube of carbon nanofibers.

Quantitative XPS was used to further study the amount of cobalt present in the inner core of the fibers. XPS is a surface sensitive technique with an escape depth of the photo-electrons for graphite of 2-3 nm, so cobalt loaded inside the

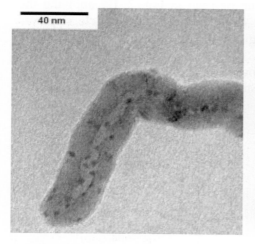

Fig. 3. TEM image of Co5, small cobalt particles are visible in and on the CNF.

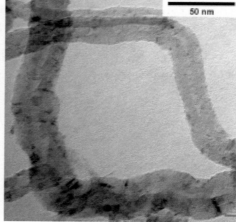

Fig. 4. TEM image of Co8, small cobalt particles are visible on the CNF.

fibers is hidden for the XPS technique. The areas of the carbon and the cobalt peak were used to calculate the Co/C ratios shown in Table 1. For Co5 this ratio is far lower than the actual loading, but also for Co8 and Co12 a higher ratio would be expected if particle sizes were taken into account. From XPS results it may be concluded that the larger the Co loading, the lower the fraction inside the tubes.

3.2 Catalytic testing

The performance of the catalysts at 1 bar is given in Table 1. The catalysts displayed ASF kinetics, with a methane selectivity varying from 46 for the lowest to 39 wt% for the highest loading. The C_{5+} selectivities varied from 29 to 34 wt%. In Fig. 5 the ASF plot of Co12 is shown. The chain growth probability (α) of the three catalysts turned out to be virtually the same: 0.61 or 0.62. After an initial deactivation of at most 30% the catalysts displayed stable activities for 70 h, which indicates that the CNF support is not gasified during catalytic operation. The activity at 1 bar, normalised on the cobalt loading (CTY), exhibited a more than two-fold increase going from the lowest to the highest loaded catalyst.

The TOF values were calculated using XRD data on the particle size and the measured CTY. A linear relation between the cobalt particle size and the TOF was found: an increase of the cobalt particle size coincides with an increase of the TOF. This relation is plotted in Fig. 6. However, this interpretation should be handled with care; it can not be excluded that this correlation is brought about, or at least affected due to the presence of a loading dependent part of the Co particles in the inner tubes of the fibers. So more research is needed to find out the cause of this correlation.

A fresh amount of Co12, the catalyst with the highest CTY was also tested at pressures closer to commercial operation. The catalysts showed stable activities during 400 h of operation. At a space-time-yield of 225 $g_{CH2} \cdot l_{cat}^{-1} \cdot h^{-1}$ the C_{5+} selectivity was as high as 86 wt%.

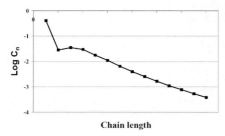

Fig. 5. ASF distribution of Co12 at 1 bar.

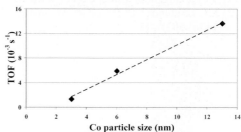

Fig. 6. Relation between cobalt particle size and TOF.

264

4. CONCLUSION

Co/CNF catalysts with small cobalt particles well dispersed on carbon nanofibers (CNF) were synthesized. It was shown that CNF is a very promising support material for the Fischer-Tropsch reaction. Stable activity of 225 $g_{CH2} \cdot l_{cat}^{-1} \cdot h^{-1}$ for 400 h was obtained at pressures of 28-42 bar syngas. A C_{5+} selectivity of 86 wt% was found, which is remarkably high for an unpromoted system.

By comparing the specific activities of the catalysts at 1 bar it was found that catalysts with smaller cobalt particles showed a lower activity. Results give an indication of a cobalt particle size effect in FT, but more research is needed to exclude other possibilities, such as the different amounts of Co in the inner tube of the fibers.

ACKNOWLEDGMENTS

The authors acknowledge J.W. Geus and C. van der Spek (TEM), A. Mens (XPS), H. Holewijn and H. Oosterbeek (high pressure FT) and the financial support of Shell Global Solutions.

REFERENCES

[1] P.J. van Berge, J. van de Loosdrecht, S. Barradas, A.M. van der Kraan, Catal. Today, 58 (2000) 321.
[2] E. Iglesia, Appl. Catal. A-Gen., 161 (1997) 59.
[3] A. Barbier, A. Tuel, I. Arcon, A. Kodre, G.A. Martin, J. Catal., 200 (2001) 106.
[4] R.C. Reuel, C.H. Bartholomew, J. Catal., 85 (1984) 78.
[5] C. Morenocastilla, F. Carrascomarin, J. Chem. Soc.-Faraday Trans., 91 (1995) 3519.
[6] K.P. de Jong, J.W. Geus, Catal. Rev.-Sci. Eng., 42 (2000) 481.
[7] E. van Steen, F.F. Prinsloo, Catal. Today, 71 (2002) 327.
[8] M.L. Toebes, J.H. Bitter, A.J. van Dillen, K.P. de Jong, Catal. Today, 76 (2002) 33.
[9] T.G. Ros, A.J. van Dillen, J.W. Geus, D.C. Koningsberger, Chemphyschem, 3 (2002) 209.
[10] M.L. Toebes, J.M.P. van Heeswijk, J.H. Bitter, A.J. van Dillen, K.P. de Jong, Carbon 42 (2004) 307.
[11] S.C. Tsang, Y.K. Chen, P.J.F. Harris, M.L.H. Green, Nature, 372 (1994) 159.
[12] J.P. Tessonnier, L. Pesant, C. Pham-Huu, G. Ehret, M.J. Ledoux, Stud. Surf. Sci. Catal., 143 (2002) 697.

Studies in Surface Science and Catalysis, volume 147
X. Bao and Y. Xu (Editors)
265

Resolving flow details in slurry bubble columns used for Fischer-Tropsch synthesis using computational fluid dynamics

Bente Helgeland Sannæs and Dag Schanke

Statoil, N-7005 Trondheim, Norway

1. INTRODUCTION

Slurry bubble columns of considerable size are being planned for commercial Fischer-Tropsch synthesis of natural gas.

Information about the flow patterns of gas and liquid is scarce. Most information has been obtained in small-scale units using air and water as the gas and liquid phases. Some researchers have used model fluids to obtain information for more realistic systems [1]. Very few studies have been conducted at increased pressure.

Computational fluid dynamics (CFD) has been used as a design tool for single-phase flows for quite some time. The use of CFD for multiphase systems still is in a developmental phase. Multiphase CFD has been used to simulate some laboratory systems where detailed information about the flow is known. This has contributed to further develop and improve the simulation models [2-4]. The main providers of CFD software are also developing their codes to be able to handle problems of commercial interest. However, there is still need for further development of the models [5].

Information about the large-scale flow in large columns cannot be readily obtained by direct measurements. Design is based on extrapolating the knowledge of the flow in smaller systems. By using CFD it is possible to gain more information about the flow. Such increased knowledge can be used to increase productivity and/or reduce costs. CFD can also be used to investigate details of the flow in regions of specific interest. Examples can be the gas distributor region, the outlet areas, and around specific internals.

The simulations reported here have been performed using the commercial codes CFX and Fluent.

2. MODELLING

The modelling of multiphase flow is based on an Eulerian formulation. The phases are treated as interpenetrating continua, and represented by a volume fraction. The pressure in the system is assumed to be the same for all phases.

2.1. Conservation of mass

The conservation of mass for each phase is given

$$\frac{\partial}{\partial t}\left(\alpha_q \rho_q\right) + \nabla \cdot \left(\alpha_q \rho_q \mathbf{u}_q\right) = 0 \tag{1}$$

2.2. Conservation of momentum

The momentum transport of each phase is governed by modified Navier-Stokes equations that account for the sharing of space between the phases.

$$\frac{\partial}{\partial t}\left(\alpha_q \rho_q \mathbf{u}_q\right) + \nabla \cdot \left(\alpha_q \rho_q \mathbf{u}_q \mathbf{u}_q\right) = -\alpha_q \nabla p + \nabla \cdot \boldsymbol{\tau}_q + \alpha_q \rho_q \mathbf{g}$$

$$+ \text{Interphase forces} \tag{2}$$

Momentum is transported between the phases via the interphase forces. The most important interphase force is the drag between bubbles and liquid. This is modelled as

$$\mathbf{M}_q = \frac{3}{4}\frac{C_D}{d}\alpha_p \rho_q |\mathbf{u}_p - \mathbf{u}_q|(\mathbf{u}_p - \mathbf{u}_q) \tag{3}$$

where C_D is the drag coefficient. Sometimes the effect of reduced liquid volume due to the presence of many bubbles is accounted for by multiplying Eq. (3) with the volume fraction of gas, α_p. Other interphase forces that may be important – depending on the flow conditions – are lift forces (transversal to the flow direction) and virtual mass forces (due the acceleration of the fluid film along with the bubble or particle). Determining the drag coefficient between bubbles and liquid is more complicated than the drag between a solid particle and a fluid due the shape distortion and possible internal flow of the bubble. Many models are available for this drag coefficient. The models used in this work are the Schilller-Naumann model, which was originally developed for solid spheres in fluid, and the Grace model, which was developed for bubbles in liquid. Due to space limitations, the reader is referred to the CFX [6] and Fluent [7] manuals for a further description of the models.

The development of improved models is an important task [8]. The size of the bubbles will also vary due to break-up of larger bubbles and coalescence of smaller bubbles. This variation is not included in the modelling reported here.

The volume fraction of each phase is governed by a transport equation. It is possible to include phase transition, but this has not been done for the simulations presented here.

2.3. Turbulence modelling

The turbulence is modelled using a $k - \varepsilon$ model that has been modified to account for the presence of two phases. The basis equations for the calculations are

$$\frac{\partial}{\partial t}\left(\alpha_q \rho_q k_q\right) + \nabla \cdot \left(a_q\left(\rho_q \mathbf{u}_q k_q - \left(\mu + \frac{u_{tq}}{\sigma_k}\right)\nabla k_q\right)\right)$$
$$= \alpha_q\left(P_q - \rho_q \varepsilon_q\right) + \text{Interaction terms}$$

$$\frac{\partial}{\partial t}\left(\alpha_q \rho_q \varepsilon_q\right) + \nabla \cdot \left(a_q\left(\rho_q \mathbf{u}_q \varepsilon_q - \left(\mu + \frac{u_{tq}}{\sigma_\varepsilon}\right)\nabla \varepsilon_q\right)\right)$$
$$= \alpha_q \frac{\varepsilon_q}{k_q}\left(C_{\varepsilon 1} P_q - C_{\varepsilon 2} \rho_q \varepsilon_q\right) + \text{Interaction terms}$$

where P_q is the production of turbulent kinetic energy which is given by a model, and the turbulent viscosity is given by

$$\mu_{tq} = C_{G\mu} \rho_q \left(\frac{k_q^2}{\varepsilon_q}\right)$$

In the simplest multiphase cases, the turbulence can be calculated for the mixture of the two phases treating them as one phase with density and viscosity calculated by volume averaging the properties of the two phases.

In many cases, the existence of a dispersed phase will cause extra turbulence in the continuous phase. One way to account for the enhanced turbulence of the continuous phase due the existing bubbles by modifying the turbulent viscosity by adding a particle induced eddy viscosity [9], which is used in the CFX simulations. Another way is to model the turbulence in each phase separately by solving $k - \varepsilon$ equations for both phases, which is the method chosen for the Fluent simulations. For a thorough description of the different models, the reader is again referred to the CFX [6] and Fluent manuals [7].

3. RESULTS AND DISCUSSION

Simulations of small-scale bubble columns have been performed using CFX-4.3. It has been shown that the wavy pattern of the gas flow that is seen in the real column is well reproduced. An example is shown in Fig. 1.

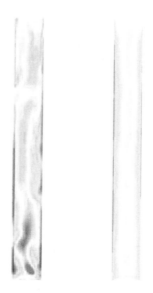

Fig. 1. Instantaneous isosurface of gas fraction in a small-scale bubble column.

Fig. 2. Instantaneous (left) and averaged over 10 seconds real time (right) gas volume fraction.

Simulations of small-scale bubble columns have also been performed using Fluent 6.1. By time averaging the instantaneous results, it is shown that the "gulf-stream" pattern – with upflow in the centre and downflow along the walls – described in the literature is also found in the simulations. Fig. 2 shows the results for the gas volume fraction in a plane through the centre of the column. By comparing the instantaneous and the time averaged results for the volume fraction it is clearly seen that the instantaneous gas volume fraction locally is much larger than the time averaged gas volume fraction.

Fig. 3 shows contours of the vertical liquid velocity in the same plane at the same time instant as the gas volume fractions in Fig. 2. A similar decrease in extreme values is seen when time averaging the instantaneous results. Both the upflow and downflow velocities are reduced.

Simulations to resolve flow details of the gas distributor region in larger columns have been performed using Fluent 6.1. Knowledge of the gas volume fraction distribution in the gas distributor region is important when different distributor designs are evaluated. A snapshot at one time instant of the gas volume fraction distribution above the distributor is shown in Fig. 2. This shows that the flow patterns are quite chaotic. The simulations show that the flow

Fig. 3. Instantaneous (left) and averaged over 10 seconds real time (right) vertical liquid velocity.

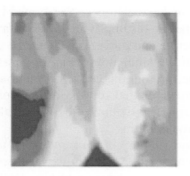 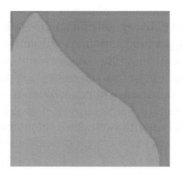

Fig. 4. Volume fraction of gas (left) and gas temperature (right) in the gas distribution region of a large-scale bubble column are shown in a plane that covers half of the reactor diameter. The centre axis of the reactor is on the left side of the figures.

varies significantly with time, showing wavy patterns and the creation and destruction of gas rich clusters. The entrance of cooler gas with respect to the slurry temperature was also investigated. The simulations show that the gas will be heated to slurry temperature shortly after introduction to the reactor. An example is shown in Fig. 4.

4. CONCLUSION

It can be concluded that CFD simulations for bubble columns and slurry bubble columns are improving. They can be used to increase the general understanding of slurry bubble columns, and are now gradually becoming a tool for aiding the design and operation of bubble columns.

REFERENCES

[1] R. Krishna and S.T. Sie, Fuel Proc. Tech., 64 (2000) 73.
[2] J. Sanyal, S. Vásquez, S. Roy, and M.P. Dudukovic, Chem. Eng. Sci., 54 (1999) 5071.
[3] J.M. van Baten and R. Krishna, Chem. Eng. Sci., 56 (2001) 503.
[4] J. Chahed, V. Roig, and L. Masbernat, Int. J. Multiphase Flow, 29 (2003) 23.
[5] J.B. Joshi and V.V. Ranade, Ind. Eng. Chem. Res., 42 (2003) 1115.
[6] CFX-4.3 Solver Manual, AEA Technology plc, Didcot, 1999.
[7] Fluent 6, User's Guide Volume 4, Fluent Inc., Lebanon, 2001.
[8] R. Kurose, R. Misumi, and S. Komori, Int. J. Multiphase Flow, 27 (2001) 1247.
[9] Y. Sato and K. Sekoguchi, Int. J. Multiphase Flow, 2 (1975) 79.

Studies in Surface Science and Catalysis, volume 147
X. Bao and Y. Xu (Editors)
271

Manganese promotion in cobalt-based Fischer-Tropsch catalysis

F. Morales Cano, O.L.J. Gijzeman, F.M.F. de Groot and B.M. Weckhuysen

Department of Inorganic Chemistry and Catalysis, Debye Institute, Utrecht University, Sorbonnelaan 16, 3584 CA Utrecht

ABSTRRACT

The effect of Mn addition on the catalytic performances of Fischer-Tropsch synthesis was investigated using a series of Co/TiO_2 catalysts prepared by two different preparation methods; incipient wetness impregnation (IWI) and homogeneous deposition precipitation (HDP). Mn was loaded using two different procedures in order to investigate its interaction with Co and TiO_2. Several characterization techniques, mainly XRD, XPS and TPR, were used to elucidate the Co oxidation state and dispersion and to estimate the Mn location. Fischer-Tropsch synthesis carried out at 1 bar of pressure showed an improvement of the catalytic performance in the Mn-promoted HDP catalyst, whereas no promotion effect was found in Mn-promoted IWI catalyst. This difference can be explained by a different Co-Mn interaction.

1. INTRODUCTION

The transformation process from syngas to hydrocarbons; i.e., Fischer-Tropsch synthesis (FTS) is gaining more and more importance due to the need for e.g. high purity diesel fuels. Overall process efficiency requires catalysts with high volume productivity, low deactivation rates, and high chain growth probability values (α), what leads to long chain hydrocarbons as the main product stream. Co-based catalysts are the most selective towards long chain hydrocarbons (mainly paraffins) what make them the catalysts of choice for low temperature FTS. In addition, water-gas shift reaction does not occur to a high extent on Co^0 [1], what means that Co-based catalysts are suitable when a rich H_2/CO feed of syngas is used e.g. from natural gas.

The use of other metals in small amounts, have been investigated over the last decades with the aim of enhancing the catalytic performances of Co-based catalysts. Several noble metals, such as Ru and Re, have been reported to enhance the Co reducibility and hence, to increase the catalytic activity [2, 3].

Other positive effects that can be achieved by using promoter elements are e.g. the inhibition of carbon deposition on Co^0 and a decrease of Co sintering what leads to the improvement of catalyst stability and a decrease of deactivation rates during FTS [4, 5]. The use of Mn as promoter for Co-based FTS catalysts can be found in the patent literature as well [6], but its exact role is unclear. In order to obtain a promotion effect, it is important to achieve a co-promotor interaction in the catalyst precursor so that a synergistic effect can be obtained [7, 8].

In the present work, four Co/TiO_2 catalysts were prepared, characterized and tested in FTS reaction in order to evaluate the promotion effect of Mn and the influence of the preparation method on the final catalytic performances. In this way, a structure-activity relationship could be approached.

2. EXPERIMENTAL

2.1. Catalyst preparation

Degussa P25 titania (surface area of 45 m^2/g, pore volume of 0.5 cm^3/g) was used as support for the preparation of cobalt FTS catalysts (Table 1). Two catalysts were prepared by incipient wetness impregnation (IWI) using aqueous solutions of cobalt nitrate and a mixture of cobalt and manganese nitrates (IWI-Co and IWI-CoMn). For this preparation method the support was pressed in pellets and sieved (0.21-0.50 mm) prior to impregnation. The other two samples were prepared by homogeneous deposition precipitation (HDP) method and the manganese was loaded after calcination by the IWI method (HDP-Co and HDP-CoMn). All catalysts were calcined at 400 °C in air for 4 hours and finally reduced at 400 °C in a H_2 flow for 6 hours followed by passivation.

2.2. Catalyst characterization

Catalyst samples were characterized using X-ray diffraction (XRD) and X-ray photoelectron spectroscopy (XPS) techniques before and after reduction. Powder XRD measurements were performed using an ENRAF-NONIUS XRD system equipped with a curved position-sensitive INEL detector and applying a Co $K\alpha_1$ radiation source ($\lambda = 1.78897$ Å). The XPS data were obtained with a Vacuum Generators XPS system, using a CLAM-2 hemispherical analyzer for electron detection. Non-monochromatic Al ($K\alpha$) X-ray radiation was used, employing an anode current of 20 mA at 10 KeV. The binding energies obtained were corrected for charge shifts using Ti $2p_{3/2} = 458.5$ eV as reference. In addition, reduction experiments were carried out by temperature programmed reduction (TPR) of calcined samples using a tubular quartz reactor flushed with 5 % H_2/Ar flow and raising the temperature at the constant rate of 10°C/min. The content of H_2 in the outflowing gas was monitored with a thermo-conductivity detector.

Table 1
Overview of Co catalysts prepared together with their metal loadings.

Sample code	Wt % Co	Wt % Mn	Preparation procedure
IWI-Co	11.5	0	IWI of Co
IWI-CoMn	11.2	1.2	IWI of Co and Mn
HDP-Co	7.5	0	HDP of Co
HDP-CoMn	7.5	2	HDP of Co + IWI of Mn

2.3. Catalytic testing

Prior to catalytic tests the samples were reduced at 300 °C for 1 hour. FTS reaction was carried out using a plug flow reactor in which the catalysts were diluted in SiC. The reaction was carried out at 1 bar of pressure, 220 °C and H_2/CO ratio of 2 (GHSV= 8100 h^{-1}). The products composition was analyzed on-line with a gas chromatograph equipped with a fused silica column of 50 meters length and a flame ionization detector.

3. RESULTS AND DISCUSSION

3.1. X-ray diffraction

XRD patterns of the calcined catalysts showed the presence of crystalline Co_3O_4 structure. Approximate Co_3O_4 particle sizes were obtained using the line broadening of diffraction peaks. The values were found to be around 13 nm for HDP catalysts and around 18 nm for IWI catalysts (Table 2). However, the catalysts prepared by the HDP method showed very low intensity diffraction peaks of cobalt species and this made it impossible to determine the particle sizes accurately. This is probably due to the presence of very small particles in amorphous state, which were not visible with XRD, and due to the lower cobalt loading contained in these catalysts. After reduction a low intensity peak due to Co^0 (HCP) was detected in the IWI catalysts, which might be the signal of only some % of the Co content due to the presence of unreduced Co in some amorphous oxidic state. The HDP reduced catalysts did not show any Co signals in XRD due to a small particle distribution.

3.2. X-ray photoelectron spectroscopy

The presence of the Co_3O_4 structure in all calcined samples was confirmed by XPS where the binding energies of $Co2p_{3/2}$ in all samples were close to 779.8 eV. The Co/Ti ratios reflected higher dispersions achieved by the HDP method since both samples prepared by this method gave higher values (Table 2). In both type of promoted catalysts Mn seemed to be already highly dispersed over the TiO_2 or on top of Co_3O_4 as revealed by the high Mn/Co ratios. This high Mn dispersion might be due to a high metal-support interaction [9]. It has also been reported that under certain conditions some metal ions can diffuse into the TiO_2 lattice forming very stable compounds [10].

Table 2
XRD and XPS results obtained for the calcined and reduced Co/TiO$_2$ catalysts.

Sample	Co$_3$O$_4$ size[a] (nm)	Co^0 size[b] (nm)	Co/Ti ratio[a]	Mn/Co ratio[a]	Co/Ti ratio[b]	Mn/Co ratio[b]
IWI-Co	17.8	12.2	0.15	0	0.23	0
IWI-CoMn	18.1	9.2	0.10	0.57	0.21	0.38
HDP-Co	13.5	-	0.28	0	0.24	0
HDP-CoMn	-	-	0.24	0.58	0.22	0.75

[a]After calcination, [b]after reduction.

Measurements of all reduced samples detected CoO as the main compound present since the catalysts were passivated. Co^{2+} oxidation state can be identified by the shake-up features present at slightly higher binding energies than the maximum and by an increase of the difference in binding energy between the Co2p$_{3/2}$ and Co2p$_{1/2}$ peaks.

After reduction, Co/Ti ratios increased in the IWI catalysts what suggests that Co was spreading over the support, whereas in the HDP catalysts the Co/Ti ratio remained similar with no significant changes in dispersion. With regard to the Mn, the Mn/Co ratio decreased for the IWI-CoMn catalyst after reduction (from 0.57 to 0.38) due to an increase of the Co signal. Hence, the Mn would be already spread over the TiO$_2$ before reduction without any close interaction with Co. In the HDP-CoMn catalyst a large increase of the Mn/Co ratio (from 0.58 to 0.75) was obtained after reduction whereas the Co/Ti ratio did not change much. Thus, this increase of the Mn signal might be due to a migration of Mn particles from the top of the cobalt particles towards the support as soon as Co$_3$O$_4$ is being reduced to Co^0. After reduction the Mn remains in the oxidic state as MnO [11], which can interact with the cobalt particles as long as these are not totally reduced.

3.3. Temperature programmed reduction

Fig. 1 shows the TPR profiles for the four catalyst precursors after calcination. Supported cobalt oxide reduction occurs in two different steps:

$$Co_3O_4 + H_2 \rightarrow 3\ CoO + H_2O$$

$$CoO + H_2 \rightarrow Co + H_2O$$

The areas of H$_2$-consumption peaks in all samples were in agreement with the stoichiometric of both reduction steps. For both unpromoted catalysts, the two peaks are attributed to the reduction from Co$_3$O$_4$ to CoO, which then reduces at higher temperatures to Co^0. However, in HDP-Co catalyst the second peak splitted in two maximum temperatures. This splitting is probably due to a different extent of metal-support interaction of the cobalt particles. The formation of CoTiO$_3$ has been reported to occur when small Co particles interact with TiO$_2$ support [12].

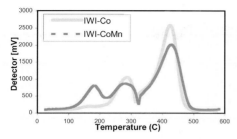 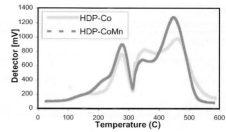

Fig 1. TPR profiles of unpromoted and Mn-promoted Co/TiO$_2$ catalysts.

The addition of Mn to both types of catalysts had a different effect on the TPR profiles. In the case of the IWI-CoMn catalyst a new reduction peak appeared at lower temperatures (~180 °C) without any effect on the reduction temperature of the Co oxides since no interaction between Co and Mn is thought to take place in this sample. The new reduction peak must be due to the reduction of Mn compounds. A different effect of Mn was found in the HDP-CoMn catalyst. Here, the presence of Mn decreased the Co reducibility since there was an increase of the maximum temperature peak area. This again points towards the existence of a Co-Mn interaction in this catalyst, which depletes Co reducibillity.

3.4. Catalytic testing

CO conversion was used as a measure of catalyst activity. In the first stage of FTS reaction deactivation occurred causing a decrease of the activity until a steady state was reached in about 24 hours. During this first stage of reaction changes in selectivity were observed, increase of α values and decrease of paraffins/olefins ratios, what may be due to structural and compositional changes in the active sites [13,14]. Table 3 summarizes the catalytic performances of the Co/TiO$_2$ catalysts under study compared at the same conversion values in order to evaluate the Mn addition.

Catalysts prepared by IWI showed better FTS performances since they yielded higher weight hydrocarbons. However, the presence of Mn promoter in the IWI catalyst did not enhance the catalytic performances, but in turn lowered the α values. HDP catalysts showed lower α values and higher P/O ratios than IWI catalysts. This means that the higher the olefinic production, the higher the chain growth probability. This is due to an increase of the α-olefin readsorption rate [15]. Nevertheless, the presence of Mn in HDP catalyst enhanced the catalytic performances increasing the C5+ yield and decreasing the methane formation, suggesting an influence of Mn on the Co active site. In addition, an improvement of catalytic stability was found for the Mn promoting catalyst as was revealed by the deactivation rate throughout long catalytic tests.

Table 3
Steady state FTS results for the four Co/TiO$_2$ catalysts under study.

Sample code	% CO conversion	Co-time yield $(10^3.s^{-1})$	% CH$_4$ select.	% C5+ select.	P/O Ratio	α values
IWI-Co	3.60	1.55	18.2	55.6	0.11	0.71
IWI-CoMn	3.60	1.56	19.2	51.4	0.15	0.67
HDP-Co	2.78	1.80	32.7	30.9	0.35	0.56
HDP-CoMn	2.78	1.80	22.3	43.5	0.22	0.63

4. CONCLUSIONS

It has been shown that the catalytic performances of Co/TiO$_2$ catalysts in FTS depend on the final Co phase distribution, which can be varied by using different preparation methods. The use of Mn as promoter element may lead to an improvement of the selectivity by decrease of the CH$_4$ formation and an increase of the α value. However, this promotion effect can only be achieved by using a suitable preparation procedure that leads to an appropriate Mn-Co interaction in the active catalyst. In the present work, an apparent Mn interaction in the HDP-CoMn catalyst led to a decrease of Co reducibility. This decrease of Co reducibility did not cause any loss of catalyst activity, but enhanced the selectivity towards higher hydrocarbons.

ACKNOWLEDGMENTS

This work is supported by a grant of Shell Global Solutions International B.V.

REFERENCES

[1] E. Iglesia, Appl. Catal. A: General, 161 (1997) 59.
[2] E. Iglesia, S.L. Soled, R.A. Fiato, J. Catal. 137 (1992) 212.
[3] J. Li, G. Jacobs, Y. Zhang, T. Das, B.H. Davis, Appl. Catal. A: General, 223 (2002) 195.
[4] S. Vada, A. Hoff, E. Adnanes, D. Schanke, A. Holmen, Top. Catal. 2 (1995) 155.
[5] E. Iglesia, S.L. Soled, R.A. Fiato, G.H. Via, Nat. Gas Conv. II, (1994) 433.
[6] M.F. Goes, P.J.M. Rek, D. Schaddenhorst, J-P. Lange, J.J.C. Geerlings, H. Oosterbeek, US5981608 Patent (1999).
[7] F.B. Noronha, M. Schmal, R. Frety, G. Bergeret, B. Moraweck, J. Catal. 186 (1999) 20.
[8] M. Roning, D.G. Nicholson, A. Holmen, Catal. Lett. 72 (2001) 141.
[9] J. Villasenor, P. Reyes, G. Pecchi, Catal. Today, 76 (2002) 121.
[10] M. Voβ, D. Borgmann, G. Wedler, J. Catal. 212 (2002) 10.
[11] L.A. Boot, M. Kerkhoffs, B. Linden, A.J.V. Dillen, J.W. Geus, F. Buren, Appl. Catal. A: General, 137 (1996) 69.
[12] Y. Brik, M. Kacimi, M. Ziyad, F. Bozon-Verduraz, J. Catal. 202 (2001) 118.
[13] H. Schulz, Z. Nie, F. Ousmanov, Catal. Today, 71 (2002) 351.
[14] G. Jacobs, P.M. Patterson, Y. Zhang, T. Das, Appl. Catal. A: General, 233 (2002) 215.
[15] E. Iglesia, S.L. Soled, R.A. Fiato, G.H. Via, J. Catal. 143 (1993) 345.

Studies in Surface Science and Catalysis, volume 147
X. Bao and Y. Xu (Editors)

The structure and reactivity of coprecipitated Co-ZrO$_2$ catalysts for Fischer-Tropsch synthesis

J.-G. Chen and Y.-H. Sun

State Key Laboratory of Coal Conversion, Institute of Coal Chemistry, Chinese Academy of Sciences, P. O. Box 165, Taiyuan 030001, China.

ABSTRACT

The co-precipitated Co-ZrO$_2$ catalysts of varying composition were investigated with BET, XRD, TPR and IR. For the low Co content catalysts, the cobalt existed as highly dispersed species, which led to low extent of reduction and favored CH$_4$ formation. For the high Co content catalysts, the reduction degree reached to 100% and improvement of Fisher-Tropsch synthesis was observed. The performance was closely related to the reducibility of catalysts.

1. INTRODUCTION

Cobalt supported on conventional supports such as SiO$_2$, Al$_2$O$_3$ or TiO$_2$ were widely investigated as catalysts for Fischer-Tropsch synthesis [1]. However, those supports inclined to form irreducible metal-support compounds, leading to both low activity for decrease of effective cobalt and poor stability under hydrothemal reaction conditions [2]. One way to avoid the problem was adopting ZrO$_2$ as support, which exhibited good resistance against H$_2$O vapor genarated in F-T synthesis [3]. Moreover, it had been found that ZrO$_2$ was beneficial for Co catalysts in F-T synthesis. Ali [4] reported that ZrO$_2$ increased the reaction rate and the chain growth probabilty as well. It was important to clarify the role of ZrO$_2$ in Co-containing catalytic systems for catalysts development. Therefore, a series of model Co-ZrO$_2$ catalysts were prepared with purpose to investigate the Co-ZrO$_2$ interaction and its influence on the structure, reducibility and performance in F-T synthesis.

2. EXPERIMENTAL

The Co-ZrO$_2$ catalysts were prepared by co-precipitation of mixed metal salts solution with Na$_2$CO$_3$. The molar ratio of Co/Zr was made by adjusting the amount of ZrOCl$_2$ and Co(NO$_3$)$_2$. The precipitate was filtered until no Cl$^-$ ion

Table 1
The structure parameters of co-precipitated Co-ZrO$_2$ catalysts

Samples	bulk ratio of Co/Zr (mol.)	S_{BET} (m^2g^{-1})	Pore. vol. (ml g^{-1})	Co$_3$O$_4$[a] size (nm)	The amounts of Co species (µmol mg^{-1})		Degree of reducion[b] (%)
					Dispersed cobalt	Separate Co$_3$O$_4$	
Co20Zr80	20:80				0.67	0	55
Co40Zr60	40:60	134.8	0.16	15	1.03	0.20	85
Co50Zr50	50:50	127.3	0.25	16	0.84	0.50	91
Co60Zr40	60:40	137.5	0.28	11	0.62	0.71	112
Co66Zr34	66:34	136.3	0.37	11	0.63	0.75	111
Co75Zr25	75:25	137.4	0.39	10	0.66	0.77	108

a: Calculated by X-ray line widths of peaks (311) of Co$_3$O$_4$ using Scherrer equation.
b: Obtained by integrating TPR area with reference to Co$_3$O$_4$.

was detected, and followed with drying at 393K, calcining at 673K for 4 hours. The nomenclature was CoxZry, where x represented the Co molar percent and y did Zr.

FT-IR spectra of calcined catalysts was obtained by scanning KBr pellets in a transmission mode with a Nicolet550 spectrometer. TPR was performed with catalyst amounts of 15mg at a heating rate of 10K/min. The carrier gas was diluted hydrogen (5%H$_2$ in Ar) and the H$_2$ consumption was detected with TCD. XRD was taken at a Dmax-γA diffractometer (Cu target, Ni filter) with a voltage of 40kV and a current of 100mA, the sample as well as reference compounds were scanned at 0.125°/min. The EXAFS was obtained using synchrotron radiation from the storage ring running at 2.2GeV and 50mA. The EXAFS data were taken in the transmission mode through a double crystal monochromator Si (111) with ion chamber as detector. The coordination shell was isolated by inverse Fourier transform of the radial structure function (RSF) over appropriate region, then the structure parameter of catalysts were obtained by phase and amplitude fitting between catalysts and reference compounds.

The evaluation of catalysts was carried out in a pressured fixed-bed reactor with a catalyst loading of 1ml. The reaction was carried out at 2MPa, 1500h^{-1}, H$_2$/CO=2 and a temperature range from 463K to 503K. The products were collected with a hot trap and a cold trap in sequence. The gaseous flow-out was analyzed with on-line chromatography. The liquid and wax were analyzed with a GC-920 chromatogram equipped with OV-101 capillary columns. Both the carbon balance and mass balance of reaction were kept among 95±5%.

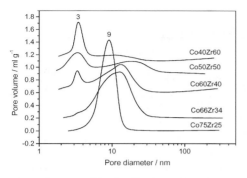

Fig.1 The pore size distribution of Co-ZrO₂

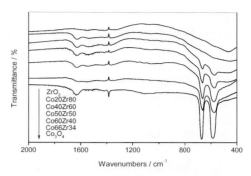

Fig.2 The IR spectra of Co-ZrO₂ catalysts.

3. RESULTS AND DISCUSSION

3.1. Texture and Structure

As shown in Table 1, the specific surface area was almost the same for catalysts of various Co content. However, the pore distribution changed considerably (see Fig.1). The fraction of wide pore (about 9nm) increased at the expense of narrow pores (3nm) with the rise of Co content, accompanied by the pore volume expanding meanwhile. It was known that the porosity of final ZrO_2 depended on polymerization degree of tetrameric ion $(Zr(OH)_2 \cdot 4H_2O)_4^{8+}$ via Zr-O-Zr [5]. In co-precipitation, the cross-polymerization between $Co(OH)_2$ and $Zr(OH)_4$ particles yields more Co-O-Zr instead of Zr-O-Zr bonds, which give rise to loose matrix upon dehydration.

XRD showed broad peaks at 36.8o in calcined catalysts, corresponding to 311 diffraction plane of Co_3O_4. The Co_3O_4 size was determined with line broadening method by referring to NaCl crystal. The Co_3O_4 particle size got smaller as the Co content increased (see Table 1). This might be that the concentrated $Co(NO_3)_2$ solution tended to form more crystallite core on contact with precipitating agent.

Co-ZrO₂ catalysts as well as pure ZrO_2 and Co_3O_4 were pressed into KBr disc and scanned with FT-IR (see Fig.2). The bands in 1000-400 cm⁻¹ were informative for solid skeleton. Co_3O_4 showed two bands at 660 and 574cm⁻¹, corresponding to metal-oxygen stretching from tetrahedral and octahedral sites, respectively [6]. Amorphous ZrO_2 did not show any remarkable adsorption. For Co-ZrO₂ catalysts, bands at 660 and 574cm⁻¹ appeared when Co content exceeded 20% and intensified with the further increase in Co content. The absence of them in Co20Zr80 implied that the cobalt existed as finely dispersed phase. The appearance of them in high Co content catalysts suggested that another Co_3O_4 phase, which was active in IR and XRD, separated. The two phases could be quantitatively determined by rating the IR bands area to pure

Table 2

The coordination parameters of Co atoms from EXAFS

Samples	The first shell				The second shell			
	N_{Co-O}	R_{Co-O}	σ^2	Quality factor	N_{Co-Co}	R_{Co-Co}	σ^2	Quality factor
Co_3O_4	5.3	1.92			4.0	2.83		
Co/SiO_2	5.8	1.92	0.0002	0.07	4.7	2.85	0	0.02
Co60Zr40	4.8	1.92	0.0003	0.148	2.7	2.87	0.0004	0.13

Co_3O_4, since the sample concentration in KBr was in control. As shown in Table 1, all cobalt in Co20Zr80 existed as dispersed species. With increase of Co content, the amounts of dispersed species reached a saturation point initially, and then decreased. Beyond Co40Zr60, the amounts leveled off.

The separate Co_3O_4 phase increased as a function of Co loadings. This was similar to $Cu-ZrO_2$ systems, on which Mohan [7] reported that 2-20 mol % Cu dissolved in ZrO_2 to stabilize fluorite structure and beyond that value isolate CuO phase generated.

The catalyst Co60Zr40 was subjected to EXAFS investigation and the local structural parameter of Co was listed in Table 2. The co-ordination number of Co-O and Co-Co was dramatically smaller than that of the pure Co_3O_4, confirming the presence of highly dispersed cobalt phase besides separate Co_3O_4 phase. It was noted that such a decline of coordination number was not observed for Co/SiO_2, though its average particle size (13 nm by XRD) was comparable with Co60Zr40. This indicated that the occurrence of finely dispersed cobalt phase, which was undetectable by XRD or IR, was proprietary to the co-precipitated $Co-ZrO_2$ system.

3.2. The reducibility of catalysts

The TPR profiles were plotted in Fig.3. For the Co20Zr80 catalyst, reduction peaks emerged at high temperatures (795 and 972K) and the reducibility was the lowest (see Table 1). For Co40Zr60 or Co50Zr50, several peaks appeared and the reducibility ranked as intermediate. For catalysts with Co content over 60%, two peaks were observed and the reducibility reached to 100%.

In-situ XRD of the typical sample Co60Zr40 was carried out to have a deep insight into the reduction of high Co content catalysts (see Fig.4). After reduction at 473K, CoO phase was observed in substitution of Co_3O_4. Rising the temperature to 573K, all diffraction peaks disappeared. But metallic Co phase was clearly observed after reduction at 673K for 4h. Dan [8] investigated the impregnated Co/ZrO_2 with XRD and disclosed the same observations. These suggested that complete reduction was achieved for Co60Zr40 upon pretreatment prior to reaction. This evidenced the mild interaction between cobalt and ZrO_2 for the lack of irreducible zirconate species. Andrea [9] found

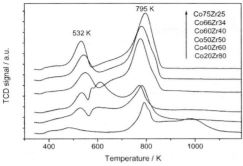 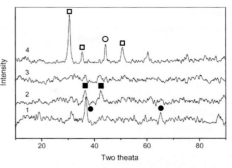

Fig.3 The TPR profiles of Co-ZrO₂ catalyst

Fig.4 The in-situ XRD of Co60Zr40 after reduction at (1) 298K; (2) 473K; (3) 573K; (4) 673K. (● Co_3O_4 ■CoO □ t-ZrO_2 ○ Co)

that a strong Co-SiO₂ interaction was displaced with a weaker Co-ZrO₂ interaction and the reducibility of Co/ZrO₂/SiO₂ catalysts improved on the addition of zirconium. However, the low Co content catalysts still showed poor reducibility as TPR revealed above. This was related to the lack of separate Co_3O_4, which otherwise would promote the reduction of highly dispersed cobalt species through H₂ spillover. So, both the mild Co-ZrO₂ interaction and the co-existence of separate Co_3O_4 phase were essential for the total reduction of cobalt in the systems.

3.3. Performance in Fischer-Tropsch synthesis

The effect of cobalt content on activity and selectivity was listed in Table 3. The CO conversion increased monotonously with the increment of Co content. The activity was also expressed as converted CO per gram cobalt per hour, and the ascending trend against Co loading meant that the utilization of cobalt improved progressively. This was a natural outcome of upgrading reducibility. Nonetheless, the TOF showed a minimum with the rise of Co loadings, suggesting the existence of different active sites.

The CH₄ selectivity declined rapidly when Co content increased from 20 to 50, and then slightly when Co content further increased. Such a change in CH₄ selectivity might be ascribed to the presence of different active sites. Taking the catalysts structure into account, one active site was supposed to be hardly reducible, finely dispersed cobalt species, which facilitated the CH₄ formation. A case in point was Co20Zr80. The other active sites should be metallic Co, which suppressed the CH₄ formation and improved the chain growth. Such sites were predominant in high Co content catalysts. The performance of Co-ZrO₂ catalysts was closely correlated with the cobalt reducibility and to two cobalt species. Similarly, Bruce [10] discovered that the addition of Ni enhanced the Co/ZrO₂ reduction and C_5^+ or 1-enes products were privileged.

Table 3
The performance of Co-ZrO$_2$ catalysts in Fisher-Tropsch synthesis[a]

Catalysts	CO Con. %	Converted CO per Co ml g^{-1} h^{-1}	TOF[b]	Hydrocabons selectivity %		
				CH$_4$	C$_2$-C$_4$	C$_5^+$
Co20Zr80	2.6	0.05	0.058	35.9	48.1	15.9
Co40Zr60	28.1	1.34	0.046	18.5	11.9	69.5
Co50Zr50	26.3	0.96	0.049	8.5	11.2	80.4
Co60Zr40	43.1	1.44	0.025	7.8	6.9	85.3
Co66Zr34	76.5	2.03	0.033	5.9	5.6	88.4
Co75Zr25	98.7	2.67	0.033	5.1	8.3	86.6

a: Reaction conditions: 473K, 1500 h^{-1}, 2MPa.
b: The H$_2$ uptake is measured with H$_2$-pulse chemisorption.

4. CONCLUSION

The co-precipitated Co-ZrO$_2$ catalysts were composed of highly-dispersed cobalt speceis and separate Co$_3$O$_4$ phase. With the rise of Co content, the amounts of latter species increased whereas the former decreased initally, and then levelled off. For the low Co content catalysts, the extent of reduction was low due to the presence of highly-dispersed cobalt species. For the high Co content catalysts, the reduction reached 100% at a mild activation condition. The performance was closely related to the reducibility of catalysts.

ACKNOWLEDGEMENT

This work is supported by State Key Foundation Program for Development and Research of China (Contract No. G19990222402) and National Science Foundation (Contract No. 20303026).

REFERENCES

[1] E.Iglesia, Appl. Catal., 161 (1997) 59.
[2] K.Gabor, E.K. Chris, J.D. Gregory, et al., J. Catal., (2003) in press.
[3] Y.H. Sun, The preparation and application of a Co-ZrO2 in Fischer-Tropsch synthesis, China Patent No. 02140487 (2002).
[4] S. Ali, B. Chen and J.G. Goodwin, J. Catal., 157 (1995) 35.
[5] D. Tichit, D.E. Alami, F. Figueras, Appl. Catal., 145 (1996) 195.
[6] B. Lefez, P. Nkeng, J. Lopitaux, G. Poillerat, Mater. Res. Bull., 31 (1996) 1263.
[7] K.D. Mohan, M.D. Avinash, V.B. Tare, et al., Solid State Ionics, 152 (2002) 455.
[8] I.E. Dan, R. Bernadette, R. Mafalie, R. Renaud, J. Catal., 205 (2002) 346.
[9] F. Andrea, C. Michael, V.S. Eric, J. Catal., 185 (1999) 120.
[10] L.A. Bruce, G.J. Hope, J.F. Mathews, Appl. Catal., 8 (1983) 349.

Studies in Surface Science and Catalysis, volume 147
X. Bao and Y. Xu (Editors)
283

Novel highly dispersed cobalt catalysts for improved Fischer-Tropsch productivity

C. Martin Lok

Johnson Matthey Catalysts, PO Box 1, Billingham, Cleveland, TS23 1LB, UK

ABSTRACT

The preparation, characterization and performance of HDC catalysts, a new generation of high-cobalt Fischer-Tropsch (FT) catalysts with dispersions at, or close to the postulated optimum dispersion of 0.15 are described. The catalysts are made in a single deposition step by deposition-precipitation of Co compounds at high pH via a cobalt ammine complex. This method leads to a uniform distribution of Co crystallites of 3-5 nm. Due to the high dispersion and Co loadings, the catalysts have a high weight and volume activity.

1. INTRODUCTION

The economy of the Fischer-Tropsch (FT) process is critically dependent on the effectiveness of the cobalt catalysts used. Therefore, the design of FT catalysts tends to focus on methods to improve the cobalt utilisation by optimising metal dispersion, improving reducibility and/or increasing stability. Iglesia has shown that the Fischer-Tropsch synthesis rate increases linearly with cobalt metal dispersion, irrespective of the chemical identity of the underlying support, at least in the dispersion range 0 - 0.1[1, see also 2-4]. As crystal structure depends only weakly on crystallite diameter over this dispersion range, it is not certain that above conclusions can be extrapolated to higher dispersions. A complication with very small cobalt crystallites might be their increased tendency to reoxidise under certain FT conditions after which the resulting cobalt oxide may react with active alumina sites resulting in inactive "cobalt aluminate" species [5, 6 see also 7]. Obviously, the more favourable cobalt-to-alumina ratio in high cobalt catalysts makes deactivation through this cobalt/support reaction less likely. Indeed it has been suggested that catalyst productivity can be increased by catalysts that maintain cobalt dispersion at or near an optimum of 0.15 while realising this dispersion at high cobalt loadings [1]. Unfortunately, the study of these catalysts is hampered by the lack of synthetic methods to prepare small Co crystallites at high metal loadings [1].

The classical method for making Co catalysts is by incipient wetness impregnation using an aqueous cobalt nitrate solution at a low pH of 2-3 [8, 9]. At this low pH only modest interaction between the positively charged alumina carrier and the similarly charged Co cations occurs and as long as the drying process has not been completed, at least part of the highly soluble cobalt nitrate remains mobile. Thus, depending on the drying conditions, Co may be deposited as relatively large clusters of crystallites of up to a few 100 nm [10]. Though these clusters break up during calcination and reduction allowing this method to yield good catalysts, dispersion is mostly only 0.05-0.10, i.e. lower than the postulated optimum of 0.15. In addition, it is difficult to realise this high dispersion at high metal loadings (>>20%) and promoters like Re are required which stabilise the cobalt phase during cobalt nitrate decomposition [9].

We report here on the newly developed so-called HDC (High-Dispersion Cobalt) catalysts [11, 12, 13] which show dispersions close to the desired 0.10-0.15 even at high metal loadings of up to 50% cobalt in the reduced state and which can be made in a single deposition step in a nitrate-free process. A feature of the process is that the cobalt is deposited at high pH thus realising a firm anchoring of the cobalt cation onto the now negatively charged support. Though the HDC method is applicable for a variety of supports, this paper focuses on alumina-based catalysts only.

2. EXPERIMENTAL

2.1. Catalyst Synthesis

A range of catalysts differing in cobalt loadings was prepared by deposition precipitation from excess solution. This was done by slurrying an alumina powder with the appropriate amount of cobalt ammine complex solution [12]. Deposition of the Co was effected by ammonia evaporation. Two aluminas were employed viz. a Sumitomo alumina (BET SA 145 m^2/g, pore volume 0.85 ml/g (0.98 des)), and a Sasol alumina (BET SA 149 m^2/g, pore volume 0.78 ml/g).

2.2. Catalyst Characterisation

Cobalt surface areas were measured using hydrogen chemisorption at 150°C in a Micromeritics ASAP 2010 after reduction at 425 °C, assuming the surface area occupied by one cobalt atom to be 0.0662 nm^2[14, 15].

Reduction behaviour was studied in the ICI Measurement Science Group, Wilton (UK) by in-situ reduction in an Anton Paar XRK reaction chamber on a Siemens D5000 X-ray diffractometer using a flow of 4 % H_2 in N_2.

Crystallite size distribution was measured at Delft University using a Philips CM30T 300 kV electron microscope.

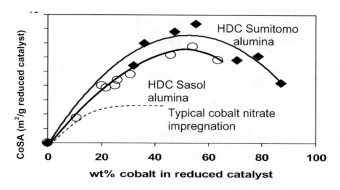

Fig. 1. Cobalt surface area as a function of cobalt content (no promoters)

2.3. Performance testing

Performance testing was done at CAER (Centre for Applied Energy Research), University of Kentucky. The catalysts were first reduced in a fixed-bed reactor at 425 °C for 4 h. They were then transferred to a 1 1 CSTR and re-reduced in-situ at 230°C for 15 h before starting the FTS reaction at 20 bar and 180°C while raising the temperature to 210°C in 3 h. The flow of the gas mixture (molar ratio H_2/CO 2.1 : 1) was then adjusted to reach 50% conversion.

3. RESULTS AND DISCUSSION

Using the HDC method the cobalt surface area increases almost linearly with cobalt content to about 30-40 m^2/g catalyst for 50 % Co in the reduced catalyst

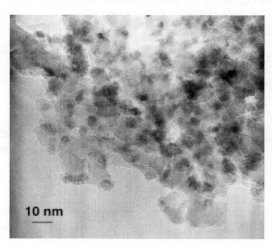

Fig. 2. TEM of a reduced 22% Cobalt HDC catalyst

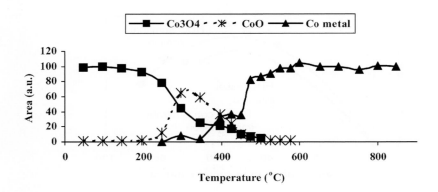

Fig. 3. Cobalt phases (XRD) during in-situ reduction of a 22% Cobalt HDC catalyst

[Fig. 1]. This compared to the 12-16 m^2/g generally attainable for a 20wt% Co catalyst made by nitrate impregnation [8]. Though the dispersion eventually declines from 0.15 for 20 wt% Co to about 0.10 for a 50 wt% Co catalyst, and subsequently to much lower values, the dispersion is better maintained than in cobalt nitrate impregnation where an earlier levelling off of the cobalt surface area or activity is observed [8, 9]. Moreover, repeated cobalt nitrate impregnations are required to reach high Co levels.

TEM showed the structure of HDC catalysts [Fig. 2] to be markedly different from that of nitrate based catalysts. The latter contain Co crystallites in large clusters [10] while in HDC the crystallites are homogeneously distributed.

In-situ XRD showed the reduction to occur in two discrete steps: at first a reduction of Co_3O_4 to CoO at 200-300 °C, followed by reduction of CoO to the metallic state at 350-450 °C [Fig. 3].

Line broadening analysis of the 22% Co catalyst yielded a crystallite size

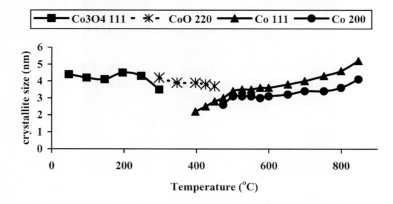

Fig. 4. Crystallite size of Cobalt HDC catalyst (22% Co) during reduction

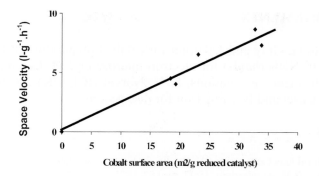

Fig. 5. Fischer-Tropsch activity (space velocity to reach 50% CO conversion) in a CSTR (after 200 h at 210 °C) as a function of cobalt surface area.

for Co_3O_4 and CoO of d = 3 and 4 nm respectively [Fig.4]. The d of 3 nm observed for the Co metal is in excellent agreement with the 3 nm determined with TEM. Similarly for the 34% catalyst a d of 4.5 and 5 nm were measured by TEM and XRD, respectively. The small crystallites are remarkably stable and only slowly lose dispersion upon raising the reduction temperature to 800°C.

The data on the crystallite size distribution indicate that the main mechanism of growth at higher reduction temperatures is by crystallite coalescence rather than "Ostwald type" ripening. This metal crystallite growth appeared to develop as a bimodal distribution. In addition, at higher temperatures the main cobalt crystal structure is face centered cubic while at lower temperatures hexagonal and other phases occur (see also [2] for nitrate based catalysts).

Fixed-bed activities of the reduced catalysts appeared to be linearly proportional to cobalt surface area per g catalyst as measured by hydrogen chemisorption confirming that the findings by Iglesia [1] are also valid for higher cobalt dispersions and loadings. Fig. 5 shows the activity results in the CSTR test displaying the same ranking as in the fixed-bed test. C_{5+} selectivity was moderate for low-Co catalysts but increased with Co loading up to the level of cobalt nitrate impregnated catalysts, possibly because of increased re-adsorption of olefins. Thus at a space velocity of 8.70 h^{-1} and 50% CO conversion, the following product composition was observed for the 42 wt% Co (oxidic) catalyst: C_{5+}: 85.1%; CH_4: 6.3%; olefins: 4.5%; C_2-C_4: 3.4%, CO_2: 0.3%, very similar to that of a cobalt nitrate impregnated catalyst. A lifetime test in the CSTR showed a slow decrease in CO conversion from 53 % after 100 h to 41% after 600 h operation.

Overall we can conclude that due to the high dispersion and cobalt loadings, HDC catalysts have a high weight and volume activity, which should have impact on the operation and/or economics of the FT process.

ACKNOWLEDGEMENTS

We are grateful to G. Gray and J. Turner for catalyst preparation, P.J. Kooyman (NCHREM, Delft, Netherlands) for electron microscopy. M. Luo and B. Davis (CAER) for performance evaluations, S. V. Norval (ICI, UK) for in-situ XRD and J. Abbott, J. Casci and B. Crewdson for discussions.

REFERENCES

[1] E. Iglesia, Natural Gas Conversion IV, Studies in Surface Science and Catalysis vol.107, Elsevier Science B.V, Amsterdam, 1997, pp 153-162.
[2] D.E. Enache, B. Rebours, M. Roy-Auberger and R. Revel, J. Catal., 205 (2002) 346.
[3] B.G. Johnson, C.H. Bartholomew and D.W. Goodman, J. Catal. 128 (1991) 231.
[4] M. Kraum, Thesis, Bochum (1999) pp 105.
[5] D. Schanke, A.M. Hilmen, E. Bergene, K. Kinnari, E. Rytter and E. Adnanes, Catal. Lett. 34 (1995) 269.
[6] G.P. Huffman, N. Shah, J. Zhao, F.E. Huggins, T.E. Hoost, S. Halvorsen, J.G. Goodwin, J. Catal. 151 (1995) 17.
[7] P.J. van Berge, J. van de Loosdrecht, S. Barradas, A.M. van der Kraan, Catal. Today, 58 (2000) 321.
[8] W.-J. Wang and Y.-W. Chen, Appl. Catal., 77 (1991 223.
[9] T. Riis, S. Eri, G. Marcelin, J. Goodwin, US Patent 4 880 763 (1989).
[10] M-C. Marion and M. Roy, US Patent No 6 235 798 (2001).
[11] R.L. Bonne, C.M. Lok, US Patent No 5 874 381 (1999).
[12] C.M. Lok, G. Gray, G.J. Kelly, WO Patent Application No 01/87480 (2001); see also Cattech 6:1 (2002) 9.
[13] C.M. Lok, S. Bailey, G. Gray, US patent 6534436 (2003).
[14] R.C. Reuel and C.H. Bartholomew, J. Catal. 85 (1984) 63.
[15] Operators Manual for the Micromeritics ASAP 2000 Chemi System V 1.00, Appendix C, Part no 200-42808-01 (1991).

Studies in Surface Science and Catalysis, volume 147
X. Bao and Y. Xu (Editors)

Reactivity of paraffins, olefins and alcohols during Fischer-Tropsch synthesis on a Co/Al$_2$O$_3$ catalyst

F. Fiore[a], L. Lietti[a], G. Pederzani[b], E. Tronconi[a], R. Zennaro[b], P. Forzatti[a]

[a]Dipartimento di Chimica, Materiali, Ingegneria Chimica "Giulio Natta", Politecnico di Milano. Piazza Leonardo da Vinci, 32 - 20133 Milano, Italy

[b]Catalysis and Process Technology Research Centre, EniTecnologie, Via Maritano, 26 - 20097 San Donato Milanese, Italy

ABSTRACT

Mechanistic aspects of the Fischer Tropsch synthesis are here investigated by chemical enrichment experiments of olefin, paraffin and alcohol molecules. The results indicate that paraffins and alcohols are terminal products, whereas olefins are hydrogenated, hydroformilated and partecipate in the chain growth.

1. INTRODUCTION

The catalytic conversion of synthesis gas into hydrocarbons (Fischer-Tropsch Synthesis, FTS) is recently receiving great attention as a powerful way of exploiting natural gas. This has determined the flourishing of many studies aiming at a better understanding of mechanistic and kinetic aspects of the synthesis. In these respects, several papers deal with the issue of the secondary reactions of olefin, and excellent papers were provided e.g. by Iglesia et al. [1], Schulz and Claeys [2], Patzlaff et al. [3]. These authors provided clear evidence in favor of olefin readsorption in the FTS, but different mechanistic implications were derived. In order to verify the occurrence of such reactions in the case of our Co-based catalyst and aiming at the development of a mechanistic-kinetic model of the reaction, an extensive research activity on the FTS was undertaken in our Labs. As a preliminary step in the ongoing kinetic analysis, a suitable reaction pathway for the synthesis was developed, based on both literature indications and direct evidences provided by mechanistic investigations. For this purpose, chemical reaction experiments were performed by addition of several "probe molecules" (paraffins, olefins and alcohols) to the CO/H$_2$ feed of a Co/Al$_2$O$_3$ catalyst operating under actual synthesis conditions.

2. EXPERIMENTAL

Chemical enrichment experiments were performed in a fixed bed reactor loaded with 3 g of a Co/Al$_2$O$_3$ catalyst (Co = 15 % w/w) operating at T = 220 °C, P = 2 MPa, H$_2$/CO = 1.6 and GHSV = 5000 h^{-1}. Paraffins (octane, n-decane), olefins (ethylene, 1- and 2-octene) and alcohols (methanol, ethanol and 1-octanol) were added to the syngas. Addition of the probe molecules does not result in significant changes in the CO conversion with respect to the "base" run.

3. RESULTS AND DISCUSSION

Paraffins and alcohols addition - In this case no significant modifications were observed under our experimental conditions in the productivity of the various reaction products with respect to the "base" case. Only a slight increase in the productivity of lower alcohols (C1-C4) was observed upon addition of methanol and ethanol. This indicates that paraffins and alcohols behave as terminal products of the synthesis, in line with other literature reports [4-6].

1- and 2-alkenes addition – Ethylene was co-fed at three different concentration levels (0.3, 0.7 and 2.6 % mol/mol). Figures 1 a-c show the results obtained in the case of the highest ethylene concentration in the feed stream. The hydrocarbons and alcohol Anderson-Schultz-Flory (ASF) distributions are reported (fig. 1 a-b), along with the olefin/paraffin (O/P) molar ratio (fig. 1c). Addition of ethylene to the syngas leads to a significant increase in the C$_2$ fraction (ethylene and ethane); the productivity of C$_{2+}$ species is also slightly enhanced, apparently up to C$_{20}$-C$_{25}$. Also, a significant increase in the 1-propanol productivity is observed (fig. 1 b), along with a minor increase in the higher alcohols formation. No significant variations in the productivity of other reaction products was noticed. Hence the results indicate that co-fed ethylene is hydrogenated to ethane, hydroformilated to 1-propanol and incorporated in the chain growth leading to higher hydrocarbons (olefins and paraffins), leading to the observed changes in the ASF distribution. The C$_{2+}$ olefins produced are in

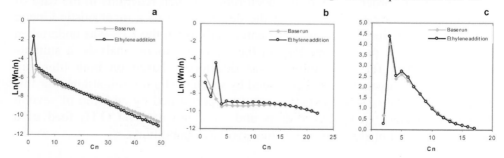

Fig. 1. Comparison of ethylene co-feed data with the "base" case. a) Hydrocarbon ASF plot; b) alcohols ASF plot; c) Olefin/paraffin molar ratio.

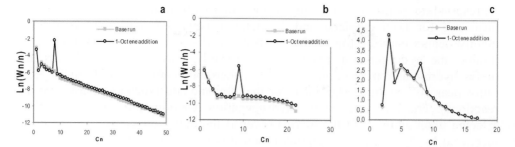

Fig. 2. Comparison of 1-octene co-feed data with the "base" case. a) Hydrocarbon ASF plot; b) alcohols ASF plot; c) Olefin/paraffin molar ratio.

turn involved in the same reactions pointed out in the case of co-feed ethylene, and this explains e.g. the slight increase in the productivity of C_{3+} alcohols along with 1-propanol. Ethylene co-feed experiments performed with lower ethylene inlet concentration (0.35 and 0.7 % mol/mol) lead to very similar results.

Similar experiments have been performed with 1-octene. Three different olefins inlet concentrations were selected (0.26, 0.55 and 1.15 % mol/mol), and results obtained in the case of the highest co-feed olefin concentration are shown in fig. 2 a-c in terms of hydrocarbons and alcohols ASF distributions (fig. 2 a and b, respectively) and of olefin/paraffin ratio in the products. The obtained results parallel those already reported for ethylene and indicate that 1-octene is hydrogenated to octane, hydroformilated to 1-nonanol and incorporated.

Fig. 3 shows the values of ethylene and 1-octene conversion and of the selectivity to the various routes depicted above for the three levels of olefin inlet concentration considered in this study. It appears that the conversion of the added olefin is high, and specifically in the case of ethylene which exhibits conversion values up to 80-90 %. 1-octene conversion is much lower, being below 40 %. The added olefins are predominantly hydrogenated to the

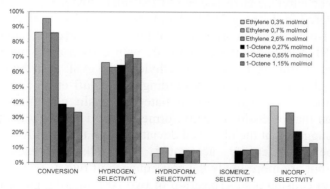

Fig. 3. Ethene and 1-octene conversion and selectivity during chemical enrichment experiments. The data refer to 3 olefin inlet concentrations.

corresponding paraffin: selectivity values to paraffins near 70 % have been measured for both ethylene and 1-octene. Likely, the olefin hydrogenation mainly occurs near the top of the catalyst bed due to the lack of water which inhibits the hydrogenation reactions [1]. Notably, during the "base" runs the olefin hydrogenation reaction is far from thermodynamic equilibrium (K_p/K_{eq} ratios near 10^{-9} could be estimated for the ethylene and 1-octene hydrogenation reactions). The K_p values of the hydrogenation reactions are not significantly affected upon addition of the olefins (even if the olefin concentration has been increased by 2 orders of magnitude), and this is characteristic of a first-order reaction in the olefin concentration. As a matter of facts, the olefin/paraffin (O/P) molar ratio is not affected by ethylene addition (fig. 1c) whereas a limited increase in the O/P ratio is evident in the case of 1-octene (fig. 2c). However, the observed deviation decreases on decreasing the inlet 1-octene concentration.

Besides hydrogenation, a significant fraction (near 10 %) of the added olefins is hydroformilated to the C_{N+1} alcohols (1-propanol in the case of ethylene and 1-nonanol in the case of 1-octene). The relevant formation of C_{N+1} alcohols from the C_N olefins clearly indicates that hydroformilation plays a role in the alcohol formation during the FTS, although a different route should be invoked to explain the formation of C_1-C_2 alcohols. It may be speculated that alcohols are formed via e.g. a CO-insertion route (as in the case of Co-based catalysts for the higher alcohol synthesis [7]), and via the previously described hydroformilation route which operates simultaneously with the CO-insertion pathway. As a matter of facts, a discontinuity is observed in correspondence of the C_4 alcohol in the alcohol ASF plot (figs. 1b and 2b). Notably, for both ethylene and 1-octene the selectivity to hydroformilation is apparently not affected by the inlet olefin concentration, hence suggesting a first-order kinetics.

The added olefin molecules are also incorporated in the chain growth, with a selectivity (estimated in this case from the olefin mass balance) which is near 30 % in the case of ethylene (irrespective of the inlet ethylene concentration) and with data scattered in the range 10-20 % in the case of 1-octene. Notably, while in the case of 1-octene the incorporation of the olefin leads to an almost parallel shift of the ASF curves for the compounds with carbon atom numbers higher than the added olefin, in the case of ethylene the ASF distributions tend to converge at high carbon number. According to Patzlaff et al. [3,8] this would suggest that the readsorbed olefin initiate the chain growth with a lower probability than that of chains directly formed in the FTS. As a matter of facts, these authors suggest that the observed deviation from the ideal ASF distribution could not be satisfactorily accounted by 1-alkenes readsorption and incorporation, but are more likely the consequence of two different mechanisms of chain growth causing a superposition of two ASF distribution. However, the observation that in the case of 1-octene co-feed experiments the incorporation of the olefin molecule leads to an almost parallel shift of the ASF curve apparently contrast with this suggestion. Finally, in the case of 1-octene, evidence for the occurrence of isomerization reactions have been collected, as pointed out by the increase in the productivity of internal octenes (selectivity close to 10 %).

The reactivity of an internal olefin (2-octene) was also investigated. Octane is the major reaction product obtained from 2-octene co-feed (hydrogenation), but minor amounts of 1-octene have also been detected (isomerization). A very small increment in the 1-nonanol productivity was also evident, likely originating from 1-octene hydroformilation, whereas indications concerning the occurrence of incorporation reactions involving 2-octene could not be obtained. Concerning the hydrogenation reaction, a significantly higher O/P ratio was observed with respect to the standard run in correspondence of the C_8 olefin/paraffin couple in the case of 2-octene addition, indicating the low reactivity of internal olefins in hydrogenation reaction.

Our data parallel those recently reported by Schulz et al. [2] and by Patzlaff et al. [3] over Co-based catalysts. Very similar results (even under a quantitative point of view) were indeed reported by these authors in spite of the different operating conditions adopted in the experiments. Notably, in the case of ethylene co-feed and in line with our data, both groups [2,3] reported conversion values in the range 85-99 % irrespective of the inlet ethylene partial pressure, with selectivities to hydrogenation and incorporation in the range 74-86 % and 14-26 %, respectively. In line with these findings, our work provides additional evidence on the independency of ethylene conversion and selectivities of hydrogenation and incorporation of the ethylene partial pressure, an indication of first-order reaction. Similar conclusions can be derived in our data from 1-octene co-feed experiments as well. Finally, it is noted that lower values for the hydroformilation reaction were reported in [2, 3], possibly due to the lower CO partial pressure used in these studies.

Mechanistic scheme of the FTS - On the basis of the results of co-feed experiments reported above a mechanistic scheme which accounts for the formation of the major reaction products was considered (fig. 4). The proposed key-intermediates in the scheme are C_N alkyl species bonded to the surface via the terminal or an internal C-atom. The terminal C-bonded species are originated upon re-adsorption of a C_N α-olefin or via C_1 addition from an adsorbed C_{N-1} alkyl species; this C_N adsorbed species may be involved in hydrogenation reactions to paraffin, chain growth via C_1 addition, hydroformilation, isomerization. The internal C-bonded species, on the other hand, may be originated upon adsorption of an internal olefin molecule and may lead upon desorption to an α-olefin via isomerization or may be hydrogenated to the corresponding paraffin; this species is apparently involved neither in hydroformilation reactions nor in the chain growth process, since branched species were not detected to a significant extent.

The proposed reaction scheme is presently considered in a parallel work as

294

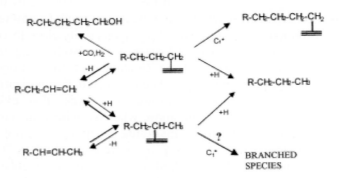

Fig. 4. Reaction scheme of the Fischer-Tropsch Synthesis.

the basis for the development of a mechanistic kinetic model of the FTS able to describe simultaneously the CO conversion and the hydrocarbon product distribution. In the model, which takes into account the olefin readsorption process and makes use of first order kinetics for both hydrogenation and chain growth of the adsorbed intermediates, a carbon number-dependent solubility for olefins was considered based on EOS predictions. The model satisfactorily describes both CO conversion and hydrocarbon distribution obtained during catalytic activity runs performed under different operating conditions in the fixed bed reactor. However, when used as a predictive tool for chemical enrichment experiments, a strong underestimation of the hydrogenation selectivity of the co-feed molecule was observed, accompanied by a symmetrical overestimation of the alkene incorporation. This may suggest that a specific catalytic site is present on the catalyst surface which does not participate in the chain growth but which is able to hydrogenate the adsorbed alkyl species, as suggested by Schulz and Claeys [9]. Notably, the model is anyway able, under a qualitative point of view, to account successfully for the observed change in the slope of the ASF distribution of C_{2+} products upon ethylene co-feed without invoking the superposition of two ASF distributions. Work is presently in progress in order to elucidate these aspects.

REFERENCES

[1] E. Iglesia, S.C. Reyes, R.J. Madon, S.L. Soled, Adv. Catal., 39(2) (1993) 345
[2] H. Schulz and M. Claeys, Appl. Catal. A:General. 186 (1999) 71
[3] J. Patzlaff, Y. Liu, C. Graffmann, J. Gaube, Catal. Today 71 (2002) 381
[4] C.N. Satterfield, R.T. Hanlon, Energy & Fuels 2 (1988) 196
[5] P.H. Emmett, J.T. Kummer, T.W. De Witt, J.Am.Chem.Soc. 70 (1948) 3632
[6] P.H. Emmett, R.J. Kores, W.K. Hall, J.Am.Chem.Soc. 79 (1957) 2989
[7] P. Forzatti, E. Tronconi, I. Pasquon, Catal. Rev.-Sci- Eng., 33 (1991) 109
[8] J. Patzlaff, Y. Liu, C. Graffmann, J. Gaube, Appl. Catal. A: General. 186 (1999) 109
[9] H. Schulz and M. Claeys, Appl. Catal. A:General. 186 (1999) 91

Studies in Surface Science and Catalysis, volume 147
X. Bao and Y. Xu (Editors)

Genesis of active sites in silica supported cobalt Fischer-Tropsch catalysts: effect of cobalt precursor and support texture

A. Y. Khodakov[a], J.S. Girardon[a], A. Griboval-Constant[a], A. S. Lermontov[b] and P.A. Chernavskii[b]

[a]Laboratoire de Catalyse de Lille, Université des Sciences et Technologies de Lille, Bât. C3, Cité Scientifique, 59655 Villeneuve d'Ascq, France

[b]Department of Chemistry, Moscow State University, 119992 Moscow, Russia

ABSTRACT

Optimization of cobalt deposition and textural properties of catalytic support represent two different but complementary approaches to enhance the performance of cobalt silica-supported Fischer-Tropsch catalysts. It is shown that low temperature decomposition of supported cobalt complexes involves formation of small Co_3O_4 crystallites. The exothermic effect due to the decomposition of cobalt precursor seems to be a major factor responsible of cobalt silicate formation. Gentle decomposition of cobalt precursor generally leads to Fischer-Tropsch catalysts with higher activity and selectivity to higher hydrocarbons. It was found that the sizes of supported cobalt particles prepared using cobalt nitrate depended on silica pore sizes. Narrow pore size distribution in periodic mesoporous silicas allows relatively high cobalt dispersion to be maintained at high cobalt loading. This leads to higher density of active sites and more active Fischer-Tropsch catalysts.

1. INTRODUCTION

Fischer-Tropsch (FT) synthesis is a part of Gas-To-Liquids (GTL) technology, which represents one of the major routes of natural gas utilization. Cobalt supported catalysts have been particularly convenient for the production of higher hydrocarbons [1, 2]. It has been largely shown that the number of active metal sites in FT catalysts and their catalytic activity are functions of both cobalt particle size and reducibility.

A weak interaction of metal and support is a particularity of silica-supported catalysts. Reduction of supported Co_3O_4 crystallites generates active sites for FT reaction. The formation of hardly reducible amorphous cobalt silicate should be avoided, since it does not produce any active sites for FT synthesis. Optimization of catalytic performance silica-supported cobalt FT catalysts would primarily require both minimizing concentrations of hardly reducible silicate species and efficient control of cobalt dispersion.

This paper focuses on two different but complementary methodologies, which have been used to handle cobalt dispersion, concentration of active sites and thus catalytic performance of silica-supported cobalt FT catalysts. The first technique involves the control of the structure of cobalt species and their catalytic performance by using different cobalt precursors and by optimizing their deposition and subsequent decomposition. The second method is based on textural properties of catalytic support.

2. EXPERIMENTAL

The catalysts were prepared via aqueous impregnation to incipient wetness. After impregnation the samples were dried at 373 K and then calcined in airflow at temperatures from 373 to 673 K. To introduce high cobalt contents (>10 wt. %), impregnation and calcination were repeated several times. Prior to catalytic measurements, the catalysts were reduced in hydrogen at 673 K for 5 h. The catalysts are labeled as xCoST, where x designates the weight percentage of cobalt, S is the impregnating cobalt salt (nitrate=N, acetate=AC or oxalate=OX) and T is the temperature of decomposition of supported cobalt salt in the flow of air.

To evaluate the effect of cobalt precursor, the catalysts were supported by Cab-o-Sil M-5 (Cabot) silica (S_{BET}=213 m^2/g, pore volume= 0.84 cm^3/g). To study the influence of support, both commercial and periodic MCM-41 and SBA-15 mesoporous silicas synthesized in our laboratories were used for supporting catalysts. Catalyst characterization was carried out using UV-visible spectroscopy, DSC-TGA, X-ray diffraction, transmission electron microscopy (TEM) and temperature programmed reduction (TPR) combined with *in situ* magnetic measurements [3]. Catalytic performance in FT synthesis was evaluated in a fixed bed reactor at 463 K at atmospheric pressure using synthesis gas with H_2/CO=2 ratio. The FT rates (in s^{-1}) are normalized by the total amount of cobalt loaded into reactor. The catalytic runs were conducted beyond 24 h in order to discard any transient behavior of the catalysts.

3. RESULTS AND DISCUSSION

3.1. Influence of cobalt precursor and conditions of its decomposition

Structure of cobalt complexes deposited on silica via impregnation. UV-visible spectra of impregnating solutions of nitrate and acetate (Figure 1) and

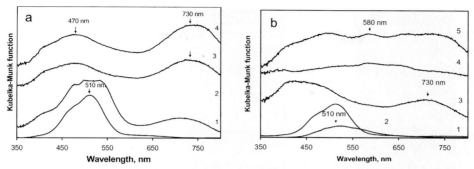

Fig. 1. UV-visible spectra of 10CoN and 10CoAC catalytic precursors: a - (1) solution of cobalt nitrate, (2) after impregnation and drying, (3) 10CoN373, (4) 10CoN673; b - (1) solution of cobalt acetate, (2) after impregnation and drying, (3) 10CoAC443, (4) 10CoAC493, (5) 10CoAC673.

oxalate suspension (not shown) present a band near 510 nm characteristic of octahedral Co^{2+} complexes [4, 5]. These octahedral complexes remain almost intact in the impregnated and dried catalysts (Figure 1, curve 2). The UV-visible spectra of impregnated CoOX were identical to the spectra of bulk salt, which indicates the presence of relatively large oxalate clusters.

Decomposition of cobalt complexes. DSC-TGA analysis showed that supported nitrate, acetate and oxalate complexes decomposed respectively in air at 423-473, 493-523 and 473-493 K. The first step of decomposition involves loss of water molecules, which is followed by disintegration of anions. Calorimetric measurements showed that the decomposition of cobalt nitrate was endothermic, whereas decomposition of acetate and oxalate was highly exothermic.

Decomposition of supported cobalt nitrate and cobalt oxalate at different temperatures led predominantly to Co_3O_4, which was identified by its XRD patterns and broad bands at 470 and 730 nm in UV-visible range [5] (Figure 1a, curves 3 and 4). The decomposition of cobalt acetate at temperatures higher than 443 K gives rise to a low intense band at 580 nm, which is likely to be attributed to Co^{2+} ions in amorphous cobalt silicate. A rather different UV-visible spectrum was observed after decomposition of supported cobalt acetate at lower temperatures (Figure 1b, curve 3). The spectrum displays only the bands attributable to Co_3O_4. It appears that Co_3O_4 is the main product of cobalt acetate decomposition at lower temperatures, while decomposition of cobalt acetate at higher temperatures leads to amorphous cobalt silicate. This suggestion is consistent with TEM data, which also show a strong impact of decomposition temperature on the structure of cobalt species in 10CoAC catalysts. Decomposition of cobalt acetate at 443 K leads to small crystallites of Co_3O_4 (Figure 2a), while no Co_3O_4 crystallites were detected after decomposition of supported cobalt acetate at higher temperatures (Figure 2b).

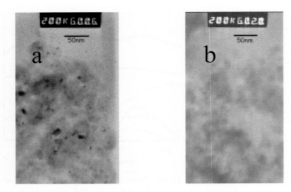

Fig. 2.TEM of 10CoAc443 (a) and 10CoAc493 (b).

It can be suggested therefore, that decomposition of supported cobalt complexes proceeds in conformity with the following schema:

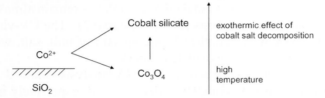

At mild conditions the decomposition primarily leads to small Co_3O_4 particles. These particles can either sinter into large Co_3O_4 crystallites or turn up into amorphous cobalt silicate when decomposition is conducted at higher temperatures. If the decomposition of cobalt salt is exothermic (CoAC and CoOX), considerable concentrations of amorphous cobalt silicate are usually produced.

Reduction of cobalt species. Typical TPR and magnetization curves of cobalt catalysts in 5%H_2/Ar (24 °/min) are shown in Figures 3. It is known that some of peaks in TPR profiles could be associated with partial reduction of metal oxides. The magnetic method, which is selectively sensitive to the metal particles, was particularly helpful in interpretation of these complex TPR data. For example on the basis of combined TPR and magnetic measurements, the low temperature peak in the TPR profile of 10CoOX673 (Figure 3) was attributed to the reduction of Co_3O_4 to CoO. Both the shape of TPR profiles and calculated extent of cobalt reduction were found to be functions of cobalt precursor and temperature of its decomposition. The extent of reduction was above 80% in CoN catalysts. The reducibility was slightly lower in CoOX (~ 60%) despite the presence of larger cobalt particles detected by XRD (>22 nm). Lower extent of reduction of CoOX seems to be related to the exothermicity of cobalt oxalate decomposition and presence of amorphous cobalt silicate. In CoAC catalysts the reducibility was a function of the temperature of cobalt acetate decomposition;

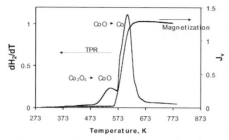

Fig. 3 Combined TPR (5%H_2 in Ar) and magnetic measurements of the reduction of 10CoOX673 catalyst.

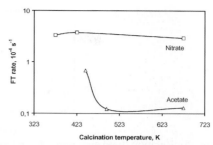

Fig. 4. FT reaction rates on 10CoN and 10CoAC catalysts as a function of calcination temperature.

the lower is acetate decomposition temperature, the higher proportion of cobalt ions can be reduced into metallic cobalt at a given temperature.

Catalytic performance. FT behavior of cobalt supported catalysts is strongly affected both by type of cobalt precursor and by its decomposition temperature. The most active were the catalysts prepared from cobalt nitrate. With the same cobalt precursor however, lower temperature of decomposition of supported cobalt salt generally results in higher FT reaction rates. The effect was more pronounced for CoAC sample; the FT rate increased by nearly 10 times (Figure 4), when the decomposition of cobalt acetate was conducted at 443 K instead of 493 K. Methane selectivity was also much lower on the catalysts calcined at lower temperatures. Low FT reaction rates were observed over Co catalysts prepared using cobalt oxalate suspensions, which was attributed to the large size of cobalt particles.

Characterization and catalytic results suggest that the temperature and exothermicity of decomposition of cobalt complexes represent major parameters, which affect cobalt dispersion, reducibility and catalytic performance of cobalt silica-supported catalysts.

3.2. Effect of support texture

Optimization of support texture represents another approach to increase the number of active sites and thus catalytic activity of cobalt catalysts. Figure 5 displays an experimental correlation between pore diameters in silica supports, sizes of supported Co_3O_4 particles and FT reaction rates. All silica supports showed monomodal pore size distribution. Cobalt dispersion seems to be essentially affected by support pore sizes [6]. Hardly reducible small cobalt clusters were observed in narrow pore MCM-41 materials (pore sizes ~ 2 nm). Easy reducible larger cobalt particles were found in large pore SBA-15 supported catalysts (pore sizes 3-10 nm). Due to better reducibility of larger cobalt particles and possibly lower oxidative deactivation during the reaction, wide pore catalysts exhibited higher FT reaction rates (Figure 5).

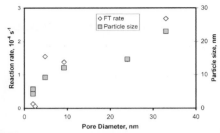

Fig. 5. Relation between pore diameters, sizes of supported Co_3O_4 particles and FT reaction rates over silica supported catalysts.

Our results also show that because of narrow pore size distribution, relatively high cobalt dispersion can be maintained in SBA-15 mesoporous silicas at high cobalt contents (up to 30 wt.%). High cobalt dispersion at high cobalt loading leads to higher density of active sites and therefore more active cobalt catalysts [7]. Over the catalysts with a high extent of cobalt reduction, hydrocarbon selectivities were only slightly affected by the type of silica support.

4. CONCLUSION

Metal dispersion, reducibility and FT catalytic performance of cobalt silica supported catalysts are strongly affected by both cobalt precursor and support textural properties. High temperature and exothermicity of decomposition of cobalt precursor usually lead to hardly reducible cobalt silicate and lower FT reaction rates. Decomposition of cobalt precursor at mild conditions seems to be essential for the design of FT catalysts with higher activity and enhanced hydrocarbon selectivity.

It was shown that the sizes of supported cobalt particles prepared using cobalt nitrate solutions depended on the support pore sizes. Large surface area and narrow pore size distribution in SBA-15 mesoporous silicas seem to prevent cobalt particles from sintering. Higher density of active sites in cobalt catalysts supported by periodic mesoporous materials results in higher FT activity.

REFERENCES

[1] E. Iglesia, Appl. Catal. A 161 (1997) 59.
[2] A. C. Vosloo, Fuel Processing Technology 71(2001) 149.
[3] P.A.Chernavskii, A.S.Lermontov, G.V.Pankina, S.N.Torbin, V.V.Lunin, Kinet.Catal., 43 (2002) 268.
[4] M.G. Ferreira de Silva, Material Research Bulletin 34(1999) 2061.
[5] Y. Okamoto, K. Nagata, T. Adachi, T. Imanaka, K. Inamura and T. Takyu, J. Phys.Chem. 95(1991) 310.
[6] A.Y. Khodakov, A. Griboval-Constant, R. Bechara and V.L. Zholobenko, J. Catal. 206(2002) 230.
[7] A.Y. Khodakov, R. Bechara and A. Griboval-Constant, Appl. Catal. A 254(2003) 273.

Studies in Surface Science and Catalysis, volume 147
X. Bao and Y. Xu (Editors)

Fischer-Tropsch synthesis on cobalt catalysts supported on different aluminas

Dag Schanke[a], Sigrid Eri[a], Erling Rytter[a], Christian Aaserud[c], Anne-Mette Hilmen[b, d], Odd Asbjørn Lindvåg[b], Edvard Bergene[b] and Anders Holmen[* c]

[a]Statoil R&D, Research Centre, Postuttak, N-7005 Trondheim, Norway.

[b]SINTEF Applied Chemistry, N-7465 Trondheim, Norway. [d]Present address: Shell Technology Norway, Drammensveien 147A, N-0277 Oslo, Norway.

[c]Dept Chemical Engineering, Norwegian University of Science and Technology (NTNU), N-7491 Trondheim, Norway.

ABSTRACT

Alumina with different surface area has been prepared by heat treatment of γ-alumina. Improved C_5^+ selectivity was obtained by using Co or CoRe supported on low surface area alumina (LSA, 15 m^2/g). LSA alumina resulted in the lowest Co dispersion and in reduced activity for secondary hydrogenation of propene.

1. INTRODUCTION

A key element in improved Fischer-Tropsch (FT) processes is the development of active catalysts with high selectivity for long chain paraffins. Supported Co catalysts are preferred in FT synthesis of waxes from natural gas due to their high activity and selectivity, low water-gas shift activity and resistance towards deactivation. Water is one of the products of the FT synthesis and water will always be present in different amounts depending on the conversion, reactor system and catalyst. It has been shown that water has a large effect on the activity and on the selectivity of C_5^+ [1, 2, 3, 4, 5]. Mass transfer effects are also very important in FT synthesis. Mass transfer can limit the productivity, but usually it is the effect on the selectivity that is of main importance.

FT synthesis is usually described as a chain growth mechanism where a C_1 unit is added to a growing chain. Linear α-olefins and paraffins are the primary products of the FT synthesis. α-olefins can also participate in secondary reactions including readsorption and further chain growth or they can be hydrogenated to paraffins. Hydrogenation to paraffins terminates the chain. It has been proposed

[6] that maximum C_5^+ selectivity is obtained by using catalyst pellets with optimum intraparticle diffusion resistance. This is achieved by increasing intraparticle diffusion resistance to the point where secondary chain building reactions of primary products (α-olefins) are maximized without inducing significant diffusion resistance to the reactants. The present work focuses on the effect of the surface area/porous structure on the catalytic performance of a series of Co catalysts supported on different aluminas. Secondary reactions on the same catalysts are studied using hydrogenation of propene as a model reaction.

2. EXPERIMENTAL

The catalysts were prepared by incipient wetness co-impregnation of alumina (Condea Puralox SCCa 45/190) using $Co(NO_3)_2 \cdot 6H_2O$ and $HReO_4$ [7]. The alumina was treated in air at different temperatures (500-1150 °C) in order to obtain supports with different surface areas. Low surface area (LSA, 15 m^2/g), medium surface area (MSA, 68 m^2/g) and high surface area (HSA, 190 m^2/g) alumina. Analysis showed that the LSA alumina contained 86 % α-Al_2O_3. The Co loading was 12 wt% and the promoted catalysts contained 0.5 wt% Re. After impregnation the catalysts were dried (3 h, 110 °C) and calcined in air (300 °C, 16 h). The catalysts were then sieved to the desired particle size, 53–90 μm. The catalysts were characterized by TPR, chemisorption, BET and XRD.

The catalysts were tested in an experimental set-up containing two isothermal fixed-bed microreactors [7,8] at 210 °C, 20 bar and $H_2/CO=2$. The reactors have an inner diameter of 1 cm and about 1.5 g catalyst diluted with SiC (1:5) were used. The catalysts were reduced in H_2 at 350 °C and 1 bar for 16 h (heating rate to 350 °C: 1 °C/min). Propene hydrogenation on the same catalysts was studied as a model reaction for secondary hydrogenation in a conventional fixed-bed microreactor made of stainless steel with 10 mm inner diameter [9,10]. The catalyst bed was diluted with SiC and 15 mg catalyst and 1185 mg SiC were used. Separate experiments have shown that SiC has no activity for propene hydrogenation. Propene hydrogenation was studied at 120 °C and 1.8 bar. The products were in both cases analyzed using online gas chromatography.

3. RESULTS AND DISCUSSION

3.1 Fischer-Tropsch syntesis

The results presented on Fig. 1 show clearly that the selectivity depends on the alumina. For unpromoted 12wt% Co catalysts the C_5^+ selectivity increases from 81 % for HSA alumina to 85 % for LSA alumina. Promoting with 0.5 wt% Re gives an increase in the C_5^+ selectivity for both LSA and HSA alumina. Similar relationships as shown in Fig. 1 between the selectivity and the alumina

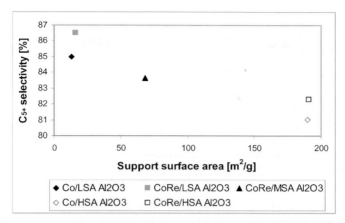

Fig. 1. C_5^+ selectivity as function of surface area for 12%Co and 12%Co-0.5%Re on low surface area alumina (LSA, 15 m²/g), medium surface area alumina (MSA 68 m²/g) and high surface alumina (HSA,190 m²/g). Fischer-Tropsch synthesis at 210 °C, 20 bar and $H_2/CO=2$.

properties also apply to the use of pore volume or pore diameter instead of surface area. In Fischer-Tropsch synthesis the selectivity of C_5^+ increases with conversion and comparisons between different catalysts should therefore be done at constant conversion. The selectivities in Fig. 1 have all been obtained at CO conversions of 49-50 %.

For Fischer-Tropsch synthesis the C_5^+ selectivity increases with the addition of water to the feed stream [11]. In fact, the increase in C_5^+ selectivity with conversion can possibly be related to the increased amount of water as the conversion increases. It was found that the improved C_5^+ selectivity obtained with LSA alumina compared to HSA alumina is observed also when water is added to the feed. For CoRe on LSA alumina the C_5^+ selectivity increases to above 91 % when water is added at the conditions of Fig. 1. Co catalysts usually have low water-gas shift activity and the use of LSA alumina reduces the CO_2 formation rate even further.

The propene/propane ratio is shown as a function of the surface area of the alumina support on Fig. 2. These results indicate that the propene/propane ratio decreases with decreasing surfaces area. The formation of propene is reduced for the low surface area (LSA) alumina support, but the reduced selectivity was not accompanied by an increase in the production of propane. The same observations were also made for the other lower olefins. These results indicate that the reduced formation of propene without any increase in the formation of propane contributes to the C_5^+ selectivity improvement of Co catalysts supported on LSA alumina. The olefin hydrogenation activity is reduced without changing the main hydrocarbon activity. It is also interesting to note that the propene/propane ratio seems to be about the same for the Co and CoRe supported catalysts. The C_5^+ selectivity, however, is higher on the Re promoted catalysts as shown on Fig 1.

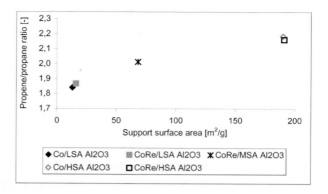

Fig. 2. Propene/propane ratio as a function of BET surface area for 12%Co and 12%Co-0.5%Re on low surface area alumina (LSA, 15 m²/g), medium surface area alumina (MSA, 68 m²/g) and high surface alumina (HSA, 190 m²/g). Fischer-Tropsch synthesis at 210 °C, 20 bar and H₂/CO=2.0.

3.1 Hydrogenation of propene

Hydrogenation of olefins was studied separately using hydrogenation of propene as a model reaction. Due to the very high hydrogenation activity of the cobalt catalysts, a very diluted propene feed was used and the temperature had to be reduced to 120 °C. Fig. 3 shows that the rate of propene hydrogenation decreases with decreasing surface area of the alumina support. The same relationship has been observed when using the pore volume of the support, the rate of propene hydrogenation increases with increasing pore volume. The rate of propene hydrogenation does not depend very much on the promoter although a slightly lower rate is observed on the CoRe supported catalysts.

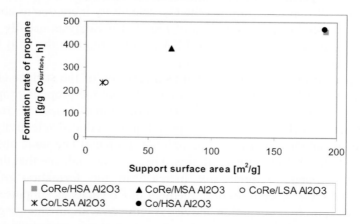

Fig. 3. Initial rate of propene hydrogenation as a function of the surface area for Co and Co-Re (12 wt% Co) on LSA (15 m²/g), MSA (68 m²/g), and HSA(190 m²/g). Propene hydrogenation at 120 °C and 1.8 bar. Feed: 1 Nml/min C₃H₆, 6 Nml/min H₂, 463 Nml/He.

The rate on Fig. 3 is given as g/g $Co_{surface}$, h, i.e. based on exposed surface atoms of cobalt. The CoRe catalysts supported on LSA alumina have lower dispersion (5.9 %) than the corresponding catalysts supported on MSA alumina (9.1 %) and HSA alumina (9.1 %) and they are therefore somewhat less active on a gram basis. The same applies also for the unpromoted Co catalysts as shown on Table 1.

Fig. 3 shows that the rate of propene hydrogenation increases with increasing surface area of the support and from Table 1 below it is evident that the dispersion increases with increasing surface area of the alumina support for the Co and CoRe catalysts. However, promoting the Co catalysts with Re increases the dispersion, but it does not affect the hydrogenation activity as shown on Fig. 3. Hydrogenation of propene on metal surfaces is usually believed to be structure insensitive reactions, but in this case the rate is different for Co and CoRe catalysts supported on different aluminas. The support itself has no catalytic activity. The low surface area alumina consists mainly of α-alumina whereas the high surface area alumina is γ-alumina indicating that the support or the interaction between the support and the metal, may play an important role for the observed effects. It could be that the observed increase in propene hydrogenation activity with increasing support surface area is due to the ability of the support to store spillover hydrogen from the metal.

The reduced activity for propene hydrogenation for the low surface area catalyst is therefore in good agreement with the observed increase in C_5^+ selectivity for the low surface area catalyst.

It has also been shown that the rate of propene hydrogenation is significantly further reduced by adding water to the reaction mixture [10]. The drop in propene hydrogenation activity was found to be reversible, i.e the rate returned to the value that was observed before water was added when the addition of water was switched off. The fact that the activity is reversible supports the idea that the decreased activity is due to competitive adsorption of water on the catalyst surface. The effect of water is also in good agreement with the observed increase in C_5^+ selectivity when water is added to the feed during Fischer-Tropsch synthesis [11].

Table 1
Dispersion of cobalt catalysts as determined by H_2 chemisorption at 40°C. HSA: High surfacea area alumina(190 m^2/g), MSA: Medium surface area alumina (68 m^2/g) and LSA: Low surface area alumina (15 m^2/g).

Catalyst:	CoRe/HSA	CoRe/MSA	CoRe/LSA	Co/HSA	Co/LSA
Dispersion:	9.1	9.1	5.9	6.9	4.2

4. CONCLUSION

Co or CoRe supported on low surface area (LSA), medium surface (MSA) and high surface area (HSA) alumina supports have been prepared by heat treatment of γ-alumina. The dispersion increases with increasing surface area of the support. The Re-promoted catalysts have higher dispersion that the corresponding unpromoted catalysts.

The rate for hydrogenation of propene (on the basis of exposed Co atoms) increases with increasing surface area of the alumina support. However, the rate of hydrogenation is independent of the promoter although the addition of promoter increases the dispersion.

Improved C_5^+ selectivity is obtained by using Co or CoRe supported on low surface area (LSA) alumina prepared by high temperature treatment of γ-alumina. The results are in accordance with reduced activity for secondary hydrogenation of olefins on LSA alumina.

ACKNOWLEDGEMENTS

The financial support from Statoil, the Research Council of Norway, SINTEF Applied Chemistry and Department of Chemical Engineering, NTNU is greatly acknowledged.

REFERENCES

[1] A.M. Hilmen, D. Schanke, K.F. Hanssen, A. Holmen, Appl. Catal. A, 186 (1999) 163.
[2] S. Krishnamoorthy, M.Tu, M.P. Ojeda, D. Pinna, E. Iglesia, J. Catal., 211(2002) 422.
[3] C.J. Bertole, C.A. Mims, G. Kiss, J. Catal., 210 (2002) 84.
[4] J. Li, G. Jacobs, T. Das, Y. Zhang, B. Davis, Appl. Catal., A, 236 (2002) 967.
[5] C.J. Kim, US Patent 5,227,4007 to Exxon Res. Eng. Co. (1993).
[6] E. Iglesia, S.C. Reyes, R.J. Madon, Adv. in Catal., 39 (1993) 221.
[7] S. Eri, K. Kinnari, D. Schanke, A.-M. Hilmen, Patent WO 02/47816 A1 (June 20 2002).
[8] D. Schanke, A.M. Hilmen, E. Bergene, K. Kinnari, E. Rytter, E. Ådnanes, A. Holmen, Catal. Lett., 34 (1995) 269.
[9] Chr. Aaserud, A.-M. Hilmen, E. Bergene, S. Eri, D. Schanke, A. Holmen, in press.
[10] Chr. Aaserud, Ph.D. Dissertation, NTNU, 2003:29.
[11] A.-M. Hilmen, O.A. Lindvåg, E. Bergene, D. Schanke, S. Eri, A. Holmen, Stud. Surf. Sci. Catal. 136 (2001) 295.

Studies in Surface Science and Catalysis, volume 147
X. Bao and Y. Xu (Editors)
307

Effect of water on the catalytic properties of supported cobalt Fischer-Tropsch catalysts

J. Li[a], B.H. Davis[b]

[a]College of Chemistry and Life Science, South Central University for Nationalities, No.5 Minyuan Road, Wuhan 430074, China

[b]Center for Applied Energy Research, University of Kentucky, 2540 Research Park Drive, Lexington, KY 40511-8410, USA

ABSTRACT

Water has been found to have a significant effect on the alumina and silica supported cobalt Fischer-Tropsch catalysts. The addition of small amounts of water slightly decreased CO conversion of Co-Pt/Al_2O_3 catalyst and effect was kinetic. Increasing the amount of added water resulted in a permanent deactivation of the catalyst. For Co/SiO2 catalyst, the addition of water in the range of 5-25% (vol.% of the feed) increased the CO conversion. Continuous addition of small amounts of water (5 vol.%) over a period of days caused an increase of CO conversion, however, for large amounts of water, continuous addition resulted in a severe catalyst deactivation.

1. INTRODUCTION

The effect of water on iron-based FT catalysts has been investigated widely [1]. The effect of water on cobalt catalysts is less well understood. Hilmen et al. [2] studied the effect of water on Co/Al_2O_3 and CoRe/Al_2O_3 catalysts by adding water to the synthesis gas feed and by model studies exposing the catalysts to H_2O/H_2. It was found that the catalysts deactivated when water was added during FT synthesis and the catalysts oxidized in H_2O/H_2 mixtures with a ratio much lower than expected for oxidation of bulk cobalt. On silica supported cobalt catalysts, the addition of water to the synthesis gas feed has been shown to slightly decrease the CO conversion [3]. Schulz et al. [4] investigated the effect of water addition on CoRu/ZrO2/aerosil and Co/MgO/ThO_2/aerosil catalysts and found that CO conversion was not influenced significantly and, remarkably, the product selectivity was improved when water was added to the feed. In this study, the effect of water on the Fischer-Tropsch catalytic properties

Table 1

Results of hydrogen TPD and percent reducibility from pulses of O_2

Catalyst	H_2 desorbed (μm)	Dispersion (%)	Cluster size (nm)	Reduction(%)
Co-Pt/Al$_2$O$_3$	141	18.4	5.6	60
Co/SiO$_2$	25.6	7.79	13.2	64

of platinum promoted Co/Al$_2$O$_3$ and Co/SiO$_2$ catalysts was investigated using two approaches: (1) changing space velocity to change CO conversion and therefore the water partial pressure and, (2) by directly adding varying amounts of water into the feed gas which was maintained at one flow rate.

2. EXPERIMENTAL

A Pt(0.5%)-15%Co/Al$_2$O$_3$ and a 12.4%Co/SiO$_2$ catalysts were used for the studies. The catalysts were prepared by incipient wetness impregnation and characterized by BET, TPR, H$_2$ chemisorption and pulse reoxidation [5]. The results are shown in Table 1. The catalytic properties for FTS were measured in a CSTR reactor. The reactor system has been described earlier [6]. A series of measurements were conducted so that the pressure of CO and H$_2$ in the feed remained constant and was 70% of the total pressure, with the remainder being argon. When water was added, a fraction of the argon was replaced by water. During the entire run the reactor temperature was 483 K, the pressure was 2.93 MPa, and H$_2$/CO was 2.0.

3. RESULTS AND DISCUSSION

3.1. Effect of space velocity / conversion on deactivation

It was found that the Pt-Co/Al$_2$O$_3$ catalyst showed good stability at a space velocity of 8 SL/gcat/h (298 K, 0.1 MPa); the deactivation rate, based on CO conversion, was only 2.1% per week (0.3% per day). Thus, the run at 8 SL/gcat/h with a deactivation rate of 0.3% per day was used as a reference. After each period at some other space velocity, the space velocity was again adjusted to 8 SL/gcat/h to compare the conversion changes. The results are shown in Fig. 1.a. For the runs at space velocities greater than 2 SL/gcat/h, the CO conversion increased almost linearly with decreasing space velocity and the catalyst did not show an apparent deactivation. When the space velocity was decreased to 2 SL/gcat/h, CO conversion did not remarkably increase and the catalyst started to deactivate. The average deactivation rate was 0.91% per day (6.4% per week), two times higher than the reference deactivation rate. However, when the space velocity changed back to 8SL/gcat/h, CO conversion had decreased from ~20 to 14.1%. The catalyst deactivated more seriously at a space velocity of 1 SL/gcat/h, and the deactivation rate was 1.77% per day (12.4% per week), After

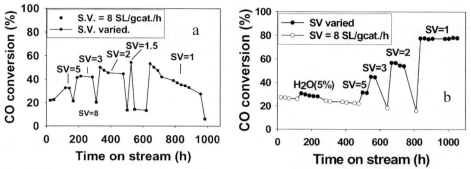

Fig. 1. CO conversion as a function of time on stream and space velocity for a: Pt-Co/Al$_2$O$_3$, b: Co/SiO$_2$ catalysts.

running at this space velocity for 320 h, the catalyst deactivated to only 5% at a space velocity of 8 SL/gcat/h. These deactivation rate patterns appear to differ for the SVs of 2 and 1. For SV = 2, the CO conversion appears to decline rapidly and then to attain a new, reasonably stable CO conversion level. But for a SV = 1, the CO conversion appears to decline continuously during the 270 h period.

The effects of space velocity on catalytic properties of Co/SiO$_2$ catalyst were investigated by the same way with Pt-Co/Al$_2$O$_3$. Firstly, the catalyst was used at a space velocity of 8 SL/gcat/h, and the catalytic activity was quite stable at this space velocity. The catalyst deactivation rate, based on CO conversion, was only 1.8% per week (0.26% per day). After several days at a space velocity of 8 SL/gcat/h, 5 vol.% water was added to the feed for 5 days. The water effect will be discussed in the next section. After the water addition was terminated, the run was continued at a space velocity of 8 SL/gcat./h for a week to re-establish the earlier rate. The catalyst was then utilized with different space velocities (5, 3, 2 and 1 SL/gcat./h). The results are shown in Fig. 1.b. At higher space velocities (8, 5 SL/gcat./h), the CO conversion increased linearly with increasing space time but deviated from linearity as the space velocity was decreased to lower values. At higher space velocities (SV= 5 and 3 SL/gcat/h), the catalyst did not show an apparent deactivation. Even when the space velocity was decreased to 1 SL/gcat/h, the catalyst showed stability; during a 2 weeks period at 1 SL/gcat/h, the catalyst did not show a measurable deactivation. This result is completely different from the performance of Pt-Co/Al$_2$O$_3$ catalyst, which displayed higher deactivation rate at the lower space velocity, and higher water partial pressure. The cobalt metal crystallite size may explain the different effects for these two catalysts. It was found that the average cobalt crystallite size of the alumina supported catalyst was 5.6 nm, much smaller than the silica supported catalyst (13.2 nm). The cobalt metal with smaller crystallite size appears to be more easily oxidized by higher partial pressure water to form cobalt oxide and the catalyst deactivates more rapidly [7].

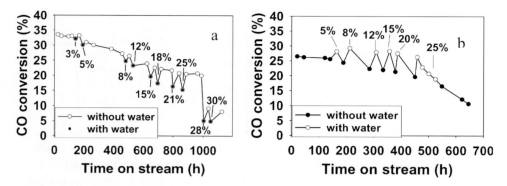

Fig. 2. Effect of water addition on CO conversion for a: Pt-Co/Al$_2$O$_3$, b: Co/SiO$_2$ catalysts.

3.2. Effect of water addition on the catalyst deactivation

3.2.1. Pt-Co/Al$_2$O$_3$

To determine the effect of water on Fischer-Tropsch synthesis, different amounts of water (3.0-30 vol.%) were added into the feed gas at a total space velocity of 8 SL/gcat./h. The results are shown in Fig. 2.a. In the range of 3-25 vol.% water added (P_{H2O}/P_{CO} = 0.6-1.2; P_{H2O}/P_{H2} = 0.3-0.6), the CO conversion decreased only slightly with increasing amounts of added water. The catalyst activity recovered after water addition was terminated. The conversion decline with time-on-stream (0.35% per day) was very close to the reference deactivation rate, showing that the catalyst was not impacted irreversibly by the brief operating periods with additional water present. This temporary decline in CO conversion when adding water may be due to the kinetic effect of water by adsorption inhibition, although most kinetic rate equations for cobalt catalyzed Fischer-Tropsch synthesis do not include the influence of water. Increasing the amount of water added to greater than 28 vol.% resulted in a severe and irreversible deactivation of the catalyst. CO conversion decreased from 15.3 to 5.0% when water was increased from 25 to 28 vol.%. More importantly, the catalyst activity was not recovered to the expected value after water addition was terminated, indicating that the catalyst had deactivated permanently. Three possibilities for the cobalt Fischer-Tropsch catalyst deactivation have been advanced: surface condensation, sulfur poisoning, and oxidation [2,8]. In this case, the catalyst permanent deactivation should be due to oxidation caused by the presence of a high partial pressure of water. From a purely thermodynamic point of view, the oxidation of bulk phase metallic cobalt to CoO or Co$_3$O$_4$ should not be possible under Fischer-Tropsch conditions [8]. Furthermore, oxidation of Co to CoAl$_2$O$_4$ is considered to be kinetically restricted during typical Fischer-Tropsch synthesis conditions. However, previous investigators have found that cobalt catalyst oxidation occurred during Fischer-Tropsch synthesis [2, 3, 8]. The extent of oxidation is dependent on the water partial pressure and the ratio of P_{H2O} to P_{H2} (or P_{CO}) [2]. In this study, the water partial pressure was 8.35 bar and the ratio of P_{H2O} to P_{H2} was 0.59, at the onset of permanent deactivation.

The effect of water addition on CO_2 selectivity of the Pt-Co/Al_2O_3 catalyst is interesting. At first, the CO_2 selectivity increased slightly to about 5% as the amount of water is increased, but when the added water was 28 vol.%, the CO_2 selectivity increased dramatically to 11.2%. This further confirmed that some surface cobalt atoms have been reoxidized to form cobalt oxide or some other form of cobalt, which was found to be active for the water-gas shift reaction. Studies in our laboratory show that an unreduced Pt-Co/Al_2O_3 catalyst gave 18.9% CO conversion for the water-gas shift reaction when a feed of CO/H_2O with a ratio of 1:1 (no added hydrogen) was introduced into the reactor at the same conditions as was used for the FTS runs described here.

3.2.2. Co/SiO₂

The effects of water addition on the Co/SiO_2 catalyst are shown in Fig. 2.b. In the range from 5 to 25 vol.% water added (P_{H2O}/P_{H2} = 0.12 to 0.8, P_{H2O}/P_{CO} = 0.24 to 1.6), the additional water increased the CO conversion. In order to determine the effect of water addition time on the catalyst activity, a smaller or a larger amount of water (5 or 25 vol.%, respectively) was added continuously into the feed for 96 h. It was found that the continuous addition of the smaller amount of water maintained the higher CO conversion in comparison to that without water addition, and did not show a significant effect on the catalyst deactivation rate (Fig. 1.b). The deactivation rate with water addition was 0.36% per day, slightly higher than the value of 0.26% per day without water addition during the initial period (0 to 120 h). The deactivation rate, based on the values at the standard condition following the addition of water (150 to 475 h. on stream) is 0.37% per day (Fig. 2.b). Thus, the deactivation during the intermittent water addition is about the same as obtained during the constant addition of 5 vol.% water (Fig. 1.b). For the addition of the larger amount of water (25 vol.%), CO conversion increased the first day, but the continuous addition of this level of water resulted in rapid deactivation of the catalyst at a rate of 1.5% per day. More importantly, the catalyst activity was not recovered to the expected value after the addition of water was terminated, indicating that the catalyst had deactivated permanently. It is likely that the cobalt was partially oxidized during the long time exposure to the higher water partial pressure. Furthermore, the activity declined about as rapidly following termination of the addition of 25 vol.% water as it did during water addition (Fig. 2.b).

The effect of water on the silica supported cobalt catalyst is in marked contrast to the alumina supported cobalt catalyst. For Pt-Co/Al2O3 catalyst, water decreased the CO conversion in the range of the water added (3-25 vol.%).

Increasing the amount of added water to 28 vol.% resulted in a permanent deactivation of the catalyst. Water also exhibited a deactivating effect on CO conversion while it was added with the feed for a CoRu/TiO$_2$ catalyst [9]. However, water had a positive effect on the CO conversion for the Co/SiO$_2$ catalyst. Some early workers [4] also found similar results, with water having an accelerating effect on the FT reaction for a silica-supported cobalt catalyst. Currently, we do not have a reasonable explanation for the support effect. A series of cobalt catalysts prepared with different pore size supports are being tested in our laboratory.

One could devise a number of possible reasons for the catalyst permanent deactivation. The situation is complex and requires much more study. However, it appears that the presence of a reasonably high P_{CO} is necessary if catalyst deactivation is to occur. This suggests that some complex that involves both P_{H2O} and P_{CO} is involved in the loss of catalytic activity. The formation of a cobalt formate like species is a potential explanation although this is very speculative with the data available. The alumina restructuring during reaction at high water conditions, i.e. boehmite formation may also be a reason for alumina supported catalyst deactivation. Another possibility is the formation of cobalt silicates or cobalt aluminates. It was shown that hydrothermal treatment of a reduced Co/SiO$_2$ catalyst with H$_2$O (and subsequent calcinations) led to the formation of both nonreducible and reducible mixed species of Co that interact with the support silica [5]. Examination of the used catalyst by XANES may provide conclusive evidence in the future. XANES has already been employed to demonstrate the presence of cobalt aluminate species in used Co/Al$_2$O$_3$ catalyst sample taken from the reactor and stored in solidified wax [7].

ACKNOWLEDGEMENTS

This work was supported by National Natural Science Foundation of China (20373090), Talented Young Scientist Foundation of Hubei (2003ABB013), EYTP of MOE and the State Ethnic Affairs Commission, P. R. China.

REFERENCES

[1] M.E. Dry, Catal. Lett. 7 (1990) 204.
[2] A.M. Hilmen, D. Schanke, K.F. Hanssen, A. Holmen, Appl. Catal. 186 (1999) 169.
[3] J.K. Minderhoud, S.T. Sie, Eur. Appl. Patent No. 83201557.2 (1988), to Shell Int. Res.
[4] H. Schulz, M. Claeys, S. Harms, Stud. Surf. Sci. Catal. 107 (1997) 193.
[5] G. Jacobs, T. Das, Y. Zhang, J. Li, B. Davis, Appl. Catal. 233 (2002) 263.
[6] J. Li, L. Xu, R. Keogh, B. Davis, Catal. Lett. 70 (2000) 127.
[7] G. Jacobs, T. Das, Y. Zhang, J. Li, B. Davis, Catal. Deactivat.. 139 (2001) 415.
[8] P.J. van Berge, J. van de Loosdrech, A.M. van der Kraan, Catal. Today, 58 (2000) 321.
[9] J. Li, G. Jacobs, T. Das, B. Davis, Appl. Catal. 233 (2002) 255.

Studies in Surface Science and Catalysis, volume 147
X. Bao and Y. Xu (Editors)
313

CeO$_2$-promotion of Co/SiO$_2$ Fischer-Tropsch synthesis catalyst

H.-B. Shi, Q. Li, X.-P. Dai, C.-C. Yu, S.-K. Shen[*]

The Key Laboratory of Catalysis CNPC, University of Petroleum Beijing, Beijing 102249, China

ABSTRACT

A series of CeO$_2$-Co (10.7 wt. %) /SiO$_2$ catalysts were prepared and evaluated for Fisher-Tropsch synthesis. The optimized ceria-promoted catalyst exhibits excellent performance in activity, C$_{5+}$ selectivity and stability. It is suggested that ceria could enhance the dispersion of metallic cobalt and increase the amount of active sites for FTS, leading to increased concentration of surface active carbon species and selectivity towards long chain hydrocarbons.

1. INTRODUCTION

The Fischer-Tropsch synthesis (FTS) is an attractive route to produce liquid hydrocarbons from abundant natural gas via synthesis gas as the result of the increasing prices of oil and environmental problems. Cobalt based catalysts have been widely used in FTS because of their have excellent activity, selectivity towards C$_{5+}$ and stability. Currently, there is significant commercial interest in producing high quality diesel and lube base stocks by FTS. Increasing the selectivity of FTS to desired products is economically attractive and attracts more and more attention.

Addition of a promoter to supported-cobalt catalyst is one of the most effective approaches for increasing the selectivity towards long chain hydrocarbons. There are two kinds of promoter: one is noble metal such as Ru or Re, and the other is oxide promoter such as ZrO$_2$ or La$_2$O$_3$ [1]. Cerium oxide because of its high oxygen storage capacity has been widely used in many fields [2], such as in catalytic conversion of light hydrocarbons, and in treatment of exhaust gas. However, previous studies on cerium oxide as a promoter of supported-cobalt FTS Catalyst are relatively deficient [3-7]. In this paper, the effect of cerium oxide addition to Co/SiO$_2$ catalyst on the performance of FTS is reported. The cerium oxide addition increases both the activity and selectivity

* Corresponding author: Shi-Kong Shen. Tel: +86 10 89733784, Fax: +86 10 69744849, Email: skshen@bjpeu.edu.cn

towards long chain hydrocarbons, and in addition the stability. The role of cerium oxide promoter playing in CO hydrogenation was investigated by means of XRD, TPR, TPSR and transient response technique.

2. EXPERIMENTAL

2.1. Catalyst preparation and evaluation

A series of CeO_2-Co (10.7 wt. %)/SiO_2 catalysts with various cerium contents were prepared by incipient wetness impregnation of cobalt nitrate and cerium nitrate mixed-solution on spherical SiO_2 of diameter 2-3 mm and BET surface area 385 m^2/g, followed by drying at 353K overnight, calcination in air for 10h, and reduction in H_2 at 700K for 8h. The evaluation of the catalysts using H_2/CO=2 synthesis gas as feed was conducted in a fixed-bed microreactor filled with 1.0g catalyst as described elsewhere [8].

2.2. TPR, H_2-TPSR and transient response experiments

H_2 temperature programmed reduction (H_2-TPR), H_2 temperature programmed surface reaction (H_2-TPSR) and transient response experiments were performed in a fixed-bed microreactor equipped with an on-line quadrupole mass spectrometer (AMTEK Q200MN). The catalyst was filled with 0.3g For TPR and 0.5g for TPSR and transient response experiments.

Transient response experiments were carried out at a gas flow rate of 15ml min^{-1}, 508K and atmospheric pressure. Prior to reaction, the catalysts were reduced at 673K for 4h, and then cooled to reaction temperature in H_2 flow. The responses of CO, H_2, CH_4, C_2H_6, C_3H_8, C_4H_{10}, and CO_2 were recorded at m/e of 28, 2, 15, 30, 29, 43 and 44, respectively. It is worth noting that nearly all linear paraffin with a carbon number larger than 4 have a fragment at m/e=43 with significant intensity, therefore the response at m/e=43 could only represent C_4H_{10} approximately.

3. RESULTS AND DISCUSSION

3.1. Optimization of ceria loading and calcination temperature

A 10.7 wt. % Co supported amount and reduction in H_2 at 700 K for 8 hours are found to be optimal for the Co/SiO_2 catalyst. The cerium oxide loading amount and calcination temperature as influencing factors on the FTS performance of the Co/SiO_2 catalysts were optimized by the CODEX method (Table 1). The optimized catalysts, which consist of Ce/Co atom ratio in the range of 0.25 and are calcined at 740K, give C_{5+} yield of 80~83%.

Table 1
Optimal design for ceria loading and calcination temperature by the CODEX method*

NO.	Ce/Co(mol/mol)	Calc. Temp.(K).	X_{CO}(%)	S_{CH_4}(%)	S_{CO_2}(%)	$S_{C_{5+}}$(%)	$Y_{C_{5+}}$(%)
1	0.05	573	74.1	8.5	0.79	83.8	62.1
2	0.3	573	90.3	7.2	1.2	87.4	78.9
3	0.05	873	62.4	9	0.67	85.8	53.5
4	0.3	873	91.2	7.7	1.3	86.4	78.8
5	0	723	80.5	8.6	1	86.1	69.3
6	0.352	723	90.1	7.1	1.3	88.3	79.6
7	0.175	511	79.5	7.5	0.89	88.4	70.7
8	0.175	935	85.9	7.8	1.2	86.9	74.6
9	0.175	723	92	7	1.6	87.5	80.5

* Reaction conditions: GHSV=500h^{-1}, P=1.2MPa, T=483K

3.2. 1000 hours stability test of the optimized CeO_2-Co/SiO_2 catalyst

A 1000 hours stability test shows that the CeO_2-Co/SiO_2 catalyst has excellent activity and stability (Fig. 1). The average result in the 1000 hours is CO conversion of 90% and C_{5+} selectivity of 81%. The product distribution in the range of C_5-C_{10}, C_{11}-C_{20} and C_{21+} are 6.8, 62.9 and 30.3 wt. % respectively. The chain growth probability of the catalyst is 0.9. After 700 hours run, a slight decrease in catalytic activity is due to covering the catalyst surface and blocking the catalyst pores by wax formed in the reaction. The activity can be regenerated easily by hydrogen sweeping at 673 K for 24 hours. The characterization of used catalysts by XRD, XPS and BET surface area prove that the chemical composition and structure of the catalyst are stable.

3.3. XRD, TPR and H_2-TPSR

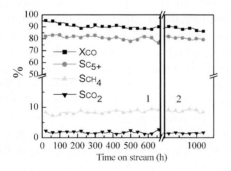
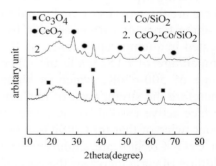

Fig. 1. The stability test of CeO_2-Co/SiO_2 catalyst at 488 K, 1.5 MPa, GHSV 500h^{-1}
1. before regeneration; 2. after regeneration

Fig. 2. XRD patterns for Co/SiO_2 and CeO_2-Co/SiO_2 catalysts

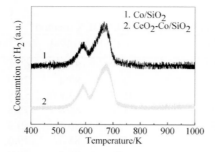

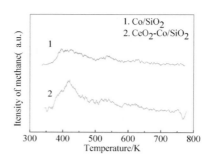

Fig. 3. TPR profiles at 30 ml min^{-1} of 11.7% H_2/Ar and heating rate of 10K min^{-1}

Fig. 4. Responses of CH_4 during H_2-TPSR at heating rate of 10 K min^{-1} and H_2/Ar=2 of 15ml min^{-1}

XRD patterns for the Co/SiO_2 and CeO_2-Co/SiO_2 are given in Fig. 2. It can be seen that the characteristic diffraction peaks of Co_3O_4 at 2θ=31.2, 36.9, 44.8, 59.4, 65.2 for CeO_2-Co/SiO_2 are lower and wider compared with that for Co/SiO_2. It suggests that the crystallite size of Co_3O_4 becomes smaller when ceria is present, namely the dispersion of cobalt is enhanced.

The TPR profiles over the fresh Co/SiO_2 and CeO_2-Co/SiO_2 shown in Fig. 3 have two reduction peaks. They correspond to the reduction of $Co_3O_4 \rightarrow CoO$ (the small one) and $CoO \rightarrow Co$ (the large one), respectively. The total area of the two peaks for CeO_2-Co/SiO_2 is 17% larger than that for Co/SiO_2. It indicates that the metallic cobalt active sites formed on CeO_2-Co/SiO_2 is more than that on Co/SiO_2.

Fig. 4 shows H_2-TPSR profiles. The both catalysts were reduced in H_2 at 673K for 4h, and then pretreated with CO/He =1/2 mixture gas at flow rate of 15 ml min^{-1} and 508K for 15 min before H_2-TPSR experiments.

During the H_2-TPSR experiments, only methane product was detected. There are three peaks of methane for the both catalysts, suggesting the formation of three types of carbon species from the chemisorbed CO. A strong peak is located in a temperature range of 373-500K, which is very close to the reaction temperature of FTS, while the other two weak peaks appear in a temperature range of 500-650K, which is higher than the reaction temperature of FTS. Therefore the strong peak of methane can be attributed to the hydrogenation of surface active carbon species. The intensity of strong peak for CeO_2-Co/SiO_2 is about twice of that for Co/SiO_2, while the intensities of another two weak peaks for CeO_2-Co/SiO_2 are slightly more than that for Co/SiO_2. It indicates that the addition of ceria increases the concentration of surface active carbon species and the activity of CO hydrogenation.

To sum up, the results from XRD, TPR and H_2-TPSR are in accord with the idea that ceria plays an important role in Co/SiO_2 catalyst, viz. enhancing cobalt dispersion and promoting the active sites for FTS.

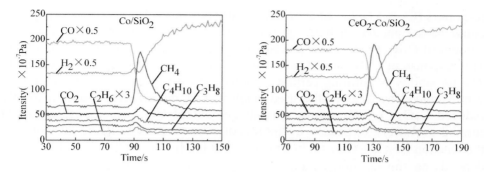

Fig. 5. Response spectra after the switch of synthesis gas (H_2/CO=2) to H_2

3.4. Transient response experiments

The responses of a step switch from synthesis gas to H_2 are shown in Fig. 5. The products of CH_4, C_2H_6, C_3H_8 and C_4H_{10} could be attributed to the hydrogenation of CH_x (x=0, 1, 2, 3), C_2H_5-, C_3H_7- and C_4H_9- species on the catalyst surface, respectively, while the product of CO_2 could be assigned to the disproportionation of CO. The total amount of C_1-C_4 paraffin products for CeO_2-Co/SiO_2 is more than that for Co/SiO_2. In particular, the response of C_4H_{10} for CeO_2-Co/SiO_2 is clearly strong compared with Co/SiO_2. The average residence time of C_4 intermediate species on the surface estimated from the response curve is about 10s for CeO_2-Co/SiO_2 and 5s for Co/SiO_2. The results indicate that the promotion of ceria not only increases the concentration of surface active carbon species and hydrogenation activity, but also benefits the selectivity towards long chain hydrocarbons.

The responses of a step switch from H_2 to synthesis gas are shown in Fig. 6. The first response product after switch was CH_4 for the both catalysts. When the intensity of CH_4 response moved up to a quite high level, the other product

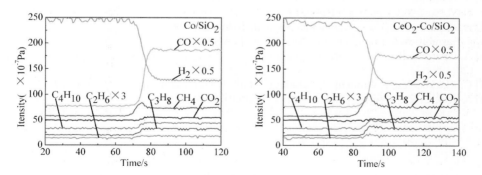

Fig. 6. Response spectra after the switch of H_2 to synthesis gas (H_2/CO=2)

responses (C_2-C_4) began to appear. This implies that CH_x (x=0, 1, 2, 3) species could be assigned to a basic group for the growing carbon chain. The total amount of C_1-C_4 products for CeO_2-Co/SiO_2 is also more than that for Co/SiO_2. The response of CO_2 for CeO_2-Co/SiO_2 is very weak compared with Co/SiO_2. The possibility of forming cerium carbonate via the reaction of CeO_2 with CO_2 was excluded by TPD of the used catalyst. CO_2 pulse reaction over a freshly reduced CeO_2-Co/SiO_2 shows a small amount of CO formed, while no CO formed over freshly reduced Co/SiO_2 (not shown). This could be attributed to the reaction $Ce_2O_3 + CO_2 \rightarrow 2\ CeO_2 + CO$. It implies that the ratio of Ce^{+4}/Ce^{+3} depends on the reaction condition. Cerium oxide could assist the catalyst to inhibit disproportionation of CO and to remove unreactive surface carbon, probably through the redox cycle of Ce^{+4}/Ce^{+3}.

4. CONCLUSION

Addition of ceria to Co/SiO_2 catalyst as a chemical promoter not only increases CO conversion and the selectivity towards long chain hydrocarbons, but also increases the stability. It is speculated that the promotion of ceria could enhance the cobalt dispersion and the active sites for FTS, consequently increasing the concentration of surface active carbon species and benefiting the selectivity towards long chain hydrocarbons.

ACKNOWLEDGEMENTS

The financial support by the Ministry of Science and Technology under grant number G1999022402 is gratefully acknowledged.

REFERENCES

[1] S. Vada, B. Chen, J.G. Goodwin Jr., J. Catal., 153 (1995) 224.
[2] A. Trovarelli, Catal. Rev. Sci. Eng., 38 (1996) 439.
[3] J. Barrault, S. Probst, A. Alouche, A. Percheron-Guegan, V. Paul-Boncour, M. Primet, Stud. Surf. Sci. Catal., 61(1991)357.
[4] B. Ernst, L. Hilaire, A. Kiennemann, Catal. Today, 50(1999) 413.
[5] J. Barrault, A. Guilleminot, J.C. Achard, et al., Appl. Catal. , 21(1986)307
[6] A. Guerrero-Ruiz., A. Seprilveda-Escribano., L. Rodriguez-Ramos., Appl. Catal., 120(1994)71
[7] C. Mazzocchia, P. Gronchi, A. Kaddouri, et al., J. Mol. Catal. , A, 165 (2001) 219
[8] X. Dai, C. Yu, S. Shen, Chin. J. Catal., 21(2000)161.

Studies in Surface Science and Catalysis, volume 147
X. Bao and Y. Xu (Editors)
©2004 Elsevier B.V. All rights reserved.

Structure-controlled La-Co-Fe perovskite precursors for higher C_2-C_4 olefins selectivity in Fischer-Tropsch synthesis

L. Bedel[a], A.C. Roger[a], J.L. Rehspringer[b] and A. Kiennemann[*a]

[a]Laboratoire des Matériaux, Surfaces et Procédés pour la Catalyse, UMR 7515, ECPM, 25 rue Becquerel, 67087 Strasbourg cedex 2, France

[b]Groupe des Matériaux Inorganiques, UMR 7504, IPCMS, 23 rue du Loess, 67037 Strasbourg cedex, France

ABSTRACT

Mixed La-Co-Fe perovskites are studied as precursors for FT catalysts. By changing the Co/Fe ratio and the La-content the precursor perovskite crystalline structure can be controlled (rhombohedral, orthorhombic or cubic), leading to different metal phases after reduction, to direct the selectivity toward C_2-C_4 olefins. The catalysts present remarkable stability under test.

1. INTRODUCTION

Fischer-Tropsch synthesis is of renewing interest over the last decade, especially for waxes formation on cobalt or iron metal catalyst [1]. However, despite an ever increasing demand on C_2-C_4 olefins, relatively few works have been initiated to study cobalt and/or iron based catalysts for light olefins production [2-4]. We have previously studied the Fischer-Tropsch reactivity of some composite $(Co,Fe)^0$ on Co-Fe spinels catalysts with both various metallic Co/Fe and metal/spinel ratios [4]. For a better control of the metal phase properties (composition, availability...) enabling a more light-olefins-oriented reactivity, cobalt and iron were integrated in a $La(Co,Fe)O_3$ perovskite-type crystalline structure in order to obtain, after partial reduction, size-controlled Co^0 or $(Co$-$Fe)^0$ particles in strong interaction with the partially reduced oxide.

2. EXPERIMENTAL

$LaCo_xFe_{(1-x)}O_3$ perovskite series (x=0, 0.25, 0.4, 0.5, 0.6, 0.75 and 1) was prepared by a Sol-gel-like method based on the thermal decomposition of mixed

La-Co-Fe propionates resin [5]. Starting materials, powdered Fe^0 and Co^{II} and La^{III} acetates, are dissolved in hot propionic acid to form exclusively the corresponding propionates. After mixing the three boiling solutions, propionic acid is evaporated until a dark pasty resin is obtained. $LaCo_xFe_{(1-x)}O_3$ perovskites result from the heating of the resin 6 hours at 600, 750 or 1100°C (2°C.min^{-1}). The $La_{(1-y)}Co_{0.4}Fe_{0.6}O_{3-\delta}$ series (y=0, 0.1, 0.2, 0.3 and 0.4) was synthesized by the same way as here above and heated at 750°C.

Thermo-programmed reduction (TPR) and dynamic magnetic measurements were performed using a 10% H_2 containing reductive flow while heating from 25 to 900°C (15°C.min^{-1}).

^{57}Fe Mössbauer spectra were recorded using a spectrometer equipped with a 50 mCi ^{57}Co in Rh source. A bath cryostat allowed measurements for all samples at liquid He temperature (4 K). The spectrometer was calibrated with reference to a sheet of α-Fe and isomer shift values are given relative to this.

Catalytic tests were performed in a fixed bed reactor under a 1 MPa syngas pressure (CO/H_2=1, GHSV=3000 h^{-1}, Q_{tot}=0.6 L.h^{-1}).

3. RESULTS AND DISCUSSION

The preparation method of the $LaCo_xFe_{(1-x)}O_3$ perovskites leads to two solid solutions: one orthorhombic for x<0.5 (isomorphic to $LaFeO_3$) and one rhombohedral for x≥0.5 (isomorphic to $LaCoO_3$).

The study of the reducibility was first carried out using TPR (see Fig. 1). All the Co-containing oxides present a two-step reduction process. The first reduction zone is centered around 450°C and is separated from the second one by a 100°C-wide steady step. Then, these oxides can be partially reduced by a 1 h long treatment at 450°C under a 10% diluted H_2 flow (50 mL. min^{-1}). The partially reduced oxides keep the initial perovskite structure. Usual temperatures in Fischer-Tropsch synthesis being about 100°C lower, no further reduction should occur during catalytic tests.

As the hydrogen consumption in the first reduction zone is conditioned by the amount of Co in the sample and increases linearly with it, this zone is assumed to be due to Co^{3+} reduction. In order to check if, during this partial reduction, metal is formed or if Co^{3+} is only reduced in Co^{2+}, dynamic magnetic measurements under reductive flow were carried out (see Fig. 2).

The magnetization of the orthorhombic perovskites, represented by $LaCo_{0.4}Fe_{0.6}O_3$ in Fig. 2, increases around 410°C in accordance with the first reduction zone observed in TPR. All the Co^{3+} and Co^{2+} oxides being paramagnetic or diamagnetic, this observed ferromagnetic behavior can only be due to the formation of Co^0 particles. The second increase of the magnetization above 600°C is caused by the reduction toward metal of the resting perovskite B-cations: Co and Fe.

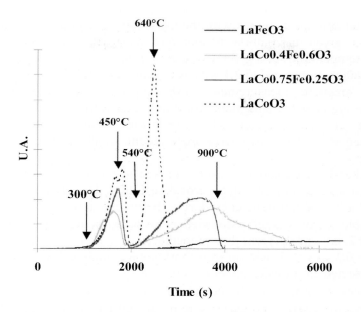

Fig. 1. TPR profiles (H$_2$ consumption) of the LaCo$_x$Fe$_{(1-x)}$O$_3$ series.

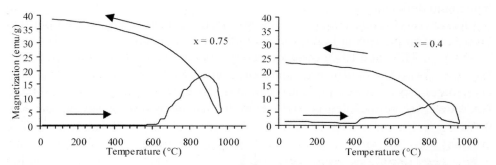

Fig. 2. Evolution of the magnetization of the LaCo$_x$Fe$_{(1-x)}$O$_3$ oxides during their reduction (10% H$_2$, 15°C.min^{-1}).

Except LaCoO$_3$ that behaves like orthorhombic perovskites, rhombohedral ones behave differently. The metal formation is only observed for temperatures higher than 600°C. No ferromagnetic metal particles are detected in the temperature range corresponding at the first hydrogen consumption measured by TPR. This hydrogen consumption is only due to the partial reduction of Co^{3+} to Co^{2+}.

Only LaCoO$_3$ and the orthorhombic perovskites (x<0.5) enable the formation of a metal phase during the partial reductive treatment at low temperature (450 °C, 1.5 h, 10 % H$_2$/He at 50 mL.min^{-1}). These partially reduced oxides will be used as Fischer-Tropsch catalysts.

The metal amount increases with the Co-loading x within the orthorhombic series for a given calcination temperature and with the decrease of the calcination temperature for a given Co-loading (max 8.6 wt% for $LaCo_{0.4}Fe_{0.6}O_3$ calcined at 600°C compared to 2.1 wt% when calcined at 750°C) [6].

Further magnetic measurements revealed that metal particles at room temperature are superparamagnetic for those generated by the reduction of $LaCoO_3$ or close to be for others (transition state between ferro and superparamagnetic). These behaviors are related to small size particles (between 5 and 10 nm).

As Fischer-Tropsch activity is often related to the available metal amount, the most Co-loaded orthorhombic perovskite $LaCo_{0.4}Fe_{0.6}O_3$ was modified by introducing a lanthanum deficiency during the preparation, in order to improve its extractable metal amount. Modified perovskites series $La_{(1-y)}Co_{0.4}Fe_{0.6}O_{3-\delta}$ (y=0, 0.1, 0.2, 0.3 and 0.4) was then prepared.

From y=0.1 to y=0.3, instead of orthorhombic, single cubic perovskite-type phase is detected on powder XRD patterns. In the diffractogram of the y=0.4 sample, traces of iron oxide (either α- or $\gamma-Fe_2O_3$) are observed. Along this $La_{(1-y)}Co_{0.4}Fe_{0.6}O_{3-\delta}$ series, the perovskite phase cell parameter a remains almost the same (from 3.902 Å for y=0.3 to 3.910 Å for y=0.2) and is not linked to the lanthanum deficiency y.

Advanced characterization by [57]Fe Mössbauer spectroscopy pointed out the presence of an increasing Fe_2O_3 amount with the lanthanum deficiency. Through magnetism, TEM and EDX complementary studies, we have demonstrated that $La_{(1-y)}Co_{0.4}Fe_{0.6}O_{3-\delta}$ oxides are made of cubic $\gamma-Fe_2O_3$ nano-cores on which crystallizes a perovskite-type $LaCo_zFe_{(1-z)}O_3$ phase. Due to its epitaxial growth, $LaCo_zFe_{(1-z)}O_3$ perovskites crystallize in the cubic system whatever the Co content z is.

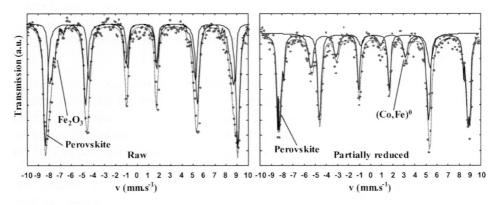

Fig. 3. [57]Fe Mössbauer spectra at 4K of $La_{0.7}Co_{0.4}Fe_{0.6}O_{3-d}$ calcined at 750°C before and after partial reduction.

These nano-composites are reducible as well as the $LaCo_xFe_{(1-x)}O_3$ series. After the reductive pre-treatment at 450°C (1 h, 2°C.min^{-1}), the γ-Fe_2O_3 nano-cores are completely reduced to form $Co_{0.5}Fe_{0.5}$ metallic alloy together with some perovskite cobalt (see Fig. 3). The metal amount increases from 2.1 wt % for y=0 up to 10.9 wt% for y=0.4. Fe_2O_3 cores seem then to act like a reservoir for reducible cations.

The catalytic activity of both $LaCo_xFe_{(1-x)}O_3$ and $La_{(1-y)}Co_{0.4}Fe_{0.6}O_{3-\delta}$ series was studied. To compare the catalysts the results are presented here at isoconversion of CO of almost 4 % (see Table 1). In addition to a very low CO_2 formation, remarkable selectivity toward desired light olefins is observed: for y=0.4, at 4.5 % of CO conversion, hydrocarbons represent 69 mol.% of the products and 48 wt% of them are in the C_2-C_4 fraction. The olefin/paraffin ratio in this fraction reaches 3. However the olefin/paraffin ratio decreases at higher CO conversion, it is of 1.4 in the C_2-C_4 fraction at a CO conversion of 15 %. The chain growth usually high for Co-based catalysts is limited by the small size of the metal particles and by the role of Fe. Metallic active phase is formed by the partial reduction of the initial material, and is in strong interaction with the partially reduced perovskite phase. That helps avoid attrition of the metal particles or excessive carbide formation. That confers these catalysts very good lifetime (see Fig. 4), neither deactivation nor change in selectivity was observed.

Table 1
Fischer-Tropsch reactivity

Catalyst	y=0 (x=0.4)	y=0.1	y=0.2	y=0.3	y=0.4
T (°C)	280	250	255	244	230
CO Conversion (%)	3.7	4.3	4.2	4.4	4.5
Selectivity to. CO_2	24.8	23.8	26.1	26.1	19.1
Selectivity to. HC	68.8	67.6	67.3	66.6	68.8
Selectivity to. Oxyg. Cmpds	6.4	8.6	6.6	7.3	12.1
Weight Distribution CH_4	52.3	46.9	47.2	49.2	31.9
Weight Distribution C_2-C_4	36.2	37.6	38.0	35.3	47.7
Weight Distribution C_{5+}	11.5	15.6	14.8	15.6	20.4
Olefin/Paraffin (C_2-C_4)	1.7	1.6	2.1	2.3	3.0
% Olefin (C_2-C_4)	63.4	62.1	67.4	69.7	75.2

Partially 450°C *in-situ* reduced $LaCo_xFe_{(1-x)}O_3$ or $La_{(1-y)}Co_{0.4}Fe_{0.6}O_{3-\delta}$ oxides (750°C-calcined), fixed bed reactor, P=1 MPa, GHSV=3000 h^{-1}, CO/H_2=1, Q_{tot}= 0.3 L.h^{-1}

324

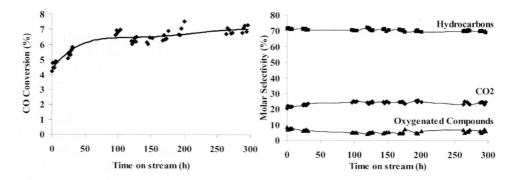

Fig. 4. Ageing of y=0.3 catalyst at 244°C

4. CONCLUSION

It has been evidenced in this work that the use of defined crystalline structure as catalyst precursor allows the control of the metal phase formation. By changing its Co/Fe ratio and the La-content, the precursor perovskite crystalline structure can be rhombohedral, orthorhombic or cubic. The reduction process of these oxides being tightly linked to their structure, it is possible to predict when preparing if the catalyst metal phase will be Co^0 or a $(Co\text{-}Fe)^0$ alloy small particles. The experimental selectivity toward C_2-C_4 olefins is particularly high while the CO_2 formation remains limited. The low extractable metal amount from pure $LaCo_xFe_{(1-x)}O_3$ perovskites can be improved by using La-deficient $La_{(1-y)}Co_{0.4}Fe_{0.6}O_{3-\delta}$ oxides, whose nano γ-Fe_2O_3 cores, in addition to driving the perovskite crystallization, act like a reducible cation reservoir.

REFERENCES

[1] H. Schulz, Appl. Catal. A, 186 (1999) 3.
[2] J.C. Hoogendorn and J.M. Salomon, Brit. Chem. Eng. (1957) 418.
[3] H. Araï, K. Mitsuishi and K. Seiyama, Chem. Lett. (1984) 1291.
[4] F. Tihay, G. Pourroy, M. Richard-Plouet, A.C. Roger and A. Kiennemann, Appl. Catal. A, 206 (2001) 29.
[5] J.L. Rehspringer and J.C. Bernier, Mat. Res. Soc. Symp. Proc. 72 (1986) 67.
[6] L. Bedel, A.C. Roger, C. Estournes and A. Kiennemann, Catal. Today 85, 207-218 (2003).

Studies in Surface Science and Catalysis, volume 147
X. Bao and Y. Xu (Editors)

Superior catalytic activity of nano-sized metallic cobalt inside faujasite zeolite for Fischer-Tropsch synthesis

Qinghu Tang, Ye Wang[*], Ping Wang, Qinghong Zhang and Huilin Wan

State Key Laboratory for Physical Chemistry of Solid Surfaces, Department of Chemistry, Xiamen University, Xiamen 361005, China

ABSTRACT

Cobalt oxide particles with a maximum of size distribution at 1.3-1.5 nm and in the state of CoO were synthesized within faujasite zeolites by a method comprising ion exchange of Co^{2+}, precipitation of exchanged Co^{2+}, and calcination. The metallic cobalt generated by the reduction of the encapsulated CoO_x at 673 K exhibited higher CO conversion in Fischer-Tropsch synthesis.

1. INTRODUCTION

Fischer-Tropsch (FT) synthesis, an important route for the transformation of the natural gas into hydrocarbon fuels via synthesis gas, has received renewed interest from the viewpoint of environment demands [1]. Supported cobalt is generally used in FT synthesis for obtaining hydrocarbons with longer carbon chains [2]. However, the development of effective catalysts with controlled product selectivity is difficult. The products of FT synthesis usually follow the Anderson-Schulz-Flory distribution. It has been reported that the mean pore diameter of the support might influence the product distributions [3]. Molecular sieves possess ordered porous structures and may possibly provide appropriate reaction spaces for regulating the product distributions. A few studies have applied microporous and mesoporous molecular sieves as catalyst supports [4, 5]. However, the impregnation method used for the loading of cobalt in most of these studies could not ensure the incorporation of cobalt into the pores of the molecular sieves, and thus it is difficult to get precise insight into the effect of pores on the activity and the product selectivities.

The preparations of nanoparticles of noble metals such as Pd and Pt within zeolites have been extensively studied [6]. A convenient method for assembling nanoparticles of noble metals in faujasite zeolites is to introduce metal ions by ion exchange and then to reduce metal ions to metal particles with a reductant.

However, in the case of non-noble transition metals such as Co, the reduction of Co^{2+} exchanged in the zeolite is difficult under temperatures lower than 800 K probably because of the strong interaction between Co^{2+} and anionic framework of zeolite [7]. Higher reduction temperatures would, however, cause the migration of a large proportion of metals out of the zeolite pores. Therefore, the development of effective methods for the synthesis of metallic cobalt inside the pores of zeolites is of significance. Recently, we have shown that the inclusion of a precipitation step after the ion exchange can result in cobalt oxide (CoO_x) in Y zeolite capable of being reduced at 673 K [8]. In this work, we report the characterizations of the CoO_x particles within the faujasite zeolites (NaX and NaY). The reduction behavior of such nano-sized CoO_x particles and the catalytic properties of the reduced samples in FT synthesis are investigated.

2. EXPERIMENTAL

For the syntheses of NaX- and NaY-encapsulated cobalt oxide samples (CoO_x-X and CoO_x-Y), Co^{2+} was first exchanged into the zeolites. After filtered, washed and dried, the Co^{2+}-containing zeolite was treated with 0.1 M NaOH. The sample was then recovered by filtration followed by washing and drying. The resultant powder was finally calcined to obtain the CoO_x-X or CoO_x-Y composites. The calcination temperature was generally 673 K and was raised to 923 K when the effect of calcination temperature was investigated.

The Co content in each sample was determined by the atomic absorption spectroscopic analysis. XRD patterns were recorded with Rigaku D/Max-C X-ray diffractometer. TEM images with EDS analysis were taken on a Phillips FEI Tecnai 30 electron microscope. XPS measurements were performed with a PHI Quantum 2000 equipment. The reduction behaviors of the CoO_x-X and CoO_x-Y were studied by H_2-TPR.

FT synthesis was carried out with a fixed-bed flow reactor operated at 2.0 MPa. The CoO_x-X or CoO_x-Y loaded in the reactor was first reduced in H_2 at 673 K. Syngas with H_2/CO ratio of 2 was then introduced and the pressure was raised to 2.0 MPa after the temperature was decreased to 523 K. Ar contained in the syngas was used as an internal standard.

3. RESULTS AND DISCUSSION

3.1. Characterizations of the cobalt oxide encapsulated in faujasite zeolites

XRD measurements of a series of CoO_x-X and CoO_x-Y samples with different Co content revealed that only diffraction peaks ascribed to NaX and NaY zeolites were observed. The peaks of crystalline Co_3O_4 did not appear for all the samples calcined at 673 K. However, as the calcination temperature was raised to 923 K, the peak at $36.8°$ assigned to the strongest line of Co_3O_4 became observable, and the Co_3O_4 particles were estimated to be ca. 29 nm in size using Scherrer's equation. These results suggest that high calcination temperature results in larger Co_3O_4 particles, which can be detected by XRD and are probably located outside the supercages of the faujasite zeolites.

TEM image of the CoO_x-Y (Co, 6.2 wt%) sample calcined at 673 K shows that the small CoO_x particles are homogeneously distributed in zeolite background (Fig. 1A). The sizes of particles were distributed in the range of 0.7-3.0 nm with a maximum of distribution at 1.3-1.5 nm (Fig. 1B). TEM measurements for the CoO_x-X (Co, 8.8 wt%) sample calcined at 673 K provided similar results. Considering the size (1.3 nm) of the supercage of the faujasite zeolites and the possible formation of a small proportion of mesopores after the treatment with NaOH, it is reasonable to conclude that most of the CoO_x particles are located within the supercages of zeolites.

On the other hand, for the samples calcined at 923 K, CoO_x particles with size larger than 20 nm could be seen on TEM micrographs, and EDS analyses confirmed that the cobalt element was not homogeneously distributed. These observations are consistent with the XRD results, supporting that higher calcination temperature causes the migration of CoO_x out of the supercages.

The oxidation state of cobalt in the CoO_x particles within the supercages of faujasite zeolites has been investigated by XPS. The binding energy of Co 2p2/3 for the sample prepared by directly calcining the Co^{2+}-exchanged X zeolite was observed at 781.7 eV, which could be attributed to Co^{2+} ions [9], indicating that no oxidation could take place in this case. For the samples prepared with the precipitation of the Co^{2+}-exchanged X or Y zeolite, the peak of Co 2p3/2 became broad and the binding energy shifted to lower value, suggesting the formation of oxidic cobalt species.

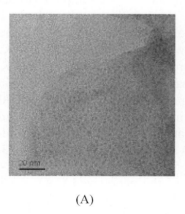

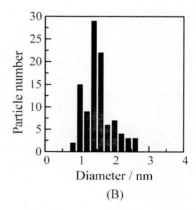

(A) (B)

Fig. 1. (A) TEM micrograph of the CoO_x-Y (Co, 6.2 wt%) sample calcined at 673 K. (B) Distribution of the diameters of the CoO_x nanoparticles in the zeolite matrix.

Table 1
Results obtained from the deconvolution of Co 2p3/2 XPS peak

Sample	Proportion of Co^{2+}	Ratio of oxidic Co^{II} to oxidic Co^{III}	Ratio of CoO to Co_3O_4
Co^{2+}-X-673 K (Co, 8.8 wt%)	∞	–	–
Co_3O_4	0	0.5	0
CoO_x-X-673 K (Co, 8.8 wt%)	0.42	3.6	3.1
CoO_x-X-923 K (Co, 8.8 wt%)	0.30	0.8	0.3
CoO_x-Y-673 K (Co, 6.2 wt%)	0.23	3.7	3.2
CoO_x-Y-923 K (Co, 6.2 wt%)	0.10	1.0	0.5

With further reference to the binding energies of oxidic Co(II) at 780.4 eV and oxidic Co(III) at 779.4 eV, we could separate the peak of Co 2p3/2 for the CoO_x-X and CoO_x-Y into three components, i.e., Co^{2+} ions, Co(II) and Co(III) both in oxidic states. The results in Table 1 show that Co^{2+} ions still remain after the precipitation and calcination. These Co^{2+} ions are probably located within the small cages such as the sodalite cages or hexagonal prisms in the faujasite zeolites, which must be difficult to be accessed by NaOH. More importantly, the CoO_x-X and CoO_x-Y samples exhibit higher ratio of oxidic Co(II) to Co(III) than Co_3O_4, suggesting that CoO co-exists with Co_3O_4 in these samples. The calculated ratios of CoO to Co_3O_4 in the CoO_x-X and CoO_x-Y calcined at 673 K were higher than 3, and decreased drastically to 0.3 and 0.5, respectively after the calcination at 923 K. These observations strongly suggest that the CoO_x particles within the supercages exist mostly in the state of CoO, and they are changed to Co_3O_4 as they migrate out of the supercage at high temperatures.

3.2. Reduction behavior of the cobalt oxide encapsulated in faujasite zeolites

Fig. 2 shows that the reduction of Co^{2+} in the Co^{2+}-X sample (without the precipitation step) occurred mainly at 1023 K along with a small peak at 823 K, while the Co^{2+} in the Co^{2+}-Y sample cannot be reduced in the whole temperature range investigated. The crystalline Co_3O_4 showed a reduction peak at 644 K. For the CoO_x-X sample calcined at 673 K, two broad reduction peaks at 590 K and 930 K were observed. The CoO_x-Y sample calcined at 673 K exhibited similar TPR profiles, and the two broad peaks were observed at 573 K and 900 K although Co^{2+}-Y did not show any reduction in the whole temperature region. It is likely that the broad reduction feature is related to the cobalt species inside the cages of zeolites. Since the reduction of CoO and Co_3O_4 takes place at 623-673 K, it is reasonable to assign the broad peak at 573 K or 590 K for the CoO_x-X or CoO_x-Y calcined at 673 K to the reduction of the nano-sized CoO_x particles, most of which are in the state of CoO, within the supercages of the zeolites. The broad peak at higher temperatures for these two samples is possibly due to the reduction of the small CoO_x clusters difficult to reduce and the remained Co^{2+}.

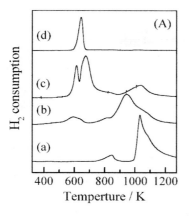 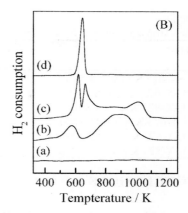

Fig. 2. TPR profiles of (A) Co-containing X (Co, 8.8 wt %) and (B) Co-containing Y (Co, 6.2 wt%) zeolites as well as Co_3O_4. A: (a) Co^{2+}-X, (b) CoO_x-X-673 K, (c) CoO_x-X-923 K, (d) Co_3O_4; (B): (a) Co^{2+}-Y, (b) CoO_x-Y-673 K, (c) CoO_x-Y-923 K, (d) Co_3O_4.

For the CoO_x-X and CoO_x-Y calcined at 923 K, TPR profiles changed significantly. Two sharper peaks appeared in the temperature range of 573-673 K and the broad peak at higher temperatures became remarkably small. This is probably related to the migration of a large part of CoO_x particles out of the supercage of the zeolites to form larger Co_3O_4 particles. The quantitative calculations showed that the reduction degrees at 673 K were 16% and 21% for the CoO_x-X and CoO_x-Y calcined at 673 K, respectively, while they were increased to 52% and 45% after the calcination at 923 K, respectively.

3.3. Catalytic results

A series of the Co-containing X and Y zeolites after reduction at 673 K were applied to Fischer-Tropsch synthesis. The Co^{2+}-X and Co^{2+}-Y samples did not show any CO conversion in Fischer-Tropsch synthesis. This is because the Co^{2+} cannot be reduced at 673 K in these two samples. On the other hand, the CoO_x-X and CoO_x-Y samples reduced at 673 K exhibited considerable CO conversions (Table 2). The product distributions were similar over these samples. The selectivity to C_5^+ hydrocarbons reached or exceeded 80%, and the value of the chain growth probability (α) was between 0.76-0.85. It should be noted that, as compared with the samples calcined at 673 K, those calcined at 923 K exhibited remarkably lower CO conversion although the latter samples possessed higher reducibility as described above. This result strongly suggests that the small cobalt particles encapsulated inside the supercage of zeolite are more active than the large ones located outside the supercage of zeolite. However, no significant differences in product distributions were observed between the samples calcined at different temperatures.

Table 2
Fischer-Tropsch synthesis over a series of Co-containing faujasite zeolites reduced at 673 K

Sample	CO conversion /%	Selectivity /%		Chain growth probability (α)
		CH_4	C_{5+}	
Co^{2+}-X-673 K (Co, 8.8 wt%)	0	–	–	–
CoO_x-X-673 K (Co, 8.8 wt%)	66	11.1	83.2	0.83
CoO_x-X-923 K (Co, 8.8 wt%)	31	10.4	80.8	0.86
Co^{2+}-Y-673 K (Co, 6.2 wt%)	0	–	–	–
CoO_x-Y-673 K (Co, 6.2 wt%)	77	17.1	72.7	0.76
CoO_x-Y-923 K (Co, 6.2 wt%)	59	8.4	86.0	0.86

4. CONCLUSIONS

The cobalt oxide particles encapsulated within faujasite zeolites could be synthesized by a procedure comprising ion exchange of Co^{2+}, precipitation of exchanged-Co^{2+} with NaOH, and calcination at 673 K. These CoO_x particles were in the size range of 0.7-3.1 nm with a maximum of distribution at 1.3-1.5 nm and were mostly in the state of CoO. Higher calcination temperature of 923 K caused the migration of the CoO_x out of the supercage to form large Co_3O_4 particles. The CoO_x particles within the supercages could partly be reduced at temperatures as low as 573 K. The metallic cobalt generated by the reduction of the encapsulated CoO_x exhibited higher CO conversion in FT synthesis.

ACKNOWLEDGEMENT

This work was supported by the Chinese Ministry of Science and Technology (No. G1999022408), and the National Natural Science Foundation of China (Nos. 20373055 and 20021002).

REFERENCES

[1] H. Schulz, Appl. Catal. A, 186 (1999) 3.
[2] E. Iglesia, Appl. Catal. A, 151 (1997) 59.
[3] B. Ernst, A. Libs, P. Chaumette, A. Kiennemann, Appl. Catal. A, 186 (1999) 145.
[4] S. Bessell, Appl. Catal. A, 96 (1993) 253; S.J. Jong, S. Cheng, Appl. Catal. A, 126 (1995) 51.
[5] A.Y. Khodakov, A. Griboval-constant, R. Bechara, V.L. Zholobenko, J. Catal., 206 (2002) 230; M. Wei, K. Okabe, H. Arakawa and Y. Teraoka, New. J. Chem, 27 (2003) 928; Y. Wang, M. Noguchi, Y. Takahashi and Y. Ohtsuka, Catal. Today, 68 (2001) 3.
[6] J.S. Feeley and W.M.H. Sachtler, J. Catal., 131 (1991) 573; L. Guczi, A. Beck, A. Horváth and D. Horváth, Top. Catal., 19 (2002) 2.
[7] L. Guczi, Catal. Lett., 7 (1991) 573.
[8] Q. Tang, Y. Wang, Q. Zhang and H. Wan, Catal. Commun., 4 (2003) 253.
[9] L. Guczi, D. Bazin, Appl. Catal. A, 188 (1993) 163.

Studies in Surface Science and Catalysis, volume 147
X. Bao and Y. Xu (Editors)

Fischer-Tropsch synthesis: effect of water on activity and selectivity for a cobalt catalyst

Tapan K. Das, Whitney Conner, Gary Jacobs, Jinlin Li, Karuna Chaudhari and Burtron H. Davis

Center for Applied Energy Research, University of Kentucky, 2540 Research Park Drive, Lexington, KY, 40511

ABSTRACT

The addition of water to the syngas ($H_2O/CO = 0.74$) feed for an unsupported cobalt catalyst utilized in a CSTR, where all of the catalyst is exposed to a common feed composition, resulted in an increase in CO conversion. Conversion remains constant during a three-day period, showing that the promotional effect was not of a transitory nature. At the same time the CO hydrogenation increased, the hydrogenation to produce methane decreased as did the secondary hydrogenation of ethene. The addition of higher fractions of H_2O/CO led to permanent deactivation of the catalyst for CO conversion but the decrease in hydrogenation to methane and of alkene products remained.

1. INTRODUCTION

Because the cobalt catalyst does not exhibit significant water-gas-shift activity, the impact of water on Fischer-Tropsch synthesis (FTS) is of concern. Water can have an impact on the catalyst by, for example, causing oxidation of the cobalt and reducing the conversion as a result of loss of active catalytic sites [e.g., 1-3 and references therein, 4,5]. Water can also have a kinetic effect by competing with the reactants for the active catalytic sites [e.g., 6-10]. In addition, some data appear to show that the impact of water depends upon the support, and even upon the cobalt loading for an alumina support [e.g., 8-10]. However, one of the early, if not the earliest, reports of the impact of water on the FTS activity and selectivity of cobalt catalysts was by Kim [11]. The addition of water (20 parts per 100 parts of 2:1 $H_2:CO$ feed) to an unsupported cobalt catalyst resulted in a greater than two-fold increase in CO conversion and a decrease in methane production by a factor of two. The experiments were carried out in a fixed-bed reactor where the concentrations of reactant and water will vary with the length of the catalyst bed, and with the unsupported cobalt catalyst the effect must have been due to cobalt [11]. If water has this effect on cobalt when it is unsupported it is anticipated that the support should not impact the water effect. However, as indicated above, the support appears to have a significant impact upon the effect of water on the supported cobalt catalyst.

In order to define the role(s) of water, our experiments were carried out in a continuous stirred tank reactor (CSTR) so that the total flow rate and partial pressures of H_2 and CO remained constant. This was accomplished by replacing increasingly larger fractions of an inert gas, Ar, with water and with a corresponding decrease in the partial pressure of Ar.

2. EXPERIMENTAL

2.1. Catalyst Preparation

A bulk cobalt catalyst was prepared using cobalt nitrate solution as a cobalt precursor. In this method a stoichiometric amount of ammonium carbonate (aqueous solution) was added to an aqueous solution of $Co(NO_3)_2.6H_2O$ (Aldrich, 98%). The precipitate was filtered and the filtered solid was washed with deionized water. The catalyst was dried overnight at 393 K and calcined at 773 K in air for 5 hours.

2.2. Water Effect

The calcined catalyst (~14.0 g; weight was accurately known) was reduced ex-situ in a fixed bed reactor with a mixture of hydrogen and helium (1:2) at a flow rate of 70 SL/h at 623K for 10 hours at atmospheric pressure. The catalyst was transferred under protection of argon to a 1 L CSTR containing 300 g of melted Polywax 3000 (start-up solvent). The catalyst was then re-reduced in-situ at 553K for 24 hours in the CSTR in a flow of 30 SL/h hydrogen at atmospheric pressure. After the activation period, the reactor temperature was decreased to 453K and synthesis gas was introduced through a dip tube while the reactor pressure was increased to 2.0 MPa (19.7 atm). The reactor temperature was then increased to 493K at a rate of 1K/min. The stirrer was operated at 750 rpm. The FT synthesis was started at a space velocity of 5.0 SL/h/gcat and H_2/CO ratio of 2.0. Different amounts of water (10-30 vol.%) were added into the feed gas at a total space velocity of 3.0 SL/h/gcat. To keep the partial pressure of hydrogen and carbon monoxide constant, 30 vol.% of argon was added to the feed. When water was added, the equivalent fraction of argon was replaced by water. Thus, the sum of water added plus argon was 30 vol% of the total feed. A high pressure ISCO syringe pump was used to add the water to the reactor. To define the impact of water addition on the activity and deactivation rate, the reaction conditions were changed back to the standard conditions with 30 vol.% argon after each run with water addition. The added water and water produced by the FT synthesis were quantified to ensure that the water balance was accurate. The flow was measured using a bubble flow meter and the composition was quantified using an on-line GC (Refinery gas analyzer, HP Quad series Micro GC).

3. RESULTS

The unsupported cobalt catalyst was active for FTS (Fig. 1). Compared to a 20 wt% cobalt catalyst supported on alumina, the supported catalyst is about 10 to 15 times as active as the unsupported catalyst when both are compared on a cobalt basis. At a space velocity of 3 SL/hr/g cat with Ar flow, the conversion approached a steady level of about 18% CO conversion. When 10% water ($H_2O/CO = 0.74$) was added, the CO conversion increased from about 18% to 24%, corresponding to a 1.3-fold increase. While the increase in conversion is not as large as was observed by Kim [11], there is a significant increase. Water addition continued for a period of three days and the conversion was essentially constant during this period. After termination of water addition, the CO conversion was about the same as before water addition. The water addition did not have a significant impact upon the catalytic stability of the unsupported cobalt. Adding 20% water ($H_2O/CO = 1.2$) caused the conversion to decline slightly; when water addition was terminated after 4 days, the catalyst activity had declined, and continued to decline during the period of normal operation.

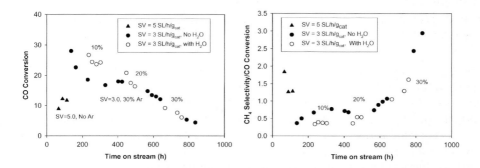

Fig. 1 (left). CO conversion during periods of normal FTS and periods of water addition.

Fig. 2 (right). Impact of water addition on methane selectivity during FTS with an unsupported cobalt catalyst.

The addition of 30% water (H_2O/CO = 1.5) did not cause the activity to increase, and the activity declined during the period following termination of water addition. Thus, the CO conversion is increased when the total catalyst inventory is exposed to a common reaction mixture in the CSTR and the only significant variable is the ratio of H_2O/CO (and H_2O/H_2). However, there is a level where a further increase in the H_2O/CO ratio ceases to cause an increase in CO conversion and, rather, causes permanent damage to the catalyst.

The methane selectivity was lower during the period of water addition than the periods before and after 10% water addition (Fig. 2). The addition of 20 or 30% water caused the methane selectivity to be slightly lower than when water was not added.

As shown in Fig. 3, the [ethene/(ethene + ethane)] fraction is larger during the periods of water addition, and this fraction seems to be larger as the amount of water added increases. At least some of the difference may be ascribed to the lower CO conversion but it still seems that the catalytic activity for secondary hydrogenation decreases in proportion to the amount of water added. For the addition of 10% water, the extent of the WGS reaction (CO_2 selectivity, Fig. 4) is the same as either before or following this period of water addition. However, the addition of 20 or 30% water causes an increase in the extent of WGS.

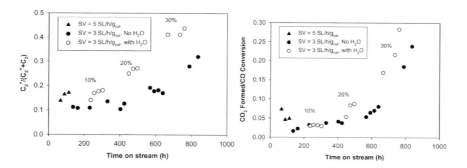

Fig. 3 (left). Impact of water addition on secondary hydrogenation of ethene (C_2^-).

Fig. 4 (right). Influence of water addition on WGS activity for an unsupported cobalt catalyst.

4. DISCUSSION

When an unsupported cobalt catalyst is exposed to lower levels of added water ($H_2O/CO \leq 0.74$) under conditions that are otherwise the same, the CO (and H_2) conversion increases. This increase is not a temporary effect but persists at the same level for more than 70 hours. This long-term effect of added water strongly suggests that water is co-adsorbed onto the catalyst surface and impacts the chemical selectivity. However, at this time it is not possible to eliminate a partial oxidation of the catalyst surface causing this effect.

The hydrogenation to produce methane is decreased with the addition of water and this persists even at water addition levels where there is a significant decline in CO conversion, and irreversible poisoning of the catalyst. Likewise, the secondary hydrogenation of ethene is lower when water is added, and this effect persists at the three levels of water addition used in this study.

In summary, these results using a CSTR, so that all catalyst particles are exposed to a constant gas composition, are in general agreement with data reported earlier for conversions using an unsupported catalyst in a fixed-bed reactor [11]. The CSTR data need to be augmented by runs with lower H_2O/CO ratios to obtain better selectivity data and to define the kinetics. Finally, the results with the unsupported catalyst raises a question that, at present, cannot be answered. If cobalt catalytic activity can be increased by the addition of water, the catalytic activity should not depend on the support. Cobalt on an alumina or titania support exhibit lower CO conversion when water is added but cobalt supported on silica exhibits a higher conversion when water is added [8-10]. These results appear to require some support-cobalt interactions, as yet unidentified, that are impacted by the presence of additional water.

REFERENCES

[1] S. Storsoeter, Chr. Aaserud, A-.M. Hilmen, O. A. Lindvag, Bergene, E.; D. Schanke, S. Eri, A. Holmen, Preprints, ACS, Div. Fuel Chem. (2002) 47, 158-159.
[2] Anne-Mette Hilmen, Odd Asbjorn Lindvag; Edvard Bergen,; Dag Schanke; Sigrid Eri; Anders Holmen, Studies in Surf. Sci. Catal. (2001) 136 (Natural Gas Conversion VI), 295-300.
[3] D. Schanke; A.M. Hilmen; E. Bergene; K. Kinnari; E. Rytter; E. Adnanes; A. Holmen, Catal. Lett. (1995) 34, 269-84.
[4] R. Zennaro, M. Tagliabue, C.H. Bartholomew, Catal. Today, (2000), 58, 309-319.
[5] P.J. van Berge, J. van de Loosdrecht, S. Barradas, A.M. van der Kraan, Catal. Today, (2000), 58, 321-334.
[6] H. Schulz, M. Claeys and S. Harms; Natural Gas Conversion IV (M de Pontes et al, eds) SSSC 107, 1997, 193.
[7] E. Iglesia, Appl. Catal. A: General, (1997) 161, 59-78.
[8] J. Li, G. Jacobs, T. Das, Y. Zhang and B.H. Davis, Appl. Catal. A: Gen., 236, 67, (2002).

[9] J. Li, X. Zhan, Y. Zhang, G. Jacobs, T. Das and B.H. Davis, Appl. Catal. A: Gen., 228, 203.

[10] J. Li, G. Jacobs, T. Das and B.H. Davis, Appl. Catal. A: General, (2002), 233 255-262.

[11] C.J. Kim, U.S. Patent 5,227,407, July 11, 1993.

Studies in Surface Science and Catalysis, volume 147
X. Bao and Y. Xu (Editors)
©2004 Elsevier B.V. All rights reserved.

Hydrogenated Zr-Fe alloys encapsulated in Al$_2$O$_3$/Al matrix as catalysts for Fischer-Tropsch synthesis

S.F. Tikhov[*][a], **V.I. Kurkin**[b], **V.A. Sadykov**[a], **E.V. Slivinsky**[b],
Yu.N. Dyatlova[a], **A.E. Kuz'min**[b], **E.I. Bogolepova**[b], **S.V. Tsybulya**[a], **A.V.
Kalinkin**[a], **V.B. Fenelonov**[a], **V.P. Mordovin**

[a]Boreskov Institute of Catalysis SB RAS, 5 Lavrentieva St., 630090,
Novosibirsk

[b]Topchiev Institute of Petrochemical Synthesis RAS, 19 Leninskii Av., 117912,
Moscow, RUSSIA

ABSTRACT

The genesis of the ZrFe intermetallides with different atomic ratios during preparation of Zr$_x$Fe$_z$H$_y$/Al$_2$O$_3$/Al catalysts and their performance in Fischer-Tropsch synthesis has been studied. The effect of structural, surface and textural properties on their activity had have been discussed.

1. INTRODUCTION

The polymetallic catalysts derived from intermetallic hydrides based upon transition metals of YIII group and the elements of the 4th period of the Periodic Table are known to be active for the hydrogenation reactions, hydrogenated ZrFe bulk catalysts demonstrating the highest output of C$_{5+}$ hydrocarbons [1,2]. However, the use of such catalysts as pure active component (AC) in the Fischer-Tropsch synthesis (FTS) is hampered due to fragility of hydride particles and a low stability towards coking [3]. A possible way of improvement of the properties of hydrogenated alloys is their encapsulation into the Al$_2$O$_3$/Al matrix via mild oxidative procedure [4]. In the work presented, the model catalysts based upon hydrogenated, pure and encapsulated Zr-Fe alloys with a varied Zr/Fe atomic ratio have been studied. Structural, textural, surface and catalytic properties of the catalysts have been elucidated in details.

2. EXPERIMENTAL

The powdered catalysts with different Zr/Fe atomic ratio ~2:1, 1:1 and 1:2 were prepared by alloying Zr and Fe at 2200-2400°C under Ar followed by hydrogenation of those alloys at a room temperature and hydrogen pressure of 5 MPa. Approximate stoichiometry of hydrides measured gravimetrically corresponds to $H_{1.5}Zr_{2.6}Fe$, $H_{0.5}ZrFe$, H_0ZrFe_2. Hydrogenated AC (fraction 0.5-0.25 mm) were mixed with the aluminum powder and subjected to hydrothermal treatment at 100°C in a special die followed by drying and calcination at 540°C. This procedure results in self-pressing of blends into compact mechanically strong metallo-ceramic monoliths [4,5]. Two types of aluminum powder (PA-4 and PA-HP) were used, the latter possessing a higher reactivity thus permitting to encapsulate more AC. The catalyst performance in FTS was evaluated for hydrogenated AC and composite catalysts (fraction 2-3 mm) in a flow type reactor at 300-310°C, 3 MPa, H_2:CO ≈ 2:1, GHSV ~6500-7400 h^{-1}. The phase composition was analyzed with URD-63 diffractometer using CuK_α radiation. The total pore volume was estimated from the values of true and apparent densities of granulated cermets [4]. The true densities of samples were estimated by using a helium Autopycnometer 1320 Micromeritics instrument. The details of microtexture were characterized by using the nitrogen adsorption isotherms measured at 77 K on an ASAP-2400 Micromeritics instrument. The crushing strength was measured using a PK-2-1 machine. The XPS spectra were recorded with an ESCA-3 spectrometer using AlK_α radiation.

3. RESULTS

As is seen in the Table 1, the following sequence of activity for pure hydrogenated alloys has been found: $Zr_{2.6}Fe$ < $ZrFe_2$ < ZrFe. As for the composite catalysts $Zr_yFe_zH_x/Al_2O_3/Al$, a large enhancement of CH_x productivity has been observed with a different sequence: $ZrFe_2$ < $Zr_{2.6}Fe$ < ZrFe, 1:1 sample being again the most effective. The same order of the C_{5+} yield for pure and encapsulated alloys has been found. The increase of the AC content

Table 1

Catalytic properties of hydrogenated Zr-Fe intermetallides in Fischer-Tropsch synthesis.

Active component		Conversion of CO, %		Hydrocarbon productivity, g/kg·h				Selec-tivity to CH_4, %
Type	Content, wt.%	Total	To CO_2	Total		C_{5+}		
				Per kg_{cat}	Per kg_{AC}	Per kg_{cat}	Per kg_{AC}	
$Zr_{2.6}FeH_x$ (2:1)	100	20.9	2.3	78	78	27	27	27,7
+ PA-HP	40.0	11.0	0.8	189	473	39	98	37.3
+ PA-4	25.6	16.6	2.5	267	1041	45	174	18.4
$ZrFeH_x$ (1:1)	100	59.1	15.0	231	231	57	57	35.2
+ PA-HP	40.0	24.4	2,6	381	953	99	248	34.9
+ PA-4	25.5	21.1	1.5	300	1172	82	320	34.9
$ZrFe_2H_x$ (1:2)	100	36.6	5.1	106	106	28	28	30.2
+ PA-4	25.4	21.5	0.6	189(188)	746	63(34)	247	27.0(46)

In brackets the data for non-hydrogenated AC are presented.

up to 40% permits to enhance the activity of 1:1 composite catalysts, but effectiveness of the AC (productivity per weight of AC) drops (Tabl. 1). The activity of non-hydrogenated AC is the same but selectivity towards CH_4 is much higher then than that for hydrogenated ones. The hydrocarbon products distribution demonstrates a non-ideal character due to enhanced CH_4 formation; the average parameter α of the Anderson-Schulz-Flory distribution being in the 0.6-0.7 range for all the catalysts studied. Comparing these data with the particle size for pure (2.5 mm) and encapsulated (0.375 mm) AC permits to conclude that, in the first approximation, the specific output (per unit wt. of Zr_yFe_z) is proportional to the AC active surface. Hence, one of the advantages of AC encapsulation into the Al_2O_3/Al matrix is the use of more dispersed and much more productive AC particles, which is not possible for the conventional fixed bed catalyst design [3].

According to XRD (Fig.1), in pure intermetallides tetragonal Zr_2Fe and $ZrFe_2$ (hexagonal for 2:1, cubic for 1:1 and 1:2 samples) phases dominate. The amount of the first phase increases, while that of the second decreases with the Fe content. For 1:2 sample a cubic β-Zr phase is also observed. After hydrogenation, tetragonal ε-Zr_2H-type phase with varied lattice parameters appears to be formed from Zr_2Fe, but in 1:2 sample this phase is absent at all. $ZrFe_2$ phase becomes cubic for all samples, the lattice parameter increasing from 2:1 to 1:2 samples. In addition, another hydride phase $Zr_xFe_yH_y$ with varied lattice parameters is observed only for 1:1 hydrogenated sample. This phase appears to be formed on the boundary between $ZrFe_2$ and ε-Zr_2H phases. After encapsulation into the Al_2O_3/Al matrix for all samples Al^0, γ-Al_2O_3 α-Fe_2O_3 and zirconia phases are observed. For the latter one, the ratio of the amount of monoclinic to tetragonal phases increases with the increasing Fe content. In 2:1 and 2:1 samples, a cubic $ZrFe_2$ phase is identified. For 1:2 sample β-Zr phase is not observed. As for hydride ε-$ZrH_{1.95}$ phase, its content is decreased in 2:1 and is almost zero in 1:1 sample.

The analysis of the texture of hydrogenated intermetallides (fraction 0.5-0.25 mm) revealed that for all samples but 1:2 one, their specific surface area is much higher than the geometric one, thus demonstrating a high roughness of their particles due to cracks and pores formation. All types of AC have mesopores (d = 390-460 Å) with a small pore volume. The hydrothermal and calcination treatments have only a weak effect on the textural characteristics of AC. As for encapsulated catalysts, the total micropore volume of the catalysts as well as the specific surface area is higher but the average micropore diameter is significantly lower as compared to hydrogenated AC. All these characteristics are close to that of a pure Al_2O_3/Al matrix and are proportional to the alumina content being higher for PA-HP aluminum powder (Tabl.2). Nevertheless, the total pore volume of the composite catalysts is much higher than that of

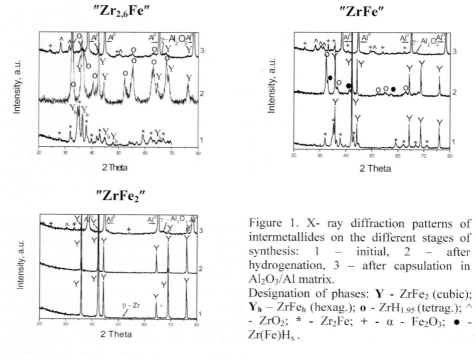

Figure 1. X- ray diffraction patterns of intermetallides on the different stages of synthesis: 1 – initial, 2 – after hydrogenation, 3 – after capsulation in Al_2O_3/Al matrix.

Designation of phases: **Y** - $ZrFe_2$ (cubic); Y_h – $ZrFe_h$ (hexag.); o - $ZrH_{1.95}$ (tetrag.); ^ - ZrO_2; * - Zr_2Fe; + - α - Fe_2O_3; ● - $Zr(Fe)H_x$.

micropores, and it increases with the AC content in the composite catalysts thus demonstrating a developed macropore (d > 1 μm) structure [4]. The crushing strength of the granulated Al_2O_3/Al matrix is higher for PA-HP derived composite due to a higher alumina content and a self-pressing in the press-form in the course of hydrothermal oxidation [6].

According to XPS, even after hydrogenation, the surface of AC is in the oxidized state: Fe^{2+}, Fe^{3+} with BE(Fe2p 3/2)~711 eV; Zr^{4+} with BE(Zr3d 5/2) ~182 eV, due to a contact with the air. The surface layer is enriched by Fe, segregation being increased after hydrogenation. The surface Fe content which is characterized by the Fe/Zr ratio is increased as compared with the bulk Fe content. Surprisingly, after encapsulation, a very low surface AC content characterized by the Fe/Al ratio < 0.013 is observed, possibly due to a shielding effect of porous Al_2O_3 partially covering the surface of AC in the course of encapsulation procedure.

4. DISCUSSION

As is seen from comparison of the catalytic, the bulk phase and the surface composition data, hydrogenation results in the enrichment and more random distribution of iron on the surface of AC. Evidently, the surface layer of AC

Table 2
Textural and mechanical properties of the catalysts $Zr_yFe_zH_x$, $Zr_yFe_zH_x/Al_2O_3$/Al, prepared from different aluminum powders (PA-HP, PA-4) and compared with Al_2O_3/Al matrix.

Active component		V_{total}, ml/g	Adsorption-desorption isotherm data			Crushing strength, MPa
Type + aluminum powder	Content, wt. %		SSA, m^2/g	V_{micro}, ml/g	d, under 2000 Å	
$Zr_{2,6}FeH_x$	0	-	0,17	0,010	390	-
+ PA-HP	40.0	0,74	15,02	0,019	50,5	1,27
+ PA-4	25.6	0,21	10,61	0,017	51	0,32
$ZrFeH_x$	100	-	0,22	0,012	440	-
+ PA-HP	40.0	0,64	42,14	0,051	49,1	2,08
+ PA-4	25.5	0,31	11,25	0,019	53	0,23
$ZrFe_2H_x$	100	-	0,03	<0,001	460	-
+ PA-4	25.4	0,24	9,09	0,015	50,8	0,58
Al_2O_3/Al, PA-HP	0	0,39	34,52	0,040	40,1	3,53
Al_2O_3/Al, PA-4	0	0,37	17,30	0,032	51,0	0,78

exists as FeC_x in the reaction media [7], but preliminary hydrogenation possibly facilitates this process positively influencing the selectivity of FeZr system. The order of activity related per the Fe surface content (g/"S_{Fe}"·h) estimated from SSA and Fe/Zr ratio was found to be passing through maximum to bulk Fe content: $Zr_{2,6}Fe$ (~2010) < ZrFe (~11550) < $ZrFe_2$ (~5480) also demonstrating a higher effectiveness of the surface active centers for 1:1 active component. However, the "real" (per mass of AC) effectiveness of catalysts also depends upon their SSA which is at maximum for 1:1 sample. Possibly, this SSA is provided by the optimum multiphase composition with hydrogenated zirconium and catalytically active iron, characterized by enhanced surface roughness and existence of microcracks.

For aluminum powder-based catalysts the porosity and the average pore diameter are closely related to the average particle size of initial Al powders [6]. Supposing a negligible blocking effect of alumina on the surface concentration of active sites due to permeation of gaseous substances through the porous layer, we found that, as compared with the known catalysts [8,9], a general level of the structural parameter χ [8] is very low ($x10^{16}$ m^{-1}): ~1.9; ~1.3 and ~0.1 for 2:1; 1:1 and 1:2 PA-4-capsulated catalysts, respectively, which is explained by a very high amount of ultramacropores in the composite catalyst [5,6]. Trends in the selectivity variation differ to methane (Tabl.1) and C_{5+} products (16.7; 27.2; 33.1). These deviations from regularities observed earlier [8,9] can be explained by the unusual pore size distribution of the composite catalysts. Also, AC is not distributed randomly on the surface of alumina, while porous alumina decorates the surface of AC by a thick (up to few μm) layer, which FTS products need to penetrate before reaching the ultramacropores.

However, all catalytic composites demonstrate a high activity which remarkably exceeds the highest level achieved by earlier described intermetallic catalysts. Thus, according to [2], on ZrFe intermetallide at 350° C and 20 MPa a C_5^+ productivity of 200 g/l$_{Cat}$·h ($\approx$60-70 g/kg$_{Cat}$·h) was obtained. In terms of the simplest FTS kinetics on Fe-contained catalysts (nearly 1st order to H_2 and 0 - ~0,2 order to CO), one can estimate that at 3 MPa and 350°C such a catalyst would produced $\approx$10-20 g/kg$_{Cat}$·h C_5^+. For composite catalysts, this value is in the range of 50-100 g/kg$_{Cat}$·h.

4. CONCLUSION

The encapsulation of hydrogenated Zr-Fe intermetallides into the Al_2O_3/Al matrix permits to preserve their main properties as well as to increase their specific activity. The multiphase composition of easy hydrogenated and FTS active component provides a non-monotonous variation of their performance with Zr:Fe ratio. The macroporous Al_2O_3/Al matrix prevents the loss of AC and permits to use highly dispersed AC while ensuring the efficient transport of reagents and products due to developed macro/microporous structure, but the selectivity towards C_{5+} needs to be improved. The composite catalysts prepared as thick coating layers on narrow tubes can ensure intensification of the heat transfer in fixed tube reactors at high linear velocities [10].

ACKNOWLEDGEMENTS

This work was financially supported by the Russian Foundation for Basic Research (Grant 02-03-32277).

REFERENCES

[1] V.V. Lunin, A.Z. Khan, J. Molec. Catal., 25 (1984) 317.
[2] A.Ya. Rozovsky, Kinet. Catal. , 40 (1999) 358.
[3] L.A. Vytnova, V.P. Mordovin, G.A. Kliger, E.I. Bogolepova, V.I. Kurkin, A.N. Shuykin, E.V. Marchevskaya, E.V. Slivinsky, Russ. J. Petrochemistry, 42 (2002) 111.
[4] S.F. Tikhov, V.A. Sadykov, Yu.V. Potapova, A.N. Salanov, S.V. Tsybulya, G.N. Kustova, G.S. Litvak, V.I. Zaikovskii, S.N. Pavlova, A.S. Ivanova, A.Ya. Rozovskii, G.I. Lin, V.V. Lunin, V.N. Ananyin, V.V. Belyaev, Stud. Surf. Sci. Catal., No 118 (1998) 797.
[5] S.F. Tikhov, Yu.V. Potapova, V.A. Sadykov, A.N. Salanov, S.V. Tsybulya, G.S. Litvak, L.F. Melgunova, React. Kinet. Catal. Lett., 77 (2002) 267.
[6] S.F. Tikhov, V.B. Fenelonov, V.A. Sadykov, Yu.V. Potapova, A.N. Salanov. Kinet. Catal., 41 (2000) 907.
[7] S. Li, G.D. Meitzner, E. Iglesia, Stud. Surf. Sci. Catal., No 136 (2001) 387.
[8] E. Iglesia, S.C. Reyes, R.J. Madon, S.L. Soled, Adv. Catal., No 39 (1992) 221.
[9] G.W. Huber, C.H. Bartolomew, Stud. Surf. Sci. Catal., No 136 (2001) 283.
[10] M.E. Dry, Catal. Today, 71(2002) 227.

Studies in Surface Science and Catalysis, volume 147
X. Bao and Y. Xu (Editors)
©2004 Elsevier B.V. All rights reserved.

A review of *in situ* XAS study on Co-based bimetallic catalysts relevant to CO hydrogenation

D. Bazin[a] and L. Guczi[b]

[a]LURE, Bât. 209 D - Université Paris-Sud, 91405 Orsay Cedex, France.

[b]Department of Surface Chemistry and Catalysis, Institute of Isotope & Surface Chemistry, CRC HAS, P. O. Box 77, H-1525 Budapest, Hungary.

ABSTRACT

The review highlights the *in situ* technique of the X-ray absorption spectroscopy (XAS) in the recent advances on understanding the structural features of solid catalysts used in the CO hydrogenation relevant to Fischer-Tropsch synthesis. The report is mostly concerned with mono- and bimetallic nanoparticles. The *in situ* XAS technique is applied for some examples illustrating the influence of the nature of the support on reducibility of Co-based mono- and bimetallic (Co-M where M = Pt, Ru or Re) systems. Based on the structural parameters a model is presented. The advantage and limitation of the X-ray absorption technique will also be discussed.

1. INTRODUCTION

The activity/selectivity of a catalyst critically depends on the preparation, composition and treatments [1, 2]. It has well been established that a thorough understanding of the physical and chemical properties is based on the precise knowledge of their structural characteristics. The importance of Co based catalysts in the Fischer-Tropsch synthesis has recently been increased and a new "gas-to-liquid" process called "Shell Middle Distillate" technology has emerged [3]. In order to prepare metallic cobalt particles with proper activity and stability, the precursor containing Co^{2+} ions must be reduced by H_2. Large Co_3O_4 particles are easy to fully reduce to metallic Co^0, but below a certain particle size (around 6 nm) the reduction of cobalt oxide is difficult [4]. Reducibility of Co^{2+} ions can be facilitated by increasing metal loading or by addition of noble metals [5].

In the present review application of the *in situ* XAS on the Co based catalysts (widely used in FT reaction), namely, on PtCo, ReCo and RuCo nanoparticles, will be illustrated. None of the industrial type of Fischer-Tropsch catalyst will be reported. We will also discuss how the information obtained, the advantages and the limitation of such an approach are handled.

2. X-RAY ABSORPTION SPECTROSCOPY (XAS)

XAS is an efficient tool to describe local organisation, such as, co-ordination numbers and interatomic distances in materials e.g. in metallic nanoparticles. The co-ordination numbers are linked to the size and the morphology of the cluster [6]. Unfortunately, this technique is insensitive to hetero-dispersion [7]. For multimetallic systems it is possible to give direct structural evidence on the presence of heterometallic bonds [8]. Such information is feasible when significant differences exist in the atomic number of the two metals. It is not the case for systems when a 5d transition metal is added to Pt as in Pt-Re. For these materials, some particular analysis procedures exist [9].

3. Co-BASED MONOMETALLIC SYSTEM

3.1 Preparation procedure

Here we are concerned with the key features of preparation resulting in the formation of nano sized cobalt particles on various supports [10-15]. For instance, on dehydrated NaY powder Shen et al. adsorbed $Co_2(CO)_8$ clusters and demonstrated by *in situ* XAS that after decarbonylation - depending on oxidative treatment - small cobalt nanoparticles are formed and they maintain its small size [16]. A new Co-doped-SiO_2-sol pillared montmorillonite was synthesised by interlayer hydrolysis and condensation of tetraethylorthosilicate in the presence of Co^{2+} ion using organic template. The results suggested that before calcination the Co species located on the SiO_2 sol in the form of $Co(OH)_2$ nano-cluster but after calcination they were transformed into cobalt oxide nanoparticles bonded to the SiO_2 pillar surface [17]. Recently, Drozdova et al. have studied the co-ordination and bonding of Co ions at cationic sites in zeolite structures through XAS. They noticed that Co^{2+} ions exchanged under condition, in which no hydrolysis occurred, were co-ordinated exclusively to framework oxygen [18]. Schwartz et al. have developed an original preparation procedure subliming $CoCl_2$ or $CoBr_2$ vapour onto the H form of MFI followed by replacing the halide ions by OH groups in subsequent heat treatments. Note that in this study, the Co sites have been characterised by XAS spectra at liquid N_2 temperature. It worth mentioning that during preparation one must avoid the formation of Co_3O_4 cluster [19] because it results in large metallic cobalt particles. As demonstrated the molecular cluster can be a proper precursor for preparation of nanoparticles.

3.2 In situ reduction of cobalt species

Reducibility of Co^{2+} ions is an important step to a working catalyst which can be studied by means of *in situ* XAS [4]. Transformation of Co_3O_4 to CoO occurs at 623-673 K, but it depends on the support and the size of cobalt oxide particles. Generally, it was established that reducibility of the Co oxide particles on silica by H_2 depends on the size of the Co_3O_4 crystallites. The ease of reducibility to metallic species decreased from larger (200-700 Å) to smaller (60 Å) particles.

The location of the metallic clusters can also be determined by XAS. For instance the porosity of mesoporous SiO_2 strongly affects the dispersion and reducibility of Co species demonstrated by *in situ* XAS [20]. Smaller particles in the narrow pores (20-50 Å) are much more difficult to reduce to metal species than larger ones located in large pores (>50 Å) of mesoporous silica.

3.3 Comparison of in situ XAF with other techniques

An important feature is to compare some physical parameters obtained from *in situ* XAS measurements with those resulting from temperature-programmed desorption (TPD), magnetic measurements (MM) and transmission electron microscopy, (TEM). For instance, H_2 TPD appears to be so complex that the reliability of chemisorption techniques is doubtful in determination of Co dispersion in the Co/SiO_2 catalysts [21]. On the other hand, the XAS showed that Co is crystallized in face centred cubic structure and small clusters are formed. The calculated average size is almost 10 times smaller than that obtained from MM and TEM, suggesting a complex morphology of Co particles. The most sensitive technique is the CO hydrogenation which shows that intrinsic activity and chain growth probability first increase with increasing size, then stabilized at ca. 6 nm. Nevertheless, the proper characterization of Co catalyst requires complementary techniques in addition to XAS.

4. BIMETALLIC SYSTEMS

4.1. PtCo bimetallic catalyst

This particular approach allows a precise determination of the electronic and structural characteristics of the metallic part of nanoparticles. For Pt-Co/NaY and Pt-Co/Al_2O_3 catalysts XAS were carried out at Pt L_{III} and Co K edges [22]. In the initial state, oxygen atoms surround Co and Pt ions in the Pt-Co/NaY sample. After *in situ* reduction Pt-O bonds are broken and replaced by platinum-metal atoms. The cobalt ions in Pt-Co/NaY are also reduced to metallic state. Note that this is also valid for the Pt-Co/Al_2O_3 sample. Metallic clusters are thus generated. The complete set of parameters implies the presence of two families of nanoparticles: monometallic Co nanoparticles and Pt-Co bimetallic clusters (Fig.1), in which only Pt-Co bonds are present (no Pt-Pt bonds).

346

4.2. The Ru-Co bimetallic system

Anchoring Co, Ru, and Ru-Co carbonyl clusters as well as their catalytic performances for CO hydrogenation have been studied by Shen et al. [23] who found significant bimetallic interaction. Also, anchoring the [RuCo₃] carbonyl clusters shows a strong reaction with the extra-framework Na$^+$ cations through involvement of the oxygen end of the bridging carbonyl ligands. On bimetallic Ru-Co catalyst CO conversion and oxygenates selectivity are higher than those on monometallic Co or Ru clusters.

Recently, Ru-Co/NaY samples in as prepared, reduced states, after exposure to air and in the form of spent catalysts have been studied by *in situ* XAS using Ru and Co K-edges [24]. In contrast to earlier XPS results [25] Ru-Co bonds did not form. The Co showed somewhat larger cluster size than ruthenium. A model based on the XAS measurements reconciles the two seemingly conflicting data and is consistent with the observed catalytic result [24].

4.3. Re-Co bimetallic catalyst

One of the most efficient EXXON processes applies this catalyst [26]. Holmen et al. also showed that in Re-Co/γ-Al₂O₃ catalyst bimetallic particles were formed after reduction at 450°C with the average particle size being less than 1.5 nm [27]. The distribution of Co in different phase is also possible. Reduction of a Re-promoted Co₃O₄/Al₂O₃ catalyst at 450°C yields a highly dispersed system with Co nanoparticles (<40 Å) a lower fraction of which is randomly dispersed over the tetrahedral vacancies of Al₂O₃ support. This occurs during reduction and not calcination.

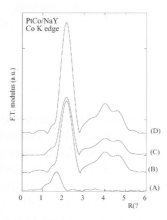
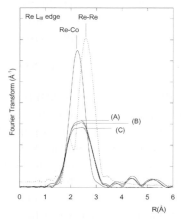

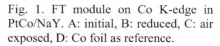

Fig. 1. FT module on Co K-edge in PtCo/NaY. A: initial, B: reduced, C: air exposed, D: Co foil as reference.

Fig. 2. FT module at the Re L$_{III}$ edge in Re-Co/NaY. In situ reduction (A), in situ air treatment (B), 2nd reduction (C) compared to the FeFF7 numerical simulations of Re-Co and Re-Re bonds

The structure and catalytic activity of the Re-Co system prepared by incipient wetness technique on Al_2O_3 and NaY zeolite have been compared [28]. The *in situ* XAS studies, XRD and TPR helped us to establish that Re promotes Co reduction and that Re-Co and Co nanoparticles are formed and the degree of reducibility depends on the support. There are two structural features (Fig. 2), that is, on Re-Co/NaY nearly all rhenium atoms are in contact with Co atoms, whereas the Co atoms are surrounded by Co atoms in the first co-ordination sphere. For the Re-Co/Al_2O_3 samples, a fraction of the rhenium in oxide form prevented the creation of the "Co Surface Phase" (CSP) which is hardly reducible.

In the CO hydrogenation the rate, α value and the olefin/paraffin ratio proves the structural architecture of the catalysts. Re promotes the Co activity and the selective formation of higher hydrocarbon. Depending the location of Re the termination/chain growth ratio is controlled influencing both the C_{5+} fraction and olefin/paraffin ratio [28].

4.4. Pd-Co bimetallic catalyst

Noronh et al. has showed that the Pd addition promoted not only the reduction of Co_3O_4 particles but also the CSP [29]. A model represented by bimetallic particles enriched with Pd and particles containing only Co has been proposed. In fact, the nature of the support as well as the preparation procedure may affect also the structural characteristics of the bimetallic systems. Recently, the limited reducibility of the Co particle size has been linked to the preparation technique. It seems that Pd acts in the bimetallic system not only as a component, facilitating Co reduction, but as an active site for H_2 dissociation to hydrogen atoms participating in the reaction [30].

5. CONCLUSIONS

We highlighted some results illustrating the usefulness of in situ XAS to study the genesis and architecture of Co-based nanoparticles as catalysts for CO hydrogenation. Accurate information on local environment around a specific element on poorly ordered systems are available using *in situ* XAS which is a major advantage versus other techniques. The structural and electronic studies have just started and the success will be complete when these synchrotron radiation related techniques will be used as standard tool.

ACKNOWLEDGEMENTS

The authors are indebted to the Hungarian Science and Research Fund (T-034920) for financial support and the COST D15/0005/99 program and the

Chemistry Department of the CNRS for supporting collaborative program with L.U.R.E synchrotron facilities.

REFERENCES

[1]   G.A. Somorjai and K. McCrea. Appl. Catal. A. 222 (2001) 3
[2]   P. Serp, P. Kalck and R. Feurer, Chem. Rev. 102 (2002) 3085
[3]   V.M.H. van Wechem, and M.M.G., Senden, Natural Gas Conversion II, (Eds.: H.E. Curry-Hyde and R.F. Howe), Stud. Surf. Sci. Catal., 81, p. 43, Elsevier Sci. Publ. Co., Amsterdam, 1994
[4]   J. Khodakov, D. Lynch, B. Bazin, N. Rebours, B. Zanier, P. Moisson, J. Chaumette, Catal. 168 (1996) 16
[5]   Z. Zsoldos, F. Garin, L. Hilaire and L. Guczi, Catal. Lett. 33 (1995) 39
[6]   I. Arcon, A. Tuel, A. Kodre, G. Martin and A. Barbier, J. Sync. Rad. 8 (2001) 575
[7]   D. Bazin, D.A. Sayers, J.J. Rehr, J. Phys. Chem. B. 101 (1997) 11040
[8]   O. S. Alexeev, B.C. Gates, Ind. Eng. Chem. Res. 42 (2003) 1571
[9]   D. Bazin, H. Dexpert, J. Lynch, J.P. Bournonville, J. Sync. Rad. 6 (1999) 465
[10]   G. Kiss, C.E. Kliewer, G.J. DeMartin, C.C. Culross and J.E. Baumgartner, J. Catal. 217 (2003) 127
[11]   G.R. Moradi, M.M. Basir, A. Taeb and A. Kiennemann, Catal. Comm. 4 (2003) 27.
[12]   B. Jongsomjit, J. Panpranot and J.G. Goodwin Jr., J. Catal., 215 (2003) 66.
[13]   D.J. Duvenhage, N.J. Coville, Appl. Catal. A., 233 (2002) 63.
[14]   F.G. Jacobs, K. Chaudhari, D. Sparks, Y. Zhang, B. Shi, R. Spicer, T.K. Das, J. Li and B.H. Davis, Fuel, 82 (2003) 1251.
[15]   E. Van Steen, F.F. Prinsloo, Catal. Today, 71 (2002) 327
[16]   G.C. Shen, T. Shido, M. Ichikawa, J. Phys. Chem. 100 (1996) 16947
[17]   J.H. Choy, H. Jung, J.B. Yoon, J. Sync. Rad., 8 (2001) 599
[18]   L. Drozdova, R. Prins, J. Ddeek, Z. Sobalik, and B. Wichterlova, J. Phys. Chem. B., 106 (2002) 2240
[19]   V. Schwartz, R. Prins, X. Wang and W.M.H. Sachtler, J. Phys. Chem. B. 106 (2002) 7210
[20]   Y. Khodakov, A. Griboval-Constant, R. Bechara, F. Villain, J. Phys. Chem. B. 105 (2001) 9805
[21]   J. Barbier, A. Tuel, I. Arcon, A. Kodre, G.A. Martin, J. Catal., 200 (2001) 106
[22]   L. Guczi , D. Bazin , I. Kovács, L. Borkó , Z. Schay, J. Lynch, P. Parent, C. Lafon, G. Stefler, Zs. Koppány , I. Sajó, Topics in Catal., 20 (2002) 129
[23]   G.C. Shen and M. Ichikawa, J. Phys. Chem. B, 102 (1998) 5602.
[24]   D. Bazin, I. Kovács, J. Lynch and L. Guczi, Appl. Catal. A., 242 (2003) 179
[25]   L. Guczi, R. Sundararajan, Zs. Koppány, Z. Zsoldos, Z. Schay, F. Mizukami and S. Niwa, J. Catal., 167 (1997) 482
[26]   B. Eisenberg, R.A. Fiato, C.H. Mauldin, G.R. Say, and S.L. Soled, Natural Gas Conversion V (Eds.: A. Parmaliana, D. Sanfilippo, F. Frusteri, A. Vaccari and F. Arena), Stud. Surf. Sci. Catal. 119, p. 943, Elsevier Sci. Publ. Co., Amsterdam, 1998
[27]   M. Rønning, D.G. Nicholson, A. Holmen, Catal. Lett., 72 (2001) 141
[28]   D. Bazin, L. Borkó, Zs. Koppány, I. Kovács, G. Stefler, L.I. Sajó, Z. Schay and L. Guczi, Catal. Lett., 84 (2002) 169
[29]   F.B. Noronh, M. Schmal, B. Moraweck, P. Delichere, and M. Brun, F. Villain and R. Frety, J. Phys. Chem. B. 104 (2000) 5478
[30]   L. Guczi, L. Borkó, Z. Schay, D. Bazin and F. Mizukami, Catal. Today 65 (2001) 51

Studies in Surface Science and Catalysis, volume 147
X. Bao and Y. Xu (Editors)
349

Fischer-Tropsch reaction over cobalt catalysts supported on zirconia-modified activated carbon

T. Wang, Y.-J. Ding*, J.-M. Xiong, W.-M. Chen, Z.-D. Pan, Y. Lu, L.-W. Lin

Natural Gas Utilization and Applied Catalysis Laboratory, Dalian Institute of Chemical Physics, Chinese Academy of Sciences, 457# Zhongshan Rd., Dalian 116023, China

ABSTRACT

An attractive Fischer-Tropsch catalyst was prepared using an activated carbon as carrier to support cobalt based catalysts. Zr promoted Co/AC catalysts remarkably enhanced the activity and the selectivity toward diesel distillates and lower the methane selectivity. This modification may be attributed to specific behavior of activated carbon with high surface area and the weak interaction between metallic cobalt active sites and activated carbon. It was emphasized that the pore size of activated carbon played a very important role in restricting the growth of carbon chain to wax.

1. INTRIDUCTION

Supported cobalt based catalysts are proven to be excellent catalysts for Fischer-Tropsch (F-T) synthesis of long-chain paraffins [1,2]. The activity of cobalt based catalysts in F-T reaction was proposed to be proportional to the area of exposed metallic Co atoms [3]. A requirement for highly active Co based catalysts is therefore a high dispersion of the cobalt metal. Cobalt interact strongly with commonly used support materials, such as alumina and silica, yielding cobalt silicates and cobalt hydrosilicates species that are non-active species for CO hydrogenation in the case of silica [4,5]. In the present work we focus on zirconia that is reported to be a promoter for F-T reaction over Co/SiO_2 [6,7], where the addition of Zr leads to improved an activity. An inert activated carbon (AC) with a high surface are employed as a carrier which has advantages of a high cobalt dispersion and non-interaction with cobalt.

---

* Corresponding Author: **Email:** dyj@dicp.ac.cn **Phone:** +86 411 4379143
Fax: +86 411 4379143

2. EXPERIMENTAL

2.1 Catalysts preparation

All catalysts used in the study were prepared by aqueous incipient wetness impregnation method. The activated carbon was sieved to 20-40 mesh and washed with distilled water for several times before being used. The activated carbon was first impregnated with a solution of $Zr(NO_3)_4 \cdot 5H_2O$, followed by drying at room temperature and calcining at 393 K for 12 h, then was impregnated with $Co(NO_3)_2 \cdot 6H_2O$ solution followed by drying at 353k for 12h, and calcining at 393k for 16 h in a flow of nitrogen at 40ml/min.

2.2 CO Hydrogenation

The testing apparatus consisted of a small fixed bed tubular reactor with an external heating system, which was made of stainless steel with 350 mm length, 8 mm inner diameter. The catalysts were in-situ reduced in a flow of H_2 before the reaction. Then the syngas ($H_2/CO = 2$) was fed into the catalyst bed and reacted for ca. 100 h under conditions of 523 K, 2.5 MPa and GHSV = $500h^{-1}$. The effluent passed through a high pressure gas-liquid separator and a low pressure gas-liquid separator in an ice-water bath. The liquid products and water including alcohols were collected by the separators. The aqueous solution containing oxygenates obtained was off-line analyzed by Varian CP-3800 gas chromatography with an FFAP column and FID detector, using 1-pentanol as an internal standard. C_4-C_{26} hydrocarbons in the oil phase of F-T products were off-line analyzed on Varian CP-3800 with SE-54 capillary column and FID as a detector. The F-T wax if it was formed was off-line analyzed on Varian CP-3800 simulated-distillation with StarSD^{TM} software. The tail gas was on-line analyzed by Varian CP-3800 GC with a Porapak QS column and TCD detector.

2.3 XRD Measurements

An X-ray diffractometer (XRD D/max-ra) was used with mono-choromatized Cu ka radiation and operated at 40kv and 30mA. The samples were scanned at a rate of 2.4 degree/min in the range $2\theta=5$-75 degrees.

2.4 Temperature-programmed reduction (TPR)

TPR experiments were conducted to determine the reducibility of catalysts. It was performed on America Micromeritics Autochem 2910. Eighty milligrams of catalyst was placed in a quartz reactor and reduced by a 10% H_2/Ar gas mixture in a flow rate of 50 ml/min. The temperature was ramped at 15 K/min and the hydrogen consumption was recorded with the thermal conductivity detector.

3. RESULTS AND DISCUSSION

Fig. 1 gives the TPR profiles of the un-promoted and Zr-promoted Co/AC catalysts. There were three major peaks at 473K, 650K and a broad one between 723K and 978K (maximum at 810K) in the TPR profile for un-promoted catalyst. The first small shoulder peak was due to the complete decomposition of the $Co(NO_3)_2$ precursor and its intensity was greatly diminished by prolonged calcination [8]. The second peak was essentially due to Co_3O_4 which was reduced in two steps $Co_3O_4 \rightarrow Co^{2+} \rightarrow Co^0$ [9], while the latter broad one was due to the methane formation at 723K-978K, which resulting from the carbon support gasification [10].

The TPR profiles of CoZr/AC catalysts were basically comparable to those of Co/AC catalysts, however, some small changes were observed in the case of CoZr/AC catalysts, the H_2 consumption peak attributed to cobalt oxide reduction was shifted to lower temperature. This might be explained due to the weak interaction between high dispersion ZrO_2 and cobalt oxides on the surface of activate carbon. It was found that almost no change appeared when Zr loadings increased, indicating that high Zr loading seems to have no significant effect on Co/AC catalysts.

Fig. 2 shows XRD patterns of Co/AC, CoZr/AC catalysts before and after F-T reaction. It could be seen that almost no diffraction peak was observed after the Co/AC and CoZr/AC samples were reduced at 673K for 4h, probably due to the formation of highly dispersed metallic cobalt or the formation of an amorphous cobalt phase (Fig. 2a and Fig.2c). While the XRD pattern of used Co/AC showed some apparent diffraction peaks, and the major peak of the cobalt metal phase was identified as the fcc form Co, with some hcp form Co. It was clear that the diffraction peaks became weaker with the addition of Zr into

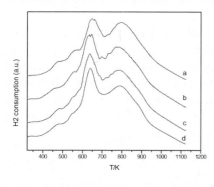

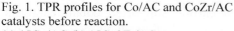

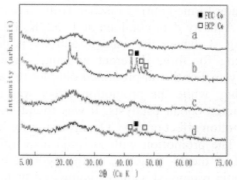

Fig. 1. TPR profiles for Co/AC and CoZr/AC catalysts before reaction.
(a) 15Co/AC (b) 15Co2Zr/AC
(c) 15Co4Zr/AC (d) 15Co6Zr/AC

Fig. 2. XRD patterns of 15Co/AC and 15Co6Zr/AC before and after F-T reaction.
(a) reduced Co/AC; (b) used Co/AC
(c) reduced CoZr/AC; (d) used CoZr/AC

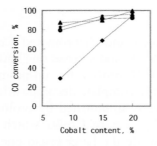

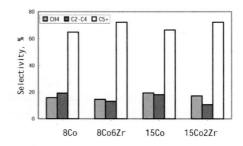

Fig. 3. CO conversion for different
catalysts as a function of cobalt-loading.
♦ 0Zr, ■ 2Zr, ● 4Zr, ▲ 6Zr

Fig. 4. Effect of promotion with zirconia
on the paraffin selectivity

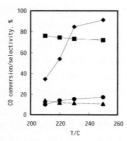

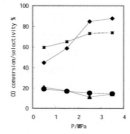

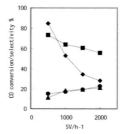

Fig. 5. Effect of temperature
on the activity and selectivity

Fig. 6. Effect of pressure
on the activity and selectivity

Fig. 7. Effect of space velocity on the activity and selectivity

▲ CH$_4$ selectivity; ● C$_2$-C$_4$ selectivity; ♦ CO conversion; ■ C$_5^+$ selectivity

Co/AC catalyst, indicating the well-dispersed metallic cobalt particles (Fig. 2d). It was worth note that the existence of the Zr promoter on the surface of activated carbon hindered the aggregation of cobalt species even in the presence of water formed during Fischer-Tropsch reaction.

Activity data obtained at T=523K, P=2.5MPa and H$_2$/CO=2 are presented in Fig. 3, where the catalyst activity is depicted as a function of the cobalt loading for three different zirconia loadings and compared to un-promoted catalysts. For the un-promoted samples the activity increased sharply with cobalt loading, while the promotion effect was not so distinctively evident for the promoted cobalt catalysts. It was found that zirconia exhibited a distinguished promotion effect on the CO conversion for low cobalt loadings.

The promotion with zirconia also influenced the product selectivity. The yield of the C$_5^+$ fraction of the hydrocarbon product increased for all of the zirconia promoted samples compared to the un-promoted catalysts (Fig. 4). It was interesting to note that the activated carbon supported cobalt catalysts displayed a novel property: very little fraction of F-T products was heavy oil parts (C$_{25}^+$), more than 60 wt. % in the liquid oil phase was diesel fuel (see Fig. 8),

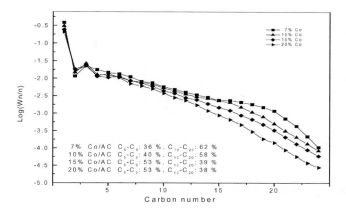

Fig. 8. Anderson-Schulz-Flory distribution over Co/AC with different Co loadings

and the methane selectivity went down with time on stream in the first hundred hours. Duration test showed little change in the activity and Hydro-carbon distribution with time on stream after the first hundred hours.

It is apparent that CO conversion increased with the reaction temperature and pressure over 15Co2Zr/AC (Fig. 5, and Fig. 6), and the optimized temperature and pressure were 503-513 K and 2.5-3.0 MPa. However, the CO conversion and C_5^+ selectivity decreased with GHSV (Fig. 7).

Fig. 8 shows the Anderson-Schulz-Flory distribution over Co/AC with different Co loadings. It was interesting to find that the selectivity towards C_{10}-C_{20} paraffins was gradually decreased with the increase of the cobalt loading. The selectivity towards C_{10}-C_{20} hydrocarbons in liquid products decreased from 62% over 7 wt. % Co/AC to 38% over 20 wt. % Co/AC catalysts, while the selectivity of C_5-C_9 hydrocarbons increased from 36% to 53% when the Co loading increased from 7 wt. % to 20 wt. %. As we know, the ratio of the cobalt active sites that located in the pores of the carrier and the others that located on the outer surface of the carrier for low cobalt loading catalyst was higher than that for high cobalt loading catalyst. These results implied that the pore structure, especially the pore size, could play a very important role in restricting growth of carbon chain during F-T reaction. More evidences from meso-pore material with different pore size from 4.0 to 8.0 nm, such as SBA-15 supported cobalt catalysts with or without promoters, confirmed powerfully the roles of the pore structure of the carrier in restricting the growth of carbon chain during F-T reaction, which will be published elsewhere. Therefore, the distribution of paraffin products of high cobalt loading was similar to that of conventional cobalt catalyst with α value of ca. 0.78, while there apparently were two α values in the curve of log(Wn/n) versus the carbon number n in the case of low cobalt loading catalyst. Duration test (more than 1000 hours) results confirmed

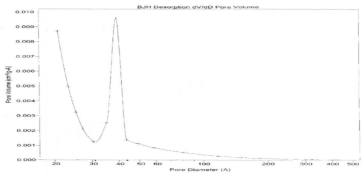

Fig. 9. Distribution of pore size of activated carbon

that the erratic effects might not result from the chromatographic effect of supports.

On the basis of the C-C band length of 0.153 nm and the C-H band length of 0.106 nm, we can estimate that the chain length of paraffin with 20 carbons is ca. 3.12 nm. So, the reasonable pore size of catalysts or carrier should be proposed in the range of 4.0-6.5 nm if the diesel distillates were synthesized. Fig. 9 shows the pore size distribution of activated carbon employed in this work. It was found that the most probable distribution of pore size fell in the range of 3.2-4.5 nm, which was surprisingly consistent with those we approximated for the diesel cut formation.

4. CONCLUSION

Activated carbon supported cobalt catalysts exhibited high activity and high selectivity toward diesel distillates and long-term stability due to its high surface area and the inert interaction between cobalt and activated carbon. The pore size of the activated carbon plays a very important role in restricting the growth of carbon chain.

REFERENCES

[1] F. Fischer, H. Tropsch and D. Dilthes, Brennstoff-chemie 6 (1925) 265
[2] M.A., Vannice, J. Catal. 50 (1977) 228
[3] E. Igleasia, S. Soled, and R.A. Fiato, J. Catal. 137 (1992) 212
[4] I. Puskas, T.H. Fleisch, J.B. Hall, B.L. Meyers R.T. Rochinski, J. Catal. 134 (1992) 615
[5] A. Feller, M. Claeys, E. van Steen, J. Catal. 185 (1999) 120
[6] S. Ali, B. Chen and J.G. Goodwin, Jr., J. Catal. 157(1995) 35
[7] F. Rohr O.A. Lindvag, A. Holmwen, E.A. Blekkan, Catal. Today 58(2000)247
[8] J.G. Haddad and J.G. Goodwin, Jr., J. Catal. 157(1995)25
[9] A.M. Hilmen, D. Schanke, and A. Holmen, Catal. Lett. 38(1996)143
[10] A. Guerrero-Ruiz, A. Sepúlveda-Escribano, I. Rodríguez-Ramos, Appl. Catal. A 120 (1994)71

Studies in Surface Science and Catalysis, volume 147
X. Bao and Y. Xu (Editors)

Effect of impregnation pH on the catalytic performance of Co/ZrO$_2$ catalyst in Fischer-Tropsch synthesis

H.-X. Zhao, J.-G. Chen, Y.-H. Sun[*]

State Key Laboratory of Coal Conversion, Institute of Coal Chemistry, Chinese Academy of Sciences, P.O. Box 165, Taiyuan 030001, China
yhsun@sxicc.ac.cn

ABSTRACT

A series of Co/ZrO$_2$ catalysts were prepared by controlling the pH value of the cobalt nitrate solution. The interaction of cobalt precursor and ZrO$_2$ support showed to be dependent on the impregnation pH and then influenced the catalytic activity of the Co/ZrO$_2$ catalysts in Fischer-Tropsch synthesis.

1. INTRODUCTION

Fischer–Tropsch synthesis (FTS) is one of major routes for converting coal-based and/or natural gas-derived syngas into chemicals and fuels [1]. Cobalt based catalysts are widely used for Fischer-Tropsch synthesis [2], especially when high chain growth probability and a low branching probability are required [3]. The focus in the development of this process is the improvement of the catalyst activity with low CH$_4$ production.

The usual supports, such as silica, titania and alumina, are easily to form surface compound with cobalt precursor, which decrease the reduction degree of the catalysts. Hence, the activity and selectivity for heavier hydrocarbons are suppressed. Zirconium oxide has attracted considerable attention recently as a catalyst support and a promoter for its high selectivity for heavier hydrocarbons in Fischer-Tropsch synthesis [2, 4]. The purpose of this paper is to study the impregnation solution pH on the catalytic performance of Co/ZrO$_2$ catalysts in Fischer-Tropsch synthesis.

2. EXPERIMENTLE

2.1 Catalyst preparation

The zirconia support was prepared by co-precipitation of zirconyl nitrate aqueous solution with ammonia hydroxide solution. Catalysts were prepared by incipient wetness impregnation using cobalt nitrate as the precursor. Nitric acid and urea were employed to regulate the solution pH. The pH value was measured by a pH meter, which was calibrated at pH values of 4 and 10 prior to each run. The catalysts were designated as Z1, Z2, Z3 and Z4 with the solution pH 1.3, 2.7, 4.4 and 6.1, respectively. The relatively higher pH of 6.1 of the cobalt nitrate solution could just keep the solution from body precipitation of cobalt species. The catalyst precursors were aged for a certain time at room temperature and then dried at 393K and calcined at 673K for 6h. The cobalt content of all the catalysts was 10 wt %. The texture properties of the catalysts are listed in Table 1.

2.2 Catalyst Characterization

The textural properties of support and catalysts were measured with ASAP-2000 Micromeritics instrument at 77 K. XRD was recorded in Rigaku D/max-γA with Cu target at 40 kV and 100 m A. Diffuse reflectance spectra (DRS) were recorded over a wavelength range from 850 to 250 nm on a double-beam Shimadzu spectrophotometer model UV-2501PC equipped with a diffuse reflectance attachment. Specpure $BaSO_4$ (Shimadzu) was the reference material for the catalysts.

Temperature programmed reduction (TPR) was carried out in a quartz reactor with a mixture of 5% H_2/N_2 as the reductive gas. The samples (0.1 g) were flushed with a N_2 flow of 40 mL/min at 393 K to remove adsorbed water and then reduced in a flow of H_2/N_2 at a rate of 10 K/min. The effluent gas was monitored by TCD. The TPR procedure was calibrated using the reduction of CuO.

The measurement of reduction degree was described elsewhere [5]. Oxygen titration was performed on a TG-151 analyzer. The catalyst (0.2 g) was reduced at 673 K, further purged with a flow of argon for 1 h to remove physically adsorbed H_2, and then maintained at 673 K for oxidation. The reducibility of metal cobalt was measured based on the weight gain after re-oxidation of the reduced sample.

Table 1
Characteristics of samples

Catalysts	Solution pH	BET/m^2/g	Pore size/nm	Reduction[a] /%
ZrO_2	---	28.0	15.2	---
Z1	1.3	27.1	15.6	100
Z2	2.7	27.5	15.4	100
Z3	4.4	28.1	16.0	100
Z4	6.1	27.9	16.1	40

[a]Extent of reduction was obtained from the weight gain during reoxidation at 673 K

DRIFTS spectra were recorded with Nicolet Magna 550 spectrometer using a spectral resolution of 8cm^{-1}. The catalyst was reduced in-situ for 6 h under atmospheric pressure by a stream of H$_2$ at 673 K. After introduction of CO for 1 h, the catalyst surface was purged with argon to remove gaseous CO and then IR spectra were recorded.

2.3 Catalytic Test

CO hydrogenation was carried out at 2 MPa, 1000 h^{-1}and a H$_2$/CO ratio of 2 in a fixed bed reactor of i.d. 10 mm. The samples were reduced in a flow of hydrogen at 673 K for 6 h at 1000 h^{-1} and then cooled down and switched to syngas. Mass balances were commenced after the reaction was on-line for 24 h. The CO, H$_2$, CO$_2$, and CH$_4$ products were analyzed on the TCD, and the gas hydrocarbons were detected on the FID. Liquid products and wax were collected in a cold trap and a hot trap respectively and then were offline analyzed on the FID which was equipped with a 35 m OV-101 capillary column.

3. RESULTS AND DISCUSSION

3.1 Surface properties of the catalysts

N$_2$ adsorption at 77 K showed that the texture hardly changed before and after the impregnation (see Table 1), and cobalt existed in the Co$_3$O$_4$ phase (see Fig. 1). Fig. 2 showed UV–Vis diffuse reflectance spectra of supported Co/ZrO$_2$ catalysts. All the catalysts showed the similar spectra, and two broad bands at about 400 and 700 nm corresponding to octahedral Co^{3+} confirmed the existence of Co$_3$O$_4$ [6]. The increase in the band intensity with the rise of the impregnation pH was due to the increase in the surface density of cobalt species [7], suggesting cobalt oxides were well dispersed on the zirconium surface in this case.

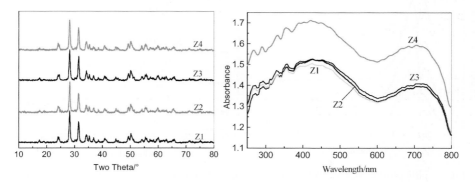

Fig. 1. XRD profiles of Co/ZrO$_2$ catalysts   Fig. 2. DRS profiles of Co/ZrO$_2$ catalysts

358

3.2 Reduction behavior of Co/ZrO₂ catalysts

The TPR profiles of Co/ZrO₂ catalysts are shown in Fig. 3. Two kinds of reduction behavior were observed. The TPR plots of samples Z1, Z2 and Z3 were almost identical to each other. Double peaks observed in samples Z1, Z2 and Z3 illustrated the step reduction of Co_3O_4. This indicated that larger Co_3O_4 particles existed on the surface of ZrO₂. The reduction of sample Z4 showed three peaks below 600℃, suggesting that different cobalt species [8] existed over ZrO₂. The zero point charge (ZPC) of ZrO₂ is reported to be about 5 [9], thus the surface of ZrO₂ was positively charged at lower pH than 5 and the adsorption of cobalt was suppressed, and consequently the interaction between cobalt and zirconium decreased. The similarity of reduction behavior of samples Z1, Z2, and Z3 implied that the same cobalt species existed on the zirconia surface. On the contrary, at higher pH than 5, the zirconia surface was negatively charged and the adsorption of positively charged cobalt ions was improved, and then the interaction between cobalt and zirconia increased. Such a different interaction directly led to the different reduction behavior of the samples.

3.3 In-situ CO adsorption on reduced catalysts

Infrared spectra of CO adsorbed on reduced catalysts are shown in Fig. 4. A group of distinct bands at 2048, 1974, 1936, 1886cm⁻¹were observed after CO adsorption on sample Z1. For Z2, the adsorption bands appeared at the same position as that of Z1 except for the relatively weaker adsorption intensity. The adsorption behavior on Z3 and Z4 was quite different from that of Z1 and Z2.

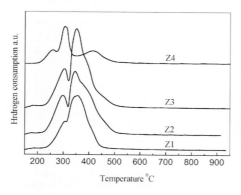

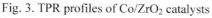

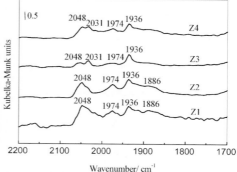

Fig. 3. TPR profiles of Co/ZrO₂ catalysts

Fig. 4. FTIR spectra of CO adsorption on reduced catalysts

The adsorption intensity at 2048, 1974, and 1936cm^{-1} became weaker and the adsorption band at 1886cm^{-1} disappeared, and at the same time, a new adsorption band appeared at 2031cm^{-1}. In addition, the adsorption intensity of sample Z4 was weaker than that of Z3. The bands above 2000 cm^{-1} are assigned to the linearly adsorbed CO, and those below 2000 cm^{-1} are attributed to the bridged-form species on Co^0 particles [5, 10]. Therefore, the adsorption bands at 2048 and 2031 cm^{-1} should be attributed to linearly adsorbed CO species, and the bands at 1974, 1936, 1886 cm^{-1} should be attributed to bridged adsorbed CO species. The strong adsorption of bridged-form CO on Z1 surface indicated that cobalt particle was larger. These facts clearly indicated that the impregnation solution pH had a significant influence upon the formation and then the adsorption properties of the surface cobalt species.

3.4 Catalytic performance for Fischer-Tropsch synthesis

CO conversion and product selectivity strongly depend on the impregnating solution pH (see Table 2). CO conversion and C$_5^+$ hydrocarbons selectivity increased and CH$_4$ selectivity decreased with the decrease of the pH. At 483 K, sample Z1 showed the highest CO conversion and C$_5^+$ hydrocarbons selectivity of 82.4% and 86.5%, respectively. Sample Z4 had the lowest catalytic performance, and the CO conversion and C$_5^+$ hydrocarbons selectivity were 33.2% and 44.4%, respectively. Such a pH dependent performance appeared to be closely related to the CO adsorption over the catalysts, especially the bridged-adsorbed CO. That might be the reason that Z1 had a better activity and selectivity.

Table 2
The performance of Co/ZrO$_2$ catalysts for Fischer-Tropsch synthesis

Catalysts	Reaction temperature /K	CO Conversion/%	CH$_4$/wt%	C$_5^+$/wt%
Z1	483	82.4	5.5	86.5
	493	94.1	5.9	82.3
Z2	483	82.6	6.3	84.8
	493	95.5	6.1	81.3
Z3	483	82.7	8.3	83.3
	493	92.5	8.5	82.0
Z4	483	33.2	33.1	44.4
	493	40.2	17.6	71.9

4. CONCLUSIONS

The interaction of cobalt precursor and ZrO_2 support showed to be dependent on the impregnation pH and then influenced the cobalt dispersion on the ZrO_2 surface. Consequently, the adsorption of CO on the catalyst surface was greatly influenced and the content of adsorbed CO increased with the increase of cobalt particle size. Meanwhile, the activity for Fischer-Tropsch synthesis increased monotonously with the increase of CO adsorption. Therefore, the impregnation solution pH should be selected lower than the ZPC of the support to weaken the interaction between cobalt and ZrO_2 and thus improve the catalytic performance of the catalysts.

REFERENCES

[1] M.E. Dry, Appl. Catal. A., 138 (1996) 319
[2] E. Iglesia, Appl. Catal. A., 161 (1997) 59
[3] H. Schulz, E.V. Steen and M. Claeys, Stud. Surf. Sci. Catal 81 (1994) 204
[4] S. Ali, B. Chen and J.G. Goodwin. Jr., J. Catal., 157 (1995) 35
[5] J.L. Zhang, J.G. Chen, J. Ren, Y.H. Sun. Appl. Catal. A., 243 (2003) 121
[6] J.Ramirez, R.Cuevas and L.Gasque, Appl. Catal. A ., 71 (1991) 351
[7] Ch. Vladov, L. Petrov and B. YtuaL. Topalova., Appl. Catal. A 94 (1993) 205
[8] E.V. Steen, G.S. Sewell, R.A. Makothe, et al. J. Catal., 162 (1996) 220
[9] J.C. Yori, C.L. Pieck, J.M. Parara, Appl. Catal. A., 181 (1999) 5
[10] M. Jiang, N. Koizumi, T. Ozaki and M. Yamada, Appl. Catal. A., 209 (2001) 59

Studies in Surface Science and Catalysis, volume 147
X. Bao and Y. Xu (Editors)
©2004 Elsevier B.V. All rights reserved.

The effect of pressure on promoted Ru and Re-Co/Niobia catalysts in the Fischer-Tropsch synthesis

Fabiana M. T. Mendes[a], Fábio B. Noronha[b], Carlos D. D. Souza[a], Mônica A. P. da Silva[c], Alexandre B. Gaspar[c] and Martin Schmal[*a,c]

[a]NUCAT/PEQ/COPPE, Universidade Federal do Rio de Janeiro, C.P. 68502, 21945-970, Rio de Janeiro, Brazil

[b]INT-Instituto Nacional de Tecnologia, Rio de Janeiro, CEP 20081-310 Brazil

[c]Escola de Química, Universidade Federal do Rio de Janeiro, C.P. 68542, CEP 21940-900, Rio de Janeiro, Brazil

ABSTRACT

Co/Nb_2O_5, $Co-Ru/Nb_2O_5$ and $Co-Re/Nb_2O_5$ catalysts were prepared and characterized by TPR. The catalysts were investigated on CO hydrogenation under 0.4, 2 and 4Mpa. The total pressure has an important effect on the product distribution. The selectivity toward diesel and C_{5+} hydrocarbons achieves a maximum at reaction pressure of 2.0Mpa. The catalysts had good stability over 50h on stream at reaction conditions.

1. INTRODUCTION

During the last decade there has been a renewed interest in the use of Fischer-Tropsch technology for the conversion of natural gas to liquids. The main factors that contributed to the interest in natural gas are: (i) An increase in the known reserves of natural gas, (ii) Environmental pressure to minimize the flaring of associated gas and (iii) Development of more active catalysts and improvement of reactor designs. This situation has driven many researches to develop more economic processes by upgrading the catalysts.

Cobalt-based catalysts are known to be effective in CO hydrogenation [1]. However, the main problem of the Fischer-Tropsch synthesis is the wide range of product distribution by the conventional FT catalysts [2]. In order to overcome the selectivity limitations and to enhance the catalyst efficiency in CO hydrogenation, several approaches have been made, such as the use of reducible support [3] and the addition of a second metal [4]. Selectivity considerations are thus extremely important and the C_{5+} formation should be as high as possible.

Bimetallic catalysts may be considered essential to many industrial catalytic processes because of the advantages such as improved stability and reducibility, slower deactivation, and noticeable increase in catalytic activity and selectivity in comparison to the monometallic constituents. The addition of noble metals to supported-cobalt catalyst has also affected the activity and selectivity of CO hydrogenation. The present authors have reported that the Re and Ru addition to a cobalt/niobia catalyst promoted the Co_3O_4 reduction and led to an interaction of Co with niobium species at the surface [3]. The Re or Ru addition increased both the activity and C_{5+} selectivity of Co/Niobia catalyst, after reduction at high temperatures and at ambient pressure conditions [3]. These results were explained by the formation of Co-NbOx interfacial sites.

There have been fewer studies in CO hydrogenation at a total pressure higher than atmosphere with bimetallic catalyst, especially with the Co-Re or Co-Ru supported catalysts [5, 6]. The main goal of this work is to characterize and investigate the total pressure effect on the activity and selectivity of promoted Co/Nb_2O_5 catalyst during long time performance of CO hydrogenation.

2. EXPERIMENTAL

2.1. Catalyst Preparation

The Nb_2O_5 support was obtained by calcination of niobic acid (CBMM) in air at 823K, for 3h. The catalysts were prepared by incipient wetness impregnation or co impregnation of the support with an aqueous solution of the cobalt nitrate, cobalt, ruthenium and rhenium chloride diluted in HCl. The catalysts contained 5wt% Co and 0.6 wt.% of Ru or Re. After impregnation, the samples were dried at 393K for 16h and calcined in air at 673K for 2h.

2.2. Characterization

TPR experiments were performed in a conventional apparatus. The samples were reduced with a mixture containing 1.74% H_2/Ar (30 cm^3/min) at a heating rate of 10K/min from 298 to 1273K.

2.3. Catalyst testing

The CO hydrogenation was performed in a flow tubular stainless steel micro reactor at 493K loaded with 2, 0.3 or 0.15g of catalyst, under 0.4, 2 and 4MPa, respectively. The total feed flow rate was held to the desired value to attain an isoconversion, using a mixture of $H_2/CO=2$ at differential conditions. Before the reaction, the samples were reduced under hydrogen flow at 773K for 16h. The reaction products were analyzed by on-line gas chromatography (Shimadzu-GC-17A), equipped with a CP-PoraBOND Q, 50m (TCD) and a CP-SIL, 50m (FID) columns.

3. RESULTS AND DISCUSSION

3.1. Characterization

Fig. 1 presents the TPR profiles of the niobia-supported catalysts. The Co/Nb$_2$O$_5$ catalyst prepared from nitrate precursor presented two peaks, at 629 and 772 K, ascribed to the reduction of Co$_3$O$_4$ and Co^{2+} species, respectively. In addition, Co/Nb$_2$O$_5$ prepared from chloride precursor presented only one peak at higher temperature (828K) and a shoulder at 879K. This TPR profile indicates the presence of Co^{2+} species, which are more difficult to reduce [7]. The presence of chlorine makes the cobalt reduction difficult and the Co^{2+} species prevails. The ruthenium addition to Co/Nb$_2$O$_5$ catalyst promotes the reduction of Co$_3$O$_4$ particles in agreement with literature [3] and the two peaks observed (530 and 691K) could be attributed to the reduction of cobalt oxide. On the other hand, the promotion of cobalt oxide reduction is not observed in the presence of rhenium oxide. In fact, the TPR profile of this catalyst was very similar to the one of the Co/Nb$_2$O$_5$ catalyst prepared from chloride precursor. These results suggest the presence of a higher amount of Co^{2+} species, which are hardly reduced.

3.2. Effect of Pressure and Catalyst Selectivity

Table 1 shows the CO conversion and selectivity for the mono and bimetallic catalysts under a reaction pressure of 0.4MPa. Co/Nb$_2$O$_5$ catalyst exhibited a low methanation activity and a high selectivity towards long chain hydrocarbons (C$_{5+}$), in agreement with previous results [3]. The ruthenium addition to Co/Nb$_2$O$_5$ catalyst changes markedly the product distribution, leading to both strong reduction of methanation and increase of C$_{5+}$ and C$_{19+}$ production. The decoration of cobalt particles by partially reduced niobium oxides species (NbOx) after reduction at 773K, which influences the cobalt adsorption properties, creates new interfacial Co-NbOx sites. The authors (3) proposed that these interfaces were responsible for the growing chain toward higher C$_{5+}$ and that the addition of the noble metals (Ru or Re) facilitate the interaction of the niobium oxide species and cobalt.

According to the literature [3, 8, 9], the increase of selectivity towards C$_{5+}$ hydrocarbons as ruthenium is added to supported Co catalysts is attributed to the higher amount of metallic cobalt. Our TPR results agree very well with that explanation since Ru strongly promoted the reduction of cobalt oxide. On the other hand, in this work, the rhenium addition led to an opposite effect. In this case, the selectivity to C$_{5+}$ decreased and CH$_4$ formation increased. In the literature [3, 9] rhenium leads to higher Co metal dispersion, which favors the C$_{5+}$ selectivity.

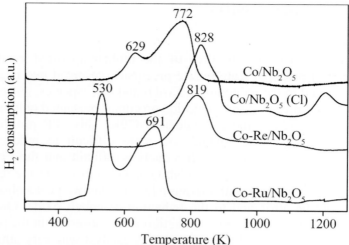

Fig. 1. TPR profiles of the catalysts supported on niobia.

Table 1

CO conversion (X) and selectivity for methane (CH_4), light hydrocarbons (C_2-C_4), gasoline (C_5-C_{12}), diesel (C_{13}-C_{18}), hydrocarbons containing more than 19 carbon atoms (C_{19+}) at 0.4 MPa and 24 h, after reduction at 773 K.

Catalyst	X (%)	Rate (μmols/g.s)	CH_4	C_2-C_4	Gasol.	Diesel	C_{19+}	C_{5+}
Co/Nb_2O_5	29.0	4.0	18.9	18.1	23.8	23.9	15.4	63.1
Co-Re/Nb_2O_5	7.9	0.7	34.0	34.9	17.5	0.8	12.9	31.2
Co-Ru/Nb_2O_5	23.1	5.9	6.3	5.6	13.6	28.6	46.1	88.2

The apparent contradictory results can be explained through the TPR analysis. Reduction profile of Co-Re/Nb_2O_5 catalyst showed a high amount of Co^{2+} species. The major presence of this Co^{2+} species could lead to a lower selectivity towards long chain hydrocarbons and instead a higher methanation activity. Similar results were obtained before [10] when CO hydrogenation reaction was performed, at the same conditions, with a Co/Nb_2O_5 (2wt%) catalyst, mainly characterized by the presence of the surface Co^{+2} species in interaction with the support.

Table 2 shows the CO conversion and selectivity for the Co/Nb_2O_5 catalyst under different reaction pressure (0.4, 2.0 and 4.0MPa). The total pressure has an important effect on the product distribution. The selectivity toward diesel and C_{5+} hydrocarbons achieves a maximum at reaction pressure of 2.0 MPa. Further increase of reaction pressure leads to the decrease of selectivity toward higher hydrocarbons fractions. Bianchi et al [11] also observed the hydrocarbon chain grown with a pressure increase from 0.5MPa to 2.0MPa during the CO hydrogenation with a Co/SiO_2 catalyst. The decreased production of higher hydrocarbons above 2.0MPa has not a suitable explanation at present.

Table 2

CO conversion (X) and selectivity for methane (CH_4), light hydrocarbons (C_2-C_4), gasoline (C_5-C_{12}), diesel (C_{13}-C_{18}), hydrocarbons containing more than 19 carbon atoms (C_{19+}) at different pressures and 48 h, after reduction at 773 K for the Co/Nb$_2$O$_5$

Pressure (Mpa)	X (%)	Rate (µmols/g.s)	CH_4	C_2-C_4	Gasol.	Diesel	C_{19+}	C_{5+}
0.4	25.9	3.5	13.1	12.0	14.6	15.8	44.5	74.9
2.0	15.2	17.3	3.3	2.3	7.8	28.7	57.9	94.4
4.0	14.6	16.7	27.6	23.4	42.3	5.4	1.5	49.0

Table 3

CO conversion (X) and selectivity for methane (CH_4), light hydrocarbons (C_2-C_4), gasoline (C_5-C_{12}), diesel (C_{13}-C_{18}), hydrocarbons containing more than 19 carbon atoms (C_{19+}) at 2.0 MPa and 48 h, after reduction at 773 K

Catalyst	X (%)	Rate (µmols/g.s)	CH_4	C_2-C_4	Gasol.	Diesel	C_{19+}	C_{5+}
Co/Nb$_2$O$_5$	15.2	17.3	3.3	2.3	7.8	28.7	57.9	94.4
Co-Re/Nb$_2$O$_5$	13.6	10.8	35.3	21.2	15.8	22.6	5.1	43.5
Co-Ru/Nb$_2$O$_5$	15.6	21.3	4.9	3.3	10.9	26.8	54.1	91.8

Table 3 shows the CO conversion and selectivity for the mono and bimetallic catalysts under 2.0MPa. The increase of reaction pressure did not change the product distribution on the Co-Ru/Nb$_2$O$_5$ and Co-Re/Nb$_2$O$_5$ catalysts. The selectivity to methane and C_{5+} practically was not altered. Tsubaki et al. [6] studied CO hydrogenation also at high total pressure condition (513K and 1Mpa) and verified that the main role of Ru itself in RuCo catalyst was reducing cobalt, not contributing FTS activity directly and the CH_4 selectivity did not show a meaningful change. Our results agree well with literature [6], where the effect of Ru or Re addition has no significant change in the catalyst performance, when worked at high pressure condition.

Fig. 2 shows the CO conversion of Co/Nb$_2$O$_5$ as a function of time on stream at the high pressure. The catalyst exhibits good stability and reaches a constant value of CO conversion (after 50 hours on stream). The catalyst stability was also very good when the reaction was performed at high-pressure conditions (2.0 and 4.0MPa).

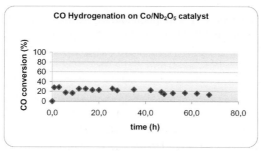

Fig. 2. CO conversion at CO Hydrogenation on Co/Nb$_2$O$_5$ Catalyst, performed at 543K and 2MPa

4. CONCLUSIONS

The Co-Ru catalyst behaved as was expected when performed at similar ambient pressure conditions (0.4Mpa), promoting the cobalt reduction, lowering the methanation activity and increasing the selectivity towards liquid hydrocarbons (C_{5+}). However, the addition of ruthenium seems not to play an important role when working at higher pressures (2.0Mpa). On the other hand, Co-Re behaved differently, being more selective to methane. The excess of chlorine used during its preparation could explain this behavior, because it makes the cobalt reduction difficult and makes the surface Co^{2+} species prevail. The total synthesis gas pressure had a great effect on the product distribution of Co/Nb_2O_5, going towards an increase in the wax formation (C_{19+}) and leading to a strong decrease in the CH_4 production, when it works at 2Mpa. Raising the pressure to 4Mpa, the selectivity shifted toward methane and light hydrocarbons, but the gasoline fraction remained at a higher value. The catalysts had good stability over 50h on stream at reaction conditions.

REFERENCES

[1] R.B. Anderson, The Fischer-Tropsch Synthesis, Academic Press Inc., Orlando, 1984.
[2] D. Schnke, in New Trends of Methane Activation, Elsevier, 1990.
[3] F. Mendes, F.B. Noronha, R.R. Soares, C.A.C. Perez, G. Marcheti and M. Schmal, Stud. Surf. Sci. Catal. (2001) 177.
[4] F.B. Noronha, A. Frydman, D.A.G. Aranda, C.A. Perez, R.R. Soares, B. Morraweck, D. Castner, C.T. Campbell, R. Frety and M. Schmal, Catal. Today, 28 (1996) 147.
[5] R.C. Everson and H. Mulder, J. Catal. 28 (1996) 147.
[6] N. Tsubaki, S. Sun and K. Fujimoto, J. Catal. 199 (2001) 236.
[7] F.B. Noronha, C.A. Perez, M. Schmal and R. Fréty, Phys. Chem. Chem. Phys. 1 (1999) 2861.
[8] E. Iglesia, Appl. Catal. 161 (1997) 59.
[9] M.P. Kappor, A.L. Lapidus and A.Y. Krylova, Proceedings of the 10th Int. Congr. Catal. (L. Guczi, F. Solymosi and P. Tetenyi, eds.) part C, p 2741-2744, Elsevier, Budapest, 1992.
[10] A. Frydman, D.G. Castner, C.T. Campbell and M. Schmal, J. Catal. 188 (1999) 1.
[11] C.L. Bianchi, F. Martini and P. Moggi, Catal. Lett. 76 (2001) 65.

Studies in Surface Science and Catalysis, volume 147
X. Bao and Y. Xu (Editors)

Synthesis of isoparaffins from synthesis gas

Xiaohong Li, Kenji Asami, Mengfei Luo and Kaoru Fujimoto

The University of Kitakyushu, Wakamatsu-ku, Kitakyushu 808-0135, Japan

1. INTRODUCTION

Synthesis of hydrocarbon mixture by Fischer-Tropsch synthesis has been already industrialized for making clean diesel fuel because of its high n-paraffin content and no sulfur. It has been well known that the product of FTS is mostly straight chain hydrocarbons, either olefin or paraffin [1, 2]. Recently, the production of hydrocarbons which is rich in iso-paraffins has collected much attention because of its excellent character for motor fuel. One of the authors have shown that the direct synthesis of branched hydrocarbons from synthesis gas using a two-component catalyst (RuKY and SO_4^{2-}/ZrO_2) [3]. In this research, the selective synthesis of C_4-C_6 iso-paraffins from synthesis gas was studied [4]. The fundamental concept is the synthesis of hydrocarbon mixtures rich in C_4-C_6 isoparaffins by hydrocracking and isomerization the primary hydrocarbons produced by Fischer-Tropsch Synthesis. Catalysts used were physical mixtures of a Co/SiO_2 FTS catalyst, zeolite or palladium-supported zeolite (Pd/zeolite). In this reaction system product hydrocarbons on the FTS catalyst diffuse to the hydrocracking catalyst either in the gas phase (volatile products) or on the surface (products such as wax, which does not vaporize under reaction conditions) and are hydrocracked or isomerized to isoparaffins. It was interesting that higher hyd rocarbons such as wax were exclusively cracked when the hydrocracking catalyst existed near the FTS catalyst.

2. EXPERIMENTAL

2.1 Catalyst preparation

The Co/SiO_2 catalyst (20wt% Co) was prepared by the incipient wetness impregnation of aqueous solutions of cobalt nitrate on the silical gel (Fujisilicia Q-15). The catalyst precursors were dried at 393 K overnight in air and then calcined in air at 573 K for 2h.

A variety of zeolites, β, H-ZSM-5, USY, H-Modenite and Na-Y, were used in this paper. Pd/zeolite (0.5 wt% loading) was prepared by ion exchange method. The ion exchange was carried out at 353 K for 6 h with $Pd(NH_3)_4Cl_2$ aqueous solution under stirring, the supported Pd/Zeolite was washed with water until no chloride ion was detected, then was dried overnight at 393 K, and followed by calcining it in air at 723 K for 2 h.

The catalysts were prepared by hybridizing (physically mixing) a variety of zeolites (β, H-ZSM-5, USY, H-Modenite, Na-Y) or Pd/zeolite (Pd/β, Pd/H-ZSM-5, Pd/USY, Pd/Na-Y) with Co/SiO$_2$ catalyst by pressure-molding the powder of Co/SiO$_2$ and zeolite or Pd/zeolite. The catalyst with between 20 and 40 mesh were selected. Another is a physically mixed catalyst with both Co/SiO$_2$ (20-40 mesh) and zeolite or Pd/zeolite (20-40 mesh). The former is named as powder hybrid catalyst; the latter is named as granular hybrid catalyst. The catalysts were reduced again 673 K for 3 h in situ before catalytic reaction.

2.2 Reaction apparatus and procedure

F-T reaction was conducted with a conventional continuous-flow fixed bed reaction apparatus (SUS tubing, 8mm I. D.). The standard condition conditions were T=513 K, P(total)=1.0 MPa, H$_2$/CO=1.9, W/F(CO+H$_2$)=5.1 g of catalyst h/mol. During the reaction, the effluent gas was cooled at a hot trap (423 K), and then was introduced to gas chromatograph equipped with NB-1 capillary column for C$_1$-C$_{19}$ (FID) and active carbon column for CO, CH$_4$, and CO$_2$ analysis (TCD). The liquid products collected in the tot trap and the products remaining in the reactor and in the catalyst were combined and analyzed by GC.

The selectivity of hydrocarbon is defined as follows:

Selectivity=$M/(C_{CO} \times \alpha_{CO})$

C_{CO} = CO flow rate in reaction gas

α_{CO} = CO conversion

M= Carbon-flow rate in iso-paraffin, n-paraffin or olefin

3.R ESULTS AND DISCUSSION

In Fig. 1, it is shown that the hydrocarbon distribution obtained on Co/SiO$_2$ only, Co/SiO$_2$+zeolite and Co/SiO$_2$+ Pd/zeolite catalysts at reaction for 4 h. CO conversion of reactions was 23.4%, 38.8%, 47.9%, respectively. CO$_2$ selectivity was only about from 0.2% to 1%. The hydrocarbon recovery is 58.6%, 83.7% and 89.1%, respectively. In the product distribution on the Co/SiO$_2$+β and Co/SiO$_2$+Pd/β hybrid catalysts, heavy hydrocarbons (C$_{12}^+$) decreased and C$_4$-C$_9$ fractions increased obviously. This fact was in good accordance with the reported findings that on the noble metal supported on an acid zeolite, hydrocarbons with longer chain were hydrocaracked more quickly than shorter ones [5]. For the conventional FTS reaction on Co/SiO$_2$ catalyst, no isoparaffins was detected. However, they were formed on the Co/SiO$_2$+β and Co/SiO$_2$+Pd/β hybrid catalysts, the selectivity to Isoparaffins on the Co/SiO$_2$+β and Co/SiO$_2$+Pd/β was 18.0% and 20.0% respectively.

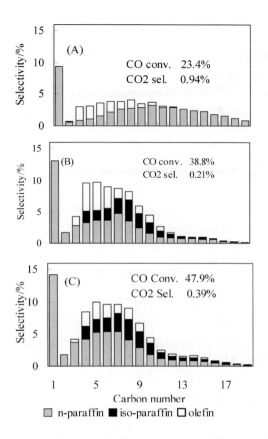

Fig.1 The hydrocarbon distribution of F-T reaction on Co/SiO$_2$ (A), Co/SiO$_2$+β (B) and Co/SiO$_2$+ Pd/β (C) catalysts. Co/SiO$_2$ catalyst: 0.8g, SiO$_2$, β or Pd/β 0.2g, time on stream: 4h.

Fig. 2 shows the time dependencies of F-T synthesis for Co/SiO$_2$ only and for a physical mixture of Co/SiO$_2$+Pd/β. When only Co/SiO$_2$ catalyst was only used, the FTS activity deactivated rather quickly during the reaction. However, on the physical mixtures of Co/SiO$_2$ FTS catalyst and Pd/β, the CO conversion increased and no deactivation was observed during the reaction. The promotional role of Pd/β for FTS activity can be attributed to these possibilities. (1) The addition of FTS activity of Pd/β catalyst, which has been ruled out experimentally. (2) The removal of was by hydrocracking, which will be proved by experimental results shown in Figs. 3 and 4. (3) the promotion of hydrogen concentration by spill-over effect from Pd to Co. Hydrogen which is activated on Pd transfer to Co site, which promote the concentration of active hydrogen on Co promote the FTS activity and CH$_4$ selectivity [6]. The latter two factors can explain the results in Fig. 2.

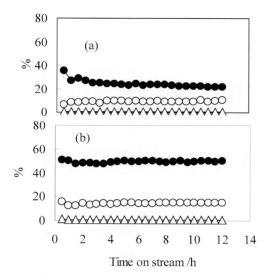

Fig.2 Time course of FTS on (a) Co/SiO$_2$, (b) Co/SiO$_2$+Pd/β. (H$_2$/CO=2 mol ratio; 1.0 MPa; W/F=5.1 g.h/mol; 513 K; 4h). ● CO conversion, ○ CH$_4$ selectivity, △ CO$_2$ selectivity

Fig. 3 shows the concept of reaction mechanism of FTS on Co/SiO$_2$ and Co/SiO$_2$ + zeolite. When used Co/SiO$_2$ catalyst only, heavy hydrocarbons (such as wax) were formed on the catalyst, and did not evaporate under reaction conditions, to retard the syngas access. Interestingly, when the hydrocracking catalyst (zeolite or Pd/zeolite) existed near the Co/SiO$_2$ catalyst, higher hydrocarbons (such as wax) diffused to the hydrocracking catalyst on which they were exclusively cracked to light hydrocarbons, to vaporize, which reduce the accumulated wax to prevent the catalyst deactivation.

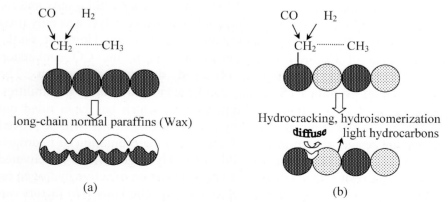

Fig. 3 The concept of reaction mechanism on (a) Co/SiO$_2$, (b) Co/SiO$_2$+zeolite.

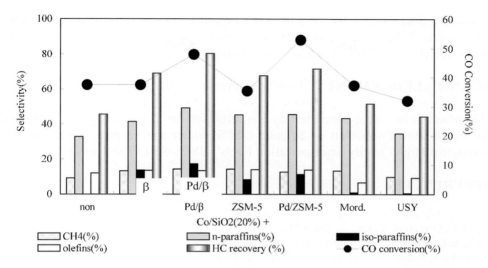

Fig. 4 The CO conversion and product selectivity in F-T reaction on various hybrid catalysts containing different Pd/zeolite. (H_2/CO=2 mol ratio; W/F=5.1; 1.0 MPa; 513 K; Reaction time 4 h)

Fig. 4 shows the CO conversion and product selectivity in FTS reaction on hybrid catalysts with different zeolite or Pd/zeolite. For FTS reaction on the Co/SiO_2 catalyst, only n-paraffins and olefins were formed, no iso-paraffin was detected. However, on the hybrid catalysts, Co/SiO_2+β, Co/SiO_2+H-ZSM-5 and Co/SiO_2+Mordenite, iso-paraffins were formed; the selectivity was 17.0%, 9.7% and 1.82% respectively. For Co/SiO_2+USY, no iso-paraffin was detected. When β and ZSM-5 were added to Co/SiO_2 catalyst, the recovery (the sum of C_1-C_{19} selectivity and CO_2 selectivity) increased, indicating that the heavy product from the conventional FTS reaction was subjected to the secondary hydrocracking effectively to form lighter hydrocarbons. This is in good agreement with the reported finding that hydrocarbons with longer chain were easier to be hydrocracked [7]. When Pd/β or Pd/ZSM-5 was used in stead of metal-free zeolite, both the CO conversion and the selectivity of iso-paraffins increased. The CO conversion was about 37%, without Pd, but it increased to 45-50% with Pd. It was hypothesized that hydrogen spillover from Pd to zeolite and Co/SiO_2 would promote the hydrocracking reaction and FTS reaction as well.

4.C ONCLUSION

It was claimed that the direct production of iso-paraffins from syngas could be realized by the hybrid reaction composed of the conventional FTS reaction and its successive hydrocracking/ hydroisomerization of the products. The addition of small amount of zeolite or palladium-supported zeolite (Pd/zeolite) to FTS catalyst could change the hydrocarbon distribution drastically which was caused by the selective cracking of higher hydrocarbons (wax).

ACKNOWLEDGEMENT

This research has been supported by Research for the Future Program of Japan Society for the Promotion of Science (JSPS) under the project "Synthesis of Ecological High Quality Transportation Fuels" (JSPS-RFTF98P01001).

REFERENCES

[1] A. A. Adesina, Appl. Catal., A, 138(1996)345
[2] M. E. Dry, Appl. Catal., A, 138(1996)319
[3] X. Song and A. Sayari, Energy and Fuels, 10, (1996)561
[4] X. Li, K.Asami, M. Luo, K. Michiki, N.Tsubaki and K. Fujimoto, Catal. Today, 84(2003)59-65
[5] K. Fjimoto, M. Adachi, H. Tominaga, Chem. Lett. (1985)783
[6] S. Sun, N. Tsubaki and K. Fujimoto, Chem. Lett.(2000)176
[7] G. Leclercq, L. Leclercq, R. Maurel, J. Catal., 44, (1976)68

Studies in Surface Science and Catalysis, volume 147
X. Bao and Y. Xu (Editors)
373

The Fischer-Tropsch synthesis over the bimetallic Ru-M/Al$_2$O$_3$ - catalysts

Larissa B. Shapovalova[*], Gauhar D. Zakumbaeva, Aliya A. Zhurtbaeva

The Institute of Organic Catalysis & Electrochemistry of the MES of the Republic of Kazakstan; 142, Kunaev str., Almaty, 480100, Kazakstan
E-mail: orgcat@nursat.kz Fax: +7 3272 915722

ABSTRACT

The structure and catalytic properties of bimetallic Ru-M /Al$_2$O$_3$ Fischer-Tropsch catalysts were studied (M = Fe, Co, V, Cr, Mn, Ce, Cu). The wide varieties of products (the hydrocarbons and the alcohols) are formed by the CO hydrogenation over bimetallic Ru-M/Al$_2$O$_3$ catalysts of cluster type. The yield and the composition of products are determined by the nature of second metals of bimetallic Ru-M/Al$_2$O$_3$ catalysts. The adsorption of CO and H$_2$ was studied using the techniques of microcalorimetry, thermal desorption, IR- spectroscopy and quantum- chemical accounts.

1. INTRODUCTION

The main task of Fischer-Tropsch synthesis is to design the effective and selective catalysts. The catalysts containing noble metals and catalysts without noble metals have been developed for obtaining different hydrocarbon fractions and oxygenates from CO by us.

In present work the Fischer-Tropsch synthesis over bimetallic Ru-M/Al$_2$O$_3$ catalysts of cluster type (M = Co, Cu, Ce, Mn, Cr, Fe, V) has been studied.

2. EXPERIMENTAL

The Fischer-Tropsch process was carried out in a flow system at 423-673K, ratio CO:H$_2$=1:2, P=0.1 MPa, V=100h^{-1}. The Ru-M /Al$_2$O$_3$ catalysts were prepared by impregnation of Al$_2$O$_3$ with mixture of Ru- and M- salts solutions. The products were determined on LKM–8MD chromatograph (column, 3.5 · 0.005m, with Separon–BD).

Quantum-chemical calculations have been made on the basis of extended Huckel method complemented with Anderson core-core repulsions as item of total electron energy. Minimal quantity of metal atoms (4-13) was taken into account. This approach seams a reasonable, because, as is known, chemisorption is highly localized phenomenon [1-4].

Adsorption of CO and H_2 on the Ru-M- catalysts was studied using the microcalorimetric technique with a Setaram-70 apparatus [5]. IR spectra were recorded on a Specord-75 UR spectrometer. Catalyst X-ray photoelectron spectra were recorded on ESCA-3 and ES-2402 instruments using MgK_{α} radiation. Operating vacuum of a physical camera was $6.6 \cdot 10^{-6}$ Pa. Electron- microscopic and diffraction characteristics were measured on a JEM-100 microscope.

3. RESULTS AND DISCUSSION

The wide varieties of products (the hydrocarbons and the alcohols) are formed by the CO hydrogenation over bimetallic Ru-M/Al_2O_3 catalysts of cluster type (Table 1). The yield and the composition of products are determined by the nature of second metals of bimetallic Ru-M/Al_2O_3 catalysts. For example, the C_1-C_3-hydrocarbons and C_1-C_4- alcohols are prevailing over Ru-Fe and Ru-Cu-catalysts. The C_2-C_4-olefines are formed on Ru-V-system ($C_2^=/C_2$= 0,36; $C_3^=/C_3$= 2,3 and $C_4^=/C_4$=2,8) The formation of high quantity of C_4-C_n n-and i-olefines and parafines take place on Ru-Ce, Ru-Co, Ru-Mn- catalysts. It should be noted that the methane is dominant interaction product formed over monometallic Ru/Al_2O_3 –catalyst in these conditions.

Incorporation of the second metals substantially affects the adsorption properties of the Ru catalyst with respect to CO and H_2 (Table 2). The maximal heats of adsoption of CO on Ru-M (1:1)/ Al_2O_3-catalysts are lower those on Ru/ Al_2O_3 and are changed in sequence: Ru-Fe/Al_2O_3>Ru-Mn/Al_2O_3≈ Ru-Ce/Al_2O_3>Ru-Cr/ Al_2O_3 >Ru-Cu/ Al_2O_3. The amount of carbon monoxide adsorbed on Ru-M/ Al_2O_3 is higher than CO_{ads} concentration on the monometallic Ru-catalyst. The maximal amount of CO ($22,32 \cdot 10^{-4}$ mol/gΣM) is adsorbed on the Ru-Cu/ Al_2O_3-catalysts. The IR data indicate that over Ru-M/ Al_2O_3 carbon oxide presents in bridging (β2, 1880-1885 cm^{-1}) and linear (β1, 2020-2080 cm^{-1} and 2170-2160 cm^{-1}) forms. The IR-spectra obtained after CO adsorption on the Ru/Al_2O_3 sample showed only one very broad band above 1980 cm^{-1}. In this case the concentration bridged forms of CO_{ads} is above 85,0%. The IR data are in a good agreement with those obtained for CO by adsorption microcalorimetry

The hydrogen adsorption heat is influenced by the nature of second components of Ru-M and varied from 101,8 kJ/mol (Ru-Ce) to 77,4 kJ/mol (Ru-Cu). The amounts of adsorbed hydrogen were changed extremely: the minimum quantity of H_{ads} = $1,25 \cdot 10^{-4}$ mol/gΣM (Ru-Cu), and the maximum quantity of H_{ads} = $9,48 \cdot 10^{-4}$ mol/gΣM (Ru-Mn).

Table 1
The CO hydrogenation on Ru-M/Al$_2$O$_3$ catalysts (P = 0.1 MPa, T = 473 K, CO:H$_2$= 1:2, V= 100 h^{-1})

Catalysts	CO-conversion, %	The products, mol.%												Alc.
		C$_1$	C$_2^=$	C$_2$	C$_3^=$	C$_3$	C$_4^=$	n-C$_4$/i-C$_4$	C$_5^=$	n-C$_5$/i-C$_5$	ΣC$_6$	ΣC$_{7-}$	ΣC$_{8+}$	C$_1$-C$_4$
		Ru:M = 9:1												
Ru-Co	15,8	8,6	0,2	8,9	7,6	14,1	2,1	11,6	2,1	17,5	16,0	3,2	8,1	–
Ru-V	30,7	12,2	2,1	5,4	27,3	11,9	27,3	9,7	4,1	–	–	–	–	–
Ru-Cu	72,3	59,8	0,8	12,5	–	13,2	–	–	–	–	–	–	–	13,7
Ru-Ce	39,7	2,5	1,0	9,4	21,4	16,4	10,6	9,8	3,1	18,6	2,5	4,8	–	–
Ru-Mn	93,2	20,4	2,2	15,6	–	15,7	9,0	20,4/6,9	4,5	3,3/2,0	–	–	–	–
Ru-Cr	25,0	27,7	2,4	16,9	–	27,4	0,9	18,5	–	4,3	1,9	–	–	–
		Ru:M = 1:1												
Ru-Fe	73,7	24,2	0,5	16,2	4,6	сл.	сл.	сл.	–	сл.	–	–	–	54,3
Ru-Co	27,2	15,2	–	4,5	–	26,6		21,3	0,2	17,6	8,6	6,0	–	–
Ru-V	21,5	2,7	1,7	3,4	3,4	7,9	13,6	19,0	13,0	34,1	–	–	–	–
Ru-Cu	35,1	36,3	8,5	31,6	–	16,8	–	6,8	–	сл.	–	–	–	сл.
Ru-Ce	7,6	8,4	0,8	4,9	22,0	1,3	9,5	5,8/18,8	7,3	12,0	9,9	–	–	–
Ru-Mn	38,1	2,6	0,9	2,8	3,0	5,0	9,8	18,5	13,8	43,6	6,4	–	–	–
Ru-Cr	35,4	29,4	–	20,9	–	20,7	2,4	5,8	–	5,7	5,1	–	–	–

Table 2
The characteristics Ru-M/ Al$_2$O$_3$ catalysts of Fischer- Tropsch Synthesis.

Catalysts Ru-M/Al$_2$O$_3$	E$_{bind}$, eV		H$_2$ Adsorption		CO$_2$ Adsorption		CO$_2$ adsorption forms	
	Ru3p$_{3/2}$	M 2p$_{3/2}$	n$_{H2}$10^{-4} mol/gΣM	q$_{H2}$ kJ/ mol	N$_{CO}$10^{-4} mol/gΣM	q$_{CO}$ kJ/ mol	B1	β2
Ru	461,5 (Ru^0)	—	4,55	146,0-32,6	5,37	199,4-19,4	7	85
Ru-Cr	464,3	578,5	2,95	101,24,1	15,8	112,7-40,6	45	39
Ru-Mn	463,6 461,7**	642,5 641,6	9,48	106,6-28,0	11,90	118,5-55,6	9	71
Ru-Ce	463,5 461,4**	3d$_{5,2}$=883,7 3d$_{5/2}$=883,2**	5,07	99,6-18,7	10,65	118,3-29,4	18	54
Ru-Cu	462,5	932,3	2,04	77,4-17,8	22,32	9,32-35,5	52	17
Ru-Fe	461,0 463,6**	711,6(Fe^{3+}) 708,7**	1,28	101,8-8,0	7,27	134,0-20,0	40	31
Ru-Co	461,2	779,6	-	-	-	-	-	-
Ru-V	461,5	517,9	-	-	-	-	-	-

**-"in situ", H$_2$, T= 773 K

The variation of the properties of Ru-M $(1:1)/Al_2O_3$ are attributed to the formation on heteronuclear Ru-M-associated clusters. It was shown that ruthenium exists mainly in the form of RuO_2 and Ru^0. The binding energy of Ru $3p_{3/2}$- electrons are changed in the next order (eV): Ru-Fe ($3d_{5/2} = 280,8$; Ru^0) > Ru-Co (461,2; Ru^0) > Ru-V (461,5; Ru^0>>Ru^{n+}) > Ru-Cu (462,5; Ru^0;Ru^{n+}) > Ru-Ce (463,5; Ru^0;Ru^{n+}) Ru-Mn (463,6; Ru^0 < Ru^{n+}) > Ru-Cr (464,3; Ru^{n+}).

The sizes of Ru partiles vary from 10 to20 Å at the surface of Ru-M/ Al_2O_3. Fine dispersed ruthenium (d $\approx$3-5 Å) produced aggregates of about 200 Å sizes. The second components of all bimetallic Ru-M- catalysts are in oxidized state (Table 2). Also the fine dispersed formations have been discovered on the surface of all bimetallic Ru-M/Al_2O_3 catalysts. Oxidised states of second metals (M = Co, Cu, Ce, Mn, Cr, Fe,V) bonded with tiny particles of Ru^0 were presented in these formations. We have assumed that these formations are Ru-M-associated clusters.

The quantum-chemical accounts and experimental methods (XPS, e-micro-copy, Messbauer spectroscopy) show that there are bimetallic Ru-M-clusters on the surface of Ru-M/ Al_2O_3-catalysts (M= Fe, Co, V, Cu, Cr, Mn, Ce). For example from the in situ XPS- spectroscopy it is seen that after the treatment with H_2 at 773 K, the binding energy of Ru$3p_{3/2}$- electrons of Ru-Fe/ Al_2O_3-catalyst shifts to higher energy region (463,6 eV) as compared with the

Table 3

Quantum-chemical accounts of 4- and 13-atomic mono- and bimetallic Ru, V, Co, Ru-V и Ru-Co-clusters

Clusters	E_{bind}*	ΔE**	Clusters	E_{bind}	ΔE
4Ru	26,77	0			
V-3Ru	39,60	12,83	Co-3Ru	32.81	26.10
2V-2Ru	39,21	12,44	2Co-2Ru	34.31	27.62
3V- Ru	29,88	3,11	3Co-Ru	34.39	27.65
4V	17,38	−9,38***	4Co	33.42	26.54
13Ru	92,52	0	13Ru	92.52	-
$1V_{(1)}$-12Ru	95,82	3,30	$1Co_{(1)}$-12Ru	101.89	9.37
$V_{(2)}$-12Ru	89,35	−2,57	$1Co_{(2)}$-12Ru	96.55	4.03
$2V_{(1,2)}$-11Ru	112,70	20,18	$1Co_{(5)}$-12Ru	96.47	3.95
$2V_{(2,3)}$-11Ru	105.43	12,91	$2Co_{(1,2)}$-11Ru	104.91	12.39
$3V_{(2,3,4)}$-10Ru	101,74	9,22	$2Co_{(2,3)}$-11Ru	99.38	6.86
$3V_{(1,3,4)}$-10Ru	110,66	18,14	$3Co_{(1,2,3)}$-10Ru	106.77	14.25
$3V_{(3,4,5)}$-10Ru	108,60	16,08	$3Co_{(2,3,4)}$-10Ru	101.24	8.72
$3V_{(4,8,12)}$10Ru	121,56	29,04	$3Co_{(3,4,5)}$-10Ru	101.97	9.39
$12V-Ru_{(1)}$	26,59	−65,93	$3Co_{(4,6,12)}$-10Ru	103.07	10.55
$12V-Ru_{(2)}$	3,48	−89,04	$12Co-Ru_{(1)}$	107.74	15.22
$12V-Ru_{(5)}$	3,25	−89,27	$12Co-Ru_{(2)}$	108.41	15.89
13V	−20,67	−116,19	13Co	106.93	10.38

E_{bind} – The total energy of all bnds in cluster, eV; ** ΔE –The binding energy of 4- and 13 atomic clusters, eV; *** the (-) show the destabilization of clusters

catalyst exposed to the air (461,5 eV). In this case Fe reduced to Fe^0-forms with $E_{bind}Fe2p_{3/2}$ =708,7 eV (in air E_{bind} $Fe2p_{3/2}$ =711,6 eV) (Table 2). Only the existence of bimetallic Ru-Fe clusters makes these oxido-reduction alterations possible. The XPS data are in a good agreement with those obtained for Ru-Fe /Al_2O_3 by Messbauer spectroscopy [5].

The quantum- chemical accounts show the formation of bimetallic Ru-M-clusters is more energetically preferably preferable than monometallic ones. For example in case Ru-Co and Ru-V- systems calculation showed that cobalt and vanadium can interact with ruthenium and form stable cluster structures. The stability of 4 or 13 atomic Co-Ru and Ru-V –clusters depends on the ratio Ru:M (M= Co, V), the locations of Ru and M in the clusters and the nature of second metals. It was shown that cobalt and ruthenium (both are crystallized on exagonal type) are able to form bimetallic Ru-Co-clusters more stable than monometallic Ru- or Co-systems. Vanadium crystallizes on cubic type, but it is able to form bimetallic Ru –V-cluster stronger than Ru-Co ones.

4. CONCLUSION

The comparison of the experimental results and the quantum- chemical accounts show that the mechanism of the Fischer-Tropsch synthesis is very complicated on Ru-M/Al_2O_3 catalysts. The analysis of the results of TPR, IRS, quantum-chemical investigation shows that the main part of CH_4 is formed on Ru^o monometallic cluster centers by dissociated mechanism [6]. The C_n-hydrocarbons (n $\geq$ 2) and alcohols are formed on the bimetallic Ru-M-clusters by undissociated mechanism. The conception and development of C-C-chains take place on the same centers. The developments of C-C-chains are realized by the implantation of the CO molecule on M-alkyl and elimination of the H_2O molecule. The α-olefines are the primary products of CO+H_2 reaction

REFERENCES

[1] R. Hoffmann, J. Chem. Phys., V.39. № 6.(1963) P.1397.
[2] A.B. Anderson, S.I. Hong, J.Z. Smialek, J. Phys. Chem., V.91. № 161.(1987) P.4245.
[3] J.U. Ammetier, H.B. Burdi, J.C. Thebeatt, R. Hoffman, J. Amer. Chem. Soc., V.100.(1978) P.3686.
[4] I.G. Efremenko, J.L. Zilberberg, C.M. Zhidomirov, A.M. Pak, React. Kinet. Catal. Lett., V.56, № 1(1995) P.77-86.
[5] G.D. Zakumbaeva, L.B. Shapovalova, J.T. Omarov, Russian Journal Appl. Chemistry , №3 (1988) P.471
[6] G.D. Zakumbaeva, L.B. Shapovalova, I.G. Efremenko, A.V. Gabdrakipov, Russian Journal of Neftechimia, № 4, (1996) 427-437

Studies in Surface Science and Catalysis, volume 147
X. Bao and Y. Xu (Editors)
379

Direct Dimethyl Ether (DME) synthesis from natural gas

Takashi Ogawa, Norio Inoue, Tutomu Shikada, Osamu Inokoshi, Yotaro Ohno

DME Development Co., Ltd, Shoro-koku Shiranuka-ch, Hokkaido, 088-0563 Japan

ABSTRACT

Dimethyl Ether (DME) is a clean and economical alternative fuel which can be produced from natural gas. The properties of DME are very similar to those of LPG. DME can be used for various fields as a clean and easy-to-handle fuel, such as power generation, transportation, home heating and cooking, etc.

An innovative process of direct synthesis of DME from synthesis gas has been developed. 5 tons per day DME production pilot plant (5TPD) project had finished very successfully in FY2001, and 100 tons per day DME production demonstration plant (100TPD) project started in FY2002. These projects are supported and funded by Ministry of Economy, Trade and Industry (METI). This JFE Direct DME Synthesis will open a new way to economical DME mass production.

1. INTRODUCTION

DME synthesis from synthesis gas (syn-gas: H_2+CO gas) has been developed for these years [1-3]. JFE (formally NKK) Corporation has been making remarkable progress in a development of direct DME synthesis from syn-gas with a liquid phase slurry reactor [4-7]. DME is the simplest ether having the chemical formula of CH_3OCH_3. It is colorless gas, and sulfur and nitrogen free. As its vapor pressure is about 0.6 MPa at 25°C, DME is easily liquefied under light pressure. DME is also a very clean diesel substitute that does no exhaust black smoke [8]. Table 1 shows physical properties and combustion characteristics of DME and other relating fuels.

DME can be distributed and stored by using LPG handling technology, which means DME does not need costly LNG tankers or LNG terminals [9]. Once natural gas is converted to DME, it will provide a competitive new alternative measure to transport natural gas.

Table 1
Physical Properties of DME and other fuel

[Properties]	DME	Propane	Methane	Methanol	Diesel
Chemical formula	CH3OCH3	C3H8	CH4	CH3OH	-
Boiling point (K)	247.9	231	111.5	337.6	180-370
Liquid density (g/cm^3 @293K)	0.67	0.49	-	0.79	0.84
Specific gravity (vs. air)	1.59	1.52	0.55	-	-
Vapor pressure (atm @293K)	6.1	9.3	-	-	-
Explosion limit	3.4-17	2.1-9.4	5-15	5.5-36	0.6-6.5
Cetane number*	55-60	(5) *	0	5	40-5
Net calorific value (MJ/Nm^3)	59.44	91.25	36.0	-	-
Net calorific value (MJ/kg)	28.90	46.46	50.23	21.1	41.86

* Estimated value

2. DIRECT DME SYNTHESIS PROCESS

Table 2 shows reactions concerning with direct DME synthesis and their reaction heats. There are mainly two overall reaction routes that synthesize DME from syn-gas, reaction (1) and (2). The reaction (1) synthesizes DME in three steps, which are reaction (3), (4) and (5). When the three reactions (3, 4, 5) take place simultaneously, the syn-gas conversion rises dramatically. Separation of by-product CO_2 from unreacted syn-gas is much easier compared with separation of by-product water (in the case of reaction (2)). Because of these reasons, JFE Direct DME Synthesis focused the development of the catalyst system and its reaction process on the reaction (1).

Typical reaction conditions are 240-280 [°C] and 3.0-7.0 [MPa]. The standard condition is 260 [°C] and 5.0 [MPa]. Catalyst loading ratio W/F is 3.0-8.0 [kg-catalyst * h/kg-mol].

Table 2
DME synthesis reaction formulas

Reaction		Reaction heat [kJ/mol-DME]	
$3CO+3H_2$	$CH_3OCH_3+CO_2$	-246	(1)
$2CO+4H_2$	$CH_3OCH_3+H_2O$	-205	(2)
$2CO+4H_2$	$2CH_3OH$	-182	(3)
$2CH_3OH$	$CH_3OCH_3+H_2O$	- 23	(4)
$CO+H_2O$	CO_2+H_2	- 41	(5)

Table 3
DME synthesis from natural gas in the direct DME synthesis plant

Unit	Reaction	
(ATR)	$2CH_4 + O_2 + CO_2 \rightarrow 3CO + 3H_2 + H_2O\downarrow$	(6)
(DME Synthesis Reactor)	$3CO + 3H_2 \rightarrow CH_3OCH_3 (DME) + CO_2$	(1)
(DME Plant Total)	$2CH_4 + O_2 \rightarrow CH_3OCH_3 (DME) + H_2O\downarrow$	(7)

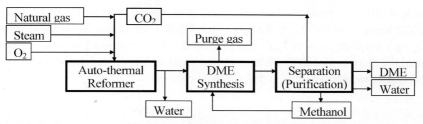

Fig. 1. Schematic process flow diagram of JFE Direct DME Synthesis process

In the process of direct DME synthesis form natural gas, the natural gas will be converted to syn-gas of $H_2/CO=1$ with by-product CO_2 (reaction (6)). Then the total process follows overall reaction (7), and eventually natural gas (methane) is converted DME and water.

The reaction of DME synthesis is highly exothermic (Table 2). It is very crucial to remove the reaction heat and to control the reaction temperature. The direct DME synthesis reactor is specialized with a liquid phase slurry reactor, which is easily remove the reaction heat. The dimensions of the slurry reactor of 5TPD were of 0.55 meter- inner-diameter and 15 meter-height.

Fig. 1 shows a schematic process flow diagram of JFE Direct DME Synthesis process. The effluent from the slurry reactor (DME and other by-products) is chilled and separated as liquid from unreacted gas. In this separation DME works as a solvent to remove CO_2 from unreacted syn-gas. CO_2 is recycled to the Auto-thermal Reformer (ATR). Methanol is recycled the DME reactor and finally converted to DME.

3. PROCESS DEVELOPMENT

During one and half years of 5TPD experimental period, six continuous plant operations were conducted. Total plant operation time reached 4,300 hours, and total DME production reached about 400 tons. Picture 1 shows the plant. Since natural gas was not available at the plant site, LPG and Coal Bed Methane were used as feedstock.

Fig. 2 shows typical 5TPD results of once-through conversion and selectivity. The once-through CO conversion reached higher than 50% at 260 °C and 5 MPa. The DME catalyst system did not generates undesirable heavier by-products such as oil, wax or higher alcohols. The degradation of the catalyst activity was confirmed to be lower than 10 points of the initial value after 1,000 hours.

Table 4
Product selectivity and DME quality of 5 TPD

a. Product C-mol ratio	: DME / (DME + Methanol)	0.91 [-]
b. Product H-mol ratio	: H_2O / (DME + Methanol + H_2O)	0.013 [-]
c. DME purity		99.9 [%]
d. Impurity of DME	: Methanol+ H_2O	< 100 pp

Table 4 shows typical product selectivity and typical product DME quality of 5 TPD (CO2 was eliminated from the products). DME once-through yield was as much as 90%. Very small amount of H_2O was produced. Fig. 3 shows a typical example of the material balance of 5TPD. When DME yield achieved 5.7 tons per, total CO conversion reached almost 95% in this material balance.

Picture 1. The 5 TPD DME pilot plant

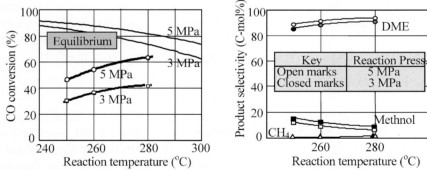

Fig. 2. Once-through CO conversion and product selectivity vs. reaction condition

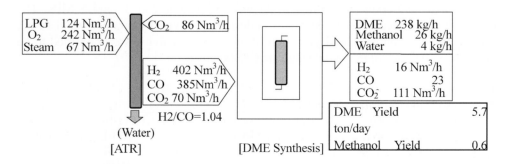

Fig. 3. Typical material balance of 5 TPD

	2002	2003	2004	2005	2006
100 TPD demo－plant					
Design					
Manufacturing,					
Construction					
Operation					
Back-up study					
Feasibility study					

Fig. 4. The schedule of 100TPD project

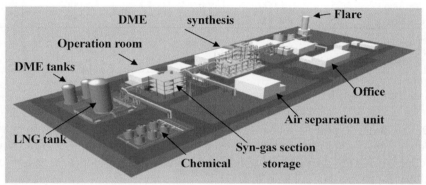

Fig. 5. Over view of the 100TPD demo-plant

4. 100TPD PROJECT

DME Development Co., Ltd, started five year DME 100TPD project with funds provided by METI in 2002. Fig. 4 shows the schedule of 100TPD project. Fig. 5 shows the overview of 100TPD. The targets of the development with 100TPD operation are as follows;

    a. To achieve targeted reactor performance.
    O Total syn-gas conversion > 95%   O Selectivity of DME > 90%
    O Purity of DME > 99%
    b. To operate the demo-plant continuously more than three months.
    c. To establish scale-up technology of the slurry reactor.
    d. To establish operation procedure.

5. ECONOMICS OF DME PRODUCTION FROM NATURAL GAS

Based on the results of 5TPD project and economic assumptions, the CIF price of DME is estimated. DME is produced at natural gas field and transported to Japan. As a result of a calorie-equivalent based sensitivity analysis, Table 4 shows the results of feasible DME production study based on natural gas feedstock, DME plant scale and transportation distance. Estimated DME CIF price (3.7-5 US$ / MMBTU) is lower than those of LPG and Diesel oil and almost similar to LNG CIF price in Japan [10].

Table 4
Study of Feasible DME Production

Area (Size of gas field) [US$/MMBtu]	Gas price [US$/MMBtu]	Plant scale [ton/d]	Production [kton/y]	FOB price (DME) [US$/MMBtu]	Transport distance [km]	CIF price (DME) [US$/MMBtu]
West Australia	1.0	2,500	830	4.4	7,000	5.0
(Large scale)	1.0	5,000	1,650	3.8	7,000	4.4
	1.0	10,000	3,330	3.3	7,000	4.0
South East Asia	1.25	5,000	1,650	4.1	5,000	4.6
(Middle scale)	1.25	10,000	3,330	3.7	5,000	4.1
Middle East	0.5	2,500	830	3.7	12,000	4.8
(Large scale)	0.5	5,000	1,650	3.1	12,000	4.2
	0.5	10,000	3,330	2.6	12,000	3.7

6. CONCLUSION

JEF Direct DME Synthesis technology successfully finished 5 TPD project with sufficient achievement. Based on these results, DME Development Co., Ltd, is now conducting a 100TPD project to establish commercial DME production technology. 100 t/d DME production will start in 2004.

The DME energy flow system would promote utilization of stranded medium and small scale natural gas field to produce an easy-to-handle fuel and to supply to the Asia-Pacific Region in the future.

ACKNOWLEDGEMENT

The authors would like to express sincere appreciation for Agency of Natural Resources and Energy, Coal Division (METI) for their long term financial support.

REFERENCES

[1] J.B. Hansen et al: SAE950063 (Feb. 1995)
[2] Air products and Chemicals: DOE/PC/90018-T7 (June. 1993)
[3] D. Romaini et al: The 2nd International Oil, Gas & Petrochemical Congress, Tehran 16-18 (May, 2000)
[4] Y. Ohno: Energy total engineering, vol.20, No.1 (1997), p.45
[5] Y. Ohno et al: Preprints of Papers Presented at the 213th ACS National Meeting, Div. Of Fuel Chemistry, p.705, San Francisco (1997)
[6] T. Ogawa et al: Proceedings of the 14th Annual International Pittsburgh Coal Conference and Workshop, 30-3004, Taiyuan, China (1997)
[7] T. Ogawa et al: The AIChE 2000 Spring National Meeting, "Oxygenated Fuels", Atlanta, UAS, (Mar.5-9, 2000)
[8] T. Fleisch: SAE950061 (Feb. 1995)
[9] Y. Kikkawa et al: Oil & Gas Journal (Apr.6, 1998), p.55
[10] Y. Ando et al: Gas 2002 Indonesia, Jakarta, Indonesia (Jan.14, 2002)

Studies in Surface Science and Catalysis, volume 147
X. Bao and Y. Xu (Editors)

Integrated synthesis of dimethylether *via* CO_2 hydrogenation

F. Arena [1,2], L. Spadaro [1], O. Di Blasi [1], G. Bonura [1] and F. Frusteri [1]

[1] Istituto CNR-ITAE, Salita S. Lucia 39, I-98126 S. Lucia (Messina), Italy

[2] Dipartimento di Chimica Industriale e Ingegneria dei Materiali, Università degli Studi di Messina, Salita Sperone 31, I-98166 S. Agata (Messina), Italy

Abstract

The direct synthesis of dimethylether (DME) *via* the integrated CO_2 hydrogenation to CH_3OH and further dehydration to DME has been carried out using a dual-functionality catalytic bed in the range 180-240°C at a pressure of 9 bar. Factors affecting the performance of Cu-Zn catalysts in the synthesis of methanol have been highlighted, while the HY zeolite results the most suitable acidic system for the *in situ* methanol dehydration. The feasibility of the *integrated* ("two-step") catalytic synthesis of DME from CO_2-H_2 streams is outlined.

1. INTRODUCTION

The pressing need to reduce the environmental impact of the technological progress imposes a continuous development of novel methodologies aimed at a severe reduction of pollutant emissions, mainly from mobile sources, to warrant a tolerable quality of life in metropolitan areas. Then, an extraordinary research effort was aimed in the last years at disclosing and synthesizing effective and *cleaner* fuels entailing lower CO_2, NO_x and SO_x emission levels [1,2]. This goal would constitute a strategic breakthrough for the raw material market, enabling the exploitation of natural gas (NG) for fuels production, instead of oil [1-3].

Amongst synthetic fuels, DME, with a calorific value of 6 Mcal/kg and a cetane number of 60, represents a promising alternative for diesel compression-ignited engines, with the major advantage of quite lower pollutants and dusts emission than gasoil [1,2]. Owing to physico-chemical properties similar to those of LPG, its use as multipurpose fuel, for chemicals' production, as refrigerant and hair spray was also devised [1-3]. Actually obtained from syngas *via* methanol, very attractive look the synthesis of DME in one step, in terms of process economy and simplicity [1-3]. In this context, further advantages would come from the exploitation of CO_2-rich syngas streams produced by the cheaper methane partial oxidation, instead of steam reforming, without any further processing [4].

This study outlines some basic results concerning the direct gas-phase synthesis of DME from CO_2-H_2 streams on multifunctional catalytic systems, shedding lights into the factors controlling activity, selectivity and productivity.

Table 1. List of catalysts

| Catalyst | Preparation method | Composition (wt%) | | | Zn_{at}/Cu_{at} | S.A.$_{BET}$ (m^2g^{-1}) |
		CuO	ZnO	Carier		
1.0 CZZ-cp	co-precipitation	28.6	29.5	46.2 [a]	1.0	76
3.0 CZZ-cb	combustion	13.5	38.0	48.3 [a]	3.0	10
1.0 CZZ-cb	combustion	26.0	25.1	48.8 [a]	1.0	24
0.3 CZZ-cb	combustion	39.0	12.4	48.4 [a]	0.3	15
0.2 CZZ-cb	combustion	42.3	9.1	48.1	0.2	14
CZ-cb	combustion	51.4	-	48.2 [a]	0.0	14
G66-A	-	35.0	33.1	30.7 [b]	1.0	105

[a] ZrO_2 as support; b) Al_2O_3 as support

2. EXPERIMENTAL

2.1. Catalysts. Several supported $Cu-Zn/ZrO_2$ catalysts were prepared both by "co-precipitation" [5], and "combustion" routes [6], varying the Cu/Zn atomic ratio in the range 0.2-3.0. A commercial $Cu-Zn/Al_2O_3$ methanol synthesis catalyst (G66-A, *Sud Chemie AG*) was used as reference system.
The list of the studied catalysts is reported in Table 1.

2.2. Temperature Programmed Reduction (TPR) analysis in the range 20–800°C was performed using a flow linear quartz microreactor fed with a 6% H_2/Ar carrier flowing at 60 *stp* cm^3·min^{-1} and a heating rate of 20°C·min^{-1}.

2.3. Metal dispersion (D) of the catalysts was evaluated by N_2O pulse chemisorption tests at 90°C, according to the procedure elsewhere described [7]. Before pulses the catalysts were reduced in flowing H_2 at 300°C for 1h.

2.4. Catalyst testing in the synthesis of methanol was carried out in a PF reactor operating at 180-240°C and 9 atm, using a stoichiometric (1:3) CO_2-H_2 reaction mixture flowing at 40 *stp* ml·min^{-1} and a catalyst sample of 0.5 g pre-reduced in *situ* at 300°C for 1h. The integrated "two-step" synthesis of DME was carried out using a physical mixture of the $Cu-Zn/ZrO_2$ catalyst and a H-ZSM5 zeolite sample (*PENTASIL-ALSI*) in the weight ratio 1:1.

3. RESULTS AND DISCUSSION

TPR measurements were carried out in order to shed lights into the effects of the preparation method and catalyst composition on the reduction kinetics of the various catalysts and settle a proper activation (reduction) protocol. The TPR profiles of the studied catalysts, with respect to the effect of the preparation method and catalyst composition (A) and Zn/Cu ratio (B) respectively, are shown in Figure 1. All the systems display a similar TPR pattern characterised by the presence of a main peak with a resolved maximum centred at 260-290°C.

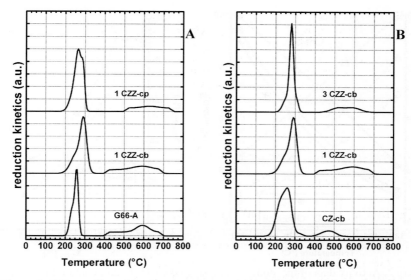

Figure 1. TPR profiles of Cu-Zn catalysts. (A) Effect of the preparation method and support; (B) Effect of the Zn/Cu atomic ratio.

Such a component monitors the reduction of the CuO phase according to a peak area corresponding to a H_2/Cu ratio close to 1 (0.93-1.09) [8]. A variable peak shape, resulting in slightly different values of the peak maximum, is then related to the reduction kinetics of differently sized CuO particles [8]. Moreover, a poorly resolved band of H_2 consumption in the range 400-700°C likely signals an incipient reduction of ZnO patches in interaction with metal Cu particles [8]. The N_2O chemisorption data of the various catalysts are summarised in Table 2 in terms of N_2O uptake, metal dispersion (D_{Cu}), MSA and θ, calculated as the normalised product of MSA_{Cu} multiplied by the difference between total and MSA_{Cu} (e.g., $4\times[MSA_{Cu}/SA_{BET}]\cdot[(SA_{BET}-MSA_{Cu})/SA_{BET}]$), taken as a measure of the interfacial $Cu^0/(ZnO\text{-support})$ area [9]. As the (high) loading of the active phase(s), Cu dispersion varies between 0.8 (CZ-cb) and 6.4% (1.0 CZZ-cp), while the MSA_{Cu} spans in the range 0.5-8.3 $m_{Cu}^2\cdot g_{cat}^{-1}$, respectively.

Table 2. N_2O chemisorption data of the catalysts

Catalyst	N_2O uptake ($\mu mol/g_{cat}$)	D_{Cu} (%)	MSA_{Cu} ($m_{Cu}^2 g_{cat}^{-1}$)	θ
1.0 CZZ-cp	100	6.4	8.3	0.389
3.0 CZZ-cb	23	2.7	1.9	0.616
1.0 CZZ-cb	69	4.2	5.7	0.724
0.3 CZZ-cb	43	1.8	3.6	0.730
0.2 CZZ-cb	38	1.3	3.1	0.690
CZ-cb	6	0.8	0.5	0.138

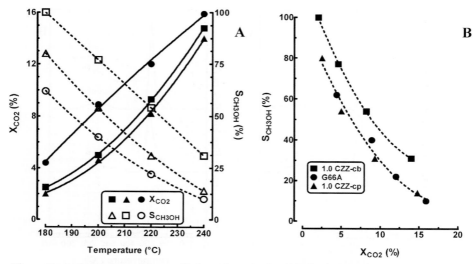

Figure 2. Activity data of the studied catalysts in the CO_2 hydrogenation to methanol in the range 180-240°C. (A) X_{CO2} and S_{CH3OH} vs. temperature of G66A($\bullet$, $\bigcirc$), 1.0 CZZ-cb ($\blacksquare$, $\square$) and 1.0 CZZ-cp ($\blacktriangle$, $\triangle$) catalysts; (B) Influence of CO_2 conversion on S_{CH3OH}.

As a consequence of rather different extents in surface area (Table 1), θ value ranges from 0.14 (CZ-cb) to 0.73 (0.3 CZZ-cb) pointing, thus, that both preparation method and Zn/Cu ratio control the development of the interfacial Cu^0/(ZnO-support) area [4-7]. It can be envisaged that different physico-chemical features could mirror in a different catalytic functionality in the CO_2 hydrogenation to methanol reaction [9]. Indeed, inspecting the trends of CO_2 conversion and CH_3OH selectivity with temperature (Fig. 2A), it emerges that the G66-A catalyst is the most reactive system, whereas ZrO_2-based systems are essentially characterised by a superior functionality for methanol formation. However, crucial is the preparation route as, in spite of a similar activity, the 1.0 CZZ-cb catalyst exhibits quite larger S_{CH3OH} values than counterpart 1.0 CZZ-cp at any T (Fig. 2A). The decisive effects of catalyst composition and preparation method become even clearer by comparing the conversion-selectivity patterns, as shown in Figure 2B. Namely, the fact that a same trend holds for catalytic data of the G66-A and 1.0 CZZ-cp systems, lying quite below that of the 1.0 CZZ-cb one, definitively proves the best performance of the latter system. Then, the effects of catalyst composition and preparation method on the CH_3OH productivity (STY_{CH3OH}) of the studied catalysts are summarised in Figure 3. For systems characterised by a Zn/Cu atomic ratio close to one (Fig. 3A), the 1.0 CZZ-cb features the highest STY_{CH3OH} with a maximum of 700g·kg$_{cat}^{-1}$·h^{-1} at 220°C, staying unchanged at 240°C. The 1.0 CZZ-cp sample attains a maximum of ca. 450g·kg$_{cat}^{-1}$·h^{-1} in the range 200-220°C, while the G66-A system features a higher STY (550g·kg$_{cat}^{-1}$·h^{-1}) at 180°C as a result of its highest specific activity.

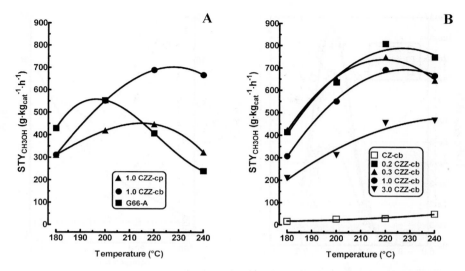

Figure 3. Activity data of the studied catalysts in the CO_2 hydrogenation to methanol in the range 180-240°C. (A) Influence of preparation method and composition on CH_3OH productivity (STY_{CH3OH}) and (B) Influence of the Zn/Cu atomic ratio on STY_{CH3OH}.

Furthermore, crucial is also the influence of the Zn/Cu ratio on the catalyst performance. In fact, for catalysts obtained by the combustion route (Fig. 3B) the system characterised by a lower Zn/Cu ratio (0.2 CZZ-cb) exhibits the largest methanol productivity with a value of ca. $800 g \cdot kg_{cat}^{-1} \cdot h^{-1}$ at 220°C.

Since neither surface area nor chemisoprtion data allow for a satisfactory rationalization of the above data, the turnover frequency of methanol formation (TOF_{CH3OH}) was plotted against θ parameter, in order to ascertain the role of the interfacial metal-oxide area on the rate of methanol formation [9]. In agreement with literature data, all the isotherms in Figure 4 feature a peculiar *volcano-shaped* trend matching with the synergetic role of Cu^0-Cu^I sites in driving the activation of H_2 and CO_2 molecules, respectively [5,7,9]. Thus, enabling a very intimate contact of the various components [6], the combustion route ensures the best compromise in terms of Cu dispersion and Cu-ZnO-ZrO_2 interaction, which results in higher methanol productivity rates [5,9].

Catalytic measurements aimed at finding a suitable acidic system for the gas phase dehydration of methanol into DME (results not reported for the sake of brevity), pointed to the H-ZSM 5 zeolite (Si/Al, 25) the one ensuring the highest (constant) CH_3OH conversion (96%) in the T range investigated [4]. Then, the results of the *integrated* two-step synthesis of DME in the range 180-240°C, on a physical mixture of the 0.3 CZZ-cb and H-ZSM-5 zeolite catalysts, are shown in Figure 5. It is immediately evident that the rate of CO_2 conversion of the 0.3 CZZ-cb catalyst is not appreciably affected by the presence of the acidic system and, as a consequence, no marked effects on the selectivity pattern are recorded.

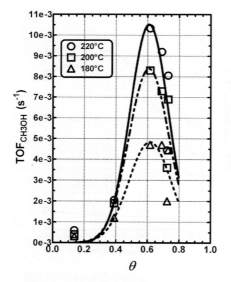

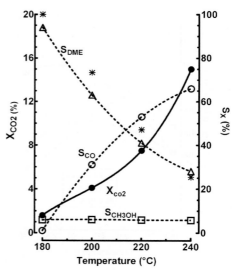

Figure 4. Relationship between TOF of methanol formation and θ at various T (Table 2).

Figure 5. Activity-selectivity data of the "0.3 CZZ-cb/H-ZSM 5" physical mixture in the range 180-240°C. For reference S_{CH3OH} data (∗) in absence of acid catalyst are included.

This can be expected on the basis of the kinetic regime of continuous gas flow catalytic tests, in contrast to batch reaction systems entailing a shift of the thermodynamic equilibrium conversion as a result of the *in situ* methanol condensation [4]. In fact, the H-ZSM 5 catalyst effectively drives an almost full conversion of the methanol produced by the 0.3 CZZ-cb catalyst to DME, as evidenced by a S_{CH3OH} keeping at a rather low level (ca. 2%) in whole range of temperature. Therefore, as the potential feasibility of the gas phase "one-step" synthesis of DME from CO_2-H_2 streams is outlined, further studies aimed at the optimisation of the dual-functionality catalytic system, reactor configuration and reaction conditions (T, P, reaction mixture composition, etc.) are in progress.

References

[1] J.B. Hansen, B. Voss, F. Joensen and I. Sigurdardottir, SAE Paper n° 950063 (1995)

[2] T. Fleisch, A. Basu, M.J. Gradassi and J.G. Masin, Stud. Surf. Sci. Catal., 107 (1997) 117

[3] T. Shikada, Y. Ohno, T. Ogawa, M. Ono, M. Mizuguchi, K. Tomura and K. Fujimoto, Stud. Surf. Sci. Catal., 119 (1998) 123

[4] M. Jia, W. Li, H. Xu, S. Hou and Q. Ge, Appl. Catal. A, 233 (2002) 7

[5] T. Fujitani, M. Saito, Y. Kanai, T. Kakumoto, T. Watanabe, J. Nakamura and T. Uchijima, Catal. Lett., 25 (1994) 271

[6] P. Bera, S.T. Arura, K.C. Patil and M.S. Hegde, J. Catal., 115 (1999) 301

[7] B. Denise, R.P.A. Sneeden, B. Beguin and O. Cherifi, Appl. Catal., 30 (1987) 353

[8] F. Arena, R. Giovenco, T. Torre, A. Venuto and A. Parmaliana, Appl. Catal. B, 45 (2003) 51

[9] T. Kakumoto and T. Watanabe, Catal. Today 36 (1997) 39

Studies in Surface Science and Catalysis, volume 147
X. Bao and Y. Xu (Editors)

A catalyst with high activity and selectivity for the synthesis of C_{2+}-OH: ADM catalyst modified by nickel

D.-B. Li, C. Yang, H.-R. Zhang, W.-H. Li, Y.-H. Sun*, B. Zhong

State Key Laboratory of Coal Conversion, Institute of Coal Chemistry, Chinese Academy of Sciences, Taiyuan 030001, China

ABSTRACT

For ADM catalyst, Ni was an effective promoter for the activity, in particular C_{2+}-OH selectivity. At a Ni/Mo molar ratio of 0.5, the highest ratio of C_{2+}-OH/C_1-OH (8.75) was obtained, which was about 10 times that of ADM catalyst in the absence of Ni (0.87). The modification of Ni showed remarkable structural effects and the great deviation of methanol from the linear A-S-F distribution indicated a different reaction route, which might be caused by the bi-functionality of nickel species, namely, the formation of corresponding precursor of alcohol and the strong ability of CO insertion.

1. INTRODUCTION

The current interest in the higher alcohol synthesis (HAS) is focused on the potential application to produce valuable gasoline enhancers in view of both utilization of resources and environmental protection. Alkali-doped molybdenum sulfide (ADM) catalyst is one of the promising systems due to its high selectivity to alcohol, high activity for water-gas shift reaction and excellent resistance to sulfur poisoning [1-3]. From a practical point of view, it is necessary to further increase the activity and selectivity to C_{2+}-OH. Previous studies showed that the addition of F-T elements (Co, Fe and Ni) could enhance the ADM catalyst's activity and selectivity to C_2+-OH due to their strong abilities to hydrogenate and promote carbon chain growth [2-3] and extensive studies was focused on the promoter of Co [4-5]. However, Ni was neglected in this field for a long time due to its strong methanation, which resulted in hydrocarbon by-products. Recent studies in syngas reactions revealed that Ni-contained catalysts showed some advantages including high activity, strong ability of CO insertion [6-9] and more important, nickel was the only metal which could keep high catalytic activity after sulfur treatment compared with other group VIII metals [10]. As a result, Ni was expected to be an ideal

---

* Corresponding author. Fax: +86-351-4041153; E-mail: yhsun@sxicc.ac.cn

promoter against to sulfur poisoning, which might enhance both activity and selectivity to higher alcohol. In the present paper, a series of nickel modified K_2CO_3/MoS_2 (ADM) catalysts were prepared and their performances for higher alcohol synthesis were investigated.

2. EXPERIMENTAL

2.1. Catalyst preparation

Ni/K_2CO_3/MoS_2 (Ni/ADM) catalysts with various molar ratios of Ni to Mo were prepared by co-precipitation method. The solutions of nickel(II) acetate $Ni(CH_3COO)_2 \cdot 4H_2O$ and $(NH_4)_2MoS_4$ were dropped to a solution of 30% acetic acid at 60℃. After stirred and reacted thoroughly for 1h, the resulted mixture was cooled to room temperature and aged over night. After filtration, the precipitates were dried and mixed with K_2CO_3 to thermal decomposition at 500℃ for 1h under N_2. ADM catalyst was prepared by thermal decomposition of a mixture of $(NH_4)_2MoS_4$ and K_2CO_3 at 500℃ under N_2. $(NH_4)_2MoS_4$ was obtained through the reaction of $(NH_4)_2S$ and $(NH_4)_2Mo_7S_{24}$ as described elsewhere [5].

2.2. Catalyst evaluation and characterization

The catalytic reactions were carried out using a stainless fixed-bed reactor with I.D. of 5mm and the catalyst was 2ml. Premixed CO and H_2 were introduced to the reactor after purifying and preheating. The products were analyzed by gas chromatographs. H_2, CO, CH_4, and CO_2 were separated by using a TDX01 column and were monitored by TCD. The liquid products were separated by using a Porapake-Q column and were monitored by FID. The mass balance was based on carbon-atom and the error of the balance of oxygen and hydrogen was within 5%. The activity of synthesis of alcohols was expressed as space-time-yield (STY), $g_{ROH}/ml_{cat.}\cdot h$ and the alcohol selectivity was based on carbon mole selectivity on a CO_2 free basis. The data were taken at steady state after 100 hr on-stream. X-ray powder diffraction (XRD) patterns of catalysts were recorded on a Rigaku D/Max 2500 diffractometer using Cu Kα radiation as the X-ray source (40KV, 100mA) for 2θ values from 5 to 70℃ at a rate of 3.0 deg/min. Scanning electron microscope (SEM) imagines of the catalysts were obtained using a LEO 438VP SEM (20KV).

3. RESULTS AND DISCUSSIONS

Table 1 gives the results of carbon monoxide hydrogenation over Ni modified ADM catalysts. Clearly, the addition of Ni increases the activity of ADM catalyst. The CO conversion and the STY of alcohol sharply increase at a Ni:Mo ratio of 3:1 The CO conversion and STY of alcohol reach maximum, which are

Table 1
Performance of CO hydrogenation over Ni/ADM catalysts with different Ni/Mo

Ni: Mo[*]	CO conv. (C-atom%)	Selectivity (C-atom -freeCO_2)		C_{2+}-OH/C_1-OH (C-atom)	STY of alcohol. (g/(l · h))	
		Alc.	HC		ROH	C_{2+}-OH
0	17.02	76.21	23.79	0.87	133	55
0.25	20.31	73.44	26.56	2.42	119	77
0.33	34.11	63.28	36.72	6.51	160	130
0.50	28.02	70.78	29.22	8.75	138	113
1.00	11.34	76.90	23.10	1.31	124	60

Reaction condition: T = 573K, P=8.0 MPa, GHSV = 2500h^{-1}, H_2/CO=1.0.
*K/Mo (molar ratio)=0.7.

34.11% and 160 kg/(ml · h), respectively. Moreover, the product distributions are greatly influenced by the addition of nickel. Ni remarkably enhanced the selectivity to C_{2+}-OH. At a Ni:Mo ratio od 1:2, the ratio of C_{2+}-OH/C_1-OH reaches a maximum (8.75), which is about 10 folds compared with that of ADM catalyst without Ni modification (0.87). From the distribution of alcohol products (see Fig.1),

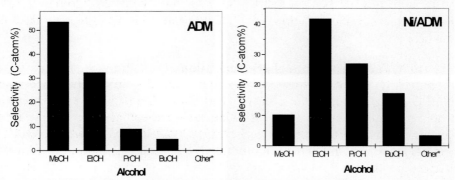

Fig. 1. Alcohol distributions over ADM and Ni/ADM (Ni/Mo=0.5) catalysts

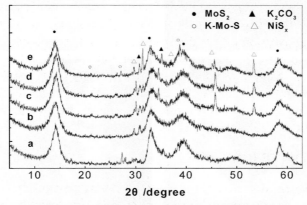

Fig. 2. XRD patterns of the catalysts: (a) ADM; Ni/ADM: (b) Ni/Mo=1/4, (c) Ni/Mo=1/3, (d) Ni/Mo=1/2, (e) Ni/Mo=1

the percentages of C_2 to C_5 alcohols all increases evidently with the percentage of methanol decreases from 53% to 11%.

The XRD patterns of Ni/ADM catalysts clearly identifies the presence of rohmbohedral Ni_3S_2 besides MoS_2, K_2CO_3 and K-Mo-S phases (see Fig. 2), and the peak intensities of Ni species become stronger with the increase of Ni/Mo ratio. Due to the complexity of multi-metal sulfide, it was difficult to fully determine the lattice phases of the samples. Usually, various crystalline or amorphous mixed phases such as NiMoS or KNiMoS phases might be present during the preparation of the sample [11].

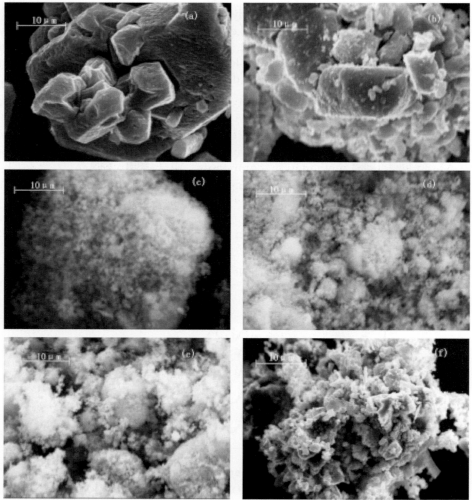

Fig.3. SEM micrographs of the catalysts: (a) MoS_2, (b) ADM, Ni/ADM: (c) Ni/Mo=1/4, (d) Ni/Mo=1/3, (e) Ni/Mo=1/2, (f) Ni/Mo=1.

The MoS$_2$ and ADM samples shows regular particles of MoS$_2$ (see Fig. 3). However, on the surface of MoS$_2$ small clusters appear which might be related to K$_2$CO$_3$ and hexagonal K-Mo-S particles. For Ni/ADM samples, morphological differences among these samples are evident. Ni/ADM sample shows an irregular shape and significant porosity, and the surface of MoS$_2$ appeared to be covered with a great deal of small-conglutinated particles. A progressive surface decoration is observed as a function of the Ni concentration. At a high ratio of Ni to Mo, the conglutination becomes weaker, and larger particles are formed on the surface of MoS$_2$. This indicates that at least two types of Ni-contained species are formed on the surface of the modified catalyst, e.g., highly dispersed and larger particulate Ni on the surface.

The A-S-F plots of alcohol products were depicted as an A-S-F distribution ($Y_n = P^{n-1} (1 - P)$), where Y_n is the mole fraction of the alcohol with a carbon number of n and P is the chain-growth probability [12] (see Fig. 4). Compared with simple ADM catalyst, Ni modified catalysts shows great deviation of methanol from the traditional linear A-S-F distribution, indicating the role of Ni in the promotion the chain propagation of C$_1$ to C$_2$. At the same time, Ni showed a strong CO insertion due to the fact of the higher chain-growth probability of C2 to C4 over Ni/ADM catalyst (0.57) compared with that of ADM catalyst (0.43).

As an excellent methanation component, Ni has strong ability for CO association. According to the reaction network based on classical CO insertion over traditional ADM catalyst [3], high C$_{2+}$-OH products was due to the formation of CHx and the strong ability of CO insertion with the presence of Ni. Studies of metal catalysts for syngas reaction revealed that the structure effects played an important role in the product distribution, and the catalysts with small metal crystallites produced alcohol while larger crystallites gave rise to methane [13]. In the present case, it is believed that for Ni/ADM catalyst, the structure

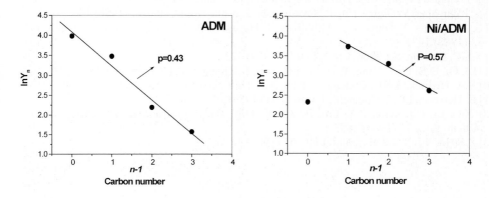

Fig. 4. A-S-F plots of alcohol over ADM and Ni/ADM (Ni/Mo=0.5) catalysts

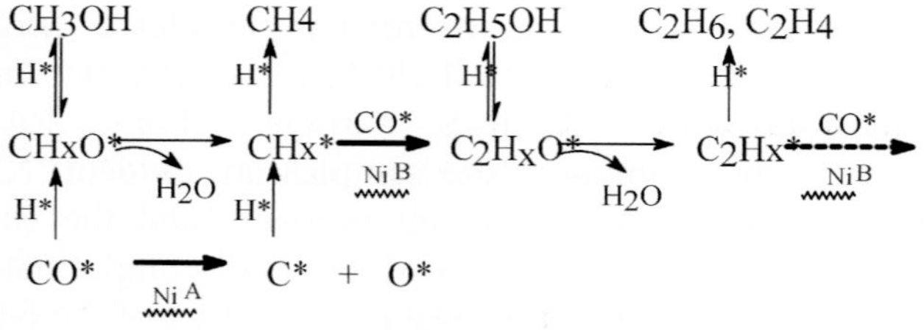
CH3OH
CH4
C2H5OH
C2H6, C2H4
H*
CO*
CHxO*
CHx*
C2HxO*
C2Hx*
H2O
Ni B
CO*
C*
O*
Ni A

Scheme1. The reaction mechanism over Ni/ADM catalyst

effect leads to the bi-functionality of Ni, namely, aggregated particle (Ni^A) and highly dispersed Ni (Ni^B) shows different functions for HAS as depicted in Scheme 1.

4. CONLUSION

The introduction of Ni to ADM catalyst can enhance both the activity and selectivity to C_2+-OH, and the optimum Ni/Mo is observed within the range of about 1/3 to 1/2. Such a promotion was found to be closely related to the structure of nickel in the catalyst, which might be due to the simultaneous presences of aggregated particles and highly-dispersed Ni.

REFERENCES

[1] R.G. Herman, Catal. Today, 55 (2000) 233.

[2] C.B. Murchison, M.M. Conway, R.R. Stevens, G.J. Quarderer, in "Proceedings, 9th International Congress on Catalysis, Cagary, 1988", Vol.2, p.626.

[3] J.G. Santiesteban, C.E. Bogdan, R.G. Herman, K. Klier, in "Proceedings,9th International Congress on Catalysis, Cagary, 1988", Vol.2, p.561.

[4] G.Z. Bian, L. Fan, Y.L. Fu, K. Fujimoto, Ind. Eng. Chem. Res., 37 (1998) 1736.

[5] J. Iranmaaaahboob, D.O. Hill, H. Toghiani, Appl. Catal., 231 (2002) 99.

[6] S. Uchiyama, Y. Ohbayashi, T. Hayasaka, N. Kawata, Appl.Catal. 42 (1988) 143.

[7] S. Uchiyama, Y. Ohbayashi, M.Shibata, T. Hayasaka, N. Kawata, T. Konishi, Appl.Catal., 42 (1988) 143.

[8] S.S.C. Chuang, S.I. Pien, Catal, Lett. 3 (1989) 323.

[9] H.J. Qi, D.B. Li, C.Yang, W.H. Li, Y.H. Sun, B. Zhong, Catal. Comm., 4 (2003) 339.

[10] S.A. Hedrick, S.S.C. Chuang, A. Pant, A.G. Dastidar, Catal. Today, 55 (2000) 247.

[11] S. Harris and R.R. Chianelli, J. Catal., 98 (1986) 17.

[12] X.G. Li, L.J. Feng, Z.Y. Liu, B. Zhong, D.B. Dadyburjor and E.L. Kugler. Ind. Eng. Chem. Res., 37 (1998) 3853.

[13] Z.R. Li, Y.L. Fu et al., Catal. Lett., 65 (1989) 43.

Studies in Surface Science and Catalysis, volume 147
X. Bao and Y. Xu (Editors)
397

On the catalytic mechanism of Cu/ZnO/Al$_2$O$_3$ for CO$_x$ hydrogenation

Qi Sun

Süd-Chemie Inc. P. O. Box 32370, Louisville, KY 40232, USA,

ABSTRACT

The structural characterization of the catalysts shows that isomorphous substitution took place between copper and zinc oxide in the catalyst. XPS analysis indicates that copper still appears as metallic copper and zinc as ZnO during CO$_x$ hydrogenation. The activity of Cu/ZnO/Al$_2$O$_3$ catalyst for methanol synthesis from CO$_x$ hydrogenation relies on both metallic copper surface area and a strong interaction (synergy) between copper and zinc oxide determine. The kinetic and *in situ* IR results reveal that methanol is formed directly from CO$_2$ hydrogenation. CO$_2$ is the primary carbon source of methanol formation. A mechanism scheme of methanol synthesis is proposed.

1. INTRODUCTION

Cu/ZnO catalyst, as an old catalyst family used in many important processes, has been attracting increasing interest for more than half a century. Recent fuel cell technology stimulates extra attention on this catalyst system since one of the production routes of the critical H$_2$ production in fuel cell involves water-gas-shift (WGS) on Cu/ZnO catalyst. The development of high activity Cu/ZnO catalysts and the investigations regarding their structures as well as catalytic properties are still the main focus because the catalytic performance directly relates to the synthesis route and the catalyst structure. However, there has long been a controversy over some important questions, such as: the nature of the active centers, the role of zinc in the catalytic process, the primary carbon source and the role of carbon dioxide in methanol synthesis as well as it's the catalytic reaction mechanism. Understanding the nature of active centers of Cu/ZnO/Al$_2$O$_3$ catalyst and catalytic mechanism of CO/CO$_2$ hydrogenation enables us effectively *design* structurally novel high performance Cu-based catalysts and *improve* the technology of CO$_2$ utilization and hydrogen production via WGS. In the present work, a systematic study on catalyst structure, active centers and catalytic mechanism for CO$_2$ or CO hydrogenation were carried out. Based on various *in situ* characterization and kinetic analysis, an active center and catalytic mechanism model were developed.

2. EXPERIMENTAL

The structure of a series of $Cu/ZnO/Al_2O_3$ catalysts with different compositions, prepared by improved co-precipitation methods, were studied utilizing XRD (X-ray diffraction), TGA (thermogravimetric analysis), TEM (transmission electron microscopy), N_2O-chemiadsorption as well as EXAFS (Extended X-ray absorption fine structure) technologies. The activity and kinetic measurements of $Cu/ZnO/Al_2O_3$ catalysts were carried out in a pressurized (2.0~5.0MPa) continuous tubular flow fixed-bed micro-reactor. The reactant composition is $CO_2/H_2=1/3$ and $CO/CO_2/H_2=20.5/5.0/74.5$ for CO_2 and CO/CO_2 hydrogenation, respectively. The surface adsorption species and methanol formation were further investigated by means of *in situ* FT-IR spectroscopy (Perkin-Elmer 2000) under real reaction conditions (2.0~5.0MPa and 100~300°C). The chemical state of copper and zinc as well as nature of active centers was characterized on a PHI-500C ESCA system (Perkin-Elmer) with Al Kα radiation (hv=1486.6eV).

3. RESULTS AND DISCUSSION

The TGA, XRD and EXAFS results indicate that most of the copper in $Cu/ZnO/Al_2O_3$ catalysts prepared with the novel oxalate gel co-precipitation method [1] is present as an *amorphous-like* structure and some of the copper is incorporated into zinc oxide lattice by isomorphous substitution. The catalytic activity for CO/CO_2 hydrogenation increases with the increase of Cu/Zn ratio

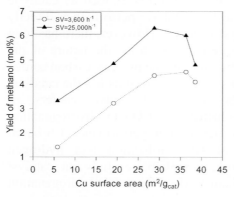
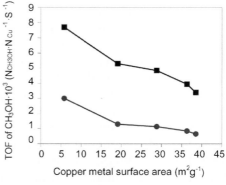

Fig. 1. The relationship between the yield of methanol and Cu specific surface area at 240°C, 2.0Mpa and space velocity of (▲)3,600h⁻¹ and (O) 25,000 h⁻¹.

Fig. 2. The correlation between the TOF of methanol formation and Cu specific surface area at 240°C, 2.0Mpa and space velocity of (●)3,600h⁻¹ and (■) 25,000 h⁻¹.

until it reaches a maximum at a Cu/Zn ratio of 1~2. The copper surface area has significant effect on the activity, but it is not a linear relationship as suggested by Chinchen et al [2]. As shown in Fig.1, the catalytic activity increases with the increase of the metallic copper surface area and showed a linear relationship in the range of 10~45% copper content (corresponding to Cu surface of 5.8~28.8m^2/g). The catalytic activity at 60% copper content (Copper surface area of 36.6m^2/g) deviates from the linear relationship and at 80% copper content the activity decreases, even though the copper surface area is the largest (38.5m^2/g). This indicates that the catalytic activity of the Cu/ZnO/Al$_2$O$_3$ catalysts depends on both the metallic copper surface and the interaction between copper and zinc. This conclusion is further clarified by the calculation of TOF (Turnover Frequency) of methanol formation over various catalysts with different copper surface areas. As shown in Fig.2, it is important to note that the TOF of methanol formation decreases with the increase of metallic copper surface area. According to Boudart's theory [3], if the catalytic activity of the catalyst only depends on the metallic copper surface area, then the plot of TOF versus metallic copper surface area would be a horizontal line. The results in Fig.2 provide us strong evidence that the interaction between copper and zinc oxide promotes the catalytic properties of the catalyst for methanol synthesis. It is demonstrated by XRD and TGA results[1,4] that when the copper content is lower, a considerable fraction of copper species are incorporated substitutionally into the ZnO lattice, resulting in a strong Cu-Zn interaction and higher TOF of methanol formation as shown in Fig.2. In this kind of structure, the Cu^{2+} in the ZnO lattice has a slightly distorted tetrahedral symmetry, which is different from the square-planar symmetry of the CuO lattice. This distortion leads to the

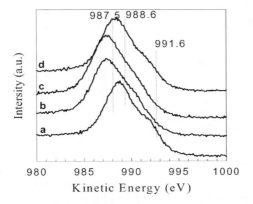

Fig. 3. XAES spectra of Zn LMM for different samples; (a) calcined, (b) reduced with H$_2$/Ar at 240°C for 4 h, (c) reacted with CO$_2$/H$_2$ flow at 0.1Mpa for 2 h, (d) exposed to pure CO$_2$ for 30min.

changes of coordination environment and electronic properties of Cu and Zn atoms in the catalyst.

in situ XPS study indicates that copper still appears as metallic copper and zinc as ZnO during reduction or reaction. Metallic copper can be partially oxidized to $Cu^{\delta+}$ species by pure CO_2. Although there is no obvious change in valence state of copper in reduction and reaction streams, a strong interaction (synergy) is observed by XAES (X-ray excited Auger electron spectroscopy) and chemical state analysis. As shown in Fig. 3, upon reduction with H_2/Ar at 240°C and working under the CO_2/H_2 atmosphere, the Auger KE (kinetic energy) of zinc shifts to lower energy and the shoulder peak at 991.6 eV disappears. However, After treatment with CO_2, the main Auger peak and the shoulder peak are restored. These reversible changes in Auger KE and shoulder peak vigorously demonstrate the *isomorphously substituted* structure and the strong interaction between copper and zinc. The formation of the isomorphously substituted structure could prevent the formation of long-range ordered zinc and copper phases. When reduced with H_2/Ar, both the copper species on the surface and in the bulk of $CuO/ZnO/Al_2O_3$ catalyst are reduced to metallic species. During this process, some oxygen atoms, which originally coordinated with the copper and zinc atoms together in *isomorphously substituted* structure, may also be removed from the ZnO lattice. As a result, the anionic (oxygen) vacancies produced in ZnO lattice would lead to the formation of ZnO micro-crystallites with serious lattice defects and structurally disordered features. The presence of anionic (oxygen) vacancies would lead to a remarkable weakening of non-local screening effect. As a result, the Auger KE of zinc shifts toward a lower level.

Fig 4 shows methanol formation rate for $CO/H_2(1/2)$, $CO_2/H_2(1/3)$ and $CO/CO_2/H_2$ (20.5/5/74.5) over the $Cu/ZnO/Al_2O_3$ catalyst at the pressure of

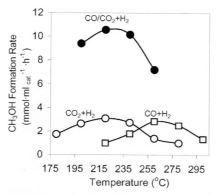

Fig. 4. Methanol formation rate for CO_2/H_2, CO/H_2 and $CO/CO_2/H_2$ over the $Cu/ZnO/Al_2O_3$ catalyst, P=2.0MPa, space velocity=4,500h^{-1}.

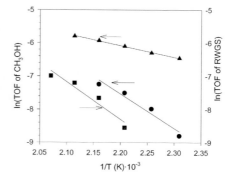

Fig. 5. Arrhenius plot ln(TOF) for methanol synthesis and RWGS reactions versus the inverse reaction temperature (1/T) over ultrafine $Cu/ZnO/Al_2O_3$ catalyst for CO_2+H_2 (●, ■) and CO/CO_2+H_2 (▲).

2.0Mpa and gas space velocity of 4,500h^{-1}. The activity measurements indicate that methanol formation rates for the individual CO and CO_2 hydrogenation are much slower than those for hydrogenation of CO/CO_2. Adding a small amount of CO_2 in $CO+H_2$ feed gas greatly promotes methanol yield and selectivity, as shown in Fig.4. Although CO_2 promotes methanol formation, the fraction of CO_2 in the outlet gas is nearly constant under all reaction conditions and the ratio of CO to CO_2 under higher space velocity conditions is higher than that under lower space velocity conditions, and this is true at various reaction temperatures (The results are not shown). An apparent activation energy could be determined from the slopes in the Fig.5, which is a plot of reaction rate (ln(TOF)) versus 1/T for CO_2+H_2 and CO/CO_2+H_2. For CO_2+H_2, the activation energy for methanol synthesis and RWGS is 20.7 kcal/mol and 22.73 kcal/mol,

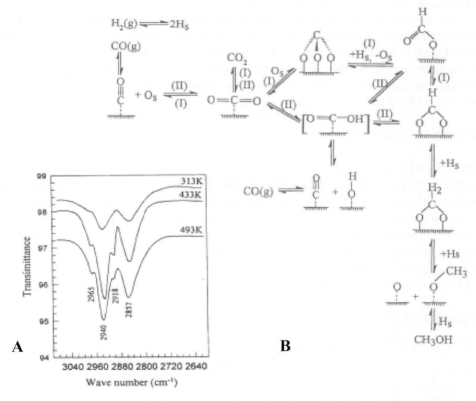

Fig. 6. (A) Formation and variation of methoxy group and methanol during temperature programmed reaction process at 2.0MPa for methanol synthesis for CO_2+H_2, b-HCOO$_s$: 2857 and 2940cm^{-1}, CH_3O_s and CH_3OH: 2918 and 2965cm^{-1}; (B) Scheme for the mechanism of methanol synthesis and RGWS/GWS reactions from CO_2/H_2 and $CO/CO_2/H_2$ over ultrafine $Cu/ZnO/Al_2O_3$ catalyst.

respectively. After the addition of CO into CO_2/H_2 feed gas, however, the activation energy for methanol synthesis is decreased to 6.52kcal/mol, although there is no significant change in the activation energy for RWGS reaction. It is clearly demonstrated that the addition of CO into CO_2/H_2 feed could lead to the decrease the activation energy for methanol synthesis. Based on above results, it is proposed that methanol is directly formed from CO_2 hydrogenation. CO_2 is the primary carbon source of methanol synthesis. The surface water species and the WGS (water gas shift) reaction play crucial roles in conversion of CO to CO_2, which continually provides active $CO_{2(s)}$ species for hydrogenation. In addition it is found in Fig.4 that the optimum reaction temperature for CO/CO_2 hydrogenation is consistent with that of CO_2+H_2, which further demonstrates that CO_2 hydrogenation is the rate determining reaction and methanol is formed from CO_2 hydrogenation. As a result, the fraction of CO_2 detected in the outlet gases keeps at a similar level and is higher than that in the feed gases.

In situ FT-IR studies directly identified that bidentate formate (b-HCOO⁻$_s$) is the necessary intermediate species for methanol synthesis and hydrogenation of b-HCOO⁻$_s$ is the rate-limiting step for methanol synthesis (Fig.6A), which is in agreement with the conclusion obtained from the kinetic testing. A mechanism scheme of methanol synthesis and the WGS/RWGS reaction was proposed (Fig.6B).

4. CONCLUSION

Both metallic copper surface area and the existence of a strong (synergetic) interaction between amorphous-like copper and zinc oxide lattice determine the activity of $Cu/ZnO/Al_2O_3$ catalysts for CO/CO_2 hydrogenation. Under the real reaction condition copper is still present as metallic copper, while pure CO_2 can partially oxidize copper to $Cu^{\delta+}$. Methanol formation is directly from CO_2 hydrogenation via a necessary intermediate species of b-HCOO⁻$_s$. The surface water species and the RWGS reaction play important roles in the conversion of CO to active $CO_{2(s)}$, which is the direct carbon source of methanol formation.

REFERENCES

[1] Q. Sun Y.L. Zhang and J. F. Deng, J. Catal., 167 (1997) 92
[2] G.C. Chinchen, K.C. Waugh, D.A. Whan, Appl. Catal., 25 (1986) 101
[3] M. Boudart and G. Djéga-Mariadassou, in "Kinetics of Heterogeneous Catalytic Reactions", Princeton University Press, 1984
[4] J.F. Deng, Q. Sun, Y.L. Zhang, Appl. Catal. A, 139 (1996) 75

Studies in Surface Science and Catalysis, volume 147
X. Bao and Y. Xu (Editors)

Solid superacids for halide-free carbonylation of dimethyl ether to methyl acetate

G.G. Volkova, L.M. Plyasova, L.N. Shkuratova, A.A. Budneva, E.A. Paukshtis, M.N. Timofeeva, V.A. Likholobov

Boreskov Institute of Catalysis, Pr. Akademika Lavrentieva 5, Novosibirsk, 630090, Russia

ABSTRACT

The production of methyl acetate directly from dimethyl ether was performed by halide-free carbonylation. The solid superacid catalysts with high density of Brønsted acid sites $Rh/Cs_xH_{3-x}PW_{12}O_{40}$ were compared to $Rh/SO_4/ZrO_2$ and $Rh/WO_x/ZrO_2$. The correlation between the nature of acid sites and activity and selectivity of the catalysts in halide-free carbonylation of DME was found.

1. INTRODUCTION

Acetic acid is an important industrial chemical which is manufactured on a large scale. The main route to acetic acid is through the carbonylation of methanol in the Monsanto process which uses a homogeneous rhodium catalyst and halide promoter [1]. Main disadvantages of this process are: a) halides are highly corrosive and are poisons for many types of catalysts, b) it is difficult to separate the products and catalyst. These problems may be overcome by developing heterogeneous catalyst that can operate effectively without halide promoter.

The first step in methanol carbonylation is activation of C-I bond (E= 240 kJ mol^{-1}) in methyl iodide. In order to exclude CH_3I it is necessary to activate C-O bond (E=350 kJ mol^{-1}) in methanol or in dimethyl ether (DME). DME is more favorable for carbonylation than methanol because it can be produced from syn-gas more effectively [2-4].

One of the main properties of solid superacids is low-temperature activation of C-C bond in alkanes that results in skeletal isomerisation of butane at room temperature [5-9]. The energy of the C-C bond (344 kJ mol^{-1}) closely matches the energy of the C-O bond (350 kJ mol^{-1}) that is why we examine solid superacids SO_4/ZrO_2, WO_x/ZrO_2 and $Cs_xH_{3-x}PW_{12}O_{40}$ in the iodide-free carbonylation of the DME to methyl acetate (MA):

$$CH_3\text{-}O\text{-}CH_3 + CO \rightarrow CH_3\text{-}CO\text{-}O\text{-}CH_3$$

Acidic cesium salts of 12-tangstophosphoric acid promoted with rhodium revealed high activity in this reaction [10-11]. Here we present performance and physical characterization of $Rh/SO_4/ZrO_2$, $Rh/WO_x/ZrO_2$ catalysts in comparison with heteropolycompounds.

2. EXPERIMENTAL

The $RhCsPW_{12}$ catalysts were synthesized as described in [11]. $Rh/SO_4/ZrO_2$, $Rh/WO_x/ZrO_2$ samples were prepared from sulfated zirconium hydroxide XZO1077/01(1), XZO999/01(2) and tungstate doped zirconium hydroxide XZO861/02 from MEL by calcination at 650-800°C and impregnation with 1%Rh using solution of rhodium chloride. Dried samples were finally calcinated at 500°C for 3 hours.

Catalysts were evaluated in flow reactor with gas analysis on line. Reaction conditions: CO/DME ratio =10/1, GHSV= 3000 h^{-1}, pressure 10 bar, temperature 473 -523K. Prior to reaction, samples were reduced at 473K in stream of $2\%H_2$-He for 2 hours. FTIR spectroscopy was applied to determine the acidity of the catalysts by monitoring the adsorption of pyridine (adsorbance band at 1540 cm^{-1} for Brønsted and 1450 cm^{-1} for Lewis acid sites). The samples were pressed into self supporting wafers containing 30-50 mg/cm^2 of material. Samples were degassed in IR cell at 250°C and then treated with pyridine at 150°C for 15 min. Excess pyridine was desorbed by evacuating of the samples at 150°C for 30 min. All the spectra were recorded at room temperature in the range of 400-6000 cm^{-1} with 4 cm^{-1} resolution, using Shimadzu FTIR-8300 spectrometer.

The number of sites of each type was calculated by fitting the data to equation: $N=A/\rho A_o$, where A (cm^{-1}) is the integrated absorbance of IR bands due to pyridine adsorbed at Brønsted or Lewis sites, ρ ($g\ cm^{-2}$) is the mass of material normalized on pellet surface, A_o ($cm\ mol^{-1}$) is the molar integrated absorption coefficient equal to 3.0 at Brønsted and to 3.5 at Lewis sites [12]. A_o was evaluated separately for Brønsted and Lewis sites by titration of pyridine on model solids possessing only Brønsted (Heteropolyacids) or Lewis (NaX) acidity at certain temperature when pyridine absorption on the walls of cell was excluded. Detailed procedure for measurement of A_o is described in [12,13].

Strength of proton sites was characterised by proton affinity (PA) calculated using the equation: PA = $[lg(3400-v_{NH})]/0.0023-51$, where PA (kJ mol^{-1}) is the energy of proton releasing, 3400 (cm^{-1}) is the wavenumber of the band of the undisturbed N-H bond of the pyridinium ion, and v_{NH} (cm^{-1}) is the wavenumber of the center of gravity of the band of the stretching vibration of the pyridinium ion which is determined from the contour of the v_{NH} line in the region of 3400-1800 cm^{-1} [12].

3. RESULTS AND DISCUSSION

IR spectra of adsorbed pyridine on acidic cesium salts of $H_3PW_{12}O_{40}$ (Fig. 1) and on sulfated and tungstated zirconia (Fig. 2) revealed bands characteristic of both Lewis (1450 cm^{-1}) and Brønsted (1540 cm^{-1}) form of adsorption. Table 1 collects the acidity data calculated from FTIR spectra of pyridine adsorption. A three-fourfold variation in Brønsted and Lewis sites is apparent across the range of samples. The maximum of Brønsted acid sites (139-118 μmol g^{-1}) was detected on $Cs_{1-2}PW$ samples, maximum of Lewis acid sites (105-110 μmol g^{-1}) on sulfated zirconia. Proton affinities provide relative scale for Brønsted acid strength. CsPW catalysts are more acidic than SO_4/ZrO_2, WO_x/ZrO_2.

Comparison of the acid strength of Cs1-1.5 samples (PA = 1120 kJ mol^{-1}) with acidity of HPW_{12} calculated by Bardin et al. (PA =1088 kJ mol^{-1}) [14] and detected by Maksimov et al. (PA = 1070 kJ mol^{-1}) [15] has shown that the parent acid is stronger than $Cs_xH_{3-x}PW_{12}O_{40}$. Total Lewis acidity of SO_4/ZrO_2 (1), (2) compares very well with those determined by FTIR spectroscopy of pyridine or CO adsorption (100-121μmol g^{-1}) in [16-18] for sulfated zirconia obtained by various synthesis routs.

The concentration of Brønsted acid sites on WO_x/ZrO_2 catalyst presented here (38 μmol g^{-1}) is somewhat higher than those obtained (34-35 μmol g^{-1}) by W. Hua and J. Sommer [19]. They used method based on the acid site titration performed by H/D exchange. However, these are the total Brønsted acid sites. Using 2,6-dimethylpyridine to selectively poison the Brønsted acid sites, Santiesteban et al. [20] reported that the strong Brønsted acid sites present over WOx/ZrO$_2$ catalyst were 2 μmol g^{-1}. C.D. Baertsch et al. [21] noted that Brønsted and Lewis acid site densities for WO_x/ZrO_2 catalysts are equal (0.04 per W-atom), we also obtained the similar values (38 and 37μmol g^{-1}) for both types of acid sites.

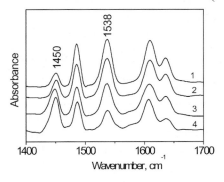

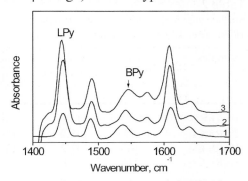

Fig. 1. FTIR spectra of pyridine adsorption on $Rh/Cs_xH_{3-x}PW_{12}O_{40}$ catalysts: 1-Cs1; 2- Cs1.5; 3- Cs2; 4- Cs2.5.

Fig. 2. FTIR spectra of pyridine adsorption on $Rh/WO_x/ZrO_2$ (1) and on $Rh/SO_4/ZrO_2$ (2,3) catalysts:

Table 1
Acidity of the catalysts from FTIR spectra of pyridine adsorption

| Samples | Brønsted acid sites | | Lewis acid sites |
	N, μmol g^{-1}	PA, kJ mol^{-1}	N, μmol g^{-1}
Rh/Cs$_1$H$_2$PW$_{12}$O$_{40}$	139	1120	25
Rh/Cs$_{1.5}$H$_{1.5}$PW$_{12}$O$_{40}$	128	1120	32
Rh/Cs$_2$HPW$_{12}$O$_{40}$	118	1150	57
Rh/Cs$_{2.5}$H$_{0.5}$PW$_{12}$O$_{40}$	48	1150	63
Rh/SO$_4$/ZrO$_2$ (1)	38	1160	105
Rh/SO$_4$/ZrO$_2$ (2)	70	1160	110
Rh/WO$_x$/ZrO$_2$	38	1180	37

Results of the DME carbonylation are presented in Table 2 and Fig.3. The best catalyst 1%Rh/Cs$_{1.5}$H$_{1.5}$PW$_{12}$O$_{40}$ revealed activity to MA=180 g l^{-1}h^{-1}, that is one order of magnitude higher than the activity of the rhodium salts of the same acid supported on silica in the iodide-free carbonylation of the DME to MA [22]. Activity dramatically changes with cesium content. Cs1.5 and Cs2 are the most active catalysts, activity of Cs1 and C2.5 is 3-5 times less. Activity and selectivity of the Cs$_x$H$_{3-x}$PW$_{12}$O$_{40}$ catalysts is higher than those of SO$_4$/ZrO$_2$ and WO$_x$/ZrO$_2$. DME conversion on sulfated and tungstated zirconia is also more than two times lower comparing with heteropolycompounds. By-products mainly consist of methane (95-97%) and ethane (up to 5%).

Low activity and poor selectivity of sulfated and tungstated zirconia in DME carbonylation comparing with alkanes isomerization may be connected with 1) lower concentration and strength of the Brønsted acid sites and 2) the difference of CO and H$_2$ chemisorption on its surface. Barton et al. reported [23] that irreversible chemisorption uptake of CO on Pt/WO$_x$/ZrO$_2$ is near two times less than H$_2$ uptakes.

Table 2
Surface area and performance of catalysts in halide-free DME carbonylation at 473K and 10 bar after 120 min on stream

| Samples | S BET m^{2}g^{-1} | DME conversion % | Rate of methyl acetate formation | | | Selectivity % |
			g l^{-1}h^{-1}	10^{-8} mol g^{-1}s^{-1}	10^{-8} mol m^{-2}s^{-1}	
1%Rh/Cs$_1$	37	11.5	60	15	0.40	91
1%Rh/Cs$_{1.5}$	57	33.0	180	45	0.78	94
1%Rh/Cs$_2$	103	31.2	170	43	0.42	95
1%Rh/Cs$_{2.5}$	216	6.4	35	9	0.042	96
1%Rh/SO$_4$(1)	115	14.7	8	3	0.026	10
1%Rh/SO$_4$(2)	127	14.7	12	5	0.039	15
1%Rh/WO$_x$	75	5.5	15	6	0.08	50

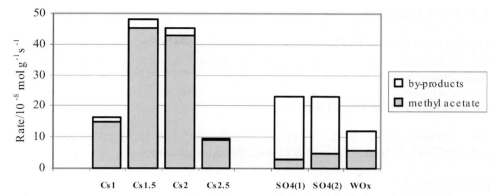

Fig. 3. The performance of $Rh/Cs_xH_{3-x}PW_{12}O_{40}$, $Rh/SO_4/ZrO_2$ and $Rh/WO_x/ZrO2$ catalysts in DME halide-free carbonylation

Relationship between selectivity to MA of the best and the worst catalysts and number of Brønsted and Lewis acid sites (Fig.4) shows that Brønsted acid sites are effective for activation of C-O bond in DME that result in high (96%) selectivity to MA. Lewis acid sites induced the cracking of DME molecule that led to the by-products formation and poor (10%) selectivity.

Correlation between catalytic performance in DME carbonylation and isomerization has been observed only for $CsPW_{12}$ catalysts. Direct comparison of the rates of DME carbonylation and alkanes isomerization [24-25] over $CsPW_{12}$ (Fig.5) has shown similar results: the catalysts with the lowest and highest Cs content are relatively inactive in both reactions. That may be attributed to at least two factors: surface area and Brønsted acidity, Cs1 catalyst has relatively low surface area, Cs 2.5 has lower density of Brønsted acid sites.

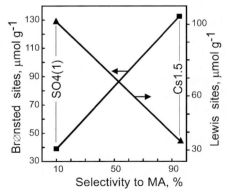

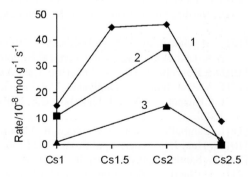

Fig. 4. Relationship between selectivity to MA of $Rh/Cs_{1.5}H_{1.5}PW_{12}O_{40}$ and $Rh/SO_4/ZrO_2$ (1) catalysts and number of Brønsted (squares) and Lewis (triangles) acid sites.

Fig. 5. Direct comparison of the rates of (1) DME carbonylation, (2) pentane isomerization [23], (3) butane isomerizaton [24] over Cs_xPW_{12} catalysts versus cesium content.

4. CONCLUSION

The $Rh/Cs_xH_{3-x}PW_{12}O_{40}$ ($1.5 \leq x \leq 2$) catalysts are the best catalysts for halide-free DME carbonylation to methyl acetate among other solid superacids due to high density of the strong Brønsted acid sites.

ACKNOWLEDGEMENTS

The authors thank MEL Chemicals Ltd. for providing the commercial XZO1077/01, XZO999/01 and XZO861/02 zirconia samples.

REFERENCES

[1] M.J. Howard, M.D. Jones, M.S. Roberts and S.A. Taylor, Catal.Today, 18 (1993) 325.
[2] T. Shikada, Y. Ohno, T. Ogawa, M. Ono, M. Mizuguchi, K. Tomura, K. Fugimoto, Stud. Sur. Sci. Catal., 119 (1998) 515.
[3] T.H. Fleisch, R.A. Sills, M.D. Briscoe, J. Natural Gas Chem., 11 (2002) 1.
[4] International DME Association Website: www.aboutdme.org
[5] K. Arata, H. Matsuhashi, M. Hino, H. Nakamura, Catal. Today, 81 (2003) 17.
[6] T. Lei, J.S. Xu, W.M. Hua, Z. Gao, Appl. Catal. A: General 192 (2000) 181
[7] D.G. Barton, S.L. Soled and E. Eglesia, Topics in Catalysis, 6 (1998) 87.
[8] T. Okuhara, N. Mizuno and M. Misono, Appl. Catal., 222 (2001) 63.
[9] I.V. Kozhevnikov, Chem.Rev., 98 (1998) 171.
[10] G.G. Volkova, L.M. Plyasova, A.N. Salanov, G.N. Kustova, T.M. Yurieva, V.A. Likholobov, Catal.Lett. 80 (2002) 175.
[11] G.G. Volkova, L.E. Pindyurina, L.S. Egorova, T.M. Yurieva, V.A. Likholobov, RP 2170724 (2001).
[12] E.A. Paukshis, IR spectroscopy for heterogeneous acid-base catalysis. Nauka, Novosibirsk, 1992 (in Russian).
[13] T.R. Hughes, H.M.White, J.Phys.Chem. 71 (1967) 2192.
[14] B. Bardin, Sh. V. Bordawekar, M. Neurock, and R.J. Davis, J.Phys. Chem. B, 102 (1998) 10817.
[15] G.M. Maksimov, E.A. Paukshis, A.A. Budneva, R.I. Maksimovskay, and B.A. Likholobov, Russian Chemical Bulletin. International Edition, 50 (2001) 587.
[16] D.J. Rosenberg and J.A. Anderson, Catal.Lett., 83 (2002) 59.
[17] E.A. Paukshtis, V.K. Duplyakin, V.P. Finevich, A.V. Lavrenov, V.L. Kirillov, L.I. Kuznetsova, V.A. Likholobov, B.S. Bal'zhinimaev, Stud. Sur. Sci. Catal., 130 (2000) 2543.
[18] L. Zanibelli, .A. Carati, C. Flego, R. Millini, Stud. Sur. Sci. Catal., 143 (2002) 813.
[19] W. Hua, J. Sommer, Appl.Catal., 232 (2002) 129.
[20] J.G. Santiesteban, J.C. Vartuli, S. Han, R. D. Bastian, C.D. Chang, J.Catal., 168 (1997) 431.
[21] C.D. Baertsch, K.T. Komala, Y-H. Chua, E.Iglesia, J.Catal., 205 (2002) 44.
[22] W.R.Wegman, European Patent 0353 722 (1989).
[23] D.G. Barton, S.L. Soled, G.D. Meitzner, G.A. Fuentes and E. Iglesia, J. Catal., 181 (1999) 57.
[24] B.B. Bardin, R.J. Davis, Topics in Catalysis, 6 (1998) 77.
[25] N. Essayem, G. Coudurier, M. Fournier, J.C. Vedrine, Catal.Lett.,34 (1995) 223.

Studies in Surface Science and Catalysis, volume 147
X. Bao and Y. Xu (Editors)
409

A new low-temperature methanol synthesis process from low-grade syngas

P. Reubroycharoen[a], T. Yamagami[a], Y. Yoneyama[a], M. Ito[b], T. Vitidsant[c], N. Tsubaki[a*]

[a]Dept. of Material System & Life Science, School of Engineering, Toyama University, Gofuku, Toyama, 930-8555, Japan

[b]Dept. of Applied Chemistry, The University of Tokyo, Tokyo, 113-8656, Japan

[c]Dept. of Chemical Technology, Chulalongkorn University, 10330, Thailand

ABSTRACT

The low-temperature methanol synthesis from syngas containing CO_2 with Cu/ZnO catalysts was carried out under 443 K and 50 bar in a semi-batch flow-type reactor with the aid of alcohol. The high one-pass total carbon conversion (61%) and methanol selectivity (>98%) were achieved without the deactivation of catalyst. The low-temperature methanol synthesis is feasibly promising for various kinds of syngas, therefore, it is facilely applied to an industrial process.

1. INTRODUCTION

Methanol, an alternative fuel, is being produced by 30 million ton per year around the world from $CO/CO_2/H_2$. It is now commercially produced by ICI process under high temperature (523-573 K) and pressure (50-100 bar), using Cu/ZnO based catalyst developed by ICI Co.. However, the efficiency of methanol synthesis, which is an extremely exothermic reaction, is severely limited by thermodynamics [1, 2]. For example at 573K and 50 bar, theoretical one - pass CO conversion is approximately 20% [3]. According to the low conversion, production cost increases as a result of the recycling of the unreacted gas. The unreacted gas is no longer recycled if one-pass conversion is high enough. In order to enhance the conversion, methanol synthesis at low-temperature has been developed by utilizing the intrinsic thermodynamic advantage. Consequently, the production cost is reduced. Moreover, syngas containing N_2 obtained from the reforming of methane using air instead of pure

O_2 can be used to reduce the production cost if one-pass conversion is high [4].

Brookhaven National Laboratory (BNL) of USA developed a low-temperature method at 373-403 K and 10-50 bar, using very strong base catalyst (mixture of NaH, alcohol and acetate) and pure syngas ($CO+H_2$). However, the trace amount of CO_2 and H_2O in the pure syngas or reaction system will deactivate the strongly basic homogeneous catalyst soon, which is the remarkable drawback of this method [5, 6]. This implies high cost from the complete purification of the syngas as well as the re-generation process of the deactivated catalyst, making the commercialization of this method impossible.

Methanol synthesis in liquid phase from pure syngas via the formation of methyl formate has been widely studied, where carbonylation of methanol and hydrogenation of methyl formate were considered as two reaction steps [7, 8]. Similar to BNL method, CO_2 and H_2O acted as poisons to the alkoxide catalyst.

A new low-temperature methanol synthesis from CO_2/H_2 on a Cu-based oxide catalyst using ethanol as a kind of catalytic solvent was developed, by which methanol was produced at 443 K and 30 bar [9]. This process consisted of three steps: 1) formic acid synthesis from CO_2 and H_2; 2) esterification of formic acid by ethanol to ethyl formate; and 3) hydrogenation of ethyl formate to methanol and ethanol. Considering the fact that water-gas shift reaction is easily conducted on Cu/ZnO catalyst [10, 11], a new route of methanol synthesis from CO/H_2 containing CO_2 is proposed, as a more practical way of methanol synthesis. It consists of the following fundamental steps:

$$CO + H_2O = CO_2 + H_2 \qquad (1)$$

$$CO_2 + H_2 + ROH = HCOOR + H_2O \qquad (2)$$

$$HCOOR + 2H_2 = CH_3OH + ROH \qquad (3)$$

$$\text{---}$$

$$CO + 2H_2 = CH_3OH \qquad (4)$$

As formic acid was not detected in the products, we suggested the reaction path as steps (2). The influences from catalyst composition, reaction time, reaction temperature and pressure, as well as reactant gas composition were investigated.

2. EXPERIMANTAL

The catalyst was prepared by the conventional co-precipitation method of an aqueous solution of copper, zinc nitrates and an aqueous solution of sodium carbonate [12].

Table 1
Reaction performances of various catalysts on methanol synthesis in batch reactor.

Catalysts	Conversion (%)			Yield (%)	
	CO	CO_2	Total carbon	Methanol	Ethyl formate
Cu/ZnO	22.8	-24.0	16.4	13.8	2.6
Cu/MnO	23.0	-40.8	14.2	0.0	14.2
Cu/MgO	15.5	23.2	16.6	5.7	10.9
Cu/CeO$_2$	11.3	27.7	13.5	0.0	13.5

443 K; initial pressure 30 bar; reaction time = 2 h; Cu/ZnO = 1.0 g; ethanol 20 ml.

2.1. Batch reactor

The reaction of activity test of various catalyst or different Cu/Zn ratio catalyst was conducted in a closed batch reactor with the inner volume of 85 ml and a stirrer was used. The composition of feed gas was $CO/CO_2/H_2/Ar$ = 32.6/5.2/59.2/3.0, except the experiments for gas composition effect.

2.2. Semi-batch flow type reactor

The reaction apparatus was a semi-batch autoclave reactor with an inner volume of 85 ml. The configuration of the reactor was reported before [13]. Passivated catalyst and alcohol of 20 ml (purity > 99.5 %) were used. The temperature of reactor exit and cold trap was controlled. The composition of feed gas was $CO/CO_2/H_2/Ar$ = 32.6/5.2/59.2/3.0.

3. RESULTS AND DISCUSSION

3.1. Batch reactor

Table 1 compared the reaction reactivity of different supported Cu catalysts. The results showed that Cu/ZnO was the best catalyst in this reaction system due to the high total carbon conversion and methanol yield. Cu/MgO also showed high reaction activity but the methanol yield was still low. No methanol was observed in the reaction performance of Cu/MnO and Cu/CeO$_2$, indicating that they had no activity for hydrogenolysis of ethyl formate to methanol.

The effect of different reactant gas composition was investigated in Table 2.

Table 2
Effect of reactant gas composition on methanol synthesis in batch reactor.

Expt. no.	Pressure (bar)			Yield (%)		
	CO	H_2	CO_2	Methanol	Ethyl formate	Total
1	13.2	15.9	0	6.0	1.3	7.2
2	9.6	19.5	0	8.0	1.8	9.8
3	9.8	17.8	1.6	13.2	2.4	15.6
4	0	21.5	7.5	11.1	2.5	13.6

443 K; initial pressure = 30 bar; reaction time 2 h; Cu/ZnO 1.0 g; ethanol 20 ml.

It is clear that syngas ($CO+H_2$) containing CO_2 exhibited the highest reaction activity. From the conversion, the rate of methanol formation from CO_2 was faster than that from CO, indicating that part of CO was converted to CO_2 at first via water-gas shift reaction, as in equation (1). Ethyl formate was the only by-product as shown in equation (2). It was also shown in Table 2 those methanols and ethyl formates were produced even if pure $CO+H_2$ were used. It is referred that H_2O formed by the reduction process of oxide surface of the passivated catalyst or the trace amount of water in ethanol solvent was involved in the process. CO possibly reacted with H_2O following the water-gas shift reaction (equation 1).

In Fig. 1, the reaction activity changed with time of reaction was shown. The conversion and yield of methanol increased with increasing in reaction time. Contrarily, the ethyl formate yield correspondingly decreased and became stable from about 2 h. This typical consecutive reaction pattern confirmed that ethyl formate was an intermediate of methanol synthesis. The reaction proceeded according to the designed route as exhibited in equations (1)-(3).

In Fig. 2, the total carbon conversion and yield of methanol increased with the increasing in reaction temperature. The selectivity of methanol gradually increased with the increasing temperature, unlike conventional gas-phase reaction where higher reaction temperature lowers methanol selectivity. As the temperatures investigated here were low, the limitation from thermodynamics did not appear.

3.2. Semi-batch flow type reactor

The time-on-stream activity change of the reaction during 20 h period was shown in Fig. 3. Although the equation (4) itself is independent of the type of alcohol solvent used, the kinetics of it, equations (2) – (3) is dependent on the type of alcohol. It was found that 2-butanol was the most effective alcoholic solvent for this low-temperature and it was employed here for continuous reaction [14]. The conversions were gradually increased and reached steady

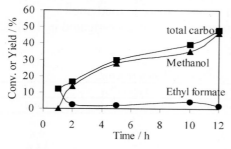

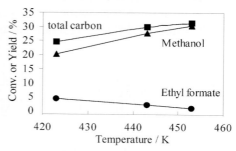

Fig. 1. The effect of reaction time on methanol synthesis. 443 K; initial pressure 30 bar; Cu/ZnO 1.0 g; ethanol 20 ml.

Fig. 2. The effect of reaction temperature on methanol synthesis. initial pressure 30 bar; time 2 h; Cu/ZnO 1.0 g; ethanol 20 ml.

at 15 h. At steady state, CO conversion and total carbon conversion were approximately 70% and 61%, respectively. More interestingly, during the initial 10 h period, the CO_2 conversion dropped to the minimum -60% and then increased to 7%. The dead volume of the reactor and the ice trap diluted the concentration of the reacted gas at the initial state. Therefore, the measured CO_2 conversion at the initial state was lower than the actual one. Similar to the results of experiments using different reactant gas in Table 2, conversion of CO_2 to methanol was faster and a part of CO was shifted to CO_2 initially, as shown in the designed reaction route, equations (1) – (3). Due to the low content of CO_2 in the feed gas, the absolute CO_2 concentration was not so high even if it was increased. It is considered that negative CO_2 conversion was not from the reaction, $2CO = CO_2 + C$, as Cu cannot cut CO bond at low temperature. As the used alcohol contained a little bit water, it seems that negative CO_2 conversion at the initial stage was due to the reaction of water contained in alcohol and CO by water-gas shift reaction, equation (1).

The relationship between the conversion and reaction pressure of the semi-batch reaction was shown in Fig. 4. It is obvious that the total carbon

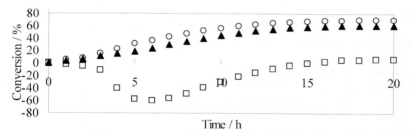

Fig. 3. Variations of conversions with time on stream. 443 K; 50 bar; Cu/ZnO 6.0 g; 2-butanol 20 ml; 20 ml/min; (○) CO%; (□) CO_2%; (▲) total carbon%.

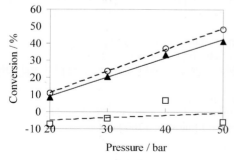

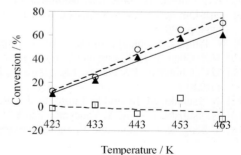

Fig. 4. Conversions with pressure for the methanol synthesis. 443 K; Cu/ZnO 3.0 g; 2-butanol 20 ml; 20 ml/min; (○) CO%; (□) CO_2%; (▲) total carbon%.

Fig. 5. Conversions with temperature for the methanol synthesis. 50 bar; Cu/ZnO 3.0 g; 2-butanol 20 ml; 20 ml/min; (○) CO%; (□) CO_2%; (▲) total carbon%.

conversion increased gradually with increasing in reaction pressure. The selectivity of methanol was consistent with the condition pressure (>98%).

The relationship between conversion and reaction temperature was shown in Fig. 5. It is interesting that the conversion continuously increased with increasing in the temperature of the reaction without changing the methanol selectivity (>98%). As the temperatures investigated here were low, the limitation from thermodynamics did not appear.

4. CONCLUSIONS

The high activity and selectivity of methanol formation was achieved at 443 K and 50 bar using 2-butanol as a solvent through a newly-designed low-temperature methanol synthesis from CO_2-containing syngas in which purification of syngas was unnecessary. The low-temperature process is feasibly promising for various kinds of syngas. Methanol synthesis at low temperature with conventional solid catalyst was a practical method, different from other low-temperature processes where a trace amount of CO_2 and H_2O remarkably deactivated the catalyst. Cu/ZnO catalyst gave the highest methanol synthesis efficiency.

REFERENCES

[1] R.G. Herman, G. W. Simmons and K. Klier, Studies in Surf. Sci. Catal., 7 (1981) 475.
[2] G.H. Graaf, P. Sijtsema, E.J. Stamhuis and G. Oosten, Chem. Eng. Sci., 41 (1986) 2883.
[3] M. Marchionna, M. Lami and A. Galleti, CHEMTECH, (1997) 27, April
[4] J. Zheng, N. Tsubaki and K. Fujimoto, Fuel, 81 (2002) 125.
[5] J. Haggin, Chem. & Eng. News, 4 (1986) 21, Aug.
[6] Brookhaven National Laboratory, U.S.Patent No. 4 614 749, 4 619 946, 4 623 634, 4 613 623 (1986), 4 935 395 (1990).
[7] V.M. Palekar, H. Jung, J.W. Tierney and I. Wender, Appl. Catal., 102 (1993) 13.
[8] Kirk-Othmer, Encyclopedia of Chemical Technology, 2nd ed., Wiley, New York, Vol. 13, 1964, p.390
[9] N. Tsubaki, Y. Sakaiya and K. Fujimoto, Appl. Catal., L11 (1999) 180.
[10] C.S. Chen, W.H. Cheng and S.S. Lin, Catal. Lett., 68 (2000) 45.
[11] G.C. Chinchen, P.J. Denny, J.R. Jennings, M.S. Spencer and K.. Waugh, Appl. Catal., 36 (1988) 1.
[12] N. Tsubaki, M. Ito and K. Fujimoto, J. Catal., 197 (2001) 224.
[13] N. Tsubaki, J. Zeng, Y. Yoneyama and K. Fujimoto, Catal. Comm., 2 (2001) 213.
[14] J. Zeng, K. Fujimoto and N. Tsubaki, Energy and Fuels, 16 (2002) 83.

Studies in Surface Science and Catalysis, volume 147
X. Bao and Y. Xu (Editors)
415

Enabling offshore production of methanol by use of an isopotential reactor

F. Daly and L. Tonkovich

Velocys, Inc., 7950 Corporate Boulevard, Plain City, OH 43064, United States

ABSTRACT

Velocys has begun developing a novel reactor that will greatly intensify the methanol synthesis reaction, thereby making the entire production process more efficient, and less costly. The concept is a microchannel isopotential reactor for methanol synthesis combined with a microchannel steam methane reformer that Velocys is also developing. This presentation will focus on methanol synthesis.

1. INTRODUCTION

The chemical industry is heavily dependent on catalytic processes. Such processes in general have lower operating costs; provide higher purity products with reduced by-product formation and less environmental impact than non-catalytic processes.

Major objectives in developing improved chemical processes typically are to use cheaper raw materials, to reduce energy demand, capital investment and operating costs. Today we are witnessing important new developments that go beyond traditional chemical engineering. Such developments have the potential to transform our concept of chemical plants leading to compact, safe, energy-efficient and environmentally friendly processes. Concepts include the use of monolith catalysts, microchannel reactors and uses of alternative forms and sources of energy, e.g., ultrasound, solar, microwave, and plasma discharges. All of these concepts offer huge potential benefits but more development work is required to bring them from bench-scale demonstrations to commercial applications to truly capture the benefits.

One application area that offers a large societal as well as economic benefit is the application of energy efficient, ultra low emissions fuel cells in a "hydrogen economy". Use of fuel cells range from the replacement of conventional internal combustion engines for transportation to displacement of batteries where long periods between re-charging is required to distributed power production in residential homes and light industry. One of the challenges to making these fuel cell applications a reality is having an inexpensive means of "storing" the hydrogen to be used for these applications.

One convenient means of storing the hydrogen fuel is in the form of methanol [1]. With methanol, the hydrogen can be released from the methane molecule by a simple, low temperature reforming reaction to produce synthesis gas containing the hydrogen to be used for the fuel cell.

Today the cost of methanol as a "fuel" compared to other hydrocarbons is expensive. If technology can be developed to lower the cost of methanol to "fuel" cost values, it will open up the use of methanol as the fuel for fuel cells.

Presently, methanol is manufactured by the hydrogenation of carbon monoxide over a copper-based catalyst, e.g., $Cu/ZnO/Al_2O_3$, in a fixed bed reactor, a basic technology that has remained unchanged for over 75 years. Average catalyst life is four years. Since the reaction is exothermic, there is a tradeoff between reaction kinetics and reaction thermodynamics. The reaction rate is greater at higher temperature, whereas equilibrium is favored at lower temperature. Operating at high temperature can increase the rate of catalyst deactivation and produce undesired side products such as ketones that can form azeotropes, making product separation more difficult.

It's well established that the intrinsic reaction rate for methanol synthesis is much faster than the heat transfer rate between the reaction vessel and reaction environment, thereby, limiting methanol production [2]. Though theoretical kinetics suggests contact times in the order of milliseconds, the slow rate of heat transfer typically necessitates contact times of seconds to minutes. If we can develop a way to utilize very short contact times, we can then dramatically increase the productivity of the reactor, i.e., much greater throughput per unit volume.

Microchannel reactors are typically defined as reactors with three-dimensional structures with inner dimensions of 10-100 micrometers [3]. Consequently, microchannel reactors in comparison to conventional chemical reactors have a high surface-to-volume ratio. This permits significantly higher heat-exchange efficiency and operation under isothermal conditions with well defined residence times [4], preventing temperature excursions and enhancing catalyst life. Consequently, a microchannel reactor yields higher throughput and better catalyst utilization.

Recently, Velocys initiated a research program to explore the feasibility of developing such a reactor system for methanol synthesis, thereby, making the entire methanol production process more efficient and less costly. The concept is a microchannel isopotential reactor combined with a microchannel steam methane reformer that Velocys is also developing. It is proposed that a microchannel isopotential reactor would have much higher heat transfer rates, catalyst forms with high effectiveness factors and temperature controlled zones.

By interleaving reaction microchannels with heat exchange channels, the isopotential reactor can rapidly remove the exothermic heat of reaction generated during methanol synthesis. Additionally, the isopotential reactor reduces the local temperature along the length of the channels. This design enables fast kinetics at the beginning of the reactor to convert the bulk of the synthesis gas to methanol, while favoring equilibrium in the latter stages to achieve superior per pass conversion. The following is a description of early results obtained using a powder $Cu/ZnO/Al_2O_3$ catalyst in both a microchannel isothermal reactor and a conventional non-isothermal tube-in-tube fixed-bed reactor.

2. RESULTS AND DISCUSSION

The performance of a $Cu/ZnO/Al_2O_3$ catalyst was compared in a single channel isothermal microreactor with a channel gap size of 60 mil and a non-isothermal tube-in-tube fixed-bed reactor. Using a feed gas with a composition of 65% H_2, 25% CO, 5% CO_2 and 5% N_2, initial experiments were performed to define the effect of contact time on CO conversion at 270°C and 51 atm. The results, presented in Figure 1, show that for contact times less than 300 ms the CO conversion is higher for the microchannel reactor versus the fixed bed reactor. However, at longer contact times CO conversion is higher in the fixed bed reactor and the difference between the two reactors increases as the contact time increases.

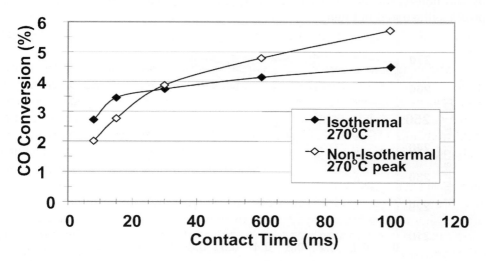

Fig. 1. CO conversions vs. contact time for isothermal microchannel and non-isothermal fix-bed reactors.

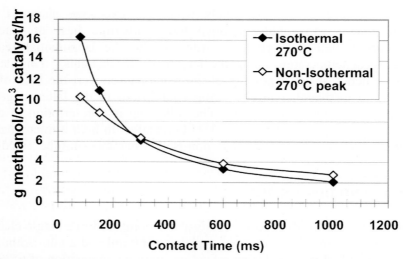

Fig. 2. Methanol yields vs. contact time for isothermal microchannel and non-isothermal fix-bed reactors at 270°C peak temperature.

The methanol yield per volume of catalyst is shown in Figure 2. At a contact time of 80 ms the microchannel reactor produced about 63% more methanol by operating isothermally than the non-isothermal, fixed-bed reactor operating at the same peak temperature.

The higher CO conversion at contact times greater than 80 ms demonstrated by the non-isothermal, fixed-bed reactor can be explained by the temperature profiles illustrated in Figure 3.

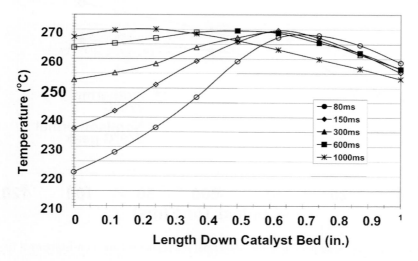

Fig. 3. External tube temperatures required to maintain 270°C peak temperature in a fix-bed reactor.

Table 1
Non-isothermal fixed-bed reactor skin temperatures versus contact times

Contact time, ms	80	150	300	600	1000
Reactor skin temperature, °C	232	241	254	263	264

As contact time is decreased, the location of the hot spot moves down the length of the reactor. To maintain the reaction temperature at 270°C peak the outside tube temperature had to be adjusted at each contact time. As would be expected, the highest heat generation (lowest contact time) required the lowest amount of external heat to maintain 270°C peak temperature. However, it can be seen (Table 1) that a large temperature gradient exists at the fastest contact times. This is believed to be the reason for the enhanced performance of the MPT reactor, which utilizes the kinetics of the full volume of catalyst more effectively by operating isothermally.

A similar microchannel reactor benefit was obtained at 250 and 230°C peak temperatures. However, at the lower temperatures, the effect is not as pronounced as that obtained at 270°C. At 250°C, both reactors are able to achieve close to the equilibrium conversion of 60% at 600ms.

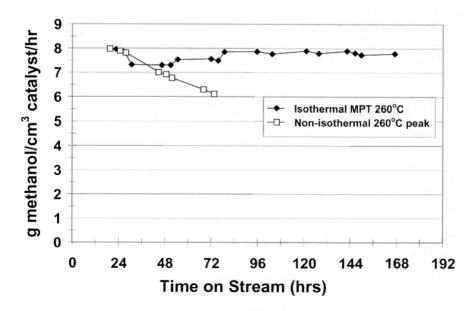

Fig. 4. Methanol yields versus time on steam for isothermal microchannel and non-isothermal fix-bed reactors at 260°C peak temperature and 300 ms contact time.

The activity of the catalyst in both reactors as a function of time on stream was also investigated. As illustrated in Figure 4, no loss in activity was observed for the microchannel reactor while a 25% reduction in methanol yield after 72 hrs on stream was observed for the fixed-bed reactor.

4. CONCLUSION

The results suggest that an isothermal microchannel reactor offers advantages over a conventional fixed-bed reactor at short contact times. However, it is proposed that significantly greater advantages can be achieved with an isopotential microchannel reactor and a catalyst more active than that presently used in commercial fixed-bed reactors. Future research endeavors will explore this proposal.

REFERENCES

[1] J.G. Highfield, Trends Phys. Chem., 5, 91 (1995).
[2] K. Kohler, British Chem. Eng., 14(10), 1548 (1969).
[3] W. Ehrfeld, V. Hessel and H. Löwe, Microreactors, Wiley-VCH, Weinheim, 2000.
[4] K. Jähnisch, V. Hessel, H. Löwe and M. Baerns, Angew. Chem. Int. Ed., 43, 406
    (2004).

Studies in Surface Science and Catalysis, volume 147
X. Bao and Y. Xu (Editors)

The promotion effects of Mn, Li and Fe on the selectivity for the synthesis of C_2 oxygenates from syngas on rhodium based catalysts

H. -M. Yin, Y. -J. Ding*, H. -Y. Luo, W. –M. Chen, L.-W. Lin

Natural Gas Utilization and Applied Catalysis Laboratory, Dalian Institute of Chemical Physics, Chinese Academy of Sciences, 457# Zhongshan Rd., Dalian 116023, China

ABSTRACT

Rh-based catalysts with the promoters of Mn, Li and Fe were prepared. The activity of the catalyst for C_2 oxygenates formation enhanced with the addition of additives. The promotion effects of these additives were investigated by means of FT-IR, CO-TPD and TPSR of adsorbed CO with H_2. IR results suggested that Fe, Mn and Li might be in close contact with Rh, and formed new active sites, which promoted CO activation on Rh surface and favored the formation of C_2 oxygenated intermediates. The results of CO-TPD and TPSR showed that only a part of adsorbed CO on the catalyst could be hydrogenated into CH_4. The ratio of CH_4 quantity in TPSR to that of desorbed CO in CO-TPD on various catalysts was calculated, and it decreased in the order: Rh/SiO_2 >$Rh-Mn/SiO_2$ >$Rh-Mn-Li/SiO_2$ >$Rh-Mn-Li-Fe/SiO_2$, which implied that non-dissociated CO increased in the reverse order. Which was consistent with the catalytic activity of C_2 oxygenates formation sequence. The improvement of the space-time yield (STY) and selectivity towards C_2 oxygenates could be explained based on that the increase of the nondissociative CO and the decrease of dissociative CO on the catalyst favored the insertion of CO into CH_x.

1. INTRODUCTION

Among metals active for CO hydrogenation, rhodium exhibits remarkable feature of selectively producing C_2 oxygenates from syngas [1-2]. In general, the mechanism of CO hydrogenation to hydrocarbons and higher oxygenates includes (a) adsorption and dissociation of H_2 and CO and the formation of CH_x blocks, (b) hydrogenation of CH_x produces alkanes; and CO insertion in the

---

* Corresponding Author: **Email:** dyj@dicp.ac.cn **Phone:** +86 411 4379143 **Fax:** +86 411 4379143

metal alkyl bond followed by H addition yields oxygenates.

We have developed an Rh-Mn-Li-Fe/SiO$_2$ catalyst and it showed unexpectedly high activity in the formation of C$_2$ oxygenates [3]. To find the nature of the promotion effect of Mn, Li and Fe, the present investigation was undertaken with the aim of (a) identifying the effects of Mn, Li and Fe on CO hydrogenation using FT-IR, temperature-programmed desorption and reaction of adsorbed CO (CO-TPD and TPSR) and (b) correlating the results of FT-IR, TPSR and CO-TPD with the catalytic activity of CO hydrogenation to elucidate the promotion effects of the additives.

2. EXPERIMENTAL

2.1 Materials and catalysts

Rhodium chloride was purchased from Johson Matthey Precious Metal Company. Catalysts were prepared by impregnating SiO$_2$ (20-40 mesh, BET surface area 260 m^2/g) with an aqueous solution of RhCl$_3$.xH$_2$O and/or Mn(NO$_3$)$_2$ and/or LiNO$_3$ and/or FeNO$_3$, and drying at 393 K for 4 hours. Rhodium loading of all the catalysts was 1.0 wt.%.

2.2 CO hydrogenation

The CO hydrogenation test was carried out at 593 K, 3.0 MPa. The catalyst was in-situ reduced in a flow of H$_2$ before reaction. The loading of catalyst was 0.4 g for each test. The testing apparatus consisted of a small fixed bed tubular reactor with an external heating system, which was made of 316 L stainless steel with 300 mm length, 4.6 mm inner diameter. The effluent passed through a condenser filled with 150 ml of cold de-ionic water. The oxygenated compounds from the product were captured by complete dissolution into the water in the condenser. The aqueous solution containing oxygenates obtained was off-line analyzed by Varian CP-3800 gas chromatography with an FFAP column and FID detector, using 1-pentanol as an internal standard. The tail gas was on-line analyzed by Varian CP-3800 GC with a Porapak QS column and TCD detector.

2.3 CO-TPD experiment

CO-TPD experiments were performed on Micromeritics Autochem 2910. After CO adsorption saturation, the catalyst bed was swept with He for 5 min. Each experiment was performed in a He flow (flow rate = 20 ml/min) with a quadruple mass spectrometer (QMS, Balzers OmniStar 300) as a detector to monitor the desorbed species under linearly increasing temperature. The heating rate was 15 K/min. MS signals at m/z = 28 (CO) were continuously recorded.

2.4 TPSR experiment

TPSR experiments were carried out as follows: after the catalyst was reduced which was the same as TPD experiment, the catalyst was swept with He for 0.5 h and then cooled down to 323 K. 5% CO-He was introduced for adsorption. After CO adsorption saturation was achieved, the catalyst was swept with He for 5 minutes; then 10% H_2-Ar was passed through the catalyst bed and the temperature was ramped at 15 K/min to the desired temperature. The signals of CH_4 (m/z=15,16) was simultaneously recorded with a Balzers Omni Star 300.

2.5 FT-IR studies

FT-IR study was conducted under 120 ml/min gas flow, H_2/CO=2. The IR spectra were recorded by a BRUKER EQUINOX 55 single-beam Fourier Transform Infrared spectrometer. The sample was reduced in flowing H_2 at 593 K for 1 h. After reduction, the catalyst was cooled down to 323 K. An FT-IR spectrum (spectrum A) of the catalyst in H_2 flow at a given reaction temperature was recorded. Subsequently, flowing H_2/CO was introduced, the temperature and pressure were increased to realize the desirable reaction conditions and maintained for 0.5 h, after relieving the pressure and purging the chamber with high-purity nitrogen for 0.5 h, spectrum B was recorded. Subtracting spectrum A from spectrum B gave a final IR spectrum.

3. RESULTS AND DISCUSSION

3.1 Effect of different promoters on performance of Rh/SiO$_2$ for CO hydrogenation

Results of CO hydrogenation on different Rh-based catalysts are shown in Table 1. Both the activity and selectivity towards C_2 oxygenates on Rh/SiO$_2$ catalyst without any promoters were poor. The activity for the formation of C_2 oxygenates increased when 1% Mn was added into Rh/SiO$_2$ catalyst, and the selectivity towards C_2 oxygenates increased greatly after 0.075% Li was introduced. The yield of C_2 oxygenates increased unexpectedly from 331.6 g/kg·h up to 457.5 g/kg·h when 0.05% Fe was added to Rh-Mn-Li/SiO$_2$ catalyst, while the selectivity towards C_2 oxygenates remained unchanged.

Table 1
Effect of different promoters on performance of Rh/SiO$_2$ for CO hydrogenation

Cat	Conv. CO %	Selectivity of products, C %							Yc$_2$+oxy g/kg·h
		C_1^+HC	MeOH	HAc	EtOH	EtOAc	HOAc	C_2^+oxy	
A	2.61	79.2	4.5	0.8	12.3	1.4	-	16.3	35.5
B	8.66	64.8	0.3	4.7	23.0	0.2	3.9	34.9	270.3
C	6.69	41.9	1.4	7.2	35.5	4.4	5.5	56.8	331.6
D	8.44	42.7	1.6	10.2	27.4	3.4	7.2	56.3	457.5

Rh=1 wt.%, T=593 K, P=3.0 MPa, H_2/CO=2 (molar ratio), GHSV=12000/h, A: Rh/SiO$_2$, B: Rh-Mn/SiO$_2$, C: Rh-Mn-Li/SiO$_2$, D: Rh-Mn-Li-Fe/SiO$_2$. Rh:Mn:Li:Fe=1:1:0.075:0.05

3.2 CO-TPD and TPSR results

TPD spectra of adsorbed CO on various Rh based catalyst samples with different promoters are shown in Fig. 1. It was clear that the amount of desorbed CO at high temperature and the total desorbed CO increased with the addition of Mn, Li and Fe, that is, the introduction of additives enhanced the amount of the strong adsorbed CO. The relative quantities of CO desorbed in CO-TPD are shown in Table 2. The dissociation of CO can be detected by determining the formation of CH_4 as shown in Fig. 2. The relative quantities of CH_4 formed in TPSR are also shown in Table 2.

As demonstrated in ref. [3] only the strongly adsorbed CO species could be hydrogenated and form CH_4, while the weakly adsorbed CO desorbed. It is generally considered that CO dissociates first and then CH_4 forms via the hydrogenation of dissociated carbon and/or CH_x species. The ratios of CH_4 produced in TPSR and CO desorbed in CO-TPD are shown in Table 2. It was found that the ratios decreased in the following order: $Rh/SiO_2 > Rh-Mn/SiO_2 > Rh-Mn-Li/SiO_2 > Rh-Mn-Li-Fe/SiO_2$, therefore the proportion of non-dissociated CO increased in the reverse order, which consisted with the increase order of space-time yield and the selectivity towards C_2 oxygenates over the

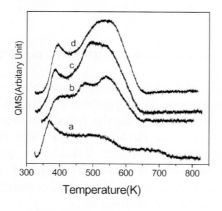

Fig. 1. CO TPD spectra of various catalysts
(a) Rh/SiO_2; (b) $Rh-Mn/SiO_2$
(c) $Rh-Mn-Li/SiO_2$; (d) $Rh-Mn-Li-Fe/SiO_2$

Fig. 2. CH_4 spectra in TPSR on catalysts
(a) Rh/SiO_2; (b) $Rh-Mn/SiO_2$;
(c) $Rh-Mn-Li/SiO_2$; (d) $Rh-Mn-Li-Fe/SiO_2$

Table 2

Relative quantities results of CO desorbed in CO-TPD and CH_4 formed in TPSR experiments on various catalysts and the corresponding CH_4/CO ratio

Catalyst	Rh/SiO_2	$Rh-Mn/SiO_2$	$Rh-Mn-Li/SiO_2$	$Rh-Mn-Li-Fe/SiO_2$
CO quantity	1.0	1.18	1.97	2.07
CH_4 quantity	1.0	1.05	1.33	1.23
CH_4/CO ratio	1.0	0.89	0.68	0.59

corresponding catalysts. From the results of CO-TPD and TPSR experiments, we could propose that the addition of Mn, Li and Fe enhanced CO dissociation and promoted CO insertion into CH_x over Rh-based catalyst simultaneously, which probably led to the enhancement of STY and selectivity to C_2 oxygenates.

3.3 FT-IR Studies

IR results after reaction at 593 K and 3.0 MPa over Rh/SiO_2 and $Rh-Mn-Li-Fe/SiO_2$ are shown in Fig. 3. It was shown that only band at 2051 cm^{-1} corresponding to linear-CO was observed [4-5] after the reaction on Rh/SiO_2 [Fig. 3(a)]. For the spectra after reaction over $Rh-Mn-Li-Fe/SiO_2$ [Fig 3(b)], band at 2038 cm^{-1} ascribing to linear-CO was detected. The red-shift of linear-CO from 2051 to 2038 cm^{-1} over $Rh-Mn-Li-Fe/SiO_2$ compared to that of Rh/SiO_2 could be attributed to the promotion effect of additives towards CO activation on Rh surface. Distinct and sharp bands at 1716, 1566, 1442, and 1343 cm^{-1} appeared in the spectra over $Rh-Mn-Li-Fe/SiO_2$. Ichikawa [6-8] assigned the bands at 1579 and 1444 cm^{-1} to bidentated acetate species, the bands at 1745, 1444and 1382 cm^{-1} to monodentated acetate and proposed the acetate species as oxygenated intermediates. Trevino [9] postulated that oxygenated precursors (a surface acetate), $C_xH_yO_z$, complex were formed at MnO sites in close proximity to the Rh-MnO interface. The role of Rh was the formation and delivery of CH_x groups and H atoms. Bands at 1716, 1566, 1442, and 1343 cm^{-1} could not be observed over un-promoted Rh/SiO_2 which indicated that the addition of Mn, Li, and Fe favored the C_2 oxygenates intermediates formation.

Extensive works have been done to investigate the unexpected effect of promoters on the activity and selectivity of Rh/SiO_2 catalyst. Sachtler and Ichikawa [5] ascribed the promotion effects by oxophilic metal promoters such as Mn, Zr, V and Ti to the formation of a tilted mode of CO adsorption on the Rh surface and enhanced the dissociation rate of CO. We have proposed in previous work [3] that promoters Fe, Mn and Li were in close contact with Rh, and formed new active sites, which were responsible for the high yield of C_2

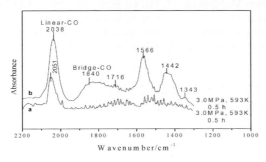

Fig. 3. IR spectra of adsorbed species on Rh/SiO_2 (a) and $Rh-Mn-Li-Fe/SiO_2$ (b)

oxygenates. Correlated with our IR results, it seemed probable that these new active sites promoted CO activation over Rh surface and favored the formation of C_2 oxygenated intermediates, thus promoted the formation of C_2 oxygenates.

4. CONCLUSION

The CO hydrogenation results showed that the introduction of Mn, Li, and Fe enhanced significantly the activity and selectivity for C_2 oxygenates formation over Rh/SiO_2 catalysts. The results of CO-TPD and TPSR indicated that these promoters could improve the amount of the strong adsorbed CO species and enhance the proportion of non-dissociated CO adsorbed on the Rh based catalysts. The increase of these non-dissociated CO might favor the insertion of CH_x into CO over Rh surface, hence lead to the enhancement of the STY and selectivity of C_2 oxygenates on the Mn, Li and Fe promoted rhodium catalysts.

IR results suggested that the addition of Mn, Li and Fe led to the red-shift of linear-CO from 2051 to 2038 cm^{-1} compared to that of Rh/SiO_2, which could be attributed to the promotion effect of additives towards the CO activation on Rh surface. Distinct and sharp bands at 1716, 1566, 1442, and 1343 cm^{-1} appeared in the spectra after the reaction over $Rh-Mn-Li-Fe/SiO_2$. The above results allowed us to propose that the promoters of Mn, Li and Fe, might be in close contact with Rh and form new active sites, which favored the acetate intermediate species formation, thus promoted C_2 oxygenates formation.

REFERENCES

[1] M. Ichikawa, T. Fukushima, and K. Shikakura, in "Proceedings, 8th International Congress on Catalysts, Berlin,"(G. Ertl. Ed.), Vol. 2, p. 69. Dechema, Frankfurt-am-Main, 1984.
[2] M. Ichikawa, and T. Fukushima, J. Chem. Soc. Chem. Commun., (1985), 321
[3] H.M. Yin, Y.J. Ding, H.Y. Luo, H.J. Zhu, J.M. Xiong, and L.W. Lin, Appl. Catal. A, 243 (2003) 155.
[4] A.C. Yang, C.W. Garland, J. Phys. Chem., 61 (1957), 1504.
[5] W.M.H. Sachtler, M. Ichikawa, J. Phys. Chem., 90 (1986), 4752.
[6] T. Fukushima, H. Arakawa, M. Ichikawa, J. Phys. Chem., 89 (1985), 4440.
[7] T. Fukushima, H. Arakawa, M. Ichikawa, J. Chem. Soc. Chem. Commun., (1985) 729.
[8] H. Arakawa, T. Fukushima, M. Ichikawa, Appl. Spectrosc., 40 (1986), 884.
[9] H. Treviño, T. Hyeon, M.H. Wolfgang, J. Catal., 170 (1997), 236.

Studies in Surface Science and Catalysis, volume 147
X. Bao and Y. Xu (Editors)
©2004 Elsevier B.V. All rights reserved.

Selective synthesis of LPG from synthesis gas

Kenji Asami, Qianwen Zhang, Xiaohong Li, Sachio Asaoka, and Kaoru Fujimoto

Faculty of Environmental Engineering, The University of Kitakyushu, Wakamatsu-ku Kitakyushu 808-0135, Japan

ABSTRACT

LPG synthesis from synthesis gas was investigated over hybrid catalysts composed of a Cu-Zn methanol synthesis catalyst and several kinds of zeolites at 523-623 K under pressure. Combination of Cu-Zn catalyst and USY or β zeolite gave high catalytic performances for LPG formation; above 70% of CO conversion and LPG selectivity were obtained at 598 K and 2.1 MPa.

1. INTRODUCTION

Liquefied petroleum gas (LPG), which is composed of propane and butanes, has environmentally benign characteristics and it has been widely used as domestic fuels or chemical feed. LPG is usually produced from two sources. The first one is from oil or gas fields, and the other is as by-products from oil refining process. Propane and butanes can be formed in other chemical reactions. Fischer-Tropsch synthesis [1, 2] is well-known to produce hydrocarbons from synthesis gas (H_2/CO), which is obtained through reforming of natural gas or gasification of coal [3]. However, a wide range of compounds such as hydrocarbons and oxygenates are produced, and distribution of hydrocarbon products follows Schulz-Flory distribution [1, 4, 5]. Thus, it is difficult to synthesize LPG selectively. On the other hand, it is reported that hydrocarbons can be produced directly from synthesis gas on hybrid catalysts composed of methanol synthesis catalyst and zeolite [6-8]. The objective of the present work is to develop novel hybrid catalysts for selective production of C_3 and C_4 hydrocarbons from synthesis gas.

2. EXPERIMENTAL

A commercial Cu-Zn catalyst was used as a methanol synthesis catalyst. Zeolite catalysts used were proton type USY, β, ZSM-5, Mordenite, and

MCM-22 zeolites, of which SiO_2/Al_2O_3 ratios were 12.2, 37.1, 14.5, 16.9, and 25.0, respectively. The hybrid catalyst was prepared by mixing Cu-Zn catalyst and zeolite physically. It was palletized to a disk, and then crushed and sieved to a size of 20-40 mesh. The hybrid catalysts were activated in-situ in a flow of 5% hydrogen in nitrogen at 573 K for 4 h. A pressurized fixed-bed flow type reaction apparatus was used for CO conversion runs. In all tests, 1 g of hybrid catalyst was charged in the stainless tube reactor with 6 mm inner diameter. Reaction temperature, pressure, and H_2/CO ratio in the feed gas were mostly 598 K, 2.1 MPa, and 2/1, respectively. All the reactants and products were analyzed by gas chromatography on line.

3. RESULTS AND DISCUSSION

3.1. Effect of zeolite type in the hybrid on LPG synthesis

Fig. 1 shows the performances of a variety of hybrid catalysts combining different zeolites with Cu-Zn methanol synthesis catalyst. Hydrocarbons and CO_2 formed with almost the same yield from CO, accompanied by a small amount of dimethyl ether (DME) over the hybrid catalysts (a). This result demonstrates that the mechanism of hydrocarbon synthesis from synthesis gas over the hybrid catalyst is different from F-T reaction mechanism. Hydrocarbons synthesis reaction proceeds through the methanol synthesis reaction, methanol dehydration to DME, and methanol/DME conversion into hydrocarbons. The equilibrium conversion of CO in the methanol synthesis under the conditions is estimated as about 2%. However, the CO conversion over each hybrid catalyst is much higher than that. The conversions of intermediate CH_3OH and DME into hydrocarbons break the equilibrium conversion of methanol formation. Production of a considerable amount of CO_2

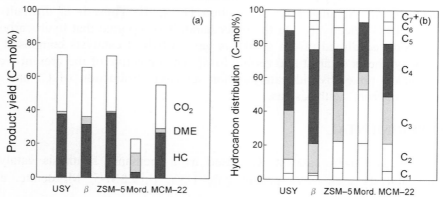

Fig. 1. Performance of the hybrid catalysts composed of Cu-Zn and various zeolites. Cu-Zn/zeolite = 50/50 (wt%), 598 K, 2.1 MPa, 4.5 g h mol^{-1}, H_2/CO=2/1.

because of the water gas shift reaction. The activity and product selectivity of the catalysts were influenced markedly by the kind of zeolite employed. USY-, β-, and ZSM-5-containing hybrid catalysts gave high activities around 70% of CO conversion, and USY- and β-containing catalysts exhibited high selectivity for LPG formation; $C_3 + C_4$ selectivity was 76% and 73%, respectively (b). The high activity of USY-containing catalyst is attributed to its strong acidity essential for formation of hydrocarbons and large size pore for easy diffusion of product hydrocarbons.

3.2. Effect of the composition on the catalytic performance

Effects of the composition of hybrid catalyst were examined over a group of Cu-Zn- and USY-containing hybrid catalysts. Fig. 2 (a) and (b) show the

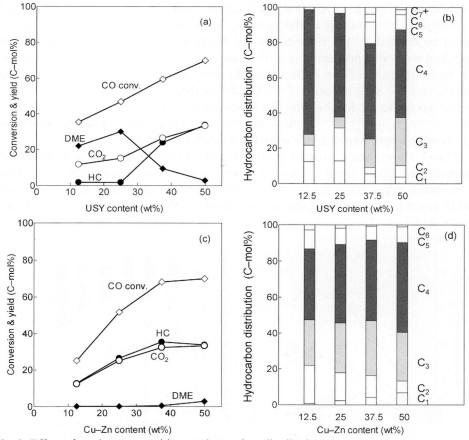

Fig. 2. Effect of catalyst composition on the product distribution.
(a), (b) Cu-Zn 50 wt%, (c), (d) USY 50 wt%, SiO_2 balance, 598 K, 2.1 MPa, 4.5 g h mol^{-1}, $H_2/CO=2/1$.

catalytic performance as a function of USY content in the hybrid catalyst, where the content of Cu-Zn catalyst is constant (50 wt%). Although CO conversion increased linearly in the whole range of the USY content, only DME and CO_2 formed with a ratio of 2/1 below 25 wt%. Above 37.5 wt%, DME decreased and almost diminished at 50 wt%. Hydrocarbons, on the other hand, increased drastically, and the yields were nearly equal to those of CO_2. Although change in hydrocarbon distribution below 25 wt% was ambiguous because of the low yields of hydrocarbons, hydrocarbon distribution shifted to lower molecular weight products between the USY content of 37.5 and 50 wt%. The highest LPG selectivity of 76% was attained at 50 wt%. Figs. 2 (c) and (d) illustrate the influence of the content of Cu-Zn catalyst in the hybrid catalyst. DME formation was very small in the range where Cu-Zn content is less than USY content. CO conversion and yields of hydrocarbons and CO_2 increased with an increase in Cu-Zn content, and leveled off above 37.5 wt%. Also, LPG selectivity increased gradually with increasing in the content. Such activity and selectivity changes would be caused by the balance among formation of methylene species from methanol or DME, propagation of the carbon chain of the intermediate olefins, and hydrogenation of the olefins into the corresponding paraffins. From these results, suitable weight ratio of the Cu-Zn methanol synthesis catalyst to USY zeolite would be 1/1 for the selective synthesis of LPG from synthesis gas.

3.3. Temperature dependence over Cu-Zn/USY catalyst

Fig. 3 shows the yields of products and CO conversion as a function of temperature. DME yield increased firstly with an increase in the reaction temperature to reach the maximum at 548K, and then decreased. CO conversion and hydrocarbon yield increased monotonously as the reaction temperature was

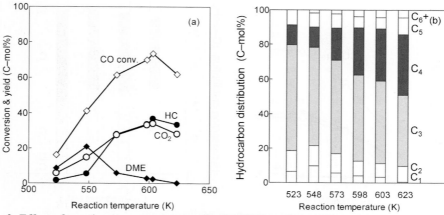

Fig. 3. Effect of reaction temperature over Cu-Zn/USY catalyst.
Cu-Zn/USY = 50/50 wt%, 2.1 MPa, 4.5 g h mol^{-1}, H_2/CO=2/1.

raised, and reached the maximum at 603K. Because the methanol synthesis reaction is exothermic, the equilibrium conversion of CO is very low at the condition of low pressure and high temperature. Low concentration of methanol and DME, as intermediates from synthesis gas to hydrocarbons, would not be in favor of the rate of hydrocarbon formation at high temperature. Distribution of the hydrocarbon products also changed with the reaction temperature. Selectivity to LPG fraction tended to increase slightly with an increase in temperature, whereas the content of propane decreased and the content of butanes increased. Furthermore, the selectivity to methane increased, and the yield of hydrocarbons decreased obviously at high temperature (623K). So the appropriate temperature of reaction was about 598K.

3.4. Influence of composition of the feed gas

H_2/CO molar ratio in the feed syngas had a great effect on the activity and selectivity of the hybrid catalyst of Cu-Zn/USY. As shown in Fig. 4, the yield of hydrocarbons and selectivity for C_3 and C_4 hydrocarbons decreased with the decrease in the H_2/CO ratio. On the other hand, the selectivity for C_2 and C_5 hydrocarbons increased obviously, and the hydrocarbon distribution became broad. The space time yield of C_3 and C_4 hydrocarbons decreased slightly with a decrease in H_2/CO ratio. According to reaction equation stoichiometry, H_2/CO ratios for LPG formation are 3/2 and 5/4 for propane and butanes, respectively. If the ratio of H_2/CO deviates from the stoichiometric value, excess reactant, H_2 or CO, should be purged in the recycle system of the unreacted reactants. This is uneconomical for the commercial application of this technology, and thus, synthesis gas with the H_2/CO ratio around 1/1 is appropriate for the efficient synthesis of LPG.

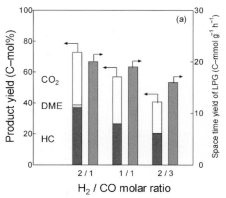 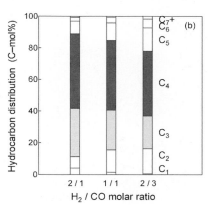

Fig. 4. Effect of H_2/CO molar ratio over Cu-Zn/USY catalyst.
Cu-Zn/USY = 50/50 wt%, 603 K, 2.1 MPa, 4.5 g h mol^{-1}.

4. CONCLUSION

Due to the synergetic effect of methanol synthesis catalyst and zeolite, hydrocarbon yield broke though the limit of thermodynamic equilibrium of methanol formation from synthesis gas. The hybrid catalyst consisted of USY zeolite and Cu-Zn methanol synthesis catalyst demonstrated a good reactivity and more than 75% selectivity for C_3 and C_4 hydrocarbons at 598K and 2.1 MPa.

ACKNOWLEDGEMENT

This work was supported by Japan Gas Synthesis LTD.

REFERENCES

[1] M.E. Dry, Catal. Today, 71 (2002) 227.
[2] H. Schulz, Appl. Catal. A: General, 186 (1999) 3.
[3] B. Jager, Stud. Surf. Sci. Catal., 107 (1997) 219.
[4] L. Fan and K. Fujimoto, Appl. Catal. A: General, 186 (1999) 343.
[5] L. Fan, K. Yoshii, and S. Yan, Catal. Today, 36 (1997) 295.
[6] C.D. Chang, W.H. Lang, and A.J. Silvestri, J. Catal., 56 (1979) 268.
[7] K. Fujimoto, Y. Kudo, and H. Tominaga, J. Catal., 87 (1984) 136.
[8] K. Fujimoto, Y. Kudo, and H. Tominaga, J. Catal. 94 (1985) 16.

Studies in Surface Science and Catalysis, volume 147
X. Bao and Y. Xu (Editors)
433

Catalytic alcohols synthesis from CO and H_2O on TiO_2 catalysts calcined at various temperatures

Dechun Ji, Shenglin Liu, Qingxia Wang and Longya Xu*

State Key Laboratory of Catalysis, Dalian Institute of Chemical Physics, Chinese Academy of Sciences, P.O. Box 110, Dalian 116023, China

ABSTRACT

The reaction performance for CO hydration on a TiO_2 catalyst under different calcination temperatures was investigated. Under reaction conditions of T = 573 K, P = 0.5 MPa, CO flow rate of 30 ml min^{-1}, TOS = 12 h, and CO/H_2O (g) = 3/2 (mol), the TiO_2 catalyst with a rutile content of 18% shows a maximum alcohols STY of 1.81 mg m^{-2} h^{-1}. In addition, the catalyst deactivation and regeneration were discussed.

1. INTRODUCTION

Conversion of CO to useful chemicals has been widely investigated by many workers. Methanol is one of the important feedstocks in chemical industries. The typical process for the production of methanol involves the reaction of carbon monoxide with hydrogen. Copper, zinc [1] and chromium oxides [2] are often selected as catalysts. In this process a large mount of H_2 (H_2/CO = 2) is required. An alternative process is that CO reacts with H_2O to produce alcohols. It is of particular advantage that in the partial oxidation of coal and typical coal gasification processes no hydrogen is required and CO can be provided as a by-product from the steel industry.

In 1984, Klinger et al. [3] first reported the reaction (3CO + 2H_2O = CH_3OH + 2CO_2). They considered that a lead oxide and an alkali metal format catalyst combination catalyzed the reaction. However, the catalyst deactivated slowly with reaction progressing because of the reduction of PbO in the presence of CO. How to keep the catalyst active sites from being be reduced in the presence of CO is the key point for the reaction. In our experiment, it is found that TiO_2 possesses such a feature in the presence of CO for the reaction and it shows much better reaction performance than PbO [4]. For the same catalyst TiO_2 and

---

* Corresponding author   e-mail: lyxu@dicp.ac.cn

solvent of tetraethylenepentamine, the reaction performance of alkali carbonate increases with its solubility (K_2CO_3 >Na_2CO_3 > Li_2CO_3) [5]. The results also indicate that the solvents show little difference among them in CO conversion. However the space time yield (STY) of CH_3OH and C_2H_5OH decreases with alkalinity of the solvent ($NH_2(CH_2CH_2NH)_3CH_2CH_2NH_2$>$NH_2(CH_2)_6NH_2$> ($HOCH_2CH_2$)$_3N$> liquid paraffin) for the same alkali carbonate of K_2CO_3 [6].

In the present work, catalytic reaction of CO and water to form alcohols on a TiO_2 catalyst under different calcination temperatures was investigated in a suspension-bed flow microreactor, together with XRD, XPS, and TPR techniques.

2. EXPERIMENTAL

2.1. Preparation of catalyst and test of reaction performance

Catalyst TiO_2 was prepared by decomposition of titanium hydroxide. The details of the process were as follows: $Ti(OC_4H_9)_4$ was dissolved in the mixture of distilled water and ethanol. Then the solution was stirred for 10 h and kept still for 10 h. Subsequently, it was treated with n-butanol azeotropic distillation method to obtain a gel powder, which was further calcined in air for 3 h at 673, 773, and 973K, et al. respectively to obtain different types of TiO_2.

Catalytic evaluation was performed in a 60 ml stainless steel reactor. The gas reactants (CO mixed with 5 vol%N_2) were saturated with water by bubbling the gas into a reservoir, which was kept at a suitable temperature to obtain the desired water concentration in the reactant flow. The flow rate of CO was 30 ml min^{-1}. A mixture of 3.0 g K_2CO_3 and 2.0 g catalyst was suspended in 30 ml of tetraethylenepentamine solvent. An internal standard method was employed to calculate the conversion of CO. The product analysis methods are similar with those in reference [4].

2.2. Characterization of catalyst

XRD patterns were taken by a Rigaku D/MAX RB diffractometer using Cu Kα radiation, operating at 40 kV and 50 mA, with a scan speed of 8 $°$ min^{-1}. XPS characterization was performed in a VG ESCA LAB MKII spectrometer.

Specific surface areas of the samples were obtained by the BET method from the adsorption isotherm at liquid nitrogen temperature of 78K, and taking a value of 0.162 nm^2 for the cross-sectional area of N_2 molecule at this temperature. Before carrying out the measurement, each sample was degassed under a reduced pressure of 10^{-5} Torr for 2 h at 573 K. H_2-TPR profiles were recorded by an on-line computer at a programmed temperature rate of 14 K min^{-1} in a 10 vol% H_2/Ar flow after the samples had been pretreated in an Ar flow at 773 K for 30 min.

3. RESULTS AND DISCUSSION

3.1. Influence of calcination temperature on the physicochemical properties of TiO$_2$

Powder diffractograms of TiO$_2$ samples were recorded from 10 to 70 °, which are compared in Fig. 1. Through changing calcination temperature, anatase and/or rutile can be obtained. At 673K, only anatase appears in XRD patterns and pure rutile at 1073K. Mixture of anatase and rutile exists between 673 and 1073K. The rutile content in the mixture calculated according to ω (R) =1/ (1+0.8I$_A$/I$_R$) [7], where ω (R) is the content of rutile phase, I$_A$ and I$_R$ are XRD intensity of anatase and rutile respectively, first increases slowly as the calcination temperature changing from 673 to 973K, then sharply rises to 100% at 1073K as shown in Fig. 2. In addition, Fig. 3 shows that the specific surface area of TiO$_2$ decreases almost linearly with the increase of calcination temperature.

3.2. Reaction performance for CO with water to alcohols on TiO$_2$ under different calcination temperatures

Fig. 3 shows the CO conversion for CO hydration on TiO$_2$ catalyst under different calcination temperatures. As the calcination temperature rises from 673 to 1073K, the specific surface area of sample decreases rapidly from 50 to 7 m^2 g^{-1}, and CO conversion decreases slowly from 8.7% to 7.8%. Since an 86% reduction in surface area produces only a 10% reduction in CO conversion, the influence of catalyst surface is almost negligible. Similarly, the specific surface area of TiO$_2$ is not related with the alcohols STY. The alcohols in the products

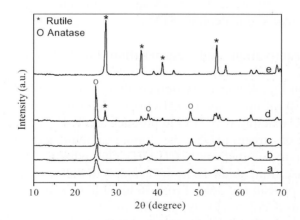

Fig. 1. XRD patterns of TiO$_2$ under different calcination temperatures (a. 673K; b. 773K; c. 873K; d. 973K; e. 1073K)

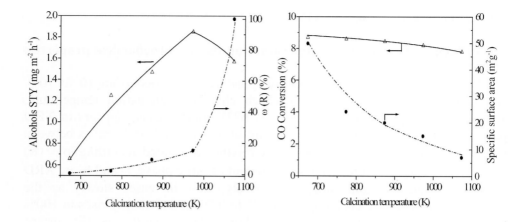

Fig. 2. (Left) Influence of calcination temperature on the rutile content ω (R) in TiO$_2$ and alcohols STY for reaction of CO with water to form alcohols

Fig. 3. (Right) Influence of calcination temperature on specific surface area of TiO$_2$ and CO conversion for reaction of CO with water to form alcohols

(Reaction conditions of Figs. 2 and 3 are: 573 K; 0.5MPa; CO flow rate of 30 ml min^{-1}; CO/H$_2$O (g) = 3/2 mol; TOS =12 h; Solvent- tetraethylenepentamine)

are methanol and ethanol. As shown in Fig.2, the alcohols STY first increases almost linearly with the increase of calcination temperature from 673 to 973K, then decreases at tempretures above 973K. The maximum alcohols STY of 1.8 mg m^{-2}h^{-1} is obtained at calcination temperature 973K, which corresponds to the rutile content of 18% due to the synergistic promotion of anatase and rutile in TiO$_2$. These results indicate that there exists an optimum rutile content in TiO$_2$ for producing the maximum alcohols STY.

3.3. Reaction deactivation and regeneration for CO hydration on TiO$_2$ calcined at 973K

As shown in Fig. 4, under reaction conditions of T = 573 K, P = 0.5 MPa, CO flow rate of 30 ml min^{-1} and CO/H$_2$O (g) = 3/2 (mol), the stability of TiO$_2$ catalyst for the CO reacting with H$_2$O to produce alcohols was performed. At the first 12 h of time-on-stream (TOS), the sample catalytic activity increases with TOS. The selectivity of methanol (CH$_3$OH) plus ethanol (C$_2$H$_5$OH) and CO conversion increase from about 12 to 46%, and 4 to 8.2%, respectively (4-12h), then both reach a plateau (12-36h). From 36 to 44 h, CO conversion and ethanol selectivity begin to decrease slowly, and the methanol selectivity changes from 33 to 27%. This means that the TiO$_2$ calcined at 973K deactivates after 36 h of

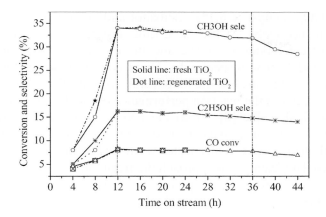

Fig. 4. Reaction performance of TiO$_2$ catalyst calcined at 973K as a function of time (573 K; 0.5 MPa; CO flow rate of 30 ml min^{-1}; CO/H$_2$O (g) = 3/2 mol; Solvent-tetraethylenepentamine)

TOS, although it shows better stability than PbO$_2$ catalyst whose reaction performance does not begin to decrease after only 20 h [4].

Klinger et al. [3] have reported that Pb(0) formed from PbO catalyst results into the catalyst deactivation for the conversion of CO with water. Here, we used XPS technique to characterize the valence states of TiO$_2$ before reaction and after the 44h life test as shown in Fig. 4. Two peaks centered at 458.3 and 464 eV respectively can be clearly observed obviously on the fresh TiO$_2$ catalyst. According to ref. [7], the two peaks belong to Ti(IV)$_{2P3/2}$ and Ti(IV)$_{2P1/2}$, respectively. The XPS spectra before and after reactions are similar, indicating that Ti^{+4} in TiO$_2$ can not be reduced to Ti(0) under such a reducing reaction atmosphere, and Ti^{+4} maybe the active centers for the reaction. This is also demonstrated by the result of H$_2$-TPR. In addition, TEM photographs show that the TiO$_2$ particle sizes before and after reactions are similar (~25.1 nm). From the above results, we can conclude that the TiO$_2$ deactivation does not result from the TiO$_2$ particle aggregation or from the reduction of TiO$_2$ for reaction of CO with water to form alcohols.

After the 44 h life test, it was found that the deactivated TiO$_2$ is partially covered by the alkali carbonate K$_2$CO$_3$. Then the mixture was treated by water wash method. The details of the process were as follows: the mixture of TiO$_2$ and K$_2$CO$_3$ was added to distilled water, then stirred for 4 h and kept still for 2 h. The resultant precipitate was washed until no CO$_3^{2-}$ ions were detected. Subsequently it was dried at 393K in N$_2$ for 2 h. Just washing the mixture of TiO$_2$ and K$_2$CO$_3$ with water, the reaction performance of the regenerated TiO$_2$ catalyst approaches to that of the fresh one as shown in Fig. 4. At the 12 h of TOS where the reaction performance is the maximum, the CO conversion of regenerated TiO_2 catalyst is 8.1%, methanol and ethanol selectivity are 32.7% and 15.5 % respectively. At the same time-on-stream the fresh TiO_2 catalyst shows the CO conversion of 8.2%, and the methanol and ethanol selectivity being 33.0% and 16.0%, respectively. There is an obvious difference between PbO and $PbTiO_3$ catalysts. For the latter catalysts, only first air combustion then water wash can restore the reaction performance of the deactivated samples to be arrived at the fresh value [8]. From the above results, the main deactivation reason is that the reaction active centers of TiO_2 catalyst are partially covered by the alkali carbonate K_2CO_3. Only water wash treatment is needed to regenerate the deactivated TiO_2 catalyst to approach the fresh one.

REFFRENCES

[1] B.J. Wood, W.E. Isakson, H. Wise, Ind. Eng. Chem. Prod. Res. Dev., 19 (1980) 197.
[2] G.B. Hoflund, W.S. Epling, D.M. Minahan, Catal. Lett., 62 (1999) 169.
[3] R.J. Klinger, J.W. Rathke, J. Am. Chem. Soc., 106 (1980) 7650.
[4] D.C. Ji, L.Y. Xu, S.L. Liu, Q.X. Wang, Chem. Lett., 11 (2002) 1160.
[5] H. Stephen, T. Stephen, Solubilities of Inorganic Compounds, Program on Pr, Oxford, 1963, Vol.1.
[6] G.X. Bai, Z.W. Zhang, J. An (eds.), Chemical Engineering Encyclopedia (in Chinese), Chemical Engineering Press, Beijing, 1990, Vol. 1, p.121.
[7] K.M. Krishna, M. Sharon, M.K. Mishra, J. Phy. Chem. Solids, 57 (1996) 615.
[8] D.C. Ji, Studies on catalytic conversion of CO and H_2O to alcohols, Ph D. Thesis., Dalian Institute of Chemical Physics, CAS, P. R. China, 2003.

Studies in Surface Science and Catalysis, volume 147
X. Bao and Y. Xu (Editors)
©2004 Elsevier B.V. All rights reserved.

Synthesis of gasoline additives from methanol and olefins over sulfated silica

Shaobin Wang[a] and James A. Guin[b]

[a]Department of Chemical Engineering, Curtin University of Technology, GPO Box U1987, Perth, WA 6845, Australia, Email: wangshao@vesta.curtin.edu.au

[b]Department of Chemical Engineering, Auburn University, Auburn, AL 36849, USA

ABSTRACT

Various sulfated solid catalysts were prepared and tested for etherification of certain C_6 olefins with methanol. Their catalytic activities were compared with the results obtained using some typical commercial ion-exchange resin catalysts. The commercial resin catalyst Amberlyst 15 showed nonselective conversion to ether, with significant accompanying isomerization, while Nafion and silica supported Nafion exhibited lower activities. Sulfated silica and montmorillonite showed comparable and sometimes higher ether yields than the commercial resin catalysts. Catalyst preparation variables influencing the catalytic activity of the sulfated silica were investigated with the objective of increasing ether yields.

1. INTRODUCTION

Oxygenated compounds such as ethers have significant potential as environmental enhancing additives for transportation fuels. Higher molecular weight oxygenates have lower water solubilities and vapor pressures than those of lower molecular weight oxygenates, e.g. MTBE, which have recently come under environmental pressure. Currently, ion exchange resins are the commercially dominated catalysts for ether production; however, they suffer from thermal stability, poor selectivity under certain conditions, and lack of regenerability. Several inorganic acid solids such as zeolites, sulfated zirconia, acid-treated clays, heteropolyoxoanions (HPA) and supported HPA catalysts have been tested for the production of ethers like MTBE [1-6]. However, few investigations have been conducted on synthesis of higher (C_5 - C_7) ethers using these types of solid acids. Indeed, most prior investigations concerning the etherification of higher olefins have focused instead on kinetic studies over the commercial resin catalysts [7-10]. In this paper, we report a preliminary investigation on the utilization of sulfated solids as catalysts and their performance in etherification of typical C_6 olefins with methanol. These particular C_6 olefins were chosen on the basis of their potential for producing oxygenated transportation fuel additives, as disclosed in a recent patent [11].

2. EXPERIMENTAL

Several commercial solids, alumina, silica, bentonite (BT), and montmorillonite (MT), were obtained from Aldrich. An activated carbon (AC) sample was obtained from Calgon. $Zr(OH)_4$ and $Fe(OH)_3$ samples were prepared by precipitation of $ZrO(NO_3)_2$ and $Fe(OH)_3$ solution with $NH_3 \cdot H_2O$ at pH = 10, followed by filtration, washing, and drying at 100 °C overnight. All these solids were then impregnated with 0.5 M sulfuric acid solution, followed by drying at 100 °C overnight and calcination at 300 °C for 3 h. A commercial ion exchange resin, Amberlyst 15 also obtained from Aldrich, was used as a reference catalyst. Two other commercial resin catalysts, Nafion NR50 and silica supported Nafion SAC-13 obtained from Fluka and Aldrich, respectively, were also evaluated as reference catalysts in this investigation.

The reactions were carried out at 80 ± 1 °C for 2 h in 25 ml stainless steel batch reactors under a pressure of 1.8 ± 0.1 MPa of dried helium and with constant agitation. 0.5 g solids were loaded in all the cases. Before reaction, all catalysts were dried in an oven at 60 °C under vacuum overnight. The reactants, consisting of 0.2 g methanol (99.93%) and 0.5 g 2,3-dimethyl-1-butene (23DM1B, 97%) or 2,3-dimethyl-2-butene (23DM2B, 99%) with an alcohol:olefin molar ratio of ca. 1:1, were mixed with heptane (4 g) as a solvent. After reaction, the reactors were quickly cooled down to room temperature. The products were then analyzed on a Varian gas chromatograph equipped with a capillary column and FID.

3. RESULTS AND DISCUSSION

Table 1 compares the catalytic activities for etherification of 23DM1B with methanol using the sulfated solid catalysts prepared in this work with some commercial resin catalysts. In the reaction, the olefin isomer 23DM2B and the ether are the major products. A small amount of 2, 3-dimethyl-2-butanol was also detected, probably due to the reaction between olefins and a trace amount of water in catalysts and reactant solution. The three commercial catalysts show varying activities with the Amberlyst 15 exhibiting the best overall conversion of 23DM1B to 23DM2B and ether of approximately 90%; however, this catalyst is nonselective for etherification having instead a high (undesirable) isomerization selectivity at 80% and a resulting 19% ether yield. Nafion NR50

Table 1
Catalytic activity of various acid-treated solids and commercial catalysts in 23DM1B etherification with methanol

Catalyst	23DM1B conversion (%)	23DM2B selectivity (%)	Ether selectivity (%)	Ether yield (%)
Amberlyst 15	91.0	78.7	20.6	18.8
Nafion NR50	29.0	44.0	53.9	15.6
Nafion SAC-13	3.2	21.7	58.9	1.9
S-BT	17.0	36.6	59.6	10.1
S-MT	44.4	46.2	52.6	23.4
S-AC	3.2	30.5	54.9	1.8
S-SiO$_2$	90.3	80.0	19.6	17.7
S-ZrO$_2$	0	0	0	0
S-Fe$_2$O$_3$	0	0	0	0
S-Al$_2$O$_3$	0	0	0	0

shows low activity and isomerization selectivity, producing an ether yield of 16%. The silica supported Nafion SAC-13 shows even lower activity than the unsupported Nafion NR50, which is somewhat different from its behavior in certain related acid-catalyzed reactions [12,13]. In those investigations, Nafion SAC-13 was found to exhibit higher activity than Nafion NR50. Among the sulfated solid catalysts prepared in this investigation, S-SiO$_2$ shows an activity similar to Amberlyst 15. S-MT gives moderate conversion (44%) but lower selectivity to isomerization (46%). Sulfated BT and activated carbon exhibit lower conversion but high ether selectivity at around 50 - 60%. Sulfated zirconia, alumina and iron oxide show no activity for this reaction. Based on the results in Table 1 in terms of ether yield, S-SiO$_2$ and S-MT are good candidates for etherification and therefore were chosen for further investigation as a function of catalyst preparation variables, with the objective of improving overall ether yield.

It is well known that etherification or isomerization of olefins generally occurs on acid sites and the overall catalytic activity is believed to be related to the acidity of the catalysts. The order of activity on resin catalysts follows the same order as their acidity [14]. Sulfation of SiO$_2$, AC and clays (bentonite and montmorillonite) can produce acid sites, which are active for etherification. Knifton and Edwards [2] have found that HF-treated montmorillonite and mineral acid treated clays showed high activity for MTBE synthesis. The inactivity of S-ZrO$_2$, S-Fe$_2$O$_3$, and S-Al$_2$O$_3$ is probably due to the strong basicity of the oxides coupled with the fact that acid treatment did not exert a strong influence on their acidity/basicity. Other investigators have found that sulfated zirconia is effective for MTBE synthesis [6]; however, the activity is dependent on the preparation method. Calcinations of sulfated zirconia at low temperatures did not produce a superacid and thus it showed a low activity. Our own investigations also indicate that sulfated zirconia calcined at 600 °C will be

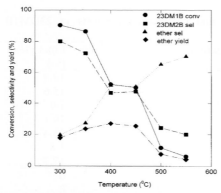

Fig. 1. Effect of calcinations temperature on activity of S-SiO$_2$.

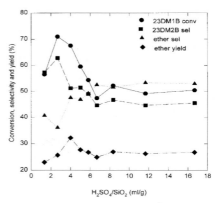

Fig. 2. Effect of loading of SO$_4^{2-}$ on activity of S-SiO$_2$.

active for etherification [14].

It is believed that calcination temperature can affect the loading of sulfur on the catalyst, thus influencing the acidity of catalyst and catalytic activity. In this regard, Fig.1 presents the catalytic behavior of S-SiO$_2$ at varying calcination temperatures. One can see that 23DM1B conversion and 23DM2B selectivity decrease with the increasing temperature while the selectivity to ether generally increases as the temperature increases. The ether yield initially increases with the increasing temperature to reach the highest value at 400-450°C and then decreases rapidly at higher temperatures. Calcination at high temperatures usually makes a greater loss in sulfate content on the surface of catalyst, resulting in the reduction of acid sites for reaction. Thermogravimetric analysis (TGA) shows that three weight loss peaks occur in the heat treatment of S-SiO$_2$. The peak below 200 °C is due to the loss of surface water while the peak at 200-300 °C can be ascribed to the decomposition of sulfate. The small peak occurring at 580-650 °C is probably the further decomposition of the remaining sulfate. This conforms the variation of activity to the calcination temperature, indicating higher activity can be obtained before 450 °C and after that the activity will decrease rapidly due to the further loss of acid sites.

The relationship between the amounts of SO$_4^{2-}$ impregnated on SiO$_2$ is illustrated in Fig.2. As shown that addition of SO$_4^{2-}$ on silica at low amount can increase 23DM1B conversion; however, the conversion decreases as additional SO$_4^{2-}$ is loaded on the solid. After the ratio reaches 7 ml/g-SiO$_2$, the conversion no longer changes, maintaining a value of 50%. The selectivity to 23DM2B exhibits a similar variation to the 23DM1B conversion while ether selectivity varies in an opposite manner. The conversion approaches the highest level of 72% at the ratio of 3 ml/g-SiO$_2$, although the best ether yield of 32% is attained at a ratio around 4 ml/g-SiO$_2$.

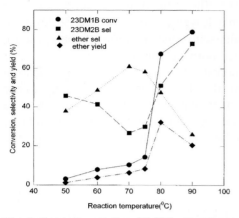

Fig. 3. Catalytic activity of S-SiO₂ at various reaction temperatures.

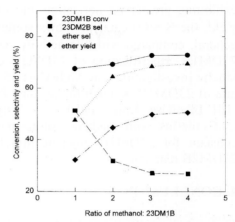

Fig. 4. The effect of sulfate loading on S-SiO₂ activity.

In order to examine activity of the prepared catalysts under varying reaction conditions, the catalytic activities of S-SiO₂ at different temperatures were investigated and are shown in Fig.3. S-SiO₂ catalysts exhibit low conversions at temperatures below 75 °C although the conversion shows an increasing trend with increasing temperature. At 90 °C the conversion reaches 80%. The ether selectivity also shows an increasing trend reaching the maximal value at 70 °C. For ether yield, the highest value, 32%, is achieved at 80 °C. This suggests that higher temperatures will not favor the etherification, in keeping with the thermodynamic equilibrium behavior expected for exothermic reactions.

The effect of methanol to olefin molar ratio on activity of S-SiO₂ was also investigated and the results are given in Fig.4. High methanol:olefin ratios are beneficial to olefin conversion, ether selectivity, and ether yield. However, the values of 23DM1B conversion, ether selectivity and yield level off at 73%, 70% and 50%, respectively, after the ratio reaches an excess around 3 - 4 mol methanol/mol-olefin.

Table 2 presents the catalytic activity of S-SiO₂ and Amberlyst 15 for etherification of the β-olefin, 23DM2B, with methanol. Similar to the results with the α-olefin, Amberlyst 15 shows a higher overall conversion while

Table 2
Catalytic activities of sulfated silica and Amberlyst 15 in 23DM2B etherification with methanol

Catalyst	23DM2B conversion (%)	23DM1B selectivity (%)	Ether selectivity (%)	Ether yield (%)
Amberlyst	27.3	33.9	64.4	17.6
S-SiO₂	17.0	20.5	76.6	13.0

exhibiting greater isomerization and lower ether selectivity. In terms of ether yield, the S-SiO$_2$ catalyst shows somewhat lower values with the β-olefin. In general, compared with the activity results with etherification of the α-olefin, 23DM1B, etherification of 23DM2B over both catalysts is much lower. Our kinetic investigations on Amberlyst 15 indicate that the methanol etherification rate of 23DM1B is much higher than that of 23DM2B at the same temperature [15]. Likewise, Rihko and Krause [16] investigated the etherification of some C$_5$-C$_7$ olefins with methanol on resin catalysts and found equilibrium and rate constant for 23DM1B etherification to be much higher than those for the 23DM2B reaction.

4. CONCLUSION

The commercial ion-exchange resin, Amberlyst 15, was an effective but nonselective catalyst in etherification of C$_6$ olefins with methanol. Nafion and silica supported Nafion were less effective for etherification. Laboratory prepared sulfated solid catalysts showed varying catalytic activities depending on the nature of solids and preparation conditions. Sulfated silica systems exhibited a comparable and sometimes higher ether yield than that of the commercial resin catalysts and thus showed potential as alternative catalysts for etherification of olefins.

REFERENCES

[1] K. Arata, Appl. Catal. A., 146 (1996) 3.
[2] J.F. Knifton and J.C. Edwards, Appl. Catal., 183 (1999) 1.
[3] F. Collignon, M. Mariani, S. Moreno, M. Remy and G. Poncelet, J. Catal., 166 (1997) 53.
[4] F. Collignon, R. Loenders, J.A. Martens, P.A. Jacobs and G. Poncelet, J. Catal., 182 (1999) 302.
[5] G. Baronetti, L. Briand, U. Sedran and H. Thomas, Appl. Catal. A., 172 (1998) 265.
[6] M.E. Quiroga, N.S. Figoli and U.A. Sedran, React. Kinet. Catal. Lett., 63 (1998) 75.
[7] R.L. Piccoli and H.R. Lovisi, Ind. Eng. Chem. Res., 34 (1995) 510.
[8] T. Zhang, K. Jensen, P. Kitchaiya, C. Phillips and R. Datta, Ind. Eng. Chem. Res., 36 (1997) 4586.
[9] M.D. Girolamo, M. Lami, M. Marchionna, E. Pescarollo, L. Tagliabue, and F. Ancillotti, Ind. Eng. Chem. Res., 36 (1997) 4452.
[10] D. Parra, J.F. Izquierdo, F. Cunill, J. Tejero, C. Fite, M. Iborra and M. Vila, Ind. Eng. Chem. Res., 37 (1998) 3575.
[11] D.E. Hendriksen, Use of Tertiary-Hexyl Methyl Ether as a Motor Gasoline Additive, United States Patent No. 5 752 992 (1998).
[12] M.A. Harmer, W.E. Farneth and Q. Sun, J. Am. Chem. Soc., 118 (1996) 7708.
[13] A. Heidekum, M.A. Harmer and W.F. Hoelderich, J. Catal., 176 (1998) 260.
[14] S. Wang and J. A. Guin, Chem. Comm., 2000, 2499.
[15] J. Liu, S. Wang and J. A. Guin, Fuel Process. Technol., 69 (2001) 205.
[16] L.K. Rihko and A.O.I. Krause, Ind. Eng. Chem. Res., 35 (1996) 4563.

Studies in Surface Science and Catalysis, volume 147
X. Bao and Y. Xu (Editors)

445

Synthesis, characterization, and MTO performance of MeAPSO-34 molecular sieves

Lei Xu, Zhongmin Liu[*]**, Aiping Du, Yingxu Wei, Zhengang Sun**

Dalian Institute of Chemical Physics, Chinese Academy of Sciences,
457 Zhongshan Road, P. O. Box 110, Dalian, 116023, China

ABSTRACT

MeAPSO-34 molecular sieves were synthesized successfully with metal ions of Ti, Cr, Mn, Fe, Co, Ni, Cu, Zn, Mg, Ca, Sr, and Ba (Me/Al$_2$O$_3$=0.05). IR, NH$_3$-TPD measurements and fixed bed MTO performances of all the samples were carried out in comparison with SAPO-34. The results illustrated that the incorporation of metal ions had great effects on structure and acidity of the molecular sieves. It was interesting that most MeAPSO-34 molecular sieves exhibited effective enhancement of the MTO performance. Especially, the incorporation of Co^{2+}, Zn^{2+} and Mg^{2+} ions showed promising modification effect for the MTO catalysts.

1. INTRODUCTION

Methanol to olefins (MTO) is a well known important reaction for non-oil route synthesis of light olefins from natural gas or coal [1]. Small pore silicoaluminophosphate (SAPO) molecular sieve was demonstrated to be effective catalysts, in which SAPO-34 has been proven to be the most attractive candidate for practical catalyst development [2]. The SAPO-34 molecular sieve after incorporating metals into the framework had been found exhibiting high selectivity to form ethylene in the MTO reaction [3-6]. The higher selectivity could be attributed to the modifications of the acidic strength, the acid distribution and the pore or channel size of SAPO-34 molecular sieve with the incorporation of metals.

The present work was to synthesize MeAPSO-34 molecular sieves with a large variety of metal ions, to research the modification effect of metal ions on physical and chemical properties of SAPO-34 and to comparatively investigate MTO performance of the synthesized samples. This would be helpful for understanding the relationship among synthesis, property and performance of the catalyst and the reaction.

* Corresponding Author: Email: zml@dicp.ac.cn Phone: +86 411 4685510 Fax: +86 411 4379289

2. EXPERIMENTAL

2.1. Synthesis

All the syntheses were carried out hydrothermally. Pseudobeohmite(72.2 wt% Al_2O_3, purissimum grade), ortho-phosphorate (85 wt%, analytical grade) and silica sol (25.0 wt% SiO_2) were used as the starting materials to get aluminum, hosphorous and silicon respectively. Triethylamine (TEA) was used as templating agent.

SAPO-34 molecular sieve was synthesized by the batch composition: 3TEA: $0.2SiO_2$: Al_2O_3: P_2O_5: $50H_2O$, and MeAPSO-34 samples were synthesized under the same condition as that for SAPO-34 apart from further addition of metal salts into the synthesis system with Me/Al_2O_3 (Me denotes metal ions) ratio equaled 0.05. The metal ions used were Cr, Fe, Ni, Cu, Ca, Sr, Ba ions (as metallic nitrate), Mn, Co, Zn, Mg ions (as metallic acetate) and Ti ion (as $Ti(SO_4)_2$).

2.2. Characterization and catalytic tests

The structure type and crystallinity of all the samples were checked by powder X-ray diffraction (XRD) on a D/max-rb diffractmeter.

Chemical composition of all the samples was determined by X-ray fluorescence analysis on a Philips Magix spectrometer.

The IR spectra were recorded on a Perkin 983G infrared spectrophotometer. The framework vibrations of the molecular sieves were examined by using 5% sample in KBr.

The measurement of acidity of all samples was performed on a homemade ammonia-TPD apparatus.

The MTO reactions were carried out on a fixed bed reactor. Reaction conditions were as follows: temperature=450°C, WHSV(MeOH)=$2h^{-1}$. On-line GC with Porapak-QS column and FID detector was used in product analysis.

3. RESULTS AND DISCUSSION

3.1. Synthesis of MeAPSO-34 molecular sieves

MeAPSO-34 molecular sieves could be synthesized successfully with Me/Al_2O_3=0.05, in which Me were Ti, Cr, Mn, Fe, Co, Ni, Cu, Zn, Mg, Ca, Sr, and Ba respectively. The relative crystallinity of all samples was calculated by the peak intensities of $2\theta \sim 9.430$, 15.940 and 20.510 in the XRD pattern on the basis of taking that of SAPO-34 as 100%. The results indicated that the relative crystallinities of most samples were higher than 85% except the cases of TAPSO-34 and CaAPSO-34 (lower than 70% without impurity). ZnAPSO-34 gave the highest relative crystallinity up to 124%, indicating that the synthesis system was favorable for the formation of CHA structure.

3.2. Chemical composition of products

The composition of the corresponding as-synthesized MeAPSO-34 samples with different metals was given in Table 1.

In case of SAPO molecular sieve, it had been concluded that there was no direct linkage between Si and P via oxygen-bridge. It was proposed that Si atoms incorporate into $AlPO_4$ structure by two different substitution mechanisms [7-8]: the first one, the Si substitution for phosphorus formed Si(4Al) structure; the second mechanism was the double substitution of neighboring aluminum and phosphorus by two silicon atoms forming Si(nAl)(n=3-0). So the ratio of (Si+P)/Al should be equal to 1 if there were no Si-O-Si (or no Si-O island) in the structure. In MeAPSO system, the formation of MeAPSO molecular sieve could be taken as the replacement of P and/or Al of the $AlPO_4$ framework by silicon and/or metal. If the ratio of (Si+P)/(Al+Me) ratio were equal to or more than 1, the incorporation should be taken as a partial replacement of Al by Me ions. If the ratio of (Si+P)/(Al+Me) ratio were less than 1, metal ions could be incorporated into the framework by replacing silicon. This could be an evidence for the incorporation of metal ions into framework structure.

In Table 1, for most samples except CuAPSO-34 and SrAPSO-34, the (Si+P)/(Al+Me) ratios were more than or equal to 1, indicating that tetrahedrally coordinated aluminum was partially replaced by Me ions. It was possible that there was also no Me-O-Al and lack of Si-O-Si linkage in the most MeAPSO-34 samples.

Table 1
Molar composition of the crystalline products of as-synthesized MeAPSO-34

Products	Molar composition of the crystalline products		
	Al: Si: P: Me	(Si+P)/(Al+Me)	Molar oxide composition
SAPO-34	1.000:0.124:0.871	0.995	$(Al_{0.501}Si_{0.062}P_{0.437})O_2$
TAPSO-34	1.000:0.152:0.873:0.018	1.007	$(Al_{0.490}Si_{0.074}P_{0.427}Ti_{0.009})O_2$
CrAPSO-34	1.000:0.150:0.881:0.004	1.027	$(Al_{0.492}Si_{0.074}P_{0.433}Cr_{0.002})O_2$
MnAPSO-34	1.000:0.135:0.912:0.031	1.016	$(Al_{0.481}Si_{0.065}P_{0.439}Mn_{0.015})O_2$
FAPSO-34	1.000:0.132:0.905:0.032	1.005	$(Al_{0.483}Si_{0.064}P_{0.438}Fe_{0.015})O_2$
CoAPSO-34	1.000:0.133:0.917:0.032	1.017	$(Al_{0.480}Si_{0.064}P_{0.440}Co_{0.016})O_2$
NiAPSO-34	1.000:0.139:0.884:0.023	1.000	$(Al_{0.489}Si_{0.068}P_{0.432}Ni_{0.011})O_2$
CuAPSO-34	1.000:0.136:0.869:0.032	0.974	$(Al_{0.491}Si_{0.067}P_{0.427}Cu_{0.015})O_2$
ZnAPSO-34	1.000:0.141:0.902:0.041	1.002	$(Al_{0.480}Si_{0.068}P_{0.433}Zn_{0.020})O_2$
MgAPSO-34	1.000:0.146:0.904:0.016	1.033	$(Al_{0.484}Si_{0.070}P_{0.438}Mg_{0.008})O_2$
CaAPSO-34	1.000:0.149:0.888:0.005	1.032	$(Al_{0.490}Si_{0.073}P_{0.435}Ca_{0.003})O_2$
SrAPSO-34	1.000:0.135:0.883:0.031	0.987	$(Al_{0.488}Si_{0.066}P_{0.431}Sr_{0.015})O_2$
BaAPSO-34	1.000:0.146:0.875:0.008	1.013	$(Al_{0.493}Si_{0.072}P_{0.431}Ba_{0.004})O_2$

448

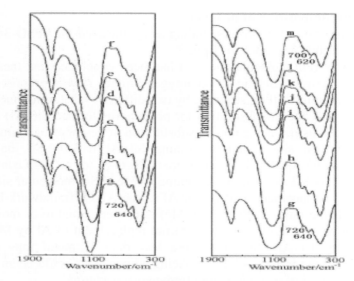

Fig. 1. IR spectra of TAPSO-34(a), CrAPSO-34(b), MnAPSO-34(c), FAPSO-34(d), CoAPSO-34(e), NiAPSO-34(f), CuAPSO-34(g), ZnAPSO-34(h), MgAPSO-34(j), CaAPSO-34(g), SrAPSO-34(k), BaAPSO-34(l), SAPO-34(m) in the framework region

3.3. IR studies in the framework region

The IR spectra of the framework vibration of as-synthesized SAPO-34, MeAPSO-34 samples were presented in Fig. 1. For SAPO-34, the bands around $1100cm^{-1}$, $620cm^{-1}$ and $700cm^{-1}$ could be observed. The bands around $1100cm^{-1}$ had been assigned to the asymmetric stretching of the TO_4 tetrahedra, the $620cm^{-1}$ was due to vibrations of the double-ring, and the $700 cm^{-1}$ could be related to T-O-T symmetric stretching. For MeAPSO-34, the framework vibration bands of all samples were similar to that of SAPO-34, but a shift of the peak positions of structure-sensitive bands could be found. For example, when metal ions were incorporated into the framework, the bands at 620, $700 cm^{-1}$ in MeAPSO-34 were shifted to around 640, $720 cm^{-1}$. Moreover, the change in the intensity of some absorption peaks could be found. For MnAPSO-34, CoAPSO-34, ZnAPSO-34 and MgAPSO-34, the intensity of peaks around $720cm^{-1}$ increased. This demonstrated that metal ions could be incorporated into the framework of SAPO-34 to form Me-O-T bond, which brought the shift of peaks and variety of intensity in the framework vibration bands.

3.4. NH₃-TPD studies

The acidity properties of the MeAPSO-34 samples were measured by NH₃-TPD technique in comparison with that of SAPO-34. TPD profiles of NH₃

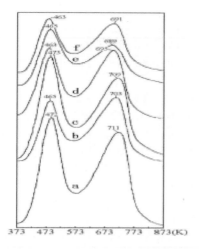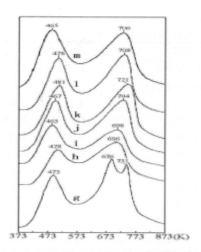

373 473 573 673 773 873(K)   373 473 573 673 773 873(K)

Fig. 2. TPD profiles of ammonia desorption from TAPSO-34(a), CrAPSO-34(b), MnAPSO-34(c), FAPSO-34(d), CoAPSO-34(e), NiAPSO-34(f), CuAPSO-34(g), ZnAPSO-34(h), MgAPSO-34(i), CaAPSO-34(j), SrAPSO-34(k), BaAPSO-34(l) and SAPO-34(m)

desorption from SAPO-34 and MeAPSO-34 are given in Fig. 2. For SAPO-34, in addition to the low temperature peak at 465K that was due to the acidity of external surface hydroxy groups, second peak is observed at 706K, which could be assigned to structure acidity. The NH_3 desorption temperatures of MeAPSO-34 molecular sieves were quite different, which was associated with the variety of intensities of both weak and strong acid sites. It was interesting that CuAPSO-34 gave two peaks of strong acid sites, and the NH_3 desorption temperatures were 676K and 731K, respectively. For FeAPSO-34, CoAPSO-34, NiAPSO-34, ZnAPSO-34 and MgAPSO-34, the amount of strong acid sites decreased. These results demonstrated that the incorporation of metal ions could adjust the acid intensity and the number of weak and strong acid sites.

3.5. MTO results of MeAPSO-34

Table 2 listed the activity and the optimum selectivities of methole to $C_2^=+C_3^=$ of MTO reaction over the SAPO-34 and MeAPSO-34 samples. It could be seen that for all MeAPSO-34 samples the initial conversion of methanol can reach 100%, which is similar to that of SAPO-34. The life-time (denoting the time on stream when conversion is lower than 100%) could be used to represent the activities of the samples, and the sequence of the life-time for the samples were: SrAPSO-34>CaAPSO-34>MnAPSO-34>BaAPSO-34>CrAPSO-34> CuAPSO-34>MgAPSO-34>FAPSO-34>TAPSO-34>SAPO-34>ZnAPSO-34> CoAPSO-34>NiAPSO-34. This indicates that the incorporation of certain metal

Table 2
Optimum product distribution in MTO over SAPO-34 and MeAPSO-34 catalysts

Samples	SAPO-34	Ti-	Cr-	Mn-	Fe-	Co-	Ni-	Cu-	Zn-	Mg-	Ca-	Sr-	Ba-
TOS (min.)	120	136	164	170	135	90	60	159	119	160	177	192	142
Products (wt%)													
CH_4	1.8	1.6	1.7	1.4	1.9	1.8	3.9	1.1	1.2	1.1	1.1	1.5	1.1
C_2H_4	53.3	50.0	54.3	56.2	58.6	60.4	55.7	52.0	59.5	54.1	52.2	53.3	50.9
C_2H_6	0.8	0.5	0.6	0.3	0.5	0.2	0.2	0.3	0.3	0.3	0.3	0.4	0.3
C_3H_6	28.9	36.7	33.8	32.1	29.8	32.5	31.5	36.4	32.1	37.5	37.7	36.6	38.5
C_3H_8	1.9	0.0	0.0	0.0	0.7	0.0	0.0	0.0	0.0	0.0	0.0	0.0	0.0
C_4H_8	7.4	7.6	6.5	5.6	5.0	3.2	4.6	6.3	4.4	4.5	5.5	5.4	6.1
C_4H_{10}	1.4	1.4	1.2	1.0	0.9	0.6	0.9	1.2	0.8	0.8	1.0	1.0	1.1
C_5^+	4.6	2.3	2.0	3.4	2.7	1.4	3.2	2.8	1.6	1.8	2.2	1.8	2.0
$C_2^=$-$C_3^=$ (wt%)	82.2	86.7	88.1	88.3	88.4	92.9	87.2	88.4	91.6	91.6	89.9	89.9	89.4

ions could yield a modification effect on activity. For example, the incorporation of four kinds of alkali-earth ions into SAPO-34 prolonged the life-time of the catalysts. It was also observed that the selectivities to form C_2+C_3 olefins of all MeAPSO-34 catalysts were superior to that of SAPO-34. Especially, the selectivities to C_2+C_3 olefins of CoAPSO-34, ZnAPSO-34 and MgAPSO-34 were more than 90%.

4. CONCLUSION

MeAPSO-34 molecular sieves had been synthesized successfully with Me/Al_2O_3=0.05. The results indicated that the incorporation of metal ions had great effects on structure and acidity of the molecular sieves. It was interesting that most MeAPSO-34 molecular sieves exhibited effective enhancement of the MTO performance. Especially, the incorporation of Co^{2+}, Zn^{2+} and Mg^{2+} ions shows promising modification effect for the MTO catalysts.

REFERENCES

[1] B.M. Lok, C.A. Messina, R.L. Patton, R.T. Gajek, T.R. Cannan, E.M. Flanigen, J. Am. Chem. Soc., 106 (1984) 6092.
[2] S. Nawaz, S. Kolboe, S. Kvisle, K.P. Lillerud, M. Stocker, H.M. Oren, Stud. Surf. Sci. Cat., 61 (1991) 421.
[3] M.A. Djieugoue, A.M. Prakash, L. Kevan, J. Phys. Chem. B, 104 (2002) 10726
[4] T. Inui, S. Phatanasri, H. Mutsuda, J. Chem. Soc., Chem. Commun., (1990) 205
[5] J. M. Thomas, Y. Xu, C.R.A. Catlow, J.W. Couves, Chem. Mater., 3 (1991) 661
[6] M. Kang, J. Mol. Catal., A Chem., 150 (1999) 205
[7] S. Ashtekar, S.V.V. Chilukuri and D.K. Chakrabarty, J. Phys. Chem. B, 98 (1994) 4878.
[8] G. Sastre, D.W. Lewis, C. Richard and A. Catlow, J. Phys. Chem. B, 101 (1997) 5249.

Studies in Surface Science and Catalysis, volume 147
X. Bao and Y. Xu (Editors)

451

Bifunctional catalysts for the production of motor fuels and aromatic hydrocarbons from synthesis gas

V. M. Mysov[*], K. G. Ione

Scientific Engineering Centre "Zeosit" SB RAS, Pr. Ak. Lavrentieva, 5, Novosibirsk 630090, Russia

ABSTRACT

The work has focused on the investigation of the one-stage synthesis gas conversion to motor fuels and the alkylation of aromatic hydrocarbons (toluene and 2-methylnaphthalene) with synthesis gas over bifunctional catalysts in a fixed-bed flow reactor. The influence of the catalysts composition, the type of the metal oxides (Zn, Cr, Fe, Co, Ru) and of acid component (ZSM-5, Beta, SAPO-5, mordenite etc.), pressure, temperature (210-440°C) and H_2/CO ratio (1.5-10) on the activity and selectivity was examined.

1. INTRODUCTION

For the last three decades, much research has been carried out on the selective synthesis of high-octane gasoline and C_4-C_6 iso-paraffins from CO and H_2 with the use of bifunctional catalytic systems composed of a Fischer-Tropsch (or Cu-Zn, Zn-Cr) component and ZSM-5 zeolite [1-5]. It is of interest to study the conversion of synthesis gas over various bifunctional catalysts aimed to produce motor fuels (gasoline, diesel), aromatic hydrocarbons and medium-weight olefins.

The use of bifunctional catalytic systems composed of a methanol synthesis catalyst and a synthetic zeolite in the syngas alkylation of aromatic hydrocarbons seems promising as regards the development of natural gas chemistry. Namba et al. [6] showed that p-xylene selectivity of 94 % at 12.1 % of toluene conversion is achieved in toluene alkylation with methanol on ZSM-5 zeolite modified by boron compounds. However, the use of the same modified zeolite in combination with Zn-Cr component as a composite catalyst for toluene alkylation with synthesis gas results in a low para-selectivity of 42.9 % at only 3.5 % toluene conversion.

[*] Corresponding Author: **Email:** mysov@batman.sm.nsc.ru **Tel/Fax:** (3832) 356251

The purpose of our paper is to investigate the one-stage synthesis gas conversion to hydrocarbons (iso-products and aromatics) and the alkylation of aromatic hydrocarbons (toluene and 2-methylnaphthalene (2-MN)) with synthesis gas to p-xylene and 2,6+2,7-dimethylnaphthalenes (2,6+2,7-DMN) respectively in a fixed bed flow reactor.

2. EXPERIMENTAL

The oxides of Zn, Cr, Fe, Co, Ru combined with synthetic zeolites were used as bifunctional catalysts. ZSM-5 type zeolites with SiO_2/Al_2O_3 = 27, 75, 100 as well as ZSM-12 (SiO_2/Al_2O_3 = 200) and β zeolite (SiO_2/Al_2O_3 = 75) were prepared through hydrothermal synthesis. Si, Mg- and Si-modified zeolites were prepared by impregnating H-form zeolites with solutions of Si, Mg-compounds. To prepare bifunctional catalysts, zeolite powder and metallic oxides powder were mixed, which was followed by pressing, crushing and sieving.

The catalytic experiments were carried out in a fixed-bed flow reactor in the gas phase at 210-440°C, 10-80 atm, $GHSV_{SYNGAS}$ = 100-20000 h^{-1}, H_2/CO = 1.5-10, $WHSV_{Tol}$ = 1-3 h^{-1} and $WHSV_{2-MN}$ = 0.1 h^{-1}. Circulation gas flow system with condensing and separation of alkylation liquid products was applied.

The reaction products were analyzed with three gas chromatographs equipped with integrators. H_2, CO, CO_2, N_2 and CH_4 were analyzed in a thermal conductivity detector (TCD) with an activated carbon column. Methanol, DME, water was analyzed with a TCD using column filled with chromosorb-102. The gaseous hydrocarbons were analyzed with a TCD using column packed with Al_2O_3. The liquid hydrocarbons were analyzed by a TCD using a 3-m-long column with Benton-34/SP-1200 on 100/120 Supelcoport or by a flame ionization detector (FID) with a 50-m-long capillary column. The identification of products was conducted by GC-MS.

3. RESULTS AND DISCUSSION

3.1. FTS+Hydrocracking reactions on bifunctional catalysts

The influence of the type of the metal oxides and of the acid component, pressure, temperature, H_2/CO ratio on the activity and selectivity of the bifunctional catalysts was examined (Table 1).

Gasoline fractions of high quality were obtained with a low content of benzene (<1 wt.%). This fact is explained by the high alkylating ability of the acid component of the bifunctional catalyst. The diesel fractions contain branched C_{10}-C_{20} hydrocarbons (40-60 wt.%) and the content of the aromatic hydrocarbons is less than 3 wt.%. The C_1-C_{20} hydrocarbons composition against carbon atom numbers differed considerably from the Schultz-Flory distribution typical for Fischer-Tropsch products. The difference is caused by the proceeding

Table 1
The investigated ranges of the reaction parameters and the performance of bifunctional catalysts in hydrocarbons (HC) synthesis from CO and H_2

Metal component	Co - Ru		Fe			Zn - Cr		
Acid component	ZSM-5, Beta, SAPO, Mordenite		ZSM-5, Beta, SAPO		ZSM-5	ZSM-5		Beta, SAPO
H_2/CO	1.5-1.8	1.7-2.0	1.7-2.0	~2	2-3	1-2		~2
Pressure, atm	10 - 20		30 - 50			60 - 80		
Temperature, °C	200-220	220-250	240-280	280-360	360-400	380-420		360-420
Selectivity, wt.%								
CH_4	7-12	8-20	4-6	5-8	1-4	4-6		3-6
C_2-C_4	5-9	6-10	7-14	8-16	16-22	20-26		26-34
C_{5+}	82-87	74-85	80-87	76-84	74-82	68-76		60-70

of some reactions on the acid component of the bifunctional catalysts, such as hydrogen redistribution, oligomerization, cracking, isomerization, and alkylation of aromatics [3, 5].

A large body of experimental data on using various bifunctional catalysts composed of a hydrogenating component (Co-Ru, Fe, or Zn-Cr) and an acid one (ZSM-5, Beta with $SiO_2/Al_2O_3 = 75$ for both zeolites) in liquid hydrocarbons synthesis was examined to define their isomerizing ability at the temperatures optimal for synthesis gas conversion.

Fig. 1 demonstrates the averaged mass ratios of the formed iso-hydrocarbons

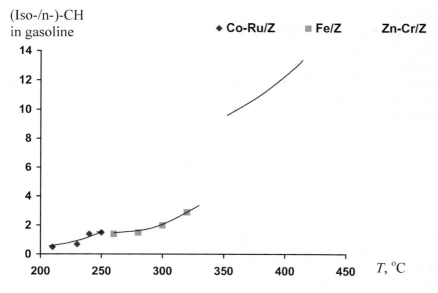

Fig. 1. Isomerizing activity of various bifunctional catalysts in gasoline synthesis from CO and H_2. Z=ZSM-5 (or Beta), $SiO_2/Al_2O_3 = 75$.

to the linear ones for three various bifunctional systems. It can be seen that Co-Ru-component catalysts, which are active in converting synthesis gas to liquid hydrocarbons at temperatures 210-250°C, scarcely isomerize the formed hydrocarbons. On the contrary, the high-temperature Zn-Cr-containing bifunctional catalysts, active in methanol synthesis, produce mainly iso-paraffins.

At temperatures under 280°C, the Fe/Z catalysts are close to the Co-Ru/Z ones in their isomerizing activity while at temperatures above 300-320°C, they tend to show a high isomerizing ability sometimes reaching iso/normal ratio in gasolines above 2 (Fig. 1). The differing isomerizing abilities of the three catalysts groups are most of all affected by the reaction temperature. An increase in the reaction temperature facilitates isomerization of the intermediate products of synthesis gas conversion over the zeolite component. Besides the temperature effect, the nature of the hydrogenating component is also important for iso-paraffins synthesis. When using Co-Ru-catalysts at temperatures 210-230°C, synthesis gas is converted mainly to n-paraffins, which in this temperature range can be scarcely isomerized by the zeolite component of the bifunctional catalyst. When using Fe catalysts, synthesis from CO and H_2 produces mainly alcohols and olefins, and at temperatures above 300°C these are converted to iso-olefins and aromatic hydrocarbons over ZSM-5 zeolites.

The high isomerizing ability of the Zn-Cr-zeolite catalysts may be probably defined by the rapidly proceeding at 360-420°C methylation of the olefins intermediately formed over the zeolite component, methanol or DME being the methylizing agent.

In the present work we are studying how the molar ratio H_2/CO in the reactor input affects the process C_{5+} productivity. Fig. 2 shows how C_{5+} productivity depends on GHSV for various H_2/CO molar ratios in the reactor input.

It can be seen that as the syngas space velocity grows, the conversion towards C_{5+} hydrocarbons increases, while the intensity of the conversion growth decreases following the increase of H_2/CO molar ratio in the reactor input.

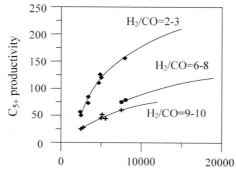

Fig. 2. Dependence of C_{5+} productivity (g/L Cat/h) on GHSV. P=80 atm, T=380°C.

Table 2
Alkylation of toluene with synthesis gas on Zn-Cr/Si-Mg-ZSM-5 catalysts

Zn-Cr/Zeolite mass ratio	30/70	30/70	30/70	70/30	70/30	70/30	70/30
Temperature, °C	380	380	400	400	400	440	440
Gas circulation	NO	YES	YES	NO	YES	NO	YES
Xylenes distribution, %							
para-	45	76	91	45	72	22	63
meta-	44	16	7	40	17	61	23
ortho-	11	8	2	15	11	17	14
Conversion, %							
Toluene	2	14	19	9	13	14	24
CO	5	11	29	20	46	12	39

3.2. The toluene alkylation with synthesis gas to p-xylene

It is known that p-xylene is a primary product in toluene alkylation with methanol [7, 8]. High p-xylene selectivity is attained by:

1) Low acidity of the external surfaces of the zeolite;

2) Short contact time of toluene with the active sites on the external surfaces of the zeolite;

3) High toluene-to-methanol ratio in feed.

These factors are also valid for toluene alkylation with synthesis gas. For the first time we have demonstrated [9, 10] that the application of gas circulation in reaction system increased the para-selectivity of Zn-Cr/Si-Mg-ZSM-5 catalysts (Table 2). It is obvious that using gas circulation with condensing and separation of alkylation liquid products permited permitted to reduce p-xylene contact time with the catalyst and to partly remove p-xylene from reacting gas. This phenomenon is due to suppression of the secondary reaction, i.e., isomerization of primarily produced p-xylene.

3.3. The 2-MN alkylation with synthesis gas to DMN

An improvement was observed in the selectivity towards 2,6- and 2,7-DMN when gas circulation was used in synthesis gas alkylation of 2-MN [11]. Therefore, the shape-selectivity check-up was carried out in circulating mode. The influence of the channel dimensions of various zeolite types on the DMN selectivity and 2-MN conversion was found (Table 3). A low 2-MN conversion and the prevailing formation of β,β-DMN were observed for medium-pore zeolites (ZSM-5). The use of large-pore zeolites, such as ZSM-12, resulted in formation of more bulky α,β-DMN and a considerable enhancement of 2-MN conversion to DMN. The combination of high acidity and large channel dimensions of β zeolite intensified the conversion of the primary β,β-DMN to α,β-DMN and then to polymethylnaphthalenes (2-MN conversion to alkylation products was 50-60%). Data shown in Table 3 can be explained by, first, increasing of the intrazeolite space for alkylation products formation and,

Table 3
Alkylation of 2-methylnaphthalene with synthesis gas

Type of zeolite		ZSM-5		ZSM-12		Beta	
Dimensions of the channels, Å		5.3×5.6		5.5×6.2		6.4×7.6	
Temperature, °C		400	440	400	440	400	440
DMN distribution, %							
2,6+2,7	β,β -	93.6	91.5	19.1	16.6	4.0	4.1
2,3+1,4	β,β - and α,α -	1.1	0.8	9.7	10.1	10.1	7.7
1,5	α,α -	–	–	–	–	traces	traces
1,7		1.6	1.1	16.3	15.8	2.7	15.7
1,3+1,6 }	α,β -	2.3	3.4	20.6	21.4	18.0	14.7
1,2		1.4	3.2	34.3	36.1	65.2	57.8
2-MN to DMN conversion, %		2	3	15	27	6	5

second, activation of alkylnaphthalenes to further alkylation with the increasing number of methyl groups in the rings.

4. CONCLUSION

Thus, the possibility is shown in the present paper to use various bifunctional catalysts in reactions of selective syngas conversion to the high-octane gasoline or diesel fractions and methylaromatics alkylation with synthesis gas syngas to the dimethylaromatics (including p-xylene, DMN etc.).

REFERENCES

[1] C.D. Chang, W.H. Lang and A.J. Silvestri, J. Catal. 56 (1978) 268.
[2] V.U.S. Rao and R.J. Gormley, Catal. Today, 6 (1990) 207.
[3] V.M. Mysov, K.G. Ione and A.V. Toktarev, in Natural Gas Conversion V (A. Parmaliana, D. Sanfilippo, F. Frusteri et al. eds.), Stud. Surf. Sci. Catal., Amsterdam, 119 (1998) 533.
[4] V.M. Mysov and K.G. Ione, RU Patent 2175960 (2001).
[5] X. Li, K. Asami, M. Luo, K. Michiki, N. Tsubaki and K. Fujimoto, Catal. Today, 84 (2003) 59.
[6] S. Namba, T. Yamagishi and T. Yashima, Nippon Kagaku Kaishi, 3 (1989) 595.
[7] A.V. Toktarev, V.M. Romannikov, G.A. Bitman and K.G. Ione, preprint of IC SB AS USSR, Novosibirsk, 1987, 1-9.
[8] J.-H. Kim, S. Namba and T. Yashima, Sekiyu Gakkaishi, Vol. 39, № 2, 1996, 79.
[9] V.M. Mysov and K.G. Ione, RU Patent 2114811 (1998).
[10] V.M. Mysov and K.G. Ione, RU Patent 2115644 (1998).
[11] V.M. Mysov and K.G. Ione, RU Patent 2119470 (1998).

Studies in Surface Science and Catalysis, volume 147
X. Bao and Y. Xu (Editors)
457

Low NOx emission combustion catalyst using ceramic foam supports

J. T. Richardson, M. Shafiei, T. Cantu, D. Machiraju, and S. Telleen

Department of Chemical Engineering, University of Houston
Houston, Texas, U.S.A. 77204-4792

ABSTRACT

Catalytic combustion has the potential to lower NO_x emissions in gas turbine applications, and BaMn hexa-aluminates have emerged as satisfactory high temperature catalysts. Ceramic foams offer advantages over monoliths as catalyst supports since the enhanced mixing avoids the formation of hot spots. This research demonstrates the feasibility of loading ceramic foams with washcoats of hexa-aluminate and confirms their superiority over monoliths.

1. INTRODUCTION

Natural gas is more efficient than coal or oil for power generation and gives less CO_2 per unit of power produced. The principal reaction is combustion of CH_4 to CO_2 and H_2O, with small emissions of CO, SO_x, and NO_x. While the first two can be reduced with improved burners and feed pretreatment, NO_x is a consequence of O_2/N_2 reactions in air at high flame temperatures (>1500). Emissions of NO_x reach 100-300 ppm and addressing this threat requires reducing emissions to < 5 ppm, especially for gas turbines whose future use will increase.

Catalytic combustion, with lean non-flame CH_4/O_2 reactions at lower temperatures, reduces emissions to 2-3 ppm [1]. Several commercial designs have emerged [2, 3], but acceptance of the technology has been slow due to extreme demands on the catalyst [4]. The main catalytic component must be active enough to ensure ignition at low temperatures (300-400°C), yet be stable enough to withstand sintering at high temperatures (1000-1300°C). The catalyst support must resist thermal shock, have high thermal stability, and operate at GHSVs up to 300000 h^{-1}. Ceramic monoliths are the preferred structures, since they comprise very small parallel channels forming a network of laminar reactors with low radial thermal conductivity and pressure drop [5]. Materials with good thermal characteristics are available (e.g. cordierite), and channel surfaces are covered with high surface area washcoats to support catalysts such as noble metals, mixed oxides, perovskites, and substituted hexa-aluminates [6].

Most designs fail to achieve good stability and lifetime due to the isolation of each channel in the monolith. Reaction inhomogeneities develop in the channels, either because of irregularities during preparation or inconsistent thermal histories [6]. Each channel behaves differently from its neighbors. Heat is not transferred from more active sites fast enough and local hot spots occur, leading to ignition of homogeneous combustion and liberation of more heat. This results in deactivation of active components and degradation of the washcoat. Regions of lower activity encourage backflashing in the channel, whereby homogeneous combustion after the bed is drawn back into the monolith.

An alternative is to use ceramic foams, which are sponge-like structures with high porosities (85 to 90%) and spherical cells connected through openings [5]. We have demonstrated certain advantages for these foams, including low pressure drop, high effectiveness factors, improved mixing and enhanced heat transfer [7]. Compared to monoliths at velocities of 1 ms^{-1}, foams have ten times the effective bed thermal conductivity, twice the areal activity and three times the pressure drop [7]. Ceramic foam-based catalytic combustion devices exhibit better mixing and have the capability to avoid hot spots and premature ignition of homogeneous combustion, since heat is quickly transported away from areas of excessive conversion. Although often treated in the patent literature, there has been very little systematic research to explore their potential [8].

In this paper, we present the results of a study on the feasibility of washcoating ceramic foams with a high temperature combustion catalyst, BaMn hexa-aluminate, and on the differences between performance of monoliths and foams under identical process conditions.

2. EXPERIMENTAL

2.1 Ceramic foams

The ceramic foams were cylindrical samples (1.27x2.54 cm) of 30-PPI α-Al$_2$O$_3$ foam , properties: pore diameter, 0.826 mm; solid density,3.94g cm^{-3}; bulk density, 0.498 g cm^{-3}; and porosity, 0.874. Other foams tested were ZrO$_2$ and cordierite, together with a 5-cm cylinder of 400 cpi cordierite monolith.

2.2 Catalyst preparation

A mixed hydrogel of BaMn hexa-aluminate (BaMn$_2$Al$_{10}$O$_{19}$), an active high temperature combustion catalyst, was prepared by carbonate co-precipitation using published procedures [4]. The slurry was aged for 3 hours at 60°C and filtered, and the cake washed and dried overnight at 110°C in air. The resulting xerogel was calcined in air from 500°C to 1400°C, and properties determined with X-ray diffraction and BET surface area measurements.

2.3 Washcoating procedures

The foam was washed with distilled water to remove loose particles and dried in air at 120°C for 2 hrs. It was then dip-coated (30 seconds) using a 10-15wt% slurry of $BaMn_2Al_{10}O_{19}$ powder (40-μ) in 95% ethanol. An aluminum nitrate binder was added to the slurry up to 20% of the $BaMn_2Al_{10}O_{19}$. Upon removal from the slurry, the foam was drained with a centrifuge and air jets. After drying at 120°C for one hour, the foam pellet was weighed to determine the loading. The dipping-drying sequence was repeated until the desired loading was achieved and the foam calcined in air at 900°C for two hours. A semi-quantitative measure of adherence was determined by immersing the washcoated foam in ether inside a sealed container, then placing it in an ultrasound bath (Branson 8200) for 30 minutes and measuring weight loss.

2.4 Activity testing

Catalytic tests were made with an automated unit, using continuous analyses of products under a wide range of air/fuel ratios and space velocities. Methane was mixed with filtered air, both regulated with Tylan FC-260V mass flow controllers to the desired fuel/air ratio. The catalyst bed comprised one ceramic foam or monolith structure or any amount of powdered catalyst loaded into a stainless steel reactor 60 cm in length and 1.5 cm I.D, heated with a 960-W Lindberg Furnace. Temperatures before and after the bed were measured with K-type thermocouples. Product gases were analyzed with on-line infrared analyzers: (California Analytical Model 300) for CO_2 and CH_4 (0 to 20 vol%) and CO (0-1000 ppm), and an Eco Physics Model CLD 700 AL Analyzer for NO and NO_x analysis up to 100 ppm..

A typical run consisted of loading the catalyst, setting the fuel and air flows (4.4% CH_4 in air), and analyzing the product gases as the furnace was heated at a programmed rate with an Omega 5508TC Temperature Controller through a sequence of heating and cooling cycles. Data were recorded in ASCII files with Excel.

A cyclic test was adopted in which the temperature was increased at a fixed rate up to 1000°C and then decreased. Each cycle was ten hours in length and the procedure was repeated to give a history of activity and stability.

3. RESULTS AND DISCUSSION

3.1 Hexa-aluminate powder properties

The xerogel was calcined from 500°C to 1400°C to establish optimum conditions for calcination (Table 1). Major changes occurred at 1100°C with the formation of $BaMn_2Al_{10}O_{19}$, and this was consistent with reported observations [6]. Calcination at 1200°C was used for subsequent preparations.

Table 1
Effect of calcination temperature on $BaMn_2Al_{10}O_{19}$ xerogel

Calcination °C	BET surface area m^2g^{-1}	Pore volume $ml\,g^{-1}$	XRD phases ( ) minor peaks
fresh	401	0.653	amorphous
500	192	0.649	amorphous, Mn oxides
900	126	0.608	$BaAl_2O_4$
1100	38.0	0.214	$BaMn_2Al_{10}O_{19}$ ($BaAl_2O_4$)
1200	22.1	0.095	$BaMn_2Al_{10}O_{19}$
1300	13.1	0.041	$BaMn_2Al_{10}O_{19}$
1400	6.9	0.031	$BaMn_2Al_{10}O_{19}$

The activity of powdered $BaMn_2Al_{10}O_{19}$ is shown in Fig. 1 at a GHSV of 10280 h^{-1} (based on catalyst volume). Conversion began at about 325°C and was complete at 650°C. The decreasing-temperature curve was slightly displaced to higher temperatures, but there was no multiplicity. Small amounts of NO_x (2-3 ppm) were detected. Analyses of the temperature profile indicated a 0.5 order for CH_4 with an activation energy of 112 kJ mol^{-1} over the complete conversion range, implying that chemical kinetics prevail with little or no transport resistance. These results are in agreement with published reports [4].

Fig. 2 shows an increase in T_{50}, the temperature for 50% conversion in each cycle, demonstrating that holding the catalyst at high temperature produced steady deactivation, suggesting that temperatures in excess of the calcination temperature (1200°C) had occurred in the bed. This could have resulted in sintering or other thermal degradation. The possibility of hot spot formation is also indicated by the presence of NO_x.

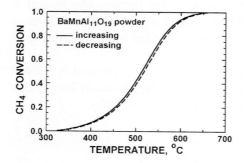

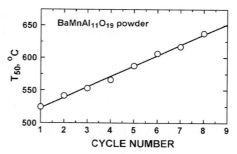

Fig. 1. Conversion curve for $BaMn_2Al_{10}O_{19}$ powder, 4.4% CH_4 in air, GHSV = 10280 h^{-1}.

Fig. 2. T_{50} for $BaMn_2Al_{10}O_{19}$ powder, same conditions as Fig. 1.

3.2 Washcoating

The ultrasound adherence test for α-Al$_2$O$_3$, ZrO$_2$ and cordierite foams without binder in the washcoat gave 5 mg loss per 30 seconds of treatment regardless of the loading level. With binder present, the loss was 0.1, 0.5 and 5 mg for ZrO$_2$, α-Al$_2$O$_3$ and cordierite respectively. Losses for unwashcoated foams were zero. These results suggest that using the binder improves adhesion in that order, but the quantitative significance is not clear.

Fig. 3, showing the effect of washcoat loading on the surface area of the foams, indicates consistent behavior. Surface area was proportional to the amount of washcoat, independent of the foam material or the presence of the binder. The average surface area for the washcoat itself was 21.9 m^2 g(cat)$^{-1}$, very close to the value for BaMn$_2$Al$_{10}$O$_{19}$ calcined at 1200°C.

Fig. 4 compares BaMn$_2$Al$_{10}$O$_{19}$ powder with 8wt% washcoated α-Al$_2$O$_3$ foam. Flow rates were adjusted to give similar space times based on the surface area of the BaMn$_2$Al$_{10}$O$_{19}$ catalyst. These were 3.00 m^{2}s cm^{-3} and 2.72 m^{2}s cm^{-3} respectively for the powder and foam. The foam not only preserved the activity of the powder but was slightly better. Equivalent performance was achieved with each of the other foam materials.

3.3 Comparison of foam and monolith

In Fig. 5, conversion for a washcoated cordierite monolith (3.33 wt% and 2.22 m^2 of washcoat) is compared on the same basis with a foam (2.14 wt% and 1.86 m^2 of washcoat) and shows a dramatic difference. The monolith was initially less active than the foam but ignited at about 725°C, rapidly reaching full conversion. A 1-ppm peak of NO$_x$ was recorded at 761°C and 684 ppm of CO at 691°C. Multiplicity was evident since decreasing the temperature gave hysteresis with extinction at 450°C and a peak of CO in excess of 1000 ppm at

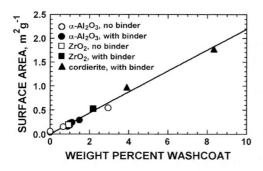

Fig. 3. Surface area of washcoated foams.

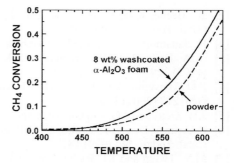

Fig. 4. Comparison of powder and foam, Space times (m^{2}s cm^{-3}): 3.00 (powder), 2.72 (foam).

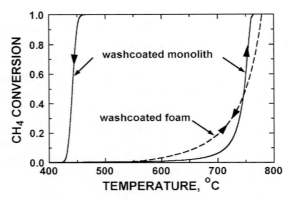

Fig. 5. Comparison of washcoated foam and monolith under similar flow condition, Space times (m²s cm⁻³): 1.90 (monolith) and 1.63 (foam).

573°C. Also, T_{50} increased about 20°C per cycle, indicating deactivation. The foam, however, displayed no hysteresis, no NO_x formation, 725 ppm of CO at 725°C, and no change in T_{50}, signifying thermally stable behavior.

4. CONCLUSIONS

This research confirms that BaMn hexa-aluminate can be washcoated onto ceramic foams that are thermally more stable than monoliths under the same test conditions for catalytic combustion. This is believed to be due to better mixing in the foams leading to the absence of hot spots. Tests to determine if this prevails for commercial gas turbine conditions are now in progress.

REFERENCES

[1] W. Beebe, K.D. Cairns, V.K. Pareek, S.G. Nickolas, J.C. Schlatter, and T. Tsuchiya, Catal. Today, 59 (2000) 95.
[2] H. Sadamori, T. Tanioka, T. Matsuhisa, Catal. Today, 26, (1995) 337.
[3] R.A. Dalla Betta, J.C. Schlatter, D.K. Yee, D.G. Loffler, T. Shoji, Catal. Today, 26 (1995) 329.
[4] P.O. Thevenin, A.G. Ersson, H.M.J. Kusar, P.G. Menon, and S.G. Jaras, Appl. Catal. A: Gen. 212 (2001) 189.
[5] J.W. Geus and J.C. van Giezen, Catal. Today, 47 (1999) 180.
[6] P. Artizzu-Duart, Y. Brulle, F. Gaillard, E. Garbowski, N. Guilhaume, and M. Primet, Catal. Today, 54 (1999) 181.
[7] M.V. Twigg and J.T. Richardson, IChemE Trans. Part A - Chem. Eng. Res. and Design. 80 (2002) 183.
[8] K. Jiratova, L. Moravkova, J. Malecha, B. Koutsky, Collect. Czech. Chem. Comm. 60 (1995) 473.

Studies in Surface Science and Catalysis, volume 147
X. Bao and Y. Xu (Editors)

Preparation of LaCoO$_3$ with high surface area for catalytic combustion by spray-freezing/freeze-drying method

Seong Ho Lee[a], Hak-Ju Kim[a], Jun Woo Nam[a], Heon Jung[b], Sung Kyu Kang[b], and Kwan-Young Lee*[a]

[a]Department of Chemical and Biological Engineering, Korea University,
1, 5-ka, Anam-dong, Sungbuk-ku, Seoul 136-701, Korea

[b]Korea Institute of Energy Research,
71-2, Jang-dong, Yusung-gu, Taejon 305-343, Korea

ABSTRACT

Perovskite-type LaCoO$_3$ oxides were prepared using citrate precursor by spray-freezing/freeze-drying. The calcination temperature for a single phase of LaCoO$_3$ could be lowered to 500°C without formation of any other oxides and the product has the highest surface area of 24 m^2/g, which has not been reported yet. It was confirmed by IR and TG analyses that the greater amount of citrate existed in the precursor dried by spray-freezing/freeze-drying than by evaporation drying, which contributed to the high surface area and well-crystallized phase of LaCoO$_3$.

1. INTRODUCTION

LaCoO$_3$ has been known to be very active catalyst for the complete oxidation of CO and hydrocarbon and it is, therefore, very suitable material for the replacement of noble catalysts used in the catalytic combustion of methane as well as for the air electrode of fuel cell [1]. However, since the high calcination temperature during the preparation reduces inevitably the surface area, many studies have been performed to overcome the problem.
Zhang et al. prepared LaCoO$_3$ with 11 m^2/g surface area by use of citrate precursor and evaporation drying method [2] and Taguchi et al. prepared LaCoO$_3$ with 12 m^2/g surface area by firing method using polyacrylate [3].

---

* Corresponding author. E-mail: kylee@korea.ac.kr.

464

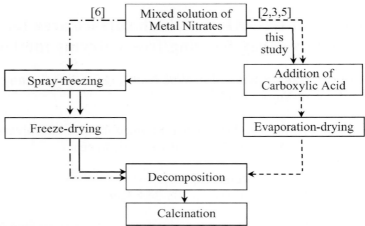

Fig. 1. Preparation methods of perovskite-type oxides reported in the references [2,3,5,6]. The solid line is a preparation method used in this study.

By Gordes *et al.*, LaCoO$_3$ with high surface area, 19 m^2/g, was obtained by drip pyrolysis method, but the LaCoO$_3$ was mixed with other oxide phases [4].

In general, to obtain the crystalline catalyst with high surface area will be possible by lowering the calcination temperature during the preparation. The lowest calcination temperature with a uniform perovskite phase of ABO$_3$ can be realized by the maximization of the homogeneity of A and B metals in the precursor before calcination.

Fig. 1 shows the preparation scheme of catalysts including the methods which have been reported. Most of them were concerned with the development of precursors and drying methods. The addition of carboxylic acids such as citric, malic, and polyacrylic acid to the mixed solution of metal nitrates [2,3,5] or the sublimation of solvent in frozen precursors for drying (spray-freezing/ freeze-drying) [6] are known to enhance the surface area of perovskite-type oxides. In this study, therefore, we have tried to combine the citrate (formed by addition of citric acid to metal nitrate solutions) precursor with spray-freezing/ freeze-drying for the preparation of LaCoO$_3$ with high surface area and compared the results with those by citrate precursor combined with the conventional evaporation drying.

2. EXPERIMENTAL

The 0.25M solutions of lanthanum nitrate and cobalt nitrate were mixed and citric acid was added. Molar ratio of an acid to total metal ions was set at 3/n, where n is the number of carboxyl groups in one molecule of acid. The mixed solutions were adjusted to various pH values by the addition of ammonia solution (28%) and then dried using a vacuum evaporator and freeze-dryer, respectively. In case of the latter, the mixed solutions were sprayed into liquid N_2 in a round-bottomed flask with N_2 carrier gas using sprayer (tip dia.: 0.3mm) before the freeze-drying. The ice powder formed was dried in freeze dryer (condenser at $-75°C$) at 0.05 torr. The dried powder was decomposed at $300°C$ for 1 hr and then calcined at 550 and $600°C$ for 5hrs. The structure of final product was confirmed by XRD (Philips XPERT MPD). BET surface area and pore volume were measured using ASAP 2010 (Micromeritics).

3. RESULTS AND DISCUSSION

3.1. X-ray diffraction and surface area analysis

Table 1 shows the XRD and BET results, which were very different depending on drying method and pH of precursors. Especially, the more remarkable difference is shown according to drying method. In case of the spray-freezing/freeze-drying, the calcination temperature for a single phase of perovskite was lower than that of the evaporation drying regardless of pH of the precursors, which resulted in the higher surface area than that of evaporation drying. Especially, in case of pH 1.4, the highest surface area, 23.7 m^2/g was obtained. To the best of our knowledge, this surface area of $LaCoO_3$ is the highest among the surface area reported.

In case of the evaporation drying, the surface area and the calcination temperature were similar to the results of Zhang *et al.* [2]. In general, evaporation drying of mixed solution may reduce the homogeneity of dried precursor after drying because the solubility of compounds in the mixed solution are different and consequently the precipitation of them cannot occur at the same time. On the contrary, the freeze-drying can prohibit this problem because the mixed solution is frozen and dried by sublimation, not by evaporation. From this point of view, the product prepared by spray-freezing/freeze-drying can have the higher surface area.

Table 1
The results of XRD and surface properties of $LaCoO_3$ prepared by various methods.

precursor	drying method	pH	XRD results and BET surface area (m^2/g)			
			500^a	550^a	600^a	650^a
nitrate added by citric acid	evaporation -drying	1.4^f	Am^b	M^c	$P^d(12.3)^e$	-
		1.7	Am	M	M	P(6.7)
		2.1^g	Am	M	P(10.2)	-
	spray-freezing/freeze-drying	1.4	P(23.7)	P(15.3)	P(7.3)	-
		1.7	M	P(18.4)	P(12.4)	-
		2.1	M	P(17.2)	P(12.0)	-

a: calcination temperature ($°C$), b: amorphous, c: mixed oxides, d: single phase of perovskite, e: BET surface area, f: pH unadjusted, g: pH just before precipitation

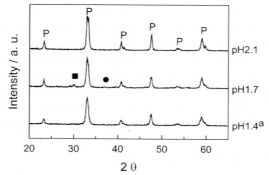

Fig. 2. XRD results of the products prepared by evaporation-drying. calcination temperature, 600°C; a, pH of precursors; ■, La_2O_3; ●, Co_3O_4; P, $LaCoO_3$.

XRD of products calcined at 600°C using the dried precursor obtained by evaporation were examined as shown in Fig. 2. Single phases of perovskite were obtained in precursors with pH 1.4 and 2.1. In case of precursor with pH 1.7, phases of La_2O_3 and Co_3O_4 are formed. It is likely that the precursor has the lower homogeneity due to pH and ammonium complexes.

XRD results of products prepared from the precursors of pH 1.4 were compared as shown in Fig. 3. To compare the extents of crystallization one another, products mixed with 50 wt% Si powder were analyzed and the intensities were standardized to that of Si[111]. In case of evaporation drying, the single oxides (La_2O_3, Co_3O_4) and amorphous phases appeared at 500 and 550°C, respectively. However, in case of spray-freezing/freeze-drying, a single phase of perovskite without any other oxides was obtained even at 500°C and the extent of crystallization was superior to that of product calcined at 550°C through the evaporation drying.

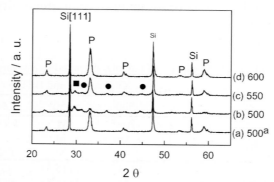

Fig. 3. XRD results of the products prepared by (a) spray-freezing/freeze-drying and (b-d) evaporation drying of precursors with pH1.4. The intensity of the results was standardized with Si[111]. a, calcination temperature; ■, La_2O_3; ●, Co_3O_4; P, $LaCoO_3$.

3.2. FTIR analysis

In order to locate different metal ions together in a molecule (malate, citrate) and then to increase the homogeneity of metal ions, organic acids like malic and citric acids are generally added to the mixed solution of metal nitrates. In this study, FTIR analysis of dried precursor was carried out to find the effect of formation of metal citrate on crystallization and surface area of the product according to the variation of drying method and pH of precursor.

Fig. 4 shows the IR spectra of dried precursors. The absorption band at 1580-1650 cm^{-1} and 1400-1500 cm^{-1} are assigned to asymmetric and symmetric C-O stretching of carboxylate anion, confirming the formation of citrate though the latter peak was superposed by a sharp peak of 1385 cm^{-1} responsible to ionic nitrate. The band at 1710 cm^{-1} is assigned to C=O stretching of free citric acid without any substitution by metal. Nitrate and ammonium ions are confirmed by NO$_3$ stretching at 1385 and 820 cm^{-1} and by asymmetric NH$_4$ stretching at 2800-3300 cm^{-1}, respectively.

As shown in Fig. 4, in case of dried precursor with pH 1.4 by spray-freezing/freeze-drying, relative intensity of C=O stretching against C-O is stronger than that of evaporation drying. It is, therefore, clear that, in case of pH 1.4, the amount of citrate in dried precursor of spray-freeze/freeze-drying is larger than that of evaporation drying. Considering the result that a single phase of perovskite was formed at the lowest calcination temperature by spray-freeze/freeze-drying with precursor of pH 1.4 as shown in Table 1, it can be known that the larger the amount of citrate in precursor is, the lower the calcination temperature for a single phase of perovskite is. Therefore, it is concluded that spray-freezing/freeze-drying is more useful method for metal citrate to cause the lower calcination temperature and the larger surface area, than evaporation drying (chemical effect on homogeneity). Evaporation drying is operated at much higher temperature compared with spray-freezing/ freeze-drying and it is likely that this difference of drying condition affected the extent of ionization and the formation of metal citrate.

The IR spectra of dried precursors with pH 1.7 and 2.1 do not show any difference between drying methods, which means that there was no chemical difference in the dried precursors. However, as shown in Table 1, all the products prepared by spray-freezing/freeze-drying showed a single phase of perovskite at lower temperature than by evaporation drying. This indicates that, as stated previously, dried precursor by spray-freezing/freeze-drying is more homogeneous than by evaporation drying, regardless of the formation of citrate. In case of the spray-freezing/freeze-drying, the precursor is dried without precipitation, while in case of the evaporation drying, the compounds in the precursor experience precipitation during the evaporation. Worse than all, the precipitation of them may not occur at the same time due to difference of solubility of the compounds. Therefore, it is reasonable that the dried precursor

468

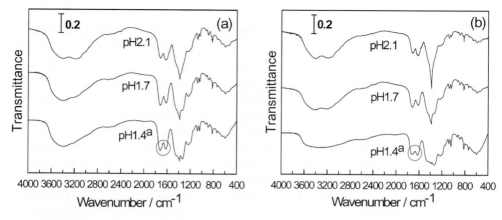

Fig. 4. IR spectra of the dried precursors by (a) evaporation-drying and (b) spray-freezing/freeze-drying. a: pH of precursors.

by spray-freezing/freeze-drying is more homogeneous than by evaporation drying (physical effect on homogeneity).

In case of precursor with pH 2.1, the intensity of C-O stretching is stronger than that of C=O stretching for both dried precursors. However, as shown in Table 1, the calcination temperature for a single phase of perovskite couldn't be lowered more than that for the precursor with pH 1.4. In this case of higher pH, it is supposed that the intensity of C-O stretching does not belong to metal citrate only but includes the band of ammonium citrate formed by addition of ammonia water.

ACKNOWLEDGEMENT

This study was supported by research grants from the Korea Science and Engineering Foundation (KOSEF) through the Applied Rheology Center (ARC), an official KOSEF-created engineering research center (ERC) at Korea University, Seoul, Korea.

REFERENCES

[1] O. Yamamoto, Y. Takeda, R. Kano, and M. Noda, Solid State Ionics, 22 (1987) 241.
[2] H. M. Zhang, Y. Teraoka, and N. Yamazoe, Chem. Lett., (1987) 665.
[3] H. Taguchi, H. Yoshioka, and M. Nagao, J. Mater. Sci. Lett., 13 (1994) 891.
[4] P. Gordes, N. Christiansen, E. J. Jensen, and J. Villadsen, J. Mater. Sci., 30 (1995) 1053.
[5] Y. Teraoka, H. Kakebayashi, I. Moriguchi, and S. Kagawa, Chem. Lett., (1991) 673.
[6] J. Kirchnerova, and D. Klvana, Int. J. Hydrogen Energy, 19 (1994) 501.

Studies in Surface Science and Catalysis, volume 147
X. Bao and Y. Xu (Editors)
469

Effect of support structure on methane combustion over PdO/ZrO$_2$

L.-F. Yang*, Y.-X. Hu, D. Jin, X. -E He, C.-K. Shi, J.-X. Cai

Department of Chemistry, State Key Laboratory of Physical Chemistry for Solid Surface, Xiamen University, Xiamen 361005, China

1. INTRODUCTION

Recently the catalytic combustion technology has been widely applied for various industrial processes [1-3]. For higher energy transfer efficiency and lower emission of air pollutants, it has long been necessary in practice to develop a catalyst with better ignition activity. Since palladium has been known to be the most active component for methane combustion, consequently investigations on catalyst support become rather attractive [4-6].

In previous studies, palladium loaded on conventional oxide support materials, such as ZrO$_2$, Al$_2$O$_3$ and SnO$_2$ demonstrated fairly good catalytic activity for methane combustion. The exclusion of SiO$_2$ as a candidate for the support reveals that the surface structure of oxide support influences on the PdO dispersion and its reactivity [7]. An appropriate interaction between PdO and surface lattice of oxide gives rise to more active sites for methane combustion. As elucidated by Juszczyk's in-situ infrared spectroscopic experiments, the active PdO phase is stabilized by insertion of the Pd^{2+} into AlO$_6$ octahedron. The extent of this interaction between support and active metal component (SMI) is apt to be affected by the kind of support, preparation parameter, thermal history, and even the palladium source. In the end, the effect of surface lattice on PdO distribution is the determinate [8].

In order to ascertain the relationship between PdO phase and support structure, Pd catalysts with Pd loaded on alumina supports with various crystal structure have been widely investigated in which the template effect of oxide surface lattice is evident. In this study, we have investigated Pd supported on zirconia support with different crystal structure for low temperature combustion of methane, and emphasis was given to the relationship between active phase and support structure.

---
* Corresponding author. Tel: +86-592-2185944; Fax: +86-592-2183047;
E-mail address: lfyang@xmu.edu.cn

2. EXPERIMENTAL

2.1. Sample preparation

Zirconia supports with monoclinic or monoclinic-tetragonal mixed lattice were synthesized by co-precipitation method from $ZrO(NO_3)_2$ aqueous solution. After calcination at an assigned temperature, these support powders were dipped into aqueous solution of palladium nitrate. The wet precursor was dried and calcined at 800℃ for 5 hours in air. The loading of palladium in the resultant catalysts was from 0.5wt.%, 1wt.% to 3wt.%. The samples are designated as [Pd wt.%]Pd/[support type (m/t+m)]-ZrO_2.

2.2. Characterization of catalysts

The phase component of calcined catalyst was analyzed by XRD method, XRD patterns were recorded with a Rigaku Rotaflex D/Max-C Diffractometer with scanning rate of 4deg./min using monochromatic CuK_a radiation (λ =0.15406 nm). Specific surface area was measured by nitrogen adsorption technique at liquid nitrogen temperature on a Carlo Elba 1900 Sorptomatic instrument. The dispersion of palladium component of the catalyst was measured by pulse-adsorption of CO in a flow of carrier gas. Prior to measuring, the catalyst was reduced by hydrogen at 400℃. The dispersion of palladium was calculated from the CO uptake by assuming a stoichiometry of $[CO]/Pd_{surface}$ =1.

2.3. Measurement of activity

Catalytic activity tests were performed using a quartz tubular fixed bed microreactor equipped with a sheathed thermocouple monitoring the reaction temperature. Combustion reaction was conducted at atmospheric pressure, and a gaseous mixture of CH_4 (2 vol.%), O_2 (8 vol.%) and N_2 (90 vol.%) was supplied at a space velocity of 48,000h^{-1}. Methane conversion in the effluent gas was analyzed by an on-line gas chromatograph (Shimadzu GC-14B) equipped with a thermal conductivity detector (TCD)

2.4. Temperature-programmed reaction

The temperature-programmed desorption (TPD) of oxygen was carried out in a flow system to observe oxygen desorption from the catalysts. About 50mg catalyst particles (d_p 0.25~0.45mm) were fixed in the middle of the quartz tubular reactor by packing quartz wool at both ends. Prior to the experiment, the sample was oxidized in an air flow (20ml/min) at 800℃ for 30min.After cooling to room temperature, the purging gas was switched to helium (impurity< 1ppm). The sample bed was heated at a rate of 20℃/min after gaseous oxygen was purged out. During this heating stage, the concentration of oxygen in the effluent gas was monitored by a quadrupole mass spectrometer (Balzers QMS

200 Omnistar).

The acidity sites on catalysts and supports were characterized by ammonia temperature-programmed desorption (NH_3-TPD) experiments, which were identical to O_2-TPD except by dilute ammonia was used in the pretreatment. The signal of fragment with m/e=15 reflected the NH_3 concentration in effluent so that it was not interfered by the water signals.

3. RESULTS AND DISCUSSION

3.1. Catalytic activity for methane combustion

XRD patterns (Fig. 1) show that the thermodynamically metastable t-ZrO_2 remains about 30 at.% when the final catalysts are acquired. On both monoclinic and monoclinic-tetragonal mixed supports, PdO presents no characteristic reflection even when the Pd loading attains 3 wt.%. This implies that PdO fails to aggregate into large crystallite on zirconia support. Further investigation by CO pulse reveals that the dispersion of Pd component on these two zirconia supports is quite different. It is shown in Table 1 that the monoclinic zirconia evidently promotes the dispersion of palladium component thereon. This is more pronounced on the lower Pd-loading samples. In the case of 0.5wt.% Pd supported on monoclinic zirconia, the dispersion is 23.8%, nearly quadruple the value for the sample supported on t+m type zirconia, of which the specific surface area is even three fold larger than that for the m-type support. The high dispersion of palladium on the less available surface implies a favorable interaction between PdO and m-ZrO_2.

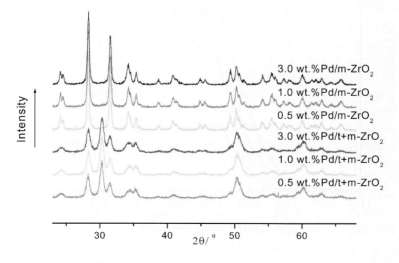

Fig. 1. XRD patterns of supported Pd catalysts

Table 1
Characteristics of Pd/ZrO$_2$ catalysts

Samples	SSA m^2/g	Pd dispersion %
3Pd/m-ZrO$_2$		5.0
1Pd/m-ZrO$_2$	14.5	14.2
0.5Pd/m-ZrO$_2$		23.8
3Pd/t+m-ZrO$_2$		2.8
1Pd/t+m-ZrO$_2$	47.7	3.1
0.5Pd/t+m-ZrO$_2$		6.3

Co-feeding of methane and air was adopted to simulate the operating condition of methane combustion. An obvious improvement in the ignition performance was acquired and a growth of activity increased rapidly by reaction temperature for the case of m-ZrO$_2$ supported catalysts, especially, for the lower Pd-loading samples. These results were shown in Fig.2. The promotion of support structure on metal dispersion undoubtedly contributes to the catalytic activity. Considering that the nature of active site for methane combustion is still obscure at present, Pd dispersion is unable to reflect directly the number of active sites; therefore, it seems unreasonable to evaluate the catalytic activity by turnover frequency.

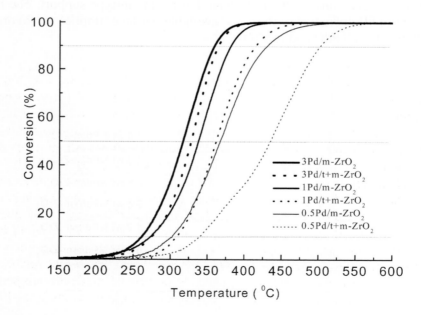

Fig. 2. Catalytic performance of Pd/ZrO$_2$ for methane combustion

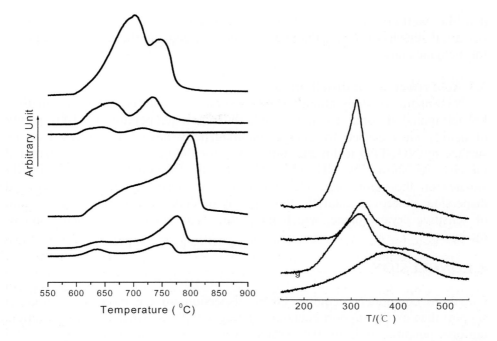

Fig. 3. O_2-TPD patterns of supported Pd catalysts in sequence from top to bottom: 3Pd/m-ZrO$_2$, 1Pd/m-ZrO$_2$, 0.5Pd/m-ZrO$_2$, 3Pd/t+m-ZrO$_2$, 1Pd/t+m-ZrO$_2$, 0.5Pd/t+m-ZrO$_2$
Fig. 4. NH$_3$-TPD patterns of supported Pd catalysts in sequence from top to bottom: 3Pd/t+m-ZrO$_2$, 1Pd/t+m-ZrO$_2$, 0.5Pd/t+m-ZrO$_2$, t+m-ZrO$_2$

3.2. PdO species on the ZrO$_2$

By means of O_2-TPD technique (Fig. 3), several PdO species that could be separated by thermal stability were revealed on the surface of zirconia support. It is evident that there is a rather stable oxygen species on the surface of PdO/t+m-ZrO$_2$. Comparative blank experiment with t+m-ZrO$_2$ excludes that oxygen desorbs from the support. So the oxygen desorption at high temperature originates from the PdO species intensively interacted with zirconia. This PdO species is unseen on the m-ZrO$_2$ supported catalyst. It is thought to be related to the decrease of Pd dispersion and catalytic activity of t+m-ZrO$_2$ supported catalysts. In the catalytic redox cycle, the reducibility of active PdO phase is predominant for the overall activity. The excessively strong Pd-O bond in the intensively interacted PdO species leads to its inertia to catalytic reaction. Combined the catalytic activity and O_2-TPD patterns, a consistent relationship between ignition activity and the oxygen desorption at moderate temperature emerges. This could be explained by the viewpoint of our previous study [9,10], that the well-crystallized PdO phase by epitaxy meet the requirement of structural sensitivity for methane activation and become the main active phase for methane combustion.

3.3. Acid effect on Pd distribution

A comparative study about these two supports would be rather informative. For tetragonal zirconia it is meta-stable under low temperature (<1100), plenty of acidity sites caused by charge inhomogeneity have been detected on its surface by NH_3-TPD technique, uniformity for PdO distribution was disrupted thereby. As shown by NH_3-TPD (Fig. 4), the Pd component preferentially situates on the acidic site. This inhomogeneity leads to the decrease of Pd dispersion; moreover, catalytic activity may be depressed by reducing the ratio of sensitive crystal plane, which meets the requirement on the activation of methane.

4. CONCLUSION

Palladium component acquires better dispersion on the monoclinic zirconia support than on the support containing tetragonal phase, since the m-ZrO_2 offers uniform potential field on the surface. And as the result of stacking the PdO layer epitaxially along the m-ZrO_2, a family of crystal planes, which was sensitive on activation of methane, is formed; therefore, an obvious improvement on catalytic activity for methane combustion is achieved.

ACKNOWLEDGEMENTS

The financial support and funding given to the G1999022400 and 20203014 projects by MoST (National Ministry of Science and Technology) and NSFC (National Natural Science Foundation) of China are gratefully acknowledged.

REFERENCES

[1] T.G. Kang, J.H.Kim, S.G.Kang and G. Seo. Catal. Today 59 (2000) 87
[2] K. Muto, N. Katada, M. Niwa. Appl. Catal. A 134 (1996) 203
[3] D. Ciuparu, M.R. Lyubovsky, and E. Altman et al. Catal. Rev., 44(2002) 593
[4] E. Garbowski, and C. Feumijantou et al. Appl. Catal. A, 87(1992) 129
[5] K. Sekizawa, H. Widjaja, S. Maeda, Y. Ozawa, K. Eguchi. Appl. Catal. A 200 (2000) 211
[6] K. Narui, K. Furuta, H. Nishita, Y. Kohtoku, T. Matsuzaki. Catal. Today 45 (1998) 173
[7] W. Juszczyk, Z. Karpinski, I. Ratajczykowa, et al. J. Catal. 120 (1989) 68
[8] R.J. Farrauto, J.K. Lampert, M.C. Hobson, E.M. Waterman. Appl. Catal. B 6 (1995) 263
[9] L. Yang, C. Shi and J. Cai, Appl. Catal. B, 38 (2002) 117
[10] C. Shi, L. Yang and J. Cai, Chem. Commun. 18(2002) 2006

Studies in Surface Science and Catalysis, volume 147
X. Bao and Y. Xu (Editors)
475

The catalytic combustion of CH_4 over Pd-Zr/HZSM-5 catalyst with enhanced stability and activity

Chun-Kai Shi, Le-Fu Yang, and Jun-Xiu Cai[*]

Department of Chemistry and State Key Laboratory for Physical Chemistry of Solid Surface, Xiamen University, Xiamen 361005, China

1. INTRODUCTION

Methane (CH_4) is a more effective greenhouse gas than CO_2 and hence CH_4 emission from natural gas engines, power plants, petroleum and petrochemical industries should be completely converted to CO_2 and H_2O by catalytic combustion [1-2]. Recently the Pd-supported zeolite catalysts are paid considerable attention due to their better low-temperature activity and high conversion efficiency for CH_4 combustion than conventional Pd/Al_2O_3 catalysts [3-5]. These Pd-supported zeolite catalysts are typically prepared by an ion-exchanged method and show a poor thermal/hydrothermal stability.

It is widely accepted that palladium oxide species play an important role as active sites during the reaction and their activities are dependent on the kinds of supports and additives used. In this work, we report the preparation of a series of Pd catalysts with different supports using the impregnation method, and the effects of these supports on the methane oxidation activity of Pd catalysts. In addition, thermal and hydrothermal stability of Pd-Zr/HZSM-5 and Pd/HZSM-5 are also evaluated at 370 °C from a practical point of view.

2. EXPERIMENTAL

2.1. Catalyst preparation

Firstly, Zr/HZSM-5 ($Si/Al_2 = 100$) support was prepared by impregnating the parent HZSM-5 powders for 12 h with aqueous solution of $Zr(NO_3)_4$, followed by drying at 120 °C for 12 h and calcining at 500 °C for 3 h. Then, the powders obtained were impregnated with $Pd(NO_3)_2$ aqueous solution. The simple and ZrO_2 modified Pd/HZSM-5 catalysts were obtained by repeating the above drying and calcining procedures. Moreover, $Pd/\gamma-Al_2O_3$ and Pd/SiO_2 samples were also similarly prepared for comparison. The desired loading of all the metals in their metallic states was 1 wt.% of the catalysts.

2.2. Catalyst characterizations

The specific surface area and micropore volume (Table 1) of the samples were measured by nitrogen-adsorption at 77 K with a Carlo 1900 Sorptomatic instrument.

Oxygen temperature-programmed desorption (O_2-TPD) tests of the catalysts were conducted in a conventional fixed-bed flow system. The samples were firstly pre-treated in air-flow (20 ml/min.) at 500 °C for 30 min, let self-cool down to room temperature (RT), and then were flushed with a high purity He at a flowing rate of 15 ml/min for 60 min to remove the physically adsorbed oxygen. Finally, the sample bed temperature was linearly increased from RT to 900 °C at a heating rate of 20 °C/min. The change of oxygen concentration from the catalysts was detected with a quadrupole mass spectrometer (Balzers QMS 200 Omnistar).

The details for methane temperature-programmed reduction (CH_4-TPR) tests of the catalysts were the same as for O_2-TPD except that a mixture of CH_4/He (4%: 96% in volume) was used as the flushing and reducing gas. Moreover, CO_2 was monitored as a product of the redox reaction between CH_4 and the PdO species on the catalysts.

2.3. Activity evaluation

The catalytic activity and lifetime tests of the catalysts were carried out using a tubular quartz microreactor of the down-flow fixed-bed type at atmospheric pressure. A powder sample of 100 mg (45-60 mesh) was placed in the reaction bed, and the reaction mixture of CH_4, O_2 and N_2 (2: 8: 90 vol.%) was fed into the reactor at a space velocity of 48000 h^{-1}. For hydrothermal stability test, the water vapor of the feed gas was provided by a glass bubbler. CH_4 conversion was analysed by an on-line gas chromatograph (Shimadzu GC-14B) equipped with an automatic sampling device and a TCD detector using a 5A sieve column at 55 °C.

3. RESULTS AND DISCUSSION

3.1. Activity and stability evaluation

Table 1

Catalytic properties of Pd-supported catalysts for the complete combustion of methane

Catalyst	Surface area $/m^2\,g^{-1}$	Micropore volume $/cm^3\,g^{-1}$	$T_{10\%}$[a]$/$ °C	$T_{50\%}$[a]$/$ °C	$T_{100\%}$[a]$/$ °C
Pd/HZSM-5	488.7[b]	0.1674[b]	315	355	425
Pd-Zr/HZSM-5	482.5[b]	0.1630[b]	292	337	380
Pd/γ-Al_2O_3	172.7[c]	0.0927[c]	352	443	490
Pd/SiO_2	197.5[c]	0.1049[c]	410	526	---

[a]Temperatures required for 10, 50 and 100% conversion of CH_4, respectively.
[b]Calculated from the Dubinin plot; [c]Calculated from the BET plot.

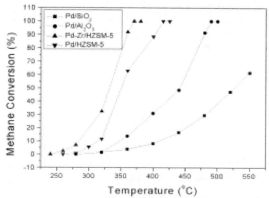

Fig. 1. Light-off curves of methane combustion over the Pd-supported catalysts.

Fig. 1 compares the catalytic activities for methane combustion over various Pd catalysts as a function of temperature. Pd catalysts with different supports display a considerable discrepancy in terms of activity. In comparison to Pd/HZSM-5, Pd-Zr/HZSM-5 shows a remarkably higher methane ignition activity and a conversion efficiency of 100%, whereas Pd/γ-Al$_2$O$_3$ and Pd/SiO$_2$ exhibit more or less lower activities for ignition and total conversion. In addition, Pd/HZSM-5 shows a higher catalytic activity than Pd/γ-Al$_2$O$_3$ and Pd/SiO$_2$, indicating that the HZSM-5 support has a better enhancing effect on the activity of loaded PdO species than the γ-Al$_2$O$_3$ or SiO$_2$ supports, and ZrO$_2$ further improves Pd/HZSM-5's activity. For further comparison, the methane conversion temperatures of all the catalysts are listed in Table 1. $T_{10\%}$, $T_{50\%}$ and $T_{100\%}$ represent the temperatures for 10, 50 and 100% methane conversions, respectively. One can clearly see that Pd-Zr/HZSM-5 exhibits the best activity among all the catalysts in terms of the three conversion levels. The $T_{10\%}$, $T_{50\%}$ and $T_{100\%}$ of Pd-Zr/HZSM-5 are 23, 18 and 45 °C lower than those of Pd/HZSM-5, respectively.

The thermal/hydrothermal stability of Pd-Zr/HZSM-5 was compared with that of Pd/HZSM-5 through monitoring the time dependence of CH$_4$ conversion over them at 370 °C (Fig. 2). In the presence or absence of 4% water vapor, Pd-Zr/HZSM-5 is found to show a remarkably lower tendency to deactivate compared to Pd/HZSM-5. Moreover, it is also worth to mention that the CH$_4$ conversion over Pd-Zr/HZSM-5 could easily be recovered to its initial value by increasing the reaction temperature to 390 °C or removing the H$_2$O in the feed gas. It is suggested that ZrO$_2$ can improve the thermal stability of the support, prevent the sintering of the supported palladium, and thus stabilize the dispersed state of palladium. As a result, with the addition of ZrO$_2$, Pd/HZSM-5 possesses remarkably enhanced thermal/hydrothermal stability.

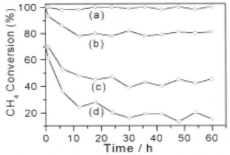

Fig. 2 Time dependence of methane conversion over Pd-Zr/HZSM-5 (a, b) and Pd/HZSM-5 (c, d) at 370 ℃ in the presence (b, d) and absence (a, c) of 4% water vapor

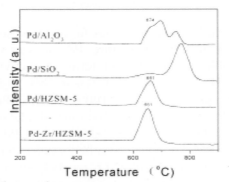

Fig. 3. O₂-TPD spectra of the samples.

3.2. O₂-TPD tests

The chemical states of PdO on the catalysts were studied by performing O_2-TPD measurements (Fig. 3). One can see that the initial O_2 evolution peak appears at ~661 °C for Pd/HZSM-5, while at ~651, 671 and 674 °C, respectively, for Pd-Zr/HZSM-5, Pd/γ-Al$_2$O$_3$ and Pd/SiO$_2$. In addition, the Pd/γ-Al$_2$O$_3$ and Pd/SiO$_2$ catalysts display obvious broad-peaks at the high-temperature region. According to Klingstedti et al. [6], the O_2 evolution peaks at lower temperature (651-674 °C) could be attributed to the decomposition of crystalline PdO, while the broad-peaks appearing at high temperatures (722-790°C) are mainly associated with the decomposition of amorphous PdO. Apparently, amorphous PdO possesses stronger Pd-O bonds than crystalline Pd-O. And ZrO$_2$ additive on Pd/HZSM-5 obviously affects the oxygen desorption properties of surface PdO species. In combination with the activity data listed in Table 1, one could come to the conclusion that the PdO species desorbing oxygen at lower temperature possess higher CH$_4$ activities, and *vice versa*.

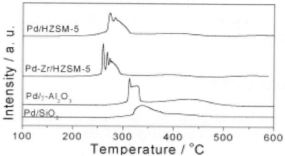

Fig. 4. CH$_4$-TPR spectra of the samples.

3.3. CH$_4$-TPR tests

In order to examine directly the CH$_4$ reducibility of PdO species on the catalysts, CH$_4$-TPR tests were carried out (Fig. 4). One can see that the CH$_4$ reduction properties of these PdO species on Pd catalysts are greatly influenced by the kinds of support and additive. All the TPR curves show two major peaks. Only the low temperature peak exhibits the reduction property of the PdO species [7]. Though the low-temperature peaks can be further split into three or more components corresponding to different PdO species reacting with methane, the initial reduction temperatures of the samples are in the order: Pd/SiO$_2$ > Pd/γ-Al$_2$O$_3$ P > Pd/HZSM-5 > Pd-Zr/HZSM-5, with Pd/SiO$_2$ being 337 °C and Pd-Zr/HZSM-5 being 267 °C. Through CH$_4$-TPR testing, Carstens et al. [2] also found that the crystalline form of PdO could be reduced more rapidly than the amorphous one. Therefore, we are tempted to ascribe the initial reduction peak to the reduction of the crystalline PdO, and the following ones to that of its amorphous state.

Crystalline PdO was shown to be more catalytically active than amorphous PdO [2,8-9]. ZrO$_2$ could act as an oxygen reservoir and is also a well-known oxygen supplier, exhibiting large oxygen mobility [2]. Therefore, ZrO$_2$ could offer oxygen atoms a relatively free pathway for approaching the Pd atoms. So the oxygen mobility of ZrO$_2$ might be a contributing factor for improving the crystallizability of PdO on Pd/HZSM-5. The crystalline PdO possesses the weak Pd-O bonds which can increase its stability and surface density of oxygen vacancies, contributing to the high turnover rate of CH$_4$ oxidation [10].

4. CONCLUSION

In a summary, a series of Pd catalysts with different supports were prepared by the impregnation method. It is found that Pd/HZSM-5 shows considerably higher CH$_4$ combustion activity than the conventional Pd/γ-Al$_2$O$_3$ and Pd/SiO$_2$. ZrO$_2$ additive on Pd/HZSM-5 can greatly improve its catalytic activity. In addition, the introduction of ZrO_2 also endows Pd-Zr/HZSM-5 with higher thermal and hydrothermal stability than those of Pd/HZSM-5. Therefore, Pd-Zr/HZSM-5 could be considered as a promising catalyst for practical low-temperature CH_4 combustion.

ACKNOWLEDGEMENT

The authors would like to thank for the financial support by the Ministry of Science and Technology of China (G1999022400).

REFERENCES

[1] R.E. Dickinson and R.J. Cicerone, Nature 319 (1986) 109.
[2] J.N. Carstens, S.C. Su and A.T. Bell, J. Catal. 176 (1998) 136.
[3] K. Muto, N. Katada and M. Niwa, Appl. Catal. A 134 (1996) 203.
[4] Y. Li and J. Armor, Appl. Catal. B 3 (1994) 275.
[5] T. Ishihara, H. Sumi and Y. Takita, Chem. Lett. (1994) 1499.
[6] F. Klingstedi, A.K. Neyestanaki, R. Bygguingsbacka, L.E. Lindfors, M. Lundén, M. Petersson, P. Tengström, T. Ollonqvist and J. Väyrynen, Appl. Catal. A 209 (2001) 301.
[7] L-F. Yang, C-K. Shi, X-E. He and J-X. Cai, Appl. Catal. B 38 (2002) 117.
[8] R. Burch and F.J. Urbano, Appl. Catal. A 124 (1995) 121.
[9] R.F. Hicks, H. Qi, M.L. Young and R.G. Lee, J. Catal. 122 (1990) 295.
[10] K.I. Fujimoto, F.H. Ribeiro, M.A. Borja and E. Iglesia, J. Catal. 179 (1998) 431.

Studies in Surface Science and Catalysis, volume 147
X. Bao and Y. Xu (Editors)

Reverse microemulsion synthesis of Mn-substituted barium hexaaluminate (BaMnAl$_{11}$O$_{19-\alpha}$) catalysts for catalytic combustion of methane

Jinguang Xu[a, b]**, Zhijian Tian**[* a]**, Yunpeng Xu**[a]**, Zhusheng Xu**[a]**, Liwu Lin**[a]

[a]Dalian Institute of Chemical Physics, Chinese Academy of Sciences, 457 Zhongshan Road, Dalian 116023, China, [*]tianz@dicp.ac.cn (Z. Tian)

[b]Institute of Applied Catalysis, Yantai University, Yantai 264005, China

ABSTRACT

The reverse microemulsion composed of 5-25% water, 3-15% Span60, 2-10% n-hexanol and 70-90% iso-octane was used as the reaction media to prepare Mn-substituted hexaaluminate catalysts. The diameters of most water droplets in the reverse microemulsion are from 10 to 80 nm. The XRD results indicated that such a reverse microemulsion is favorable to the formation of hexaaluminates phase. The heat stability and CH$_4$ combustion activity of the BaMnAl$_{11}$O$_{19-\alpha}$ catalysts prepared with Ba(NO$_3$)$_2$ as a precursor is as good as the ones with Ba(i-OC$_3$H$_7$)$_2$.

1. INTRODUCTION

Recently catalytic combustion has attracted considerable attention due to its potential for burning natural gas and methane in turbines with extremely low levels of NO$_X$ emissions [1]. The main problem in the development of a catalytic combustion process is the seeking of suitable catalysts with high stability and high activity [2]. Pd-based catalysts are known to be active for methane combustion, but these catalysts lose activity at high temperatures when PdO decomposes to Pd [3]. Various heat-resistant metal oxides have been tested as catalysts and supports [4]. Among all the catalysts, hexaaluminate catalysts developed by Arai et al. show high stability and high activity for methane combustion [5]. Many researchers have made great efforts to enhance the heat stability and catalytic activity of hexaaluminate catalysts [6-8]. Arai et al. prepared the barium hexaaluminates using a sol-gel method by hydrolysis of aluminium and barium composite alkoxides in a mixture of water and alkanols [5]. Zarur et al. prepared a barium hexaaluminate support, which had the largest surface area ever reported in the literature, in a reverse microemulsion media also by the hydrolysis of aluminium and barium alkoxides [6].

Considering the costliness of barium alkoxides, we report here a new route to a production of Mn-substituted barium hexaaluminate ($BaMnAl_{11}O_{19-\alpha}$) catalysts in a reverse microemulsion media by using $Ba(NO_3)_2$ instead of $Ba(i-OC_3H_7)_2$. The effects of the composition of reverse microemulsion on the properties of $BaMnAl_{11}O_{19-\alpha}$ catalysts were also investigated in this work.

2. EXPERIMENTAL

The reverse microemulsion system consisted of Span60, n-hexanol, iso-octane and the solution of manganese nitrate and barium nitrate, which were employed as the surfactant, cosurfactant, continuous oil phase and dispersed aqueous phase, respectively. Typical microemulsion composition consisted of 5-25% water or brine (the solution of manganese nitrate and barium nitrate), 3-15% Span60, 2-10% n-hexanol and 70-90% iso-octane. Calculated amounts of $Al(i-OC_3H_7)_3$ was dissolved in 2-propanol to form the precursor solution, which was then slowly added to the reverse microemulsion at 50ºC. The hydrolyzed mixture was then allowed to age for 24h. After aging, the materials generated were subjected to supercritical drying to remove iso-octane, Span60, n-hexanol, 2-propanol and the residual water. The details of the supercritical drying were illustrated elsewhere [9]. The materials were then calcined at 1200ºC for 4h.

The size of the water droplets in the reverse microemulsion was measured with a Malvern Zetasizer 1000 laser light scattering meter. The BET specific surface areas were measured by physisorption of nitrogen at 77K using a Micrometrics ASAP 2010 instrument. Powder XRD patterns of the catalysts were recorded by a Rigaku D/max-rB X-ray diffractometer with CuKα radiation. The XRD instrument was operated at 40 KV and 50 mA. The spectra were scanned between 5º and 70º at a rate of 5º/min.

The catalytic activity for the combustion of methane was tested by a fixed-fed quartz reactor system. Electronics mass flow controllers (MFCs) were used to control the gas flow rates. For each experiment, 1ml of the catalyst (grain size 20 - 40 mesh) was loaded to the reactor (i.d=15 mm). 800 mL/min (STP) of 1% CH_4 in air was used as the feed. The gas hourly space velocity (GHSV) was monitored to 48000 h^{-1}. Effluent gases were analyzed on-line using a GC equipped with a carbon molecular sieve column and a thermal conductivity detector (TCD). T_{10} and T_{90} were used to represent the temperatures required for 10% and 90% CH_4 conversion, respectively.

3. RESULTS AND DISCUSSION

3.1 Effect of reaction media on phase composition

The details of preparation and the specific surface areas of the BaMnAl$_{11}$O$_{19-\alpha}$ catalysts are shown in Table 1. Ba and Ba^{2+} in the Catalyst ID represent Ba(i-OC$_3$H$_7$)$_2$ and Ba(NO$_3$)$_2$ as the precursor of the BaMnAl$_{11}$O$_{19-\alpha}$ catalysts, respectively. WP refers to a mixture of water and 2-propanol, which was used as the reaction medium for the hydrolysis of Ba(i-OC$_3$H$_7$)$_2$ and Al(i-OC$_3$H$_7$)$_3$. RM implies that reverse microemulsion was applied as the reaction medium.

The phase compositions of BaMnAl$_{11}$O$_{19-\alpha}$ catalysts prepared in different reaction media are shown in Fig.1.

Ba/RM-35 sample consisted of a single β-Al$_2$O$_3$ phase while Ba/WP sample contained β-Al$_2$O$_3$, α-Al$_2$O$_3$ and BaAl$_2$O$_4$ phase. The BET surface area of Ba/RM-35 sample is 32.1 m^2/g, which was larger than that of Ba/WP sample. The decrease of the BET surface area of Ba/RM-35 sample was due to the presence of α-Al$_2$O$_3$ and BaAl$_2$O$_4$ phase. A single β-Al$_2$O$_3$ phase also appeared in the Ba^{2+}/RM-35, Ba^{2+}/RM-70 and Ba^{2+}/RM-140 samples, which were prepared with Ba(NO$_3$)$_2$ as precursors.

Table 1

Preparation and properties of the BaMnAl$_{11}$O$_{19-\alpha}$ catalysts

Catalyst ID	Ba precursor/Reaction media	R^a	S$_{BET}$^{b} m^2/g
Ba/WP	Ba(i-OC$_3$H$_7$)$_2$/H$_2$O+ i-C$_3$H$_7$OH	35	21.3
Ba/RM-35	Ba(i-OC$_3$H$_7$)$_2$/Reverse Microemulsion	35	32.1
Ba^{2+}/RM-35	Ba(NO$_3$)$_2$/ Reverse Microemulsion	35	16.0
Ba^{2+}/RM-70	Ba(NO$_3$)$_2$/ Reverse Microemulsion	70	31.5
Ba^{2+}/RM-140	Ba(NO$_3$)$_2$/ Reverse Microemulsion	140	20.0

a: R=H$_2$O/Al(iso-OC$_3$H$_7$)$_3$ (molar ratio); b: after Calcinations at 1200 °C for 4h

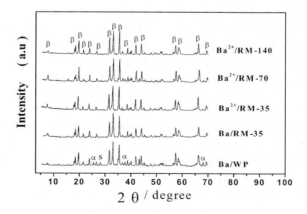

Fig. 1. XRD spectra of the BaMnAl$_{11}$O$_{19-\alpha}$ catalysts. β: β-Al$_2$O$_3$; α: α- Al$_2$O$_3$; S: BaAl$_2$O$_4$

A mixture of water and alkanols is usually used as the reaction medium for the hydrolysis of alkoxides [5]. Zarur et al. reported that the barium hexaaluminate (BHA) and CeO_2-BHA synthesized with reverse microemulsion as the reaction media displayed excellent methane combustion activity, owing to their high surface area [6]. Our results indicate that the reverse microemulsion is favorable to the formation of hexaalumintes phase.

3.2 Replacement of $Ba(i\text{-}OC_3H_7)_2$ by $Ba(NO_3)_2$

Barium alkoxide is usually applied as the precursor to prepare barium hexaaluminate or substituted barium hexaaluminate. It was claimed that the uniform mixing of components at a molecular level could be attained by using barium and aluminum alkoxides as precursors [5]. In our work, we tried to use $Ba(NO_3)_2$ to replace $Ba(i\text{-}OC_3H_7)_2$ as the precursor for the synthesis of Mn-substituted barium hexaaluminate ($BaMnAl_{11}O_{19-\alpha}$) catalysts. As shown in Fig. 1, a single β-Al_2O_3 phase appears in the Ba^{2+}/RM-35, Ba^{2+}/RM-70 and Ba^{2+}/RM-140 samples. The BET surface areas of the three samples varied with the H_2O/Al(iso-$OC_3H_7)_3$ ratio (R). As R increases from 35 to 70, the BET surface area of the sample increases from 16.0 to 31.5 m^2/g. However, the BET surface area decreases while R increases from 70 to 140. It is worthy to note that the BET surface area of Ba^{2+}/RM-70 sample is comparable to those of Ba/RM-35 sample.

Two reactions appear in the hydrolysis of $Al(i\text{-}OC_3H_7)_3$: hydrolysis (Equation 1) and polycondensation (Equation 2),

$$Al(i\text{-}OC_3H_7)_3 + 3H_2O \quad \rightarrow \quad Al(OH)_3 + 3i\text{-}C_3H_7OH \tag{1}$$

$$Al(OH)_3 + Al(OH)_3 \quad \rightarrow \quad (OH)_2Al\text{-}O\text{-}Al(OH)_2 + H_2O \tag{2}$$

The hydrolysis reaction is responsible for creating new seeds and is favored by high H_2O/Al(i-$OC_3H_7)_3$ ratios (R). In contrast, the polycondensation reaction gives rises to particles growth of the existing seeds and is promoted by low H_2O/Al(i-$OC_3H_7)_3$ ratios (R) [10].

Machida et al. reported that R must be lower than the stoichiometric ratio in order to promote the polycondensation reaction [11]. However, with $Ba(NO_3)_2$ to replace $Ba(i\text{-}OC_3H_7)_2$ as precursors, the R must be higher than 35 due to the solubility of $Ba(NO_3)_2$. As shown in Table1, the BET surface area of Ba^{2+}/RM-70 sample is larger than that of Ba/WP sample. This indicates that the reverse microemulsion used as the reaction medium play a key role in the hydrolysis and polycondensation of $Al(i\text{-}OC_3H_7)_3$. Zarur et al. reported that nanometer-size aqueous micelles in a reverse microemulsion could be used to control hydrolysis and polycondensation of Ba and Al alkoxides [10]. We determined the size distribution of water droplets in our reverse microemulsion. The results are shown in Fig. 2.

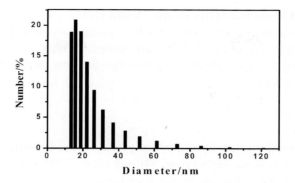

Fig. 2. The size distribution of the water droplets in the reverse microemulsion
H_2O: 15.3 vol.%; H_2O/SPAN60 = 140 (molar ratio)

The size of water droplets in the reverse microemulsion was below 120 nm and the diameters of most water droplets were from 10 to 80 nm. It is obvious that such nanometer-sized water droplets controlled the hydrolysis and polycondensation of $Al(i-OC_3H_7)_3$ although the R was much higher than the stoichiometric ratio. However, the BET surface area of the resulting materials is not increased with increasing the R, as reported by Zarur et al. [10].

3.3 Methane combustion activities of catalysts

The combustion activity of methane over $BaMnAl_{11}O_{19-\alpha}$ catalysts prepared by different methods is presented in Fig. 3.

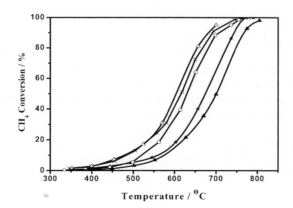

Fig. 3. CH_4 combustion over $BaMnAl_{11}O_{19-\alpha}$ catalysts
($\bigcirc$):Ba/RM-35;($\triangle$):Ba^{2+}/RM-70;($\times$):Ba/WP;($\bigstar$):Ba^{2+}/RM-140; ($\blacktriangle$): Ba^{2+}/RM-35

The CH_4 combustion activity of the $BaMnAl_{11}O_{19-\alpha}$ catalysts prepared with $Ba(NO_3)_2$ as precursors depends on the water/ aluminum isopropoxide ratio. The Ba^{2+}/RM-70 sample showed high CH_4 combustion activity, which is higher than that of Ba/WP sample. This results illustrate that the new route with $Ba(NO_3)_2$ as a precursor is practical.

4. CONCLUSION

The preparation medium has an important effect on the formation, heat stability and methane combustion activity of Mn-substituted hexaaluminate catalysts. With the reverse microemulsion as the preparation medium, the nanometer-size water droplets in the reverse microemulsion control the hydrolysis and polycondensation of $Al(i-OC_3H_7)_3$. Mn-substituted barium hexaaluminate ($BaMnAl_{11}O_{19-\alpha}$) catalysts prepared by reverse microemulsion synthesis method with $Ba(NO_3)_2$ as a precursor show high heat stability and methane combustion activity.

ACKNOWLEDGMENT

The authors would like to thank the financial support from Ministry of Science and Technology of the People's Republic of China (G1999022401).

REFERENCES

[1] L.D. Pfefferle and W.C. Pfefferle, Catal. Rev-Sci. Eng., 29 (1987) 219.
[2] E.M. Johansson, K.M.J. Danielsson, E. Pocoroba, E.D. Haralson and S.G. Järäs, Appl. Catal. A, 182 (1999) 199.
[3] M.F.M. Zwinkels, S.G. Järäs, P.G. Menon and T.A.Griffin, Catal. Rev-Sci. Eng., 26 (1993) 319.
[4] Ben W.-L. Jang, R.M. Nelson, J.J. Spivey, M. Ocal, R. Oukaci and G. Marcelin, Catal. Today, 47 (1999) 103.
[5] M. Machida, K. Eguchi and H. Arai, J. Catal., 103 (1987) 385.
[6] A.J. Zarur and J.Y. Ying, Nature, 403 (2000) 65.
[7] E.M. Johansson, M. Berg, J. Kjellstom and S.G. Järäs, Appl. Catal. B, 20 (1999) 319.
[8] P. Artizzu-Duart, J.M. Millet, N. Guilhaume, E. Garbowski and M. Primet, Catal. Today, 59 (2000) 163.
[9] J. Xu, Z. Tian, J. Wang, Y. Xu, Z. Xu and L. Lin, Korean J. Chem. Eng., 20 (2003) 217.
[10] A.J. Zarur, H.H. Hwu and J.Y. Ying, Langmuir, 16 (2000) 3042.
[11] M. Machida, K. Eguchi and H. Arai, Bull. Chem. Soc. Jpn., 61 (1988) 3659.

Studies in Surface Science and Catalysis, volume 147
X. Bao and Y. Xu (Editors)
©2004 Elsevier B.V. All rights reserved.

Effect of drying method on the morphology and structure of high surface area BaMnAl$_{11}$O$_{19-\alpha}$ catalyst for high temperature methane combustion

Zhijian Tian[*]**, Jinguang Xu, Junwei Wang, Yunpeng Xu, Zhusheng Xu, Liwu Lin**

Dalian Institute of Chemical Physics, The Chinese Academy of Sciences, Dalian 116023, China, [*]tianz@dicp.ac.cn (Z. Tian)

ABSTRACT

Conventional oven drying (CD) and supercritical drying (SCD) methods were applied to the preparation of BaMnAl$_{11}$O$_{19-\alpha}$ catalysts. The effect of drying methods on specific surface area, pore structure, morphology and combustion activity of the samples was investigated. The catalysts obtained by SCD have higher surface area, narrower pore size distribution, and higher combustion activity than those obtained by CD.

1. INTRODUCTION

Catalytic combustion of natural gas is a promising technology conversion process because of more efficiently burning in a wider air-to-fuel ratios and lower NO$_x$ and CO emissions than conventional thermal combustion. However, at high temperature, e.g. in the gas turbine (>1200 °C), the complete combustion of residual hydrocarbons is diffusion controlled rather than reaction controlled, so that the reaction rate is dependent on not only the number of active sites but also the available surface area. Therefore, thermally stable catalysts with high surface area and advanced pore structures are very desirable [1-4]. As the more pronounced change in texture of catalysts is induced by synthesis conditions, in this paper, we used a sol-gel process to prepare Ba-Mn-Al hydroxide hydrogel and dried it by different methods. The effect of drying method on the pore structure, morphology and catalytic activity of obtained BaMnAl$_{11}$O$_{19-\alpha}$ hexaaluminate catalyst for methane combustion was investigated.

2. EXPERIMENTAL

Ba-Mn-Al hydroxide hydrogel was synthesized using a sol-gel process by drop-wise addition of an aqueous barium and manganese nitrate salt solution to a warm aluminum isopropoxide isopropanol solution under vigorous stirring. The gel/sol so formed was allowed to age for 12h. The hydrogel obtained was then divided into two parts. One part was dried by supercritical ethanol drying (8.0MPa, 260°C) (SCD) and an aerogel was obtained. The other part was dried in an oven at 110°C for 12h in air (CD), and converted to a xerogel. The aerogel and xerogel samples were calcinated in a muffle furnace at 1200°C for 4h. The corresponding $BaMnAl_{11}O_{19-\alpha}$ catalysts were labeled as SCD and CD.

The specific surface area (S_{BET}) and the mean pore diameter (d_P) of the samples were determined by nitrogen physisorption at 77K using a Micromertics ASAP 2010 instrument. Pore size distributions were calculated by applying the Barrett-Joyner-Halenda (BJH) method to the desorption branch of the isotherms. Powder XRD characterization was performed with a Rigaku D/MAX-RB X-ray diffractometer using a copper target at 40KV and 50mA. Transition electron microscopy (TEM) images were taken using a JEOL JEM-1200EX microscope.

The catalytic evaluation was performed by a fixed-bed quartz reactor system under atmospheric pressure with an on-line GC. 1% CH_4 in air was used as the feed, and the gas hourly space velocity (GHSV) was monitored to 48,000h^{-1}. The detail was described elsewhere [5].

3. RESULTS AND DISCUSSTION

3.1 Pore structure, texture and specific surface area

The preparation and morphological properties of the Ba-Mn-Al samples are summarized in Table.1. The XRD spectra reveal that after calcination at 1200°C in air both the aerogel and xerogel samples transformed from amorphous materials to well crystalline Ba-β-Al_2O_3 hexaaluminate phases with trace of $BaAl_2O_4$ phases.

The adsorption-desorption isotherms of the aerogel and xerogel samples are shown in Fig.1.The aerogel sample shows a type IV isotherm with a type H3 desorption hysteresis loop according to IUPAC classification [6]. The shapes of

Table 1
Preparation and Morphological Properties of $BaMnAl_{11}O_{19-\alpha}$

Catalyst	Drying	Calcination °C	Crystalline bulk phase	S_{BET} $m^2 \cdot g^{-1}$	V_P^b $cm^3 \cdot g^{-1}$	d_P^c nm
aerogel	SCD	-	-	260.5	1.36	20.9
xerogel	CD	-	-	370.2	0.32	3.5
$BaMnAl_{11}O_{19-\alpha}$	SCD	1200	Ba-β-Al_2O_3	72.2	0.35	19.5
$BaMnAl_{11}O_{19-\alpha}$	CD	1200	Ba-β-Al_2O_3	21.3	0.14	25.5

the hysteresis loops have often been identified with specific pore structures. The type H3 hysteresis loop is associated with particles giving rise to slit-shaped mesopores. The xerogel sample shows a type-IV isotherm with a type H2 desorption hysteresis loop. According to the suggestion of IUPAC, this type of hysteresis loop is resulted from crossed pores with narrow necks and wide bodies (often referred to as 'inkbottle' model mesopores).

The morphology of the aerogel and xerogel samples and the BaMnAl$_{11}$O$_{19-\alpha}$ catalysts were investigated by TEM and the results are shown in Fig.2. The particles in the aerogel sample are fibroid (Fig.2a), which are obviously different from the agglomerates in the xerogel sample (Fig.2b).

Fig.3 shows the pore-size distributions of the aerogel and xerogel samples and corresponding catalysts after calcination at 1200°C. The aerogel sample shows a bimodal pore-size distribution, with maxima around 2.59 nm and 30.0 nm (Fig.3a). The mean pore diameter (d_p) is 20.9 nm. The xerogel sample displays a unimodal pore-size distribution, with a maximum around 3.52 nm, and the mean pore diameter (d_p) is 3.48 nm (Fig.3b).

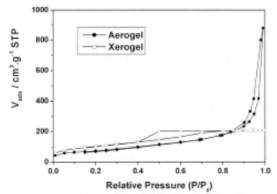

Fig.1. The adsorption-desorption isotherms of the aerogel and xerogel samples.

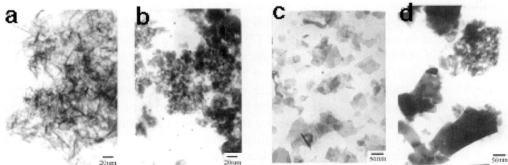

Fig.2. TEM images of BaMnAl$_{11}$O$_{19-\alpha}$ catalysts before and after calcination
(a) aerogel, (b) xerogel, (d) BaMnAl$_{11}$O$_{19-\alpha}$ (SCD), (d) BaMnAl$_{11}$O$_{19-\alpha}$ (CD)

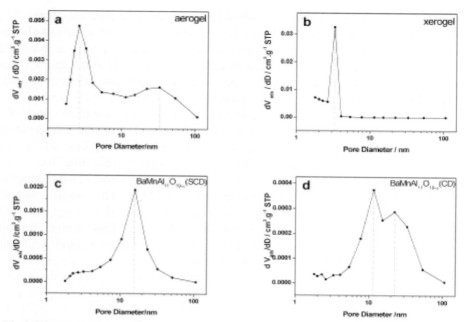

Fig.3. The pore-size distribution of the aerogel, xerogel and corresponding catalysts
(a) aerogel, (b) BaMnAl$_{11}$O$_{19-\alpha}$, (SCD) (c) xerogel, (d) BaMnAl$_{11}$O$_{19-\alpha}$ (CD)

The wet solid network in a gel consists of strands of 'polymeric substances' or chains of particles joined at nodes and forming a three-dimensional cross-linked structure. During the drying of the gel, the capillary forces, originated by the surface tension of the liquid phase, tend to densify the solid by closely packing together the elementary particles, destroying the most fragile part of the three-dimensional network and reducing the porous volume of the resulting solid. With supercritical drying, the gel collapse was suppressed due to disappearance of the capillary pressure. The fibroid structure of the gel was maintained and large pores were present in the aerogel after drying. In contrast, with conventional oven drying, the capillary pressure is in an order of ~100MPa [7]. Such high a pressure made the gel to shrink and collapse and the pores in the xerogel after drying is smaller.

After calcination at 1200°C, the SCD catalyst consists of thin plate-like crystals (50-100nm), whereas the CD catalyst contains rather thick and large (100-200nm) agglomerates (Fig.2c and Fig.2d). The isotherms and the hysteresis loops of the SCD and CD catalysts are found to be very similar and akin to type IV and type H3 of the IUPAC classification, respectively. The platelet geometry of primary particles observed by electron microscopy is associated with the nature of the hexagonal facet of hexaaluminate crystals and is consistent with the presence of slit-shaped micropores revealed from the N_2 adsorption analyses.

The $BaMnAl_{11}O_{19-\alpha}$ catalyst (SCD) gives a unimodal pore-size distribution, with the maximum around 14.4 nm, and the mean pore diameter (d_p) is 19.5 nm (Fig.3c). The $BaMnAl_{11}O_{19-\alpha}$ catalyst (CD) shows a bimodal pore-size distribution centered at 11.5 nm and 21.0 nm, and the mean pore diameter (d_p) is 25.5 nm (Fig.3d). It is obvious that a collapsing of the smaller mesopores (around 3.0 nm) occurred and larger pores formed in the calcination process. Larger particles were also formed due to particle aggregation caused by the collapsing of pores, and the BET surface area decreased simultaneously.

Ishikawa et al [8] reported that the contact area was associated to the pore volume. A large pore volume corresponded to a small contact area. The sintering first takes place at the contact area between particles. Mutual diffusion of ions through the contact area follows as the crystallites grow. As shown in Table 1, the pore volume of the aerogel sample is about 4 times larger than that of the xerogel samples. So the contact area between fibroid particles in the aerogel sample is much smaller than that in the xerogel sample. During calcination, sintering of the particles in the aerogel sample was weaker than that in the xerogel sample due to the smaller contact area. Such a weaker sintering made the platelet crystals in the SCD catalyst powders thinner than that in the CD catalyst powders. It was reported that the surface area of hexaaluminate depends on the thickness of the hexagonal facet of hexaaluminate crystals [9]. The thinner the hexagonal facet of hexaaluminate crystals gets, the larger the surface area of hexaaluminates would be.

As shown in Table 1, the BET surface area of the xerogel sample was $370.2 m^2 \cdot g^{-1}$. After calcination at 1200°C, it decreased to $21.3 m^2 \cdot g^{-1}$. Although the BET surface area of the aerogel sample was $260.5 m^2 \cdot g^{-1}$, it reduced to $72.2 m^2 \cdot g^{-1}$ after calcination, which is much higher than that ($21.3 m^2 \cdot g^{-1}$) of the xerogel sample. Therefore, the crossed inkbottle pores obtained by conventional oven drying are much easier to collapse than the slit-shaped pores by supercritical drying.

3.2 The combustion activity of methane

The catalytic activity for methane combustion over the $BaMnAl_{11}O_{19}$ catalysts obtained by different drying methods was investigated and the results are shown in Fig.4. The ignition temperature (T_{10}) of the SCD catalyst is ca. 450°C, which is 20°C lower than that of the CD catalyst. This is attributed to a more homogenous distribution of active phase in the $BaMnAl_{11}O_{19}$ catalyst obtained by the supercritical drying than that of the CD catalyst [5]. At high temperature, the mass transfer phenomena on the surface or inside the catalyst particle are emphasized in the reaction of catalytic combustion of methane. The pore size and porosity affect the effective diffusivity of reaction gas in the catalyst. The T_{90} of the SCD catalyst is 50°C lower than that of the CD catalyst.

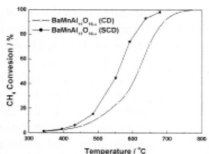

Fig. 4 CH$_4$ combustion over Mn-substituted hexaaluminate catalysts

The high activity of the SCD catalyst can be attributed to the high effective gas diffusivity.

4. CONCLUSION

Drying methods have a strong effect on the phase composition, surface area, and pore structure and combustion activity of the Mn-substituted hexaaluminate catalysts synthesized by the sol-gel method. Compared with the samples obtained by CD, those by SCD have a higher surface area, a narrower distribution of pore sizes and a higher combustion activity.

ACKNOWLEDGMENT

The authors would like to thank the financial support from Ministry of Science and Technology of the People's Republic of China (G1999022401).

REFERENCES

[1] J.G. McCarty, Nature, 403 (2000) 35
[2] L.D. Pfefferle and W.C. Pfefferle, Catal. Rev-Sci. Eng., 29 (1987) 219
[3] M. Machida, K. Eguchi and H. Arai, J. Catal., 123 (1990) 477
[4] M.F.M. Zwinkels, S.G. Järäs, P.G. Menon and T.A. Griffin, Catal. Rev-Sci. Eng., 35 (1993) 319
[5] J. Wang, Z. Tian, J. Xu, Y. Xu, Z. Xu and L. Lin, Catal. Today, 83 (2003) 213
[6] K.S.W. Sing, D.H. Everett, R.A.W. Haul, L. Moscou, R.A. Pierotti, J. Rouquerol and T. Siemieniewska, Pure Appl. Chem., 57 (1985) 603
[7] L.L. Hench and J.K. West, Chem. Rev., 90 (1990) 33
[8] T. Ishikawa, R. Ohashi, H. Nakabayashi, N. Kakuta, A. Ueno and A. Furuta, J. Catal., 134 (1992) 87
[9] H.Arai, T. Yamada, K. Eguchi and T. Seiyama, Appl. Catal., 26 (1986) 265

Studies in Surface Science and Catalysis, volume 147
X. Bao and Y. Xu (Editors)
493

Synthesis of nano-sized $BaAl_{12}O_{19}$ via nonionic reverse microemulsion method: I. Effect of the microemulsion structure on the paticle morphology

Fei Teng, Zhijian Tian[*], Jinguang Xu, Guoxing Xiong[*], Liwu Lin

Dalian Institute of Chemical Physics, Chinese Academy of Sciences, Dalian 116023, China. tianz@dicp.ac.cn (Z. Tian), gxxiong@dicp.ac.cn (G. Xiong)

ABSTRACT

Nano-sized $BaAl_{12}O_{19}$ particles with different morphologies, i.e. spherical, cylindrical, spindly, and acicular, were prepared in the reverse microemulsion. The effects of microemulsion composition and drying method on the particle morphology were investigated. It was found that the acicular $BaAl_{12}O_{19}$ catalyst has higher thermal stability and catalytic activity for methane combustion than the spherical one.

1. INTRODUCTION

Catalytic combustion of natural gas has many advantages over the conventional flame combustion in suppressing emissions (NO_x, CO) and improving energy efficiency [1]. However, the catalysts with high surface areas are required for methane combustion at high temperatures. Due to the high thermal stability of hexaaluminate, it has become the focus of the researches. To improve its stability, various preparation methods are explored [1-3]. Recently, J. Y. Ying et al. [4] have prepared the $BaAl_{12}O_{19}$ catalyst with the largest surface area up to now by the reverse microemulsion method. The reverse microemulsion method has been proved effective to control the morphology of the particles [5-8]. In present article, the $BaAl_{12}O_{19}$ catalysts were synthesized in a nonionic reverse microemulsion. The influences of water content and drying method on the particle morphology, as well as the activity for methane combustion over the acicular and spherical $BaAl_{12}O_{19}$ catalysts, were investigated.

2. EXPERIMENTAL

2.1. Preparation of the $BaAl_{12}O_{19}$ catalyst
The reverse microemulsion systems used in the present study consist of

polyoxyethylene (6) tridecyl alcohol ether ($C_{13}EO_6$, MW=464), n-hexanol, cyclohexane and aqueous solution (1.2M mixture solution of $Ba(NO_3)_2$ and $Al(NO_3)_3$ (Ba:Al=1:12 mole ratio) or 2.5M $(NH_4)_2CO_3$ solution), which are employed as the surfactant, cosurfactant, continuous oil phase and dispersed aqueous phase, respectively. To prepare the reverse microemulsion, the accounted amounts of $C_{13}EO_6$, n-hexanol, cyclohexane, and the aqueous solution were mixed under stirring until it became transparent. The weight ratio of aqueous solution in the microemulsion was designed as water content (wt%). The typical microemulsion compositions are given in Table.1, and water content was altered from 8wt%, 10wt% to 25wt%.

The precursor of the $BaAl_{12}O_{19}$ catalyst was synthesized by rapid mixing the same volume of nitrates-containing and $(NH_4)_2CO_3$-containing microemulsion under vigorous stirring at room temperature. The microemulsion was stirred continually for 5h and then allowed to age for 24h. After aging, the acetone was added to demulsify the system. The precursor was recovered by centrifugation. The recovered precursor was washed sufficiently with acetone, deionized water, and ethanol in sequence. Then it was divided into two parts and dried by supercritical ethanol drying (8.0MPa, 260 °C) (SCD) and conventional oven drying (CD), respectively. The details of the drying method were illustrated elsewhere [9]. The dried samples were calcined at 1200 °C for 5h under flowing air in a muffle. The corresponding $BaAl_{12}O_{19}$ catalysts were labeled as SCD and CD.

2.2. Characterization and Activity Test for Methane Combustion

The morphology, crystal structure, surface area and pore structure of the $BaAl_{12}O_{19}$ catalysts were characterized by TEM, XRD and nitrogen physisorption, respectively. The morphology of the Ba-Al precursor in microemulsion was also observed with TEM by directly subjecting several drops of microemulsion to photographing. The electric conductivity of microemulsion was measured with a DDS-11A model conductometer.

The reaction of methane combustion was carried out in a conventional flow system under atmospheric pressure. A mixture gas of methane (1 vol.%) and air (99 vol.%) was fed into the catalyst bed at a gas hourly space velocity (GHSV) of 48000 h^{-1}. The inlet and outlet gas compositions were analyzed by an on-line gas chromatography (GC) with a packed column of carbon molecular sieve and a thermal conductivity detector (TCD).

3. RESULTS AND DISCUSSIONS

3.1. Effect of water content on the particle morphology

The preparation conditions are summarized in Table 1. Fig. 1 presents the TEM images of the Ba-Al precursor in microemulsion. In a microemulsion with a low water content (8wt%), a few spherical particles are observed. At 15wt%

Table 1
Preparation and morphologies of the Ba-Al samples after calcination at 1200°C for 5h

Sample	W: CS: S:O (wt:wt)	Drying	Particle shape [*]	Particle size (nm) [*]
(A)	1.6:1.2:3.5:13.5	SCD	Spherical	15~20
(B)	3.2:1.2:3.5:13.5	SCD	Cylindrical	15×40
(C)	6.1:1.2:3.5:13.5	CD	Spindly	(50~100)×(400~500)
(D)	6.1:1.2:3.5:13.5	SCD	Acicular	(5~10)×(100~200)

W: water, CS: n-Hexanol, S: $C_{13}E_6$, O: Cyclohexane; * After calcination at 1200°C for 5h.

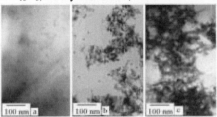

Fig. 1. TEM images of the precursors prepared in the microemulsions with different water contents. (A) 8wt%. (B) 15wt%. (C) 25wt%

water content, cylindrical particles are formed, accompanying by few spherical particles. At a higher water content (25wt%), a network of the rodlike precursor particles appears.

TEM images of the calcined particles are shown in Fig. 2. It is found that the particle morphology varies from spherical, cylindrical, to acicular or spindly, with increasing of water content. The spherical particles of 15-20nm in diameter were prepared in the microemulsion of 8wt% water content (Fig. 2A). The small cylindrical particles of 15nm in diameter and 40nm in length (Fig. 2B) were prepared at 15wt% water. Fig. 2C and D present the images of the particles obtained at 25wt% water but via different drying methods. The spindly particles obtained by CD are of 50-100nm in diameter and 400-500nm in length (Fig. 2C), and the acicular particles obtained by SCD are of 10nm in diameter and 100-200nm in length (Fig. 2D). Comparing Fig.1 with Fig. 2, it could be found that the morphology of the precursor particles is retained to some extent in the calcined particles. It seems that the particles have some capability of shape

Fig. 2. TEM images of the calcined $BaAl_{12}O_{19}$ prepared in the microemulsions with different water contents. (A) 8wt%. SCD (B) 15wt%. SCD (C) 25%. CD (D) 25wt%. SCD.

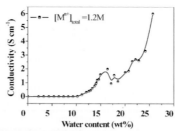

Fig. 3. The variations of electric conductivity with water content for the nitrates-containing reverse microemulsion.

memory. A kind of shape memory effect was also observed by Yan et al. on the preparation of MgO nanorods from rodlike $Mg(OH)_2$ nanocrystallites [10].

The morphology of the precursor particles might be related to the original microstructure of microemulsion. It was reported that with the increasing of water content in microemulsion, the connectivity of the aqueous phase increases, and the micelles could vary from spherical, rod-like, to thread-like (bicontinuous structure). The connectivity of aqueous phase in a nonionic microemulsion could be revealed by electric conductivity [7, 8]. The electric conductivity of our microemulsion system was also determined at different water content (Fig. 3). The electric conductivity is close to zero as the water content is lower than 10wt%, indicating that the water droplets are highly discrete in the continuous oil phase. The conductivity gradually increases with further increasing of water content, which arises from percolation phenomenon [11]. Above 20wt% water content, there is a steep increase of conductivity with increasing of water content, which indicates that the bicontinuous structure forms in microemulsion, and the water channels act as electric conducting channels [12].

At lower water contents, the interaction between the micelles is very weak, and the spherical micelles are retained; with the increasing of water content, the interaction of the micelles becomes so strong that regular spherical micelles could not be maintained. The spherical micelles would probably change into rodlike or cylindrical. With further increasing water content, bicontinuous structure (interconnected water channels) would form [13]. The formation of the particles with different morphologies may be related to the varieties of microemulsion microstructures, which acts as a soft template. The particles nucleate, grow, and assembly within the restricted space enclosed by surfactant and cosurfactant molecules, therefore the micelle morphology would be imprinted into the precursor particles. After calcination, the $BaAl_{12}O_{19}$ catalysts particles retain their precursor morphology due to the shape memory effect.

Mann et al. [14] have synthesized filament-like barium sulphate in an ionic microemulsion. They attributed the particle morphology to the strong electrostatic interactions between the inorganic phase and the ionic head group of AOT. In our study, because there only existed weak hydrogen bonds or van de waals force between the inorganic phase and $C_{13}EO_6$, the morphology of the particles more likely resulted from that of the micelles. Zurur et al. [4] also prepared the particles with different morphologies by altering water content of the microemulsion.

The drying method also has an influence on the shape formation of the $BaAl_{12}O_{19}$ particles. Comparing the Fig. 2C and 2D, it could be found the spindly particles obtained by CD are much larger than the acicular particles (SCD). The supercritical drying could effectively relieve the aggregation of the particles caused by capillary tension of water. Thus, the microstructure morphology imprinted in the precursor particles could be effectively retained during the supercritical drying. The large spindly particles could be probably considered as a result of aggregation of the acicular particles under capillary tension of water.

3.2. Catalytic combustion of methane

The texture and crystal structure properties of the acicular and the spherical $BaAl_{12}O_{19}$ catalysts are given in Table 2, and their light-off curves for methane combustion are shown in Fig. 4.

The ignition temperature (T_{10}) and complete combustion temperature (T_{90}) of the acicular sample (Curve 2) are ca. 690 and 770°C, which are 65 and 98°C lower than those of the spherical one (Curve 1) respectively. The higher activity of the acicular $BaAl_{12}O_{19}$ catalyst in methane combustion could be attributed to the higher surface areas, larger pore volume and pore size than those of the spherical one. At low temperature, the large surface area is favorable for the adsorption and activation of more reactive species. While at high temperature, the reaction is controlled mainly by the mass transfer process, and the large pore volume and pore size could favor the effective gas diffusivity.

Since the contacting points between particles are considered the sintering sites, the decrease in the number of contact points could suppress the particles sintering [15-17]. Compared with the spherical particles, the higher resistants to

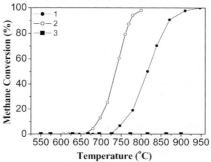

Fig. 4. Activity of methane combustion over the $BaAl_{12}O_{19}$ catalysts: 1-Sample A, 2-Sample D, 3- Empty reactor packed with quartz beads.

Table 2
The texture and crystal structure properties of the $BaAl_{12}O_{19}$ catalysts activity for methane combustion.

Sample	Particle shape	SSA $(m^2 g^{-1})$	V_p (ml g-1)	d_p (nm)	Crystal phase
A	Spherical	46.60	0.09065	7.702	β-Al_2O_3
D	Acicular	47.08	0.1483	17.74	β-Al_2O_3

SSA: Specific surface area, V_p: Average pore volume, d_p: Average pore size.

the sintering for the acicular particles could be primarily ascribed to their acicular shape. Therefore, the acicular $BaAl_{12}O_{19}$ catalyst has larger surface area, and their pore volume and size than the spherical one after calcination at 1200°C. Barry et al. [15] and Ishikawa et al. [16] have prepared fibrous alumina with large surface area after high temperature calcination, they also attributed it to the fibrous shape of the particles.

4. CONCLUSIONS

The results show that the particle morphology of $BaAl_{12}O_{19}$ could be controlled by the microemulsion microstructure, which varies with the microemulsion compositions. The particle morphology could be improved by the supercritical drying. The acicular $BaAl_{12}O_{19}$ catalyst possesses higher thermal stability and catalytic activity for methane catalytic combustion than the spherical one.

ACKNOWLEDGEMENT

The financial support from the Ministry of Science and Technology of the People's Republic of China is gratefully acknowledged (G1999022401).

REFERENCES

[1] L.D. Pfefferle and W.C. Pfefferle, Catal. Rev. Sci. Eng., 29 (1987) 219.
[2] M. Machida, K. Eguchi and H. Arai, J. Catal., 123 (1990) 477.
[3] G. Groppi, M. Bellotto, C. Cristiani, P. Forzatti and P. L. Villa, Appl. Catal. A: General, 104 (1993) 101.
[4] A.J. Zarur and J.Y. Ying, Nature, 403 (2000) 65.
[5] L. Qi, J. Ma, H. Cheng and Z. Zhao, Colloids Surf. A, 108 (1996) 117.
[6] D. Walsh, J.D. Hopwood and S. Mann, Science, 164 (1994) 1576.
[7] J. Tanori and M.P. Pleni, Adv. Mater., 7 (1995) 862.
[8] K. Kandori, K. Kon-no and A. Kitahara, J. Colloid Interface Sci., 11 (1987) 5579.
[9] J. Wang, Z. Tian, J. Xu, Y. Xu, Z. Xu and L. Lin, Catal. Today, 83 (2003) 213
[10] L. Yan, J. Zhuang, X. Sun, Z. Deng and Y. Li, Mater. Chem. and Phys., 76 (2002) 119.
[11] D.O. Shah and R.M. Jr. Hamlin, Science, 171 (1971) 483.
[12] H. Hoffmann, G. Platz and W. Ulbricht, J. Phys. Chem., 85 (1981) 3160.
[13] A.J. Zarur, N.Z. Mehenti, A.T. Heibel and J.Y. Ying, Langmuir, 16 (2000) 9168.
[14] J.D. Hopwood and S. Mann, Chem. Mater., 9 (1997) 1819.
[15] H. Zhu, J.D. Riches and J.C. Barry, Chem. Mater., 14 (2002) 2086.
[16] T. Ishikawa, R. Ohashi, H. Nakabayashi, A. Ueno and A. Furuta, J. Catal., 134 (1992) 87.
[17] T. Horiuchi, T. Osaki, T. Sugiyama and T. Mori, J. Non-crystalline Solids, 291 (2001) 187.

Studies in Surface Science and Catalysis, volume 147
X. Bao and Y. Xu (Editors)

Selective oxidation of CH$_4$ to CH$_3$OH using the Catalytica (bpym)PtCl$_2$ catalyst: a theoretical study

X. Xu[*][a, b], G. Fu[a], W. A. Goddard III[*][b] and R. A. Periana[c]

[a]State Key Laboratory for Physical Chemistry of Solid Surfaces; Department of Chemistry, Center for Theoretical Chemistry, Xiamen University, Xiamen, 361005, China

[b]Materials and Process Simulation Center (139-74), Division of Chemistry and Chemical Engineering, California Institute of Technology, Pasadena, California 91125, USA

[c]Department of Chemistry, University of Southern California, Los Angeles, California 90089, USA

1. INTRODUCTION

Selective alkane oxidation by transition metal complexes in solution has been the focus of substantial effort since the 1970s [1-19]. However the low activity of alkanes and the typically higher activity of the desired products make this process a great challenge.

In 1998, Periana et al. of Catalytica Inc. reported a breakthrough in developing an effective catalyst for high yield selective oxidation of CH$_4$ to CH$_3$OH [10]. The initial catalyst was dichloro-(η^2-{2,2'-bipyrimydl}) platinum (II) complex, (bpym)PtCl$_2$. In well-dried sulfuric acid (80 ml at 102%) they found that 72% of 115 mmol CH$_4$ at 3400 kPa was converted by 50 mmol of catalysts to product (mixture of CH$_3$OSO$_3$H + CH$_3$OH) with selectivity of 81% in 2.5 hours at 220 °C. They also reported another system, (NH$_3$)$_2$PtCl$_2$, which was more active, but less stable.

The Catalytica process is promising, since it has high yield at relatively low temperatures. On the other hand, it suffers from several drawbacks such as (1) catalyst rate is too low; (2) separation cost is too high; (3) solvent is too corrosive and (4) SO$_2$ reoxidation is too expensive. What is more, (5) the catalyst is severely inhibited by water.

We have embarked on a project to elucidate the fundamentals of the bpym and ammine systems, indicating stability, activity and selectivity [16, 17]. Our objective is to obtain an improved understanding of how the current catalyst

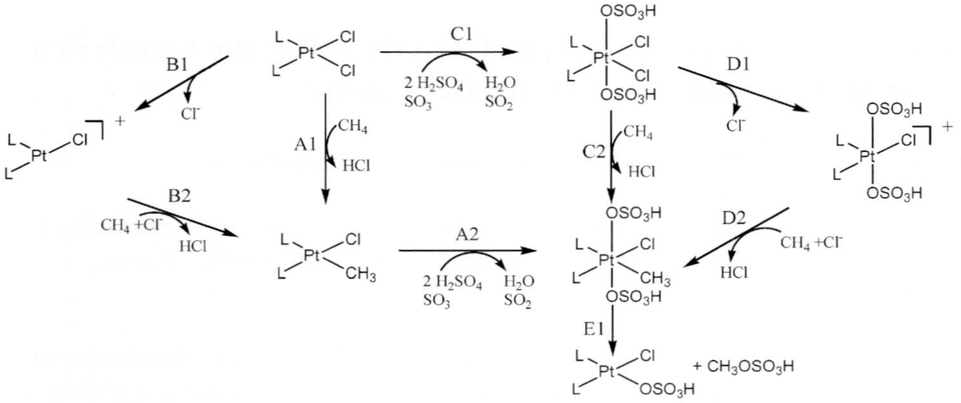

500

Fig. 1. Reaction pathways. Pathway 1: A1, A2, E1; Pathway 2: B1, B2, A2, E1; Pathway 3: C1, C2, E1; Pathway 4: C1, D1, D2, E1.

works such that we might suggest new catalysts with improved performance.

Here we report the overall reaction thermodynamics as depicted in Fig. 1. Reaction potential energy profiles for C-H activation are shown in Fig. 2.

2. COMPUTATIONAL DETAILS

All quantum mechanical calculations were carried out using the B3LYP flavor of density functional theory (DFT) with Jaguar [20]. The basis set used is LACVP(+)** [20]. Analytic Hessian was calculated at each optimized geometry to ensure that each minimum is a true local minimum (LM, only positive frequencies) and that each transition state (TS) has only a single imaginary frequency (negative eigenvalue of the Hessian). All calculations used the Poisson-Boltzmann continuum approximation to describe the effect of sulfuric acid solvent [16, 17, 20]. The reaction thermodynamics has been reported as the free energy changes in solution at the experimental reaction temperature of 453 K [10], denoted as $\Delta G(s, 453\ K)$; while the reaction potential energy profiles were reported as the enthalpy changes in solution at 453 K, $\Delta H(s, 453\ K)$.

3. RESULTS AND DISCUSSION

It is believed that the Catalytica process involves three steps: C-H activation, oxidation and functionalization [10]. As shown in Fig. 1, Pathways 1 and 2 are activation-before-oxidation processes; while Pathways 3 and 4 are activation-after-oxidation processes. Pathways 1 and 3 may be considered as associative pathways, where the departure of ligand Cl⁻ is accompanied by the coordination of CH_4; while Pathways 2 and 4 are dissociative pathways,

Table 1

Reaction thermodynamics. Free energies (in kcal/mol) in sulfuric acid solution at 453 K, G (s, 453 K).

	$(NH_3)_2$	bpym	bpymH$^+$	bpymH$_2$$^{2+}$
A1 (C-H activation)	25.3	24.6	21.7	18.8
A2 (Oxidation)	-8.0	-2.3	2.3	20.7
B1 (Ligand dissociation, T-complex)	9.4	21.8	25.6	35.4
B2 (C-H activation)	15.9	2.8	-3.9	-16.6
C1 (Oxidation)	16.8	23.4	33.1	44.0
C2 (C-H activation)	0.6	-1.6	-9.1	-4.5
D1 (Ligand dissociation)	29.8	35.4	40.5	53.1
D2 (C-H activation)	-29.2	-36.6	-49.7	-58.8
E1 (Functionalization)	-13.7	-18.1	-18.0	-25.4

involving the formation of a T-complex as an intermediate [10]. Previous calculations suggested that protonation of bpym ligand is a favorable process; while the replacement of Cl$^-$ with HSO$_4^-$ is an favorable process in dry sulfuric acid [16, 17]. Hence we used L$_2$PtCl$_2$ (L$_2$ = (NH$_3$)$_2$, bpym, bpymH$^+$ and bpymH$_2$$^{2+}$) as the initial catalysts. The calculated reaction free energies are summarized in Table 1.

The formation of L$_2$PtCl(CH$_3$) from L$_2$PtCl$_2$ (reaction A1) is endothermic for all ligands. The (NH$_3$)$_2$ ligands are the most endothermic (ΔG(s, 453 K) = 25.3 kcal/mol). The bpym ligands are less endothermic than the (NH$_3$)$_2$ ligands, leading to 24.6, 21.7 and 18.8 kcal/mol as the number of protons on the bpym ligands is increased from 0 to 2.

The oxidation step (reaction A2) involves a two-electron redox reaction: oxidation of Pt(II) to Pt(IV) couples with the reduction of S(VI) in the form of SO$_3$ to S(IV) in the form of SO$_2$. Experimentally, the oxidation step is rate-determining [10]. We find that the formation of the Pt(IV) octahedral complex is the most exothermic for the NH$_3$ ligand (ΔG(s, 453 K) = -8.0 kcal/mol); it is mildly exothermic (-2.3 kcal/mol) for the bpym ligand and mildly endothermic (2.3 kcal/mol) for the singly protonated form; the doubly protonated form is particularly unfavorable, being endothermic by 20.7 kcal/mol. If the activation barrier to oxidation follows the same trend, the NH$_3$ ligand complex should be the most active, having the lowest barrier.

The functionalization step involves the removal of CH$_3$OSO$_3$H from the Pt(IV) complex via reductive elimination to regenerate a Pt(II) complex (reaction E1). We find this step to be exothermic for all ligands with the trend that (most exothermic first)

$$bpymH_2^{2+} > bpymH^+ \approx bpym > (NH_3)_2.$$

Experimental evidence suggests that E1 is a fast step [10].

In the dry sulfuric acid solvent, it is possible that oxidation step proceeds before C-H activation. We find that oxidation of L$_2$PtCl$_2$ (reaction C1) is around

502

20 kcal/mol more endothermic than that of $L_2PtCl(CH_3)$ (reaction A2). This is reasonable as CH_3 is a better electron donor than Cl such that electron density in Pt is higher in $L_2PtCl(CH_3)$ than that in L_2PtCl_2. We find that oxidation of (bpym)$PtCl_2$ (23.4 kcal/mol) is less favorable than oxidation of $(NH_3)_3PtCl_2$ (16.8 kcal/mol) and protonation of bpym further disfavors the oxidation step (G(s, 453 K) = 33.1, 44.0 kcal/mol for L2 = bpymH+ and bpymH22+, respectively).

The calculations show that C-H activation on Pt (IV) (reaction C2) is thermo-neutral (ΔG(s, 453 K) = 0.6 kcal/mol) for L_2 = $(NH_3)_2$ and mildly exothermic for the bpym ligands (-1.6 ~ -9.1 kcal/mol). As compared to reaction A1, it is indicated that higher oxidation state of Pt is in favor of the C-H activation.

Based on the thermodynamics, we infer that the oxidation-before-activation pathway (Pathway 1: C1, C2, E1) is preferable to the oxidation-after-activation pathway (Pathway 3: A1, A2, E1) for L_2 = $(NH_3)_2$; while Pathway 1 and Pathway 3 are competitive for L_2 = bpym. We deduce that protonation of bpym disfavors Pathway 1 but favors Pathway 3.

Periana et al. suggested that C-H activation takes place via electrophilic substitution, going through a dissociative step to form the 14-electron T-complex (reaction B1). Our previous calculations showed that C-H activation can involve either electrophilic substitution or other pathways such as oxidative addition

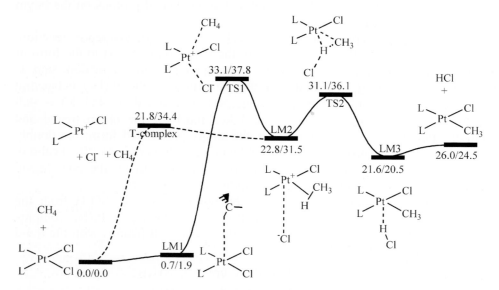

Fig. 2. Reaction energy profiles: associative pathways (solid line) versus dissociative pathways (dashed line) for the C-H activation step via electrophilic substitution. H (sol, 453 K) for the $(NH_3)_2PtCl_2$/(bpym)$PtCl_2$ systems.

depending on the nature of the ligands [16]. Here we compare the dissociative pathway with the associative pathway (reaction A1) where the incoming CH_4 occupies the open coordination site as one Cl^- is leaving without going through the T-complex.

As shown in Fig. 2, we find that the methane complex (LM2) is a distinct stable intermediate. Take the free reactants $L_2PtCl_2 + CH_4$ as references (ΔH(s, 453 K) = 0.0 kcal/mol), LM2 is endothermic by 22.8 and 31.5 kcal/mol; while TS1 in the associative pathway is uphill by 33.1 and 37.8 kcal/mol for $L_2 = (NH_3)_2$ and bpym, respectively. In LM2, Cl^- remains associated as an ion pair and can be involved in TS2 to break the C-H bond. TS2 is uphill by 31.1and 36.1 kcal/mol for $L_2 = (NH_3)_2$ and bpym, respectively. Hence C-H activation should be easier for $L_2 = (NH_3)_2$ than that for $L_2 =$ bpym via the associative pathway.

For the dissociative pathways, formation of the T-complex is endothermic by 21.8 and 34.4, which are 11.3 and 3.4 kcal/mol lower than those of TS1 in the associative pathways for $L_2 = (NH_3)_2$ and bpym, respectively. Hence the dissocative pathway outweighs the associative pathway in the ammine system; while the associative pathway is competitive to the dissociative pathway in the bpym system. We find

$$(bpymH_n^{n+})PtCl_2 \rightarrow [(bpymH_n^{n+})PtCl]^+(\text{T-complex}) + Cl^-$$

ΔH(s, 453 K) = 39.7, 50.3 kcal/mol for n=1, 2, respectively,
ΔG(s, 453 K) = 25.6, 35.4 kcal/mol for n=1, 2, respectively.

Thus, protonation of bpym strongly disfavors the dissociative pathways. Preliminary calculations show that TS1 and TS2 in the associative pathway would decrease by ~ 4 kcal/mol as bpym is protonated.

4. CONCLUSION

We performed computational studies of the thermodynamics of the Catalytica Pt(II) catalysts for selective oxidation of CH_4 to CH_3OH. All calculations were carried out at the B3LYP/LACVP(+)** level, including the Poisson-Boltzmann continuum approximation to describe salvation effects. We examined thermodynamics of four possible reaction pathways, involving dissociative versus associative pathways; and oxidation-after (before)-activation pathways. We present the reaction energy profiles for the C-H activation via electrophilic substitution. Our results can be summarized as follows.

(1) The dissocative pathway outweighs the associative pathway in the ammine system; while the associative pathway is competitive with the dissociative pathway in the bpym system.

(2) The ammine system should be more active than the bpym systems, as both activation and oxidation steps are easier in the ammine system, in agreement with the experimental observation.

(3) The oxidation-after-activation pathways are preferable to the oxidation -before-activation pathways for the bpym systems; whereas the ammine ligands are in favor of the oxidation-before-activation pathways.

(4) Protonation of bpym disfavors the dissociative pathway but favors the associative pathway.

(5) Protonation is preferable in the activation step, but unfavorable in the oxidation step.

ACKNOWLEDGEMENTS

This work was supported by NSFC (20021002, 29973031), NSF-FJ (2002F010), TRAPOYT from MOE, and funding from Chevron.

REFERENCES

[1] S.E. Bromberg, H. Yang, C.M. Asplund, T. Lian, K.B. McNamara, K.T. Kotz, J.S. Yeston, M. Wilkens, H. Frei, R.G. Bergman and C.B. Harris, Science, 278 (1997) 260.
[2] J.A. Davies, P.L. Watson, J.L. Liebman and A. Greenberg (eds.), Selective Hydrocarbon Activation, Wiley-VCH, New York, 1990.
[3] C. Hall and R.N. Perutz, Chem. Rev., 96 (1996) 3125.
[4] A. Sen, Acc. Chem. Res., 21 (1988) 421.
[5] A.E. Shilov and G.B. Shul'pin, Chem. Rev., 97 (1997) 2879.
[6] J. Sommer and J. Bukala, Acc. Chem. Res., 26 (1993) 370.
[7] K.M. Walktz and J.F. Hartwig, Science, 277 (1997) 211.
[8] G.S. Hill, G.P.A. Yap and R.J. Puddephatt, Organometallics, 18 (1999) 1408.
[9] R.A. Periana, D.J. Taube, E.R. Evitt, D.G. Loffler, P.R. Wentrcek, G. Voss and T. Masuda, Science, 259 (1993) 340.
[10] R.A. Periana, D.J. Taube, S. Gamble, H. Taube, T. Satoh and H. Fujii, Science, 280 (1998) 560.
[11] L. Johansson and M. Tilset, J. Am. Chem. Soc., 123 (2001) 739.
[12] H. Heiberg, L. Johansson, O. Gropen, O.B. Ryan, O. Swang and M. Tilset, J. Am. Chem. Soc., 122 (2000) 10831.
[13] H. Heiberg, O. Swang, O.B. Ryan and O. Gropen, J. Phys. Chem., A103 (1999) 100004.
[14] K. Mylvaganam, G.B. Bacskay and N.S. Hush, J. Am. Chem. Soc., 122 (2000) 2041.
[15] K. Mylavaganam, G.B. Backsay and N.S. Hush, J. Am. Chem. Soc., 121 (1999) 4633.
[16] J. Kua, X. Xu, R.A. Periana and W.A. Goddard III, Organometallics, 21 (2002) 511.
[17] X. Xu, J. Kua, R.A. Periana and W.A. Goddard III, Organometallics, 22 (2003) 2057.
[18] T.M. Gilbert, I. Hristov and T. Ziegler, Organometallics, 20 (2001) 1183.
[19] I. Hristov and T. Ziegler, Organometallics, 22 (2003) 1668.
[20] Jaguar 4.0, Schrodinger, Inc., Portland, Oregon, 2000.

Studies in Surface Science and Catalysis, volume 147
X. Bao and Y. Xu (Editors)
©2004 Elsevier B.V. All rights reserved.

Micro-plasma technology - direct methane-to-m ethanol in extremely confined environment -

Tomohiro Nozaki, Shigeru Kado, Akinori Hattori, Ken Okazaki, Nahoko Muto

Mechanical and Control Engineering, Tokyo Institute of Technology
2-12-1 O-okayama, Meguro, Tokyo, JAPAN 152-8551

ABSTRACT

Flow-type, non-equilibrium, micro-plasma reactor achieved 17% one-pass methanol yield through selective oxidation of methane at 25°C and 1 bar within 280 ms. The breakthrough involved is that the least reactive methane was excited by reactive non-equilibrium plasma, while successive destruction of reactive methanol was minimized simultaneously within the extremely confined environment: highly reactive and quenching environment, paradoxical conditions in conventional thermochemical processes, are established simultaneously in a single reactor.

1. INTRODUCTION

The great benefit of direct methanol synthesis from natural gas has attracted substantial effort over the past decade. However, methanol yield has generally been lower than 5% [1-3]. More promising results that claimed 10% yield (including formaldehyde) were difficult to reproduce [4-5]. Periana et al. synthesized a methanol derivative (CH_3OSO_3H) with 70% yield at 220°C and 35 bar using a platinum catalyst in concentrated H_2SO_4-SO_3 [6]. Unfortunately, this process requires large amounts of toxic chemicals, such as H_2SO_4, and 2.5 hours reaction time. Bjorklund et al reported 25% methanol yield with multi-recycling reaction system; however, one-pass methanol yield could not exceed 4% [7]. One promising solution to achieve a commercial target (greater than 10% methanol yield) is to provide extreme quenching conditions within a single reactor even at elevated temperature and pressure (450-500°C and 30-60 bar) to avoid successive destruction of highly reactive methanol [3,8-9].

2. EXPERIMENTAL

2.1. Micro-plasma reactor

Micro plasma reactors, here we report, can satisfy such a special requirement [10-11]. It consisted of a Pyrex thin glass tube (I.D. 1.0 mm × 60.0 mm) and a twisted metallic wire inside the tube (0.2 mm × 100.0 mm). The reactor was secured in a heat reservoir to maintain constant reaction temperature. A high-voltage sine wave (2 kV at 75 kHz) was applied between twisted metallic wire and grounded heat reservoir. Power consumed by micro-plasma reactor was 5 W. The principal of generating plasma is similar to barrier discharge, which is characterized by large numbers of filamentary discharges with 1-10 ns duration. Creation of a highly pulsed ionized atmosphere is able to activate the least reactive methane at room temperature by suppressing excess discharge current that contributes to significant gas temperature increase. Methane conversion in barrier discharges has been studied before, and the comprehensive investigation has been provided by Larkin et al. [12]. They investigated partial oxidation of methane using a co-axial barrier discharge reactor equipped with a water jacket. However, the result showed little impact on methanol yield: one-pass methanol yield was 4% with 60 seconds reaction time. Large reactor space, in either radial or lateral, seemed not to restrict further oxidation of methanol. Creation of non-equilibrium plasmas within an extremely confined environment is therefore strongly required. The different quenching effect between the micro-plasma reactor and conventional barrier discharge reactors was clearly distinguished in terms of electric characteristics [13].

2.2. Gas analysis

Figure 1 shows micro-plasma reactor and gas analysis set-up. Methane and oxygen mixture was fed into micro-plasma reactor. Nitrogen was premixed into methane cylinder as a reference for gas analysis (10vol%). We monitored

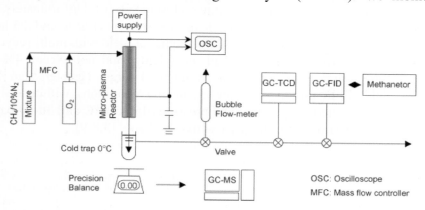

Fig. 1. Micro-plasma reactor and measurement set-up

emission from N_2 (second positive band) and NO (γ band) [14-15] during plasma processing. However, emission from either N_2 or NO was absent. Chemical reactions involving N_2 or NO_X, which may act as catalytic agent, were reasonably negligible. Liquid products such as H_2O, CH_3OH, HCHO, $HCOOCH_3$, HCOOH, and C_2H_5OH were collected by cold trap (0°C). Total weight of liquid mixture was measured after 60-90 minutes processing. Each component except H_2O was then quantitatively analyzed by GC-MS (Shimadzu QP-5000: Poraplot Q, carrier gas He). CH_4, O_2, N_2, and H_2 were analyzed with gas chromatograph equipped with TCD detector (Shimadzu GC-8A: Molecular sheave 5A, carrier gas Ar). CO, CO_2, CH_4, C_2H_6, and C_2H_4 were introduced to gas chromatograph via methanizer (FID detector: Shimadzu GC-8A: Porapack Q, carrier gas N_2). We used gas tight syringes for gas sampling. The syringes were rinsed with Acetone at least 10 times and then dried by air-blow before gas sampling. Each measurement was carried out twice to make sure that there was not previous residues remained.

3. REQUIREMENTS FOR MICRO-PLASMA REACTOR

Reactor diameter must be smaller than at least quenching diameter of CH_4/O_2 flame in order to create a highly quenching environment to avoid further oxidation of methanol: quenching diameter is the minimum distance that a flame can propagate through a combustible gas mixture [16]. On the other hand, according to the Paschen's law, discharge space must be wider than 10 m to ignite discharge with moderate applied voltage [17]. These points imply that the glass tube must have an inner diameter of between 10 m and 2000 m at

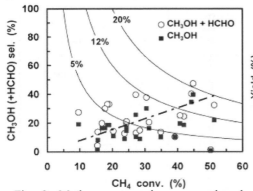

Fig. 2. Methane conversion vs. methanol (+formaldehyde) selectivity. Discharge power was 5 W. Glass inner diameter: 1 and 2 mm. Total gas flow rate: 5-20 sccm. CH_4/O_2 ratio: 1.0-5.0. Temp.: 25-120°C. Pressure: 1 - 3 bar. Twisted metallic wires: Mo, Ni, Pt, and stainless steel.

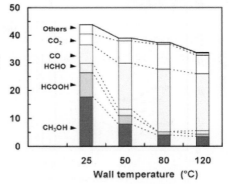

Fig. 3. Effect of reaction temperature on product yields. The uppermost line corresponds to methane conversion. 'Others' was mostly ethanol. Discharge power: 5 W. Pressure: 1 bar. Glass inner diameter: 1 mm. Total gas flow rate: 10 sccm. Initial CH_4/O_2: 2.0. Twisted wire was stainless steel.

atmospheric pressure. Heat produced by plasma and exothermic reaction was removed effectively with this dimension. Therefore, low temperature synthesis was maintained even though high methane conversion was achieved.

4. RESULTS AND DISCUSSIONS

4.1. General aspects of direct methanol synthesis

Figure 2 shows one-pass methane conversion and methanol selectivity for all experiments. Wide ranges of experimental conditions were examined as listed in the figure legend. One-pass methanol yield reached commercial target (>10%): maximum yield even reached 17%. Formaldehyde was also produced; however, unlike other methods, the yield was much smaller than other oxygenates and methanol was selectively produced. The most interesting and positive aspect is that methanol selectivity is proportional to methane conversion. We also used Ni, Mo, and Pt metallic wires as catalyst, but the result was independent of the metal variety in this set up. Methane conversion and methanol selectivity were mostly affected by oxygen concentration.

Figure 3 presents oxidative products in terms of reaction temperature. The uppermost line corresponds to methane conversion. Further reduction in initial oxygen mainly produced ethane and ethylene, while higher initial oxygen usually led to carbon monoxide. At low temperature where thermal reaction is negligible small, methane conversion depends strongly on specific input energy, i.e. W/Q (J/m^3). Here, $W=f{\times}E$ is input power (W), where f is frequency and E is discharge energy per cycle, and Q is total gas flow (Nm^3/s). Operating frequency

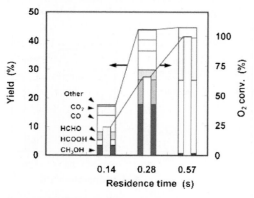

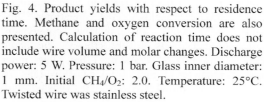

Fig. 4. Product yields with respect to residence time. Methane and oxygen conversion are also presented. Calculation of reaction time does not include wire volume and molar changes. Discharge power: 5 W. Pressure: 1 bar. Glass inner diameter: 1 mm. Initial CH$_4$/O$_2$: 2.0. Temperature: 25°C. Twisted wire was stainless steel.

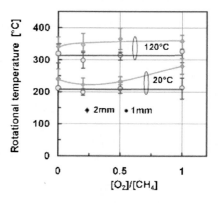

Fig. 5. Rotational temperature, which corresponds to plasma gas temperature. Reactor wall temperature: 20°C and 120°C, and reactor size: 1 mm and 2 mm.

induces the crucial effect of achieving high methane conversion in a miniaturized reactor within a reasonable time. Methanol selectivity got extremely high when reaction temperature was lower than its boiling point. In fact, methanol yield attained 17% when wall temperature was 25°C. Methanol seemed to condense on the reactor wall, then get removed from the oxidizing plasma region efficiently, avoiding further destruction either by electron impact or thermal decomposition. Selectivity was suddenly decreased at 50°C, and then it kept almost constant between 80°C and 120°C. We propose that high-pressure synthesis up to 60 bar would improve the current process to a greater degree than further miniaturization. High-pressure synthesis is desirable for methanol synthesis from an equilibrium perspective. The high-pressure situation also enhances the quenching effect by gas phase molecular collisions. Changes in boiling and melting point of methanol solution allow us to lower reaction temperature even below zero.

Fig. 4 shows product yields and methane conversion with respect to reaction time. Fig. 4 also indicates oxygen conversion as blank columns. Methane conversion and methanol selectivity increased monotonically with reaction time. However, if oxygen was totally consumed, methane conversion would stop and all oxygenates were ready to decompose. Although existence of non-equilibrium plasma is essential to trigger the process at room temperature, effective methane excitation that engenders various oxygenates was only achieved in the presence of oxygen. According to Otsuka et al. [3], active ionic oxygen species, such as O^-, O_2^-, and O_2^{2-}, are believed to initiate dissociation of alkanes' C-H bond. Not only dissociative excitation by electron impact, but also production of ionic oxygen species must play an important role for moderate oxidation of methane in low temperature plasma catalysis. Oxygenates were not produced if oxygen was totally consumed, while they are continuously decomposed by plasma. Even though most of oxygenates were trapped on the reactor wall, they were still exposed to plasma and further decomposed to CO and CO_2 when residence time was excessively long. Oxygenates must be trapped on the wall, then exhausted from the reactor as soon as possible.

4.2. "Real" reaction temperature

We also characterized rotational temperature of CH (431.5 nm) from emission spectroscopy, which reasonably represents plasma gas temperature of corresponding emission region. The detailed procedure how to determine rotational temperature has been published by Nozaki et al. [18-20]. Rotational temperature is plotted on Fig. 5 with respect to O_2/CH_4 ratio. Rotational temperature was about 200°C higher than wall temperature because microdischarges are localized within narrow filamentary regions. However, rotational temperature was low enough so as not to initiate thermal reactions. It is also interesting to note that rotational temperature was constant with respect to

O_2/CH_4 ratio owing to efficient heat removal through reactor wall, although it seems to be slightly affected when the glass diameter was 2 mm. When wall temperature increased from 20°C to 120°C, then rotational temperature also increased 100°C higher.

5. CONCLUDING REMARKS

1. The micro-scale flow-type non-equilibrium plasma generated within the extremely confined environment realized 17% one-pass methanol yield at 25 °C and 1 bar within 280 ms. The CH_4/O_2 mixture could be processed with no dilution, carbon/wax deposits, or explosion.
2. Highly reactive and quenching environment, paradoxical conditions in conventional thermochemical processes, are established simultaneously in a single reactor. Liquid condensation due to low temperature synthesis is crucial for high methanol yield, while adequate reaction rate was maintained even at room temperature due to non-equilibrium plasma catalysis.
3. Methanol selectivity was proportional to methane conversion.
4. Methanol was selectively produced rather than formaldehyde. New radical species, such as various ions and vibrational state species are anticipated as promising key radicals at low temperature plasma catalysis.

REFERENCES

[1] P.S. Casey, T. McAllister, K. Foger, Ind. Eng. Chem. Res. 33 (1994)1120.
[2] K. Omata, N. Fukuoka, K. Fujimoto, Ind. Eng. Chem. Res. 33 (1994) 784.
[3] K. Otsuka, Y. Wang, Appl. Catal. A-Gen. 222 (2001) 145.
[4] N.R. Hunter, H.D. Gesser, L.A. Morton, P.S. Yarlagadda, Appl. Catal. 57 (1990) 45.
[5] W. Feng, F.C. Knopf, K.M. Dooley, Energy & Fuels 8 (1994) 815.
[6] R.A. Periana et al., Science 280 (1998) 560.
[7] M.C. Bjorklund, R.W. Carr, Ind. Eng. Chem. Res. 41 (2002) 6528.
[8] W.C. Danen et al., Preprints Petro. Chem. Div. Published by ASC, 36 (1991) 166.
[9] Y. Sekine, K. Fujimoto, Energy & Fuels 10 (1996) 1278.
[10] T. Nozaki, Japan Patent Application 2003-078489 (2003).
[11] T. Nozaki, A. Hattori, S. Kado, K. Okazaki, 225th ACS National Meeting, CD-ROM: Fuel Chem. Div., Plasma Techno. & Catal. CATL-5 (2003).
[12] D.W. Larkin, L.L. Lobban, R.G. Mallinson, Catal. Today 71 (2001) 199.
[13] T. Nozaki, A. Hattori, S. Kado, K. Okazaki Proc. 3rd Int. Workshop on Basic Aspects of Non-equilibrium Plasmas Interacting with Surfaces 1 (2003) 8.
[14] G. Herzberg, Molecular Spectra and Molecular Structure, (London 1950), p32.
[15] N. Gherardi, S. Martin, F. Massines, J. Phys D: Appl. Phys 33 (2000) L104.
[16] B. Lewis, Combustion, Flames and Explosions of Gases (Academic press, INC, 1987).
[17] von Engel, Ionized Gases (Oxford University Press, 1965).
[18] T. Nozaki, Y. Unno, Y. Miyazaki, K. Okazaki, J Phys. D Appl. Phys., 34 (2001) 2504.
[19] T. Nozaki, Y. Miyazaki, Y.Unno, K . Okazaki, J Phys. D Appl. Phys., 34 (2001) 3383.
[20] T. Nozaki, Y. Unno, K. Okazaki, Plasma Sources Sci. Technol. 11 (2002) 431.

Studies in Surface Science and Catalysis, volume 147
X. Bao and Y. Xu (Editors)
©2004 Elsevier B.V. All rights reserved.

Towards a better understanding of transient processes in catalytic oxidation reactors

Renate Schwiedernoch[a], Steffen Tischer[a], Hans-Robert Volpp[b], Olaf Deutschmann[a]

[a]Institute for Chemical Technology, University of Karlsruhe, Engesserstr. 20, 76313 Karlsruhe, Germany, deutschmann@ict.uni-karlsruhe.de

[b]Institute of Physical Chemistry, Heidelberg University, INF 253, 69120 Heidelberg, Germany

ABSTRACT

Light-off of the partial oxidation of methane to synthesis gas on a rhodium-coated cordierite honeycomb monolith at short contact times is studied experimentally and numerically. The objective of this investigation is a better understanding of transient processes in catalytic oxidation reactors. The numerical simulation predicts the time-dependent solid-temperature distribution of the entire reactor and the two-dimensional flow fields, species concentrations, and gas-phase temperature profiles in the single channels. A detailed reaction mechanism is applied, and the surface coverage with adsorbed species is calculated as function of the position in the monolith. During light-off, complete oxidation of methane to water and carbon dioxide occurs initially. Then, synthesis gas selectivity slowly increases with rising temperature. The numerically predicted exit temperature, conversion, and selectivity agree well with the experimentally derived data.

1. INTRODUCTION

Several promising production routes in the area of natural gas conversion are based on high-temperature (approximately 1000°C) catalysis at short contact times [1-3]. These processes are often carried out in monolithic structures coated with the catalytic material, in which the explosive gas/oxygen mixture is fed. The understanding of the reactor behavior at transient conditions is crucial for the overall performance, scale-up, and, in particular for reactor safety. Reliable models and computational tools can help in the identification of parameter regions (temperature, mixture composition, pressure, reactor geometry), at which unsafe reactor operation occurs and assist in the development of a safe reactor design and operation.

In this paper, we apply a new approach for modeling the transient states of catalytic monoliths such as light-off and shut-down of the reactor as well as oscillating behavior and temporal as spatial variations of the feed. Exemplarily, we will discuss the light-off of the catalytic partial oxidation (CPO) of methane to synthesis gas,

$$CH_4 + \frac{1}{2} O_2 \rightarrow CO + 2 H_2,$$

over a rhodium coated cordierite honeycomb monolith at short contact times. In the experiment, exit gas-phase temperatures and species concentrations are monitored by thermocouples and mass spectroscopy, respectively. In the numerical study, the recently developed computational tool DETCHEM^{MONOLITH} [4-6] is applied, which permits the three-dimensional simulation of the temperature distribution of the entire monolith coupled with two-dimensional laminar reactive flow field simulations of a representative number of single monolith channels. The latter predicts the gaseous velocity, species concentrations, temperature profiles, and surface coverage based on a multi-step heterogeneous reaction mechanism.

2. EXPERIMENT

The reactor set-up is described in our recent study on CPO of methane on a rhodium/alumina monolith [5]. In contrast to this work, we now use cordierite as support material. The reactor consists of a 40 cm long quartz tube with a diameter of 8 mm. The Rh coated honeycomb monolith, placed inside the tube, is of 10 mm length and 6 mm in diameter. The hydraulic diameter of the single channels, which have a rectangular cross-section, is 0.7 mm. A 10 mm long inert monolith in front of the catalyst serves as radiation shield. A ceramic cloth is wound around both monoliths preventing gas bypass.

The exit gas phase temperature is monitored using a type N thermocouple, which is placed in a quartz nose behind the monolith. The product composition is determined by mass spectroscopy (QMS) as function of time.

In the example discussed, a constant molar inlet CH_4/O_2 ratio of 1.6 with 78 vol.%, Ar dilution is chosen. The mixture flows at 298 K and atmospheric pressure with a total mass flow of 1.7 slpm into the quartz tube reactor. This flow rate corresponds to a single channel inlet velocity of 2.0 m/s and a residence time of few milliseconds.

A furnace is applied to ignite the reaction by heating up the monolith using a temperature ramp of 5 K per minute. As soon as light off occurs, the furnace was turned off. Here, light-off is defined as the rapid increase of conversion detected by the QMS, which occurs at 850 K for the conditions chosen.

3. MODELING

The model couples the heterogeneous reactions on the catalytic surface, the reactive flow field in the single channels, and the heat balance of the solid structure of the monolith as well as the thermal insulation, quartz tube and monolithic heat shield in front of the reactor. The laminar reactive flow in the single channels is modeled by the steady-state, two-dimensional (axi-symmetric) boundary layer equations with transport coefficients that depend on composition and temperature [7]. Surface and gas phase reactions are modeled by elementary-step like reaction mechanisms. The catalytic reaction on the wall takes a varying surface coverage of adsorbed species along the channel wall into account.

The residence time of the reactants inside the single channel is much smaller than the time scale for temperature variation in the monolith solid due to the larger heat capacity of the solid monolith compared to that of the gas flow. Therefore, the simulation of the flow through a single channel can be treated as steady-state at during one time step of the transient simulation of the heat balance of the solid structure. The heat flux from the gas phase into the monolith bulk, due to convection and chemical reactions, is then calculated for the solid temperature profiles and initial flow conditions by simulation of a representative number of single channels at each time step. Hence, a transient heat conduction equation is solved leading to time-varying temperature profiles for the single channel simulations as well. Heat losses due to radiation and convection at the exterior boundary/insulation of the monolith are taken into account. Based on these models, the recently developed computational tool DETCHEM[MONOLITH] predicts the two/three-dimensional temperature profiles, species concentrations, and velocity distribution of the entire structured catalyst as function of time. For a more detailed description of the model and the software it is referred to [4-6].

A detailed surface reaction mechanism is applied to model the partial and complete oxidation of methane on rhodium. Since this mechanism also takes intermediate steps and species into account, further global reactions such as steam and CO_2 reforming and the water gas shift reactions are automatically included. The detailed surface reaction mechanism was taken from our former study [5] and consists of 38 reactions among 6 gas phase species and 11 surface species. The simulation also includes a very detailed gas-phase reaction mechanism of up to 414 reactions among 35 species. However, gas phase reactions are not significant at the conditions chosen and will, therefore, not be discussed further in the current paper.

In order to describe the catalyst loading, the ratio of active catalytic to geometrical surface area has to be specified, which can be derived from

Table 1
Properties of the solid and insulation material [8]

physical properties	cordierite monolith	insulation
heat conduction λ_{rad} [W/(m K)]	2	1.36
density ρ [kg/m^3]	2100	2205
heat capacity c_p [J/(kg K)]	800	730

chemisorption measurements. The surface reaction mechanism applied was derived for a Rh/alumina catalyst [5] and it was found that the now used catalyst based on cordierite has only a tens of the active catalytic surface area.

The material properties of the support and the insulation are given in Table 1. The total porosity of the monolithic structure is 0.495. Heat losses due to conduction, convection, and thermal radiation at the exterior walls of the monolith are included in the simulation.

4. RESULTS AND DISCUSSION

Figure 1 (left) shows the numerically predicted temperature distribution of the solid structure during light-off. The time is set to zero when the conversion of methane rapidly increases. At this time the furnace, which was used to heat up both the inert and the catalytic monolith, is switched off. The cold inflowing gas is heated up (Fig. 1 middle) and an almost homogeneous temperature distribution occurs.

It can clearly be seen how the reaction starts at the front end, there methane is oxidized, mainly to CO_2 and H_2O (Fig. 2), and the temperature increases. Further downstream, only little conversion occurs. During ignition radial temperature profiles are established over the entire monolith due to external heat losses (the furnace is turned off now) as well as in the single channel, where the exothermic heterogeneous reactions lead to higher temperatures on the catalytic wall. After approximately 2s, a hot spot occurs in the first section of the catalytic monolith. After 50 s, the solid temperature profile is smoothed out due to heat conduction the solid structure and the steady state is reached.

In our former study [5], in which an alumina support was used, the reaction started at the back end of the monolith and no hot spot was observed. It should be noted that not only the different heat conduction of the support matters but also the total catalyst loading, monolith porosity and geometry. The detailed simulation offers a useful tool to study the impact of these different external parameters on reactor performance, in particular at transient conditions.

The product profiles in the single channel (Fig. 2) reveal that total oxidation is the main reaction during light-off leading to a the fast temperature increase. The only slightly exothermic formation of synthesis gas begins slowly with increasing temperatures. It can also be seen that a large part of the formed hydrogen comes from a consecutive reaction, in which the steam, mainly formed in the entrance region of the catalyst where a large amount of oxygen is still available, reacts with methane. This endothermic reaction causes the slight temperature decrease further downstream aside from heat losses.

The much faster diffusion of hydrogen in comparison to the other species leads to the different shape of the radial profiles, which also means that the processes in the single channel need to be modeled by a two-dimensional description as done here; the overall reaction is transport-limited after ignition.

It is also worth to mention that the overall conversion reaches a maximum a few seconds after ignition, which seems to be caused by the interaction of different heat transport modes. Thus, the ambient temperature, which is still close to the initial furnace temperature, the initially very high total oxidation rate releasing a large amount of reaction heat and leading to the temperature peak discussed, and the still relatively warm heat shield have a large impact on the chemical processes during light-off.

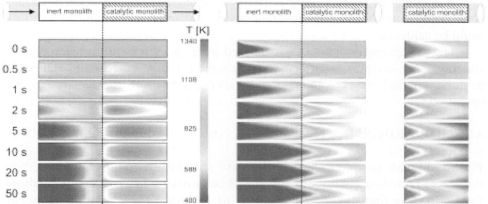

Fig. 1. Temperature of the solid structure of the monoliths (left), gas-phase temperature (middle) and methane mole fraction (right, range: 0.0-0.137) in a single channel in the center of the monolith during light-off; the radial coordinates are stretched differently.

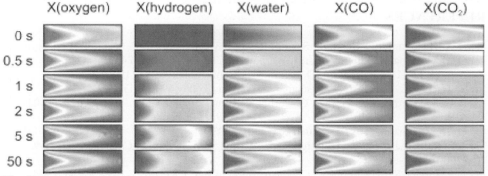

Fig. 2. Gas-phase mole fractions in a single channel in the centre of the monolith during light-off; the radial coordinates is stretched. (ranges: $O_2 = 0 - 0.08$, $H_2 = 0 - 0.05$, $H_2O = 0 - 0.07$, $CO = 0 - 0.05$, $CO_2 = 0 - 0.03$

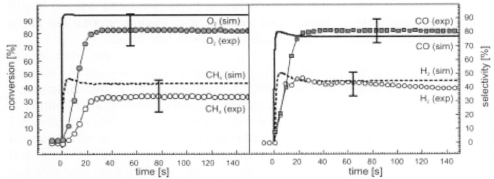

Fig. 3. Comparison of experimentally derived and numerically predicted conversion and selectivity.

A qualitatively good but not perfect agreement between experimentally derived and numerically predicted conversion and selectivity as function of time could be achieved.

The model and computer code applied offers a useful tool to reach a better understanding of transient processes of catalytic oxidation reactors. They can now be applied to explore safe parameter regions for reactor design and operation.

ACKNOWLEDGEMENTS

We would like to thank Jürgen Warnatz and Jürgen Wolfrum (Heidelberg University) for their continuous support of our work on heterogeneous catalysis. This work was financially supported by the Deutsche Forschungsgemeinschaft (DFG), J. Eberspächer GmbH & Co (Esslingen/Germany), and ConocoPhillips Inc. (Ponca City/OK, USA).

REFERENCES

[1] D.A. Hickman, L.D. Schmidt, Science 259 (1993) 343.
[2] O. Deutschmann, L.D. Schmidt, J. Warnatz, AIChE J. 44 (1998) 2465.
[3] A.S. Bodke, D.A. Olschki, L.D. Schmidt, E. Ranzi, Science 285 (1999) 712.
[4] S. Tischer, C. Correa, O. Deutschmann, Catalysis Today 69 (2001) 57.
[5] R. Schwiedernoch, S. Tischer, C. Correa, O. Deutschmann, Chem. Eng. Sci. 58 (2003) 633.
[6] O. Deutschmann, et al., www.detchem.com, 2003.
[7] L.L. Raja, R.J. Kee, O. Deutschmann, J. Warnatz, L.D. Schmidt, Catalysis Today 59 (2000) 47.
[8] D.R. Lide, CRC Handbook of Chemistry and Physics, CRC Press, Boca Raton, 1994.

Studies in Surface Science and Catalysis, volume 147
X. Bao and Y. Xu (Editors)
517

Direct oxidation of methane to oxygenates over supported catalysts

S.A. Tungatarova, G.A. Savelieva, A.S. Sass, and K. Dosumov

D.V.Sokolsky Institute of Organic Catalysis and Electrochemistry. Republic of Kazakhstan. 142, D.Kunaev st., Almaty 480100.

ABSTRACT

Optimal compositions of thermostable catalysts over carriers on the base of Mo heteropoly compounds for the oxidation of CH_4 to CH_2O were developed. It was established the correlation for yield and productivity by CH_2O with the content of acidic centers and the I and II forms of structural reactive oxygen. It was detected the positive role of water vapor on the yield and selectivity of products. Regeneration of the low-percentage catalysts was developed.

1. INTRODUCTION

Light alkanes are the great part of natural and oil gases and have narrow practical application. Processing of gases is one of the problems of oil-gaseous industry. Alternative methods for utilization of alkanes by catalytic partial oxidative conversion to industrial useful products are interesting. Scientists try to find a single-stage catalytic method for oxidative conversion of alkanes to important products of industrial organic synthesis (alcohols, aldehydes, olefins) [1-4]. Supported complexes of polyoxide catalysts were used for the processes of partial oxidative conversion of methane to oxygen-containing compounds and the products of oxidative dehydrogenation. Exploration of the new compounds of catalysts over carriers on the base of Mo heteropoly compounds (HPC) for the process of oxidative conversion of methane, determination of optimal conditions of reactions, investigation of surface properties of catalysts by the complex of physical-chemical methods were carried out.

2. EXPERIMENTAL

Experiments were carried out in a flow type system with the removal of products in the cooling zone. Tubular quartz reactor (length - 0,2 m, diameter - $1,9 \times 10^{-2}$ m) with fixed bed catalysts was used for investigation. The catalysts were prepared by impregnation of granulated carriers: aluminosilicate, alumina, cordierite, active carbon, zeolites, clinoptilolite. The 12th series Mo HPC were used as active component of catalysts. Block cordierite carriers with different secondary carriers (cordierite, silica gel, zeocar, Al_2O_3, «Sph» silica gel, aluminosilicate) and protective layers are used too.

Experiments were realized in reaction mixtures, containing methane, oxygen, inert gas with or without water vapor at the reaction temperatures 873-1173K and atmospheric pressure. The content of reaction components was widely varied.

Physicochemical investigations were carried out with using of IR, ESDR, EPR, TPR and isotope methods. Analysis of reaction products was carried out by chromatographic and spectrophotometric methods.

3. RESULTS AND DICUSSIONS

It was detected the influence of content of HPC over support on yield and selectivity of CH_2O, C_2 hydrocarbons in the process of oxidative conversion of methane for 0,5-20% HPC/aluminosilicate. Curves had polyextreme character for low-percentage catalysts (< 5% mass.). It was shown that polyextreme change of catalytic properties was determined by formation of compounds in the system of HPC - carrier by IR-spectroscopy and TPR methods [5]. Change of binding strength, reactivity of structural oxygen of catalyst and acid-base properties were determined by formation of these compounds. The presence of four types of oxygen, which take part in the interaction with H_2 at 573 (I), 743 (II), 943-948 (III) and 973-1273 K (IV) were registered during investigation of interaction of structural oxygen of initial heteropoly acid (HPA) with H_2 from Ar-H_2 flow by TPR method. The effect of HPA supporting over aluminosilicate on structural characteristics, type of structural oxygen and reactivity in oxidation of CH_4 to CH_2O were also tested. It was demonstrated by IR and XRD methods that supporting of HPA over carrier in the quantity of 5-20 mass % results in forming of Keggin structure over carrier as in the case of compact acid. Uniformity of observable IR, XRD, ESDR and TPR spectra for compact HPA and 5-20 mass % supported catalysts indicate on this fact. Decreasing of HPA content from 5 to 0.1 mass % results in displacement of all four peaks interaction of structural oxygen with H_2 in TPR regime into region of higher temperatures. This fact indicates on increasing of binding strength and change of reactivity concerning H_2 in the result of change structural characteristics of supported acid. The data, obtained by TPR method, were confirmed by IRS data: certain typical absorption bands of HPA disappeared - Mo-O-Mo, but the Mo=O and Si-O-Mo bands preserved for supported catalysts with content of HPA < 5 mass %. However, their shift by 10-30 sm^{-1} toward the higher frequencies region of valence and deformation vibration of aluminosilicate bands was registered. Combination of TPR and IRS data indicate on formation of a new type of structure, caused by interaction of HPA fragments with carrier (for example, Si-O-Mo) together with amorphous structure of HPA in low-percentage supported catalysts.

It was investigated the influence content of HPA on concentration variation of acidic centers with different acid strength. These acidic centers were detected by EPR method on the base of adsorption of test molecules of aromatic compounds (benzene, naphthalene, anthracene) with certain ionization potential. The acidity of the HPC varied in direct correlation with the amount of the weak-acid centers of Brønsted type (7.4 - 8.2 eV) and catalytic activity in the reaction of the formation of CH_2O.

Influence of the substitution of proton in HPA on activity and selectivity in reaction was studied. Yield of CH_2O was decreased in next series depending upon cation: H $\geq$ Ba > Mg > Li > Sr > Na > Bi $\geq$ Ca (for $H_4SiMo_{12}O_{40}$). Substitution of proton on cations expands the temperature range of raised productivity.

It was detected that addition of water vapor to reaction mixture increased the yield of CH_2O, as show in Fig.1, and stabilized the structure of catalyst. It is known that HPC is destroyed in the case of loss the water of crystallization. The investigation on the role of the concentration of water vapor in reaction mixture allows to develop the optimal conditions for selective action of catalysts in the process of partial oxidation of methane to formaldehyde and oxidative dimerization of methane (ODM). The influence of the concentration of water vapor in reaction mixture on direction of oxidative conversion of CH_4 is presented on Table 1. The optimal conditions for formation of CH_2O and parallel carrying out the process of ODM with insignificant forming of C_2H_6 and C_2H_4 were observed at action the excess of water vapor ($CH_4:H_2O=1:90$). Lowering the concentration of water vapor ($CH_4:H_2O=1:54$ and than 1:18) decreased the yield of CH_2O and increased the yield of C_2-hydrocarbons.

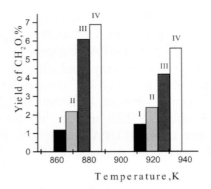

Fig.1. Influence of the content of water vapor in reaction mixture on yield of CH_2O over 2,3% $H_3SiMo_{12}O_{40}$/aluminosilicate. CH_4-0,5%, O_2-10,0%, $(Ar\pm H_2O)$-89,5%. CH_4:H_2O ratio: I-1:0; II-1:18; III-1:54; IV-1:90.

The positive influence of water vapor was studied by using of isotopic labeled water (D_2O and $H_2{}^{18}O$) instead of normal water vapor. Irrespective of chemical composition of supported HPC and reaction temperature (873-1023K) in the product of oxidative conversion of methane the deuterium enters both to initial methane, water and other products - formaldehyde, C_2-hydrocarbons. The ^{18}O oxygen atoms were detected in oxygen containing reaction products (CH_2O, CH_3OH, CO_2, and H_2O). These data are argument of participation of water vapor as soft oxidant in reaction of partial oxidation of methane to formaldehyde across formation of H- and OH-groups in structure of supported catalysts and stabilization the heteropoly acid structure over the carrier.

Table 1
The influence of content of the water vapor in reaction mixture on direction of oxidative conversion of CH_4 over 2,3% $Na_4PFeMo_{11}O_{39}$/aluminosilicate
CH_4 - 0,5%, O_2 - 10,0%, $(Ar+H_2O)$ - 89,5%, $W = 7840h^{-1}$

Ratio	T,K	Selectivity, %				Conversion*,%
CH_4:H_2O (vol.)		CH_2O	C_2H_4	C_2H_6	C_3H_6	
	773	35,0	-	58,3	-	1,2
1:18	873	30,5	22,6	40,3	2,4	12,4
	923	12,8	24,7	56,7	4,5	17,8
	773	75,6	-	-	-	0,9
1:54	873	27,5	9,8	29,4	27,5	10,2
	923	20,0	5,7	27,8	45,0	20,9
	773	95,2	-	-	-	2,1
1:90	873	81,7	-	8,3	-	8,2
	923	3,4	18,0	48,6	-	1,1

* - by-products - CO_2, H_2, traces of CO.

It was detected the optimal ratio of reaction components in mixtures lean by methane - ($CH_4 : H_2O : O_2 = 1 : 90 : 20$, vol) and rich by methane - ($CH_4 : H_2O : O_2 = 1 : 0,53 : 0,45$, vol) [6,7]. The content of water vapor was varied from $CH_4:H_2O=1:0,25$ (9,8% H_2O) to $CH_4:H_2O=1:0,60$ (23,4% H_2O) over 2,3% $H_3SiMo_{12}O_{40}$/aluminosilicate. It was shown in Table 2 that optimal conditions for formation of CH_2O were installed at high concentration of water vapor ($CH_4:H_2O=1:0,53$). However subsequent raising the concentration of water vapor was undesirable in connection with washing the active phase from the surface carrier.

By TPR method, it was detected that treatment of catalysts in mixture (O_2, water vapor) at 923K after reaction both at 923K and 1173K regenerated all forms of structural oxygen, which are specific for initial supported HPA. Thus, our method for oxidation of CH_4 for production of CH_2O at 873-923K with periodical regeneration of catalyst by mixture (O_2, water vapor) at 923K results in preservation all types of structural oxygen. Positive effect of this fact was installed in case of use of cascade reactor [7].

Table 2

Influence of reaction mixture composition on yield and productivity in the partial oxidation of methane to formaldehyde over 2,3% $H_4SiMo_{12}O_{40}$/aluminosilicate. Space velocity (W) = 8000h^{-1}, T = 923K

$CH_4:H_2O:O_2$	Composition of mixture, (vol.%)	Yield, (gHCHO/nm^{3}CH$_4$)	Productivity, (gHCHO/kgHPC·h)
1:0,25:0,45	CH$_4$ - 39,0; O$_2$ - 17,6; H$_2$O - 9,8; Ar - 33,6	13,8	1690,0
1:0,27:0,45	CH$_4$ - 39,0; O$_2$ - 17,6; H$_2$O - 10,5; Ar - 32,9	20,8	2711,0
1:0,53:0,45	CH$_4$ - 39,0; O$_2$ - 17,6; H$_2$O - 20,7; Ar - 22,7	22,3	2897,0
1:0,55:0,45	CH$_4$ - 39,0; O$_2$ - 17,6; H$_2$O - 21,5; Ar - 21,9	23,3	2850,0
1:0,59:0,45	CH$_4$ - 39,0; O$_2$ - 17,6; H$_2$O - 23,0; Ar - 20,4	16,8	2120,0
1:0,60:0,45	CH$_4$ - 39,0; O$_2$ - 17,6; H$_2$O - 23,4; Ar - 20,0	13,8	1810,0

The yield of CH_2O was decreased in series of carriers from 2959.0 to 1345.0 g CH_2O/kg HPC·h: clinoptilolite (+10%HCl) ~ aluminosilicate > silica gel KSM-5 ~ CaA > clinoptilolite > KSM-2,5 ~ fluorinated Al_2O_3 [7]. Block catalysts showed sufficiently high activity too. Over the Mg-contacts, containing silica gel as secondary carrier, were obtained the similar high yields. Thus, using of Si-containing oxides as carriers with high acidity and coordinating of hydroxide groups over surface, which is necessary for HPC to stabilize its structure, generates prerequisites for forming of active and selective catalysts as the activity of catalysts correlates with power and type of acidic centers over supported catalysts.

4. CONCLUSION

It was detected the positive role of water vapor on yield and productivity of the process. The presence of water vapor stabilized the HPC structure in catalysts, reactivated of structural oxygen in reaction moment. The water was not only diluent, but participant of reaction too. Activity of the catalysts increased with reduction the content of HPC over carrier from 20 to 1%, correlated with the capacity of weak acid centers in catalysts and content of I and II forms of oxygen, which are typical for HPC. The low-percentage catalysts are characterized by two states over the carrier: a crystal phase with Keggin's structure, part of HPA occurs in amorphous state and interacts with carrier. It was developed the method for regeneration of catalysts by periodical thermo-steam-air treatment.

REFERENCES

[1] T.Sugino, A.Kido, N.Ueno, and Y.Udagawa, J. Catal., 190 (2000) 118.
[2] A.Parmaliana, F.Arena., J. Catal., 167 (1997) 57.
[3] S.A.Tungatarova, G.A.Savelieva, and K.Dosumov, TOCAT 4 Fourth Tokyo Conference on Advanced Catalytic Science and Technology, July 14-19, 2002, Tokyo, Japan, 223.
[4] Yang Chao, Xu Nanping, Shi Jun, Ind. Eng. Chem. Res., 37 (1998) 2601.
[5] S.A.Tungatarova, G.A.Savelieva, and N.M.Popova, Kinetika i kataliz, in press.
[6] G.A.Savelieva, S.A.Tungatarova, A.S.Sass, and K.Dosumov, Ninth International Symposium on Heterogeneous Catalysis, September 23-27, 2000, Varna, Bulgaria, 877.
[7] S.A.Tungatarova, G.A.Savelieva, and K.Dosumov, Sustainable Strategies for the Upgrading of Natural Gas: Fundamentals, Challenges, and Opportunities, NATO Advanced Study Institute, July 6-19, 2003, Villamoura, Algarve, Portugal, 345.

Studies in Surface Science and Catalysis, volume 147
X. Bao and Y. Xu (Editors)

Synthesis of methanesulfonic acid and acetic acid by the direct sulfonation or carboxylation of methane

Sudip Mukhopadhyay, Mark Zerella, and Alexis T. Bell

Department of Chemical Engineering, University of California-Berkeley, CA
4720-1462, U.S.A.

ABSTRACT

Methane can be converted with high selectivity to either methanesulfonic acid or acetic acid by sulfonation or carboxylation of methane using a peroxide initiator at relatively low temperatures.

1. INTRODUCTION

The discovery and development of catalytic processes for the conversion of methane to chemicals and liquid fuels is the subject of ongoing research. Current approaches for the conversion of methane start by reforming methane to synthesis gas and then converting this mixture to hydrocarbons and/or oxygenated products (e.g., aldehydes, alcohols, carboxylic acids). Since 65-75% of the capital cost of this indirect approach is associated with methane reforming, there is an interest in identifying strategies for the direct conversion of methane to products. The present work was undertaken to investigate novel approaches for the functionalization of methane with SO_3 and CO_2 in order to produce methanesulfonic acid (MSA) and acetic acid, respectively.

2. SYNTHESIS OF METHANESULFONIC ACID

The current commercial process for the synthesis of methanesulfonic acid (MSA) involves the chlorine oxidation of methylmercaptan and produces six moles of HCl per mole of MSA, resulting in a coupling of the demand for the primary product and the byproduct [1]. Recent studies show that MSA can be synthesized with high yield from methane and SO_3 in the presence of a free radical initiator using fuming sulfuric acid as the solvent [2].

2.1 Experimental Method

Reactions were carried out in a 100 cm^3 Parr autoclave. $K_2S_2O_8$ and 3 mL of fuming H_2SO_4 were added to a glass liner containing a stir bar. The reactor was pressurized with 650 psig CH_4, and heated to 65°C for up to 3 h. Following reaction, the reactor was cooled, vented, and the liquid product was added slowly to 0.5-1.0 cm^3 of H_2O to convert any unreacted SO_3 to H_2SO_4. Reaction products were characterized by ^1H and ^{13}C NMR.

2.2 Results and Discussion

The reaction of methane and SO_3 was initiated by the addition of a free radical initiator, namely $K_2S_2O_8$. $K_4P_2O_8$, Urea-H_2O_2, CaO_2, and KO_2 were also used in as initiators. Triflic acid (TFMSA) and $RhCl_3$ were used as promoters.

Table 1 illustrates typical conversion of SO_3, the limiting reagent, measured after 3 h of reaction. In all cases the selectivity to MSA is virtually 100%, with only trace amounts of methyl bisulfate being observed as a byproduct.

Fig. 1 demonstrates that the conversion of SO_3 is a strong function of the solvent. Conversion was highest in the most acidic solvent, TFMSA, and it was least in non-acidic DMSO. Only slightly lower conversion from that in pure TFMSA was achieved using H_2SO_4 containing 0.2M TFMSA. For longer reaction times, e.g. > 2 h, the conversion of SO_3 to MSA is significantly higher (87%) in this solvent mixture than in pure TFMSA (50%).

Failure to achieve 100% SO_3 conversion suggests that unreacted SO_3 is consumed in secondary reactions or that the initial charge of initiator, $K_2S_2O_8$, is fully consumed prior to the completion of the reaction. The latter explanation can be eliminated, since, the addition of a new charge of initiator, after attainment of the plateau in SO_3 conversion, did not further increase the conversion of SO_3. The more probable interpretation is that a portion of the MSA formed reacts with SO_3 to form $CH_3(SO_3)_2H$. To confirm the possibility of forming such a product, samples of pure MSA containing increasing amounts of SO_3 were prepared. ^1H NMR spectra of these samples are shown in Fig. 2. The ^1H peak for protons

Table 1

Effect of different promoters on the methane sulfonation reaction[a]

Initiator, mmol	Promoter, mmol	% SO$_3$ Conv. to MSA
$K_4P_2O_8$, 0.20	None	20
$K_2S_2O_8$, 0.20	None	60
$K_2S_2O_8$, 0.20	TFMSA[b]	90
Urea-H_2O_2, 0.74	None	23
Urea-H_2O_2, 0.74	$RhCl_3$, 0.33	86
CaO_2, 0.74	None	63
KO_2, 0.74	$RhCl_3$, 0.33	17

Reaction condition : methane, 650 psig; solvent, 3 mL, time, 3 h; temperature, 65°C; [b]0.2M TFMSA in fuming sulfuric acid is used.

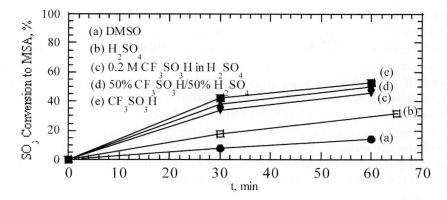

Fig. 1. Effect of solvents on MSA synthesis. Reaction condition: methane, 650 psig; solvent, 3 mL, time, 3 h; temperature, 65°C; [b]0.2M TFMSA in fuming sulfuric acid is used.

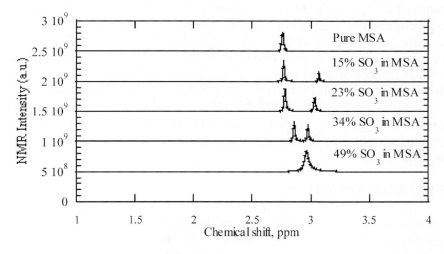

Fig. 2. *In-situ* ^{1}H NMR of mixtures of MSA with SO_3

associated with the methyl group of MSA appears at 2.5 ppm. Upon addition of SO_3, a new peak appears downfield of that for pure MSA at 3.1 ppm which is assigned to $CH_3(SO_3)_2H$.

The conversion of SO_3 to MSA passes through a maximum with increasing temperature. The decreases in SO_3 conversion at temperatures above about 70°C is attributed to the decomposition of the initiator, which releases O_2, an inhibitor of reaction [3]. Since SO_3 is an essential component of MSA, the rise in SO_3 conversion to MSA with increased initial concentration of SO_3 is not surprising. The decrease in SO_3 conversion to MSA for SO_3 concentrations above 30% is possibly due to the removal of dissolved SO_3 through the formation of $H_2S_2O_7$,

$H_2S_3O_{10}$, $H_2S_4O_{13}$, etc. [4]. The decrease in the conversion of SO_3 to MSA may also be due to formation of methanedisulfonic acid (2-3%), and $CH_3(SO_3)_2H$ (3-5%), which are observed at SO_3 concentrations above 40%. The effects of $K_2S_2O_8$ concentration on the conversion of SO_3 to MSA in the absence and presence of TFMSA were studied. Here again the addition of $K_2S_2O_8$ increases the conversion of SO_3 by a significant amount. However, above a certain $K_2S_2O_8$ concentration the conversion of SO_3 to MSA decreased.

MSA formation is believed to proceed via free radical mechanism. At reaction temperature, an initiator, $I\bullet$, is generated (i.e., $SO_4^-\bullet$, $PO_4^-\bullet$, $OH\bullet$, $CaO\bullet$, $KO\bullet$, $RhO\bullet$, or $RhOO\bullet$), which can abstract an H atom from methane to produce a $CH_3\bullet$ radical ($CH_4 + I\bullet \rightarrow CH_3\bullet + IH$). $CH_3\bullet$ then reacts with SO_3 to generate $CH_3SO_3\bullet$ ($CH_3\bullet + SO_3 \rightarrow CH_3SO_3\bullet$), which in turn reacts with methane to form MSA and $CH_3\bullet$ ($CH_3SO_3\bullet + CH_4 \rightarrow CH_3SO_3H + CH_3\bullet$). This mechanism is supported by the observation that a free radical scavenger such as O_2 inhibits the reaction. Appearance of C_2H_6 in the head-pace gas also supports the proposed mechanism.

The acceleration of MSA formation in more acidic solvents is ascribed to an increase in the concentration of $SO_4^-\bullet$, since the dissociation of $S_2O_8^{2-}$ is known to be more extensive under acidic conditions [5]. Consistent with this, we have observed that the rate of MSA formation increases with increasing acidity of the medium (see Fig. 1), i.e., DMSO $< H_2SO_4 <$ TFMSA. The effect of solvent acidity on the maximum conversion of SO_3 to MSA is due to the lowering of SO_3 consumption through the formation of $CH_3(SO_3)_2H.2CH_3SO_3H$ complex. Therefore, the net effect of increasing solvent acidity is to increase the maximum possible yield of MSA.

3. ACETIC ACID SYNTHESIS

Acetic acid is an important commodity chemical that is currently synthesized from methane or coal in a three-step, capital- and energy-intensive process. Here we show that acetic acid can be formed by the reaction of methane and CO_2 in an acid solvent [6].

3.1 Experimental Method

The carboxylation of methane with CO_2 was carried out in fuming sulfuric acid and in the presence of $K_2S_2O_8$ and $VO(acac)_2$. The solvent was chilled to 5-8°C during these additions to minimize the thermal decomposition of $K_2S_2O_8$. The reactor was then purged with N_2 to expel the air from the system and then pressurized with 120 psig of CO_2 and then with 80 psig of CH_4. The reactor was heated to 85°C under stirring and maintained at this temperature for 16 h. After the stipulated period of time, the reactor was quenched with ice. Two grams of water were slowly added to hydrolyze any unreacted SO_3. The reaction mixture was then filtered and analyzed by ^1H NMR. Acetic acid was distilled from the aqueous sulfuric acid solution, and identified by ^{1}H and ^{13}C NMR, and Raman spectroscopy.

3.2 Results and Discussion

Under standard reaction conditions, approximately 7% of the methane was converted to acetic acid. Only a small amount of the methyl ester of the acid was observed as a byproduct. ^{13}C-labeling experiments using ^{13}CH$_4$ and ^{13}CO$_2$ shows that acetic acid is produced exclusively from methane and CO$_2$ supplied to the reactor. With trifluoromethanesulfonic acid and its anhydride as the solvent, the yield increased to 13%. When trifluoroacetic acid was used as the solvent, the yield was 16%. However, ^{19}F NMR studies shows that trifluoroacetic acid reacts with CH$_4$ to form CH$_3$COOH and CHF$_3$, as was first suggested by Spivey and coworkers [7].

Fig. 3a and b show ^{1}H NMR spectra of the reaction product mixture in fuming sulfuric acid prior to the addition of water. The peak in Fig. 3a is assigned to the mixed anhydride of acetic acid and sulfuric acid, CH$_3$C(O)-O-SO$_3$H. Upon hydrolysis the mixed anhydride was converted to acetic acid (Fig 3b). Thus, it appears that the formation of acetic acid involves the following stoichiometric reactions:

$$CH_4 + CO_2 + H_2SO_4 \rightarrow CH_3C(O)\text{-}O\text{-}SO_3H + H_2O \quad \Delta G^\circ_{rxn} \cong +60 \text{ kJ/mol} \qquad (1)$$

$$H_2O + SO_3 \rightarrow H_2SO_4 \qquad\qquad\qquad \Delta G^\circ_{rxn} = -79 \text{ kJ/mol} \qquad (2)$$

$$CH_3C(O)\text{-}O\text{-}SO_3H + H_2O \leftrightarrow CH_3COOH + H_2SO_4 \quad \Delta G^\circ_{rxn} \cong -5 \text{ kJ/mol} \qquad (3)$$

The direct reaction of CH$_4$ with CO$_2$ to produce acetic acid is thermodynamically unfavorable (CH$_4$ + CO$_2$ → CH$_3$COOH $\Delta G^\circ_{rxn} \cong$ +55 kJ/mol); however, the combination of the reactions 1, 2, and 3 is thermodynamically favorable. VO(acac)$_2$ and K$_2$S$_2$O$_8$ are both essential to the formation of mixed acid anhydrides. While the specific functions of VO(acac)$_2$ and K$_2$S$_2$O$_8$ have not been identified, it is possible to suggest a role for each component. S$_2$O$_8^{2-}$ cleaves thermolytically to SO$_4^-\bullet$ then reacts with methane to generate CH$_3\bullet$. This radical reacts with CO$_2$ to form CH$_3$CO$_2\bullet$. However, since CH$_3$CO$_2\bullet$ is thermodynamically unstable, it is likely stabilized as CH$_3$CO$_2^-$ by reaction with V(IV) in VO(acac)$_2$. This hypothesis is supported by the observation that peracetic acid (CH$_3$COOOH) decomposes to CH$_4$ but not when VO(acac)$_2$ is present. In fuming sulfuric acid, the formation of CH$_3$C(O)-O-SO$_3$H may proceed via the reaction of CH$_3$CO$_2^-$ with SO$_3$ and H$^+$.

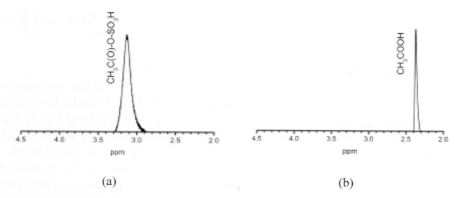

(a)   (b)

Fig. 3. ^{1}H NMR spectra of the reaction mixture prior to hydrolysis (a) and after hydrolysis (b). Reaction conditions: H_2SO_4 (10.0 g), SO_3 (3.0 g), $K_2S_2O_8$ (3.7 mmol), $VO(acac)_2$ (0.16 mmol), CH_4 (80 psig), CO_2 (120 psig), 16 h, 85°C.

4. CONCLUSION

In summary, the present work shows that methane can be converted with high selectivity to either methanesulfonic acid or acetic acid by sulfonation or carboxylation of methane using a peroxide initiator at relatively low temperatures.

ACKNOWLEDGMENT

This work was supported by grants from Atofina North America and BP.

REFERENCES

[1] (a) J.I. Kroschwitz, M. Howe-Grant. Kirk Othmer Encyclopedia of Chemical Technology, Wiley: New York, 1991. (b) R. Guertin; U.S. Patent 3,626,004, 1971.
[2] (a) N. Basickes; T.E. Hogan; A. Sen, J. Am. Chem. Soc. 118 (1996) 13111.; (b) L.J. Lobree; A.T. Bell, Ind. Eng. Chem. Res., 40 (2001) 736.; (c) S. Mukhopadhyay; A.T. Bell, Ind. Eng. Chem. Res., 41 (2002) 5901.; (d) S. Mukhopadhyay; A.T. Bell. Org. Process Res. Dev., 7 (2003) 161.; (e) S. Mukhopadhyay; A.T. Bell. Angew. Chem. Int. ed., 42 (2003) 1019.; (f) S. Mukhopadhyay; A.T. Bell. Angew. Chem. Int. ed., 2003, in press; (g) S. Mukhopadhyay; A.T. Bell. Org. Process Res. Dev., 2003, in press; (h) S. Mukhopadhyay; A.T. Bell. J. Am. Chem. Soc., 2003, (i) S. Mukhopadhyay; A.T. Bell. Chem. Comm., 2003,
[3] D.A. House, Chem. Rev., 62 (1962) 185.
[4] (a) R.J. Gillespie; E.A. Robinson. Non-Aqueous Solvent Systems (Waddington, T.C. ed.), Academic Press: London, 1965, p. 162; (b) R.J. Gillespie, J. Chem. Soc., (1950) 2516.
[5] S. Patai, The chemistry of Functional Groups, Peroxides, Wiley, New York, 1983.
[6] M. Zerella; S. Mukhopadhyay; A.T. Bell. Org. Lett., 5 (2003) 3193.
[7] G.M. Wilcox; G.W., Roberts; J.J. Spivey; J. Appl. Catal., 226 (2003) 317.

Studies in Surface Science and Catalysis, volume 147
X. Bao and Y. Xu (Editors)
529

New study of methane to vinyl chloride process

Christophe Lombard and Paul-Marie Marquaire[*]

Département de Chimie Physique des Réactions, UMR 7630 CNRS - INPL
ENSIC, 1 rue Grandville, BP 451, 54001 Nancy Cedex, France.

1. INTRODUCTION

Several studies concern the direct conversion of natural gas to higher values, transportable fuels as C_2 compounds, using the chlorine pathway [1,4]. The first common step is the mono-chlorination of methane. In fact, in Benson's process [1,2], methyl chloride produced in a flame methane/chlorine was pyrolysed immediately in this flame to form ethylene and acetylene. In the Chlorine-Catalysed Oxidative-Pyrolysis (C.C.O.P.) process presented by Senkan [3,4], methyl chloride is produced in a first step from methane using well-known process [5]. Then, methyl chloride is pyrolysed with small amounts of oxygen to form acetylene, ethylene and vinyl chloride without significant formation of soot. More recently, we proposed a new process [6,7], the Methane to Vinyl Chloride (M.T.V.C.) process: methyl chloride, produced in a first step, is pyrolysed, in a second step, with small amounts of chlorine to form vinyl chloride, acetylene and ethylene. Experimental results and simulations were presented neglecting, at first approximation, the products from methylene chloride. In this paper, we report on new experimental and mechanistic studies about M.T.V.C. process also called "chloro-pyrolysis" of methyl chloride. This new study takes into account the chemical consequences of the formation of methylene chloride.

2. EXPERIMENTAL AND RESULTS

A preliminary study is needed in order to estimate the influence of experimental parameters on the issue of the reaction. It is well-known that the thermal reaction CH_3Cl/Cl_2 (about $400\,°C - 600\,°C$) is a long chain reaction of chlorination of which major products are CH_2Cl_2, $CHCl_3$, CCl_4 and HCl [5]. It depends on residence time and initial chlorine ratio. Minor products are chloroethanes and traces could be chloroethylenes. The main radicals of this long chain reaction are $CH_2Cl^•$ and $Cl^•$. At low conversion, the reactive system could be described by the following primary mechanism [5]:

Initiation steps:

CH_3Cl		$\rightarrow CH_3^{\bullet}$	$+ Cl^{\bullet}$	(1)
Cl_2	$+ M$	$\rightarrow 2 Cl^{\bullet}$	$+ M$	(2)

Transfer steps

$CH_3^{\bullet}$	$+ CH_3Cl$	$\rightarrow CH_4$	$+ CH_2Cl^{\bullet}$	(3)
$CH_3^{\bullet}$	$+ Cl_2$	$\rightarrow CH_3Cl$	$+ Cl^{\bullet}$	(4)

Propagation steps:

$CH_2Cl^{\bullet}$	$+ Cl_2$	$\rightarrow CH_2Cl_2$	$+ Cl^{\bullet}$	(5)
$Cl^{\bullet}$	$+ CH_3Cl$	$\rightarrow HCl$	$+ CH_2Cl^{\bullet}$	(6)

Termination steps:

$2 CH_2Cl^{\bullet}$	$\rightarrow CH_2Cl\text{-}CH_2Cl$	(7)

We can introduce a primary chain length λ which is defined by the ratio between the propagation rate r_p Eq. (5), and the main termination rate r_t Eq. (7). For a temperature lower than 750 °C, the following relationship (8) for λ is achieved:

$$\lambda = \frac{k_5}{\sqrt{k_2 k_7}} \sqrt{\frac{[Cl_2]}{[M]}} \qquad (8)$$

The chain length λ depends on two main parameters: chlorine ratio via $[Cl_2]$ and temperature. λ increases with $[Cl_2]$ and since the apparent "activation energy" E_λ is negative ($E_\lambda \approx -22$ kcal/mol), λ decreases when the temperature increases. The chain length λ provides only a general trend but it seems that two parameters play a major role in the issue of reaction: temperature and chlorine ratio.

The CH_3Cl/Cl_2 thermal gas phase reaction has been studied in a plug flow reactor at temperatures comprised between 600 and 1100 °C, atmospheric pressure with helium dilution and a residence time in the range 10 – 30 ms. The Cl_2/CH_3Cl molar input ratio is in the range 0 – 50 % and the operating partial pressure of methyl chloride is between 76 Torr and 127 Torr, helium making balance to atmospheric pressure. Methyl chloride, chlorine and helium are premixed before the reactor but not preheated. The chosen reactor is an annular cylindrical quartz reactor. The thickness of the annular zone is equal to 0.5 mm. The reactor's heating is achieved by spiral electric wires, which are coiled around the external tube. This heating zone is directly followed by an indirect external water-quenching zone.

Using mass spectrometry (CPG-MS), the following products have been identified in the major part of experiments: C_2H_3Cl, CH_2Cl_2, C_2HCl, CH_4, C_2H_2, C_2H_4, $CH_2=CCl_2$, $CHCl=CHCl$ (cis and trans isomers), C_2HCl_3, C_4H_4. All of these products and CH_3Cl have been quantified in all experiments, except C_2HCl because we didn't find pure sample or mixture of the latter. The first exploratory working shifts allowed defining attractive operating conditions called "operating point". The operating point is featured by a temperature of 950 °C, a Cl_2/CH_3Cl input ratio of 15 % and a residence time of 10 ms. The partial pressure of CH_3Cl is close to 0,1 atm. Then, the conversion of methyl chloride is around 30 % and the selectivities of products are: C_2H_3Cl: 22.4 %, CH_2Cl_2: 34.6 %, C_2H_2: 11.9 %, C_2H_4: 4.9 %, CH_4: 3.6 %, E-$C_2H_2Cl_2$: 3.6 %, Z-$C_2H_2Cl_2$: 3 %, C_4H_4: 2.5 %, $CH_2=CCl_2$: 1.8 % and C_2HCl_3: 1.1 %. With a conversion of CH_3Cl about 30 % and a selectivity of C_2 (C_2H_3Cl, C_2H_2, C_2H_4) about 40 %, these results are cheering but CH_2Cl_2's selectivity remains high.

2.1. Effect of temperature

The conversion of methyl chloride increases very slightly between 600 °C and 800 °C (Fig. 1.). The reaction consists of only the single chlorination of CH_3Cl, a long chain reaction but whose chain length decreases with increasing temperature (see above). When the temperature increases, the reaction becomes a short chain reaction and the propagation Eq. (5) becomes minor respect to the termination Eq. (7). In fact, the yield of CH_2Cl_2 decreases with temperature when, in the same time, the yield of C_2H_3Cl increases. CH_2Cl-CH_2Cl, from reaction Eq. (7), leads very rapidly to C_2H_3Cl and HCl for a temperature above 800 °C. But for a temperature above 950 °C, C_2H_3Cl also leads to C_2H_2, whose yield increases strongly with temperature. Then, the yield of C_2H_3Cl presents an optimum for approximately 950 °C (Fig. 1.).

For temperatures higher than 900 °C, the chlorination and pyrolysis of CH_3Cl take place simultaneously leading to a strongly increasing of conversion with temperature. It is the "chloro-pyrolysis" of CH_3Cl.

2.2. Effect of chlorine ratio

The conversion of CH_3Cl increases near linearity with respect to chlorine ratio (Fig. 2.). But, in pure pyrolysis condition (without chlorine), the conversion is only equal to 4.4 %. With a chlorine ratio of 5 %, the conversion is 14.7 %. It seems that chlorine plays the role of a chemical inductor of the pyrolysis of CH_3Cl. The yields of CH_2Cl_2, C_2H_3Cl and C_2H_2 increase with the increase of the chlorine ratio, especially the yield of CH_2Cl_2 (Fig. 2.). But, the yields of these products are less and less influenced by chlorine ratio when the latter is high (above 30 %). In the same time, the yields of C_4H_4, $CH_2=CCl_2$, CHCl=CHCl and C_2HCl_3 increase strongly with chlorine ratio (not presented in this study), even at high chlorine ratio. Chlorine leads to a great formation of CH_2Cl_2, C_2H_3Cl and C_2H_2, but, at the same time, it could also involves a great consumption of these same products, especially at high chlorine ratio, leading to the formation of C_4H_4, $CH_2=CCl_2$, CHCl=CHCl and C_2HCl_3 (see paragraph 3).

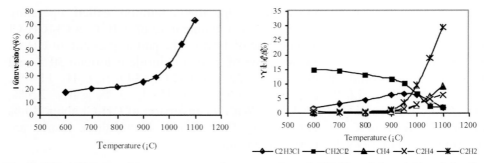

Fig. 1. The influence of temperature on the conversion of CH₃Cl and the yields of the five main products.

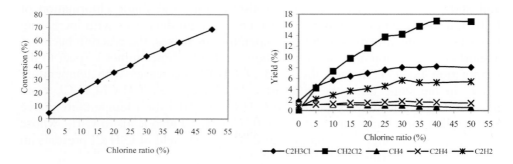

Fig. 2. The influence of chlorine ratio on the conversion of CH₃Cl and the yields of the five main products.

3. MECHANISM AND SIMULATIONS

A detailed chemical mechanism involving 99 species including C_1 to C_4 compounds (62 molecules and 37 free radicals) and 504 elementary reactions (mainly non-reversible) has been developed using available mechanisms from literature [2, 8-10] to model the "chloro-pyrolysis". This mechanism is as complete as possible. The Chemkin-II codes [11] were used to calculate concentrations of species. The comparison of the calculated concentrations with experimental values shows a good agreement when temperature, chlorine ratio and residence time are modified, and we claim that the mechanism is validated by experiment. In Fig. 3. calculated conversion of CH₃Cl and calculated partial pressures profiles of C_2H_3Cl and CH_2Cl_2 (indicated by lines) are compared to those determined experimentally (indicated by symbols).

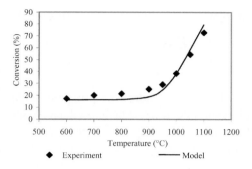 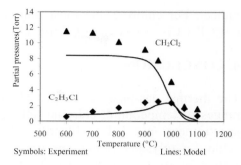

Fig. 3. Comparison of experiment and model for conversion and partial pressures of CH_2Cl_2 and C_2H_3Cl.

As seen in Fig. 3., the agreement between the model predictions and the experiment is pretty good, which suggests that the major steps involved in the "chloro-pyrolysis" process are reasonably well described by the mechanism. The main results of the analysis of the mechanism were presented in prior work [7]: at low temperature (T < 700 °C), the thermal gas-phase reaction CH_3Cl/Cl_2 is a long chain reaction producing mainly CH_2Cl_2 and HCl. At high temperature (T > 900 °C), the reaction becomes a short chain reaction producing mainly attractive C_2 compounds (C_2H_3Cl, C_2H_2 and C_2H_4) by recombination of C_1 radicals ($CH_2Cl^•$ and $CH_3^•$) but also undesirable CH_2Cl_2. The analysis of this new mechanism shows that CH_2Cl_2 is mainly consumed by metathesis with the major radicals of reactive system as $Cl^•$, $CH_3^•$ and $CH_2Cl^•$ to produce the $CHCl_2^•$ radical. The last one can react with chlorine to form chloroform $CHCl_3$. But, simulations show that $CHCl_2^•$ is mainly involved in recombination's reactions with $CH_2Cl^•$ or with itself to form $CH_2Cl\text{-}CHCl_2$ or $CHCl_2\text{-}CHCl_2$. In fact, $CH_2Cl^•$ and $CHCl_2^•$ are the most concentrated radicals. $CH_2Cl\text{-}CHCl_2$ and $CHCl_2\text{-}CHCl_2$ can undergo some reactions of HCl elimination (deshydrochlorination) to form isomers of dichloroethylene ($CH_2=CCl_2$, Z $CHCl=CHCl$ and E-$CHCl=CHCl$) or trichloroethylene (C_2HCl_3). The deshydrochlorination of isomers of dichloroethylene leads to acetylene chloride C_2HCl.

At the operating point specified above, simulations show that the major part of chlorine is consumed in the first part of the reactor. Then, the single pyrolysis of CH_3Cl follows its "chloro-pyrolysis" and the conversion of CH_3Cl is slower in the second part of reactor. In fact, the most important reaction channel for CH_3Cl is $CH_3Cl + Cl^• \rightarrow CH_2Cl^• + HCl$, Eq. (6). Then, the presence of chlorine in the system accelerates the rate of conversion of CH_3Cl because the easy initiation of chlorine and the metathesis involving chlorine liberate the $Cl^•$

534

radical. For chlorine, the most important reaction channel of consumption is $CH_2Cl^{\bullet} + Cl_2 \rightarrow CH_2Cl_2 + Cl^{\bullet}$, Eq. (5).

4. CONCLUSION

The thermal gas-phase reaction CH_3Cl/Cl_2 has been studied to prospect wide ranges of temperature, chlorine ratio and residence time. The gas-phase mass spectrometry has led to identify all the chemical compounds produced by "chloro-pyrolysis". Some specific operating conditions lead to interesting formation of C_2 compounds (C_2H_3Cl, C_2H_2 and C_2H_4) with a significant conversion of CH_3Cl. The presence of chlorine induces an increase of methyl chloride's conversion and C_2 compounds' yields. But, it also induces the formation of CH_2Cl_2, itself leading to the formation of dichloroethylene and trichloroethylene.

A detailed radical mechanism is proposed and it explains qualitatively and quantitatively the formation of all the major compounds in agreement with experimental results. Especially, the analysis of the mechanism explains the responsibility of CH_2Cl_2 in the formation of $C_2H_2Cl_2$, C_2HCl and C_2HCl_3.

A chlorine multistage supply in the reaction zone could decrease the formation of CH_2Cl_2 with the same conversion and C_2 compounds' yield.

REFERENCES

[1] S.W. Benson, U.S. Patent N° 4,199,533 (1980).
[2] M. Weissman and S.W. Benson, Int. J. Chem. Kinet., 16 (1984) pp. 307-333.
[3] S.M. Senkan, U.S. Patent N° 4,714,796 (1987).
[4] A. Granada, S.B. Karra and S.M. Senkan, Ind. Eng. Chem. Res., 26 (1987) pp. 1901-1905.
[5] B.E. Kurtz, Ind. Eng. Chem. Process Des. Develop., 11 (1972) pp. 332-338.
[6] P.M. Marquaire, M. Al Kazzaz, Y. Muller and J. Saint Just, Studies in Surface Science and Catalysis, 107 (1997) pp. 269-274.
[7] P.M. Marquaire and M. Al Kazzaz, Studies in Surface Science and Catalysis, 119 (1998) pp. 259-264.
[8] M. Weissman and S.W. Benson, Prog. Energy Combust. Sci., 15 (1989) pp. 273-285.
[9] T.J. Mitchell, S.W. Benson and S.B. Karra, Combust. Sci. And Tech., 107 (1995) 223-260.
[10] S.B. Karra and S.M. Senkan, Ind. Eng. Chem. Res., 27 (1988) pp. 1163-1168.
[11] R.J. Kee, F.M. Rupley and J.A. Miller, CHEMKIN II. A Fortran Chemical Kinetics Package for the Analysis of Gas-Phase Chemical Kinetics, Sandia Laboratories Report, SAND 89-8009B (1993).

Studies in Surface Science and Catalysis, volume 147
X. Bao and Y. Xu (Editors)

Structure-activity relationships in the partial oxidation of methane to formaldehyde on FeO$_x$/SiO$_2$ catalysts

F. Arena[a], G. Gatti[b], S. Coluccia[b], G. Martra[b] and A. Parmaliana[a]

[a] Dipartimento di Chimica Industriale e Ingegneria dei Materiali, Università di Messina, Salita Sperone 31, 98166 S. Agata (Messina), Italy

[b] Dipartimento di Chimica IFM, Università di Torino, Via P. Giuria 7, 10125 Torino, Italy

ABSTRACT

The effects of the preparation method on structures and dispersion of FeO$_x$/SiO$_2$ catalysts have been investigated by DR-UV-Vis, FTIR and TPR measurements. The efficiency of the preparation route based on "adsorption-precipitation" of Fe^{2+} (ADS/PRC), in enhancing the oxide dispersion is ascertained. The performance of ADS/PRC catalysts in the partial oxidation of CH$_4$ to HCHO (MPO) is compared with that of conventional systems obtained by "incipient wetness" of silica with aqueous solutions of Fe^{III} (INC/WET). The ADS/PRC method, enabling a higher dispersion of the active phase, provides very effective MPO catalysts.

1. INTRODUCTION

An impressive amount of research work has been devoted during the last decades to the investigation of supported iron-based systems as highly effective catalytic materials for selective redox reactions [1-8]. Many catalyst formulations include zeolites as carrier, their behaviour pattern being strongly related to the structure of the Fe sites stabilised into the matrix framework [1-3]. Likewise, the structure of active sites controls the functionality of FeO$_x$/SiO$_2$ catalysts in the MPO reaction [4-8]. In fact, both theoretical considerations and experimental findings suggest that a proper design of MPO catalyst entails the dispersion of a suitable active phase over an "inert" silica matrix [8]. Then, "isolated" Fe^{3+} species assist selective oxidation paths leading to HCHO, whereas *Fe$_2$O$_3$ nanoparticles* mostly yield CO$_x$ because of a high availability of lattice oxygen promoting the further oxidation of HCHO [4-8]. Then, a preparation route enabling a high dispersion and a superior MPO functionality of FeO$_x$/SiO$_2$ catalysts was disclosed [5-8].

This paper is aimed at highlighting the efficiency of the preparation method based on "adsorption-precipitation" of Fe^{II} ions on silica carriers [5-8] to promote

Table 1. List of SiO$_2$ and FeO$_x$/SiO$_2$ catalysts

Code	SiO$_2$ support	Preparation method - Fe precursor	Fe loading (wt%)	S.A.$_{BET}$ (m$^2\cdot$g^{-1})	Surface Fe loading (Fe$_{at}\cdot$nm^{-2})
Si 4-5P	-	-	0.020 [a]	385	0.006
F5	-	-	0.015 [a]	600	0.003
1-FS	Si 4 5P	INC/WET – Fe^{III}	0.095	400	0.025
3-FS	Si 4-5P	INC/WET – Fe^{III}	0.430	400	0.116
FS-1	Si 4 5P	ADS/PRC – Fe^{II}	0.350	400	0.095
FS-2	F5	ADS/PRC – Fe^{II}	0.095	600	0.017
FS-3	F5	ADS/PRC – Fe^{II}	0.370	600	0.067

[a] Fe as impurity.

the dispersion of FeO$_x$/SiO$_2$ catalysts. The influence of the structure of active sites on the catalytic pattern of the title system in MPO is addressed.

2. EXPERIMENTAL

2.1. Catalysts. FeO$_x$/SiO$_2$ catalysts (**FS-x**) were prepared by the "adsorption-precipitation" (ADS/PRC) route [5-7], using powder silica carriers (Si 4-5P and F5, *Akzo Nobel*) and FeSO$_4$ precursor. For comparison, another series of catalysts (**x-FS**) was prepared by "incipient wetness" (INC/WET) of the above silicas with aqueous solutions of Fe(NO$_3$)$_3$. After impregnation, all the catalysts were dried at 100°C and then calcined at 600°C (6h) [5-7].
The list of catalysts is reported in Table 1.

2.2. UV-Vis Diffuse Reflectance (DR-UV-Vis) spectra were acquired in reflectance mode, using a *Perkin-Elmer Lambda 900* instrument equipped with an integrating sphere. The spectra were processed by the "Kubelka-Munk" function.

2.3. IR measurements were carried out with a *Bruker Equinox 55* instrument equipped with a DTGS detector and running at 4 cm^{-1} resolution. Spectra of adsorbed NO are reported in absorbance after subtraction of the spectrum of the sample before NO adsorption ("blank" spectrum).

2.4. TPR measurements were performed using a quartz microreactor fed with a 6% H$_2$/Ar carrier flowing at 30 *stp* cm$^3\cdot$min^{-1} [7]. All the experiments were run in the range 200–800°C with a heating rate of 20°C$\cdot$min^{-1}, while deconvolution of TPR profiles was made by the "PeakFit" software (*Jandel Scientific*).

2.5. Catalyst testing in MPO was performed by a batch reactor at 650°C and 1.7 bar with a flow rate of 1.0 *stp* L$\cdot$min^{-1} (He:N$_2$:CH$_4$:O$_2$=6:1:2:1) [5-7].

3. RESULTS and DISCUSSION

The iron oxide dispersion can be comparatively evaluated from the DR UV-Vis spectra of the samples in air, normalised to the relative maximum intensity, as shown in Fig. 1. The spectrum of the FS-2 (a) sample, loaded with 0.095 Fe wt% exhibits a band with maximum at ca. 45000 cm^{-1} due to ligand to metal charge transfer (LMCT) transitions typical of isolated Fe^{3+} ions in octahedral coordination. In the case of FS-3 (b), the higher Fe loading results in the presence of additional components in the 40000-30000 cm^{-1}, monitoring the presence of di- or oli-gomeric FeO_x species [9]. For the similarly loaded FS-1 catalyst (0.35%), obtained by the ADS/PRC route using a lower surface area silica (e.g., Si 4-5P) carrier, the DR UV-Vis spectrum (c) also outlines a tail in the range 30000-15000 cm^{-1}, typical of larger bi-dimensional FeO_x patches and/or *Fe₂O₃ nanoparticles*

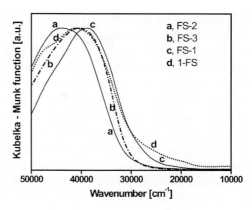

Fig. 1. DR-UV-Vis spectra of FeO_x/SiO_2 catalysts in air.

[9]. In spite of a lower Fe content (0.095%), the adsorption in this region is even more intense for the 1-FS sample (INC/WET), definitely proving that the ADS/PRC gives a much better dispersion of the active phase than INC/WET. Likely, a selective electrostatic interaction with the negatively polarised hydroxyl groups of silica prevents the mobility of Fe^{II} ions and the consequent formation of *Fe₂O₃ nanoparticles* during calcination [5-7, 10].

The coordination of Fe ions on the silica surface was probed by IR analysis of adsorbed NO further to a pretreatment of samples at 400°C under vacuum (Fig. 2). The spectrum of the FS-2 catalyst (Fig. 2A) shows that the admission of the first dose of NO (a) produces a single band at 1818 cm^{-1} due to mononitrosylic Fe^{II}-NO adducts [11], while a rise of the NO pressure (b-h) enhances this band leading to the appearance of the NO gas phase roto-vibrational pattern (1950-1830 cm^{-1}). Then, Fe^{II} sites with only one coordinative vacancy are present in this system.

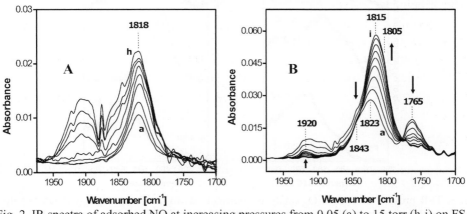

Fig. 2. IR spectra of adsorbed NO at increasing pressures from 0.05 (a) to 15 torr (h-i) on FS-2 (A) and FS-3 (B) catalysts pretreated *in situ* at 400°C under vacuum.

By contrast, the spectrum of the FS-3 sample at the lowest NO coverage (Fig. 2B, a), besides to the mononitrosylic band (1823 cm^{-1}) displays a shoulder at 1843 cm^{-1} and a band at 1765 cm^{-1} typical of the symmetric and antisymmetric stretching modes of dinitrosylic Fe^{II}-(NO)$_2$ species [11]. Rising the NO pressure (b-i), these two components decrease progressively transforming into two other signals; one quite weak at 1920 cm^{-1}, and the other, at *ca.*1805 cm^{-1}, appearing as a shoulder of the main peak at 1815 cm^{-1}. These monitor the symmetric and antisymmetric stretching modes of trinitrosylic Fe^{II}-(NO)$_3$ adducts, respectively, formed by the insertion of a NO molecule on dinitrosyl adducts [11]. Moreover, the increased coverage yields a higher intensity of the mononitrosylic band, the maximum of which shifts to lower frequency as a result of the disappearance of the shoulder at 1843 cm^{-1} and growth of that at 1805 cm^{-1}. The concomitant occurrence of mono- and poly-nitrosylic adducts signals that, at higher loading, Fe^{II} centres with higher degree of coordinative unsaturation, besides to sites with only one coordinative vacancy, are present [9,10].

The redox features of such surface species were highlighted by TPR analysis of the various FeO$_x$/SiO$_2$ samples. The effects of loading (A) and preparation method (B) on the "normalised" TPR pattern of FeO$_x$/SiO$_2$ catalysts are shown in Fig. 3. Because of the similar *constitutional* Fe content, the F5 and Si 4-5P silicas feature an analogous reduction pattern characterised by a H$_2$ consumption band spanned in the range 300-750°C with a broad maximum centred at 550-570°C, accounting for a H/Fe ratio of 0.48 and 0.45, respectively [7]. Apart of preparation method, Fe addition enhances the reduction kinetics yielding a shift of the overall TPR pattern to lower T, though no changes in the H/Fe ratio (0.49-0.57) are recorded [7,10].

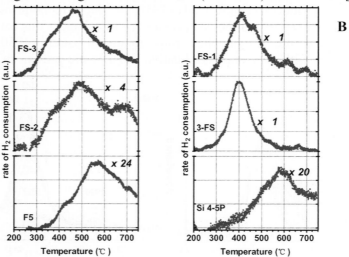

Fig. 3. TPR profiles of FeO$_x$/SiO$_2$ catalysts "normalised" to the Fe content: A) Effect of the loading on F5 supported ADS/PRC catalysts; B) Effect of the preparation method on Si 4-5P supported catalysts with similar (0.35-0.43%) Fe loading.

Table 2. Fitting of TPR spectra parameters of FeO_x/SiO_2 catalysts.

Sample	Aggregated Species				Isolated Species					
	M_1/fwhm$_1$ (°C)	A_1 (%)	M_2/fwhm$_2$ (°C)	A_2 (%)	M_3/fwhm$_3$ (°C)	A_3 (%)	M_4/fwhm$_4$ (°C)	A_4 (%)	M_5/fwhm$_5$ (°C)	A_5 (%)
F5	-	-	425/95	11.1	510/95	19.0	586/117	41.2	700/124	28.7
Si 4-5P	-	-	411/99	8.4	508/92	23.4	596/101	40.0	700/118	28.2
FS-2	-	-	385/106	21.6	484/98	27.6	573/118	24.6	696/124	26.2
FS-3	309/85	4.9	385/106	26.7	474/98	33.4	575/118	21.2	695/124	13.9
3-FS	305/97	7.0	395/106	75.9	493/99	10.5	587/118	3.8	699/124	2.9
FS-1	325/93	8.8	410/100	37.7	491/99	20.6	578/118	14.1	694/126	12.2

Assuming that the reduction of Fe species would not be affected by loading and carrier [10], the concentration of the various surface species was estimated by the deconvolution of TPR spectra into five Gaussian peaks with same centre (M, °C) and width at half-maximum (fwhm, °C) [10]. The deconvolution parameters reported in Table 2 provide a very good fit ($r^2>0.98$) of all the experimental TPR profiles (results not reported), suggesting that the 1st and 2nd peaks are associated to easily reducible "aggregated" moieties and the last three components to that of "isolated" species with a different degree of coordinative unsaturation [10]. Then, a quantitative evaluation of the relative abundance (A_x) of the various species (Table 2) reveals that both preparation method and Fe loading concur to affect the surface structures of FeO_x/SiO_2 systems determining, thus, marked differences in the catalytic behaviour [6-8,10]. Activity data in MPO are reported in Table 3 in terms of hourly CH_4 conversion, selectivity, reaction rate, HCHO productivity (STY_{HCHO}) and Fe atom specific activity (σ_{Fe}), calculated on the basis of the analytical loading (Table 1). The superior performance of ADS/PRC FeO_x/SiO_2 catalysts (FS-x) is immediately evident. The bare Si 4-5P and F5 silicas, owing to their (very low) Fe content, exhibit the lowest reaction rates and STY_{HCHO} values of 310 and 200 $g \cdot kg_{cat}^{-1} \cdot h^{-1}$, respectively. As the largest concentration of "isolated" Fe moieties (Table 2), yet, both systems are characterised by the highest σ_{Fe} value (1.0 s^{-1}). Addition of ca. 0.08 wt% of Fe to F5 silica by INC/WET (1-FS) allows a ca. threefold rise in activity and no changes in S_{HCHO} (62%) resulting in a STY_{HCHO} of 600 $g \cdot kg_{cat}^{-1} \cdot h^{-1}$. With respect to the F5 carrier, yet, such figures mirror a decrease in σ_{Fe} to 0.52 s^{-1}. By contrast, at the same loading, the FS-2 sample (ADS/PRC) exhibits a much higher activity resulting in a σ_{Fe} of 0.88 s^{-1}, and an even higher S_{HCHO} (84%) implying a marked rise in STY_{HCHO} to 1,350 $g \cdot kg_{cat}^{-1} \cdot h^{-1}$. The worst impact on activity is found adding by INC/WET 0.4 wt% of Fe to Si 4-5P carrier, as the 3-FS sample features a σ_{Fe} of only 0.14 s^{-1} and a comparatively low S_{HCHO} (48%) resulting in a STY_{HCHO} of only 590 $g \cdot kg_{cat}^{-1} \cdot h^{-1}$. These data signal a very poor MPO functionality evidently linked to a drop in dispersion [4-8,10]. Indeed, despite of the comparable loading (0.35%), the FS-1 sample (ADS/PRC) displays an extraordinary activity, more than four times larger than that of 3-FS catalyst, resulting in a σ_{Fe} of 0.76 s^{-1}, which coupled to a rather high S_{HCHO} (55%), provides the utmost STY_{HCHO} value of 2,800 $g \cdot kg_{cat}^{-1} \cdot h^{-1}$.

Table 3. Activity data of FeO_x/SiO_2 catalysts in MPO at 650°C

Sample	w_{cat} (mg)	$X_{CH4,\cdot h}$ (%)	S_{HCHO} (%)	S_{COx} (%)	reaction rate ($\mu mol_{CH4}\cdot s^{-1}\cdot g^{-1}$)	STY_{HCHO} ($g\cdot kg_{cat}^{-1}\cdot h^{-1}$)	σ_{Fe} (s^{-1})
Si 4-5P	50	3.5	78	22	3.6	310	1.01
F5	50	2.7	65	35	2.8	200	1.04
1-FS	25	4.3	62	38	8.8	600	0.52
FS-2	**10**	**2.9**	**84**	**16**	**14.9**	**1,350**	**0.88**
3-FS	25	5.4	48	52	11.1	590	0.14
FS-1	**5**	**4.6**	**55**	**45**	**47.3**	**2,800**	**0.76**
FS-3	**5**	**4.4**	**52**	**48**	**43.7**	**2,480**	**0.66**

Confirming the negligible influence of the carrier on the efficiency of the preparation method [5-7], the FS-3 catalyst, bearing a similar Fe content (0.37%) by ADS/PRC on F5 silica, features a catalytic pattern like that of the FS-1 sample (Table 3). Noteworthy, the match of activity and characterization data is fairly evident plotting σ_{Fe} and ε_{HCHO} (e.g., $\sigma_{Fe}\times S_{HCHO}$) vs. "$A_3+A_4+A_5$" values, this last corresponding to the concentration of isolated species on the various catalysts (Table 2). As shown in Fig. 4, the occurrence of two linear relationships confirms the main role of isolated species in driving the MPO reaction, in contrast to easily reducible *Fe₂O₃ nanoparticles* enabling the formation of huge amounts of *lattice oxygen ions* unreactive to CH_4 but prompting an easy conversion of HCHO to CO_x [4-8,10].

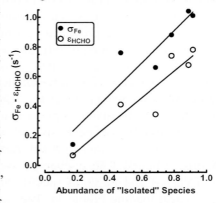

Fig. 4. Specific activity (σ_{Fe}) and rate of HCHO formation (ε_{HCHO}) of Fe atoms vs. the abundance of "isolated" species (e.g., "$A_3+A_4+A_5$") from TPR modeling.

REFERENCES

[1] R.W. Joyner and M. Stockenhuber, Catal. Lett., 45 (1997) 15
[2] X. Feng and W.K. Hall, J. Catal., 166 (1997) 368
[3] N. Mimura and M. Saito, Catal. Lett., 58 (1999) 59
[4] T. Kobayashi, Catal. Today, 71 (2001) 69
[5] A. Parmaliana, F. Arena, F. Frusteri and A. Mezzapica, German Patent GEM 17 (2000) SÜD CHEMIE AG
[6] F. Arena, F. Frusteri, T. Torre, A. Venuto, A. Mezzapica and A. Parmaliana, Catal. Lett., 80 (2002) 69
[7] F. Arena, F. Frusteri, L. Spadaro, A. Venuto and A. Parmaliana, Stud. Surf. Sci. Catal., 143 (2002) 1097
[8] F. Arena and A. Parmaliana, Acc. Chem. Res., 36(12) (2003) 867 and references therein
[9] S. Bordiga, R. Buzzoni, F. Geobaldo, C. Lamberti, E. Giamello, A. Zecchina, G. Leofanti, G. Petrini, G. Tozzola and G. Vlaic, J. Catal., 158 (1996) 486
[10] F. Arena, G. Gatti, S. Coluccia, G. Martra, and A. Parmaliana, J. Catal. (2004) submitted
[11] G. Spoto, A. Zecchina, G. Berlier, S. Bordiga, M.G. Clerici and L. Basini, J. Mol. Catal. A 158 (2000) 107

Studies in Surface Science and Catalysis, volume 147
X. Bao and Y. Xu (Editors)
541

Methanol from oxidation of methane over MoO$_x$/La-Co-oxide catalysts

X. Zhang, D.-H. He*, Q.-J. Zhang, B.-Q. Xu, Q.-M. Zhu

Key Lab of Organic Optoelectronics & Molecular Engineering of Ministry of Education，Department of Chemistry, Tsinghua University, Beijing 100084, China

ABSTRACT

The selective oxidation of methane to methanol with O$_2$ as an oxidant over MoOx/La-Co-O catalysts was investigated. Comparatively high CH$_3$OH selectivity (*ca.* 60%) and yield (*ca.* 6.7%) were achieved over 7wt% Mo/La-Co-O catalyst. The properties of the catalysts have also been characterized with, XRD, XPS and H$_2$-TPR. The proper reducibility and O$^-$/O^{2-} ratio of the catalysts enhanced the production of methanol in the selective oxidation of methane.

1. INTRODUCTION

Selective oxidation of methane is one of the most attractive potential industrial processes for the use of abundant natural gas resources. The supported molybdenum-based catalysts have been considered to be active in this reaction [1]. SiO$_2$, Al$_2$O$_3$, Al$_2$O$_3$-SiO$_2$, MgO, TiO$_2$, TiO$_2$-SiO$_2$, SnO$_2$, ZnO, HZSM-5 and US-Y etc. have been used as the supports of MoO$_3$ [1]. The reducibility and the participation of O^{2-} and/or O$^-$ species play important roles in the formation of C$_1$-oxygenate [1]. In this paper, we report the catalytic performance of novel MoO$_x$/La-Co-oxide catalysts in the selective oxidation of methane. The properties of the catalysts have also been characterized with BET method, XRD, XPS and H$_2$-TPR so as to establish their relation to catalytic performance.

2. EXPERIMENTAL

La-Co-oxide powder was prepared by citric acid sol-gel method as described in the literature [2]. The MoO$_x$/La-Co-oxide catalysts were prepared by impregnation of La-Co-oxide powder with different amount of (NH$_4$)$_6$Mo$_7$O$_{24}$·4H$_2$O (Tianjing Chemical Co., A.R.) aqueous solution. The catalyst precursors were dried at 110 $^\circ$C for 8 h, and then calcined at 500 $^\circ$C for 8 h. The catalysts were called n-MLC, in which "n" indicates %wt of the Mo-loading.

The selective oxidation of methane was carried out in a specially designed high-pressure continuous-flow micro reactor and controlled with MRCS800-4B micro-reaction apparatus [3]. The effluent was analyzed with an on-line gas chromatograph (SHIMADZU GC-8A, TCD, H_2 carrier), which was equipped with two columns: GDX-403 for the separation of CH_3OH, $HCHO$, C_2H_4, C_2H_6 and H_2O, and TDX-01 for the separation of N_2, O_2, CH_4, CO and CO_2. The addition of nitrogen in the feed gas acted as both diluent and internal standard.

The BET specific surface areas of the samples were measured on a Micrometric ASAP-2010C equipment. The phase structures of the catalysts were identified with a BRUKER D8 Advance powder X-ray diffractometer with Cu K_α (λ=0.15406 nm). The XPS spectra were collected with a PE PHI-5300 ESCA spectrometer, Al K_α (1486.6 eV, 10.1 kV). The quantitative TPR experiments of the samples were carried out in home-built TPR apparatus [3].

3. RESULTS AND DISCUSSIONS

The blank experiment using the reactor packed with quartz wool was carried at 420 °C and 5.0 MPa. The result showed that 4.9% CH_3OH selectivity at 0.9% CH_4 conversion was obtained at 420 °C, indicating that the gas phase oxidation of methane had no great influence on the formation of C_1-oxygnate under such reaction conditions. In the catalytic oxidations, the volume up and down the catalyst bed (0.5 g, 20~40 mesh) was filled with quartz wool. MoO_3 appeared low CH_4 conversion (1.6%) and low formaldehyde selectivity (6.9%). The La-Co-oxide mainly resulted in the total oxidation of methane to CO and CO_2. However, MLC catalysts showed an activity for the formation of CH_3OH. The conversion of CH_4 and products distribution closely depended on the Mo-loading.

Fig. 1 shows the influence of Mo-loading on the catalytic performance of MLC catalysts. A slight increase of CH_4 conversion was observed with the

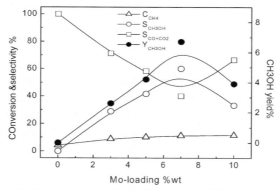

Fig. 1. Effect of Mo-loadings on catalytic performance of MLC catalyst in selective oxidation of methane. Reaction conditions: 420°C, 4.2 MPa, $CH_4/O_2/N_2$=9/1/1 and 14400 ml/(g.cat).h^{-1}.

increase of Mo-loading. The maximal values of CH_3OH selectivity and yield were obtained over 7-MLC catalyst. The results imply that molybdenum oxide could be the active sites in the selective oxidation of methane to methanol. Furthermore, a synergetic effect might occur between the molybdenum oxide and the La-Co-oxide.

The reaction temperature had distinct influence on the catalytic performance of 7-MLC catalyst (Fig. 2). As the reaction temperatures increased from 380 to 460 °C, CH_4 conversion gradually increased from 5.0% to an almost steady level of ca. 11% as a result of nearly complete consumption of oxygen in the feed gas. At the temperature of 420 °C, the highest ca. 60% CH_3OH selectivity and 6.7% CH_3OH yield were obtained. On further increasing temperature, both CH_3OH selectivity and yield decreased while CO_x selectivity increased. The results indicate that high temperature accelerated deep oxidation of CH_4 and CH_3OH.

The BET surface areas of the MLC catalysts slightly increased from 10.8 to 11.8 m^2/g with the increase of Mo-loading from 3 to 10%wt. The XRD results showed that both La-Co-oxide and MLC catalysts mainly composed of $LaCoO_3$ and Co_3O_4. The relative quantity of both $LaCoO_3$ and Co_3O_4 in these samples was estimated according to the corresponding diffraction peak intensity. The results showed that the ratio of $LaCoO_3/Co_3O_4$ nearly kept constant with the increase of Mo-loading. Therefore, it is deduced that amorphous and well-dispersed molybdenum oxide located on the catalysts.

The XPS results are presented in Table 1. The binding energy (BE) of $Co_{2p3/2}$ on the well crystal Co_3O_4, Co_2O_3, CoO and $LaCoO_3$ are 779.8, 779.6, 780.4 and 779.1 eV, respectively [4]. The BE of $La_{3d5/2}$ on La_2O_3 and $LaCoO_3$ is 833.7 and 833.2 eV, respectively [4]. The La-Co-oxide in present study showed lower BE of La_{3d} (833.0 eV) and lower BE of Co_{2p} (779.3 eV) than La_2O_3 and Co_2O_3. This suggests the existence of deficient sites around Co and La.

When Mo was loaded on La-Co-oxide, the binding energy varied with Mo-loading. Both the binding energy of Co and La slightly increased with

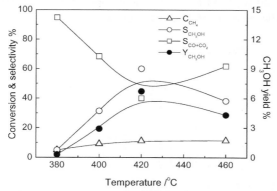

Fig. 2. Effect of reaction temperatures on catalytic performance of 7-MLC catalysts in selective oxidation of methane. Reaction conditions: 4.2MPa, $CH_4/O_2/N_2$=9/1/1 and 14400ml/(g.cat).h^{-1}.

Table 1
XPS results of the catalysts

Catalysts	Co (eV)		La (eV)		Mo (eV)	
	Co 2p3/2	Co 2p1/2	La 3d5/2	La 3d3/2	Mo 3d5/2	Mo 3d3/2
La-Co-oxide	779.3	794.5	833.0	850.1	-	-
3-MLC	779.3	794.5	833.4	850.5	232.0	235.2
7-MLC	779.5	794.6	833.6	850.6	232.1	235.3
10-MLC	779.9	795.4	833.6	850.6	232.4	235.6
MoO₃	-	-	-	-	233.5	236.7
Co₃O₄*	779.8					
Co₂O₃*	779.6					
CoO*	180.4					
LaCoO₃*	779.1		883.2			
La₂O₃*			833.7			

* The data are quoted from reference [4]

Mo-loading. The BE of Mo$_{3d}$ on MLC catalysts was lower than that on MoO₃. The results imply the occurrence of interaction between molybdenum oxide and La-Co-oxide, resulting in deficient sites around Mo. With the increase of Mo-loading, the BE of Mo$_{3d}$ slightly increased, due to that the decrease of the deficient sites around the Mo.

The catalysts showed asymmetrical O$_{1s}$ XPS peaks at ca. 529.8 eV. These O$_{1s}$ XPS spectra were fitted with Lorentzian function and analyzed in detail [2]. The results reveal that the existence of oxygen species in mixed oxidation state on the catalysts surface. The involved oxygen species were lattice oxygen species (O$^{2-}_{1s}$) at 528.8-529.0 eV, adsorbed oxygen species (O$_2{}^{2-}$/O$^-_{1s}$) at 50.1-530.2 eV and O$_{1s}$ of oxygen-containing contamination OH$^-$/CO$_3{}^{2-}$ at 532.1-532.5 eV. The relative content of the three kinds of oxygen species can be estimated from the relative area of sub-peaks. It is worth to note that the ratio of O$^-$/O^{2-} gradually decreased with the Mo-loading (Fig. 3a).

The TPR results are summarized in Table 2. MoO₃ exhibited two TPR peaks at 889 and 1006 K with respect to the step-wise reduction of Mo^{6+}⇒Mo^{4+}⇒Mo^0 [5]. La-Co-oxide showed TPR peaks near 664, 728 and 912 K. The H₂ consumption of TPR peak at 664 K was about one half of that of TPR peak at

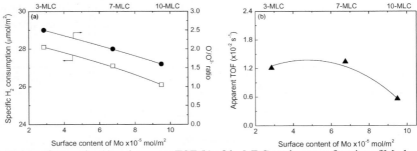

Fig. 3. Variety of properties (a) and apparent TOF (b) of the MLC catalysts as a function of Mo-loading.

912. Since La is non-reducible under the TPR conditions, such TPR peaks are belong to the reduction of Co. Lago et al. [6] reported that $LaCoO_3$ showed TPR peaks at 633 and 844 K, corresponding to the reduction of $Co^{3+} \Rightarrow Co^{2+} \Rightarrow Co^0$. Co_3O_4 also appeared two TPR peaks near 530 and 730 K, which are respectively due to the reduction of $Co^{3+} \Rightarrow Co^{2+}$ and $Co^{2+} \Rightarrow Co^0$ [7]. Hence, the TPR peaks near 664 and 912 K of the La-Co-oxide were the reduction of $Co^{3+} \Rightarrow Co^{2+}$ and $Co^{2+} \Rightarrow Co^0$ in $LaCoO_3$. The TPR peak at 728 K might be the reduction of $Co^{2+} \Rightarrow Co^0$ of Co_3O_4.

The TPR profiles and the reduction temperature (T_p) of the MLC catalysts were different from the TPR profile of MoO_3 and La-Co-oxide, further indicating the occurrence of interaction between molybdenum oxide and La-Co-oxide. The T_P of MLC catalysts shifted toward the high temperature with Mo-loading. The results indicate that the reducibility of MLC catalysts gradually reduced with Mo-loading. The H_2 consumption of MLC catalysts slightly increased with Mo-loading, probably due to the increase of the reducible oxide (Table 2).

The results obtained in this work clearly show that molybdenum oxide loading on the carrier La-Co-oxide induced some change of properties, which determined the catalytic performance. The changes of properties of MLC catalysts might be due to the occurrence of interaction between molybdenum oxide and La-Co-oxide. The catalysts with low Mo-loading showed strong interaction compared to the catalysts with high Mo-loading. Therefore, the catalysts with low Mo-loading had higher reducibility and more deficient sites than the catalysts with high Mo-loading. The deficient sites and the reducibility influenced the activation and transform of oxygen species as the sequence [2]: $O_{2\ (gas)} \Leftrightarrow O_{2\ (surface)} \Rightarrow O^{2-}_{\ (surface)} \Rightarrow O_2^{2-}_{\ (surface)} \Leftrightarrow O^-_{\ (surface)} \Rightarrow 2O^{2-}_{\ (lattice)}$, so the O^-/O^{2-} ratio varied with Mo-loading.

In order to elucidate the influence of the properties on the activity and selectivity of MLC catalysts, several means have been employed. The catalytic performance is presented as apparent turnover frequency (TOF), which is defined as the number of moles of CH_3OH formed per mole of molybdenum. The TOF values reported here are apparent TOF values since all the supported molybdenum atoms are assumed to be active in the selective oxidation of methane. The reducibility of the catalysts is expressed by the specific H_2

Table 2

TPR peaks temperature and H_2 consumption

Catalyst	TPR peaks temperature (K)			H_2 consumption ($\mu mol/(g.cat)$)			
	I	II	III	I	II	III	total
La-Co-oxide	664	728	912	60	91	151	302
3-MLC	740	937	-	92	210	-	304
7-MLC	745	949	-	88	216	-	304
10-MLC	748	973	-	78	230	-	308
MoO_3	889	1006	-	-	-	-	-

consumption, which is defined as H_2 consumption per m^2 of the catalyst. The relationships of the apparent TOF and the reducibility properties as function of Mo-loading are shown in Fig. 3a and 3b. It is evidence that an increase in Mo-loading generally resulted in a decrease in the specific H_2 consumption and O^-/O^{2-} ratio, and in the case of the apparent TOF showing the maximal value on 7-MLC catalyst.

As the above results showing, the specific H_2 consumption and O^-/O^{2-} ratio varied with Mo-loading, furthermore, such properties had close relationship with the catalytic performance. It is well known that both O^- species and O^{2-} species are active in the selective oxidation of methane to methanol [1]. And, it seems that O^- species are more active than O^{2-} species [1]. However, O^- species also have high activity in the oxidation of methanol [2]. Likewise, much higher reducibility of the catalysts can result in deep oxidation of methane and methanol. Hence, the catalysts with the proper reducibility and O^-/O^{2-} ratio, such as 7-MLC catalyst, enhanced the production of methanol in the selective oxidation of methane.

4. CONCLUSIONS

The interaction between loaded Mo and carrier La-Co-oxide leads to the modification of the properties of MoO_x/La-Co-oxide catalysts, which in turn determined the catalytic performance in the selective oxidation of methane to methanol. The high activity and selectivity of CH_3OH were obtained on 7-MLC catalyst. The proper reducibility and O^-/O^{2-} ratio of the catalysts enhanced the production of methanol in the selective oxidation of methane.

ACKNOWLEDGEMENTS

Thanks for the financial support from the National Key Fundamental Research Project Foundation of China (G1999022406).

REFERENCES

[1] K. Tabata, Y. Teng, T. Takemato, E. Suzuki, M.A. Banares, M.A. Pena, J.L.G. Fierro, Catal. Rev., 44(1) (2002) 1.
[2] W. Wang, H.B. Zhang, G.D. Lin, Z.T. Xiong, Appl. Catal. B., 24 (3-4) (20002) 219.
[3] X. Zhang, D.H. He, Q.J. Zhang, Q. Ye, B.Q. Xu, Q.M. Zhu, Chin. J. Catal., 24(4)(2003)305.
[4] J.F. Moulder, W.F. Stickle, P.E. Sobol, K.D. Bomben, Handbook of X-ray Photoelectron Spectroscopy, edited by J. Chastain, Perkin-Elmer Corporation, Eden Prairie, MN, 1992.
[5] A. Parmaliana, F. Arena, F. Frusteri, G. Martra, S. Coluccia, V. Sokolovskii, Stud. Surf. Sci. Catal., 110 (1997) 347.
[6] R. Lago, G. Bini, M.A. Pena, and J.L.G. Fierro, J. Catal., 167 (1997) 198.
[7] J.L. Li and N.J. Coville, Appl. Catal. A., 208(1-2) (2001) 177.

Studies in Surface Science and Catalysis, volume 147
X. Bao and Y. Xu (Editors)

SbO$_x$/SiO$_2$ catalysts for the selective oxidation of methane to formaldehyde using molecular oxygen as oxidant

Haidong Zhang, Pinliang Ying, Jing Zhang, Changhai Liang, Zhaochi Feng and Can Li[*]

State Key Laboratory of Catalysis, Dalian Institute of Chemical Physics, Chinese Academy of Sciences, P. O. Box 110, Dalian 116023, China
[*] Corresponding author. Fax: +86 411 4694447, Email: canli@dicp.ac.cn

ABSTRACT

SbO$_x$ and SbO$_x$/SiO$_2$ catalysts were prepared and investigated for methane selective oxidation to HCHO. HCHO selectivity up to 41% can be obtained on Sb$_2$O$_5$/SiO$_2$ catalyst at 873 K and just drop gently to 18% with temperature up to 923 K. HCHO selectivity for SbO$_x$/SiO$_2$ catalysts decreases gently with reaction temperature, so considerable value of HCHO selectivity can still be obtained at high temperatures.

1. INTRODUCTION

Methane selective oxidation to C$_1$-oxygenats has attracted much attention in the past several decades [1-3]. The conversion of methane to C$_1$-oxygenats still remains a major challenge in heterogeneous catalysis because of the large gap between chemical reactivity and thermal stability of methane and its transform products. Many supported single metal oxides or multi-component metal oxides have been employed as the catalysts for the selective oxidation of methane [4-7], but most oxides are not selective in this reaction. It was found that silica-supported vanadium and molybdenum oxides were relatively effective catalysts in this reaction, but the HCHO yield is still poor for practical application [2, 8-9]. Searching for more selective catalysts is always crucial for the methane selective oxidation. Antimony oxide has been used as a compound of some selective oxidation catalysts [10-12]. But, to our knowledge, antimony oxides have never been used as the major active component in a catalyst for methane selective oxidation and its catalytic performance in this reaction has never been reported. An interesting work about the design of selective oxidation catalysts done by Hutchings et al. reported that the temperature required for the methanol decomposition on Sb$_2$O$_3$ is higher than that on many other oxides [13-14], indicating that methanol, an possible product of methane selective oxidation, is quite stable on Sb_2O_3. Therefore, it might be possible to achieve a higher selectivity to C1-oxygenates in the methane selective oxidation on antimony oxides. In this contribution, a series of SbO_x/SiO_2 catalysts were prepared and investigated for methane selective oxidation using O_2 as oxidant under atmospheric pressure. Our primary results show that high HCHO selectivity can be obtained on SbO_x/SiO_2 catalysts, especially on Sb_2O_5/SiO_2 catalysts.

2. EXPERIMENTAL

1-20 wt.% Sb_2O_5/SiO_2 and Sb_2O_3/SiO_2 catalysts were prepared by impregnation of silica (specific surface area is 391 m^2 g^{-1}) with $SbCl_5$ and $SbCl_3$ (Arcos) - ethanol solution accompanying the hydrolysis with an aqueous ammonia solution. Additionally, a 10 wt% Sb_2O_3/SiO_2 catalyst and a 5 wt% Sb_2O_3/SiO_2 catalyst were prepared by impregnation with Sb_2O_3 (Arcos)-water slurry and Sb_2O_3-tartaric acid solution respectively. A 5 wt% MoO_3/SiO_2 catalyst was prepared by impregnation of silica with aqueous solution of ammonium heptamolybdate (Arcos). Sb_2O_5 were prepared by hydrolysis of $SbCl_5$ with aqueous ammonia. All impregnates and precipitates were dried at 393K for 12h and then calcined in air at 923K for 6h. The catalytic reactions were carried out in a quartz tubular reactor and an Agilent 6890N GC equipped with an FID and a TCD and two packed columns (Porapak Q and TDX-01) was employed for on-line analysis of products. Methane conversion and HCHO selectivity were calculated using an external standard method based on following equations:

$$CH_4 \text{ conversion} = (M_{CH4, in} - M_{CH4, out})/M_{CH4, in} \times 100\%$$

$$HCHO \text{ selectivity} = M_{HCHO}/(M_{CH4, in} - M_{CH4, out}) \times 100\%$$

where $M_{CH4, in}$ and $M_{CH4, out}$ are the amount (mol) of CH_4 feed in and remaining in the effluent; M_{HCHO} is the amount (mol) of HCHO detected. On average, the estimated error in replica experiments was within ± 5% and carbon balance was better than 95%. UV Raman spectra were recorded on a homemade UV Raman spectrograph using 325 nm line of a He-Cd laser with 25 mW output as the excitation source. XRD were taken on a Rigaku Rotaflex (RU-200B) X-ray diffractometer with Cu-Kα radiation (40 kV, 40 mA).

3. RESULTS AND DISCUSSION

Fig. 1 shows methane conversions and the selectivity to HCHO for Sb_2O_3/SiO_2 and Sb_2O_3 catalysts at 873 K. For Sb_2O_3/SiO_2 catalysts prepared from $SbCl_3$, the selectivity to HCHO increases with loading from 13.5% (1 wt.%) to 25.8% (20 wt.%) although methane conversion drops from 1.8% (1 wt%) to

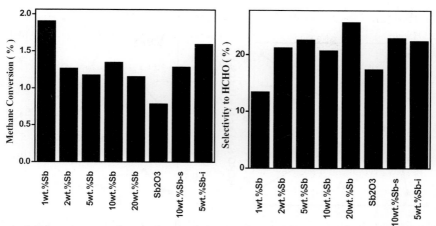

Fig. 1. Methane conversion and selectivity to formaldehyde for Sb_2O_3/SiO_2 and Sb_2O_3 catalysts at 873 K. 1-20 wt.% Sb are Sb_2O_3/SiO_2 catalysts prepared by impregnation with $SbCl_3$-ethanol solution.10 wt.% Sb -s and 5wt.% Sb-i are Sb_2O_3/SiO_2 catalysts prepared by impregnation with Sb_2O_3 -water slurry and Sb_2O_3 -tartaric acid solution respectively. 0.2 g catalyst was used in each test. GHSV=3750 h^{-1}, $CH_4/O_2/He$ = 17:5:3 (volume ratio).

1.2% (20 wt.%). For the catalysts prepared from Sb_2O_3 by slurry method and impregnation method respectively, methane conversions are 1.4% and 1.6% while HCHO selectivity are 23% and 22.5%. The selectivity to HCHO for Sb_2O_3 is 17.5%, which is lower than those for Sb_2O_3/SiO_2 catalysts prepared by different methods with different precursors except the lowest loading one (1 wt.%). Sb_2O_3 shows the lowest methane conversion value, 0.79%, among all catalysts in Fig. 1. This suggests that Sb_2O_3/SiO_2 catalysts can show better activity and selectivity than bulk Sb_2O_3 in methane selective oxidation. Fig. 2A gives methane conversions and HCHO selectivity for Sb_2O_5/SiO_2 catalysts at 873 K. The selectivity to HCHO increases from 18% to 41% with increasing Sb_2O_5 loading from 1 wt.% to 20 wt.% while methane conversion drops from 1.0% to 0.47%. A similar trend has been observed for methane conversion and HCHO selectivity at 923 K. The selectivity to HCHO also increases from 7.6% to 18% with increasing Sb_2O_5 loading from 1 wt.% to 20 wt.% while methane conversion drops from 4.6% to 2.1%. The methane conversion for bulk Sb_2O_5 at 873 K and 923 K are 0.48 % and 1.7 % while HCHO selectivity are 10.8 % and 8.3 %, respectively. The HCHO yields for bulk Sb_2O_5 at 873 k and 923 K are 0.05 % and 0.14 %, respectively. It turns out that Sb_2O_5/SiO_2 catalysts are more selective and active than bulk Sb_2O_5 in methane selective oxidation. Fig. 2B shows the one pass yield of HCHO for Sb_2O_5/SiO_2 catalysts vs. Sb_2O_5 loading at 873 K and 923 K, respectively. The HCHO yield increases with Sb_2O_5 loading and reaches its maximum value for 5 wt.% Sb_2O_5/SiO_2 at either 873 K or 923 K. Fig. 2C gives HCHO selectivity vs. CH_4 conversion for 5 wt% Sb_2O5/SiO_2 catalyst. HCHO selectivity increases from 25 % to 30 % while CH_4 conversion

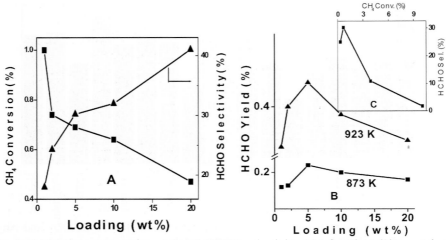

Fig. 2 (A) Methane conversions (■) and HCHO selectivity (▲) for Sb₂O₅/SiO₂ catalysts at 873 K; (B) One pass HCHO yield for Sb₂O₅/SiO₂ vs. Sb₂O₅ loading at 873 K (■) and 923 K (▲); (C) HCHO selectivity vs. CH₄ conversion for 5 wt% Sb₂O₅/SiO₂ catalyst. 0.2 g catalyst was used in each test. GHSV=3750 h⁻¹, CH₄/O₂/He = 17:5:3 (volume ratio).

increases from 0.33 % to 0.70 % and then drops to 1.8 % with increasing CH_4 conversion up to 10.1 %. From Fig. 1 and Fig. 2, we also can see that Sb_2O_5/SiO_2 catalysts are more selective than Sb_2O_3/SiO_2 catalysts.

Well-dispersed metal oxides catalysts are generally believed to be helpful for achieving high selectivity and activity in this reaction. Fig. 3 shows the XRD patterns for the silica used as support, 1-20 wt% Sb_2O_5/SiO_2 catalysts and Sb_2O_5 prepared by hydrolysis of $SbCl_5$. No diffraction peak of Sb_2O_5 has been observed in the XRD patterns of Sb_2O_5/SiO_2 catalysts, even for the loading of 20 wt.%. This suggests that Sb_2O_5 is highly dispersed on the surface of Sb_2O_5/SiO_2 catalysts. In fact, the XRD patterns of Sb_2O_3/SiO_2 catalysts prepared from either $SbCl_3$ or Sb_2O_3 by slurry and impregnation methods also show no diffraction peak of bulk Sb_2O_3. This may be the reason why the supported Sb_2O_3 and Sb_2O_5 on SiO_2 catalysts show better activity and HCHO selectivity than bulk Sb_2O_3 and Sb_2O_5. Fig. 4 gives the UV Raman spectra of silica, 1-20 wt% Sb_2O_5/SiO_2 catalysts and Sb_2O_5 prepared by hydrolysis of $SbCl_5$. The UV Raman spectrum of bulk Sb_2O_5 gives major bands at 440, 515 and 693 cm⁻¹. No bands at 577 and 782 cm⁻¹ were detected for bulk Sb_2O_5. The bands at 577 and 782 cm⁻¹ begin to appear in the spectrum of 1 wt% Sb_2O_5/SiO_2 catalyst. These two bands become more intensive with increasing Sb_2O_5 loading till it reaches 5 wt% and then weaken. The bands at 577 and 782 cm⁻¹ should be ascribed to the low oligomeric antimony species. In the UV Raman spectrum of 5 wt% Sb_2O_5/SiO_2 catalyst, the bands at 440 and 515 cm⁻¹ appear and their intensity rise as Sb_2O_5 loading increases. The bands at 440 and 515 cm⁻¹ should be ascribed to the formation of

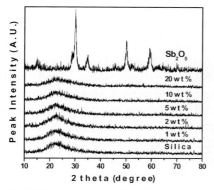

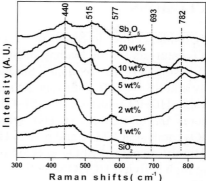

Fig. 3. XRD patterns for the silica used as support, 1-20 wt% Sb$_2$O$_5$/SiO$_2$ catalysts and bulk Sb$_2$O$_5$ prepared by hydrolysis of SbCl$_5$.

Fig. 4. UV Raman spectra of Sb$_2$O$_5$/SiO$_2$ catalysts (1-20 wt%), Silica and Sb$_2$O$_5$ prepared by hydrolysis of SbCl$_5$. Laser excitation, 325 nm

aggregated antimony entities on the surface of silica. The agglomeration of the oxidic species probably begins to take place while Sb$_2$O$_5$ loading reaches 5 wt%. For Sb$_2$O$_5$/SiO$_2$ catalysts with Sb$_2$O$_5$ loading higher than 5 wt%, the amount of polymeric antimony species increases and the antimony species with the structure like that of bulk Sb$_2$O$_5$ are dominant on the surface. This may be the reason why the optimum Sb$_2$O$_5$ loading for HCHO yield is 5 wt% and also suggests high dispersion of Sb$_2$O$_5$/SiO$_2$ catalysts is necessary for achieving high HCHO yield. The HCHO yield obtained for each Sb$_2$O$_5$/SiO$_2$ catalyst increases with temperature from 873 K to 923 K. Additionally, for 5 wt% Sb$_2$O$_5$/SiO$_2$ catalyst, the maximum HCHO yield value was observed to increase from 0.21% (873 K) to 0.43% (923K) and HCHO selectivity rises with CH$_4$ conversion at first and then drops. These are very interesting performance observed in methane selective oxidation for the Sb$_2$O$_5$/SiO$_2$ catalysts. In the methane selective oxidation on heterogeneous catalysts such as MoO$_3$/SiO$_2$, using O$_2$ as oxidant, the selectivity to HCHO drops dramatically with temperature. For the 5 wt% MoO$_3$/SiO$_2$ catalyst tested under the same conditions as SbO$_x$/SiO$_2$ catalysts, the HCHO selectivity is 10% at 873 K and drops dramatically to 0.66% at 923 K with 4.1% and 18% methane conversions, respectively. For Sb$_2$O$_3$/SiO$_2$ and Sb$_2$O$_5$/SiO$_2$ catalysts, higher HCHO selectivity than that for 5 wt% MoO$_3$/SiO$_2$ catalyst was achieved. HCHO selectivity for SbO$_x$/SiO$_2$ catalysts decrease gently with temperature so considerable value of HCHO selectivity can still be obtained at high temperatures. For 5 wt% Sb$_2$O$_5$/SiO$_2$ catalyst, HCHO selectivity drops from 30.2 % to 10.9 % with temperature from 873 K to 923 K and the HCHO yield increases from 0.10 % (823 K) to 0.43 % (923 K). The high HCHO selectivity for SbO$_x$/SiO$_2$ catalysts make it possible to combine SbO$_x$/SiO$_2$ catalysts with some reactors, such as a continuous recycle reactor separator [15], to achieve high HCHO yield with less CO$_x$ generated.

4. CONCLUSIONS

Antimony oxides, either in Sb_2O_5 or Sb_2O_3, can be highly dispersed on SiO_2 in SbO_X/SiO_2 catalysts for the loading of SbO_X up to 20 wt%. In methane selective oxidation using O_2 as oxidant, SbO_X/SiO_2 catalysts show good HCHO selectivity and Sb_2O_5/SiO_2 catalysts are more selective than Sb_2O_3/SiO_2 catalyst. Considerable HCHO selectivity for Sb_2O_5/SiO_2 catalysts can be still obtained at temperatures up to 923 K. HCHO selectivity up to 41% was obtained for Sb_2O_5/SiO_2 catalyst at 873 K and a 18% HCHO selectivity was achieved even at 923 K. Highly dispersed SbO_X species on SbO_X/SiO_2 catalysts is positive for achieving high activity, HCHO selectivity and HCHO yield.

ACKNOWLEDGMENTS

This work was financially supported by the State Key Project of Ministry of Science and Technology (Grant 1999022407), the BP-China project for Clean Energy and the National Natural Science Foundation of China (NSFC Grant 20273069).

REFERENCES

[1] J.H. Lunsford, Catal. Today, 63 (2000) 165.
[2] K. Tabata, Y. Teng, T. Takemoto, E. Suzuki, M. Bañares, M.A. Peña and J.L.G. Fierro, Catal. Rev.-Sci. Eng., 44 (2002) 1.
[3] R.H. Crabtree, Chem. Rev. 95 (1995) 987.
[4] J.A. Barbero, M.C. Alvarez, M.A. Bañares, M.A. Peña and J.L. Fierro, Chem. Commun., (2002) 1184.
[5] Kiyoshi Otsuka and Ye Wang, Appl. Catal. A: General, 222 (2001) 145.
[6] V.D. Sokolovskii, N.J. Coville, A. Parmaliana, I. Eskendirov and M. Makoa, Catal. Today, 42 (1998) 191.
[7] A. de Lucas, J.L. Valverde, L. Rodriguez, P. Sanchez and M.T. Garcia, Appl. Catal. A: General, 203 (2000) 81.
[8] T.J. Hall, J.S.J. Hargreaves, G.J. Hutchings, R.W. Joyner, S.H. Taylor, Fuel Proc.Tech., 42 (1995) 151.
[9] A. de Lucas, J.L. Valverde, P. Cañizares, L. Rodrigues, Appl. Catal. A: General, 172 (1998) 165.
[10] T. Inoue, K. Asakura, Y. Iwasawa, J. Catal., 171 (1997) 457.
[11] J. N. Al-Saeedi, V.V. Guliants, O.G. Pérez and M.A. Bañares, J. Catal., 215 (2003) 108.
[12] Y. Moro-oka, Catal. Today, 45 (1998) 3.
[13] G.J. Hutchings and S.H. Taylor, Catal. Today, 49 (1999) 105.
[14] S.H. Taylor, J.S.J. Hargreaves, G.J. Hutchings and E. W. Joyner, Appl. Catal. A: General, 126 (1995) 287.
[15] Ioannis C. Bafas, Ioannis E. Constantinou, Constantinos G. Vayenas, Chem. Eng. J., 82 (2001) 109-115

Studies in Surface Science and Catalysis, volume 147
X. Bao and Y. Xu (Editors)
553

Highly stable performance of catalytic methane dehydro-condensation to benzene and naphthalene on Mo/HZSM-5 by addition and a periodic switching treatment of H₂

Ryuichiro Ohnishi, Ryoichi Kojima, Yuying Shu, Hongtao Ma, and Masaru Ichikawa[*]

Catalysis Research Center, Hokkaido University, Kita-Ku, N-11, W-10, Sapporo 060-0811, Japan.

1. INTRODUCTION

The catalytic dehydroaromatization of methane towards benzene and naphthalene is of great scientific importance and industrial benefit for the effective utilization of natural gas and biogas to provide alternative petrochemical feed stocks and bulky hydrogen production for fuel cells. In the last decade, a lot of research work on methane dehydroaromatization has been conducted,[1–6] but the problem of serious decrease in catalytic activity and selectivity with time on stream is still unsolved and vital drawback from the viewpoint of industrial application of this process. Effective methods to solve the problems by adding a few percent of CO2 to the methane feed on HZSM-5 supported Mo and Re catalysts [9–11]. However, these oxidative gases produce CO as by-products which make it di fficult to separate hydrogen from the reaction products. Here we present a novel and improved method to achieve excellent stabilit y and catalytic performance of methane dehydrocondensation towards benzene and naphthalene at 1023–1073 K by a few % H2 addition to methane and a periodic switching operation of hydrogen, owing to the effective removal of poisoning coke on Mo/HZSM-5 catalyst.

2. EXPERIMENTAL

The Mo/HZSM-5 catalyst (6wt% Mo loading) was prepared by the wet impregnation technique with an NH4ZSM-5 zeolite (SiO2/Al2O3 = 40) and an aqueous solution of ammonium heptamolybdate. Honeycomb type ZSM-5 (Meidensha Co., Japan) was heated at 723 K for 5h to remove the adsorbed water in ZSM-5 channel, cooled in vacuum and impregnated with aqueous solution containing proper amount of ammonium molybdate for 30min. The

554

catalyst was dried in vacuum and calcined at 773K for 5h. About 10g (
20mm, height 130mm) of 6wt% Mo loaded honeycomb ZSM-5 catalyst was
charged into a high-pressure stainless steel reactor (20mm i.d.). After
carburization with methane stream at 823K for 30 min, the methane
dehydroaromatization reaction with and without H_2 addition were performed
under the conditions of 1023K, 0.3MPa and methane space velocity of 2520
ml.g-MFI^{-1}.h^{-1} in unit weight of MFI used. The reaction was carried out in a
fixed-bed continuous-flow system at 1023–1073K by a periodic switching with
methane and H_2 for 30 min each. The products were analyzed by two on-line
GCs equipped with FID and TCD detectors.

TG-TPO measurement was conducted on a TG/DTA/MASS System (Mac
Science Co., TG-DTA 2020S). The O_2/He stream (flow rate: O_2=15 ml/min
and He=100 ml/min) was fed on 30 mg of Mo/HZSM-5 catalyst after being used
in the reaction for 10h. Then, the temperature was increasing at a rate of 10 K
min^{-1} and H_2O, CO, CO_2 were continuously monitored with a mass spectrometer
(ThermoLab VG Gas) at m/e=18, 28 and 44 respectively. Mo K-edge XAFS
(X-ray absorption fine structure) measurements were conducted on Mo/HZSM-
5 samples at the Photon Factory of the National Laboratory for High Energy
Physics(KEK-PF, Tsukuba, Japan) to characterize the Mo/HZSM-5 catalyst
before and after a few% H2 addition and a periodic switching operation with H2
flow at 1023-1073K.

3. RESULTS AND DISCUSIION

Figure 1 shows the formation rate of benzene on 6wt% Mo/HZSM-5 catalyst in
the methane dehydroaromatization reaction with and without H_2 addition into

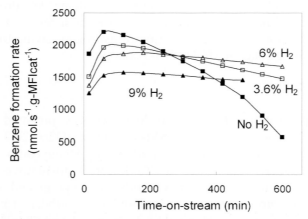

Fig. 1. Formation rate of benzene in the methane dehydroaromatization reaction on 6wt% Mo
loaded honeycomb ZSM-5 catalyst under flowing methane with 0, 3.6, 6 and 9% H_2 and
under the conditions of 1023 K, 0.3 MPa and 2520 ml.g^{-1} h^{-1} of CH_4 space velocity.

methane feed gas. At 1023K and 0.3 MPa, the catalytic activity without hydrogen addition in the feed decreased quickly after 10h of the reaction to ca. 26% of the initial activity because of high coke formation. In contrast, with 3.6, 6 and 9% (mole/mole) H_2 addition, catalytic activities for benzene kept 74, 88 and 93% of the highest activities, respectively, after 10h of time on stream in the methane dehydroaromatization reaction. Although the methane conversion was suppressed by the addition of H_2 in the early stage of the reaction, total amount of benzene formed in 10h with 3-6% H_2 in the feed still increased. As shown in Table 1, the formation rates of other products such as hydrogen and naphthalene were similarly stabilized by the addition of H_2 in the feed. It is obvious from TG-TPO measurements in figure 2 and 3 that the coke formation is suppressed by a small amount of H_2 addition to methane feed. After 10 h of the reaction under pure methane flow, ca. 2.4 mg of coke was formed on the catalyst. In contrast, only 1.7 and 1.1 mg of coke were produced on the catalyst during the methane dehydroaromatization reaction for 10h under flowing methane with 3.6 and 6% H_2, respectively, as shown in figure 2.

Table 1
Effect of H_2 addition in methane feed in the methane dehydroaromatization reaction on 6wt% Mo loaded honeycomb ZSM-5 catalysts

H_2 amount in feed (%)	CH_4 conversion (%)	Product distribution in hydrocarbons (%)			
		Benzene	Toluene	Naphthalene	C_2+C_2'
0	12.7(10.4)	75(67)	4(6)	9(5)	10(21)
3.6	11.0(9.9)	77(75)	5(5)	9(5)	8(13)
6	9.5(8.7)	76(77)	4(5)	10(6)	9(11)
9	7.6(6.7)	75(77)	4(5)	9(6)	10(12)

[a]Reaction temperature: 1023K, reaction pressure: 3atm, methane specific velocity:2520ml/g.h

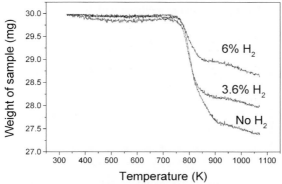

[b]Data were taken at 120min (480min) of time one stream.
Fig. 2. Weight changes in TPO experiments of catalysts being used in the methane dehydroaromatization for 10h under flowing methane with 0, 3.6 and 6% H_2

Figure 3 shows the corrected sum of peak height profiles of $CO+CO_2$ (CO_x) in TG-TPO experiments. The CO_x profile from Mo/ZSM-5 being used in pure methane stream consisted of two peaks at 790 and 870K, which were attributed to the carbon associated with molybdenum and to carbonaceous deposits on the Brønsted acid sites of the zeolite, respectively [11, 16]. In contrast, the other two CO_x profiles of catalysts used under methane with hydrogen only gave one peak at the low temperature side. The result indicates that H_2 in feed suppresses the formation of coke formed on Brønsted acid sites. This is very important because it makes possible to regenerate the used catalyst at lower temperature, where the sublimation of molybdenum oxide and deterioration of zeolite framework may be avoided.As shown in Fig. 4, at 1073 K, 0.3 Mpa by the switching operation between CH4 and H2 for 30 min a stable and high formation rate of benzene(10300 nmol g-cat-1 s-1) and H_2 (20800 nmol g-cat-1 s-1; benzene/H2 molar ratio =12) at 15% conversion of methane, close to thermodynamically

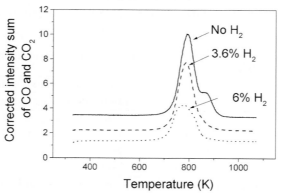

Fig. 3. CO + CO_2 profiles in TPO experiments of catalysts being used in the methane dehydroaromatization for 10h under flowing methane with 0, 3.6 and 6% H_2

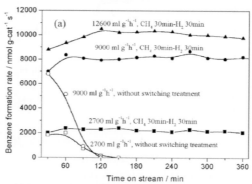

Fig.4. Formation rat of benzene in methane dehydrocondensation reaction on 6%Mo/HZSM-5 at 1073K, 0.3Mpa and 2700-12600mlg-1h⁻¹ with a periodic switching treat-ment between CH4 for 30 min and H2 for 30 min(solid symbols) and without the treatment (open symbols)

limited conversion of methane at 1073 K, was obtained in a prolonged time-on-stream for 100 h. In contrast, without the sitching treatment, the benzene formation rate decreased drastically and reached zero after only 2 h on stream. Again, a serious decreasing trend for benzene formation without the switching treatment, i.g., in a contenious mode of methane flow, and the distinct promotion effect on the catalyst stability with the switching operation between CH4 and 100% H2 could be observed.

Note that coke formation was substantially suppressed by the periodic switching operation between methane and H2 as compared with that using a methane flow alone.

TPO profiles of used Mo/HZSM-5 catalysts by addition of6%v/v H_2 to methane flow and by the switching operation of methane and H_2 showed a suppression of inactive coke peak at higher temperature of 850K, demonstrating their effectiveness for removal of poisoning coke formed on the Brønsted acid sites.[8-10] Mo K-edge EXAFS experiments were conducted to reveal the change in the active phase of the Mo/HZSM-5 catalyst on the Mo/HZSM-5 by 6% H2 addition to methane at 1023K, the Fourier transform (FT) spectra of which are shown in Fig. 5. As has been described in ref.15, the FT function of the sample that reacted with methane at 1023K for 10hr showed a characteristic peak at 2.7 Å assigned to Mo-Mo in Mo_2C as an active phase for methane dehydrocondensation towards benzene and naphthalene [11]. The features of the FT spectrum were unchanged and its intensity relatively increased after the H_2 addition experiment. These results suggest that the contribution of Mo_2C as the active phase dispersed in HZSM-5 is restored, while a poisoning carbon is effectively removal by the 100% H2 switching treatment or 6% v/v addition of hydrogen at 1023 K. The excellent thermal stability of Mo_2C in the hydrogen atmosphere is at the exceeding temperatures over 1023 -1273K.

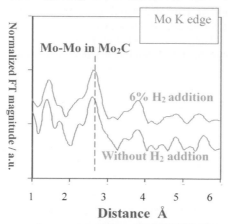

Fig. 5. Mo K-edge Fourier transform spectra of used Mo/HZSM-5 samples under CH4 sream at 1023K, 0.3MPa and and 6% H2 addition to CH4 feed at 1023K and 0.3MPa, respectively.

4. CONCLUSIONS

A small amount of H_2 addition into methane feed is a very effective to suppress a coke on Brønsted acid sites the Mo/HZSM-5 catalyst during the methane dehydroaromatization reaction at 1023 - 1073K. Furthermore, a periodic switching treatment with hydrogen achieved a high and excellent stability of catalytic performance in the methane dehydrocondensation towards benzene and naphthalene substantially effective to regenerate the active Mo/HZSM-5 catalyst due to the efficient removal of coke, while a thermal resistance of Mo_2C as an active phase for the reaction towards benzene and naphthalene on Mo/HZSM-5 catalyst at 1023-1073K.

ACKNOWLEDGMENTS

We appreciate the financial support from the New Energy and Industrial Technology Development Organization (NEDO) of Japan through the 2000 - 2002FY Regional Consortium R&D Program (12S1001).

REFERENCES

[1] L. Wang, L. Tao, M. Xie, G. Xu, J. Huang and Y. Xu, Catal. Lett., No. 21 (1993) 35.
[2] F. Solymosi, A. Erdohelyi and A. Szoke, Catal. Lett., No. 32 (1995) 43.
[3] S. Liu, Q. Dong, R. Ohnishi and M. Ichikawa, Chem. Commun., (1997) 1455.
[4] W. Ding, G.D. Meitzner and E. Iglesia, J. Catal., No 206 (2002) 14
[5] Y. Shu, D. Ma, L. Xu, Y. Xu and X. Bao, Catal. Lett., No. 70 (2000) 67.
[6] Y. Shu and M. Ichikawa, Catal. Today, No. 71 (2001) 55.
[7] L. Wang, R. Ohnishi and M. Ichikawa, J. Catal., No. 190 (2000) 276.
[8] Y. Shu, R. Ohnishi and M. Ichikawa, J. Catal., No. 206 (2002) 134.
[9] Y. Lu, D. Ma, Z. Xu, Z. Tian, X. Bao and L. Lin, Chem. Commun., (2001) 2048.
[10] Y. Shu, R. Ohnishi and M. Ichikawa, Catal. Lett., No. 81 (2002) 9.
[11] S. Liu, L. Wang, R. Ohnishi and M. Ichikawa, J. Catal., No. 181 (1999) 175.
[12] Z. Liu, M.A. Nutt and E. Iglesia, No 81(2002)271
[13] Y. Shu, H. Ma, R. Ohnishi and M. Ichikawa, Chem. Commun.(2003)86.

Studies in Surface Science and Catalysis, volume 147
X. Bao and Y. Xu (Editors)

Reactions of propane and n-butane on Re/ZSM catalyst

F. Solymosi[*], **P. Tolmacsov and A. Széchenyi**

Institute of Solid State and Radiochemistry, the University of Szeged and Reaction Kinetics Research Group of the Hungarian Academy of Sciences P.O. Box 168, H-6701 Szeged, Hungary

1. INTRODUCTION

A great effort is being made to convert methane into more valuable compounds. One of the most attractive results was obtained on $Mo_2C/ZSM-5$ catalyst, which transfers methane into benzene with 80% selectivity at 10% conversion at 973 K [1-3]. Several studies have been performed to improve the catalytic performance of $Mo_2C/ZSM-5$ and to develop a more active and stable catalyst for the methane – benzene conversion. The most successful test was made by Ichikawa et al.[4], who found that Re/ZSM-5 exhibits similar catalytic behavior as $Mo_2C/ZSM-5$ for the aromatization of methane. Re/ZSM-5 proved also effective in the aromatization of ethane [5].

In the present paper the reactions of propane and butane – which are also formed in the aromatization of methane – are investigated on supported Re. We hope that the evaluation of the reaction pathways of these hydrocarbons over Re/ZSM-5 helps us to understand better the formation of benzene from methane on the same catalyst.

2. EXPERIMENTAL

The gases used were of commercial purity (Linde). NH_4-ZSM-5 was a commercial product (Zeolite Intern.), which was calcined to produce H-ZSM-5 in air at 868 K for 4 h. SiO_2 was the product of Aerosil 380. Re/H-ZSM-5 samples with different Re loadings were prepared by impregnation (incipient wetness) of H-ZSM-5 with the aqueous solution of $(NH_4)_2ReO_4 \cdot 4H_2O$. The resulting materials were dried at 393 K, calcined in air at 773 K for 4 h; and then reduced at 523 K for 30 min. Catalytic reactions were carried out at 1 atm of pressure in a fixed-bed, continuous flow reactor consisting of a quartz tube (8 mm i.d.) connected to a capillary tube. The flow rate was 12 ml/min. The carrier gas was Ar. The hydrocarbon content was 12.5 %. Generally 0.3 g of loosely compressed catalyst sample was used. Reaction products were analyzed gas chromatographically using a Hewlett-Packard 4890 gas chromatograph with a Porapack QS+S column and a HP-PLOT Al_2O_3 column. The catalyst has been characterized by XPS.

3. RESULTS AND DISCUSSION

3.1. Reactions of propane on H-ZSM-5

First the catalytic behavior of ZSM-5 of different compositions was tested. The reaction was observed above 823 K. As shown in Fig 1A, the SiO_2 /Al_2O_3 (further Si/Al) ratio dramatically influenced the catalytic performance of ZSM-5. At 873 K the highest conversion (87%) was obtained for the sample with Si/Al = 30, which decreased to 66% for Si/Al = 50, to 37% for Si/Al = 80 and 7% for Si/Al = 280. There were no or only little changes in the rate of formation of various products in time on stream (5-160 min). Whereas methane was produced with 31-36% selectivity on all samples, the selectivity of other main products changed drastically with the composition of ZSM-5. Whereas the selectivity of benzene was high 28% on ZSM-5 with Si/Al = 30, it gradually decayed with the increase of this ratio and attained almost a zero value on ZSM-5 of high silica content. At the same time the selectivity of ethylene grew from 15% to 57%. Besides these products, ethane, propylene and toluene were also formed with low selectivities. Traces of butane and pentane were also detected. Variation of the reaction temperature showed that higher temperature is favorable for the formation of benzene, while at lower temperature the selectivity to the propylene and ethylene significantly increased (Fig. 1B). No appreciable change occurred in the selectivity of methane formation.

3.2. Reactions of propane on Re/ZSM-5

To identify the surface species formed during the interaction of propane with the Re/HZSM-5 catalyst, detailed IR spectroscopic measurements were performed. Propane adsorbed weakly and nondissociatively on Re catalysts below 300 K. At higher temperatures, the characteristic vibrations of π-bonded, di-σ-bonded propylene and propylidyne were identified by FTIR spectroscopy.

In the next experimental series Re was deposited on the most effective ZSM-5 support (Si/Al ratio = 30). Besides the enhanced conversion benzene formed with higher rate at the expense of the formation of methane and ethylene. This was well exhibited in the values of selectivities. As shown in Fig. 2A an increase in the loading of Re from 2% to 5% somewhat increased the conversion, but it did not lead to the further improvement of the catalyst, as regards the benzene formation. Further increase of the Re content to 10% caused decay in the conversion, but only slightly affected the product distribution.

Re exerted a more profound influence on the catalytic performance of the less effective ZSM-5 (Si/Al ratio = 280); at 873 K the conversion increased from

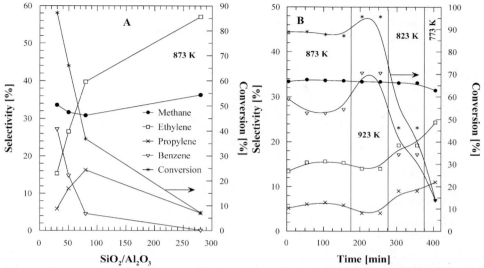

Fig. 1. Effects of SiO₂/Al₂O₃ ratio at 873 K (A) and the temperature (B) on the reaction of propane on ZSM-5(30)

12.0% to 47-48% (Fig. 2B). The formation of propylene dramatically increased at the expense of that of ethylene. Whereas on pure ZSM-5 benzene was detected only traces, in the presence of 2% Re it formed with 5-10% selectivity. At higher Re content (10%), the conversion of propane was less than on 2% Re/ZSM-5, but the selectivities of the main products changed only slightly.

An examination of the Re/ZSM-5 (Si/Al = 30) catalyst after reaction revealed the deposition of coke. TPR studies showed that the hydrogenation of surface carbons on 2% Re/ZSM-5 starts at 600 K and gave two peaks at 710 and 820 K. The amount of carbon, as determined in the form of methane, was 3, 46 x 10^5 nmol. According to the calculation this means that about 1, 2% of propane that passed through the catalyst bed in 155 min decomposed to carbon.

The effect of space velocity on the reaction of propane was determined by the variation of flow rate of the gas mixture. We obtained that with the increase of the flow rate, the velocity of the formation of all products increased. On the basis of the selectivities, the higher flow rate was favorable for the formation of ethylene and propylene. In contrast, the selectivity of benzene and ethane significantly, while that of methane slightly decreased.

The catalytic efficiency of Re has been also investigated on silica, which behaved practically inactive towards the reaction of propane even at 873 K. Re/SiO₂, however, was an active catalyst: the main process was the dehydrogenation of propane. The selectivity of propylene was 40-45% at the conversion of 9-10%. Other products were ethylene and methane. Benzene was detected only in a small amount with 5-7% selectivity.

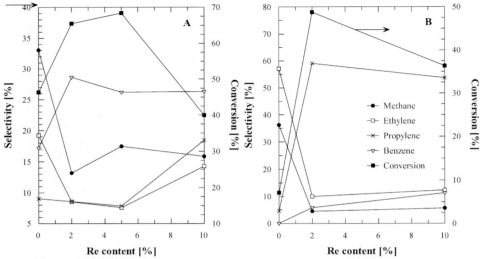

Fig. 2.Effects of Re content of ZSM-5(30) at 823 K (A) and ZSM-5(280) (B) on the reaction of propane at 873 K

3.3. Reactions of propylene on ZSM-5 and Re/ZSM-5

As the key compound in the conversion of propane is propylene it was necessary to exploit as to how the Re on ZSM-5 affects the reactions of propylene. On the most effective ZSM-5 (Si/Al = 30) the conversion of propylene was 72-75% even at 773 K, which remained constant for several hours. The main products of the reaction are as follows: benzene (S=31%), toluene (S=26-27%), other aromatics (12%), methane (S=11-12%), ethylene (7.5%) and ethane (~7%).

The deposition of Re on this ZSM-5 improved its catalytic performance. The conversion of propylene increased up to 85% at 2% Re and to 91% at 5% Re contents. The selectivities of the aromatic compounds, however, decreased with the rise of the Re content. The situation was different on the less active ZSM-5 (Si/Al = 280). Adding 2% Re to this zeolite increased the conversion from 57% to 72-65%, and also improved the capability to aromatization. The selectivity of the fomation of benzene changed from 17% to 25-35 %, and that of toluene from 12% 18-15%.

3.4. Reaction of n-butane

The reaction was followed at 823 K. As in the previous case, the product distribution sensitively depended on the composition of ZSM-5. Highest conversion, 96%, was obtained for the sample with Si/Al ratio = 30 and the lowest, 22%, with Si/Al ratio of 280. There was no or only a slight decay in the rate of formation of various products in time on stream. Whereas the cracking of

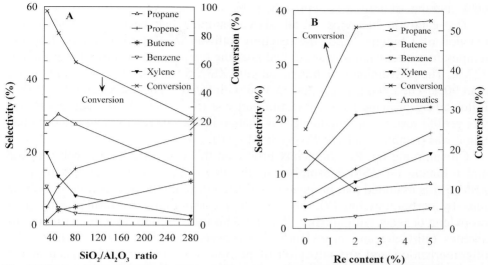

Fig. 3.Effects of SiO$_2$/Al$_2$O$_3$ ratio (A) and Re content of ZSM-5(280) (B) on the reaction of n-butane at 823 K

butane was the dominant reaction pathway, yielding propane, propene, ethylene, ethane and methane in decreasing amounts, the production of aromatics (benzene, toluene and xylenes) occurred on all ZSM-5 samples. The selectivity of aromatics was the highest on ZSM-5 with high alumina content, and decreased to low values at Si/Al ratio of 280 (Fig. 3A).

Although ZSM-5 samples alone exhibited relatively high activity, an enhancement of its catalytic performance with Re was experienced on all samples. Results obtained for ZSM-5 (Si/Al ratio=280) are plotted in Fig.3B. The conversion of n-butane was doubled even at 2% Re content. This zeolite exhibited very little aromatizing tendency, the presence of 2 and 5% Re, however, promoted the formation of both benzene and xylene. The selectivity of aromatics attained a value of 10-16%. A significant increase was observed in the selectivity of butenes, whereas the formation of lower hydrocarbons has been reduced. Detailed analysis of butenes showed that it consists of 1-butene, trans 2-butene, iso-butene and cis 2-butene. From among them iso-butene was produced with the highest selectivity. Note that the effect of Re in the dehydrogenation and aromatization of n-butane approached that of Mo$_2$C catalyst on the same ZSM-5 samples [6].

The important role of Re in the conversion of n-butane into other compounds is clearly demonstrated by the results obtained for Re dispersed on SiO$_2$, which exhibited only slight catalytic activity. On Re/SiO$_2$ not only the cracking and dehydrogenation of n-butane were catalyzed, but the aromatization reaction was also promoted. At 2% Re content the selectivity of butene attained a value of

60%, and that of aromatics approached 5%.

The catalytic behavior of all above samples has been examined in the reaction of 1-butene. Since the reaction pathway is strongly influenced by the temperature, the reaction was studied near the temperature of former experiment, 723 K. The conversion of 1-butene on pure ZSM-5 samples was very high, more than 90%. The main products on ZSM-5 with Si/Al = 280 are xylene (S = 20%), followed by propane (S = 16%) and benzene (S = 3%). Depositing 2% Re on this zeolite resulted in practically no change in the conversion and selectivities. Same was obtained on Re/ZSM-5 (Si/Al = 80).

On summing up, Re effectively promoted the aromatization of both propane and n-butane on ZSM-5 samples. In the description of this process on pure ZSM-5 it is assumed that propane is activated through the abstraction of hydride ions by carbonium ion [7]. The dimethyl carbonium ions formed may give propylene or react with alkenes. Parallel with this reaction route the cracking of alkenes (ethylene and propylene) also proceeds to yield lower alkanes. The oligomerization and aromatization of propylene occur on the Brönsted acidic sites in the zeolite cavities. The same picture is valid for the reaction of n-butane on ZSM-5. The favourable effect of Re can be explained by its high activity towards the dehydrogenation of propane and n-butane. This was well exhibited by using an inactive silica support, which contains no Brönsted sites. The fact that aromatic compounds were also produced over Re/SiO_2 catalysts may suggest that the Re created some active acidic sites on SiO_2 or at the Re/SiO_2 interface which facilitated the coupling of reactive hydrocarbon species formed.

The study of the chemistry and reaction pathways of different hydrocarbon fragments (CH_3, C_2H_5, C_3H_7, C_4H_9) over Re (001) surface in UHV are in progress with the use of XPS, HREELS and TPD spectroscopic methods. It is expected that the results may help to establish the finer mechanism of the above reactions.

ACKNOWLEDGEMENTS

This work was supported by the Hungarian Academy of Sciences and by the OTKA TS 40877 and 38233.

REFERENCES

[1] L. Wang, L. Tao, M. Xie and G. Xu, Catal. Lett. 21 (1993) 35.
[2] D. Wang, J.H. Lunsford, and M.P. Rosynek, Topics in Catal. 3 (1996) 299.
[3] F. Solymosi, A. Szőke and J. Cserényi, Catal. Lett. 39 (1996) 157.
[4] S. Liu, Q. Dong, R. Ohnishi, and M. Ichikawa, J.C.S. Chem. Commun. 1445 (1997).
[5] F. Solymosi and P. Tolmacsov, to be published.
[6] F. Solymosi, R. Németh and A. Széchenyi, Catal. Lett. 82 (2003) 213.
[7] Y. Ono, Catal. Rev.-Sci. Eng. 34(3) (1992) 179.

Studies in Surface Science and Catalysis, volume 147
X. Bao and Y. Xu (Editors)
©2004 Elsevier B.V. All rights reserved.

Dehydro-aromatization of CH$_4$ over W-Mn(or Zn, Ga, Mo, Co)/HZSM-5(or MCM-22) catalysts[*]

L.-Q. Huang, Y.-Z. Yuan, H.-B. Zhang[**], Z.-T. Xiong, J.-L. Zeng, G.-D. Lin

Department of Chemistry & State Key Laboratory of Physical Chemistry for the Solid Surfaces, Xiamen University, Xiamen 361005, China

ABSTRACT

Several promoted W/HZSM-5(or MCM-22)-based catalysts for DHAM reaction were developed. Over the W-Zn-Ga(or Mo-Co)/MCM-22 catalyst, 70% selectivity of benzene (S$_{Ben}$) at 14~17% conversion of methane (X$_{CH4}$) could be achieved at 1073 K. XPS and NH$_3$-TPD measurements demonstrated that heavy deposition of carbon was the main reason leading to deactivation of the catalyst. Reoxidation by air can regenerate the deactivated W-based catalyst effectively.

1. INTRODUCTION

Catalytic dehydro-aromatization of methane (DHAM) to aromatics and H$_2$ in absence of O$_2$ has drawn increasing attention [1-4]. For practical uses, W-based catalyst is one of the promising DHAM catalysts [5, 6]. The main advantage is its highly heat-resistive performance. The high reaction temperature is beneficial to attaining high CH$_4$-conversion (thermodynamic equilibrium conversion of 11.3% and 21% was predicted for: 6CH$_4$ = C$_6$H$_6$ + 9H$_2$, at 973 K and 1073 K, respectively [4]), yet do not lead to loss of W by sublimation. Recently, highly active, selective and heat-resistive W/HZSM-5(or MCM-22)-based catalysts promoted by Mn(or Zn, Ga, Mo, Co) were studied and developed in our lab. The results shed light on the design of practical catalysts for DHAM process.

2. EXPERIMENTAL

Unpromoted catalysts, W/HZSM-5(or MCM-22), were prepared according to the method described previously [5]. Promoted W-based catalysts were prepared by first impregnating a certain amount of HZSM-5 (Si/Al molar ratio at 38) or MCM-22 (Si/Al=30) zeolite carrier with a calculated amount of nitrate of Mn

*This work was supported by the National Priority Fundamental Research Project (No: G1999022400) of China. **Corresponding author, Email: hbzhang@xmu.edu.cn

(or Zn, etc.) in aqueous solution, followed by drying at 393 K and calcining at 673 K for 4 h, and subsequently impregnating with a calculated amount of H_2SO_4-acidified $(NH_4)_2WO_4$ aqueous solution (pH=2~3), and drying at 393 K and calcining at 773 K in air for 5 h. The catalyst assay was carried out in a fixed-bed down-flow reactor-GC combination system. Conversion of CH_4 and selectivity of hydrocarbon products were evaluated with an internal standard analyzing method. The amounts of coke on catalyst were determined by TG-DTA method.

3. RESULTS AND DISCUSSION

3.1 DHAM reactivity of W/HZSM-5(or MCM-22)-based catalysts

The results of activity assay of a series of catalysts for DHAM reaction were shown in Table 1. Over the Mn (or Zn, Mo)-promoted W/HZSM-5(or MCM-22) catalysts, a X_{CH4} of 14~17% with the S_{Ben} at a value of 56~72% could be achieved in the initial 2~3 h of the reaction at 1073 K.

Analogous to the cases over Mo/HZSM-5-based catalysts reported by Ichikawa et al [3], addition of a small amount ($\leq 1.8\%$) of CO_2 to the feed-gas was found to significantly enhance X_{CH4} and S_{Ben} and to improve the coke-resistive performance of the catalysts. Such a result was shown in Fig. 1-A. Over the 3%W-1.5%Mn/HZSM-5 catalyst, a 3% increase of X_{CH4} accompanied by 4% increase of S_{Ben} was observed in the initial 3 h of reaction, and after 6 h of reaction, X_{CH4} still maintained at ~10%, with the corresponding S_{Ben} at ~28%. This demonstrates the greatly improved durability of the catalyst. The promoting effect by the added CO_2 originated most probably from its role as scavenger of functioning surface of the catalyst via Boudouart reaction: $CO_2 + \underline{C}$ (deposited C) $\rightarrow 2CO$ (as implied by emergence of CO with gradually increasing concentration in the products), thus in favor of alleviating the deposition of carbon and deferring deactivation of the catalyst due to coking to a greater extent.

Table 1

DHAM reactivity over catalyst systems with different promoters [a]

Catalyst (metal component in wt%)	CH_4-conv. / % [b]	Selectivity / % (C-basis) [b]				Durability / h
		Ben.	Tol.	C_2H_4	C_2H_6	
3%W/HZSM-5	20.8	51.1	4.1	6.3	1.7	3
3%W-1.5%Mn/HZSM-5	17.0	55.8	3.8	6.9	1.1	5
8%W/MCM-22	13.3	52.0	3.2	2.4	1.0	5
8%W-0.6%Zn/MCM-22	14.0	61.2	3.4	3.7	0.9	6
8%W-0.5%Mo/MCM-22	15.4	60.9	3.4	4.3	0.9	6
8%W-0.6%Zn-0.1%Ga/MCM-22	15.3	71.1	5.9	4.2	1.0	≥ 7.5
8%W-0.4%Mo-0.1%Co/MCM-22	14.5	72.2	5.1	4.3	0.8	≥ 7.5

a) Reaction conditions: 0.1MPa, 1073 K, GHSV of feed-gas CH_4+10%Ar at 960 h^{-1}.
b) Data taken when the selectivity of benzene reached maximum.

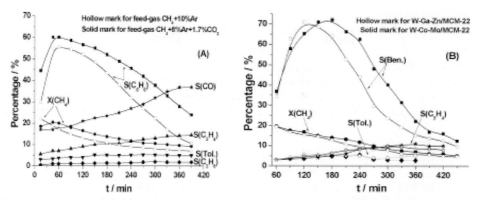

Fig. 1. DHAM reactivity over: (A) 3%W-1.5%Mn/HZSM-5 and (B) 8%W-0.6%Zn-0.1%Ga/ MCM-22 (hollow mark) and 8%W-0.4%Mo-0.1%Co/MCM-22 (solid mark) at: 0.1MPa, 1073 K, GHSV of feed-gas CH_4+10%Ar at 960 h^{-1} (unless otherwise specified).

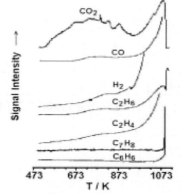

Fig. 2. CH_4-TPR/TPSR-MS profiles of the oxidative precursor of 3%W/HZSM-5 catalyst.

The CH_4-TPR/TPSR-MS profiles (Fig. 2) of the oxidative precursor of 3%W/HZSM-5 catalyst showed that its reduction-activation by CH_4 started from ~500 K. As temperature rising, the detected gas-phase products included successively CO_2, CO, H_2, C_2H_6 and C_2H_4, while benzene and toluene did not appear until the TPR/TPSR peaks of CO_2 and CO came to an end (implying that the reduction-activation of the catalyst was completed). In essence, such a CH_4-TPR/TPSR process corresponded quite well to the initial induction period in the catalyst test, from which it was shown that temperature as high as ~1073 K was requisite for the reduction-activation of the W/HZSM-5 catalyst.

3.2 Carbon deposition and catalyst regeneration

The XPS spectra of W(4f) and C(1s) of the 3%W/HZSM-5 catalyst with different time on the reaction stream are shown in Fig. 3.

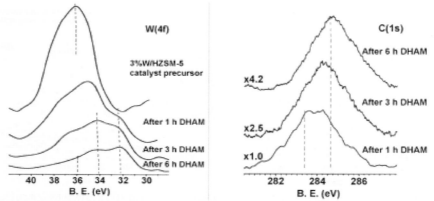

Fig. 3. XPS spectra of W(4f) and C(1s) of the 3%W/HZSM-5 catalyst with different time on stream of the reaction.

With reference to Refs [7, 8] and assuming the $W(4f)_{7/2}$ binding energy (B.E.) of W^{6+}, W^{5+}, W^{4+}, W^{2+}, WC at 36.0, 35.3, 34.5, 33.2, 32.2 eV and $\Delta E_b = E_b(W(4f)_{5/2}) - E_b(W(4f)_{7/2}) = 2.1 \pm 0.1$ eV (B.E.) and $I(W(4f)_{5/2}) / I(W(4f)_{7/2})$ (intensity ratio of the peaks) $= 0.78 \pm 0.03$ for each kind of the W^{n+} species in the same valence-state, analysis and fitting of these W(4f)-XPS spectra were done by computer. The results were shown in Table 2. It was evident that there co-existed W-species with mixed valence-states at the surface of functioning W/HZSM-5 catalyst, and after 6 h of the reaction, WC was the dominant W-containing surface species.

The C(1s)-XPS spectra of the functioning catalyst in different stages of the reaction provided another evidence for the formation of the WC species. It can be seen from Fig. 3 that the observed C(1s)-XPS peaks appeared at 283.3 and 284.7 eV (B.E.) in the initial 1 h of reaction, which may be attributed to C(1s) of WC and amorphous C-deposits, respectively [9]; while after 6 h of the reaction, the amorphous carbon became predominant C-containing species at the surface of the functioning catalyst, most probably due to the WC being covered by heavy carbon deposits.

Table 2

Composition of W-species at the surface of functioning 3%W/HZSM-5 catalyst

Time on stream of reaction (h)	Percentage (%) of different W-species in total surface W			
	W^{6+}	W^{5+}	W^{4+}	WC
1	19.1	32.0	28.8	20.1
3	12.5	24.2	32.1	31.2
6	7.5	18.2	20.1	54.2

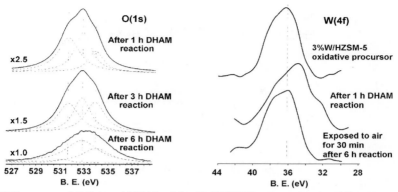

Fig. 4. XPS spectra of O(1s) and W(4f) of the 3%W/HZSM-5 catalyst in different states.

The argument of the WC species formation could also be supported by the O(1s)-XPS spectra of these samples shown in Fig. 4. According to Leclercq et al. [9], the observed O(1s)-XPS features may resolve into contributions from three types of surface O-species: metal-oxide O(1s) at 531.6 eV (B. E.), surface hydroxylic O(1s) at 532.8 eV (B. E.), and O(1s) of oxygen-doped/contaminated tungsten carbides at 533.9 eV (B. E.). With reaction time prolonged, the sub-peak at 531.6 eV greatly weakened, even vanished, and the sub-peak at 533.9 eV correspondingly enhanced, implying that most of WO_X turned into WC.

Through the analysis and fitting of the XPS spectra, it was indicated that the molar ratio of the component elements at the surface of 3%W/HZSM-5 catalyst was Al/C/O/Si/W = 1.2/~0/63.1/34.4/1.3 for the catalyst in oxidative state, and 1.5/41.7/35.5/20.8/0.5 for the functioning catalyst after 1 h of reaction. The ratio changed to 0.7/75.0/14.3/9.7/0.3 after 3 h of reaction, and to 0.6/82.7/10/6.6/0.1 after 6 h of reaction, indicating that most portion of the surface of the functioning catalyst was covered by carbon deposits after 6 h of reaction.

NH_3-TPD investigation revealed that ca.80% of strong and medium-strong acidic sites were lost after 5 h of reaction. N_2-BET measurements demonstrated that, after 1 h reaction, specific surface area of the functioning catalyst decreased to 302 m^2g^{-1} from 368 m^2g^{-1} for the oxidative precursor of catalyst. It further came down to 75 m^2g^{-1} after 5 h of reaction, implying that carbon deposits blocked most of the zeolite pores, thus resulting in the loss of a greater part of the surface. In fact, heavy deposition of carbon on the surface of functioning catalyst was the main reason leading to deactivation of the catalyst.

Reoxidation by air can regenerate the deactivated 3%W/HZSM-5 catalyst effectively, with the Al/C/O/Si/W molar ratio at the renewed surface recovered to 1.8/~0/64.7/31.7/1.9 (Fig. 4). Results of activity assay of the 3%W-1.5%Mn /HZSM-5 catalyst after a series of operation/regeneration (by air-oxidation each time for 1 h at 873 K) showed that, it is not until after ten times of regeneration that a small decrease in the catalyst activity was observed (Fig. 5). It seems that

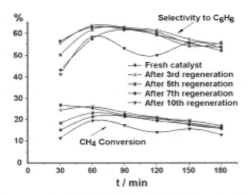

Fig. 5. DHAM reactivity over the 3%W-1.5%Mn/HZSM-5 catalyst after undergoing a series of operation/air-regeneration. Reaction conditions: 0.1MPa, 1073 K, GHSV at 960 h^{-1}.

the reason leading to the decrease in the catalyst activity was probably due to dealuminization to some degree of the framework structure of HZSM-5 zeolite caused by aqueous vapor resulting from the coke combustion in air-regeneration processes repeatedly.

3.3 Action of promoter

As shown by the previous NH$_3$-TPD studied [5], the promoters may play an important role in the optimized adjustment of surface acidity of the catalyst so as to alleviate coking and in inhibiting aggregation of the W-species and improving their dispersion at the surface and channels by doping of them into WC$_X$O$_Y$ matrix. Over the doubly promoted catalyst 8%W-0.4%Mo-0.1%Co/MCM-22, a 1.2% increase of X$_{CH4}$ accompanied by the 20% increase of S$_{Ben}$ in comparison to that of the unpromoted catalyst 8%W/MCM-22 was observed in the initial 5 h of the reaction (see Table 1 & Fig. 1-B). After 7 h of the reaction, X$_{CH4}$ still maintained at 6.2%, with the corresponding selectivity to benzene, toluene, C$_2$H$_4$ and C$_2$H$_6$ at 16.3%, 2.9%, 9.9% and 1.7%, respectively. The durability of the catalyst was greatly improved.

REFERENCES

[1] L. Wang, L. Tao, M. Xie, G. Xu, J. Huang, Y. Xu, Catal. Lett. 21 (1993) 35.
[2] D.J. Wang, J.H. Lunsford, M.P. Rosynek, J. Catal. 169 (1997) 347.
[3] R. Ohnishi, S. Liu, M. Ichikawa, J. Catal. 182 (1999) 92.
[4] J.L. Zeng, Z.T. Xiong, H.B. Zhang, G.D. Lin, K.R. Tsai, Catal. Lett. 53 (1998) 119.
[5] Z.T. Xiong, H.B. Zhang, J.L. Zeng, G.D. Lin, Catal. Lett. 74 (2001) 233.
[6] W. Ding, G.D. Meitzner, D.O. Marler, E. Iglesia. J. Phys. Chem. B 105 (2001) 3928.
[7] J. Haber, J. Stoch, L. Ungier, J. Sol. St. Chem., 19 (1976) 113.
[8] W. Grenert, E.S. Shpiro, R. Feldhaus, K. Anders, et al., J. Catal., 107 (1987) 522.
[9] G. Leclercq, M. Kamal, J.M. Giraudor, P. Devassine, et al., J. Catal., 158 (1996) 142.

Studies in Surface Science and Catalysis, volume 147
X. Bao and Y. Xu (Editors)
©2004 Elsevier B.V. All rights reserved.

Detailed mechanism of the oxidative coupling of methane

Y. Simon[*] , F. Baronnet, G.M. Côme and P.M. Marquaire

Département de Chimie Physique des Réactions, UMR 7630 - CNRS,
ENSIC-INPL, 1 rue Grandville - BP 451, 54001 NANCY Cedex (France).

ABSTRACT

To determine the relative importance of gas-phase and surface reactions in the oxidative coupling of methane (OCM), experimental investigations were performed. Our experimental results were compared to simulated values derived from a kinetic model taking into account heterogeneous and gas-phase reactions. We propose an original approach derived from Benson's techniques to estimate the kinetic parameters of surface reactions.

1. INTRODUCTION

Whereas it is well established that the OCM is a homogeneous-heterogeneous reaction, several points are still open on the reaction mechanism, especially the relative importance of gas-phase and surface reactions.

Experiments in a continuous flow reactor over La_2O_3 as a catalyst were performed in a large temperature range (1023 K to 1173 K) and on a variable quantity of catalyst. The results were compared to simulated values.

2. EXPERIMENTAL

The experimental setup is a catalytic jet-stirred reactor and has been previously described [1-4]. The reactor is made of quartz and has a gas-phase of constant volume $V = 113$ cm^3.The catalyst, La_2O_3, made from lanthanum carbonate, has a B.E.T. specific area equal to 1.2 $m^2.g^{-1}$. The particles have a diameter of around 60 µm and are compacted into pellets (diameter : 12.6 mm, thickness : 1 mm, mass : 0.45 g). Catalyst pellets are laid on a flat surface inside the reactor. The outlet gas stream was analysed by on-line gas

---
* Corresponding Author: **Email**: Yves.Simon@ensic.inpl-nancy.fr **Phone**: (33) 383.17.51.22 **Fax**: (33) 383.37.81.20

chromatography. The reaction was carried out in the following conditions : inlet composition mixture : $CH_4/O_2 = 4$, temperatures from 1023 to 1173 K and number of catalyst pellets from 0 to 8.

3. KINETIC MODELLING

Experimental results were compared to simulated values. These simulations were performed by means of Chemkin Surface Program on the basis of a kinetic model which takes into account elementary heterogeneous and gas-phase reactions. The mechanism used for simulation includes a heterogeneous part and a homogeneous one. The homogeneous part is described by a set of over 450 elementary reactions. It takes into account all the elementary reactions between molecules and free radicals including less than three carbon atoms. This gas- phase mechanism is well known and has been confirmed by a large amount of experimental data for different hydrocarbon reactions, particularly for the homogeneous oxidation of methane [5].

The heterogeneous mechanism takes into account the results from the mechanistic studies reported in the literature [6-12]. The experimental studies of oxygen chemisorption have shown that this reaction leads to the dissociation of diatomic O_2 molecule to form two active atomic oxygen centres.

$$O_2 + \lozenge \quad \underset{\longleftarrow}{\longrightarrow} \quad O_2(s) + \lozenge \quad \longrightarrow 2\,O(s)$$

The rate determining step is assumed to be the O_2 dissociation reaction. Heterogeneous mechanism is written assuming that reactive surface oxygen species interact with CH_4 and all major reaction products. We presumed that oxygen surface reactions are Eley – Rideal reactions. The first step is the activation of methane.

$$CH_4 + O(s) \rightarrow CH_3. + OH(s)$$

Estimation of the kinetic parameters of surface reactions.

In this work, we have chosen a method of estimation based on an original approach derived from BENSON's techniques [13]; it can be an alternative to the ab-initio methods which are more complex to handle.

The activation energy of elementary surface reactions is assumed to be the same as that of the equivalent gas phase reaction. The preexponential factors are estimated by using partition functions of reactants and transition state. For example, in a reaction such as:

A + B(s) → AB$^{\neq}$(s) → products

the preexponential factor is given by the equation :

$$A = \frac{kT}{h} \frac{Nq_{AB}^{\neq}}{q_A q_B}$$

k : Planck constant

h : Bolzman constant

T : Temperature (K)

N : Avogadro number

where q_A, q_B, $q_{AB}^{\neq}$ are the partition functions of A, B and the activated complex AB$^{\neq}$. Partition functions of gas-phase molecules can be calculated. On the other hand, the determination of the partition function of absorbed species [B(s), AB$^{\neq}$(s)] requires some assumptions. Therefore, we suppose that the difference between the partition functions of these two species is only due to the vibrational part q_v. Hence, a vibrational analysis of adsorbed species was carried out to estimate the vibrational partition function q_v. The frequency values used are those given by Benson [13]. Finally, this mechanism is a set of 20 elementary reactions.

4. RESULTS AND DISCUSSION

The formation of different reaction products (CO, CO_2, C_2H_4, C_2H_6) and CH_4 conversion were simulated at different experimental conditions (temperature, number of catalyst pellets) using the Chemkin Surface Program. In our model the specific area and the site density are not adjustable parameters. The value of the specific area was determined by using B.E.T. method and that of site density was taken in the literature [14]. Only the frequency factors have been optimized to adjust calculated and experimental results. The correlation between experimental and simulated values is satisfactory (see Fig. 1 and 2). As a result of the kinetic sensitivity analysis, the set of 20 heterogeneous reactions could be reduced to 6 reactions (Table 1).The production rate analysis of the main species has allowed to write the reaction scheme of OCM (Fig. 3).

Table 1

OCM reduced surface mechanism (◊ symbolizes surface site and (s) adsorbed species)

Reactions	A(mol, cm, s)	E (cal. mol^{-1})
$O_2 + 2◊$ → $2\,O(s)$	$6.1\ 10^{16}$	-26700
$CH_4 + O(s)$ → $CH_3. + OH(s)$	$2.8\ 10^{08}$	8400
$2\,OH(s)$ → $H_2O + O(s) + ◊$	$3.0\ 10^{23}$	100
$CO + O(s)$ → $CO_2 + ◊$	$2.7\ 10^{09}$	3000
$C_2H_4 + O(s)$ → $C_2H_3. + OH(s)$	$2.1\ 10^{10}$	5900
$C_2H_6 + O(s)$ → $C_2H_5. + OH(s)$	$2.5\ 10^{10}$	5800

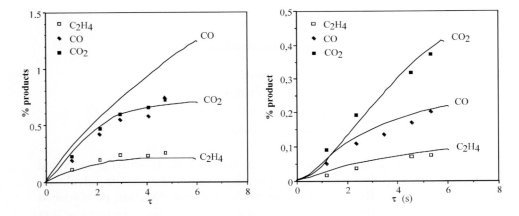

Fig. 1. 1023 K and 1 catalyst pellet   Fig. 2. 1173K and 3 catalyst pellets

Experimental/caculated production of CO, CO_2 and C_2H_4 (molar fraction vs gas space time)

According to this scheme, the initiation is exclusively the surface reaction:

$$CH_4 + O(s) \rightarrow CH_3. + OH(s)$$

This heterogeneous step is the main source of the free radicals involved in the chains of the gas-phase reaction. This mechanism shows that the catalytic reactions are also important for CO_2 formation from CO and for C_2H_4 formation from C_2H_6. We can note that the mechanism of OCM includes two distinct reaction pathways. The first pathway leads to the formation of oxygenated compounds Ox (CO, CO_2, HCHO...) and the second one leads to the formation of hydrocarbons C_2+ (C_2H_6, C_2H_4, C_2H_2...). Therefore, when the conversion is low, C_2+ and Ox selectivities are controlled by the rates of formation of C_2H_6 and HCHO respectively, through the reactions:

$$CH_3. + O_2 \rightarrow HCHO + OH. \qquad r = k(CH_3.)(O_2)$$

$$2\ CH_3. \rightarrow C_2H_6 \qquad r = k\ (CH_3.)^2$$

The C_2+ selectivity increases when O_2 concentration decreases and when the initiation rate increases. On the other hand, when the conversion is high, a third reaction pathway decomposes C_2H_4 into oxygenated species and, as a consequence, C_2+ selectivity decreases.

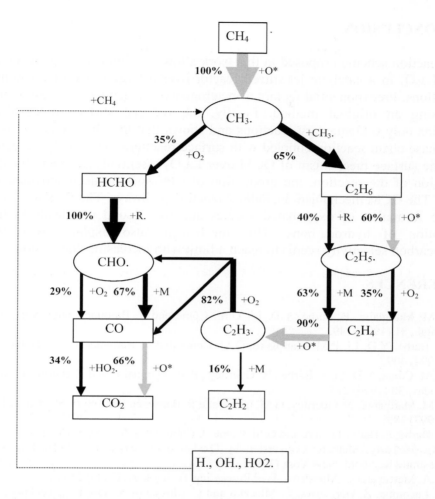

Fig. 3. Mechanism of the oxidative coupling of methane over La_2O_3 (1023 K, CH_4 conversion : 10%). Black arrows are gas-phase reactions and grey arrows are surface reactions.

The presence of La_2O_3 increases the C_2+ selectivity when the gas space time is low by introducing a new surface initiation reaction. Then, for higher gas space time, the secondary reaction of C_2H_3. oxidation decreases C_2+ selectivity. This analysis explains that many authors have found that C_2+ selectivity decreases drastically with increasing conversion of methane [15-17].

5. CONCLUSION

The reaction scheme proposed in this work allows to simulate OCM reaction over La_2O_3 in a catalytic jet-stirred reactor over a large range of operating conditions. Preexponential factors of heterogeneous reactions were estimated by using an original method. Finally, the reduced surface mechanism contains only 6 elementary reactions and shows that the OCM reaction is a gas-phase chain reaction coupled with surface reactions. This study suggests that the surface mechanism of OCM over La_2O_3 essentially accounts for the initiation of the reaction, the production of CO_2 and the decomposition of C_2H_6. The OCM mechanism includes 2 reaction pathways. The first one leads to the formation of oxygenated species and the second one leads to the formation of hydrocarbons. This mechanism also explains why the hydrocarbon selectivity seems to reach a limit with increasing conversion.

REFERENCES

[1] P.M. Marquaire, P. Barbé, Y.D. Li, G.M. Côme and F. Baronnet, Stud. Surf. Sci. Catal., 81 (1994) 149.

[2] P. Barbé, Y.D. Li, P.M. Marquaire, G.M. Côme and F. Baronnet, Catal. Today, 21 (1994) 409.

[3] G.M. Côme, Y.D. Li, P. Barbé , N. Gueritey , P.M. Marquaire and F. Baronnet, Catal. Today, 30 (1996) 215.

[4] P.M. Marquaire, N. Gueritey, G.M. Côme and F. Baronnet, Stud. Surf. Sci. Catal. 119 (1997) 383.

[5] P. Barbé, F. Battin-Leclerc and G.M. Côme, J. Chim. Phys. 92 (1995) 1666.

[6] J.G. McCarty, Methane Conversion by Oxidative Processes, E.E. Wolf Ed., Van Nostrand Reinhold, New York (1992).

[7] G.A. Martin and C. Mirodatos, Fuel Processing Technology 42 (1995) 179.

[8] S. Lacombe, Z. Durjanova, L. Mleczko and C. Mirodatos, Chem. Eng. Technol., 42 (1995) 216.

[9] P.M. Couwenberg, Qi Chen and G.B. Marin, Ind. Eng. Chem. Res., 35 (1996) 3999.

[10] D. Wolf, M. Slinko, M. Baerns and E. Kurkina, Appl. Catal. A : General, 166 (1998) 47.

[11] K. Coulter and D.W. Goodman, Catal. Letters, 20 (1993) 169.

[12] C. Shi, M. Xu, M.P. Rosynek and J.H. Lunsford, J. Phys. Chem., 97 (1993) 216.

[13] S.W. Benson, Thermochemical Kinetics, 2nd Edition, Wiley Interscience (1976).

[14] S.-J. Huang, A.B. Walters, M.A. Vannice, Appl. Cat. B, 17 (1998) 183.

[15] Z. Kalenik and E.E . Wolf, Methane Conversion by Oxidative Processes, E.E. Wolf Ed., Van Nostrand Reinhold, New York (1992).

[16] T. Le Van, C. Louis, M. Kermarec, M. Che, J.M. Tatibouët, Catal. Today, 13 (1992) 321.

[17] T. Wakatsuki, M. Yamamura, H. Okado, K. Chaki, S. Okada, K. Inabu, S. Susuki and T. Yoshinari, Natural Gas Conversion IV, Studies in Surface Science and Catalysis, Vol. 107, Elsevier (1997).

Studies in Surface Science and Catalysis, volume 147
X. Bao and Y. Xu (Editors)
©2004 Elsevier B.V. All rights reserved.

High performance methane conversion into valuable products with spark discharge at room temperature

S. Kado[a], Y. Sekine[b], K. Urasaki[b], K. Okazaki[a] and T. Nozaki[a]

[a]Department of Mechanical and Control Engineering, Tokyo Institute of Technology, 2-12-1 O-okayama, Meguro-ku, Tokyo 152-8552, Japan

[b]Department of Applied Chemistry, School of Science and Engineering, Waseda University, Okubo, Shinjuku-ku, Tokyo 169-8555, Japan

ABSTRACT

Non-catalytic methane activation with spark discharge showed high energy efficiency and high selectivity to acetylene by the collision of electrons possessing high energy enough to dissociate methane into atomic carbon. The application of spark discharge to steam reforming of methane was effective for hydrogen production at low temperature. The selectivity to desired product was improved by the combination with catalyst.

1. INTRODUCTION

From a point of view of chemical utilization of natural gas, methane conversion into valuable products such as hydrogen, synthesis gas, C_2 and higher hydrocarbons has been examined under various conditions. The activation of methane requires a high reaction temperature due to its thermal stability, but such condition is not suitable for the selective formation of ethane/ethylene and oxygenates by OCM (Oxidative Coupling of Methane) and POM (Partial Oxidation of Methane). That is one of the reasons why high yield of these products, sufficient for the commercialization, has not been obtained [1].

We have paid attention to low-temperature plasma, which has a property of high electron temperature and low gas phase temperature. Although the application of this technique to chemical reaction makes it possible to develop a process, which is not restricted to thermal equilibrium, the problems are low energy efficiency and low reaction selectivity. In this study, it was found that spark discharge and combined system with catalyst have a possibility to overcome these problems.

2. EXPERIMENTAL

The reactivity and energy efficiency for methane activation was compared among three types of cold plasmas such as dielectric barrier discharge (DBD), corona discharge and spark discharge using methane of 99.9999% purity as a feed gas. A flow type reactor composed of a quartz tube with 16.5 mm i.d. and 19.5 mm o.d. (1.5 mm thickness) was used under the conditions of atmospheric pressure and room temperature without any catalysts. In DBD, the electrodes configuration was the coaxial type with copper electrode of high voltage cathode andaluminum ground electrode. AC 75 kHz high voltage was applied between the electrodes. The quartz tube of the reactor played a role of dielectric material. The gap distance was 1.0 mm, and the supplied discharge power was 17 W. In corona and spark discharge, the electrodes configuration was needle to needle type with a gap distance of 5.0 mm. DC negative high voltage was applied between 1.0 mm diameter stainless steel electrodes. The supplied discharge power was 4.6 W in corona discharge and1.4 W in spark discharge. In spark discharge, the pulse frequency was 50 Hz, and the pulsed current reached 20-30 A in 300-500 ns.

Hydrogen production by steam reforming of methane was examined using spark discharge without catalyst at atmospheric pressure and 393 K in order to prevent the condensation of water. The discharge power was varied from 0.4 to 2.1 W by controlling pulse frequency from 35 to 275 Hz under the conditions of 2 mm gap distance and $H_2O/CH_4 = 5/5$ cm^3min^{-1}.

Catalytic spark discharge was also investigated for the selective formation of ethylene with Lindlar catalyst (PbO-$PdO/CaCO_3$) purchased from N.E. Chemcat Corporation and for CO_2 reforming of methane with $Ni_{0.48}/C_{0.52}$ catalyst prepared by Miura et al. in Kyoto University [2] under the conditions of atmospheric pressure, room temperature and 10 mm gap distance. The catalyst bed was located between the electrodes supported by stainless steel mesh.

The waveforms of applied voltage and current were measured with a digital oscilloscope. The gaseous products were analyzed by gas chromatography equipped with FID (G-950 column and Porapak N column) and TCD (SHIN-Carbon column). The amount of deposited carbon was estimated from the total amount of CO and CO_2 in thecombustion of deposited materials in the air at 923 K. The emission from the discharge region was continuously recorded by an ICCD camera with 50 ms exposure, and a 500 mm spectrometer equipped with 1200 grooves mm grating was used.

3. RESULTS AND DISCUSSION

3.1. Reactivity of low temperature plasmas

Table 1 shows the products selectivity and methane conversion when pure methane was fed. In DBD, the oligomerization to hydrocarbons higher than C_3 was the main reaction (others in Table 1 were considered to be the liquid hydrocarbons). The main component in C_2 hydrocarbons was ethane formed by the coupling of CH_3 radicals. In corona discharge, although the formation of C_2 hydrocarbons was dominant, the consecutive dehydrogenation of ethane to ethylene and acetylene was observed. In addition, a large amount of carbon was deposited on the electrodes with the increase in the conversion. On the other hand, in spark discharge, acetylene was produced with the selectivity of 85% with small amount of carbon deposition.

From a point of view of the energy efficiency, methane conversion rate (mol/J) was estimated. Methane conversion rates of DBD and corona discharge were almost the same at 0.2μmol/J, while the spark discharge reaction showed the high efficiency of 1.7μmol/J. Up to the present, the highest efficiency of 2.4μmol/J was obtained by the optimization of conditions. In this case, the specific energy requirement (SER) of acetylene was 11.7 kWh/kg-C_2H_2, which was as same as that of Huels process (12.1 kWh/kg-C_2H_2) [3].

The gas temperature determined by Boltzmann plot method of CH rotational band [4] was 420 K in spark discharge, which was lower than that of DBD (530 K) and corona discharge (960 K). This indicated that plasma chemical reaction was dominant, and high acetylene selectivity was not due to the thermal reaction in spark discharge. We have observed strong emission due to atomic carbon species (C^+ and C^{2+}) and H Balmer series from spark discharge channel, which meant that methane was highly dissociated into atomic carbon and hydrogen by the collision of electrons possessing high energy [5].

3.2. Steam reforming of methane with spark discharge

The excitation energy of the dissociation of water molecule into OH and H radicals is the same level as that of methane into CH_3 and H radicals (~9 eV). From the results mentioned above, the efficient production of OH radical from

Table 1
Results of methane activation with DBD, corona and spark discharge

Discharge	Flow rate / cm^3min^{-1}	CH_4 conv. (%)	Selectivity (Carbon based %)						Efficiency / $\mu molJ^{-1}$
			C_2H_6	C_2H_4	C_2H_2	C_3-C_5	Others	Carbon	
DBD	30	18.4	32.6	3.1	2.4	29.5	29.7	2.7	0.2
	15	24.9	33.2	2.9	2.6	34.1	23.7	3.4	0.1
Corona	30	4.2	45.8	24.6	15.1	4.8	4.8	4.9	0.2
	15	10.2	32.0	25.9	18.9	7.4	0.4	15.4	0.2
	6	24.8	17.1	18.2	19.8	6.3	4.1	34.4	0.2
Spark	30	11.8	3.9	4.5	85.5	3.4	0.7	1.9	1.7
	15	19.1	3.6	5.1	84.0	4.2	0.0	3.1	1.4

Conditions: pure methane flow, room temperature, atmospheric pressure.

water is expected by the use of spark discharge. When the spark discharge was applied to steam reforming of methane at $H_2O/CH_4=1$, the production of CO (reforming) proceeded with 65-70 % selectivity as well as the formation of C_2H_2 (coupling) with 30% selectivity. The energy efficiency ($\Delta H_{(reforming + coupling)}/W$) reached about 55%, and the SER of 1 Nm^3 H_2 was 2.2 kWh/Nm^3-H_2 [6], which is as same as that of gliding arc process [7] only for steam reforming of methane, in which much more improved SER of 0.11 kWh/Nm^3-H_2 was obtained in POM.

3.3. Catalyst-spark discharge combined system

In spark discharge channel, methane was highly dissociated into atomic carbon, so it is suitable for the selective formation of acetylene or reforming, but not preferable for the useful products such as ethylene and methanol. To show the advantage of low gas phase temperature, the selective formation of ethylene was examined using Lindlar catalyst, which is suitable for selective hydrogenation of acetylene compounds to olefin compounds. From a point of view of the material used as a catalyst, CO_2 reforming of methane was also attempted using Ni/C catalyst. The experiments were conducted in the same way as reported previously [8].

3.3.1. Selective formation of ethylene

Fig. 1 shows the effect of the height of catalyst bed at 10 mm spark gap and $CH_4/H_2=1/4$. The height of 10 mm catalyst bed means that the gap was filled with catalyst particle, and that of 0 mm was the result without catalyst. With the decrease in the height of catalyst bed, methane conversion and ethylene selectivity increased slightly, but the main product remained to be ethane. It was considered that the existence of hydrogen promoted the further hydrogenation from ethylene to ethane.

The effect of hydrogen concentration was examined at 2.5 mm height of catalyst bed as shown in Fig. 2. With the decrease in hydrogen content, ethylene

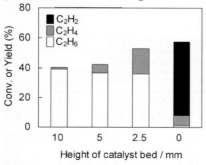

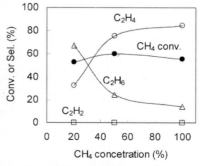

Fig. 1. Effect of height of Lindlar cat. bed at $CH_4/H_2=1/4$, 10 cm^3min^{-1} total flow rate, 10 mm gap, 100 Hz frequency.

Fig. 2. Effect of CH_4 concentration at 10 cm^3min^{-1} total flow rate.

selectivity increased dramatically, and reached 85% when pure methane was fed. In this case, ethylene yield was 47%, which was much higher than that reported in catalytic OCM [1]. The higher yield is expected by increasing the input power, but it is doubtful whether the production of ethylene with spark discharge is economic or not.

3.3.2. CO₂ reforming of methane with Ni₀.₄₈/C₀.₅₂ catalyst

We have reported that NiMgO solid solution catalyst enhanced CO selectivity in spark discharge at room temperature [8]. Without catalyst, it was necessary to dilute methane with carbon dioxide up to $CH_4/CO_2=1/4$ in order to obtain 90% CO selectivity, while 84.5% selectivity to carbon monoxide was obtained with NiMgO catalyst without dilution ($CH_4/CO_2=1$). It was also found that NiMgO catalyst without pretreatment of H_2 reduction at 1123 K did not influence the reaction selectivity.

Fig. 3 shows profiles of the mass decrease against the increase in temperature in hydrogen flow. NiMgO catalyst was hard to be reduced. On the other hand, $Ni_{0.48}/C_{0.52}$ catalyst was reduced at 410 K. Although Ni/Carbon catalyst was easily reduced, the gasification of carbon also proceeded at 760 K.

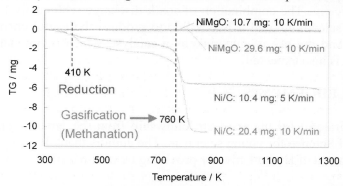

Fig. 3. Results of thermo-gravimetry of NiMgO catalyst and $Ni_{0.48}/C_{0.52}$ catalyst in H_2 flow.

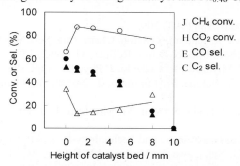

Fig. 4. Effect of height of $Ni_{0.48}/C_{0.52}$ catalyst bed on CO_2 reforming of methane at $CH_4/CO_2=1$, 20 cm³min⁻¹ total flow rate, 10 mm gap, 100Hz frequency.

So, it is impossible to use Ni/Carbon catalyst for the reforming of methane, which needs high reaction temperature. In spark discharge, the gas temperature was as low as 420K as mentioned above, it is possible to use Ni/Carbon catalyst for the reforming of methane without the gasification of carbon.

Fig. 4 shows the results of CO_2 reforming of methane with Ni/Carbon catalyst without pretreatment at $CH_4/CO_2=1$ when the height of catalyst bed was varied from 10 mm to 1 mm. When the spark gap was filled with Ni/Carbon catalyst (10 mm height), the reaction did not proceed because carbon is a good electric conductor. Under other conditions, Ni/Carbon catalyst showed the activity to reforming reaction, and higher selectivity to carbon monoxide was obtained that that of non-catalytic reaction. At 1 mm height of catalyst bed, the highest CO selectivity of 87% was obtained.

4. CONCLUSIONS

For methane activation, the spark discharge showed the high energy efficiency and selectivity to acetylene without catalyst in spite of low gas temper ature of 420 K. The production of hydrogen with spark discharge by steam reforming was also efficient. The combination of spark discharge with catalyst influenced the reaction selectivity, and it is possible to produce the useful chemicals with high yield. And the use of materials as a catalyst, which is not stable at high temperature, is also expected.

ACKNOWLEDGEMENTS

S. Kado is grateful for his research fellowship from the Japan Society for the Promotion of Science for young scientists. The authors thank Prof. K. Miura and Dr. H. Nakagawa of Kyoto Univ. for providing us with $Ni_{0.48}/C_{0.52}$ catalyst.

REFERENCES

[1] O.V. Krylov, Catal. Today, 18 (1993) 209.
[2] K. Miura, H. Nakagawa, R. Kitaura, T. Satoh, Chem. Eng. Sci., 56 (2001) 1623.
[3] H. Gladisch, Hydrocarbon Process. Petrol. Refiner, 41 (1962) 159.
[4] T. Nozaki, Y. Unno, Y. Miyazaki, K. Okazaki, J. Phys. D: Appl. Phys. 34 (2001) 2504.
[5] S. Kado, K. Urasaki, Y. Sekine, K. Fujimoto, T. Nozaki, K. Okazaki, Fuel, 82 (2003) 2291.
[6] Y. Sekine, K. Urasaki, S. Kado, S. Asai, M. Mtsukata, E. Kikuchi, ISPC-16, Taormina, 2003.
[7] A. Czernichiwski, Oil & Gas Science and Technology-Rev. IFP, 56 (2001) 181.
[8] S. Kado, K. Urasaki, H. Nakagawa, K. Miura and Y. Sekine, ACS Symposium Series 852, Utilization of Greenhouse Gases, C.J. Liu et al. Eds., (2003) 302.

Studies in Surface Science and Catalysis, volume 147
X. Bao and Y. Xu (Editors)

583

Combined single-pass conversion of methane via oxidative coupling and dehydroaromatization: a combination of La$_2$O$_3$/BaO and Mo/HZSM-5 catalysts

Y.-G. Li, H.-M. Liu, W.-J. Shen, X.-H. Bao, Y.-D. Xu[*]

State Key Laboratory of Catalysis, Dalian Institute of Chemical Physics, Chinese Academy of Sciences, 457 Zhongshan Road, Dalian 116023, China. xuyd@dicp.ac.cn

ABSTRACT

Combined single-pass conversion of CH$_4$ via oxidative coupling and dehydroaromatization over La$_2$O$_3$/BaO and 6Mo/HZSM-5 catalyst system was evaluated. The improvement in the stability is mainly attributed to the high activities of the Mo carbide species in the gasification of cokes and reverse Boudouard reactions.

1. INTRODUCTION

Recently, methane dehydroaromatisation (MDA) over Mo/HZSM-5 catalysts under non-oxidative conditions has received considerable attention [1-3]. Heavy formation of cokes is one of the most serious obstacles for its industrial application. It was reported that the addition of CO$_2$ (less than 3%) and/or proper amount of H$_2$O to the CH$_4$ feed can improve and enhance the catalyst stability [4, 5]. Along this direction, we have already reported that combined single-pass conversion of methane via oxidative coupling and dehydroaromatization is a proper way to improve the reaction stability by a combination of SrO-La$_2$O$_3$/CaO and Mo/HZSM-5 catalysts [6]. In this work, we put the La$_2$O$_3$/BaO catalysts in the top layer and the Mo/HZSM-5 catalysts in the second layer in a fixed bed reactor, and the CH$_4$/O$_2$/N$_2$ reactants (N$_2$ was used as the internal standard for products analysis) was introduced in downstream mode, so that the OCM and MDA reactions can happen in turn to give a combined single-pass conversion. The improvement mechanism on the catalyst system is discussed based on the high activities of the Mo carbide species in the gasification of cokes and the reverse Boudouard reactions.

2. EXPERIMENTAL

The 6wt%Mo/HZSM-5(denoted as 6Mo/HZ) catalyst was prepared by the conventional impregnation method as previously described [7]. The 14wt% La_2O_3/BaO (denoted as LB) catalysts were prepared by the coprecipitation method. An $(NH_4)_2CO_3$ solution was added dropwise into a mixed solution of La and Ba nitrates in a desired ratio, and then filtrated. It was then dried at 393 K for 4 h and calcined at 1073 K for 4 h. The samples were pressed, crushed and sorted into sizes of 20-40 meshes and stored in a desiccator before use.

The reaction tests were carried out in a quartz tubular fixed-bed reactor at atmospheric pressure and 1003 K as reported in ref. [6]. The charge of the LB catalyst was 0.15 g and that of the 6Mo/HZ catalyst was 1.0 g. The tail gas was analyzed by a Varian 3800 online gas chromatograph and calculated as described elsewhere [8]. The calculation of H_2 concentration in tail gas and oxygen balance was based on GC data.

3. RESULTS AND DISCUSSION

3.1. Reaction performance on LB-6Mo/HZSM-5

Table 1
The combined single-pass conversion of methane via LB and 6Mo/HZ at 1003K.

Catalyst	CH_4 Conv., %	Product Yield, %					H_2 concen in tail gas, Vol%
		CO/CO_2	C_2	BTX	C_{10}	Coke	
6Mo/HZ[a,d]	17.7	0.1/0	0.4	10.7	2.1	4.4	17.1
6Mo/HZ[b,d]	0.9	0/0	0.2	0.1	0.0	0.6	1.1
LB-quartz[a,e]	4.2	0/0.6	3.6	-	-	-	-
LB-6Mo/HZ a,e	20.1	2.5/0	0.3	8.9	2.0	6.3	18.6
LB-6Mo/HZ b,e	15.9	2.8/0	0.5	8.3	0.6	3.7	14.5
LB-6Mo/HZ c,e	14.9	2.9/0	0.6	7.7	0.5	3.2	13.6
6Mo/HZ[a,f]	21.0	2.8/0	0.4	10.2	2.1	5.5	17.9
6Mo/HZ[c,f]	13.6	2.8/0	0.7	6.9	0.4	2.8	10.8
6Mo/HZ[a,g]	19.5	1.6/0	0.4	10.8	1.9	4.8	17.8
6Mo/HZ[b,g]	10.5	1.3/0	0.9	5.8	0.3	2.2	9.2

Data were taken after running the reaction for (a) 60 min; (b) 960 min; (c)1380min.
The feed composition was (d) $CH_4:N_2$ =9:1 and (e) $CH_4:O_2:N_2$=9:0.11:1.
(f) $CH_4:CO_2:(N2+He)$=9:0.12:1.1   (g) $CH_4:CO_2:(N2+He)$=9:0.05:1.1.

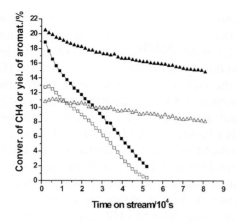

Fig. 1. The changes of the CH_4 conversion (solid symbols) and aromatics yields (empty symbols) over different catalyst system. On LB-6Mo/HZ ($\blacktriangle$, $\triangle$); On 6Mo/HZ ($\blacksquare$, $\square$).

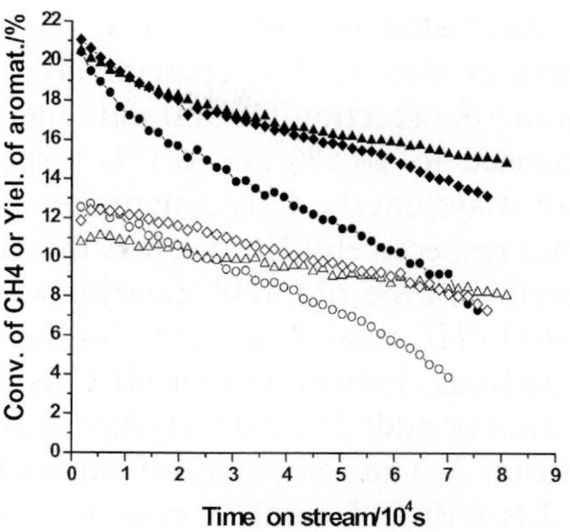

Fig. 2. The effect of CO_2 on the changes of CH_4 conversion (solid symbols) and aromatics yields (empty symbols) on LB-6Mo/HZ with 1.2% O_2 ($\blacktriangle$, $\triangle$); on 6Mo/HZ with 0.6% ($\bullet$, $\bigcirc$) and 1.3% CO_2($\blacklozenge$, $\diamondsuit$).

The results are listed in Table 1 and shown in Fig. 1. After running the reaction for 60 min, the CH_4 conversion and the yield of aromatics on the 6Mo/HZ were 17.7% and 12.8% and on the LB-6Mo/HZ catalysts were 20.1% and 10.9% respectively. However, the catalytic performances change with time on stream on the 6Mo/HZ and the LB-6Mo/HZ catalyst system quite differently. The deactivation of the 6Mo/HZ catalyst was very serious. The CH_4 conversion dropped from 17.7% to 0.9% and the yield of aromatics from 12.8% to 0.1%, respectively, after running the reaction for 960 min. On the other hand, for the LB-6Mo/HZ catalyst system the deactivation was suppressed significantly, and after 960 min on stream the yield of aromatics was 8.9% at a methane conversion of 15.9% as shown in Fig. 1.

3.2. Effect of CO_2 added in the feed and formed from OCM

It was reported that the addition of CO_2 (less than 3%) to the CH_4 feed could enhance the catalyst stability [4]. Both CO_2 and H_2O are the by-products of the OCM reaction. The concentration of CO_2 is about 0.6% in the OCM tail gas produced on the LB catalyst during the combined reaction process, which was estimated on the basis of oxygen balance.

In order to investigate the promotional effect of CO_2 formed from the OCM on the LB catalyst, A CH_4-CO_2 co-feed gas with the CO_2 concentration of 0.6%, was introduced on the 6Mo/HZ and the corresponding catalytic performances are also listed in Table 1. The CH_4 conversion was 19.5% and the yield of aromatics was 12.7%, respectively, at the initial stage of the reaction. After running the reaction for 960 min, the CH_4 conversion and the yield of aromatics decreased to 10.5% and 6.1%, respectively. Meanwhile, for the LB-6Mo/HZ catalyst system the CH_4 conversion and the yields of aromatics were 14.9% and 8.2%, respectively. Even when the content of CO_2 was doubled in the co-fed experiment, the 6Mo/HZ catalyst was not as stable as it was in the combined LB-6Mo/HZ catalyst system. On the other hand, only CO could be detected in the tail gas, indicating that all CO_2 produced on the LB catalyst in the OCM reaction or added in the CH_4 feed was transformed into CO as shown in Table 1 and Fig. 3. The selectivity of carbonaceous deposits was greatly suppressed on the LB-6Mo/HZ catalyst system as shown in Fig. 4. After 6h on stream the selectivity of cokes increased remarkably on the 6Mo/HZ catalyst while it remained unchanged on the LB-6Mo/HZ catalyst system. This again illustrates that the addition of CO_2 in the feed is an effective way to remove the coke via the reverse Boudouard reaction, i.e. $CO_2+C=2CO$. However, it appears that the addition of CO_2 in the feed is not the sole factor to give better stability of the LB-6Mo/HZ catalyst system.

3.3. The deactivation rate constants, k_d, on the different catalysts

The deactivation rate constants, k_d, of various catalysts, which are calculated from the deactivation curve by plotting the natural logarithm of the BTX formation rate versus the reaction time on stream for the LB-6Mo/HZ and 6Mo/HZ according to ref. [9], are shown in Fig. 5 and the data are listed in Table 2.

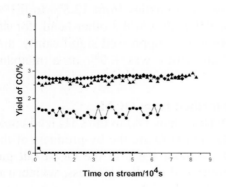

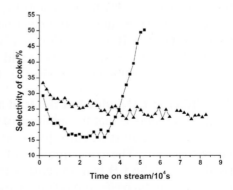

Fig. 3. The variation of the CO yields with time on stream on 6Mo/HZ (■); on LB-6Mo/HZ with 1.2% O_2 (▲); on 6Mo/HZ with 0.6% (●) and 1.3% CO_2 (◆).

Fig. 4. The variation of the selectivity to coke with time on stream over 6Mo/HZ (■), and LB-6Mo/HZ(▲).

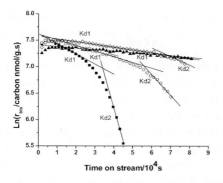

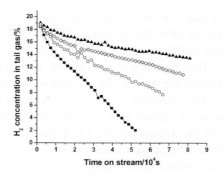

Fig. 5. The deactivation curves on LB-6Mo/HZ with O_2 1.2%(▲); on 6Mo/HZ with 0.6%CO_2(○) and 1.3%CO_2(◇); on 6Mo/HZ (■).

Fig. 6. The variation of H_2 concentration in tail gas with time on LB-6Mo/HZ with O_2 1.2%(▲); on 6Mo/HZ with 0.6%CO_2(○) and 1.3%CO_2(◇); on 6Mo/HZ (■).

Table 2

The deactivation rate constants, k_d, on various catalysts.

Catalyst	6Mo/HZ	6Mo/HZ[a]	6Mo/HZ[b]	LB-6Mo/HZ[c]
k_{d1}*	1.84	1.07	0.53	0.28
k_{d2}*	12.8	3.98	1.67	-

*The dimension of the deactivation rate constants, k_d, is $10^{-5} s^{-1}$;
a, CO_2:CH_4=0.6%; b, CO_2:CH_4=1.3%;c,O_2:CH_4=1.2%.

The addition of CO_2 into the CH_4 feed with the CO_2 concentration of 0.6% on the 6Mo/HZ catalyst showed higher activity and yield of aromatics than the LB-6Mo/HZ catalyst system in the time of 500 min on stream, but the value of its k_{d1} was 1.07, which is larger than that of the LB-6Mo/HZ catalyst system of 0.28. This indicates that another factor also plays a role in stabilizing the catalytic behaviors of the LB-6Mo/HZ catalyst system. Even when the CO_2 concentration was increased to 1.3%, the value of k_{d1} on the 6Mo/HZ was 0.53, which is still much higher than that on the LB-6Mo/HZSM-5 catalyst. The results implied that the combined reaction of OCM-MDA on the LB-Mo/HZSM-5 catalyst system is a more efficient way to stabilize the MDA reaction than the addition of CO_2 in CH_4 feed when the available oxygen amount is the same.

H_2O is another component containing oxygen atom in the tail gas after the OCM reaction on the LB catalyst. Recently, the promotional effect of water for the MDA on Mo/HZSM-5 catalyst was reported [5]. The authors claimed that addition of 1.7% H_2O to the CH_4 feed displays the best stability at 998K. A reasonable explanation on these results is that the gasification of the coke, i.e. $H_2O+C=CO+H_2$ would also play a role in eliminating carbonaceous deposits. In our case, it would produce about 52% of the amount of CO in the tail gas formed after the combined OCM-MDA reaction if all of the water reacted with coke in this way.

The gasification of cokes with H_2O probably plays an important role in eliminating carbonaceous deposits. This was confirmed by further analyzing the H_2 concentration in the tail gas formed from the combined single-pass reaction. Figure 4 shows that the change of the hydrogen concentration in tail gas with time on stream on the LB-6Mo/HZ catalyst system, which is higher than those when CO_2 was added in the CH_4 feed on the 6Mo/HZ catalyst. The main contribution to hydrogen formation, of course, is from the MDA reaction $(6CH_4=C_6H_6+9H_2)$. However, it decreased sharply due to serious deactivation. The addition of CO_2 into the CH_4 feed stabilized the 6Mo/HZ catalyst via the reverse Boudouard reactions on the Mo carbide species. The gasification of coke by water, which was formed from the OCM reaction on the LB catalyst, also made a contribution to the coke elimination and H_2 production.

This work shows that the combined single-pass conversion of methane via oxidative coupling and dehydroaromatization over the LB-6Mo/HZSM-5 catalyst system is an effective way for enhancing the activity and stability of the 6Mo/HZSM-5 catalysts. Such an enhancement is mainly due to the proper formation of CO_2 and H_2O in the OCM reaction over LB catalysts, which react with coke formed in the MDA reaction on 6Mo/HZSM-5 catalysts via the reverse Boudouard reaction and the gasification reaction. At the same time, more hydrogen and CO are produced in the tail gas. However, the reaction condition and the catalyst system must be tuned very carefully.

ACKNOWLEDGMENTS

Financial supports from the Ministry of Science and Technology of China, the Natural Science Foundation of China, and the BP-China Joint Research Center are gratefully acknowledged.

REFERENCES

[1] Y. Xu, and L. Lin, Appl. Catal. A, 188 (1999) 53.
[2] Y. Shu and M. Ichikawa, Catal. Today, 1 (2001) 86.
[3] Y. Xu, X. Bao, L. Lin, J. Catal., 216 (2003) 386.
[4] S. Liu, L. Wang, Q. Dong, R. Onishi and M. Ichikawa, Chem. Commun.,11 (1998) 1217.
[5] S. Liu, R. Onishi and M. Ichikawa, J. Catal., 220 (2003) 57.
[6] Y. Li, L. Su, H. Wang, Y. Xu, X. Bao, Catal., Lett. 89 (2003) 275.
[7] D. Ma, Y. Shu, X. Bao, Y. Xu, J. Catal., 189 (2000) 314.
[8] Y. Shu, D. Ma, X.Bao, Y. Xu, Catal. Lett., 66 (2000) 161.
[9] O. Levenspiel, J. Catal., 25 (1972) 265.

Studies in Surface Science and Catalysis, volume 147
X. Bao and Y. Xu (Editors)

Formation of acetaldehyde and acetonitrile by CH₄-CO-NO reaction over silica supported Rh and Ru catalysts

Nobutaka Maeda, Toshihiro Miyao and Shuichi Naito[*]

Department of Applied Chemistry, Kanagawa University, 3-27-1, Rokkakubashi, Kanagawa-ku, Yokohama, 221-8686, Japan

ABSTRACT

The exposure of carbon-deposited Rh and Ru surfaces prepared by the decomposition of methane to a flow of CO was found to facilitate C-C bond formation of surface carbonaceous species toward higher hydrocarbons. It was also revealed that the adsorption of NO on the carbon-deposited surface prior to the admission of CO caused the selective formation of acetaldehyde and acetonitrile in the subsequent CO-TPR step. Furthermore, the addition of NO during CH₄-CO reaction stimulated the insertion of CO into surface methyl group to form acetaldehyde.

1. INTRODUCTION

The conversion of methane into valuable chemicals such as higher hydrocarbons and oxygenates has become much more important in recent chemical industries where an alternative resource of petroleum is strongly desired. A great number of studies have been reported on the methane conversion processes such as syntheses of syngas in steam or dry reforming, the partial oxidation to form methanol or formaldehyde, and the non-oxidative dehydroaromatization to form benzene. In 1990's Amariglio et al. and van Santen et al. reported that the hydrogenation of Pt or Co surface after the decomposition of methane brought about various higher hydrocarbons [1, 2]. In the case of Pt/SiO₂, the admission of CO instead of H₂ in the second step was found to be effective for the oligomerization of methane [3]. On the contrary, very few successful studies have been reported on the carbonylation of methane to form oxygenates, which is still one of the most challenging targets in modern catalytic chemistry [4-6].

We have already reported the selective formation of benzene in CH₄-CO reaction at lower temperatures and a marked effect of CO which accelerates the formation of C-C bond of surface carbonaceous species over Rh/SiO_2 catalysts [7-9]. Recently, we have turned our efforts to elucidate the reactivity of surface carbonaceous species provided from methane and to establish a new reaction pathway of methane over group VIII metal catalysts. We have found the novel interaction of surface carbonaceous species with adsorbed CO and NO, and revealed the possibility of heterogeneous catalytic carbonylation of methane with CO in the presence of NO.

2. EXPERIMENTAL

5wt% silica supported Rh and Ru catalysts were prepared by impregnating silica into an aqueous solution of corresponding metal chlorides. After drying the resulting powder was placed in a quartz reactor connected with a closed gas circulation system and reduced by 26.7 kPa of hydrogen at 673K (Rh/SiO_2) or 723K (Ru/SiO_2) for 6 hrs. The measurement of hydrogen adsorption at room temperature was employed to estimate the metal dispersions of the catalysts, which resulted in 37.1 and 29.5% dispersion of Rh and Ru atoms, respectively. The TPD and TPR experiments were carried out as follows: the reduced catalyst was exposed to 1.3kPa of CH_4 at 523K for 1hr in the first step and then the reactor was cooled down to the room temperature. 1.3kPa of NO was introduced at room temperature in the case of NO-present experiments. Then, NO in the gas-phase was substituted for He (He-TPD) or CO (CO-TPR) and the catalyst temperature was raised slowly at a rate of 1.7K/min. Liquid N_2 cold trap was employed during these procedures to gather the primary products and to shift the reaction equilibrium. Products distribution was analyzed by TCD and FID-GCs, MASS and GC-MS. The adsorbed species present on the catalyst surface during these processes were observed by means of FT-IR spectroscopy.

3. RESULTS AND DISCUSSION

The effect of adsorbed CO on the reactivity of surface carbonaceous species formed by the decomposition of methane was studied over 5wt%Rh/SiO_2 catalyst. Figs.1 (a) and (b) display He-TPD and CO-TPR profiles over the methane decomposed surfaces. Each plot in these figures represents the total molar amounts of products gathered between the indicated temperature and the previous one. In the He-TPD experiment only methane was desorbed, implying little formation of C-C bond during the first decomposition process of methane. On the contrary, in the case of CO-TPR (Fig.1 (b)) various hydrocarbons were released with the decrease of methane, exhibiting a sharp contrast with the TPD experiment where only methane was desorbed. These results clearly indicate the promotion effect of adsorbed CO on the formation of C-C bond from lower fragments such as CH_X (a) or C_2H_X (a) toward higher hydrocarbons.

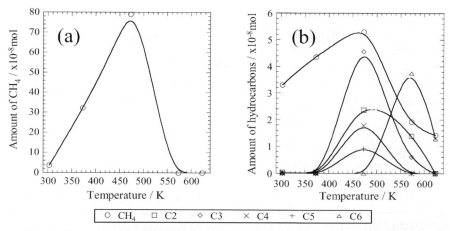

Fig.1. Temperature dependence of the amount of released hydrocarbons during He-TPD;(a) and CO-TPR;(b) processes after CH$_4$ decomposition (523K, 1hr.) over 5wt%Rh/SiO$_2$

The dependence of CO partial pressure upon the product distribution during CO-TPR step was investigated. Only methane was released under lower CO partial pressure. Drastic changes of product distributions were observed at around 3 Torr of CO, and further increase of CO-pressure brought about an excellent selectivity to C$_{2+}$ hydrocarbon products. The selectivity toward benzene reached 33.6% in 10 Torr of CO. It is reasonable to consider that saturated adsorption of CO leads to extinguish the vacant Rh sites and to stabilize carbonaceous species in keeping hydrogen-rich conformation. These effects of adsorbed CO ultimately resulted in the oligomerization of surface carbonaceous fragments into higher hydrocarbons.

When CH$_4$-decomposed Rh surface was exposed to 1.3kPa of NO gas at room temperature prior to the subjection to a flow of CO, subsequent CO-TPR experiment triggered a prominent change of products distribution compared with the case of NO absent experiment (see Fig.1 (b)) where only hydrocarbons were desorbed. As seen from Fig. 2, higher selectivity to acetaldehyde and acetonitrile was realized at 473K, although a simultaneous reduction of nitric oxide by CO took place to form considerable amounts of N$_2$ and CO$_2$. Owing to a fact that no acetaldehyde and acetonitrile were released in the absence of adsorbed NO, it is reasonable to suppose that the admission of NO brings about a specific ability for surface Rh atoms and adsorbed species. Fig.3 displays the infrared spectra of adsorbed species during this CO-TPR process. At room temperature two typical peaks were observed at 1667 and 2070cm^{-1},which can be assigned to Rh-NO$^-$ and Rh^0-CO adsorbed species, respectively. It was noticed that adsorbed NO disappeared with raising the temperature, while new peaks assignable to

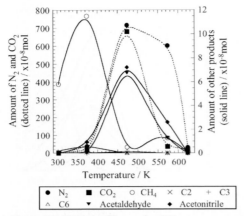

Fig.2. CO-TPR profile on the CH_4-decomposed and subsequent NO-adsorbed surface of 5wt%Rh/SiO$_2$

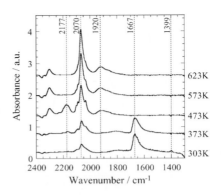

Fig.3. FT-IR spectra during CO-TPR on the CH_4-decomposed and subsequent NO-adsorbed surface

gm-dicarbonyl Rh^+ emerged gradually up to 473K, which situated in the shoulder position(seen at approximate 2030 and 2090cm^{-1}) of Rh^0-CO. However, at higher temperatures these shoulder peaks disappeared again due to the reduction and aggregation of Rh^+ clusters. These phenomena can be reasonably explained in terms of the formation of cationic rhodium species with oxygen atoms generated by the dissociation of NO at appropriate temperatures. Since reduced Rh^0 site is active only for the formation of hydrocarbons (see Fig.1 (b)), Rh^+ site would be essential for the formation of acetaldehyde. At the same time a characteristic band at 2177 cm^{-1} emerged at 473K, which can be assigned to surface cyano groups from which acetonitrile may be formed. Isotopic tracer experiments revealed that acetaldehyde was formed through the insertion of CO into surface methyl groups formed by the decomposition of methane. Accordingly, Rh^+ site is active for CO insertion, as reported by some authors such as S. S. C. Chuang et al. [10]. And the associative desorption of methyl group with cyano group was proposed as a mechanism of acetonitrile formation because all carbon atoms of acetonitrile were supplied from only methane.

The interaction of surface carbonaceous species with CO and NO over Ru/SiO$_2$ catalyst was also investigated by the similar experiments with Rh catalyst described above. The results of CO-TPR on the Ru surface in the absence and the presence of adsorbed NO are summarized in Table 1. Although various kinds of hydrocarbons were formed in the absence of NO similar to Rh catalyst (see Fig.1 (b)), a small amount of acetaldehyde was formed without adsorbing NO. The results of XPS and IR studies revealed the existence of a small amount of residual chlorine even after the H_2 reduction at 723K. These cationic ruthenium species may enhance the CO-insertion process, which

Table 1. Total amounts of products formed during CO-TPR(up to 623K) in the absence as well as in the presence of pre-adsorbed NO after the decomposition of methane for 1hr. at 523K over 5wt%Ru/SiO$_2$ (x10^{-8}mol)

	CH$_4$	C2	C3	C4	C5	C6	CH$_3$CHO	CH$_3$CN
NO-absent surface	33.99	0.99	0.41	0.55	0.13	0.06	0.24	0
NO-present surface	10.20	0.47	0	0	0	0.07	2.41	16.64

provides a small amount of acetaldehyde. Detailed study on the effect of residual chlorine with the variation of reduction temperatures and ruthenium precursors are underway and will be reported in the near future. On the other hand in CO-TPR experiments, large amounts of acetaldehyde and acetonitrile were selectively released from NO-preadsorbed surface of Ru/SiO$_2$ catalyst same as the case of Rh catalyst. A ten-fold increase in the molar amounts of acetaldehyde compared to NO-absent surface was achieved by the function of oxygen atoms generated from NO. The ratio of acetonitrile to acetaldehyde over Ru catalyst was six times as high as that over Rh catalyst, implying the facile formation of C-N bond on the Ru surface.

The possibility of the continuous catalytic reaction of methane with CO and NO was studied over Rh/SiO$_2$ catalyst at various temperatures to yield acetaldehyde and acetonitrile. The results are summarized in Table 2, which clearly demonstrates that the reaction at 403K is more active for CO-insertion process toward acetaldehyde than those at other temperatures. A negligible amount of acetonitrile was formed at all temperatures, while N$_2$ and CO$_2$ were the main products. These results rationally demonstrate that higher temperatures give rise to undesirable NO-CO reaction as a predominant reaction pathway in

Table 2. Rh-catalyzed CH$_4$-CO-NO reaction after 2hr.; reaction condition: P$_{CH4}$=P$_{CO}$=1.3kPa, P$_{NO}$=0.01kPa

Reaction temp. / K	Amount of products / x10^{-8}mol			
	CH$_3$CHO	CH$_3$CN	N$_2$	CO$_2$
403	6.9	0	13.9	25.5
473	3.7	0.1	444.8	912.5
523	0	0	969.8	2013.3

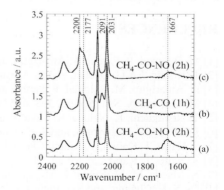

Fig.4. FT-IR spectra in the reaction of CH$_4$ with CO and NO at 403K (see text for (a) ~ (c).)

preference to CO-insertion process and associative desorption of CN(a) with CH$_3$(a). Fig. 4 shows the result of IR study during CH$_4$-CO-NO reaction. In the reaction of methane with CO and NO at 403K (see the spectrum (a)), geminate type-adsorbed CO species were observed at 2031 and 2091cm^{-1}, indicating the existence of Rh$^+$ species which may enhance the insertion of CO into methyl group. The removal of NO from gas-phase, however, caused the appearance of linear type-CO at 2070cm^{-1} indicating the formation of Rh^0 species on the surface (in spectrum (b)). It is worth noticing that under this condition the rate of acetaldehyde formation was depressed considerably with the increase of ethane formation rate. Reintroduction of NO into the reaction mixture caused the transformation of Rh^0 sites into Rh$^+$ ones and the insertion pathway toward acetaldehyde was opened again (spectrum(c)). This is the first report demonstrating the mechanistic clarification of successful heterogeneous carbonylation of methane to acetaldehyde. The most important discovery in this work is that heterogeneous insertion of CO into CH$_3$(a) takes place over cationic metal surface and that the stabilization of cationic sites over the metal particle will realize the successful synthesis of acetaldehyde from methane and CO in the absence of oxidants such as nitric oxide and oxygen.

4. CONCLUSION

(1) Acetaldehyde was formed for the first time through the insertion of CO into surface methyl group formed by the dissociation of methane in CH$_4$-CO-NO reaction over Rh/SiO$_2$ and Ru/SiO$_2$ at 403K.
(2) Cationic Rh$^+$ species formed by NO is responsible for the insertion process.
(3) Acetonitrile was also formed by the associative desorption of CH$_3$ (a) and CN (a) species formed by the reaction of CO with NO.

REFERENCES

[1] M. Belgued, P. Pareja, A. Amariglio, and H. Amariglio, Nature, 352, 789 (1991).
[2] T. Koerts, J.A.G. Deelen, and R. A. van Santen, J. Catal., 138, 101 (1992).
[3] H. Amariglio, M. Belgued, P. Pareja, and A. Amariglio, Catal. Lett., 31, 19 (1995).
[4] M. Lin, and A. Sen, Nature, 368, 613 (1994).
[5] T. Nishiguchi, K. Nakata, K. Takaki, and Y. Fujiwara, Chem. Lett., 1141 (1992).
[6] J.C. Choi, Y. Kobayashi, and T. Sakakura, J. Org. Chem., 66, 5262 (2001).
[7] S. Naito, T. Karaki, T. Iritani, Chem. Lett., 887 (1997).
[8] S. Naito, T. Karaki, T. Iritani, M. Kumano, Stud. Surf. Sci. Catal., 119, 265 (1998).
[9] N. Maeda, M. Miyauchi, T. Miyao, and S. Naito, Catal. Lett., 84, 245 (2002).
[10] S. S. C. Chuang, and S. I. Pien, J. Catal.,135, 618 (1992).

Studies in Surface Science and Catalysis, volume 147
X. Bao and Y. Xu (Editors)

Methane dehydroaromatization over alkali-treated MCM-22 supported Mo catalysts: effects of porosity

L.-L. Su, Y.-G. Li, W.-J. Shen, Y.-D. Xu[*], X.-H. Bao[*]

State Key Laboratory of Catalysis, Dalian Institute of Chemical Physics, Chinese Academy o f Sciences, 457 Zhongshan Road, Dalian 116023, China
xuyd@dicp.ac.cn; xhbao@dicp.ac.cn

ABSTRACT

Methane dehydroaromatization (MDA) was carried out on 6Mo/MCM-22 and 6Mo/alkali-treated MCM-22 catalysts. The results of catalytic evaluation, MAS NMR and pore distribution experiments demonstrated that the alkali treatment of MCM-22 could create some mesopores on the parent MCM-22 zeolite. Appropriate amount of mesopores coexisted with its inherent micropores would benefit the diffusion of the reactants and products, thus increase the stability of the MDA reaction on the 6Mo/alkali-treated MCM-22catalyst.

1. INTRODUCTION

Since Wang et al. reported that Mo/HZSM-5 catalyst is active and selective for methane dehydroaromatization (MDA) in 1993, many studies on this topic have been carried out [1-4]. It is well recognized that the Mo/HZSM-5 is a bi-functional catalyst; The MoOx species are reduced by CH_4 in the induction period and probably are in the form of Mo_2C and/or MoO_xC_y, which are responsible for methane dehydrogenation and for the formation of C_2 species, while the Brönsted acid sites in the zeolite channels are associated with the oligomerization of ethylene and the formation of benzene and other aromatics. Due to the obvious difference in the molecule sizes of the reactant (CH_4) and products (mainly C_6H_6 and $C_{10}H_8$) in MDA, pore size distribution and channel structure of the zeolites may play an important role from the point of view of shape-selectivity and mass transport in the zeolite channels. It was reported that a zeolite with a pore diameter of ca 0.6nm, which is consistent with the dynamic diameter of benzene molecules, is a good catalyst component for MDA reaction [5]. Recently, it has been reported that appropriate alkali treatment of the MFI zeolite is an effective method to change its pore size distribution and to create a kind of secondary mesopores [6-8] and the approach has been approved to be effective for the MDA reaction on Mo/HZSM-5 catalysts [9]. In addition, MCM-22 has been reported to be a good catalyst component for MDA reaction [10]. This kind of zeolite has two independent pore systems: two-dimentional 10MR intralayer channels, and 12MR interlayer supercages with a depth of 18.2 Å, both accessible through 10 MR apertures. It seems to be interesting to study if the alkali treatment of MCM-22 zeolite can also create some mesopores on the parent zeolite and thus is also beneficial for the MDA reaction on Mo/MCM-22 catalysts.

2. EXPERIMENTAL

MCM-22 was synthesized by referring to ref. [11]. The alkali treatment of MCM-22 zeolite was performed as follows: MCM-22 was added to an aqueous NaOH solution and stirred for appropriate time under a given temperature condition. The conditions of alkali treatment of MCM-22(P), i.e., the concentration of sodium hydroxide solution/M, temperature/K, time/h were set at 0.1, 323, 1 denoted as AT1 and 0.2, 323, 1 denoted as AT2, respectively. After filtering and drying, the zeolite was ion-exchanged with NH_4NO_3 solution to obtain the NH_4^+ form zeolite. Mo-containing catalysts (Mo wt% = 6%) were prepared by impregnation method described in ref. [10]. The reaction was carried out in a quartz tubular fixed-bed reactor at atmospheric pressure and 973 K, 1500ml/g.h as described in previous work [10]. N_2 adsorption and desorption experiments were carried out using an AUTOSPRB-1 Micromeritics equipment. Multinuclear MAS NMR experiments were carried out at 9.4 T on a Bruker DRX-400 spectrometer using 4 mm ZrO_2 rotors.

3. RESULTS AND DISCUSSION

3.1 The effect of alkali treatment on framework structure and acidity of MCM-22

Figure 1 shows the ^1H MAS NMR spectra of MCM-22 before and after alkali treatment and the corresponding deconvolution results of the ^1H MAS NMR spectra are listed in Table 1. Four peaks at ca. δ = 1.7, 2.2, 3.7 and 5.8 ppm, can be clearly resolved, which are almost the same as reported in our previous work [12]. Accordingly, the bands at ca. δ = 1.7 and 2.2 ppm are attributed to silanol group and Al-OH species, respectively. The band at ca. δ = 3.7 ppm is associated with bridging OH groups (Brönsted acid sites). The resonance at ca. δ = 5.8 ppm is commonly assigned to another kind of Brönsted acid sites, i.e. the restricted Brönsted acid sites. After mild alkali treament of MCM-22 the amount of Brönsted acid sites at 3.7 ppm was not affected noticeably, while for the MCM-22(AT2) sample the amount of Brönsted acid sites decreases as we can see from Table 1. On the other hand, the peak intensity at 1.7 ppm increases with the severity of alkali treatment process, indicating that there are more disfigurement sites on the severe by alkali treated MCM-22. At the same time, the Si/Al ratio determined by XRF experiments decreases after alkali treatment as shown in Table 1.

^{27}Al MAS NMR spectra of the MCM-22 before and after the alkali treatment are shown in Fig. 2. The peak at 55ppm is attributed to the tetrahedral framework aluminum and the peak at 0 ppm is ascribed to the extra framework aluminum. Before and after the alkali treatment the ratio between the framework and extra framework Al does not change obviously, demonstrating that the Al species of the MCM-22 has not been seriously influenced by the alkali treatment. However, the Si/Al ratio of the zeolite determined by XRF experiments decreased after the alkali treatment. Since the Al species is not expelled from the zeolite as shown in Fig. 2 and there is not any Al species being introduced during the alkali treatment process, the decrease in the Si/Al ratio after alkali

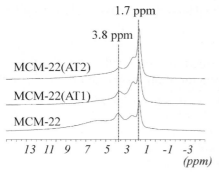

Fig. 1. ^{1}H MAS NMR spectra of MCM-22 before and after alkali treatment

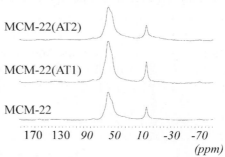

Fig. 2. ^{27}Al MAS NMR spectra of MCM-22 before and after alkali treatment

Table 1
The deconvolution result of ^{1}H MAS NMR spectra for MCM-22 before and after alkali treatment

	Number of hydroxyls/u.c.		SiO_2/Al_2O_3[a] of zeolite before Mo loading	SiO_2/Al_2O_3[a] of zeolite after Mo loading	Mo%[a]
	Si-OH-Al	Si-OH			
MCM-22	3.5	2.4	19.7	21.2	6.2
MCM-22(AT1)	3.2	3.4	16.8	17.2	7.0
MCM-22(AT2)	2.2	4.2	15.1	15.4	6.0

a: determined by XRF experiment

treatment was resulted from the dissolution of siliceous species on the zeolite. It seems that the alkali solution with appropriate concentration would preferentially attack the siliceous species and lead to its extraction from the zeolite. The inertness of the alkali treatment towards Si-O-Al bond in the framework preserves the specific Brönsted acid sites.

3.2 Effect of the alkali treatment on the pore distribution of the MCM-22

The change of pore distribution is shown in Fig. 3 and the physical properties of the MCM-22 zeolte before and after the alkali treatment are listed in Table 2. It is clear that with the alkali treatment condition becoming more and more severe, the more obvious change in pore distribution could be observed and the mesopores part possessed more proportion all over the MCM-22 zeolite. The dissolution of siliceous species, caused by the alkali treatments, probably starts preferentially on the positions where the crystillization is poor, such as at the boundaries or defects of the MCM-22 zeolite crystallites. And this kind of dissolution of siliceous species may lead to the formation of some mesopores in the zeolite. At the same time, the external surface areas increase after the alkali

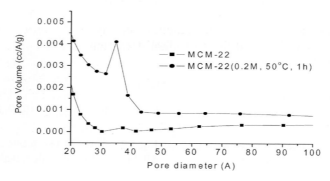

Fig. 3. Pore distribution in the range of mesopore scope for MCM-22 and MCM-22(AT2)

Table 2
Physical properties of MCM-22 before and after alkali treatment

	S_{BET} $m^2 \cdot g^{-1}$	S_{micro} $m^2 \cdot g^{-1}$	$S_{ext.}$ $m^2 \cdot g^{-1}$	S_{meso} $m^2 \cdot g^{-1}$	V_{micro} $cm^3 \cdot g^{-1}$	V_{meso} $cm^3 \cdot g^{-1}$	Pore size (A) micro	Meso
MCM-22(P)	515	422	93	76	0.21	0.25	5.3	--
MCM-22(AT1)	505	361	144	145	0.18	0.50	5.4	52
MCM-22(AT2)	485	296	188	192	0.14	0.47	5.5	49

treatment. This may be attributed to the fact that the siliceous species residing along the boundaries of the crystallites or in the framework has been dissolved during the alkali treatment, and such a kind of dissolution will induce the formation of some small crystallites and the secondary mesopores form between crystallites boundaries.

3.3 The catalytic performance of MDA on Mo/MCM-22 and Mo/MCM-22(AT) catalysts

Figure 4 and Figure 5 show the reaction results on the alkali treated MCM-22 supported Mo catalysts. The depletion rate of methane and the formation rate of aromatics are higher on Mo/ MCM-22(AT1) catalyst than those on the untreated Mo/MCM-22 catalyst. Particularly, the naphthalene yield is much higher on Mo/ MCM-22 (AT1). Since the result of ^{1}H MAS NMR demonstrates that the Brönsted acid sites have not been seriously affected by the alkali treatment Mo/MCM-22(AT1) catalyst, the enhancement of the reaction performance after the alkali treatment is mainly resulted from the change of the pore structure. On the Mo/MCM-22(AT1) catalyst, being coexisted the micropores and mesopores, not only the aromatics product can be selectively formed, but also they can diffuse out of the zeolite channels timely due to the existence of the mesopores, thus the reaction activity and stability are enhanced.

However, when the conditions of alkali treatment are too severe such as in the case of the Mo/MCM-22(AT2) catalyst, its catalytic performance decreases sharply. Firstly, the alkali treatment conditions used on the Mo/MCM-22(AT2) catalyst are too severe, so that the amount of Brönsted acid sites decrease obviously as shown in Figure 1 and Table 1. This indicates that the Brönsted acid sites have been seriously changed for the Mo/MCM-22(AT2) catalyst. Secondly, because of the severe conditions of the alkali treatment there are too

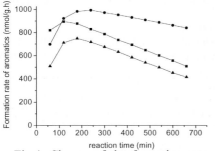

Fig.4. Change of the formation rate of aromatics with time on stream on 6Mo/MCM-22(■), 6Mo/MCM-22(AT1)(●), and 6Mo/MCM-22(AT2)(▲) catalysts.

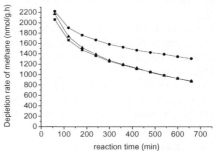

Fig.5. Change of the depletion rate of CH₄ with time on stream on 6Mo/MCM-22(■), 6Mo/MCM-22(AT1)(●), and 6Mo/MCM-22(AT2)(▲) catalysts.

many mesopores created and too many micropores reduced on the Mo/MCM-22(AT2) catalyst, The ratio between micropores and mesopores on it is not suitable for the MDA reaction. Appropriate amount of mesopores coexisted with inherent micro-pores would be beneficial for the diffusion of the reactant and products and thus increase the stability of the catalyst.

The alkali treatment of MCM-22 zeolite mainly changed the porosity and channel structure. A proper alkali treatment could create proper amount of mesopores, which coexisted with the inherent micropores of the MCM-22 zeolite due to the dissolution of siliceous species of the zeolite framework. The 6Mo/MCM-22 catalysts prepared by using a proper alkali treated MCM-22 zeolite, with the coexistence of mesopores and micropores on it, showed higher depletion rate of methane and formation rate of aromatics as well as reaction stability when compared with the conventional untreated 6Mo/MCM-22 catalysts. The present work again demonstrates that pore-size distribution and channel structure of the zeolite play an important role in catalyst performance due to the shape selectivity and mass transfer in the zeolite channels.

ACKNOWLEDGMENTS

Financial supports from the Ministry of Science and Technology of China, the Natural Science Foundation of China, the Chinese Academy of Sciences and the BP-China Joint Research Center are gratefully acknowledged.

REFERENCES

[1] L. Wang, T. Li, M. Xie, Y. Xu, Catal. Lett. 21 (1993) 35.
[2] Y. Xu, L. Lin, Appl. Catal. A: 188（1999）53.
[3] Y. Shu, M. Ichikawa, Catal. Today 71（2001）55.
[4] Y. Xu, X. Bao, L. Lin, J. Catal. 216（2003）386.
[5] C. Zhang, S. Li, Y. Yuan, W. Zhang, T. Wu, and L. Lin, Catal. Lett. 56 (1998) 207.
[6] M. Ogura, S. Shinomiya, J. Tateno, Y. Nara, E. Kikuchi, M. Matsukata, Chem. Lett. (2000) 882
[7] T. Suzuki, T. Okuhar, Micro. Meso. Mater., 43 (2001) 83
[8] M. Ogura, S. Shinomiya, J. Tateno, Y. Nara, M. Matsukata, Appl. Catal., 219 (2001) 53
[9] L. Su, L. Liu, J. Zhuang, H. Wang, Y. Li, W. Shen, Y. Xu, X. Bao, Catal. Lett. 91 (2003) 155.
[10] Y. Shu, D. Ma, L. Xu, Y. Xu, X. Bao, Catal. Lett. 70 (2000) 67.
[11] A. Cizmek, B. Subotic, I. Smit, A. Tonejc, R. Aiello, Micro. Mater., 8 (1997) 159.
[12] D. Ma, Y. Shu, X. Han, Y. Xu and X. Bao, J. Phys. Chem. B 105 (2001) 1786.

Studies in Surface Science and Catalysis, volume 147
X. Bao and Y. Xu (Editors)

Middle pressure methane conversion into C_2 hydrocarbons on supported Pt-Co catalysts

L. Borkó and L. Guczi

Department of Surface Chemistry and Catalysis, Institute of Isotope and Surface Chemistry, CRC, Hungarian Academy of Sciences, P. O. Box 77, H-1525, Budapest, Hungary

ABSTRACT

Pt-Co bimetallic catalysts were prepared by incipient wetness technique on alumina or NaY zeolite and were characterized by XRF, XRD, TPR, CO chemisorption and X-ray photoelectron spectroscopy. Methane conversion to higher hydrocarbons was investigated in a flow system by high temperature one-step, constant flow mode at 750°C and by two-step low temperature pulse mode at 250°C under non-oxidative conditions. Catalytic activity in oxygen-free transformations at atmospheric pressure is stable only for a short time. The simultaneous use of higher pressure and high temperature weakens the poisoning effect and makes the process possible to form primarily ethane (and ethene) which is predicted from thermodynamic calculations. The nature of support also significantly affects the amount of C_{2+} products, which can be attributed to structural differences of metal particles (location, reducibility, dispersion).

1. INTRODUCTION

Direct oxygen-free methane conversion into C_{2+} hydrocarbons takes place typically at relatively high temperatures [1, 2]. In our earlier works [3-7] we demonstrated the advantage of the bimetallic Co-based catalysts compared to monometallic ones. In two-step non-oxidative low temperature conversion of methane the catalytic activity decreased in the sequence of Pt-Co > Pd-Co > Ru-Co > Re-Co > Co. Among the samples Pt-Co bimetallic catalysts proved to be the best. The role of platinum was to facilitate reduction thereby modify the bulk and the surface structure of Pt-Co samples [8]. The co-ordination numbers, the repartition of the two metals inside a bimetallic cluster can be established by in situ EXAFS experiments. In fact, the complete set of parameter points to the presence of two kinds of metallic clusters. Monometallic Co nanoparticles and bimetallic Pt-Co nanoparticles coexist, for the latter almost only Pt-Co bonds exist. The presence of monometallic Pt particles is negligible. The majority of particles are bimetallic. By addition of platinum to cobalt the reduction of Co^{2+} ions is facilitated and monometallic Co clusters and bimetallic Pt-Co bimetallic nanosized particles are formed. The extent of reducibility can be indicated by TPR and XPS measurements.

Catalytic activity in oxygen-free transformations at atmospheric pressure is maintained for a short period of time, because a major part of the catalyst surface becomes poisoned by graphitic carbon. The use of higher pressures and high temperature makes it possible to run the process to form primarily ethane (and ethene) which is predicted from thermodynamic calculations. Increasing pressure does not affect the equilibrium in the case of methane-ethane conversion reaction while it has negative effect on the other reaction routes including the coke formation. The subject of the present study is to reveal the role of middle pressures in methane transformation processes on supported cobalt catalysts modified by platinum.

2. EXPERIMENTAL

2.1. Sample preparation and chemical characterization

The comparison of the impregnation preparation method with the ionexchange technique for the Pt-Co/NaY system was reported before in ref. [9]. In the present work we wanted to use the same type of preparation for both the alumina and the zeolite supported catalysts that is why we applied the impregnation. For the preparation of bimetallic samples by incipient wetness method Al_2O_3 90 (70-230 mesh ASTM Merck, fraction 0.2-0.5 mm) and NaY 14-8960 (Strem Chemicals LOT 123032-S) were used. Pt-Co bimetallic catalysts on alumina or NaY zeolite were prepared by incipient wetness technique. Al2O3 and NaY supports were added into the 4 cm^3 solution of $Co(NO_3)_2 \times$ 6 H_2O (Merck) and $[Pt(NH_3)_4](NO_3)_2$ (Aldrich Chemical Co.) with constant stirring. Samples were dried for 24 h at 383 K and analysed with X-ray fluorescence technique (XRF), X-ray diffraction (XRD), X-ray photoelectron spectroscopy (XPS), temperature programmed reduction (TPR) and CO chemisorption. The physical-chemical properties of samples are listed in Table 1.

Table 1
Physical-chemical properties of the supported $Pt_{10}Co_{90}$ samples [8]. In parenthesis the atomic compositions of the metals are shown.

Support	Composition (wt.%) (XRF)		CO(ads) (μmol/g)	Extent of reduction		Kind of particles (EXAFS)
	Pt	Co		TPR	XPS Co 2p peak	
Al_2O_3	1.93 (8)	6.38 (92)	54	0.46	0.55	Co and Pt-Co
NaY	1.81 (8)	5.86 (92)	42	0.98	0.83	Co and Pt-Co

2.2. Catalytic reaction

The reaction for methane conversion was investigated in a flow system in high temperature one-step, constant mode and in low temperature two-step pulse mode under non-oxidative conditions. The one-step constant mode experiments were pursued using 100 mg catalyst at 750°C. Methane-helium mixture was introduced into the system with flow rate of 25 cm^3 min^{-1} by atmospheric and middle pressures 1-10 bar. The products were analyzed by means of a gas chromatograph type CHROMPACK CP 9002 using a 50 m long plot fused silica column (0.53 mm ID) with stationary phase of CP-Al$_2$O$_3$/KCl with a temperature programmed mode. Conversion was characterized by the ΣC_{2+} /$C_{CH4}\times100$; Selectivity was calculated by (C_i/ΣC_{2+}) for i = 2-10; (C_{2+}: C_2-C_{10}; C_{3+}: C_3-C_{10} saturated, unsaturated, cyclic, aromatic hydrocarbons; C_{CH4}: initial methane concentration).

In the two-step mode, detailed elsewhere [3], 100 mg catalyst was used at 250°C. Methane pulses (0.5 cm^3, 24 μmol) was introduced into a stream of helium with flow rate of 100 cm^3 min^{-1}.

After the methane was chemisorbed the system was carefully flushed with helium, then the hydrogenation started and the products were trapped in a loop of 3 cm^3 at liquid nitrogen temperature. Finally the loop was closed, allowed to warm up to ambient temperature and the products were analyzed by gas chromatograph. Total activity was characterized by the amount of C_{1+} and (or) C_{2+} products in micromoles (in methane equivalents) related to 1g catalyst. Specific activity was calculated from C_{1+}/M_{surf} = C_{1+}/(CO)$_{ads}$; C_{2+}/M_{surf} = C_{2+}/(CO)$_{ads}$; (M_{surf} -calculated from CO adsorption data). C_{1+}: C_1-C_{10}; C_{2+}: C_2-C_{10} saturated, unsaturated, cyclic, aromatic hydrocarbons. Selectivity was calculated by (C_i/ΣC_{1+}) $\times100$ for i = 1-8.

3. RESULTS AND DISCUSSION

3.1. Character of bimetallic particles

It is well known that Co^{2+} ions in Co/SiO$_2$ system are hardly reducible [10, 11]. For example, cobalt loading and alumina porosity also influences cobalt reducibility [12]. The difference between the $Pt_{10}Co_{90}/Al_2O_3$ and $Pt_{10}Co_{90}/NaY$ bimetallic systems is reflected in the TPR and XPS measurements. Extent of Co^{2+} reduction in the $Pt_{10}Co_{90}/NaY$ sample is in the range of 83 and 98 % that is higher than in the Al_2O_3 supported one. This high extent reducibility can be ascribed to two factors: (i) the metal particles are in the zeolite supercage despite the lack of ion exchange process, and/or (ii) Cobalt Surface Phase (CSP) is not present which is non reducible material. For nanometer scale metallic particles, oxidation conditions lead generally to a complete cleavage of intermetallic bonds. The oxidation, however, was observed only for alumina supported bimetallic sample, which is also an indication that metallic particles are inside the zeolite.

X-ray diffraction study, the comparison of XRD trace of the $Pt_{10}Co_{90}/NaY$ and NaY support and the curve obtained by modeling indicated that the $Pt_{10}Co_{90}/NaY$ sample contains nanosize particles. The amounts of larger crystallites are less than 1%, the prevailing part of the crystallites is X-ray amorphous. This result also supports the presence of mainly nanoparticles in zeolites. On the basis of calculations from EXAFS data using a cubo-octahedral model, we can estimate the maximum diameter of these particles is about 1 nm.

3.2. One-step high temperature mode: effect of support and pressure

The nature of support and total pressure affects the catalytic properties of the samples as shown in Table 2. Change of the support facilitated the conversion of reaction on $Pt_{10}Co_{90}$ catalyst from 0% to 0.86% on Al_2O_3 and NaY zeolite, respectively, at 1 bar total pressure. The corresponding values at 10 bar pressure are 0.24% and 2.0%. Alumina supported catalyst is not active at atmospheric pressure, while at 10 bar pressures the reaction yields C_2 (ethane and ethene) products with selectivity of 100%. On NaY supported Pt-Co catalyst the 75% C_2 selectivity characteristic at atmospheric pressure increased to 100% at 10 bar.

Table 2
Methane conversion in plug flow reactor: flow rate 25 cm^3 min^{-1}, methane/helium concentration 4%, catalyst temperature750 °C.

Parameter	$Pt_{10}Co_{90}/Al_2O_3$		$Pt_{10}Co_{90}/NaY$	
Pressure (bar)	1	10	1	10
Conversion (%)	-	0.24	0.86	2.0
Ethane selectivity (%)	-	72	67	54
Ethene selectivity (%)	-	28	8	46
Aromatics selectivity (%)	-	-	21	-
C_{3+} selectivity (%)	-	-	25	-

3.3. Two-step low temperature mode: total and specific activity

For the sake of comparison the two-step non-oxidative methane conversion is presented. The structure revealed by the EXAFS, TPR, XPS studies, is well reflected in two-step methane conversion process.

In Table 3 is shown that the C_{1+} selectivity is 100% on both catalysts, no coke remained on the surface. In our former studies [3] it was established, that after chemisorption of small amount of CH_4 on the supported $Pt_{10}Co_{90}$ catalyst surface, in a second, hydrogenation step the surface CH_x species formed could be completely removed in form of C_{2+} hydrocarbons. NaY supported PtCo catalyst performs much better activity both in CH_4 activation and in C-C bond formation, than Al_2O_3 supported sample. The specific activities related to the accessible metal atoms are smaller in case of Al_2O_3 supported samples than in case of $Pt_{10}Co_{90}/NaY$ bimetallic catalysts.

From the comparison it looks obvious that in the one-step constant flow mode the temperature increase alone is not enough but the simultaneous enhancement of pressure gives satisfactory results. At atmospheric pressure in the high temperature one step regime the catalyst surface becomes poisoned in a fast process. However, at 10 bar pressure the transformation mainly into ethane and ethylene starts, as well, decreasing the poisoning effect, in accordance with the thermodynamic expectations.

Wolf et al. [13] has put forward a detailed surface kinetics for the methane conversion in non-oxidative reaction. The detailed model gives an interpretation why the catalytic activity is maintained in the long run despite the high carbon concentration at the Pt surface. According to the model after the formation of CH_x species and methane chemisorption ethylidyne species is set up on the surface. This ensures the progress of the process even at low concentration of CH_3 species. Pressure/flow rate modifies the ethane formation through the variation of surface carbon concentration, and ethane formation can be optimized by changing the parameters. The higher pressure makes it possible to run the process at temperature where at atmospheric pressure no catalytic activity exists due to the formation of graphite on the surface.

Table 3

Two-step mode methane activation: effect of support on the amount of products and selectivity of $Pt_{10}Co_{90}$ catalysts

Support	Conversion (%)	Selectivity (%)		Total activity (µmol/g)		Specific activity	
		methane	C_{2+}	C_{1+}	C_{2+}	C_{1+}/M_{surf}	C_{2+}/M_{surf}
Al_2O_3	0.8	28	72	2.6	1.9	0.05	0.04
NaY	2.1	6	94	5.0	4.7	0.12	0.11

I-step: CH_4 activation (at 250°C, 100 cm^3 min^{-1} He, 22.3 µmol CH_4 pulse);
II-step: CH_x hydrogenation (at 250°C, 100 cm^3 min^{-1} H_2, 10 min exposure time).

The model evaluated and applied as a simulation to high-pressure (10 bar) to improve C_{2+} yields has not been experimentally tested. Regarding to the predictions of [13] the enhancement of pressure from 1 to 10 bars in methane conversion leads to enhance the reaction rate by two orders of magnitude. In our experiments the enhancement of pressure from 1 to 10 bars leads to enhance the reaction rate on the 23 folds that is in a fair agreement with theoretical predictions of [13].

4. CONCLUSIONS

The effect of supporting material and the reaction pressure on the catalytic properties of $Pt_{10}Co_{90}$ catalyst in non-oxidative methane conversion was studied. Supporting on NaY results in significantly improved activity and selectivity compared to supporting on Al_2O_3, which can be attributed to the structural differences of metal particles (location, dispersion and reducibility). The use of middle pressure enhances the methane conversion, C_2 selectivity and suppresses the poisoning of the surface in accordance with the thermodynamic predictions.

ACKNOWLEDGEMENTS

The authors are indebted to the Hungarian Science and Research Found (grants # T-030343 and T043521) for financial support of the research.

REFERENCES

[1] J.H. Lunsford, Catal. Today, 63 (2000) 165.
[2] T.V. Choudhary, E. Aksoylu and D.W. Goodman, Catal. Rev.-Sci. Eng., 45 (2003) 151.
[3] L. Guczi, K.V. Sarma and L. Borkó, Catal. Lett., 39 (1996) 43.
[4] L. Guczi, K.V. Sarma and L. Borkó, J. Catal., 167 (1997) 495.
[5] L. Guczi, L. Borkó, Zs. Koppány and F. Mizukami, Catal. Lett., 54 (1998) 33.
[6] L. Guczi, G. Stefler, Zs. Koppány and L. Borkó, React. Kinet. Catal. Lett., 74 (2001)259.
[7] D. Bazin, L. Borkó, Zs. Koppány, I. Kovács, G. Stefler, I. Sajó, Z. Schay and L. Guczi, Catal. Lett., 84 (2002) 169.
[8] L. Guczi, D. Bazin, I. Kovács, L. Borkó, Z. Schay, J. Lynch, P. Parent, C. Lafon, G. Stefler, Zs. Koppány, and I. Sajó, Topics in Catalysis, 20 (2001) 129.
[9] L. Guczi, G. Lu and Zs. Koppány, in: J. Weitkamp, H. G. Karge, H. Pfeifer, and W. Hölderich (Eds.), Stud. Surf. Sci. Catal., 84 (1994) 949.
[10] A.Kogelbauer, J. G. Goodvin and Oukaci, J. Catal., 160 (1996) 125.
[11] A. R. Belambe, R. Oukaci and J. G. Goodwin, J. Catal., 166 (1997) 8.
[12] L, Guczi, R. Sundararajan, Zs. Koppany, Z. Zsoldos, Z.Schay, F. Mizukami and S. Niva, J. Catal., 167 (1997) 482.
[13] M. Wolf, O. Deutschmann, F. Behrendt and J. Warnatz, Catal. Lett., 6 (1999) 15

Studies in Surface Science and Catalysis, volume 147
X. Bao and Y. Xu (Editors)
©2004 Elsevier B.V. All rights reserved.

607

The critical role of the surface WO$_4$ tetrahedron on the performance of Na-W-Mn/SiO$_2$ catalysts for the oxidative coupling of methane

S.-F. Ji[* a, b, c], T.-C. Xiao[c], S.-B. Li[b], L.-J. Chou[b], B. Zhang[b], C.-Z. Xu[b], R.-L. Hou[b] and M.L.H. Green[c]

[a]The Key Laboratory of Science and Technology of Controllable Chemical Reactions, Ministry of Education, Beijing University of Chemical Technology, 15 Beisanhuan Dong Road, P.O.Box 35, Beijing 100029, China
Email: jisf@mail.buct.edu.cn

[b]State Key Laboratory for Oxo Synthesis and Selective Oxidation, Lanzhou Institute of Chemical Physics, Chinese Academy of Sciences, 342 Tianshui Road, Lanzhou 730000, China

[c]Wolfson Catalysis Centre, Inorganic Chemistry Laboratory, University of Oxford, South Parks Road, Oxford, OX1 3QR, U.K.

ABSTRACT

A series of M-W-Mn/SiO$_2$ catalysts (M = Li, Na, K, Ba, Ca, Fe, Co, Ni, and Al) have been prepared and their catalytic performances for OCM were evaluated. The roles of the WO$_4$ tetrahedra and the WO$_6$ octahedra on the catalyst surface have been investigated using X-ray photoelectron spectroscopy (XPS) and laser Raman spectroscopy (LRS).

1. INTRODUCTION

A 5(wt)%Na$_2$WO$_4$-2(wt)%Mn/SiO$_2$ catalyst system was first reported by Li and his co-workers [1-2], and has been one of the most effective catalysts for the oxidative coupling of methane [3-5]. The full trimetallic Na, W, and Mn formulation on α-cristobalite is very important for higher CH$_4$ conversion and C$_2$ hydrocarbon selectivity. However, the role of WO$_4$ tetrahedron on the catalyst surface, which is formed during the OCM reaction in the presence of Na and Mn in the catalyst, has not been studied. In this work, the M-W-Mn/SiO$_2$ catalysts (M = Li, Na, K, Ba, Ca, Fe, Co, Ni, and Al) have been prepared with the same molar content as a 5(wt)%Na$_2$WO$_4$-2(wt)%Mn/SiO$_2$ catalyst and their catalytic performances for OCM were evaluated. The roles of the WO_4 tetrahedron and the WO_6 octahedron on the catalyst surface have been investigated using X-ray photoelectron spectroscopy (XPS) and laser Raman spectroscopy (LRS).

2. EXPERIMENTAL

2.1. Catalyst Preparation

The catalysts were prepared by the incipient wetness impregnation method. The silica gel support (32-53 mesh) was first impregnated with an aqueous solution of $Mn(NO_3)_2$ at 90°C, and for 4-5 h at 120°C; then with aqueous solutions containing an appropriate amount of $(NH_4)_5H_7(WO_4)_6$ and $M(NO_3)$ (M = Li, Na, K, Ba, Ca, Fe, Co, Ni, and Al) at 90°C, then dried for 4-5 h at 120°C. Finally, the catalysts were calcined in air for 6 h at 830-850°C. Herein, the amounts of the various components are expressed in the same molar weight as $5wt\%Na_2WO_4$-$2wt\%Mn/SiO_2$ catalyst.

2.2. Catalytic Activity Test

The catalysts were evaluated at 800°C, and 1 bar with a GHSV of 25,400 h^{-1}. 0.2g catalysts were loaded in a quartz fixed-bed microreactor (i.d. 6mm). The reactants, CH_4 and O_2 at a ratio of 3.2:1, were used without further purification. The reaction products were analyzed with an on-line SC-6 GC equipped with a TCD, using a Porapak Q column for the separation of CH_4, CO_2, C_2H_4, and C_2H_6, and a 5A molecular sieve column for the separation of O_2, CH_4, and CO.

2.3. Catalyst Characterization

The XPS analysis of the catalysts was performed with a VG ESCALAB 210 spectrometer. A Mg target was used as the anode of the X-ray source. The binding energies were calibrated using the Si (2p) line at 103.4 eV as the reference. The Raman spectra of the catalysts were recorded in a Yuon Jobin Labram spectrometer with a resolution of 2 cm^{-1}. It is equipped with a CCD camera enabling microanalysis on a sample point. A 514.5 nm Ar^+ laser source was used and the spectra acquired in a back scattered confocal arrangement. The scanning range was set from 90 to 3000 cm^{-1}.

3. RESULTS

3.1. Catalytic Reaction Results

The results of CH_4 conversion and product selectivity of the M-W-Mn/SiO$_2$ catalysts (M = Li, Na, K, Ba, Ca, Fe, Co, Ni, and Al) for OCM are listed in Table 1. Over the Na- and K-W-Mn/SiO$_2$ catalysts, CH_4 conversion and C_2H_4 selectivity are approximately 30% and 40% respectively. After 5 h time on stream, the performance of the catalysts was essentially unchanged. With the

Table 1
The catalytic performance of the M-W-Mn/SiO$_2$ catalysts [a]

Catalysts [b]	Conv. of CH$_4$(%)	Selectivity(%)				Yield(%)	
		C$_2$H$_4$	C$_2$H$_6$	CO	CO$_2$	C$_2$+	COx
Li$_2$WO$_4$-Mn/SiO$_2$	20.5	35.4	28.7	17.8	18.1	13.1	7.4
Li$_2$WO$_4$-Mn/SiO$_2$ [c]	18.2	25.7	27.6	25.2	21.5	9.7	8.5
Na$_2$WO$_4$-Mn/SiO$_2$	29.5	42.6	23.8	11.5	22.1	19.6	9.9
Na$_2$WO$_4$-Mn/SiO$_2$ [c]	28.9	43.9	22.7	11.6	21.8	19.3	9.6
K$_2$WO$_4$-Mn/SiO$_2$	29.6	39.6	22.8	12.7	24.9	18.5	11.1
K$_2$WO$_4$-Mn/SiO$_2$ [c]	29.8	39.7	22.7	13.3	24.3	18.6	11.2
BaWO$_4$-Mn/SiO$_2$	21.3	18.7	13.8	45.5	22.0	6.9	14.4
BaWO$_4$-Mn/SiO$_2$ [c]	18.6	13.2	12.5	51.0	23.3	4.8	13.8
CaWO$_4$-Mn/SiO$_2$	19.0	12.1	13.2	55.2	19.5	4.8	14.2
CaWO$_4$-Mn/SiO$_2$ [c]	18.2	11.4	13.6	54.3	20.7	4.6	13.6
FeWO$_4$-Mn/SiO$_2$	18.3	7.5	7.1	67.3	18.1	2.7	15.6
FeWO$_4$-Mn/SiO$_2$ [c]	17.7	7.5	7.6	67.1	17.8	2.7	15.0
CoWO$_4$-Mn/SiO$_2$	18.3	11.9	11.1	56.1	20.9	4.2	14.1
CoWO$_4$-Mn/SiO$_2$ [c]	18.1	11.1	12.7	56.4	19.8	4.3	13.8
NiWO$_4$-Mn/SiO$_2$	18.4	12.1	11.7	52.3	23.9	4.4	14.0
NiWO$_4$-Mn/SiO$_2$ [c]	17.9	10.6	9.9	56.7	22.8	3.7	14.2
Al$_2$(WO$_4$)$_3$-Mn/SiO$_2$	17.9	8.6	8.0	68.1	15.3	3.0	14.9
Al$_2$(WO$_4$)$_3$-Mn/SiO$_2$ [c]	17.1	8.4	8.3	69.3	14.0	2.9	14.2

[a] Reaction conditions: T = 800℃; CH$_4$:O$_2$ = 3.2:1; GHSV=25,400ml h^{-1}; 0.2g catalyst; Stream time = 0.5h. [b] The molecular formula of the metal tungstate was identified by XRD. [c] Reaction conditions: T = 800℃; CH$_4$:O$_2$ = 3.2:1; GHSV=25,400ml h^{-1}; 0.2g catalyst; Stream time = 5h.

Li-W-Mn/SiO$_2$ catalyst, the CH$_4$ conversion was only 20%, and the C$_2$H$_4$ selectivity 35.4%. After 5 h time on stream, there was a small decrease for CH$_4$ conversion, but the C$_2$H$_4$ selectivity is decreased dramatically and C$_2$H$_6$ selectivity is decreased slightly. For the Ba-W-Mn/SiO$_2$ catalyst, the CH$_4$ conversion and C$_2$H$_4$ selectivity were 21.3% and 18.7% after 0.5 hour test; however, CO selectivity was up to 45.5%. After testing for 5 hours, CH$_4$ conversion and C$_2$H$_6$ selectivity decreased to 18.6% and 13.2%. However, CO selectivity was increased.

Over the Ca-, Fe-, Co-, Ni-, and Al-W-Mn/SiO$_2$ catalysts, both CH$_4$ conversion and C$_2$ selectivity were low, with a high selectivity to CO. After an activity test of 5 hours, the CH$_4$ conversion and product selectivity did not change significantly.

3.2. X-Ray Photoelectron Spectroscopy

Table 2 shows the observed binding energies and near-surface elemental compositions of the fresh and the post-catalysts. In the fresh catalysts, the near-surface concentration of Mn is between 1.1% and 2.3%. After the catalyst was tested for 5 hours, the near-surface concentrations of Mn in all of the catalysts are higher than these in the fresh catalysts. In the fresh catalysts, the near-surface concentrations of W are between 1.1% and 2.2%, and those of the metal ions are between 0.4% and 12.8%, significantly varying with different catalysts. The near-surface concentrations of Li, Na, K, and Ba are relatively higher than the stoichiometric composition of metal tungstate, while those of Ca, Fe, Co, Al and Ni are lower than the stoichiometric tungstate composition. These results indicate that the Li, Na, K, and Ba are enriched, and Ca, Fe, Co, Al and Ni are deficient on the catalyst surface.

The binding energy of W4f also changes with the different additives. Generally speaking, catalysts modified with alkline and alkline earth metals have the higher binding energy of W4f, and the binding energy of W4f over the fresh and used Na-W-Mn/SiO$_2$ are 35.8 eV, while the Ca modified catalyst has the lowest W4f binding energy among the alkline and alkline earth metal modified catalyst, which is only 35.1eV. Compared the alkline and alkline earth metal modified catalysts, the transition metal modified W-Mn/SiO$_2$ catalysts have the lower binding energy of W4f, they are no higher than 35.2eV.

Table 2
Observed XPS binding energies (eV) and near-surface compositions (at.%) of the M-W-Mn/SiO$_2$ catalyst components

Catalysts	M [a]		W(4f)		Mn(2p)		Si(2p)		O(1s)(SiO$_2$)		O(1s)(MOx) [b]	
	BE (eV)	at. %	BE (eV)	at. %	BE (eV)	at. %	BE (eV)	at. %	BE (eV)	at. %	BE (eV)	at. %
Li$_2$WO$_4$-Mn/SiO$_2$[c]	55.7	5.7	35.9	2.2	641.7	1.5	103.4	20.9	532.8	51.1	530.9	18.6
Li$_2$WO$_4$-Mn/SiO$_2$[d]	55.9	5.2	35.6	2.0	641.6	1.8	103.4	20.8	532.7	52.5	530.6	17.7
Na$_2$WO$_4$-Mn/SiO$_2$[c]	1072.2	12.8	35.7	1.6	641.5	2.3	103.4	19.9	532.9	49.7	530.8	13.7
Na$_2$WO$_4$-Mn/SiO$_2$[d]	1072.1	12.6	35.8	1.5	641.4	2.4	103.4	19.2	532.9	49.5	530.9	14.8
K$_2$WO$_4$-Mn/SiO$_2$[c]	293.3	7.1	35.6	1.5	641.7	1.4	103.4	21.1	532.8	51.0	531.0	17.9
K$_2$WO$_4$-Mn/SiO$_2$[d]	293.4	7.5	35.7	1.6	641.5	1.7	103.4	21.5	532.9	50.2	530.9	17.5
BaWO$_4$-Mn/SiO$_2$[c]	780.1	4.8	35.7	2.2	641.5	1.1	103.4	20.0	532.9	49.8	530.9	22.1
BaWO$_4$-Mn/SiO$_2$[d]	779.9	4.3	35.3	2.3	641.3	1.5	103.4	20.1	532.8	50.3	530.5	21.5
CaWO$_4$-Mn/SiO$_2$[c]	347.5	0.5	35.1	1.1	641.8	1.2	103.4	24.7	532.7	56.8	531.6	15.7
CaWO$_4$-Mn/SiO$_2$[d]	347.3	0.7	35.1	1.2	641.7	1.3	103.4	24.9	532.9	56.5	531.4	15.4
FeWO$_4$-Mn/SiO$_2$[c]	710.4	0.4	35.2	1.1	641.6	1.1	103.4	25.1	532.8	50.6	531.5	21.7
FeWO$_4$-Mn/SiO$_2$[d]	710.3	0.5	35.1	1.2	641.6	1.5	103.4	25.2	532.8	50.1	531.5	21.5
CoWO$_4$-Mn/SiO$_2$[c]	781.3	1.5	35.0	1.7	641.2	1.5	103.4	20.7	532.9	55.8	530.4	18.8
CoWO$_4$-Mn/SiO$_2$[d]	781.5	1.4	35.0	1.7	641.3	1.8	103.4	20.5	532.8	55.6	530.5	19.0
NiWO$_4$-Mn/SiO$_2$[c]	855.8	0.8	35.0	2.0	640.9	2.1	103.4	19.7	532.8	55.5	530.2	19.9
NiWO$_4$-Mn/SiO$_2$[d]	855.6	0.9	35.0	2.1	640.8	2.3	103.4	19.3	532.7	55.3	530.2	20.1
Al$_2$(WO$_4$)$_3$-Mn/SiO$_2$[c]	73.9	1.2	35.0	1.8	640.8	1.8	103.4	21.5	532.8	59.8	530.2	13.9
Al$_2$(WO$_4$)$_3$-Mn/SiO$_2$[d]	74.0	1.1	35.1	1.8	640.8	1.9	103.4	21.0	532.8	59.5	530.1	14.7

[a] Li$_2$WO$_4$-Mn/SiO, Na$_2$WO$_4$-Mn/SiO, K$_2$WO$_4$-Mn/SiO, BaWO$_4$-Mn/SiO, CaWO$_4$-Mn/SiO, FeWO$_4$-Mn/SiO$_2$, CoWO$_4$-Mn/SiO$_2$, NiWO$_4$-Mn/SiO$_2$ and Al$_2$(WO$_4$)$_3$-Mn/SiO$_2$ catalysts are Li(1s), Na(1s), K(2p), Ba(3d), Ca(2p), Fe(2p), Co(2p), Ni(2p), and Al(2p), respectively. [b] MOx represents the metal oxides except SiO$_2$. [c] Fresh catalysts. [d] After testing 5 h (T = 800 ℃; CH$_4$:O$_2$ = 3.2:1; GHSV=25,400 h^{-1}; 0.2g catalyst).

3.3. Laser Raman Spectroscopy

The Raman spectra of the fresh and the post-catalysts are shown in Fig. 1. The laser Raman spectrum of unpromoted 3.1%W-2%Mn/SiO$_2$ was measured and listed as the reference to compare with the addition of the different metal modifiers on the vibration of the catalyst surface. Comparison with the previous literature results suggests that the Raman bands between 952 cm^{-1} and 905 cm^{-1} can be assigned to tetrahedral WO$_4$, and bands between 889 cm^{-1} and 862 cm^{-1} to octahedral WO$_6$. The bands at 701 cm^{-1} and 574 cm^{-1} are attributed to manganese oxide, and those below < 500 cm^{-1} to α-cristobalite and to quartz SiO$_2$, respectively.

It can be seen from Fig. 1 that the fresh Na-W-Mn/SiO$_2$ catalyst has two bands at 947 and 905 cm^{-1} due to the tetrahedral WO$_4$, and there are four bands at 683, 615, 574, and 509 cm^{-1} due to manganese oxide species, while no Raman band assignable to octahedral WO$_6$ were observed (Fig. 1A-d). The fresh K-W-Mn/SiO$_2$ catalyst has two bands at 952 and 909 cm^{-1} (for tetrahedral WO$_4$), and four bands at 696, 623, 579, and 515 cm^{-1} (for manganese oxide). Again no Raman bands for octahedral WO$_6$ were observed (Fig. 1A-f). After an activity test for 5 hours, no change was observed in the Raman bands assigned to the tetrahedral WO$_4$ and manganese oxide.

The fresh Li-W-Mn/SiO$_2$ catalyst has two bands at 941 and 917 cm^{-1} due to tetrahedral WO$_4$, one at 862 cm^{-1} for octahedral WO$_6$, and one strong band at 467 cm^{-1} and one weak band at 206 cm^{-1} for quartz SiO$_2$; no Raman bands due to the manganese oxide can be seen (Fig. 1A-b). After an activity test for 5 hours,

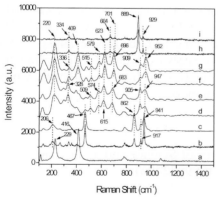 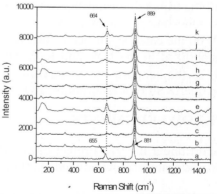

Fig.1A. The Raman spectra of a)α-cristobalite, b)Li-W-Mn/SiO$_2$(fresh), c)Li-W-Mn/SiO$_2$ (post), d)Na-W-Mn/SiO$_2$ (fresh), e)Na-W-Mn /SiO$_2$ (post), f)K-W-Mn/SiO$_2$ (fresh), g)K-W -Mn/SiO$_2$ (post), h)Ba-W-Mn/SiO$_2$ (fresh), and i)Ba-W-Mn/SiO$_2$ (post) catalysts

Fig.1B.The Raman spectra of a)3.1%W-2%Mn /SiO$_2$, b)Ca-W-Mn/SiO$_2$ (fresh), c)Ca-W-Mn /SiO$_2$(post), d)Fe-W-Mn/SiO$_2$ (fresh), e)Fe-W -Mn/SiO$_2$(post), f)Co-W-Mn/SiO$_2$ (fresh), g)Co -W-Mn/SiO$_2$(post), h)Ni-W-Mn/SiO$_2$ (fresh), i)Ni-W-Mn/SiO$_2$ (post), j)Al-W-Mn/SiO$_2$ (fresh), and k) Al-W-Mn/SiO$_2$ (post) catalysts.

the band at 941 cm^{-1} for the tetrahedral WO_4 decreased greatly while the band at 862 cm^{-1} (octahedral WO_6) increased significantly (Fig. 1A-c), indicating that the surface structure of the catalyst had changed. Comparison with the catalytic activity of the Li-W-Mn/SiO$_2$ catalyst (entry 1, 2, Table 1) the fresh catalyst has a good initial CH_4 conversion activity and C_2 selectivity but, after 5 hour test, the CH_4 conversion and C_2H_4 selectivity decreased greatly, suggesting the WO_4 tetrahedron on the catalyst surface might play an important role in the CH_4 conversion and the C_2H_4 selectivity.

The fresh Ba-W-Mn/SiO$_2$ catalyst has one weak band at 929 cm^{-1} for the tetrahedral WO_4, one strong band at 889 cm^{-1} for the octahedral WO_6, and two weak bands at 701 and 664 cm^{-1} for the manganese oxide (Fig. 1A-h). This catalyst exhibits a high activity for CH_4 conversion and C_2H_4 selectivity (entry 7, 8, Table 1). After an activity test of 5 hours, the Raman band at 929 cm^{-1} for tetrahedral WO_4 decreased (Fig. 1A-i) and the CH_4 conversion decreased slightly, while the C_2H_4 selectivity dropped rapidly. This further suggests that the tetrahedral WO_4 plays a key role for the CH_4 conversion and C_2H_4 selectivity.

Each of the Ca-, Fe-, Co-, Ni-, and Al-W-Mn/SiO$_2$ catalysts has a band at 890 cm^{-1} for octahedral WO_6, and no Raman bands due to the tetrahedral WO_4. After an activity test for 5 hours, the activity of the catalysts is almost the same and the Raman spectra are not changed significantly (Fig. 1B-c, -e, -g, -i, and -k), suggesting that the surface structure of these catalysts is stable.

4. CONCLUSIONS

In all the trimetallic catalysts containing alkali metal ions or alkaline-earth metal ions, tungsten, and manganese, supported on SiO$_2$, the presence of tetrahedral WO_4 on the catalyst surface results in active and selective OCM catalysts. It can be stabilized in the Na-, and K-W-Mn/SiO$_2$ catalysts, which has α-cristobalite SiO$_2$ as the support, but is unstable in Li-, or Ba-W-Mn/SiO$_2$ catalysts, which have quartz SiO$_2$ or amorphous SiO$_2$ as the support.

ACKNOWLEDGMENTS

Support under the 973 Project of the Ministry of Science and Technology, P. R. China, is gratefully acknowledged (Grant G1999022406).

REFERENCES

[1] X. Fang, S. Li, J. Gu, and D. Yang, J. Mol. Catal., (China). 6 (1992) 225.
[2] X. Fang, S. Li, J. Lin, and Y. Chu, J. Mol. Catal., (China). 6 (1992) 427.
[3] J.H. Lunsford, Angew. Chem. Int. Ed. Engl., 34 (1995) 970.
[4] J.H. Lunsford, Catal. Today, 63 (2000) 165.
[5] S. Li, Chin. J. Chem., 19 (2001) 16.

Studies in Surface Science and Catalysis, volume 147
X. Bao and Y. Xu (Editors)

613

Hydrogenolysis of methyl formate to methanol on copper-based catalysts and TPR characterization[#]

Y. Long, W. Chu[*], Y. H. Ye, X. W. Zhang, X. Y. Dai

Department of Chemical Engineering, Sichuan University, Chengdu 610065, China

ABSTRACT

It was found that Zr and B were effective promoters for Cu/Cr/Si catalyst with low Cr content used in the model reaction of hydrogenolysis of methyl formate to methanol. TPR, XRD results and the catalytic performance of catalysts were analyzed. It was indicated that there was an optimum Cu^+/Cu^0 ratio in the catalysts. The catalyst with 6 wt.% of Zr had lower reduction temperature and obtained the maximum yield of methanol at a lower temperature.

1. INTRODUCTION

Methanol is more important than ever for the variety of energy-related markets. Interests in methanol have stimulated researches to lower the cost of its production. Methyl formate (MF), a useful intermediate of many important chemicals, has been more noticed recently. Methanol synthesis is a highly exothermic reaction and it was thermodynamically favorable at low temperatures. In recent years, liquid-phase methanol synthesis at low temperature, where high equilibrium conversion was expected, has been proposed to overcome the above-mentioned problems in the conventional methanol production process [1-4]. The two-step mechanism was proposal for the methanol synthesis at low temperature (MSLT), and the second-step, the hydrogenolysis of MF, was worth studying. Copper chromite catalysts are frequently employed in industrial hydrogenolysis. However, a problem arises from environmental pollution due to the large content of chromium (Cr wt.% > 35%) in the copper chromite catalysts. In this work, the investigation on the hydrogenolysis of methyl formate to methanol on Cu/Cr/Si catalyst and those promoted with Zr or B have been

Supported by NNSFC（No.29903011）and the sub-division of 973 key project by MST of China (G1999022406)
* Corresponding Author. Email: chuwei65@yahoo.com.cn; Phone: +86 28 85400591

performed in order to decrease the content of chromium and to develop new efficient catalysts at low temperature.

2. EXPERIMENTAL

2.1. Catalyst preparation
The catalyst was prepared by the complex co-precipitation method [2]. The precipitation was filtered and washed after ageing for 4h. Finally, the precursors were dried at 373-393K overnight, calcined at 623K for 4h, and reduced at 538K.

2.2. Reaction test
Hydrogenolysis of methyl formate (MF) was carried out in a conventional fixed-bed reactor system working at atmospheric pressure. The catalyst (100mg) was held in place by a glass-wool plug sandwich. A MF/H_2 mixture was used as reactant gas. The reaction temperature ranged from 408K to 438K. The reactor effluent was analyzed with an on-line gas chromatograph equipped with a GDX-103 column. Thermal conductivity detector was used.

2.3. Catalyst characterization
The experimental apparatus for TPR included a gas chromatograph, a microreactor with 40-100 mesh catalyst and a temperature program controller [5,6]. The temperature program reduction of catalyst was carried out with 30ml min^{-1} mixing gas of H_2 and N_2 (H_2/N_2=6.9/93.1,v/v) at a rate of 5 K min^{-1}.X-ray diffraction patterns of the samples were recorded on a Philips D/max-rA diffractometer with a filtrated Cu Kα radiation, 2θ angles from 10 to 70.

3. RESULTS AND DISCUSSION

3.1. Effect of promoter and its content in the catalyst
The reaction results of the catalysts, in which we added different additives as the forth component, were shown in Table 1. The experimental results showed that the catalytic activities of the catalysts promoted by B or Zr were higher than that of $CuCr_{22}Si$ catalyst, but MeOH selectivity decreased a little. According to MeOH yield, we concluded that it was useful to enhance MeOH yield by adding B or Zr to the Copper-based Catalysts with low Cr content.

B and Zr were efficient promoters to Cu/Cr/Si catalyst for the hydrogenolysis of methyl formate, so we investigated the effect of different promoter content on catalytic activity. From Table 1, we could also find that $CuCr_{22}Si-B_2$(wt%) catalyst was more active at a lower temperature (408K), while $CuCr_{22}Si-B_{10}$(wt%) sample was more active at a higher temperature (438K).

Table 1
Effect of different additive on the performance of Cu/Cr/Si catalyst

Catalyst	T/K	MF Conversion	MeOH Selectivity	CO Selectivity	CH$_4$ Selectivity	MeOH Yield
	408	3.31	69.93	6.25	23.82	2.31
CuCr$_{22}$Si	423	24.14	87.46	9.31	3.22	21.12
	438	52.08	84.85	13.48	1.67	44.56
	408	18.20	82.77	4.2	13.03	15.11
CuCr$_{22}$Si -B$_2$	423	48.31	81.34	15.89	2.77	39.34
	438	71.53	73.92	23.64	2.44	52.91
	408	11.57	83.04	4.73	12.24	9.62
CuCr$_{22}$Si -B$_{10}$	423	47.66	85.64	11.86	2.50	40.85
	438	70.39	83.78	14.33	1.89	58.97
	408	22.08	78.25	5.36	16.39	17.29
CuCr$_{22}$Si -Zr$_2$	423	54.84	83.52	12.54	3.94	45.80
	438	68.19	80.29	16.99	2.71	54.77
	408	39.17	87.25	10.69	2.06	34.19
CuCr$_{22}$Si -Zr$_6$	423	63.44	83.37	14.89	1.75	52.88
	438	71.04	78.37	19.66	1.97	55.68
	408	37.71	78.29	7.58	14.13	29.52
CuCr$_{22}$Si -Zr$_{15}$	423	55.67	80.72	15.85	3.43	44.94
	438	73.50	80.67	17.26	2.07	59.31

Reaction conditions: 100mg; space velocity 4000h^{-1}; H$_2$/MF =2.

The maximum yield was obtained when 6 wt.% of Zr was added to Cu/Cr/Si catalyst. The addition of Zr showed a stronger influence on the MF conversion than that on the methanol selectivity. But when the reaction temperature increased to 438K, the catalyst with 15 wt% Zr was more active than the one with 6 wt % Zr. Therefore, it was suggested that the addition of Zr caused some changes on the surface of copper-chromite catalysts, adjusted the ratio of Cu$^+$/Cu^0 on the surface of catalyst, and enhanced the specific activity for MF hydrogenolysis[7].

3.2. Enfluence of reaction temperature

The effect of reaction temperature on the performance of MF hydrogenolysis has been investigated. The experiments were carried out at 408-468K. The results of the reactions were shown in Fig. 1. Under the conditions studied, MF conversion increased markedly with the rise of reaction temperature below 438K, and then increased by inches when the temperature was higher than 438K. MeOH selectivity increased till 423K, and then decreased with the rise of reaction temperature. CH$_4$ and CO were produced as by-products. A maximum MeOH yield was obtained at 438K. The former research suggested that this reaction was affected by thermodynamics and dynamics corporately. Kinetics

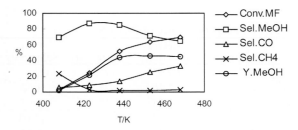

Fig. 1. The effect of the reaction temperature on reaction performance

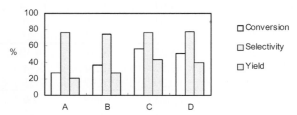

Fig. 2. Comparison of catalytic activity of catalysts prepared by different methods
A- the mixing solution of $(NH_4)_2Cr_2O_7$ and sodium silicate(S1) was added into ammonia copper solution(S2), then promoter solution(S3) was added; B- S1 and S2 were added together into a small amount of distilled water, then S3 was added; C- S2 was added into S1, then S3 was added; D- the precipitation gained by method C (S3 wasn't added) was infused in S3 for 24h

control at a low reaction temperature, while thermodynamics properties controlled over a certain temperature [8-11].

3.3. Role of catalyst preparation method

The relative activities of catalysts prepared by different methods were shown in Fig. 2. It was shown that the best catalyst was prepared by method C, followed by method D, B and A. We considered that the active components were more dispersed in the catalysts prepared by method C and D, and promoter Zr worked better in the samples to affect the surface character of Cu/Cr/Si catalysts.

3.4. TPR characterization

3.4.1. TPR results for catalyst of different sample amount

Patterns #1A–#3C in Fig. 3 represented the TPR patterns of 50, 25 and 15mg of Cu/Cr/Si-B$_{10}$ catalyst, respectively. At the same conditions, the reduction temperature of pattern #1A was 534K; those of pattern #2B, #3C were 521K and 516K, respectively. We could also see that the peak area was decreased with the decrease of catalyst amount. According to the results, we choose 15mg as the experimental sample amount.

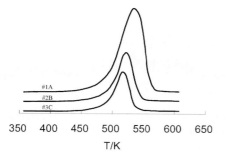

Fig. 3. H₂-TPR profiles for Cu/Cr/Si-B₁₀ catalyst of different amount

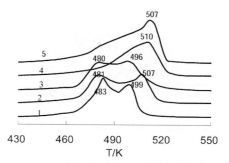

Fig. 4. H₂-TPR profiles of catalysts with different Zr promoter content (#1=0, #2=2%Zr, #3=6%Zr, #4=10%Zr, #5=15%Zr)

3.4.2. TPR results for catalysts with different Cr content

From the results of TPR spectra of catalysts with different Cr content, we could see that $Cu/Cr_{42}/Si$ catalyst exhibits a single peak at 471K, and when the Cr content decreased, the peak shifted to higher temperature and its height became shorter. This indicated that there was an interaction between Cr and Cu.

3.4.3. TPR results for catalysts with different Zr promoter content

The TPR profiles for Cu/Cr/Si-Zr catalysts with various Zr contents were presented in Fig. 4. Patterns 1–5 represented the samples containing 0, 2%, 6%, 10% and 15wt.% Zr, respectively. Two peaks were identified for catalysts having Zr contents of 0, 2 and 6wt.%, respectively. The first peak represented the reduction of CuO to Cu^+. The other peak indicated that Cu^0 was formed from Cu^+. At the same time, the ratio of the two peaks' height changed when Zr was added. The changes might be brought about by the dissolution of zirconium ions in copper oxide phase [12,13].

3.4.4. TPR results for catalysts before reaction and after reaction

Fig. 5 showed the TPR patterns of the Cu/Cr/Si-Zr₆ catalyst after calcination and after reaction. It could be seen that the peaks' area were reduced obviously after reaction and their reduction temperature became lower. The ratio of the two peaks' area first over second also changed from 1.08 to 0.38, which was suggested to be the suitable Cu^+/Cu^0 ratio in catalysts.

3.5. XRD characterization

Fig. 6 showed the XRD pattern of Cu/Cr/Si-B catalyst, only the diffraction peaks of $CuCrO_2$ and CuO were observed, it indicated that the main part of Cu/Cr/Si-B catalyst existed at a non-crystal state [14]. The intensity of $CuCrO_2$ and CuO phases decreased when B was added into Cu/Cr/Si catalyst, which demonstrated the increase of copper dispersion.

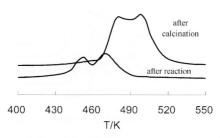

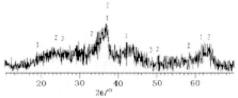

Fig. 6. XRD pattern of Cu/Cr/Si-B catalyst
1—CuCrO$_2$ 2—CuO   3—SiO$_2$

Fig. 5. H$_2$-TPR profiles of catalysts before and after reaction

4. CONCLUSION

The hydrogenolysis of methyl formate to methanol was investgated over novel Cu/Cr/Si catalysts promoted by B or Zr. It was found that Zr and B were effective promoters for Cu/Cr/Si catalyst with low Cr content used in methanol production from methyl formate. The catalyst with 6 wt.% of Zr had lower reduction temperature and obtained the maximum yield of methanol at a lower temperature. TPR results indicated that there was a optimum Cu$^+$/Cu^0 ratio in the catalysts. The results of TPR spectra of catalysts with different Cr content indicated that there is an interaction between Cr and Cu.

REFERENCES

[1] M. Marchionna, M. Lami, A. Galletti, Chemtech. April, (1997) 27
[2] W. Chu, T. Zhang, C.H. He, et al., Catal. Lett. 79 (2002) 129
[3] S. Ohyame, Appl.Catal. A. 180 (1999) 217
[4] W. Chu, Y.T. Wu, L.W. Lin, et al., Chemistry Progress 13(2) (2001) 128
[5] Y Zhang, W Chu, C.R. Luo, Q. Zhou, Plasma Chem. Plasma Process, 20, (2000) 137
[6] W Chu, R Kieffer, A Kiennemann, J.P. Hindermann, Appl. Catal. A, 121 (1995) 95
[7] E.D. Guerreiro, O.F. Gorriz, G. Larsen, et al., Appl Catal. 204 (2000) 33
[8] X.Q. Lin, Y.T. Wu, Z.L. Yu, et al., Stud. Surf. Sci. Catal., 119 (1998) 557
[9] W.K. Chen, X.Q. Liu, S.Z. Luo, et al., J. Nat. Gas Chem., 9(2) (2000) 139
[10] X.Q. Liu, Y.T. Wu, W.K. Chen, et al., J. Nat. Gas Chem., 9(1) (2000) 50
[11] Y. Choi, K. Futagami, T. Fujitani, et al. Appl Catal A: General, 208 (2001) 163
[12] W.S. Ning, H.Y. Shen, H.Z. Liu, Appl Catal A: General, 211 (2001) 153
[13] J.P. Lang, Catalysis Today, 64 (2001) 3
[14] B.J. Liaw, Y.Z. Chen, Appl Catal A: General, 206 (2001) 245

Studies in Surface Science and Catalysis, volume 147
X. Bao and Y. Xu (Editors)

Role of SnO$_2$ over Sn-W-Mn/SiO$_2$ catalysts for oxidative coupling of methane reaction at elevated pressure

Lingjun Chou, Yingchun Cai, Bing Zhang, Jian Yang, Jun Zhao, Jianzhong Niu, and Shuben Li[*]

State Key Laboratory for Oxo Synthesis and Selective Oxidation, Lanzhou Institute of Chemical Physics, Chinese Academy of Sciences, Lanzhou 730000, China

ABSTRACT

The SnO$_2$-doped W-Mn/SiO$_2$ catalyst for oxidative coupling of methane (OCM) to high hydrocarbons has been studied in a micro- stainless-steel reactor at elevated pressure. Under 750°C, 1.0×10^5 h^{-1} GHSV and 0.6MPa, 20.3% yield of C$_2^+$ hydrocarbons and 29.7% CH$_4$ conversion with 33.5% selectivity of C$_3 \sim$ C$_4$ is obtained. The storage-oxygen capability in W-Mn/SiO$_2$ catalyst is enhanced with adding SnO$_2$, SnO$_2$ promotes the migration of Na, W and Mn to the catalyst surface; W^{5+} possibly existed over the surface of SnO$_2$ doped W-Mn/SiO$_2$ catalyst.

1. INTRODUCTION

In the last two decades, most of the attention on the direct conversion of methane has been devoted to the oxidative coupling reaction (OCM), which involves the catalytic conversion of methane with oxygen to C$_2$+ hydrocarbons and water [1-2]. One of the most effective catalysts for the OCM is W-Mn-/SiO$_2$ [3], several groups leaded by Li, Lunsford and Lambert have studied for the OCM over the catalyst on the elementary reaction and role of active composition [4-6]. Most of the researcher's have focused their attention on the OCM at atmospheric pressure, however, the studies on the OCM under elevated pressure are less [7-9]. Recently, we reported that a comparatively high yield of higher hydrocarbons was achieved over SnO$_2$ promoted W-Mn/SiO$_2$ catalysts under lower pressure as 0.6MPa in the stainless- steel reactor for oxidative conversion of methane [10，11]. In this paper, we reported the effects of SnO$_2$ on crystalline phase; the relation between catalyst structure and catalytic activities was discussed.

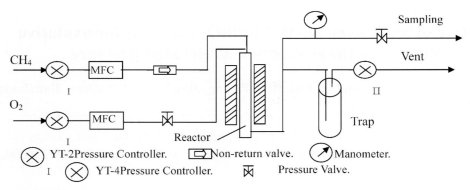

Fig. 1. The schematic diagram of the apparatus on OCM at elevated

2. EXPERIMENTAL

2.1. Catalyst preparation

The SnO_2 modified W-Mn/SiO_2 catalysts were prepared by the equal-volume impregnation method [6]. Silica particles were washed with water several times and dried. The pretreated silica particles were impregnated with the aqueous solutions of $Mn(NO_3)_2$ and $Na_2WO_4 \cdot 2H_2O$ in appropriate concentrations at 50℃ respectively. The impregnated silica particles were evaporated to dryness, and impregnated by the toluene solutions of 5% di-n-butyltin dilaurate at 50℃ and dried overnight at 100℃; the resultant were calcined in air at 850℃ for 8 hrs. No carbon deposit was detected over the fresh catalyst surface by XRD.

2.2. Analysis

The major products for oxidative coupling of methane at elevated pressure were detected by two on-line gas chromatographs using TCD with 5A column which separated O_2, CH_4, CO and with Poropak Q column which separated CH_4, CO_2, C_2H_4, C_2H_6, another GS using FID with a Al_2O_3 capillary tube column separated higher hydrocarbons. All data were taken after 30 minutes of reaction.

2.3. Reactor system

In all experiments, the reactant gases, methane and oxygen without diluent, were co-fed into the reactor. The catalytic runs were carried out in a vertical-flow fixed-bed reactor system (In Figure 1) made by a stainless-steel tube (I.D.=8 mm); 0.4 g catalysts were loaded in the reactor, while the remaining space of the reactor was filled with ceramic particles (40 mesh).

2.4. Catalyst characterization

XRD patterns of the catalysts were obtained with a D/MAX-RB diffractmeter using Cu kα radiation X-ray at 50 kV and 60 mA. XPS analysis of the catalysts was performed with a VG ESCALAB 210 spectrometer, near-surface compositions were calculated using sensitivity factors, which were provided in the software of the instrument.

Temperature programmed desorption (TPD) of O_2 was carried out in Y-type quartz reactor with 0.1g catalysts using AM-100 catalyst character system. The fresh catalysts was treated in O_2 (30ml min^{-1}) at 800℃ for 30 min, followed by decreased to room temperature with O_2 (30 ml min^{-1}) flow, then sequentially flushed by He (50 ml min^{-1}). Each run was conducted between room temperature and 850℃ at a heating rate of 30℃ min^{-1} by He at a flow rate of 50 ml min^{-1}.

3. RESULTS AND DISCUSSION

The significant reaction on the SnO_2 modified W-Mn/SiO$_2$ catalysts for OCM was found, 20.3% yield of C_2^+ hydrocarbon and 29.7% CH$_4$ conversion with 33.5% selectivity of $C_3 \sim C_4$ could be obtained at a CH$_4$/O$_2$ ratio of 4 and $1.0 \times 10^5 h^{-1}$ GHSV (Fig.2); through investigation of reaction conditions, the 2.5-4.0 of CH$_4$/O$_2$, 1.0-$1.5 \times 10^5 h^{-1}$ of GHSV and 720-780℃ as the optimal operated condition was established. As SnO_2 addition in W-Mn/SiO$_2$ catalyst, CH$_4$ conversion and selectivity of C_2^+ were enhanced obviously. But SnO_2-Na-Mn/SiO$_2$ catalyst exhibits lower C_2^+ selectivity; addition of Na$_2$WO$_4$, the C_2^+ selectivity increases markedly because Na$_2$WO$_4$ is responsible for selectivity of higher hydrocarbons [1]. Thus a kind of synergistic reaction occurs positively among Sn, Mn and W over the SnO_2-promoted W-Mn/SiO$_2$ catalysts.

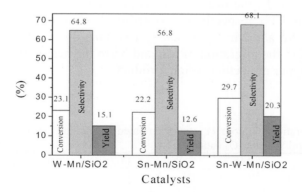

Fig. 2. Performance of SnO_2 promoted W-Mn/SiO$_2$ catalysts for the oxidative coupling of methane without diluent at elevated pressure under *P=0.6MPa*、 *CH$_4$/O$_2$=4*、 *T=750 ℃,GHSV=1.0×10^{5}h^{-1}*.

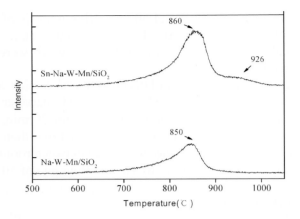

Fig. 3. Comparison between the O_2-TPD results of 8.0%SnO_2/W-Mn/SiO_2 and W-Mn/SiO_2 catalysts.

The comparison between the O_2-TPD results of 8.0%SnO_2/W-Mn/SiO_2 and W-Mn/SiO_2 catalysts were shown in Fig. 3. As may be seen, the oxygen storage capability in W-Mn/SiO_2 catalyst was enhanced with addition of SnO_2 so as to 1.5 times than that of W-Mn/SiO_2 catalyst. So the surface lattice oxygen in SnO_2 doped W-Mn/SiO_2 catalyst was responsible for methane activation, catalytic activity for OCM over Sn-W-Mn/SiO_2 catalyst (Fig.2) is increased the contribution due to enhancing oxygen storage capability.

Near-surface concentration and related binding energy of SnO_2/W-Mn/SiO_2 and W-Mn/SiO_2 catalysts are shown in Table 1. As can be seen, the near-surface concentration of Na and Mn are 5.8% and 1.3% over W-Mn/SiO_2 catalyst, after adding SnO_2 in W-Mn/SiO_2, the near-surface concentration of Na and Mn are increased to 8.2% and 3.1% respectively. So adding SnO_2 promotes obviously that the migration of Na and Mn to the catalyst surface due to the interaction between SnO_2 and SiO_2; the migration of Mn and Na to the near surface resulted in a marked shift to higher molecular weight products.

Table 1
Near-surface concentration (atom%) and related binding energy of 8%SnO_2/-W-Mn/SiO_2 and W-Mn/SiO_2 catalysts.

Cat.	Na_{1s}		Mn_{2p}		W_{4f}		$O_{1s}(SiO_2)$		$O_{1s}(MO_x)$	
	BE	In*	BE	In	BE	In	BE	In	BE	In
Na-W-Mn	1071.8	5.8	641.4	1.3	35.1	2.4	532.5	51.4	530.1	16.1
Sn-Na-W-Mn	1072.0	8.2	641.1	3.1	35.3	2.3	532.7	46.0	530.2	19.2

*In= Near-surface concentration (atom%).

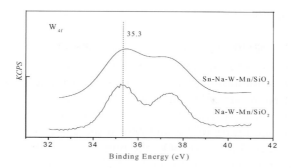

Fig. 4. XPS spectra in W_{4f} region of $SnO_2/W-Mn/SiO_2$ and $W-Mn/SiO_2$ catalysts.

When adding SnO_2 in $W-Mn/SiO_2$, the peaks of $W4f_{7/2}$ and $W4f_{5/2}$ had overlapped (Fig. 4); W^{5+} possibly existed over the surface of SnO_2-doped $W-Mn/SiO_2$ catalyst. The structural, catalytic performance and spectroscopic results indicated that SnO_2 played a dual role as chemical promoter, it was that promoted Mn and Na_2WO_4 migrate to near surface. A kind of transmittal mode about molecular oxygen was existed possibly between W and Mn or among W, Mn and Sn; the transmittal mode is due to existence of metal oxide being changeable valence in a series of W-Mn catalysts. The mechanism for OCM over SnO_2 doped $W-Mn/SiO_2$ catalysts should be described about equal to which of $W-Mn/SiO_2$ catalyst by using the Redeal-Redox mechanism [1].

The fresh SnO_2-W-Mn/SiO_2 catalyst and the catalysts after reaction at 0.6MPa and 3.5MPa were subjected to XRD measurement (Fig.5). The crystalline phase of the fresh catalyst is SnO_2, Mn_2O_3, Na_2WO_4 and α-cristobalite. $Na_2W_2O_7$ and tridymite are detected after reaction 8.0 hrs at 0.6MPa and do not change during further pressure and longer reaction time. It shows that the catalyst is steadier under high pressure.

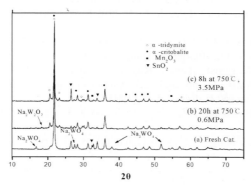

Fig. 5. XRD patterns of used and fresh $8\%SnO_2/W-Mn/SiO_2$.

4. CONCLUSION

A CH_4 conversion of 29.7% with C_2+ selectivity of 68% is obtained in a micro-stainless-steel reactor at elevated pressure. Lower temperature and lower CH_4/O_2 were more suitable for oxidative coupling of methane to C_{2+} formation over the SnO_2-doped W-Mn/SiO_2 catalysts. This reaction is especially sensitive to operating conditions, particularly CH_4/O_2 ratio and total pressure. Adding SnO_2 promotes the migration of Na and Mn to the catalyst surface, it is due to the interaction between SnO_2 and SiO_2; the migration of Mn and Na to the near surface resulted in a marked shift to higher molecular weight products; and the oxygen storage capability in W-Mn/SiO_2 catalyst was enhanced with addition of SnO_2. So the surface lattice oxygen in SnO_2 doped W-Mn/SiO_2 catalyst was responsible for methane activation, catalytic activity for OCM over Sn-W-Mn/SiO_2 catalyst is increased the contribution due to enhancing oxygen storage capability. When adding SnO_2, the peaks of $W4f_{7/2}$ and $W4f_{5/2}$ had overlapped; W^{5+} possibly existed over the surface of SnO_2 doped W-Mn/SiO_2 catalyst.

ACKNOWLEDGEMENT

The authors gratefully acknowledge financial support of this work by the Ministry of Science and Technology (Grant 1999022406), P.R. China, and by the Natural Science Foundation of China (Grant 20373081).

REFERENCE

[1] S.B. Li, Chin. J. Chem., 19 (1) (2001) 16.
[2] J.H. Lunsford, Angew. Chem. Int. Ed., 34 (1995) 970.
[3] X.P. Fang, S.B. Li, and J.Z. Lin, J. Mol. Catal. (China), 6 (1992) 225.
[4] Sergei Pak, P. Qiu, J.H. Lunsford, J. Catal., 179 (1998) 222.
[5] A. Palermo, J.P.H. Vazquez, A.F Lee., M.S. Tikhov, R.M. Lambert, J. Catal., 177 (1998) 259.
[6] D.J. Wang, M.P. Rosynek and J.H. Lunsford, J. Catal., 155 (1995) 390.
[7] A. Ekstrom, R. Regtop and S. Bhargava, Appl. Catal., 62 (1990) 253.
[8] Z.Q Yu, X.M. Yang, J.H. Lunsford and P. Rosynek, J.Catal., 154 (1995) 163.
[9] M. Pinabian-Carlier, A.B. Hadid and C.J. Cameron, in "Natural Gas Conversion" (A.Holmen, K. J. Jens and S. Kolboe, Eds.), (1991) 183.
[10] L.J. Chou, Y.C. Cai, B. Zhang, J.Z. Niu, S.F. Ji and S.B. Li, React. Kinet. Catal. Lett., 76(2) (2002) 311.
[11] L.J. Chou, Y.C. Cai, B. Zhang, J.Z. Niu, S.F. Ji and S.B. Li, Appl. Catal. A: General , 238 (2003) 185.

Studies in Surface Science and Catalysis, volume 147
X. Bao and Y. Xu (Editors)

Infrared study of CO and CO/H$_2$ adsorption on copper catalysts supported on zirconia polymorphs

Z.-Y. Ma, R.-Xu, C.-Yang, Q.-N. Dong, W.-Wei, Y.-H. Sun[*]

State Key Laboratory of Coal Conversion, Institute of Coal Chemistry, Chinese Academy of Sciences, Taiyuan 030001, China

ABSTRACT

Cu/ZrO$_2$ catalysts prepared with different zirconia polymorphs as supports were characterized by FT-IR technique to determine the effect of zirconia polymorphs on methanol synthesis from CO hydrogenation. The results showed that the zirconia polymorphs impacted greatly influence on copper state and behavior of CO adsorption. Simultaneously, the formation of intermediate species and methanol were also affected by zirconia polymorphs.

1. INTRODUCTION

Cu/ZrO$_2$ showed good performance for methanol synthesis from CO/CO$_2$ hydrogenation [1.2]. Mechanistic studies demonstrated that zirconia played an active role in the synthesis of methanol. Daniel [3] discovered that there existed a spillover of absorbed hydrogen and carbon monoxide in the interface of zirconia and copper. Fisher [4] also proposed that methanol synthesis mainly performed on zirconia by studying the surface intermediate species as CO/H$_2$ adsorption on Cu/Zr/SiO$_2$ catalyst. On the other hand, zirconia possesses three polymorphs, namely, monoclinic, tetragonal and cubic, which had different performance for catalytic reactions. Our previous studies revealed that zirconia polymorphs performed dramatically different capacities for CO adsorption and conversion [5]. In this paper, three Cu/ZrO$_2$ catalysts prepared by dispersing copper on amorphous, monoclinic and tetragonal zirconia polymorphs were characterized by in-situ FT-IR with the main goal to determine the effect of zirconia polymorphs on CO adsorption and CO/H$_2$ reaction.

2. EXPERIMENTAL

Cu/ZrO$_2$ catalysts were prepared by impregnation cupric nitrate with amorphous

---

[*] Corresponding author. Tel: +86 351 4049612; E-mail: yhsun@sxicc.ac.cn

(am-), monoclinic (m-) and tetragonal (t-) zirconia followed with drying and calcination at 623 K for 3 h. FT-IR spectra of Cu/ZrO$_2$ were collected using an in-situ cell on a Nicolet Magna-550-II equipped with TGS detector at a resolution of 4 cm^{-1}. 15 mg sample was pressed into a Φ15 mm self-supporting wafer and pre-reduced at 573 K under H$_2$ atmosphere for 4 h and then evacuated (<10^{-5} Pa) at the same temperature for 1 h. After cooling to the ambient temperature, the reference spectrum was collected. Thereafter, the CO or CO/H$_2$ (1:2 molar radio, 60 Torr) was introduced at gradually increasing temperature for 20 min and then the sample was evacuated at ambient temperature to purify the gas CO. The spectrum was obtained by collecting 64 scans at ambient temperature.

3. RESULTS AND DISCUSSION

3.1. CO adsorption

The FT-IR spectra of CO adsorbed on copper containing zirconia polymorphs recorded at different temperatures were shown in Fig. 1. A band at about 2127 cm^{-1} was detected when CO adsorbed on Cu/am-ZrO$_2$ at 298 K. It shifted to 2117 cm^{-1} and the peak intensity reached the maximum when the adsorption temperature increased to 373 K, and then the peak intensity decreased continuously along with the temperature increasing from 473 K to 623 K. The analogical behavior was observed when CO adsorbed on Cu/m-ZrO$_2$ catalyst with the different peaks located at 2113 and 2123 cm^{-1} with temperatures increasing. As reported by Kohler [6] that the band at 2127 cm^{-1} or 2123 cm^{-1} could be ascribed to CO adsorption on Cu$^+$ sites, which then shifted to lower frequencies for the change of copper electronic state by CO reduction at high temperature. But for Cu/t-ZrO$_2$, only a band located at 2113 cm^{-1} appeared and the position hardly changed with adsorption temperature increas, suggesting that metal copper phase existed on the catalyst surface after reducing by H$_2$. Moreover, CO disproportionation was reported to occur on the surface of the catalyst, i.e. 2CO $\rightarrow$ C+CO$_2$, and the reaction was activated at high temperatures. The graphite carbon deposited on the active sites, which then enshrouded the CO adsorption and resulted in the decrease of band intensity.

Corresponding to CO adsorption and reaction on three Cu/ZrO$_2$ catalysts, the intermediate species were observed in the region of 1800-1000 cm^{-1} at different temperatures. Several bands at 1608, 1444, 1328, and 1224 cm^{-1} with a shoulder at 1573 cm^{-1} appeared on Cu/am-ZrO$_2$ when CO adsorbed at 298 K. The band at 1608, 1328, 1224 cm^{-1} could be ascribed to bicarbonate species [7], which could be proved as the bidentate forms by the position of v_{as}C-O at 1608 cm^{-1}. While the band situated at 1444 cm^{-1} was the characteristic of ion carbonate, and the shoulder at 1573 cm^{-1} was attributed to formate with another band overlapped by band at 1328 cm^{-1} [7]. When adsorption temperature increased to 373 K, the

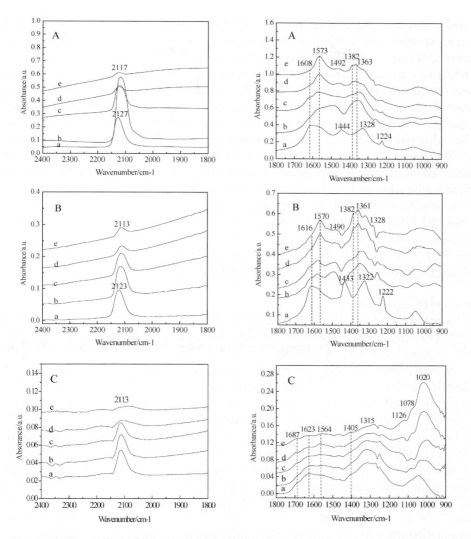

Fig. 1. FT-IR profiles of CO adsorbed on Cu/am-ZrO$_2$ (A), Cu/m-ZrO$_2$ (B) and Cu/t-ZrO$_2$ (C) at different temperatures. (a: 298 K, b: 373 K, c: 473 K, d: 573 K, e: 623 K)

band at 1608, 1328, 1224 cm^{-1} weakened and the ion carbonate disappeared completely, but two bands located at 1492 and 1356 cm^{-1} appeared, indicating the transformation of unidentate carbonate from bidentate bicarbonate or ion carbonate. The band at 1573 cm^{-1} became stronger when the temperature increased to 573 K due to more surface hydroxyl involved to form formate. And the formate species was almost stable even at 623 K as well as unidentate carbonate. For Cu/m-ZrO$_2$, ion carbonate and bidentate bicarbonate species also presented at 298 K. And the unidentate carbonate could be observed with the intensity decreas along with the rise of temperature. It could be found that formate species was formed at low temperature on Cu/m-ZrO$_2$ and Cu/am-ZrO$_2$ catalysts. However, weak formate band located at 1564cm^{-1} was observed at 573 K when CO adsorbed on Cu/t-ZrO$_2$, and the carbonate and bicarbonate species could be detected even up to 623 K, which was different from that on Cu/am-ZrO$_2$ and Cu/m-ZrO$_2$ catalysts. Copper introduction did not change the properties that t-ZrO$_2$ possessed weaker capacity of formation formate compared to the other two supports [5]. It was interested to note that when CO adsorbed on Cu/t-ZrO$_2$ several new bands situated at 1020, 1078 and 1128 cm^{-1} emerged and became obvious with the temperature increasing, which could be ascribed to the different vibration of C-O of the methoxide species [4]. The hydrogen, required for the hydrogenation of formate species to methoxide, could be provided as Schild [8] revealed that CO could react with the surface hydroxyl to form carbonate and hydrogen on the zirconia surface at higher temperature: $CO+2OH^- \rightarrow CO_3^{2-} +H_2$, The phenomenon might be due to the predominance of carbonate and bicarbonate species for Cu/t-ZrO$_2$ catalyst.

3.2. CO+H$_2$ adsorption

The behavior of CO/H$_2$ reaction on three Cu/ZrO$_2$ catalysts was illustrated in Fig. 2. For Cu/am-ZrO$_2$ catalyst, the behavior of CO adsorption was similar to those without hydrogen. At the same time, the CO$_2$ bands appeared for the sake of water-gas shift (WGS) and the bands intensity increased with the rise of adsorption temperature. On the other hand, when syngas adsorbed below 473 K, similar intermediate species could be detected as those of CO adsorption on Cu/am-ZrO$_2$ catalyst. However, when adsorption temperature was above 473 K, the formate became the major species and bidentate bicarbonate was hardly observed, while the intensity of formate and bicarbonate was almost equal when CO adsorbed on Cu/am-ZrO$_2$ catalyst at the same temperature. As adsorption temperature increased further, the intensity of formate decreased and a broad new band located at 1012 cm^{-1} appeared. Keeping it in mind that H$_2$ could be dissociated to atomic hydrogens on copper surface and then spilled over to zirconia surface [4], it could be concluded that Cu enhanced the formation of formate and methoxide species by providing efficient atomic hydrogen to the carbonaceous species presented on the supports.

In the case of CO/H$_2$ adsorption on Cu/m-ZrO$_2$ catalyst, bi dentate bicarbonate, unidentate carbonate and formate species could be detected below 473 K, which was also similar to those of CO adsorption. As adsorption temperature increased further, the intensity of formate bands weakened gradually, which was hardly detected till 623 K. Simultaneously, several new bands located at 1025, 1076 and 1118 cm^{-1} attributed to the gas methanol were observed, which was distinctive to that of CO adsorption on Cu/m-ZrO$_2$. Schild [8] considered that π-bond formaldehyde, which formed on Cu/ZrO$_2$-based catalysts in the catalytic

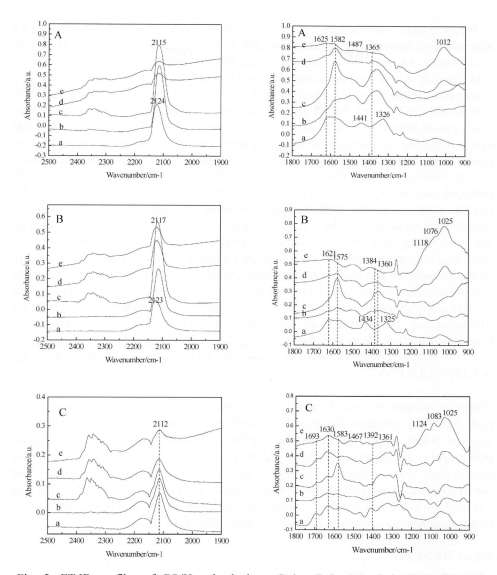

Fig. 2. FT-IR profiles of CO/H₂ adsorbed on Cu/am-ZrO₂ (A), Cu/m-ZrO₂ (B) and Cu/t-ZrO₂ (C) at different temperatures. (a: 298 K, b: 373 K, c: 473 K, d: 573 K, e: 623 K)

reaction of CO hydrogenation, was the key intermediate species, and the subsequent reduction yielded surface-bond methanol. But in our present work no formaldehyde species could be detected. Then the conclusion could be drawn that the presence of hydrogen reduced the formate to methoxide, this was in line with the formate-to-methoxide mechanism postulated by Sun [1.2].

When syngas adsorbed on Cu/t-ZrO$_2$ catalyst, the CO band shifted little with adsorption temperature increase from 298 K to 623 K, and gas CO was almost disappeared. Similar to CO adsorption at low temperature, no formate species was formed when syngas adsorption on Cu/t-ZrO$_2$. It was noted that when adsorption temperature increased to 473 K, formate species with bands at 1583 and 1361 cm^{-1} could be obviously detected, which was formed at lower temperature than that of CO adsorption. These differences were also attributed to the presence of hydrogen in syngas current, which made the formate species easily formed at lower temperatures from carbonate species hydrogenating. Also, it was similar to the other two catalysts that gas methanol was obviously detected and reached maximum when adsorption temperature elevated to 623 K. On the other hand, the formate species almost disappeared at 573 K and the carbonate species increased slightly. These indicated that the formate species was completely converted to methanol species at 573 K, and the formation of methanol was easier on Cu/t-ZrO$_2$ than on Cu/am-ZrO$_2$ and Cu/m-ZrO$_2$ catalysts. With the increase of reaction temperature the peak intensities of bicarbonate and bidentate carbonate species decreased slightly in company with the enhancement of methanol species, which made it clear that the carbonate species could be hydrogenated to formate in the presence of hydrogen following the formate-to-methoxide mechanism.

4. CONCLUSION

Zirconia polymorphs greatly affected the copper state and CO adsorption as well, and then also affected the formation and transformation of the intermediate species during CO hydrogenation. Cu/ZrO$_2$ catalysts with different zirconia polymorphs as supports showed different capacity of intermediate species conversion to methanol in the presence of hydrogen, while the formate-to-methoxide mechanism was followed for three Cu/ZrO$_2$ catalysts. The Cu/t-ZrO$_2$ exhibited higher properties for methanol synthesis from CO hydrogenation than that of Cu/am-ZrO$_2$ and Cu/m-ZrO$_2$ catalysts.

REFERENCES

[1] Y. Sun and P.A. sermon, J. Chem. Soc. Chem. Commun. (1993) 1242.
[2] Y. Sun and P.A. sermon, Catal. Lett., 29, (1994) 361
[3] Daniel Bianchi, Tarik Chfik, Jean Teichaner et al., Appl. Catal., 112 (1994) 57
[4] Ian A. Fisher, Alexis T. Bell, J. Catal., 178 (1998) 153
[5] Z. Ma, R. Xu, C. Yang, W. Wei, and Y. Han, Preprints of ACS Div. of Fuel Chem. 45 (2003) 193
[6] M.A. Kohler, N.W. Cant, M.S. Wainwright, and D.L. Trimm, J. Catal., 117 (1989) 188.
[7] Daniel Bianchi, Tarik Chafik, Mohamed Khalfallah, et al. Appl. Catal., 105 (1993) 223.
[8] C. Schild, A, Wokaun, and A. Barker, J. M. Catal., 63 (1990) 223

Studies in Surface Science and Catalysis, volume 147
X. Bao and Y. Xu (Editors)

Nonoxidative aromatization of methane over Mo/MCM-49 catalyst

Qiubin Kan, Dongyang Wang, Ning Xu, Peng Wu, Xuwei Yang, Shujie Wu and Tonghao Wu*

Department of Chemistry, Jilin University, Changchun 130023, China

ABSTRACT

MCM-49 with perfect hexagonal shape and about 300~500 nm in diameter and 25~50 nm in thickness has been hydrothermally synthesized, which was used as support to prepare Mo/MCM-49 catalyst. Methane aromatization over Mo/MCM-49 catalyst was carried out in the absence of oxygen, which shows high activity, benzene yield and long catalyst life. Effects of Mo content, reaction temperature and feed space velocity on the catalytic performance of Mo/MCM-49 catalyst was investigated.

1. INTRODUCTION

The conversion and utilization of natural gas have become worldwide hotspot as the petroleum resources have been reducing. Non-oxidative aromatization of methane is an important catalytic reaction in the ways of methane conversion. Since the Mo/ZSM-5 catalyst with high activity and selectivity for methane dehydro-aromatization in the absence of oxygen was firstly reported by Dalian Institute of Chemical Physics, Chinese academy of Science [1], much attention have been paid to this catalyst system [2-4]. Up to now the best result achieved over Mo/ZSM-5 catalysts is ~11% of methane conversion, ~75% of benzene selectivity and less than 10 hour of catalyst life. Clearly improving the stability and selectivity of catalyst is a difficult subject because of seriously coking and Mo leaching during reactions. Many research works have been done by changing different active component and zeolite support. By comparing studies over catalysts of the different molecular sieve supports, such as ZSM-5, ZSM-8, ZSM-11, ZSM-22, Beta, Y, HM, SAPO-34, MCM-41, etc, the most beneficial support structures were understood. They should contain two dimensional 10-member ring cross channels [5,6]. In addition, the support should also have suitable acidity. Although a lot of research works have been well done, it is pity that no other catalyst systems better than Mo/ZSM-5 were found before the year

2000. Based on the previous conclusion of the relations between the catalytic performance and catalyst support structure, we considered MCM-22 zeolite family with MWW structure including MCM-22, MCM-49, MCM-56, ITQ-1, ITQ-2 as suitable catalyst support for methane conversion into aromatics[7,8] because they have 10-member ring cross channels slight less than ZSM-5 and thermal stability and acidity similar with ZSM-5. Moreover, considering the effect of the channel length on the diffusion of reactants and products, we decided to prepare nano-size MCM-49 as Mo-based catalyst support and to carry out the non-oxidative aromatization reactions of methane under the same conditions as used over Mo/ZSM-5 catalysts.

2. EXPERIMENTAL

2.1. Synthesis and Characterization of MCM-49

Synthesis mixture with composition of $1.0SiO_2 : 0.03\sim0.05Al_2O_3 : 0.05\sim0.09Na_2O : 0.05\sim0.35HMI : 12\sim35H_2O$ was prepared using sodium aluminate (Al_2O_3 44.7%, Na_2O 39.2%), sodium hydroxide, hexamethyleneimine (99%, Sigma) as the template, and silica gel (Na_2O 0.23 wt% and SiO_2 25 wt%) as the silica source. A typical synthesis procedure to prepare highly crystalline MCM-49 was as follows: 4.56 g of sodium aluminate and 20 ml of HMI were dissolved in 80 ml of distilled water. To this solution 120 g of silica gel was added with stirring. After becoming homogeneous, the slurry was then transferred to Teflon-lined stainless-steel autoclaves and heated at 443 K under the condition of 60 rpm rotation for 72 hrs. The resulting solids were recovered by filtration and washed with distilled water until the filtrate was neutral. The wet products were dried overnight in air at 393 K to obtain the as-synthesized samples. The occluded organic was removed by heating the as-synthesized samples in a quartz tube at 811 K for 3 hrs under a continuous flow of nitrogen before a flow of air was passed at 823 K for 6 hrs to obtain the calcined samples.

As-synthesized and calcined samples were examined by using XRD for phase identification. XRD measurements were carried out using CuKα radiation on the SHIMADZU XRD-6000 diffractometer in the 2θ range of 5~40°. The morphology and size of MCM-49 crystals were determined using a HITACHI H-8100 transmission electron microscope.

2.2. Preparation of Mo/MCM-49 catalysts and catalytic test

H-MCM-49 was obtained by three times repeated ion exchange of the calcined MCM-49 with an aqueous solution of 1 mol L^{-1} ammonium nitrate and calcination in air at 773 K for 3 hrs. The Mo/MCM-49 catalysts with different Mo loadings were prepared by mechanical mixing of MoO_3 and H-MCM-49 and calcination in air at 773 K for 3 hrs. Catalytic tests were carried out under atmospheric pressure in a continuous flow fixed bed quartz reactor of 1 cm ID, which was charged with 0.5 g of catalysts of 40~60 mesh. The feed gas mixture of 92.46% methane and 7.54% nitrogen was introduced into the reactor at 1500 ml g^{-1} h^{-1} through a mass flow controller and the temperature was raised to 973 K within 45 mins and kept at the temperature during reaction. An on-line GC-17A gas chromatograph was used, equipped with a 6m×3mm HayeSep D 80/100 column to analyze H$_2$, N$_2$, CO, CO$_2$, CH$_4$, C$_2$H$_4$, C$_2$H$_6$ by on-line TCD and a CBP1-M50-025 quartz capillary to analyze aromatics such as benzene, toluene, naphthalene by on-line FID. The methane conversion and the product selectivity were calculated on a carbon number base by inner standardization method.

3. RESULTS AND DISCUSSION

3.1. Synthesis and characterization of MCM-49

XRD patterns of typical MCM-49 samples (Fig not shown here) show that the as-synthesized sample has the structure of MCM-49 zeolite and several peaks in the XRD spectra are broad and overlap. Upon calcination these peaks become sharper and some new peaks appear. A comparison of the line positions and the relative intensities of the peaks between our samples and the ones reported in literature [1] have been made and fairly identical results are obtained.

Transmission electron micrographs of standard sample of MCM-49 are showed in Fig. 1. Indeed the most reports in literature indicted that MCM-49 crystals were in form of round lamellar particles or spherical aggregates composed of aggregated platelets. However, the MCM-49 synthesized here is the crystals of very thin regular hexagonal sheets, about 300~550 nm in diameter and 20~50 nm in thickness. This perfect hexagonal shape is a reflection of P6/mmm symmetry of the framework structure. Differing from the description by other reports, our samples are the uniform separate single crystals and have not aligned further into larger lamellar particles. Viewed perpendicular to the c direction, the single crystal consists of several layers and a TEM image of MCM-49 is able to distinguish the position of the sinusoidal channels and the supercages.

Fig. 1. TEM photos of as-synthesized MCM-49

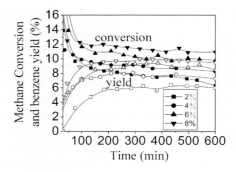
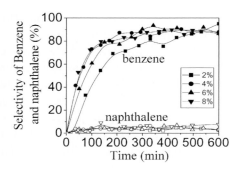

Fig. 2. Effect of Mo loading on methane
conversion and benzene yield

Fig. 3. Effect of Mo loading on selectivity of
benzene and naphthalene over Mo/MCM-49

3.2. Catalytic conversion of methane over Mo/MCM-49 catalysts

3.2.1. Effect of Mo contents on the catalytic performance of Mo/MCM-49 catalysts

Fig. 2 and Fig. 3 give the results of methane aromatization in the absence of oxygen at 973 K over the Mo/MCM-49 catalysts with different Mo contents. Being a highly selective catalyst for the benzene and naphthalene, Mo/HMCM-49 produces trace amount of C_2 hydrocarbons and toluene. The amount of coke deposition is important information, which can be obtained by GC analysis through adding nitrogen as standard. Here we only discuss methane conversion and the production of benzene and naphthalene. The effects of MoO_3 loading on the catalytic performance were examined on 10 h time on stream. It was found that methane conversion significantly increases after the introduction of Mo to the MCM-49. However, the optimum Mo content in the catalysts is about 6~8 wt% for the dehydro-aromatization reaction of methane. Further addition of Mo results in a decrease in the activity and selectivity of the catalyst for methane conversion to benzene and naphthalene. Benzene formation reaches a maximum value after running for 200~400 mins. The yield of benzene is obviously enhanced on the 8 wt% loading catalyst and the highest value is 10%. When the Mo content is less than 4%, the methane conversion does not exceed 9% after methane is on stream for 100 mins. It is obvious that with the decrease in Mo loading, the selectivity for benzene is markedly decreased.

3.2.2. Effect of reaction temperature on the catalytic performance of Mo/MCM-49 catalyst

As expected from kinetic and thermodynamic considerations, high temperature will be advantageous to methane conversion. The influence of

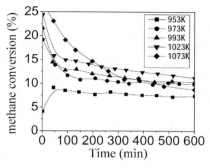

Fig. 4. Effect of reaction temperature on
methane conversion over Mo/MCM-49

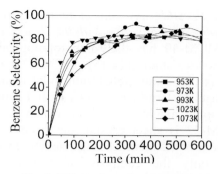

Fig. 5. Effect of reaction temperature on
benzene selectivity over Mo/MCM-49

temperature on the reaction over 6 wt% (MoO$_3$) Mo/MCM-49 catalyst is shown
in Fig. 4 and Fig. 5. A significant increase of methane conversion can be
observed when raising the reaction temperature from 953 K to 1073 K. However,
if the temperature is too high, for example 1073K, serious carbon deposition
occurs, and therefore the methane conversion quickly decrease with the reaction
times and the selectivity for benzene becomes lower than that at lower reaction
temperatures. A proper temperature range for the methane aromatization is
973~1023 K.

3.2.3. Effect of feed space velocity on the catalytic performance of Mo/MCM-49 catalyst

The feed space velocity of methane exerts significant influences on the
reaction of methane. Fig. 6 and Fig. 7 reveal that with the increase in the space
velocity of methane (e.g. >1200ml/g.h), methane conversion substantially
decreases, while benzene yield has not big change. In fact a proper scope of the
space velocity is between 750 ml g^{-1} h^{-1} and 1200 ml g^{-1} h^{-1}.

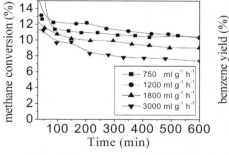

Fig. 6. Effect of feed space velocity on
methane conversion over Mo/MCM-49

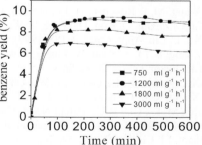

Fig. 7. Effect of feed space velocity
on benzene yield over Mo/MCM-49

3.2.4. Possible association of the activity and selectivity of Mo/MCM-49 for methane aromatization with the structure of MCM-49 support

MCM-49 has the same topological structure as MCM-22. According to Leonowicz'studies[14,15], MCM-22 possesses two independent pore systems, one is two-dimensional sinusoidal channels of 10-member ring and the other consists of large supercages with 12-member ring inner diameters of 0.71 nm and inner height of 18.2 nm which access through 10-member ring apertures. MCM-22 is sheet crystal, on the exterior surface of which there exist large amount of 12-member ring pockets. In previous investigation it has been found that only the catalysts of zeolite supports containing two dimensional 10-member rings cross channels in Mo/zeolite catalysts had good catalytic performances for non-oxidative aromatization of methane[5,6] and the catalysts of zeolite supports with large pores or one-dimensional pore channels show poor activity and selectivity. In this point of view, it is considered that high activity and selectivity for methane aromatization over Mo/MCM-49 should be related to the cross channels of 10-member ring. Because nano-sized MCM-49 as support of Mo/MCM-49 has positive effect on aromatization reaction of methane in comparison with larger size of MCM-49 crystals, it is concluded that shorter channels of nano-sized MCM-49 may be beneficial to the diffusion of products of aromatics out the pores of the catalyst and/or to the increment of Mo active sites, which also implies that diffusion of aromatic products in the pore channels of MCM-49 should influence overall reactions.

ACKNOWLEDGEMENTS

This work is supported by NSFC (20273025) and the Ministry of Science of China (1999022400).

REFERENCES

[1] L. Wang, L. Tao, M. Xie, G. Xu, J. Huang, Y. Xu, Catal. Lett. 21 (1993) 35.
[2] D. Wang, M. P. Rosynek, H. Lunsford, J. Catal. 169(1997) 347.
[3] S. Li, C. Zhang, Q. Kan, D. Wang, T. Wu, L. Lin, Appl. Catal. A: 187(1999) 199.
[4] Y. Xu, L. Lin, Appl. Catal. A 188(1999) 53.
[5] C. Zhang, S. Li, Y. Yuan, W. Zhang, T. wu, L. Lin, Catal. Lett. 56 (1998) 207.
[6] S. Li, Doctor Thesis, 2000, 84.
[7] N. Xu, Q. Kan, X. Li, Q. Liu, T. Wu, Chem. J. Chinese U. 21 (2000) 949.
[8] N. Xu, Q. Kan, J. Zhang, X. Li, L. Ji, T. Wu, Chem. J. Chinese U. 21 (2000) 1113.
[9] Rubin M. K.; Chu P. US Patent No. 4 954 325 (1990).
[10] Shang Li, Chun-Lei Zhang, Yi Yuan et al.. Chem. J. Chinese U., 12 (1998) 2005.
[11] Shuang Li, Chunlei Zhang, Qiubin Kan et al., Appl. Catal. A: 187 (1999) 199-206
[12] Shuang Li, Ding Ma, Qiubin Kan et al., Reat. Kinet. Catal. Lett., 70(2000) 349.
[13] Shuang Li, Chunlei Zhang, Qiubin Kan et al., Stud. Surf. Sci. Catal, 130 (2000) 3609.
[14] M.E. Leonowicz, J.A. Lawton, et al, Science, 264 (1994) 1910.
[15] A. Lawton, S.L. Lawton, M.E. Leonowscz, et al, Stud. Surf. Sci. Catal., 98 (1995) 250.

Studies in Surface Science and Catalysis, volume 147
X. Bao and Y. Xu (Editors)

Selective oxidation of methane to C_1-products using hydrogen peroxide

T.M. Nagiev, L.M. Gasanova, E.M. Mamedov, I.T. Nagieva, Z.Y. Ramasanova, and A.A. Abbasov

Institute of Chemical Problems, National Academy of Sciences of Azerbaijan, 29 H. Javid av., 370143 Baku, the Republic of Azerbaijan

ABSTRACT

There were presented the results of experimental investigations on synchronization between natural gas selective oxidation and hydrogen peroxide's decomposition. It was shown that methane synchronous gas-phase oxidation by hydrogen peroxide is flexibly controlled towards both formaldehyde selective formation and hydrogen-containing gas or methanol production. The usage of heterogeneous biomimetic catalyst $PPFe^{3+}OH/AlSiMg$ allowed carrying out the process of methane direct selective synchronous oxidation to methanol by hydrogen peroxide under mild conditions with a high yield of methanol (46.5%) at a selectivity of 97%.

1. INTRODUCTION

Single-carbon molecules are gaining importance for modern science and industry because of their accessibility and the possibility to synthesize a variety of organic compounds on their basis.

There was studied a chemical system comprising synchronized reaction of hydrogen peroxide decomposition and methane oxidation to C_1-products. In such system methane oxidation, depending on process parameters, goes in different directions to form respective products. Hydrogen peroxide is applied as an oxidant, because it embraces high inducing activity along with efficiency and specificity of its action.

The core of hydrogen peroxide's inducing activity consists in the fact that being transformed to the active form – $\cdot OH$ and $HO_2\cdot$ radicals (primary reaction), hydrogen peroxide is selectively consumed in another conjugated reaction forming respective products. The way these two conjugated brutto-reactions take is of a synchronized character:

$$H_2O_2 + H_2O_2 \rightarrow [\dot{O}H, HO_2\dot{}] \rightarrow 2H_2O + O_2 \qquad \text{(primary reaction)} \qquad (1)$$

$$H_2O_2 + CH_4 \longrightarrow \begin{cases} CH_2O + H_2O \\ H_2, CO, CO_2 \\ CH_3OH + H_2O \end{cases} \qquad \text{(secondary reaction)} \qquad (2)$$

since a minimum of molecular oxygen accumulation is in agreement with a maximum of methane conversion [1].

2. EXPERIMENTAL

Experimental investigations that revealed chemical conjugation in oxidation reactions by hydrogen peroxide in the gas phase have been carried out in a through-flow integral quartz glass reactor, the construction of which ensured the introduction of H_2O_2 into the reaction zone in the undecomposed form. By special experiments it was shown that under H_2O_2's adding into the reaction zone through the quartz tube at a velocity of 6-40cm/s, the concentration of H_2O_2 stays constant until its reaching the reaction zone.

Via injection syringe pressed down by mechanical device (electromotor-driven) water solution of hydrogen peroxide (15÷25w%) was added directly to pre-reaction zone through quartz tube separately from the substrate being oxidized. Conical quartz tube was 95mm long with a diameter of 5÷10mm at the input and 4mm at the output. High linear velocity of H_2O_2 in the tube (6-40cm/s) excluded any possibility of its decomposition until it is added into the reaction zone. Under these conditions one could keep concentration of H_2O_2 as nearly as it was initially adjusted. Any possibility of molecular oxygen's participation as an initial oxidant in the process has been excluded.

Through the quartz tube, separately from H_2O_2, a natural gas (96% CH_4) has been added to the reactor, and then during the reaction it reacted with H_2O_2.

Hardening zone adjoined the reaction zone (at a distance of 15÷20mm). In that place the tube tapered critically. Such a hardening system excluded any possibility of reagents and products being oxidized beyond the reaction zone.

Oxidation of methane by hydrogen peroxide was carried out at low contact times in a through-flow system (τ=0.8÷2.35s), which to a high extent excluded possibility of products further transformation during secondary reactions.

As to the participation in the reaction of O_2 formed as a result of hydrogen peroxide's decomposition, it should be noted that the reaction medium residence time in the through-flow reactor, in our case, did not allow O_2 participating in the process [2].

3. DISCUSSION AND RESULTS

There was studied a single-stage process for formaldehyde production via conjugated homogeneous oxidation of methane by hydrogen peroxide [3]. The given process featured the fact that no methanol was formed as a semiproduct. It was shown a high efficiency of hydrogen peroxide as a generator of highly reactive intermediates capable of oxidizing methane selectively to formaldehyde and under optimum conditions (T=520°C, τ=1.2s, $CH_4 : H_2O_2 = 1 : 1$, $C_{H_2O_2} =$ 25w%) the yield of formaldehyde made up about 40% at a high selectivity of 94% (see Table 1). Formaldehyde was formed as a result of methane direct oxidation by-passing the stage of methanol formation. To support this viewpoint the results of experiments are shown in Table 1.

Table 1

Conjugated oxidation of CH_4 and CH_3OH by H_2O_2 (T=520°C, $C_{H_2O_2}$ = 25w%)

Experiment	n	v_{CH_4} L/h	$v_{H_2O_2}$ mL/h	v_{CH_3OH} mL/h	Yield of products, w%				c_{CH_4} %	c_{CH_3OH} %	N, %
					CH_2O	CH_3OH	CO_2	CO			
	$CH_4:H_2O_2$				$CH_4\text{-}H_2O_2\text{-}H_2O$						
1	1:0.24	1.2	1.44	-	6.8	-	-	-	6.8	-	100
2	1:0.5	"-"	3.2	-	14.0	1.0	-	-	15.3	-	97.5
3	1:1	"-"	5.74	-	39.0	2.4	-	-	41.4	-	94.0
4	1:1.2	"-"	7.2	-	35.5	7.5	-	3.2	45.5	-	78.0
5	1:1.6	"-"	9.5	-	30.0	13.0	2.4	5.6	50.0	-	60.0
	$CH_3OH: H_2O_2$				$CH_3OH\text{-}H_2O_2\text{-}H_2O$						
6	1:0.25	-	1.44	1.44	4.5	-	2.0	-	-	6.5	69.0
7	1:0.5	-	2.32	"-"	9.2	-	4.0	1.6	-	14.0	67.5
8	1:1	-	4.72	"-"	7.9	-	5.8	2.5	-	16.2	48.8
	$CH_4:CH_3OH:H_2O_2$				$CH_4\text{-}CH_3OH\text{-}H_2O_2\text{-}H_2O$						
9	0.7:0.4:1	0.8	5.74	0.72	2.5	-	5.9	0.5	4.3	16.1	-
10	0.8:0.2:1	0.82	"-"	0.36	3.7	-	4.3	1.6	6.2	13.4	-

n – molar ratio; c – conversion; N – selectivity; in experiment 8 the yield of CHOOH – 1.3w%; in experiments 9 and 10 the yield of C_2H_6 – 0.9 and 0.7w%; C_3H_6 – 0.17 and 0.1w%; C_3H_8 – 0.25 and 0.2w%.

In the course of these experiments under the conditions of formaldehyde selective formation methanol was added into the reaction system in various amounts. Given the results of these experiments it follows that the formation of formaldehyde in the presence of methanol is of nonselective character, i.e. it features a big number of by-products - C_3H_6, C_3H_8, C_2H_6, CO and CO_2. The formation of formaldehyde in small amounts is caused by the presence of methanol in the initial mixture in the amount up to 40w%, because under these conditions H_2O_2 is probably consumed on methanol's deep oxidation.

First ever under homogeneous conditions by chemical induction without applying any heterogeneous catalysts it came possible to effectively carry out high-temperature oxidation of methane to molecular oxygen and carbon dioxide by appreciable reduction of reaction temperature (the yield of H_2 is up to 74 mol %, CO_2 - 23%, CO -1.5%) [4].

The experimental data on the effect of contact time on methane oxidation process (Fig. 1) show that an increase in the contact time (τ) raises the extent of methane conversion. With an increase in τ afteroxidation of CO to CO_2 occurs, that leads to a decrease in CO content down to quite small amount even at τ=0.6s, whereas the yield of H_2 reaches its maximum at 74mol%.

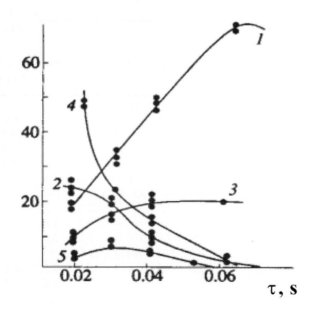

Fig. 1. Dependence of gas content (mol%) change in reaction gaseous mixture on the contact time (τ) at T=880°C, $C_{H_2O_2}$ = 15w%, CH_4: H_2O_2=1:3; 1 – H_2, 2 – CO_2, 3 – CO, 4 – CH_4, 5 – O_2.

The chemical system being developed features flexibility, which, by intelligent control of reaction rate, enables one, depending on the purpose, to get quantitatively varied composition of products such as H_2, CO and CO_2.

For each of these oxidation reactions there were proposed free radical mechanisms. From the mechanisms proposed it is seen that the relation between indicated synchronous reactions Eq. (1) and Eq. (2) is set through elementary stages by means of intermediary radicals such as $\cdot OH$ and $HO_2\cdot$.

There was studied the process of direct oxidation of methane by hydrogen peroxide to methanol under pressure, which allowed, compared to other existing processes, to reduce the temperature and pressure, increase conversion and selectivity: at $T=400°C$, $p=7$atm, the yield of methanol – 17.6%, selectivity – 99.5%.

It was studied and developed detail mechanism of methanol production. It was shown that the synchronization between the reactions of hydrogen peroxide decomposition and methane oxidation in the gas phase, inducing gas-phase homogeneous oxidation, takes place in the case of heterogeneous catalyzed oxidation, too.

The biomimetic catalyst synthesized by us and mimicking the mechanism of monooxygenase enzyme's (cytochrome P-450) function allowed developing a new highly selective catalytic system. The given system enabled gas-phase oxidation of such inert gas as methane under mild conditions featuring high efficiency: the yield of methanol reached 46.5% at $T=180°C$.

The given works led to the development of extremely active monooxygenase mimic – $PPFe^{3+}OH/AlSiMg$ [5]. The studied mimic is capable of catalyzing two interrelated synchronous reactions – catalase and monooxygenase. Synchronization mechanism is realized through common reactive intermediate – $PPFe^{3+}OOH/AlSiMg$, which is consumed in both reactions but with different rate.

REFERENCES

[1] T.M. Nagiev, Russian J. Phys. Chem. 74 No.11 (2000)1853.
[2] T.M. Nagiev, L.M. Gasanova, S.Z. Zulfugarova, Ch.A. Mustafaeva, and A.A. Abbasov, Chem. Eng. Comm., 190 No. 5-8 (2003) 726-748.
[3] T.M. Nagiev and L.M. Gasanova, Khim. Fizika, 3 No. 10 (1984) 1455-1461.
[4] T.M. Nagiev, L.M. Gasanova, and Z.U. Ramasanova, Zhurnal Fiz. Khim. 68 No.1 (1994) 23-28
[5] T.M. Nagiev and M.T. Abbasova, Russian J. Phys. Chem. 71 7 (1997) 1088-1092.

Studies in Surface Science and Catalysis, volume 147
X. Bao and Y. Xu (Editors)

Methane activation on alumina supported platinum, palladium, ruthenium and rhodium catalysts

Ruth L. Martins[a]**, Maria A. Baldanza**[a]**, Mariana M.V.M. Souza**[a] **and Martin Schmal**[*a,b]

[a]NUCAT/PEQ/COPPE, Universidade Federal do Rio de Janeiro, C.P. 68502, 21945-970, Rio de Janeiro, Brazil. [*]E-mail: schmal@peq.coppe.ufrj.br

[b]Escola de Quimica, Universidade Federal do Rio de Janeiro, C.P. 68542, 21940-900, Rio de Janeiro, Brazil.

ABSTRACT

Methane activation was conducted on Pt, Pd, Ru and Rh alumina supported catalysts. Simultaneously with the gas chemisorption, H_2, CO_2 and CO were evolved from all the catalyst surfaces. The H_2 evolution resulted from the association of hydrogen atoms provided from dissociative chemisorption of methane. CO and CO_2 evolution was explained by the partial migration of carbon adspecies, from metal surface to the metal-support interface, followed by their interaction with the alumina O^{2-} ions.

1. INTRODUCTION

The direct methane conversion into more valuable chemicals remains a challenge for catalyst researchers. By far, the most promising route has been the oxidative coupling [1, 2] although the parallel formation of carbon oxides still results into poor selectivities to C_2 products. Another approach is the two-step reaction sequences, involving transition metal catalysts and hydrogen. The dissociative chemisorption of methane occurs above 373K and produces three forms of carbon, C_α, C_β and C_γ, which are distinguishable by temperature they react with hydrogen. High flow rates of methane [3-6], membrane-assisted process [7] and hydrogen acceptors [8] are alternatives used experimentally to ensure the metal surface free from hydrogen adspecies, but indeed, result in a very low conversion. In this paper it was reported a comparative study about the interaction of CH_4 with 5% of transition metal loaded catalysts supported on alumina. In an attempt to go more deeply into the mechanism of the methane interaction with the catalyst surfaces, two types of experiments were performed:

"in situ" Infrared measurements of methane adsorption and surface reaction of hydrogen with carbon adspecies formed when methane pulses were chemisorbed at several temperatures.

2. EXPERIMENTAL

2.1. Catalyst Preparation
The Pt, Pd and Ru catalyst, with 5% (w/w) of metal, were supplied by Acros Organics and were used as such. The Rh catalyst was prepared by wet point impregnation of an alumina provided by Davison Grace, with a $RhCl_3.xH_2O$ solution (Aldrich). The dried catalyst was calcined up to 673K in a flow of synthetic air (50 cm^3 min^{-1}).

2.2. Methane chemisorption and reaction of carbon adspecies with hydrogen
Infrared experiments were conducted using a Fourier Transform spectrometer, Perkin Elmer 2000, and self supported wafers (9.8 mg cm^{-2} thickness) by using a Pyrex cell with CaF_2 windows. During the sample pretreatment and gas adsorption the cell was attached to a vacuum glass system. The spectral domain was between 4000 and 1000 cm^{-1} with a 4 cm^{-1} resolution. The samples were reduced "in situ" with pure hydrogen at 673 K for one hour, followed by evacuation up to 10^{-5} Torr at the same time and temperature. After cooling down to room temperature the infrared spectrum was recorded and used as background for the other adsorption experiments. Adsorption studies were conducted by exposing the reduced wafers to 10 Torr of methane at room temperature, followed by heating the wafer to 373, 473 and 573K. Experiments with CO_2 chemisorption were also conducted on the reduced wafers, by exposing them to 30 Torr of CO_2, followed by evacuation up to 10^{-5} Torr.

The activated chemisorption of methane followed by reaction of the carbon adspecies with hydrogen were conducted in a TPD/TPR 2900 Micromeritics equipment using a Balzers quadrupole mass spectrometer as detector. All the catalysts (100 mg) were reduced "in situ" with hydrogen (50 cm^3 min^{-1}) in a ramp of 2K min^{-1} up to 673K, staying at this temperature overnight. The hydrogen was swept from the surface by flowing He at 673K for one hour, then the sample temperature was reduced to the reaction temperature, and the catalyst surface was exposed to three pulses of methane (44,6µmols each) in flowing He, followed by three pulses of hydrogen (44,6 µmols each). Each plotted point in Figures 2-5, represents the average of the three pulses.

3. RESULTS AND DISCUSSION

3.1. Infrared experiments

Infrared spectra of methane chemisorbed on reduced wafer of Pt/Al_2O_3 at different temperatures, are exhibited in the Fig. 1. At 228K, a band at 1304cm^{-1} was observed and assigned as symmetric CH$_3$ bending of methane, in the gas phase [9]. As the temperature increased, bands at 1652, 1437 and 1229 cm^{-1} were formed and were interpreted as asymmetric stretch of unidentate and bidentate carbonate species [9]. The spectrum **e**, obtained by exposition of the just reduced catalyst to 30 Torr of CO_2, at room temperature, followed by vacuum up to 10^{-5} Torr, gave support to the above interpretation.

Similar spectra were obtained for the other alumina metal loaded catalysts, suggesting that methane was dehydrogenated at the metal surface, and the resulted carbon adspecies migrate to the metal-support interface, interacting with O^{2-} ions of the support and forming carbonated adspecies. In a previous work [10] it was reported similar behavior when Pt/ZrO_2 catalyst was exposed to methane at different temperatures. Bands at 1613, 1564, 1441 and 1229 cm^{-1} were interpreted as bicarbonate and carbonate species formed when carbon adspecies provided from dehydrogenation of methane at Pt^{o}, migrate toward metal-support interface and interact with the zirconia OH and O^{2-} ions.

3.2. Methane Chemisorption and reaction of C adspecies with hydrogen

Simultaneously with the methane pulses, H_2, CO_2 and CO were evolved from all the catalyst surfaces as detected by the mass spectrometer. The H_2 evolution resulted from the association of hydrogen atoms provided from dissociative chemisorption of methane, as indicated by reactions (1) and (2):

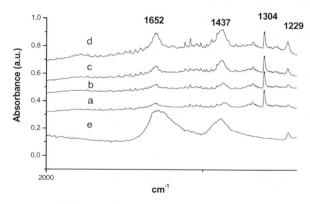

Fig, 1. Pt/Al_2O_3 reduced at 773K: **a**, CH_4 chemisorption at 228K; wafer heated to: **b**, 373K; **c**, 473K; **d**, 573K; **e**, CO_2 chemisorption at 228K on reduced wafer.

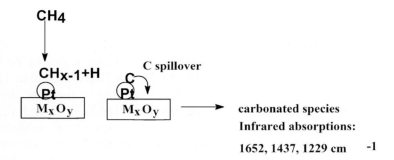

Scheme 1. Spiltover of C adspecies from metal surface to the metal-support interface, followed by interaction with OH and/or O^{2-} ions of the support.

$$CH_4 + 2^* \rightarrow CH_{3\ ads} + H_{ads} \tag{1}$$

$$H_{ads} + H_{ads} \rightarrow H_2, \text{ where * represents a } Pt^\circ, Pd^\circ, Rh^\circ \text{ or } Ru^\circ \text{ sites.} \tag{2}$$

CO_2 and CO evolutions can be understood by the partial migration of carbon adspecies, formed at metal surface, to the metal-support interface followed by their interaction with the alumina O^{2-} ions, as proposed in the Scheme 1. Part of this carbonated species still remains attached to the alumina surface, as was seen in the Infrared spectra exhibited at Fig. 1.

The carbon adspecies which remain attached to the metal surfaces, were hydrogenated when the catalyst was submitted to pulses of H_2. Only methane was observed, in amount detectable by the mass spectrometer. Figures 2-4 represent the methane activated chemisorptions followed by the hydrogenation of the carbon adspecies, exhibited at Fig. 5, from the different metal loaded catalysts.

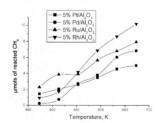

Fig. 2. μmols of reacted CH_4

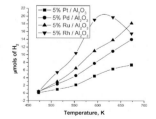

Fig. 3. μmols of H_2 evolved during CH_4 chemisorption.

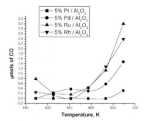

Fig. 4. μmols of CO evolved during CH_4 chemisorption.

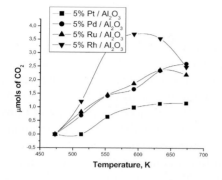 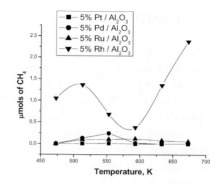

Fig. 5. μmols of CO_2 evolved during methane chemisorption.

Fig. 6. μmols of CH_4 evolved by reaction of CH_x adspecies with H_2.

Despite their low reactivities, Rh and Ru were the most active catalysts, as observed by the data of Fig. 2. Pt loaded catalyst was the less active. In general, the methane conversion increases as the temperature of the reaction increases for all catalysts. The H_2 evolution also increases with the temperature of the reaction except for Rh, at 673K. However, it cannot be ruled out the possibility of hydrogen spillover from metal particles onto the support, when the temperature of the reaction was greater then 613K [11].

The profiles of CO evolution show the same tendency for increasing with the temperature of reaction. Ru and Rh loaded alumina showed a minimum for CO evolution at 553 K while Pd loaded alumina showed a little maximum at the same temperature, and since 599K, Pd, Ru an Rh catalysts exhibited a linear increase of CO evolution with the temperature of the reaction. Pt loaded alumina showed a different behavior, being the less effective for CO production, which was detected, in the gas phase, only at 673K.

For CO_2 evolution, Rh loaded catalyst exhibited a maximum at 593K, with a similar profile for H_2 evolution showed in Fig. 2 Ru loaded catalyst exhibited two maxima at 553 and 633K, respectively. Pd and Ru loaded catalysts presented similar behavior with respect to CO_2 evolution, and Pt loaded catalyst was the less selective. The presence of carbon oxides species were supported by Infrared experiments where carbonated species were observed at support surface.

Rhodium catalyst was also the best metal for hydrogenation of carbon adspecies. Probably, for this catalyst, two types of carbonaceous material were formed during methane chemisorption, being hydrogenated at different temperatures (at 513K and above 600K). Pt and Ru catalysts exhibited similar behavior with respect to hydrogenation of carbon adspecies, and Pd showed a little maximum for CH_4 evolution at 553K.

4. CONCLUSIONS

Activated chemisorption of methane takes place at Pt, Pd, Ru and Rh supported catalysts with simultaneous evolutions of H_2, CO and CO_2. Carbon oxides were believed to result by the migration of part of carbon adspecies from the active metal sites toward the support where their interaction with OH and/or O^{2-} ions takes place. Pt catalyst was the less selective for the carbon oxides evolutions and carbonated adspecies formation. Rh was the most active metal for methane activation and also for stabilizing carbon adspecies in different structures (carbidic and amorphous carbon) suitable for hydrogenation around 500K and at temperatures higher than 599K. Pt and Ru catalysts showed similar behavior with respect to hydrogenation of carbon adspecies being much less active than Rh catalyst. The same is truth for Pd catalyst, which presented a little maximum for methane evolution at 553K. As no higher hydrocarbon species were detected it can be concluded that the operation conditions used in this work (pulses of methane in flowing helium) were not suitable for carbon nucleation needed for the formation of higher species.

REFERENCES

[1] G.E. Kelleer and M.M. Bhasin, J. Catal. 73 (1982) 9.
[2] J.H. Lunsford, Angew Chem. Int. Ed. Engl. 34 (1995) 970.
[3] M. Belgued, A. Amariglio, P. Paréja, H. Amariglio, J. Catal. 159 (1996) 441.
[4] M. Belgued, A. Amariglio, P. Paréja, H. Amariglio, J. Catal. 159 (1996) 449.
[5] P. Paréja, S. Molina, A. Amariglio, H. Amariglio, Appl. Catal. 168 (1998) 289.
[6] M. Belgued, A. Amariglio, L. Lefort, P. Paréja, H. Amariglio, J. Catal. 161 (1996) 282.
[7] O. Garnier, J. Shu, B.P.A. Grandjean, Ind. Eng. Chem. Res. 36 (1997) 553.
[8] P. Paréja, M. Mercy, J.C. Gachon, A. Amariglio, H. Amariglio, Ind. Eng. Chem Res. 38 (1999) 1163.
[9] L.H. Little in Infrared Spectra of Adsorbed Species, Academic Press, London, New York, 1966.
[10] R.L.Martins, M.M.V.M. Souza, M.A. Baldanza, M. Schmal, Proceedings of the 18th North American Catalysis Society Meeting, P-314, June 2003.
[11] J.M. Hermann, P. Pichat in Studies in Surface Science and Catalysis, vol 17, Spillover of Adsorbed Species, G.M. Pajonk, S.J. Teichner, J.E. Germain, Editors, Elsevier Amsterdam, Oxford, New York, Tokyo, 1983.

Studies in Surface Science and Catalysis, volume 147
X. Bao and Y. Xu (Editors)
©2004 Elsevier B.V. All rights reserved.

Improvement of the selectivity to propylene by the use of cyclic, redox-decoupling conditions in propane oxidehydrogenation

N. Ballarini[a], F. Cavani[a], A. Cericola[a], C. Cortelli[a], M. Ferrari[a], F. Trifirò[a], R. Catani[b] and U. Cornaro[c]

[a]Dipartimento di Chimica Industriale e dei Materiali, Viale Risorgimento 4, 40136 Bologna, Italy. *cavani@ms.fci.unibo.it*

[b]Snamprogetti SpA, Viale De Gasperi 16, 20097 San Donato Milanese MI, Italy.

[c]EniTecnologie SpA, Via Maritano 26, 20097 San Donato Milanese MI, Italy.

ABSTRACT

Catalysts based on silica-supported vanadium oxide were tested in the oxidative dehydrogenation of propane, under both co-feed and redox-decoupling conditions. It was found that it is possible to achieve higher selectivity to the olefin by separation of the two steps of the redox mechanism. The performance was affected by the amount of vanadia loaded on silica.

1. INTRODUCTION

The oxidative dehydrogenation (ODH) of light alkanes represents one possible alternative to the current industrial productions of olefins via energy-intensive, endothermal catalytic dehydrogenation (DH) or steam cracking [1]. Of particular interest for a possible industrial application is the ODH of propane and of ethane. Several papers have been published in recent years, which examine a variety of catalytic systems for this reaction [1]. In the case of propane ODH, however, the selectivity to propylene, sometimes very high for low hydrocarbon conversion, falls when propane conversion increases, due to consecutive combustion reactions. It has been proposed that one method to limit the undesired combustion in the selective oxidation of hydrocarbons is to run the reaction under cyclic conditions (redox-decoupling operation), in which the two steps of the redox mechanism are carried out separately: (a) during the "reducing step", the hydrocarbon is put in contact with the oxidized catalyst; the alkane is transformed to the products, and the catalyst progressively becomes more reduced along with the increasing reduction time; (b) during the "oxidizing" step, the reduced catalyst is put in contact with air, to restore the original oxidation state. An analogous approach has been developed in a two-zone, fluidized-bed reactor proposed by Lopez Nieto et al. [2]. In previous papers [3,4] we have investigated the catalytic properties of V/Nb mixed oxides and of V/Si/O cogels in propane ODH under both co-feed and redox-decoupling conditions. In the present work we examine the effect of vanadia loading on catalytic performance of silica-supported vanadium oxide.

2. RESULTS AND DISCUSSION

Catalysts prepared and characterized are listed in Table 1. Commercial silica spheres (Sud Chemie, T4359E1) were impregnated with an aqueous solution of $(NH)_4VO_3$, and calcined at 550°C. Before catalytic tests, all samples were treated in flowing air inside the reactor, at 550°C for 0.5 h.

Fig. 1(left) plots the propane and oxygen conversions as functions of the reaction temperature under co-feed operation for sample 4, while Fig. 1(right) reports the selectivity to the products, i.e., propylene, CO, CO_2, and oxygenates (mainly acetone and acetic acid). The H_2 produced is given as molar percentage in the outlet stream. Fig. 1(left) also gives the C balance, as calculated from the ratio between the sum of yields to the different products and the propane conversion. Under the conditions employed, total oxygen conversion was reached at 450°C. Propane conversion increased in the temperature range 390-to-450°C, and slightly also above 450°C, notwithstanding total oxygen conversion. This indicates that at high temperature a non-oxidative reaction contributed to propane conversion under anaerobic conditions.

At temperature between 400 and 450°C, prevailing products were oxygenates and propylene, the selectivity of which however decreased for increasing temperature, with a corresponding increase in the formation of carbon monoxide and carbon dioxide. Under these conditions, C balance was initially close to 70%, and then became higher than 90% at 450°C. The H_2 formation was either nil, or negligible (less than 0.1 mol. % in the gas stream at 450°C).

Table 1
Main characteristics of samples

Sample n,	wt.% V_2O_5	Surface area, m^2/g
1	2.0	280
2	6.8	274
3	10.0	259
4	15.0	254
5	100	8

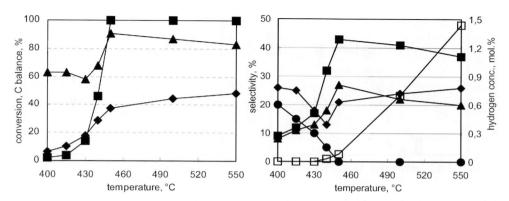

Fig. 1. Left: conversion of propane (◆), of oxygen (■) and C balance (▲) as functions of the reaction temperature. Right: selectivity to propylene (◆), to carbon monoxide (■), to carbon dioxide (▲), to oxygenates (●), and hydrogen concentration (□), as functions of the reaction temperature. Co-feed operation. Catalyst: sample 4. Residence time 2 s. Feed: 20% propane, 20% oxygen, remainder helium.

At temperatures higher than 450°C, the selectivity to propylene increased from 13 to 25%, with a corresponding lower formation of carbon oxides. The C balance worsened (from 91 to 83%), indicating an unsteady accumulation of coke. The amount of H_2 produced increased considerably, and well corresponded to the higher propylene concentration. Therefore, data indicate that at high temperature propylene is produced by two different mechanisms, (i) an ODH mechanism in the fraction of reactor in which O_2 is still present in the gas phase, and (ii) a DH one in the fraction of reactor in which the catalyst operates under anaerobic or almost-anaerobic conditions.

The behaviour of samples 1-5 in co-feed operation is compared in Fig. 2, which plots the propane conversion and the selectivity to propylene as functions of temperature. Sample 1 reached total O_2 conversion at around 550°C, while samples 2-4 at approximately 450°C. Therefore samples 2-4 had comparable activity, notwithstanding the different vanadia loadings, and were more active than sample 1. Surprisingly, unsupported vanadia (sample 5) turned out to be the most active sample, despite the very low surface area. Considerable differences were also found in propylene selectivity; samples 1 and 2, having the lowest vanadia content, were more selective at low temperature, because of the absence of oxygenates. Differences were smaller at high temperature.

V_2O_5 (sample 5) was unselective, at both low and high temperature. At total oxygen conversion, the highest yield to propylene was obtained with sample 1 (yield 13.5% at 550°C). Samples 1 and 2 also yielded light hydrocarbons (methane, ethane and ethylene), but with a selectivity lower than 2%. Other differences between catalysts concerned the formation of H_2 in the

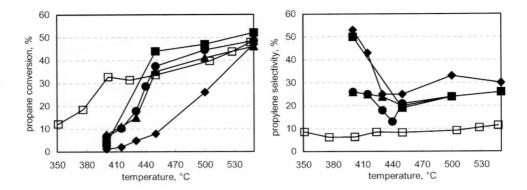

Fig. 2. Conversion of propane (left) and selectivity to propylene (right) as functions of temperature. Catalysts: samples 1(◆), 2(■), 3(▲), 4(●) and 5 (□). Co-feed operation. Residence time 2 s. Feed: 20% propane, 20% oxygen, remainder helium.

temperature range 500 to 550°C; hydrogen production was the lowest in sample 1 (0.7 mol.% at 550°C), due to the larger availability of O_2 in the gas phase. The hydrogen concentration obtained with samples 2-4 at 550°C was in the range between 1 and 2 mol.%.

Fig. 3 plots the effect of the half-cycle reduction time on propane conversion and on selectivity to the products, for sample 4, under redox-decoupling conditions (tests were carried out following the protocol described in detail in a previous work, [3]). Increasing reduction times led to (i) a decrease of propane conversion, which however did not become nil, indicating the presence of a DH contribution to propane transformation, (ii) a decline of the selectivity to carbon oxides, while the selectivity to propylene only slightly changed, and (iii) a worsening of the C balance, due to the higher formation of coke. Hydrogen concentration also was higher, as a consequence of the occurrence of propane DH and of coke formation.

Fig. 4 compares the catalytic performance of samples 1, 3 and 4 under cyclic conditions. For all catalysts the conversion achieved was low, even on the oxidized samples (i.e., for shorter reduction times). As for operation under co-feed conditions, sample 1 was the least active catalyst. However, on the opposite of that observed under co-feed conditions, best selectivity to propylene was with samples having the highest vanadia content (samples 3 and 4). This was due to the lower coke formation, as inferred from the C balance, while the selectivity to carbon oxides was very low for all catalysts. The formation of coke was either due to a contribution of silica (more relevant in samples having lower coverage of the support), or to the reduction of vanadium, with development of a V species over which the formation of coke was accelerated. Reduction of vanadium for sample 1 was achieved in correspondence of shorter reduction times than for samples 3 and 4, due to the lower loading of active phase. Aerobic conditions (i.e., co-feed operation) are thus necessary to limit the formation of coke in samples having low support coverage.

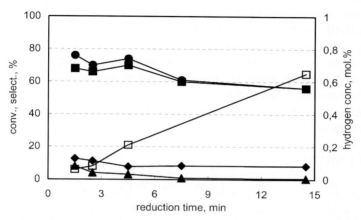

Fig. 3. Effect of the half-cycle reduction time on propane conversion (◆), on selectivity to propylene (■) and to carbon oxides (▲), on C balance (●), and on hydrogen concentration (□). T 500°C; gas residence time 2 s; feed: 20% propane in helium. Catalyst: sample 4.

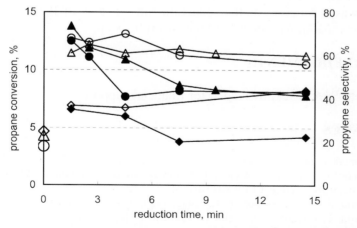

Fig. 4. Effect of the half-cycle reduction time on propane conversion (full symbols) and on selectivity to propylene (open symbols). Catalysts: samples 1 (◆), 3 (▲), 4 (●). Temperature 500°C; gas residence time 2 s; feed: 20% propane in helium. Values of selectivity at zero reduction time are those achieved under co-feed operation.

The amount of hydrogen formed confirmed differences between samples. In the case of sample 1, hydrogen production was remarkable even for very short reduction times, and then slightly increased, while with samples 3 and 4 the formation of H_2 was negligible for shorter reduction times, and became relevant only for longer times. This also suggests a different relative contribution of ODH and DH in propylene formation under anaerobic conditions, depending on the amount of vanadia loaded in catalysts.

Selectivity to propylene achieved with samples 1, 3 and 4 under redox-decoupling conditions for shorter reduction times (i.e., for yet oxidized catalysts), is compared in Fig. 4 with the selectivity achieved under co-feed conditions at the same propane conversion level, in the range between 7 and 14%. Cyclic conditions led to a better selectivity to propylene, as a consequence of the lower formation of oxygenates and of carbon oxides than under co-feed, aerobic conditions.

4. CONCLUSIONS

The performance of silica-supported vanadium oxide catalysts in propane ODH was tested under both co-feed and cyclic, redox-decoupling conditions. Cyclic operation led to a better selectivity to propylene than co-feed, due to the lower formation of carbon oxides. The performance was affected by the amount of vanadia loaded in catalysts. Higher vanadia contents led to a higher instantaneous conversion and to a better selectivity as well, due to the lower formation of coke. The relative contribution of ODH and of DH on propylene formation was a function of the reduction time, that is of the reduction level achieved for vanadium oxide during the anaerobic step.

ACKNOWLEDGEMENTS

Snamprogetti SpA is acknowledged for financial support.

REFERENCES

[1] E.A. Mamedov and V. Cortes-Corberan, Appl. Catal. A 127 (1995) 1.
[2] M.L. Pacheco, J. Soler, A. Dejoz, J.M. Lopez Nieto, J. Herguido, M. Menendez, J. Santamaria, Catal. Today, 61 (2000) 101.
[3] N. Ballarini, F. Cavani, M. Ferrari, R. Catani and U. Cornaro, J. Catal., 213 (2003) 95.
[4] N. Ballarini, F. Cavani, C. Cortelli, C. Giunchi, P. Nobili, F. Trifirò, R. Catani and U. Cornaro, Catal. Today, 78 (2003) 353.

Studies in Surface Science and Catalysis, volume 147
X. Bao and Y. Xu (Editors)

Oxidative activation of light alkanes on dense ionic oxygen conducting membranes

M. Rebeilleau[a], A.C. van Veen[a], D. Farrusseng[a], J.L. Rousset[a], C. Mirodatos[a], Z.P. Shao[b], G. Xiong[b].

[a]Institut de Recherches sur la Catalyse, 2. Av. A. Einstein 69626 Villeurbanne Cedex France

[b]State Key Laboratory of Catalysis, Dalian Institute of Chemical Physics, Chinese Academy of Sciences, P.O. Box 110,Dalian 116023, People's Republic of China

ABSTRACT

The oxidative dehydrogenation of ethane to ethylene (ODHE) has been studied in a catalytic membrane reactor (CMR) using a dense mixed ionic oxygen and electronic conducting perovskite membrane $Ba_{0.5}Sr_{0.5}Co_{0.8}Fe_{0.2}O_{3-\&}$. At 1080K, an ethylene yield of 66% was obtained with the bare membrane. After Pd cluster deposition, the ethylene yield reached 76% at 1050K. Ni cluster deposition led to a decrease of ethane conversion compared to the bare membrane without changing ethylene selectivity.

1. INTRODUCTION

Extensive efforts are currently devoted to the chemical conversion of light alkanes, i.e. C_2H_6 and C_3H_8, to valuable chemicals, both being besides methane main constituents of natural gas. Oxidative dehydrogenation (ODH) of alkanes to alkenes is considered as an attractive alternative to the existing thermal steam cracking processes due to the lower temperature required and decreased coke formation. Unfortunately, yields obtained up to now in conventional reactors still remain by far too low for industrial developments. At low temperature (<873K), the best ethylene yield currently reported is below 40% and was obtained with a mixed oxide $Mo_{0.73}V_{0.18}Nb_{0.09}O_x$ [1]. The best results for ODHE at high temperatures were close to 44-49% over mixed oxide catalysts [2, 3]. In fact, the reaction of ethane with gaseous oxygen, over all known catalyst is accompanied by the thermodynamically favored formation of carbon oxides, thereby decreasing selectivity at a given conversion.

In this way, dense catalytic membrane reactors (CMR) are regarded as promising candidates for enhanced reactor configurations in partial oxidation reactions. Here, dense ionic oxygen conducting membranes (IOCM) perform the selective separation of oxygen from air on one side of the membrane while providing activated oxygen to the other side being fed with the alkane. Thus, dense CMR combines in the same unit oxygen separation and alkane conversion, leading with high probability to breakthroughs in cost efficiency when mass transfer and reaction rate are tuned adequately.

Recently, promising results using IOCM were reported for oxidative dehydrogenation of ethane (ODHE) [4, 5]. However, the temperature range used remains still at a too high level for an industrial application and so, it appears necessary to increase the yield of ethylene at lower temperature, keeping as reference the yield of 16% obtained at 973K as reported in [4]. The absence of catalyst on top of the membrane, on the reaction side, is certainly a feature to change when opting to increase conversion without decreasing selectivity. In fact, even if perovskites are known to catalyze many oxidation reactions, like oxidative coupling of methane or oxidative dehydrogenation of light hydrocarbons, the contact surface, i.e. the geometric surface of the membrane, is quite low to play a significant role taking parallel proceeding gas-phase reactions into account. In order to increase the surface activity and consequently the alkane conversion at lowered temperature, we decided to deposit a catalyst on the surface. Pd and Ni metallic clusters, known to be active for alkane activation, were deposited by laser vaporization of corresponding metallic rods to obtain a catalytic surface modification with high metal dispersion and uniform cluster size.

For this purpose, we used a bare $Ba_{0.5}Sr_{0.5}Co_{0.8}Fe_{0.2}O_{3-\&}$ membrane comparable to that employed in [4], which showed good results in oxygen permeation and in partial oxidation of methane [6-8]. The performances in ODHE of the bare and surface modified membranes are compared in this paper, in order to i) demonstrate that the concept of CMR with dense IOCM is validated for ODHE, and ii) evaluate the influence of surface modification of the membrane on ethane conversion and ethylene selectivity.

2. EXPERIMENTAL

The $Ba_{0.5}Sr_{0.5}Co_{0.8}Fe_{0.2}O_{3-\delta}$ (BSCFO) perovskite powder was prepared using an adapted variant of the so-called citrates method. In this method, stoichiometric amounts of $Ba(NO_3)_2$, $Sr(NO_3)_2$, $Co(NO_3)_2 \cdot 6H_2O$ and $Fe(NO_3)_3 \cdot 6H_2O$ (purity >99.5%) were fully dissolved in a small amount of distilled water, followed by addition of EDTA and Citric Acid with a molar ratio of Perovskite: EDTA: Citric Acid equal to 1:1.5:3. The obtained purple solution was heated to 373K, and a gel-like material was formed by evaporation of water after about three hours, which transformed into a brown foam by a heat treatment at 575K for three hours. This foam was calcined at 1173K for four hours in air yielding a perovskite powder, that was further homogenized by grinding in a mortar. Raw membrane discs were pressed by applying a pressure of 140 MPa for one minute. The complete densification of the raw discs was accomplished by sintering in dense alumina boats at 1425K for eight hours.

A Nd-YAG laser was focused onto a metallic rod, which was driven in a slow screw motion. Short helium bursts, synchronized with laser pulses, cooled the plasma and a supersonic expansion occurred at the exit of the source, inducing the formation of a beam of clusters, containing neutral and ionized species. The masses of the charged clusters could be analyzed by a time of flight mass spectrometer, hence allowing for reproducible cluster size adjustment. These clusters were sent to a vacuum chamber and deposited either on the surface of the dense ionic oxygen conducting membrane or on amorphous carbon to allow an easy characterization by electron microscopy. The typical deposition rates monitored by a quartz microbalance, were 5 nm/cm²/min (6.1 µg/min) for the case of the non-porous materials used and the thickness of the deposition was close to 8 Å. Two rods were used with the Pd_{100} and Ni_{100} composition.

The formation of a pure perovskite structure was checked by X-ray diffraction using a Bruker D5005 system in the 2θ range of 3 to 80°, a step width of 0.02°, a counting time of 1s and Cu $K_{\alpha1+\alpha2}$ radiation (1,54184 Å). The bulk elementary composition was determined by ICP-OES analyzing a sample dissolved by heating at 523 to 573 K in a mixture of H_2SO_4 and HNO_3. TEM micro graphs were taken with a JEOL JSM-840A electron microscope.

The oxidative dehydrogenation of light hydrocarbons was studied using a reactor as described previously [8]. Discs (about 1 mm thick) were sealed in between two dense alumina tubes using gold rings as chemically inert sealant. Furthermore, the side wall of the disc was extensively covered with gold paste suppressing radial contributions to the oxygen flux permeating through the active cross-section of 0.5 cm². The sealing was carried out in the beginning of experimental runs by heating to 1073 K for one night. The oxygen-rich side was fed at a constant total pressure adjusted to 120 kPa and a constant total flow rate of 50 ml·min^{-1} using a mixed stream of O_2 and N_2 (evaporated liquid N_2) individually controlled by mass flow controllers. The reaction side was fed by a mixture of ethane and He individually controlled by mass flow controllers. On both membrane sides the introduced gases were preheated while passing the void space between the alumina tube and an inserted quartz tube and left afterwards rapidly via the tube center. Two pressure transducers were installed allowing to record the total pressure on each membrane side of the reactor. Ethane oxidation experiments were conducted within the temperature range of 973K to 1100K, with a flow of 35mL·min^{-1} C_2H_6/He and a C_2H_6 partial pressure equal to 0.25atm. The reactant gases (O_2, N_2, C_2H_6) and the product gases (H_2, CH_4, CO_2, C_2H_4, C_2H_6 and H_2O) were analyzed by two on-line gas chromatographs, GC, connected to a PC based automated data collection and analysis. The first GC was equipped with a TCD and a molecular sieve 13X column, allowed the detection of O_2, N_2, CH_4, CO and H_2O. H_2, CH_4, CO_2, C_2H_4, C_2H_6 and H_2O concentrations were deduced from the second GC equipped with a Haye-Sep D column and TCD. Moreover, Ar was introduced as internal standard with the reactants in order to determine the flow increase caused by the molar expansion during the reaction. The carbon balance could be closed within 4%. Gas leakage due to an insufficient sealing or an incomplete densification of the membrane could in case be detected by monitoring the N_2 concentration.

3. RESULTS

The effect of temperature on the ethane conversion and ethylene selectivity is represented in Fig. 1 for the bare membrane and those modified by Pd and Ni deposition, respectively.

3.1 ODHE performances on bare membrane

The ethane conversion increased from 11% to 72% within the temperature range of 975 - 1080K. Selectivity decreased slightly from 96% to 92%, and the yield of ethylene reached a maximum value of 68% at 1080K. These results fully agree with those reported in [4]. During the experiment, we did not observed any oxygen in gaseous phase, indicating a total consumption of oxygen permeating through the membrane. The selectivity of the other carbon containing products (i.e. CH_4, CO_2, CO) was always below 5%. However, a large amount of hydrogen was detected (around 12 vol.%), indicating that direct dehydrogenation of ethane ($C_2H_6 \rightarrow C_2H_4 + H_2$) occurs together with ODHE ($C_2H_6 + \frac{1}{2} O_2 \rightarrow C_2H_4 + H_2O$). By comparing i) the amount of oxygen that ODHE would require for the observed ethylene yield, and ii) the amount of oxygen effectively permeating through the membrane (from side permeation measurements), we estimated the contribution of the direct dehydrogenation of ethane to 80% and that of oxidative dehydrogenation to 20%. Furthermore, since no coke formation was observed during all the experiments, one major role of permeating oxygen is also to react with carbon deposits leading to the observed side CO_x products.

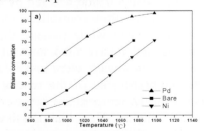 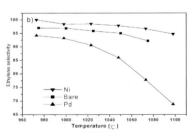

Fig. 1 Ethane conversion (a) and ethylene selectivity (b) as a function of temperature for a bare BSCFO membrane and both Pd and Ni modified BSCFO membrane. Conditions: Air side F_{air}=50ml/min, P_{air}=1.2 atm; reaction side: $F_{ethane/helium}$=37mL/min, P_{ethane}=0.25atm.

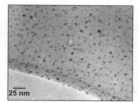

Fig. 2. TEM image of Pd deposited on amorphous carbon (to allow imaging)

3.2 Effects of surface modification by Ni and Pd deposition.

Laser deposition allowed to obtain an excellent dispersion of Ni and Pd particles with a homogeneous particle size (d $\cong$ 4 nm) as shown in Fig. 2. In this case, the surface coverage with metal was close to 10 % (1μg/cm^2).

The effects of Pd and Ni cluster deposition at the surface of the BSCFO membrane on both ethane conversion and ethylene selectivity are shown in Fig. 1. In the case of Pd deposition, the ethane conversion increased fourfold at 973K leading to an ethylene yield of about 38%, to be compared with the reference value of 16% obtained with the bare membrane. The selectivity reached almost 100% at 1100K. A slight decrease of ethylene selectivity was observed at high temperature with a minor increase in CO$_2$ formation. The maximum value in yield of ethylene was close to 76% at 1050K (with ethylene selectivity 86%), to be compared with the maximum value of 68% obtained with the bare membrane at 1080K. In this way, the Pd deposition on the surface membrane allowed to increase significantly the ethylene yield above the level obtained for an unmodified membrane which was investigated at even higher temperatures. On the other hand, in the case of Ni deposition the conversion was lowered compared to that obtained with the bare membrane within all the temperature range. The ethylene selectivity was not affected by this modification, but the yield of ethylene reached only 37% at 1050K and 67% at 1100K. In both cases, like for the bare membrane, there was no oxygen detected in the effluent gas which indicates that all the oxygen crossing the membrane was consumed. Furthermore, no coke formation was observed during membrane reactor operation, while reference experiments in the absence of oxygen using a conventional tubular configuration indicated massive coke formation.

4. DISCUSSION

The evaluation of the intrinsic activity of membrane and catalyst is not easily accessible due to many parameters. Firstly, in temperature range of 973-1173K the gas phase reactions (leading to direct dehydrogenation) are likely to play important role. Secondly, the oxygen flux crossing the membrane is a function of temperature and of the oxygen partial pressure gradient, and so, the effective oxygen flux crossing the membrane during the reaction cannot be not directly measured. However, from the hydrogen amount detected we could estimate in the case of the bare membrane the contribution of oxidative dehydrogenation was about 20%. Concerning the experiments with the two catalysts deposited, we observed minor changes in oxygen permeation (as calculated from oxygenated products) which cannot explain the changes in ethane conversion. Thus, it seems that Ni plays an inhibitor role for the ethane activation while Pd allows to increase it. An explanation to the observed differences most likely relates to the different properties of metals used. Under the pertaining conditions, Pd is likely to stay predominately in metallic form favoring the direct dehydrogenation, or allowing for enhanced ODHE by surface oxygen supplied from the membrane. In the case of nickel, kinetics and thermodynamics would favor an oxidation of Ni to nickel oxide in the presence of oxygen crossing the membrane. Due to the expected mobility of the oxide phase at high temperature, a solid/solid reaction with the perovskite could have formed a kind of surface layer decreasing in turn the accessibility of the hydrocarbon to the active membrane surface.

5. CONCLUSION

The dehydrogenation of the ethane to ethylene was studied in a dense membrane reactor made of $Ba_{0.5}Sr_{0.5}Co_{0.8}Fe_{0.2}O_{3-\&}$ before and after modification of the surface on the reaction side. In agreement with Wang et al. [4], a per pass ethylene yield of 68% with an ethylene selectivity of 92% was obtained at 1080K. After deposition of Pd clusters on top of the membrane, a major improvement of the CMR performance was achieved with a maximum per pass ethylene yield of 76% with 86% ethylene selectivity at 1050K. In contrast, a decrease of ethane conversion was observed over all range of temperatures after the Ni deposition without change in ethylene selectivity. The concept of dense membrane reactor as an alternative to the conventional thermal dehydrogenation process seems therefore validated. High and stable productivity is established by the achieved formation rate of ca. 7×10^2 $kg_{C_2H_4}/g_{Pd}/h$ and an overall system efficiency of ca. 4×10^{-1} $kg_{C_2H_4}/kg_{membrane}/h$. Further works work on mechanism understanding and process modeling are subjects of ongoing work.

ACKNOWLEDGEMENTS

Financial support by the EC program CERMOx (G55RD-CT-2000-0035).

REFERENCE

[1] S.J Conway, D.J. Wang, J.H. Lunsford, Appl. Catal., 79 (1991) L1.
[2] O.J. Velle, A. Andersen, K.-J. Jens, Catal. Today, 6 (1990) 567.
[3] L. Ji., J. Liu, Chem. Commun, 1996, 1203.
[4] W. Wang, Y. Cong, W. Yang, Catalysis Letters 84 (2002) p. 101-106.
[5] F.T. Akin, Y.S. Lin, J. Membr. Sci., 209 (2002), 457.
[6] Z. Shao, W. Wang, Y. Cong, H. Dong, J. Tong, G. Xiong, J. Membr. Sci., 172 (2000) 177.
[7] Z. Shao, G. Xiong, H. Dong, W. Yang, L. Lin, Separation and Purification Technology, 25 (2001) 97.
[8] A.C. van Veen, M Rebeilleau, D. Farrusseng, C. Mirodatos, Chem. Commun. 2002, 32.

Studies in Surface Science and Catalysis, volume 147
X. Bao and Y. Xu (Editors)
661

Effect of support on performance of MoVTeO catalyst for selective oxidation of propane to acrolein

C. J. Huang, W. Guo, X. D. Yi, W. Z. Weng and H. L. Wan[*]

State Key Laboratory for Physical Chemistry of Solid Surfaces, Department of Chemistry, Xiamen University, Xiamen, 361005, China

ABSTRACT

Selective oxidation of propane to acrolein over supported $MoV_{0.2}Te_{0.1}O_x$ catalysts has been investigated. It was found that supports affected the structure, reducibility and O_2-desorption behaviour of the catalysts, which were closely related to catalyst performance. Among the catalysts studied, $MoV_{0.2}Te_{0.1}/SiO_2$ showed the best performance for oxidation of propane to acrolein.

1. INTRODUCTION

Selective transformation of propane to acrolein is a challenging task in both economic and scientific terms. Most catalysts for this reaction reported in the literature are unsupported multi-component oxides. Among them, AgBiVMoO catalyst shows the best catalytic performance under conditions where, however, primary gas-phase dehydrogenation of propane occurred [1]. The modification of this catalyst by coprecipitating the main components in the presence of support led to an increase in propane conversion, but no improvement in acrolein yield [2]. Recently, K-modified Fe/SiO_2 was reported active for oxygenates formation from propane oxidation [3]. Over this catalyst, acrolein selectivity was 22% with a propane conversion of 10% at 748 K. However, the systematic studies on the effect of supports have been scarcely reported.

In this paper, we report an investigation on the effects of the supports (CaO, CeO_2, SiO_2, ZrO_2 and $\gamma-Al_2O_3$) on the catalytic performance of $MoV_{0.2}Te_{0.1}O_x$ catalyst for selective oxidation of propane to acrolein.

2. EXPERIMENTAL

2.1 Catalyst preparation

Supported $MoV_{0.2}Te_{0.1}O_x$ catalysts with MoO_3 content of 11 wt% were

---

* Corresponding author

Table 1
Surface areas and XRD phases of catalysts

Catalysts	Code	S.A.$_{BET}$ (m^2 g^{-1})	Phases observed
MoV$_{0.2}$Te$_{0.1}$O$_x$	M	7.53	MoO$_3$
MoV$_{0.2}$Te$_{0.1}$/CaO	MCa	19.12	CaMoO$_4$, CaO
MoV$_{0.2}$Te$_{0.1}$/CeO$_2$	MCe	17.34	CeO$_2$
MoV$_{0.2}$Te$_{0.1}$/SiO$_2$	MSi	60.16	MoO$_3$, SiO$_2$
MoV$_{0.2}$Te$_{0.1}$/ZrO$_2$	MZr	6.42	ZrMo$_2$O$_8$, ZrO$_2$
MoV$_{0.2}$Te$_{0.1}$/Al$_2$O$_3$	MAl	139.81	γ-Al$_2$O$_3$

prepared by incipient wet impregnation of commercial oxides (i.e., CaO, CeO$_2$, SiO$_2$, ZrO$_2$ and γ-Al$_2$O$_3$) with an aqueous solution (pH ~ 6) of heptamolybdate, metavanadate and tellurium acid. The resulting precursors were dried at 110 °C overnight and then calcined in air at 600 °C for 5 h. For the unsupported catalyst, the above solution was evaporated and the resulting solid was dried and calcined under the same conditions as the supported samples. In all cases, the atomic ratio of Mo/V/Te = 1.0/0.2/0.1 was fixed. The catalyst samples, along with their notations and some physico-chemical properties such as specific areas and phases, are shown in Table 1.

2.2 Catalytic tests

Catalytic tests were carried out in a fixed bed flow reactor (quartz tube, 5 mm i.d.) under the following conditions: T_R, 500 °C; P_R, 1 bar; W_{cat}, 0.27 g. The reactant gas mixture with a molar composition of C$_3$H$_8$: O$_2$: He = 1.2 : 1.0 : 1.2 was passed through the catalyst bed at a space velocity of 3900 ml•g-cat^{-1}•h^{-1}. The products were analysed by two gas chromatographs (an FID-GC with GDX-103 column and a TCD-GC with squalane/Al$_2$O$_3$ and carbosieve columns). The data were collected after 0.5 h on stream.

2.3 Catalyst characterization

XRD patterns were recorded on a Rigaku Rotflex D/max-C X-ray powder diffractometer (30 kV, 20 mA) scanning at 8° min^{-1} using Cu K_a radiation.

Laser Raman spectra were collected under ambient conditions using Renishaw UV-Vis Raman 1000 System equipped with a CCD detector and a Leica DMLM microscope. A green laser (λ = 514.5 nm) was used for excitation.

TPR experiments were performed on a flow apparatus using a 5% H$_2$/Ar mixture flowing at 20 ml/min. The heating rate was 10°C/min. Hydrogen consumption was monitored by a TCD detector after removing the water formed.

O$_2$ -TPD was carried out in a flow system connected to a quadruple mass spectrometer (Balzers QMS 200 Omnistar). In each experiment, 0.2g catalyst was used. The sample was pretreated by heating under flowing air at 600°C for 30 min then cooling to room temperature. After purging with helium (20 ml/min) for 30 min, the sample was heated at 20°C/min from room temperature to 800°C.

3. RESULTS and DISCUSSION

3.1 Catalytic performance

From Table 2, significant influence of support on the activity and selectivity of $MoV_{0.2}Te_{0.1}O_x$ catalyst is observed. All of the supported catalysts, except MCa, show higher activities than the unsupported ones. The oxidative conversion of propane decreases in the following order: MAl > MZr > MSi ~ MCe >> M ~ MCa. The improvement in activity cannot be simply attributed to the increase in catalyst surface area (S.A.). For example, both M and MZr have almost the same S.A. (see Table 1), however, their activities are quite different. In comparison with unsupported catalyst, MCa shows a lower selectivity to CO_x and a higher selectivity to propylene, while MCe, MZr and MAl display higher selectivities to CO_x and lower selectivities to propylene and acrolein. Among these catalysts only MSi exhibits a marked functionality towards the formation of acrolein (30%). The products over other catalysts are mainly C_3H_6 (16-60%), CO_x (39-82%) and a small amounts (0.6-5.7%) of CH_4, C_2H_4 and oxygenates.

3.2 XRD results

Fig. 1 shows XRD patterns of the catalysts. Phases identified by XRD are shown in Table 1. Both M and MSi catalysts present XRD peaks assignable to MoO_3, although the peaks of the latter are very weak and accompanied by a broad peak due to amorphous SiO_2. In the case of MCa and MZr catalysts, the strong interactions between supports and Mo species lead to the formation of new phases, i.e. $CaMoO_4$ and $ZrMo_2O_8$, respectively. For MCe and MAl catalysts, only peaks from the oxide supports are observed, indicating that the loaded oxides highly disperse over the supports. In all the catalysts, no phase containing V or Te is detected.

3.3 Laser Raman spectra

Fig. 2 shows the Raman spectra of unsupported and supported $MoV_{0.2}Te_{0.1}O_x$ catalysts. MCa presents the bands from $CaMoO_4$ at 793, 848 and 878 cm^{-1} [4].

Table 2
Catalytic performance of the catalysts

Catalyst	C_3H_8 conversion (%)	Selectivity (%)				Acrolein yield (%)
		C_3H_6	CO_x	Acr	Others *	
Blank	~ 0	—	—	—	—	—
M	3.9	56.0	39.0	4.4	0.6	0.2
MCa	2.4	60.7	33.6	1.1	4.7	0.3
MCe	21.7	16.0	81.9	0.5	1.5	0.1
MSi	22.1	25.5	42.2	30.1	2.1	6.7
MZr	25.0	28.6	68.5	1.2	1.7	0.3
MAl	32.4	34.6	59.4	0.1	5.7	~ 0

* $C_2H_4 + C_2H_4O + C_3H_6O + CH_4$

On MZr, three distinct bands at 748, 944 and 998 cm^{-1} are observed. Based on the results of XRD analysis, the bands are tentatively attributed to $ZrMo_2O_8$. On the other hand, MAl displays only one broad band at around 960 cm^{-1}, which has been assigned to polymolybdate by many authors [5]. Generally, the Raman bands in the region of 870 ~ 960 cm^{-1} are characteristic of amorphous phase such as surface molybdate and polymolybdate species [5]. The bands due to the amorphous phase are also observed in the spectra of MSi (877, 960 cm^{-1}) and MCe (898, 960 cm^{-1}) catalysts. For MCe, the weak band at 818 cm^{-1} is ascribed to a trace amount of MoO_3 [6]. These observations suggest that the supports in the MAl, MCe and MSi catalysts are responsible for the high dispersion of Mo species, especially in the case of MAl.

Among the catalysts investigated, only M and MSi present intense bands due to crystalline MoO_3 at 666, 818 and 994 cm^{-1}. However, their Raman spectra are actually different. The most striking difference is the intensity ratio of the 818-cm^{-1} band to the 994-cm^{-1} band, which is higher for MSi than for M. This observation is significant since the 818-cm^{-1} band and 994-cm^{-1} band have been attributed to the Mo-O-Mo and Mo=O stretching vibrations, respectively [7]. It was proposed that over transition metal oxide surface the bridging oxygen species are more active than the terminal ones [8], which is consistent with the result obtained over MoO_3 [9]. For MoO_3 crystal, Bruckman et al. [10] found that the (010) face, on which Mo-O-Mo sites are

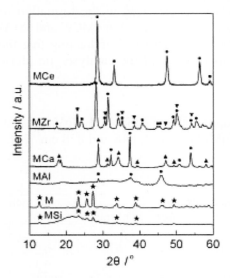

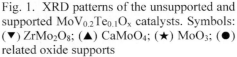

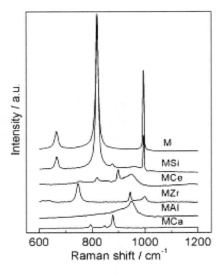

Fig. 1. XRD patterns of the unsupported and supported $MoV_{0.2}Te_{0.1}O_x$ catalysts. Symbols: ($\blacktriangledown$) $ZrMo_2O_8$; ($\blacktriangle$) $CaMoO_4$; ($\star$) MoO_3; ($\bullet$) related oxide supports

Fig. 2. Raman spectra of the unsupported and supported $MoV_{0.2}Te_{0.1}O_x$ catalysts

preferably located [11], is responsible for oxygen insertion of allyl to form acrolein. Based on the above analyses, it is suggested that the unique catalytic performance of MSi in the selective oxidation of propane to acrolein is probably related to its abundant Mo-O-Mo sites over the crystalline MoO_3.

3.4 H_2-TPR results

The H_2-TPR profiles of various catalysts and related oxide supports are presented in Fig. 3, in which $T_{o,red}$ indicates the onset temperature of reduction and T_{max} the temperature of the main reduction peak. CeO_2 shows a reduction profile with a weak peak at about 510 °C and a peak at temperatures higher than 750 °C. Other supports present TPR profiles with no reduction peak up to 750 °C as expected. However, a pronounced effect of support on the reduction behaviour of the catalysts is found for all the supported ones. In the case of M, the reducibility is very weak at temperature lower than 700 °C. When M is loaded on the supports, $T_{0,red}$ and T_{max} shift to lower temperatures. This may be related to the increased dispersion and/or the structure modification of the loaded oxides caused by the supports. Based on the $T_{0,red}$ and the T_{max} values shown in Fig. 3, the reducibility sequence of various catalysts can be given as follows: MZr > MAl > MSi ~ MCe > MCa > M. The sequence is almost the same as that of activity of the catalysts except MAl. This reveals that the oxidative conversion of propane is closely related to the reducibility of the catalysts. Compared with MZr, the higher activity of MAl may be attributed to its much larger S_{BET} value (Table 1) and higher dispersion, as evidenced by the results of XRD and Raman.

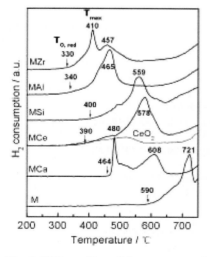

Fig. 3. TPR profiles of the unsupported and supported $MoV_{0.2}Te_{0.1}O_x$ catalysts

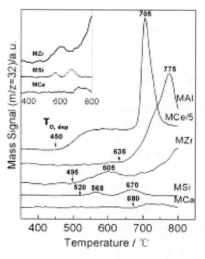

Fig. 4. O_2-TPD profiles of the supported $MoV_{0.2}Te_{0.1}O_x$ catalysts

3.5 O_2-TPD measurements

Fig. 4 shows the O_2-TPD profiles of supported $MoV_{0.2}Te_{0.1}O_x$ catalysts. Based on the onset temperature of O_2 desorption ($T_{o,dep}$), the mobility scale of oxygen species over the catalysts can be evaluated as follows: MCe > MZr > MSi > MAl >MCa. This sequence is almost the same as that of the CO_x selectivity as shown in Table 1. Among the catalysts investigated, MCe catalyst shows the strongest peak with the lowest $T_{o,dep}$, which could explain its highest selectivity to CO_x formation. In contrast, MCa catalyst displays two very weak peaks at temperatures higher than 700 °C. On this catalyst, not only fewer CO_x but also little oxygenates are produced. The selectivity of C_3H_6, which is generally assumed as an intermediate in the conversion of propane to acrolein, is as high as 60%, while the selectivity of acrolein is fairly low. This may be due to the poor insertion ability of oxygen species on catalysts for transformation of C_3H_6 to oxygenates. On the other hand, MSi exhibits an intermediate behaviour of O_2 desorption, as evidenced by its TPD profile with $T_{o,dep}$ higher than that of MZr but lower than that of MCa. The moderate inserting function of the oxygen species on MSi would be beneficial to the formation of acrolein.

In conclusion, the selective oxidation of propane over $MoV_{0.2}Te_{0.1}O_x$ catalysts is profoundly affected by the nature of supports. The supports influence the structure, reducibility and O_2-desorption behaviour of the catalysts, which are closely related to their catalytic performance. Among the catalysts investigated, $MoV_{0.2}Te_{0.1}/SiO_2$ shows the best performance for the oxidation of propane to acrolein.

ACKNOWLEDGEMENT

This project is supported by the Ministry of Science and Technology of China (grant No. G1999022408)

REFERENCES

[1] M.M. Bettahar, et al., Appl. Catal. A: Gen., 145 (1996) 1.
[2] M. Baerns, O.V. Buyevskaya and M. Kubik, Catal. Today, 33 (1997) 85.
[3] Y. Teng and T. Kobayashi, Chem. Lett., (1998) 327
[4] C.C. Williams, J.G. Ekerdt and J.M. Jehng, J. Phys.Chem., 95 (1991) 8781.
[5] G. Mestel, Catal. Rev.-Sci. Eng., 40 (1998) 451
[6] M.A. Py, Ph. E. Schmid, J.T. Vallin, Nuovo Cimento, 1977, 38B, 271
[7] I.R. Beattie and T.R. Gilson, J. Chem. Soc. A. (1969) 2322
[8] R.S. Weber, J. Phys. Chem., 98 (1994) 2999
[9] M.R. Smith and U.S. Ozkan, J. Catal., 141 (1993) 124
[10] K. Bruckman, J. Haber, and T. wiltowski, J. catal., 106 (1987) 188
[11] R.A. Hernandez, and U.S. Ozkan, Ind. Eng. Chem. Res., 29 (1990) 1454

Studies in Surface Science and Catalysis, volume 147
X. Bao and Y. Xu (Editors)

Infrared study of the reaction pathways for the selective oxidation of propane to acrolein over MoPO/SiO₂ catalyst

X. D. Yi, W. Z. Weng*, **C. J. Huang, Y. M. He, W. Guo and H. L.Wan***

State Key Laboratory for Physical Chemistry of Solid Surfaces, Department of Chemistry and Institute of Physical Chemistry, Xiamen University, Xiamen 361005, China

ABSTRACT

To elucidate possible reaction pathways for propane selective oxidation to acrolein over MoPO/SiO₂ catalyst, the surface species formed by adsorption of 2-Br-propane, isopropanol, propane/O₂ and propene on the catalyst and the transformation of these species at elevated temperature were studied by IR spectroscopy. The results suggest that isopropoxy species is one of the intermediates for the selective oxidation of propane to acrolein over the catalyst. The isopropoxy species can either convert to acetone by dehydrogenation or to propene by β-hydrogen elimination, and the latter can further convert to acrolein through allylic process.

1. INTRODUCTION

The partial oxidation of light paraffins has attracted much attention because it represents a route to obtain more valuable organic compounds from low cost saturated hydrocarbons. However, it is usually accompanied with many difficulties as a result of the high reactivity of the oxygenated products compared to the low reactivity of alkanes under the same reaction conditions [1-3]. The selective oxidation of propane to acrolein and acrylic acid has been studied over various catalysts, including VPO catalyst, heteropolyacids and other mixed oxide catalysts, but only a few showed good catalytic performances [4-7].

Current research on propane selective oxidation to acrolein concerns mainly fundamental aspects. A debate still exists on the reaction mechanism and little attention has been given to the study of the nature of the adsorbed species or possible intermediates formed by interaction of reactants with the catalyst and their chemistry of surface transformation. Since methylene C-H bonds in propane are weaker than those of the methyl groups, propane activation has been proposed to occur by the initial cleavage of these weaker C-H bonds to form isopropoxy species [8]. Recently, we found that MoPO/SiO$_2$ (Mo/P/Si = 3.5/5/100, molar ratio) catalyst was active and selective in the propane selective oxidation to acrolein, with a 19.8% selectivity to acrolein at the propane conversion of 30.2% [9]. By using a conventional fixed bed reactor, we have studied the oxidative conversion of propane and a series of probe molecules (e. g. isopropanol, 1-propanol, propene, acetone and propanal) related to the intermediates and products of the selective oxidation of propane to acrolein over MoPO/SiO$_2$ catalyst. The results suggest that isopropoxy species is likely to be the intermediate of propane oxidation over the catalyst. In this paper, the surface species formed by adsorption of 2-Br-propane, isopropanol, propane/O$_2$ and propene on the catalyst and the transformation of these species at elevated temperature were studied by IR spectroscopy. It is expected that these experiments will provide us with some useful information to understand the reaction mechanism of propane oxidation to acrolein over the catalyst.

2. EXPERIMENTAL

The catalysts were prepared by impregnation of silica (294 m^2/g) with a solution containing the desired quantities of ammonium heptamolybdate and NH$_4$H$_2$PO$_4$. Impregnated samples were dried overnight at 393 K and calcined in air at 653 K for 4 h followed by at 823 K for a further 6 h. The IR spectra were recorded in transmission mode with a Perkin-Elmer Spectrum 2000 FTIR spectrometer using an *in situ* IR cells (BaF$_2$ windows) connected to an evacuation-gas manipulation apparatus. The catalyst powder was pressed into self-supporting discs that were calcined in O$_2$ at 773 K for 1 h, followed by evacuation at the same temperature from 10 min and cooled under vacuum to certain temperature to record the background spectra of the cell and the catalyst wafer. The selective molecule was then introduced into the cell at 323 K and the gas/solid interface was maintained for 10 min to 4 h. The IR spectra of adsorbed species were recorded at 323, 373, 423, 473 or 573 K, respectively, after the gas phase components in the IR cell were removed by evacuation. The spectrum of the adsorbed species was referenced to the background spectrum of the pretreated catalyst recorded under vacuum at the same temperature before the admission of adsorbates.

3. RESULTS AND DISCUSSION

3.1. IR spectra arising from 2-Br-propane and isopropanol adsorptions

It was believed that isopropyl species were formed in the rate-determining step for the oxidative dehydrogenation of propane to propene on the catalyst surface [10]. The study of kinetic isotopic effects for the oxidative dehydrogenation of propane to propene on MoO$_x$/ZrO$_2$ catalyst also showed that the methylene C-H bond activation was the rate-determining step for propane dehydrogenation reaction [11]. Formation of isopropyl species during propane oxidation requires the abstraction of a hydrogen atom from the methylene group. Since the strength of C-Br bond in 2-Br-propane is weak than that of C-H bond in the methylene group of propane, the isopropyl species should be formed more easily from 2-Br-propane than from propane. Therefore, we can use 2-Br-propane as a probe molecule to investigate the conversion of isopropyl species on the catalyst. The adsorption of 2-Br-propane on MoPO/SiO$_2$ catalyst at 323 K is characterized by a spectrum with bands at 2983, 2943 and 2885 (v_{CH}) cm^{-1}, 1469 and 1461 ($\delta_{as, CH3}$) cm^{-1}, 1389 and 1378 ($\delta_{s, CH3}$) cm^{-1}, and 1343 (δ_{CH}) cm^{-1} (Fig. 1a). It is interesting to see that the IR bands shown in Fig. 1a are very similar to those arising from the adsorption of isopropanol over the same catalyst at the same temperature (Fig. 2a). This observation suggested that an isopropoxy species was readily formed by adsorption of 2-Br-propane on the catalyst at 323 K. Since the MoPO/SiO$_2$ catalyst itself was not IR transparent in the region below 1300 cm^{-1}, we are unable to observed the band of C-O

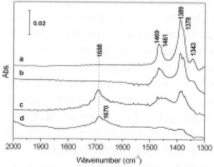

Fig. 1. IR spectra of adsorbed species arising from interaction of MoPO/SiO$_2$ catalyst with 2-Br-propane at 323 K for 10 min followed by evacuation at (a) 323 K and heating under vacuum to (b) 373 K, (c) 423 K and (d) 473 K.

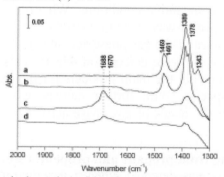

Fig. 2. IR spectra of adsorbed species arising from interaction of MoPO/SiO$_2$ catalyst with isopropanol at 323 K for 10 min followed by evacuation at (a) 323 K and heating under vacuum to (b) 373 K, (c) 423 K and (d) 473 K.

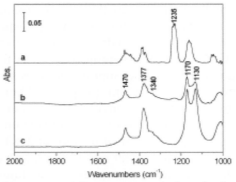

Fig. 3. IR spectra of (a) gaseous 2-Br-propane and the adsorbed species arising from interaction of (b) 2-Br-propane and (c) isopropanol with Al$_2$O$_3$ at 323 K for 10 min followed by evacuation at 323 K.

stretching located at ~1100 cm^{-1} [12].However, by comparing the IR spectra arising from the adsorptions of 2-Br-propane (Fig. 3b) and isopropanol (Fig. 3c) over Al$_2$O$_3$, it can be concluded that an isopropoxy species is readily formed at 323 K. This is also evidenced by the observation of intensity decrease of the band at 1235 cm^{-1} for R$_2$BrC-H deformation (Fig. 3a) after the adsorption accompanied by the appearance of the bands at 1130 and 1170 cm^{-1} due to the strongly coupled C-O/C-C stretchings [12]. A rational explanation to the observation is that the carbon-oxygen interaction is stronger than the carbon-metal interaction. Therefore the isopropyl species interact most probably with a lattice oxygen atom rather than with a surface metal cation [2]. As the samples containing the adsorbed isopropoxy were heated to 423~473 K under vacuum, two new bands indicating the formation of acetone (1688 cm^{-1}) and acrolein (1670 cm^{-1}) were observed (Fig. 1 and 2, spectra c and d). The assignments of these two bands are based on the comparison of the above spectra with the IR spectra arising from acetone and acrolein adsorptions over the catalyst under the comparable conditions. When the sample containing the adsorbed isopropoxy species was heated under 5% O$_2$/He atmosphere from 323 K to 673 K, the IR spectra of the gas phase acetone (1738, 1722 cm^{-1}, $v_{C=O}$) and propene (~1650 cm^{-1}, $v_{C=C}$) were detected at 473 K. This observation indicates that acetone and propene are the main products for the oxidative conversion of surface isopropoxy species. With further increasing the temperature of the sample to 573~673 K, the bands of gas phase acrolein (1733, 1710 cm^{-1}, $v_{C=O}$) were also detected.

3.2. IR spectra arising from propane and propene adsorptions

No surface species was detected when a gas mixture of propane/O$_2$ was introduced to the catalyst until the temperature of the sample was raised to 473 K. After the gas phase components were removed by evacuation, a weak band at

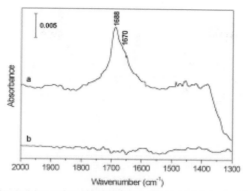

Fig. 4. IR spectra of adsorbed species arising from interaction of MoPO/SiO$_2$ catalyst with propane at 473 K for 4 h followed by evacuation at (a) 473 K and heating to (b) 523 K.

1688 cm^{-1} and a shoulder at 1670 cm^{-1} assignable to adsorbed acetone and acrolein, respectively, were observed at 473 K (Fig. 4a). The formation of acetone strongly suggests that propane is activated by abstracting a hydrogen atom from –CH$_2$– group. This is followed by bonding of the lattice oxygen near by over the catalyst with the secondary carbon of the isopropyl species leading to the formation of an isopropoxy species. However, the isopropoxy species can not be detected under the experimental conditions of propane adsorption (Fig. 4a), since the IR spectra shown in Fig. 1 and 2 clearly indicated that the conversion of isopropoxy species to acetone and acrolein take place at 423 K.

Since propene is one of the main products of the oxidative conversion of surface isopropoxy species, the adsorption and conversion of propene over the catalyst is also studied. The IR spectrum of the surface species arising from interaction of MoPO/SiO$_2$ catalyst with propene at 373 K is shown in Fig. 5a. Compared to the spectra shown in Fig. 1 and 2, the bands at 2983, 2943, 2885, 1469, 1461, 1389, 1378 and 1346 cm^{-1} can be assigned to isopropoxy species. However, the appearance of additional bands at 3091 and 3023 ($v_{=CH}$), 1863 ($2w_{CH2}$), 1649 ($v_{C=C}$) and 1426 ($\delta_{=CH2}$) cm^{-1} indicates the coexistence of another adsorbed species. These bands are observed after adsorption of allyl alcohol over the catalyst and can therefore be assigned to an allyl alcoholate species [12]. As the temperature of the sample was increased to 423 K, two new bands, ascribable to the adsorbed acetone (1688 cm^{-1}) and acrolein (1671 cm^{-1}) species were observed. These results indicate that two major routes may be involved in the activation of propene over MoPO/SiO$_2$ catalyst. The first one involves breaking of the C-H bond of the methyl group with subsequent oxygen insertion to give allyl alcoholate, which then converts to acrolein; the second route is associated with the electrophilic attack of the Brønsted acid sites on the catalyst at the propene double bond, giving rise to an isopropoxy species, which is then oxidized to acetone.

672

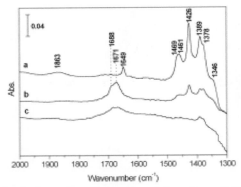

Fig. 5. IR spectra of adsorbed species arising from interaction of MoPO/SiO₂ catalyst with propene at 373 K for 0.5 h followed by evacuation at (a) 373 K and heating under vacuum to (b) 423 K and (c) 473 K.

4. CONCLUSIONS

The above results suggest that isopropoxy species is one of the intermediates for the selective oxidation of propane to acrolein over MoPO/SiO₂ catalyst. The isopropoxy species can either convert to acetone by dehydrogenation or to propene by β-hydrogen elimination, and the latter can further convert to acrolein through allylic process.

ACKNOWLEDGEMENTS

This project is supported by MSTC (grant No. G1999022408) and NSFC (grant No. 20021002).

REFERENCES

[1]   C. Batiot, Appl. Catal. A: Gen., 137 (1996) 179.
[2]   M.M. Bettahar, G. Costentin, L. Savary and J.C. Lavally, Appl. Catal. A: Gen., 145 (1996) 1.
[3]   M.M. Lin, Appl. Catal. A: Gen., 207 (2001) 1.
[4]   J.C. Vedrine, Top. Catal., 21 (2002) 97.
[5]   A. Kaddouri, C. Mazzocchia and E. Tempesti, Appl. Catal. A: Gen., 180 (1999) 271.
[6]   M. Ai, J. Catal., 101 (1986) 389.
[7]   Y.C. Kim, W. Ueda and Y. Moro-oka, Appl. Catal. A: Gen., 70 (1991) 175.
[8]   H.H. Kung, Adv. Catal., 40 (1994) 1.
[9]   X.D. Yi, W.Z. Weng, C.J. Huang, Y.M. He, W. Guo and H.L. Wan, Chn. J. Catal., 24 (2003) 755.
[10] P.M. Michalakos, M.C. Kung, J. Jahan and N.H. Kung, J. Catal., 140 (1993) 226.
[11] K. Chen, E. Iglesia and A.T. Bell, J. Phys. Chem. B. 105 (2001) 646.
[12] G. Busca, E. Finocchio, V. Lorenzelli, G. Ramis and M. Baldi, Catal. Today, 49 (1999) 453.

Studies in Surface Science and Catalysis, volume 147
X. Bao and Y. Xu (Editors)
©2004 Elsevier B.V. All rights reserved.

Oxidative ethane activation over oxide supported molten alkali metal chloride catalysts

S. Gaab, J. Find, R. K. Grasselli and J. A. Lercher

Institute of Chemical Technology, TU München, Lichtenbergstrasse 4, 85747 Garching, Germany

ABSTRACT

Oxidative dehydrogenation of ethane was studied with Li/Dy/Mg/Cl mixed oxides with various Cl⁻ loadings as catalyst. Ethylene yields of up to 77% were obtained with catalysts containing above 8 wt% Cl⁻. Based on kinetic data and *in situ* high-temperature XRD measurements as well as TG-DSC analysis, a new reaction model is proposed. The active species of the catalytic system are postulated to be dissolved in molten LiCl, supported on Dy_2O_3/MgO. Atomic oxygen in the LiCl melt is speculated to form the catalytically active [OCl⁻] species. The decomposition of [OCl⁻] leads to the formation of O• + Cl⁻ or O⁻ + Cl•. Both radical species are highly oxidizing and readily activate alkanes via hydrogen abstraction.

1. INTRODUCTION

Ethene and propene are important starting materials for the production of petrochemicals and fine chemicals. In recent years the increasing demand of ethene and propene has spurred substantial interest in the development of alternatives routes other than FCC and naphtha cracking [1-3]. Catalytic oxidative dehydrogenation (ODH) of alkanes represents a promising option. However, alkene yields in the ODH over mixed oxide catalysts are well below 40% with substantial formation of carbon oxides [4].

Among many mixed metal oxides studied for the ODH of ethane, Li_2O/Dy_2O_3/MgO mixed oxides have been found of some interest in the past [5] and have more recently experienced a renaissance through incorporation of halides and specific preparation procedures [6-9]. Ethylene yields up to 77% were obtained with such catalysts. For this reason Li/Dy/Mg mixed oxides containing various Cl⁻ levels were investigated to explore the relations between the catalytic material and catalytic activity and selectivity. A tentative mechanism is presented and discussed.

2. EXPERIMENTAL

Li/Dy/Mg/O/Cl precursors were prepared by wet impregnation using MgO, Dy_2O_3, $LiNO_3$, aqueous HCl (37%) and NH_4Cl as starting materials. Details are given elsewhere [6]. The precursors were calcined for 12 hours at 973 K.

The concentration of Li, Mg, and Dy was analyzed by AAS (UNICAM 939 AA-Spectrometer). The concentration of Cl⁻ was analyzed using a Metrohm 690 Ion Chromatograph.

XRD phase analysis was performed by means of a Philips X`Pert instrument. *In-situ* high-temperature XRD experiments were done in an environmental chamber (HTK Paar, 1200) using synthetic air flow. The TG-DSC measurements were performed using a modified Setaram TG-DSC thermoanalyzer.

Kinetic measurements were performed in a tubular fixed bed reactor. The weight hourly space velocity was 0.8 h⁻¹ and the ethane to oxygen ratio was one. The composition of the products was analyzed with a Hewlett Packard 6890 Series gas chromatograph equipped with an FID and TCD detector. A Pora Plot Q and a Molsieve column were used for product separation.

3. RESULTS AND DISCUSSION

3.1. Material properties

The elemental composition, BET surface area, and the XRD-phase composition of the Li/Dy/Mg/O/Cl materials are compiled in Table 1. In all cases, MgO was the dominating crystalline phase. Crystalline $LiDyO_2$ and Dy_2O_3 were observed. The distribution between the $LiDyO_2$ and Dy_2O_3 phases strongly depends on the Cl⁻ content of the catalyst. At low Cl⁻ concentration $LiDyO_2$ dominates among these two phases, while at high Cl⁻ concentration Dy_2O_3 take the place of the $LiDyO_2$ phase. A Li_2O was observed by Raman spectroscopy (not shown) for catalysts having a Li to Cl ratio above one. LiCl is

Table 1
Catalyst composition and characterization

Code	MgO [wt%]	Dy_2O_3 [wt%]	Li_2O [wt%]	Cl [wt%]	Li/Dy/Cl atomic ratio	BET [m²/g]	XRD phases at 298 K
C01	85.2	7.7	6.2	0.9	0.20/0.02/0.01	3.37	MgO; $LiDyO_2$
C02	87.1	7.9	3.6	1.4	0.11/0.02/0.02	3.36	MgO; $LiDyO_2$
C04	84.7	7.7	4.3	3.3	0.14/0.02/0.04	3.03	MgO; $LiDyO_2$
C06	81.9	7.4	5.6	5.1	0.19/0.02/0.07	6.29	MgO; $LiDyO_2$; Dy_2O_3
C08	82.4	7.5	4.9	5.2	0.16/0.02/0.07	2.71	MgO; $LiDyO_2$; Dy_2O_3
C12	81.1	7.3	3.1	8.5	0.10/0.02/0.12	8.47	MgO; Dy_2O_3
C16	76.5	6.9	4.6	12.0	0.16/0.02/0.18	11.3	MgO; Dy_2O_3

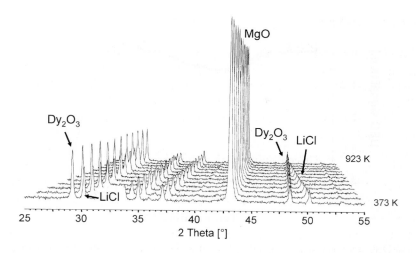

Fig. 1. XRD diffractograms of preactivated catalyst C16 at various temperatures in synthetic air flow (50 K steps).

not observed as XRD crystalline phase at room temperatures because the water attached caused long range disorder. It appears as crystalline phase after heating the samples in flowing N_2 above 373 K.

To describe the phase transformations under reaction conditions, X-ray diffractograms of preactivated catalysts were taken *in situ* between 373 and 923 K. The diffractograms show that MgO is present over the entire temperature range studied, $LiDyO_2$ (with samples having less than 8 wt% Cl⁻) between 373 and 573 K, while Dy_2O_3 appears at 473 K and gains in intensity with rising temperature at the expense of $LiDyO_2$ (not shown). LiCl appeared as a crystalline phase at 373 K and persisted as such up to 823 K (Fig. 2). The disappearance of its XRD peaks is attributed to the transformation into a melt. This conclusion is based on several arguments: (1) The Dy_2O_3 (known to form mixed oxides with Li⁺) diffraction lines did not change in intensity, position or half-width. (2) Before disappearance the XRD lines of LiCl shifted to higher *d*-values. The thermal expansion coefficient α deduced from this shift (44.8×10^{-6} K⁻¹) is in excellent agreement with the literature value (46.0×10^{-6} K⁻¹) [10]. (3) The melting process could be directly followed by TG-DSC measurements. Fig. 2 shows the TG-DSC analysis of a Cl⁻ free and a Cl⁻ containing catalyst. Both catalysts showed broad endothermic signals between 573 and 673 K due to $MgCO_3$ decomposition and water release as it was verified by MS. Additionally, a weak signal was observed at 869 K for the Cl⁻ free catalyst which results from CO_2 release from the catalyst. In contrast, Cl⁻ containing catalysts showed strong endothermic signals with a maximum at 877 K without any changes of the catalyst mass, which is in perfect agreement with the transformation of LiCl into

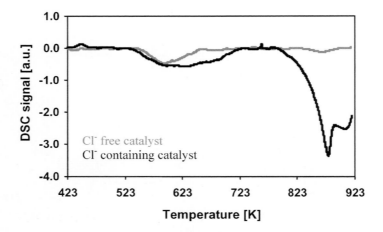

Fig. 2. TG-DSC analysis (Cl⁻ free catalyst grey line; Cl⁻ containing catalyst black line).

a melt. It is interesting to note that in contrast to pure LiCl (m.p. = 878 K) melting/ spreading already occurred at 800 K.

This leads to an overall picture of the catalyst consisting of a MgO core decorated with crystallites of Dy_2O_3. This two phase particle is covered with ~ 20 monolayer of LiCl able to sorb oxygen [11].

3.2. Catalytic properties

The ethane conversion and the ethene yield as a function of temperature are illustrated in Figs. 3A and 3B, respectively. The activity of the samples increases dramatically with the concentration of Cl⁻ in the material. The higher activity is accompanied by a higher selectivity if the LiCl is maintained in the molten state. Within the samples containing molten LiCl the selectivity hardly varied as long as least a monolayer of LiCl was available.

The existence of molten LiCl under reaction conditions in the Li/Dy/Mg oxide catalysts studied is concluded to be a key factor for the effectiveness in the chloride-mediated ODH of paraffins [6]. Depending on the reaction conditions and catalyst composition two regimes, one of high apparent activation energy (~ 246 kJ/mol) and one with lower activation energy (~ 147 kJ/mol), were observed (Fig. 4). The former regime corresponds to catalysts with high Cl⁻ loadings which exhibited high activity at low temperatures. The latter corresponds to catalysts with low Cl⁻ content. The abstraction of hydrogen (dissociative adsorption) is speculated to be involved in the rate determining step in regime with the higher energy of activation. This regime is mostly affiliated with catalysts of low Cl⁻ concentration. Catalysts with high concentration are much more active and seem to become limited in the availability of [OCl⁻] diffusing within the molten layer.

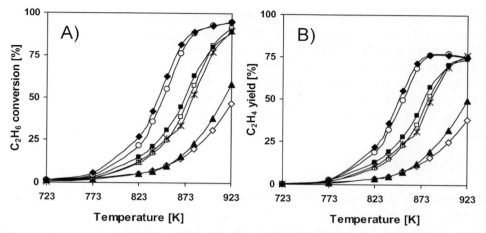

Fig. 3. Catalytic performances of C01 to C16 catalysts at different temperatures: A) C_2H_6 conversion and B) C_2H_4 yield ($\lozenge$ C01, $\blacktriangle$ C02, $\square$ C04, $\blacksquare$ C06, * C08, $\circ$ C12 and $\blacklozenge$ C16).

Based on these observations we speculate that the active site for the ODH of ethane must be associated with the generation of oxidized chloride (OCl^-) or the decomposition product of such ion. The steady state concentration of this species or any precursor must be extremely small, as the absence of oxygen in the feed led to the immediate decrease of the catalytic activity. Note also that the reaction is true oxidative dehydrogenation with H_2 not being detected under any circumstance. Tentatively we speculate that the active site for the ODH of

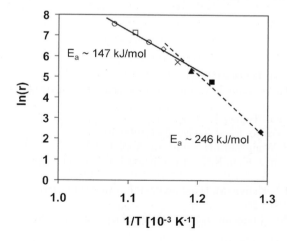

Fig. 4. Arrhenius plot: The slope of isothermal datapoints for catalysts with different Cl loadings is used (Fig 3B) to calculate the apparent activation energy.

ethane reside in molten LiCl, supported on Dy_2O_3/MgO. Oxygen is dissolved dissociatively in the melt forming the catalytically active [OCl⁻]. At higher temperatures the [OCl⁻] ions decompose to O• + Cl⁻ or O⁻ + Cl•. Both radicals can easily activate ethane by homolytic hydrogen abstraction forming C_2H_5• and HCl (i.e., when starting with Cl•). HCl then reacts with O⁻ to form •OH and Cl⁻. The so formed C_2H_5• radicals react further with •OH to form C_2H_4 and H_2O, thereby closing the reaction cycle. The cycle starting from O• proceeds in analogy.

4. CONCLUSIONS

The catalytic performance of novel chlorinated Li/Dy/Mg mixed oxides with varying chloride content was explored with respect to the ODH of ethane. In these materials a LiCl melt supported on the Dy_2O_3 decorated MgO particles exists under reaction conditions. It is speculated that the catalytically active species is affiliated with the OCl⁻ ion or one of its decomposition products. The increase in Cl⁻ concentration and specific surface area leads to an increase in the amount of available LiCl melt and a corresponding increase of active species. The effectiveness of this species, i.e., [OCl⁻], lies in its ability to disproportionate to: O• + Cl⁻ or O⁻ + Cl•, whereby both, O• and Cl•, are highly oxidizing and capable of activating and abstracting hydrogen homolytically. At present we cannot unequivocally conclude whether unique performance of the catalyst lies in the absence of highly coordinatively unsaturated sites or due to the presence of a mobile phase offering oxidation sites without the possibility of strong sorption.

REFERENCES

[1] Oil and Gas Journal, January (1997), 12.
[2] M.M. Bhasin, J.H. Mc Cain, B.V. Vora, T. Imai and P.R. Pujadó, Appl. Catal. A., 221 (2001) 397.
[3] M.A. Banares, Catal. Today, 51 (1999) 319.
[4] F. Cavani and F. Trifirò, Catal. Today, 24 (1995) 307.
[5] S.J. Conway, D.J. Wang and J.H. Lunsford, Appl. Catal. A, 79 (1991) L1.
[6] S. Gaab, M. Machli, J. Find, R.K. Grasselli and J.A. Lercher, Topics in Catalysis, 23 (2003) 151.
[7] M.V. Landau, M.L. Kaliya, M. Herskowitz, P.F. van den Oosterkamp and P.S.G. Bocqué, Chemtech, 26 (1996) 24.
[8] M. Herskowitz, M.V. Landau and M.L. Kaliya, German Patent No. DE 19502747 C1 (1996).
[9] A. Hartung, S. Gaab, J. Find, A. Lemonidou and J.A. Lercher in: Proceedings of the DGMK-Conference (2000) 139.
[10] A. Ievins, M. Straumanis and K. Karlsons, Z. Phys. Chem. B40 (1938) 146.
[11] S. Gaab, J. Find, R.K. Grasselli and J.A. Lercher, prepared for publication

Studies in Surface Science and Catalysis, volume 147
X. Bao and Y. Xu (Editors)

Effects of additives on the activity and selectivity of supported vanadia catalysts for the oxidative dehydrogenation of propane

Hongxing Dai[*], **Larry Chen**[*], **T. Don Tilley**[**], **Enrique Iglesia**[*], **and Alexis T. Bell**[*]

Chemical Sciences Division, Lawrence Berkeley National Laboratory and Departments of Chemistry** and Chemical Engineering*, Berkeley, CA 94720, U.S.A.

ABSTRACT

The activity and selectivity of alumina- and magnesia-supported vanadia for propane oxidative dehydrogenation can be enhanced by can be enhanced by interactions of vanadia with molydena for the alumina-supported catalyst and by phosphorous doping for the magnesia-supported catalyst.

1. INTRODUCTION

Oxidative dehydrogenation (ODH) of propane to propene is an attractive alternative to non-oxidative dehydrogenation, because ODH is favored by thermodynamics even at low temperatures and does not lead to the formation of coke and lower molecular weight products. Vanadia supported on Al_2O_3 and MgO have emerged as the most active and selective catalysts among those that have been investigated [1-8]. Investigations of propane ODH on VO_x/Al_2O_3 have shown that the highest specific activity is achieved at near-monolayer coverages of polyvanadate species, 7.5 V/nm^2 [2-8]. When vanadia is supported on MgO, neither polyvanadate nor V_2O_5 domains are formed and, instead, small domains of $Mg_3(VO_4)_2$ are formed [1, 2 (and references therein), 9]. The specific rate for ODH increases with increasing apparent surface density of V_2O_5, reaching a plateau at a surface density of about 4 V/nm^2, which corresponds closely to the surface density of VO_x in $Mg_3(VO_4)_2$. The objective of the present studies was to investigate whether even higher ODH activities and propene selectivities could be achieved by modification of alumina and magnesia supported vanadia. The influence of using a molybdena monolayer between the vanadia and the alumina was examined in the case of VO_x/Al_2O_3 [10], whereas doping the magnesia with phosphorous was examined in the case of VO_x/MgO [11].

2. Experimental

$VO_x/MoO_x/Al_2O_3$ was prepared as follows. Alumina was first impregnated with an aqueous solution of ammonium heptamolybdate in order to deposit an equivalent monolayer of MO_x on to the support. The deposited material was dried and then calcined in air at 773 K. Vanadia was then deposited on to this support via impregnation from a isopropanol solution of vanadyl isopropoxide. The dried material was then calcined a second time at 773 K. The BET surface area of the $VO_x/MoO_x/Al_2O_3$ materials was 190 m^2/g. Raman spectroscopy revealed that the molybdena layer decreased the tendency of vanadia to form V_2O_5 domains and enhanced the formation of polyvandate species. TPR experiments showed that the molybdena layer increased the reducibility of the dispersed vanadia relative to that observed for vanadia dispersed on pure alumina.

High surface area MgO ($\sim$200 m^2/g) was doped with P by adding an appropriate quantity of $(O)P(O^tBu)_3$ and then calcining this material in air at 823 K. Vanadia was then dispersed on the P-doped MgO by contacting the support with $(O)V(O^tBu)_3$ and then calcining the solid mixture in air at 823 K. The resulting material has a BET surface area of $\sim$ 190 m^2/g. The only species observed by Raman spectroscopy is magnesium orthovanadate, irrespective of the degree of P-doping.

Propane ODH rates were measured in a quartz microreactor. The partial pressure of propane was 13.5 kPa and that of oxygen, 1.7 kPa. The reaction products were analyzed by gas chromatograph. The propane and oxygen conversions were kept below 2% and 20%, respectively, in all experiments.

3. RESULTS AND DISCUSSION

3.1 Effects of MoO_x on alumina-supported VO_x

Figure 1a shows initial propene formation rates normalized per V-atom as a function of V surface density. The specific rate of ODH reached a maximum at a surface density of $\sim$7-8 VO_x/nm^2. At a given V surface density, propene formation rates are 1.5-2.0 times higher when VO_x is dispersed on MoO_3/Al_2O_3 than on pure Al_2O_3. The trend shown for high surface densities in Fig. 1 is consistent with the expected loss of accessibility of V centers as three dimensional V_2O_5 structures form at apparent VO_x surface densities above 7.5 V/nm^2.

The increase in the specific rate of propane ODH with V surface density below 7.5 V/nm^2 seen in Fig. 1a for vanadia dispersed in alumina and alumina

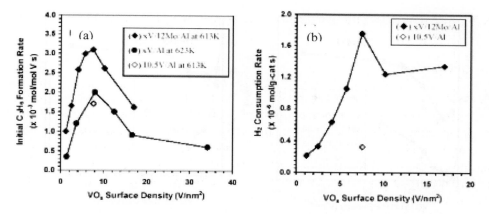

Fig. 1. Effect of VO_x surface density on (a) the specific activity of VO_x/Al_2O_3 and $VO_x/MoO_x/Al_2O_3$ and (b) the rate of H_2 reduction at 613 K.

covered by a monolayer of molybdena is closely associated with an increase in the reducibility of VO_x domains [8]. This relationship is clearly observed in Fig. 1b. The higher propane ODH activity and reducibility of vanadia dispersed on alumina containing a monolayer of molybdena is believed due to the formation of V-O-Mo bonds between the two dispersed oxide layers, and may reflect the formation of polymolybdovanadate structures [10].

Previous kinetic and mechanistic studies of propane ODH on VO_x and MoO_x catalysts have shown that primary and secondary reactions can be accurately described by the scheme [2, 3, 5-7, 8]:

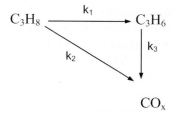

Each reaction is accurately described using pseudo-first-order dependencies on propane and propene reactants and zero-order in O_2 [3, 6, 7]. The k_2/k_1 and k_3/k_1 ratios reflect relative rates of C_3H_8 combustion and dehydrogenation, and of C_3H_6 combustion and dehydrogenation, respectively. Figure 2a show k_2/k_1 ratios as a function of VO_x surface density. These ratios decrease with increasing VO_x surface density up to ~2 V/nm^2 and then increase monotonically for higher surface densities. The k_2/k_1 ratios increased with increasing V surface density on both supports, but they were significantly lower on MoO_x/Al_2O_3 than on pure Al_2O_3 supports at all surface densities.

682

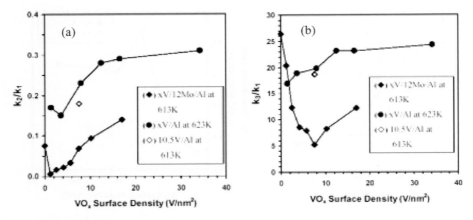

Fig. 2. Effect of VO_x surface density on (a) k_2/k_1 and (b) k_3/k_1.

The influence of vanadia surface density on k_3/k_1 ratios is shown in Fig. 2b. The k_3/k_1 ratio decreases with increasing surface density up to 7.5 VO_x/nm^2, and then increase gradually as V surface densities increases beyond this value. The values of k_3/k_1 are much lower for VO_x domains supported on MoO_x/Al_2O_3 than for domains supported at similar surface densities on Al_2O_3. These data indicate that the molybdenum interlayer suppresses propene combustion rates relative to propane ODH rates. On Al_2O_3, the k_3/k_1 ratio increases monotonically with increasing vanadia surface density. The high initial value of k_3/k_1 on MoO_x/Al_2O_3 reflects a significant catalytic contribution from MoO_x domains, which show much higher k_3/k_1 ratios than VO_x domains (13, 14). As vanadia covers an increasing fraction of the exposed MoO_x surface, it reduces the contributions from the latter to combustion rates, because the specific activity of vanadia is much higher than that of molbydena [12, 13]. A minimum k_3/k_1 value is achieved at a surface density of 7.5 VO_x/nm^2, which corresponds to a theoretical polyvanadate monolayer. Above a surface density of 7.5 V/nm^2, however, k_3/k_1 increases with increasing surface density as in the case of pure Al_2O_3 supports. In spite of some residual contributions from less selective MoO_x domains, VO_x domains dispersed on MoO_x/Al_2O_3 give much lower k_3/k_1 ratios than when dispersed on Al_2O_3 at all VO_x surface densities. This appears to reflect the higher dispersion of VO_x species on MoO_x/Al_2O_3 supports than on Al_2O_3, which minimizes unselective reactions prevalent in V_2O_5 crystallites. These unselective reactions lead to the observed increase in k_3/k_1 ratios with increasing VO_x surface densities on both MoO_x/Al_2O_3 and Al_2O_3 supports.

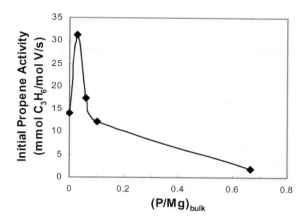

Fig. 3. Effect of P/Mg ratio on the specific activity of VO_x/P-MgO (VO_x surface density = 3.4 V/nm^2) at 773 K.

3.2 Effects of P on magnesia-supported VO_x

The rate of propane ODH on P-doped VO_x/MgO increased with increasing P doping as shown in Fig. 3, reaching a factor of two relative to the rate observed on pure MgO, and then decreased for P-doping levels above 5 wt% P_2O_5. This trend exactly parallels that observed for the effects of P-doping on the reducibility of the vanadia. However, P-doping had no effect on the either the selectivity of propane ODH to propene or the secondary combustion of propene. TPR shows that with increasing P-doping of MgO, the reducibility of vanadia passes through a maximum at doping level of 5 wt% P_2O_5.

4. CONCLUSIONS

In summary, the present studies have shown that the rate of propane ODH to propene over alumina- and magnesia-supported vanadia can be enhanced by increasing the reducibility of the supported vanadia. This effect is consistent with the finding that the first step in the ODH of propane involves the reductive addition of C_3H_8 to a surface VO_x species to form a surface propoxide species. While the selectivity of propane ODH to propene and the secondary combustion of propene to CO_x are unaffected by P-doping of VO_x/MgO, a significant enhancement in both properties is observed when a layer of molybdena is introduced between highly dispersed vanadia and alumina.

684

ACKNOWLEDGEMENTS

This work was supported by the Office of Basic Energy Sciences, Chemical Sciences Division, of the U.S. Department of Energy under contract DE-AC03-76SF00098.

REFERENCES

[1] H. Kung, Adv. Catal., 40 (1994) 1.
[2] S. Albonetti, F. Cavani, and F. Trifiro, Catal. Rev. Sci. Eng., 38 (1996) 413.
[3] A. Khodakov, J. Yang, S. Su, E. Iglesia, and A.T. Bell, J. Catal., 177 (1998) 343.
[4] A. Khodakov, B. Olthof, A.T. Bell, and E. Iglesia, J. Catal., 181 (1999) 205.
[5] K. Chen, S. Xie, E. Iglesia, and A.T. Bell, J. Catal., 189 (2000) 421.
[6] K. Chen, A. Khodakov, J. Yang, A.T. Bell, and E. Iglesia, E., J. Catal., 186 (1999) 325.
[7] K. Chen, E. Iglesia, and A.T. Bell, J. Phys. Chem. B, 105 (2001) 646.
[8] K. Chen, A.T. Bell, and E. Iglesia, J. Catal., 209 (2002) 35.
[9] C. Pak, A.T. Bell, and T.D. Tilley, J. Catal., 206 (2002) 49.
[10] H. Dai, A.T. Bell, and E. Iglesia, J. Catal., in press.
[11] L. J.-Y. Chen, MS thesis, Department of Chemical Engineering, University of California, Berkeley, 2002.
[12] K. Chen, S. Xie, A.T. Bell, and E. Iglesia, J. Catal., 198 (2001) 232.
[13] M.D. Argyle, K. Chen, E. Iglesia, and A.T. Bell, J. Catal., 208 (2002) 139.

Studies in Surface Science and Catalysis, volume 147
X. Bao and Y. Xu (Editors)

Oxidative dehydrogenation of ethane over Pt-Sn catalysts

B. Silberova, J. Holm and A. Holmen[*]

Department of Chemical Engineering, Norwegian University of Science and Technology (NTNU), Sem Sælands vei 4, Trondheim, N-7491, Norway

ABSTRACT

The catalytic activity of Pt and Pt-Sn monolithic catalysts has been examined for the production of ethene by short contact time oxidative dehydrogenation of ethane. The experiments were performed in a continuous flow reactor. High yields of ethene (~63%) were obtained over Pt-Sn catalyst at 850 °C when hydrogen was added to the feed.

1   INTRODUCTION

Light hydrocarbons are important feedstocks for the production of olefins such as ethene and propene. Although large-scale steam cracking is the main process for production of olefins, several other possible routes have received considerable attention. Catalytic dehydrogenation and oxidative dehydrogenation at short contact times are examples of approaches that have been studied recently by several groups [1-5]. High yields of ethene and propene have been observed during oxidative dehydrogenation on Pt-coated foam monoliths at extremely short contact times (1 – 10 ms).

It has also been shown [4] that enhancement of the ethene yield can be obtained by addition of hydrogen to the reaction mixture and by promoting the Pt catalyst by Sn. Addition of a second metal such as Sn to Pt is well known and used in other processes like catalytic reforming [6] and dehydrogenation [7].

The present study focuses on the addition of hydrogen during oxidative dehydrogenation of ethane over Pt and Pt-Sn monolithic catalysts. The experiments were performed over a large temperature range. The effect of hydrogen addition was tested by varying the H_2/O_2 ratio in the feed. Furthermore, Pt catalysts with various Sn loading were prepared and tested under the same reaction conditions.

2   EXPERIMENTAL

Cordierite ceramic monolith ($2MgO*2Al_2O_3*5SiO_2$) with cell density of 62.2 cells/cm^2 was used as support. Cylindrical pieces of the monolith (10 mm long, 15 mm diameter) were washcoated by a dispersion of Disperal P2 (BET = 286 m^2/g). The washcoated monoliths were dried at 120 °C for 4 h and the washcoating step was repeated until the weight increase was 15% of the primary monolith weight. After calcination in a flow of air at 550 °C for 4.5 h, the washcoated monoliths were impregnated with $Pt(NH_3)_4(NO_3)_2$ to obtain a Pt loading of about 1 wt.%. The impregnated monoliths were then dried at 80 °C for 4 h and calcined at 550 °C in a flow of air for 4.5 h. The Pt/monoliths were subsequently impregnated by $SnCl_2*2H_2O$, dried at 80 °C and calcined in a flow of air at 100 °C for 0.5 h and at 700 °C for 1.5 h. Pt/monoliths with three different Sn loadings (1, 3 and 7 wt.%) were prepared and tested for oxidative dehydrogenation of ethane. The Pt-Sn monoliths were reduced in the reactor prior to the experiments in a flow of hydrogen at 700 °C for 1.5 h.

Experiments were carried out at atmospheric pressure and in the temperature range of 400 – 950 °C in a conventional flow apparatus consisting of a quartz reactor with inner diameter of 15 mm. The Pt or Pt-Sn coated monolith was placed between two inert monolith pieces acting as radiation shields. At the reactor outlet, two water-cooled condensers were installed for removal of water. The amount of water was calculated based on the oxygen material balance and expressed as O-selectivity. The total flow rate consisting of ethane, air and argon was 2000 Nml/min. In some experiments, hydrogen was added to the reaction mixture. In these cases, the flow of argon was reduced to keep the total flow rate constant. The ethane/oxygen ratio in the feed was 2/1 and the ratio of hydrogen/oxygen was 3/1 if not otherwise mentioned. The contact time through the coated monolithic piece was estimated to 0.04 s at standard conditions. "Dry" samples of the product stream were analyzed by two on-line gas chromatographs.

3   RESULTS AND DISCUSSION

3.1   Effect of H$_2$ addition and Pt-Sn catalysts

Oxidative dehydrogenation of ethane was performed over Pt and Pt-Sn monolithic catalysts without and with addition of hydrogen. At lower temperatures the lowest conversions of ethane were obtained using Pt-Sn catalyst with addition of hydrogen. However, at high temperatures the conversion of ethane was the same for all the experiments performed (Fig. 1A). Reactor temperatures measured inside the reactor were higher with addition of hydrogen to the ethane/oxygen mixture during the experiments over the Pt catalyst and even more using the Pt-Sn catalyst (Fig. 1A). If less ethane is converted but the reactor temperatures are increased when hydrogen is added, other exothermic reactions must occur. Looking at the production of water during the experiments over the Pt catalyst (Fig. 1B), a large amount of water was formed when hydrogen was present in the feed stream. Using the Pt-Sn catalyst, even more significant amounts of water was produced (Fig. 1B). It is obvious that oxygen reacts with hydrogen leading to the formation of large amounts of water. Using Pt catalyst, the conversion of oxygen was already complete at 600 °C when hydrogen was added to the feed. Without addition of hydrogen, not all oxygen was converted at the same temperature. Higher temperatures (< 850 °C) were necessary to achieve total conversion of oxygen over the Pt-Sn catalyst.

As shown on Fig. 2A, the addition of hydrogen obviously contributed to higher selectivities of ethene and lower selectivities of carbon oxides for the Pt catalyst (Figs. 2B and 2C). The selectivities of ethene were even more enhanced by adding Sn to the Pt catalyst (Fig. 2A). This effect can be assigned firstly to the reaction of oxygen with hydrogen, which is preferred to the reaction of oxygen with ethane and secondly to high activity of the Pt-Sn catalyst for dehydrogenation of light alkanes [7]. A high activity of Pt-Sn catalysts for this type of reaction is ascribed to the fact that Sn inhibits the formation of highly dehydrogenated surface species and Sn also decreases the size of surface Pt ensembles [8]. The Pt-Sn catalyst also showed a significant effect on the formation of carbon monoxide (Fig. 2B) and carbon dioxide (Fig. 2C), which were lowered in the whole temperature range studied.

Methane is also one of the components formed during oxidative dehydrogenation of ethane (Fig. 2D).

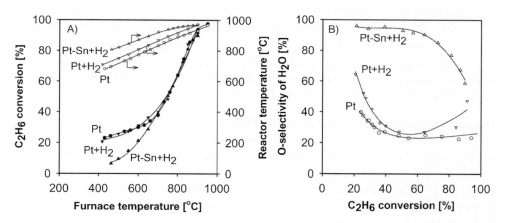

Fig. 1. Oxidative dehydrogenation of ethane over Pt and Pt-1wt.%Sn catalysts with and without addition of H_2. Feed [Nml/min]: C_2H_6 (308), air (733), H_2 (0 - 462) and Ar (959 - 497). Total flow rate: 2000 Nml/min. A: Conversion of ethane and reactor temperature as a function of the furnace temperature. B: O-selectivity of water as a function of the conversion of ethane.

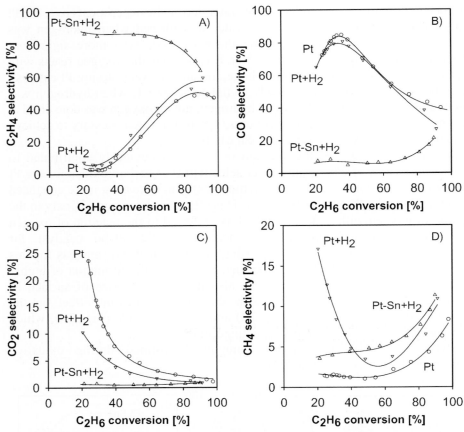

Fig. 2. Effect of addition of H_2 on Pt and Pt-1wt.%Sn monolithic catalysts on the selectivity of ethene (A), carbon monoxide (B), carbon dioxide (C) and methane (D). Feed [Nml/min]: C_2H_6 (308), air (733), H_2 (0 - 462) and Ar (959 - 497). Total flow rate: 2000 Nml/min.

Higher formation of methane was obtained over the Pt catalyst with hydrogen addition compared to the one without hydrogen addition at low furnace temperature. The addition of hydrogen might thus contribute to the formation of methane via CH_y species, which are formed at low furnace temperatures on the Pt catalysts [9-11]. Using the Pt-Sn catalyst, the formation of methane was even higher compared to the Pt catalyst, which is probably due to higher activity of the Pt-Sn catalyst for hydrogenolysis of light alkanes [12].

Higher hydrocarbons such as C_3, C_4 and C_5 components were also formed during oxidative dehydrogenation. The addition of hydrogen to the ethane/oxygen mixture enhanced slightly the yield of C_3 hydrocarbons (from 0.47% to 0.50%) but a more significant effect was observed by addition of Sn to the Pt catalyst (the yield of C_3 hydrocarbons increased to 0.83%). The highest yields of C_4 and C_5

hydrocarbons (1.94% and 0.06%, respectively) were obtained over Pt-Sn catalysts with addition of hydrogen.

3.2 Effect of Sn loading

The addition of 1wt.%Sn to the Pt catalyst and with addition of hydrogen showed a significant effect on the selectivity of ethene. However, further addition of Sn to the Pt catalyst did not reveal any large enhancement in the selectivity of ethene. The highest yield of ethene (~ 63%) was obtained over the Pt-7wt.%Sn catalyst at 850 °C, which is only a few percent higher value compared to using Pt-Sn catalysts with Sn loadings of 1wt.% and 3wt.%. The yields of ethene obtained are higher than the ones reported by Henning and Schmidt (45-57% at 920-925°C) [4] or Buyevskaya et al. (34-46% at 855-867 °C) [5]. The selectivities of carbon monoxide were the same for all three different catalysts at high temperatures. Even though the selectivity of carbon monoxide was highest using the Pt-3wt.%Sn at low temperatures, no obvious effect of the Sn loading on the carbon monoxide production can be deduced. The selectivities of methane and carbon dioxide were hardly influenced by higher Sn loading on the Pt catalyst.

3.3 H_2 output

The consumption and the production of hydrogen are demonstrated by H_2/O_2 ratios as a function temperature in Fig. 3A. It is obvious from Fig. 3A that the amount of hydrogen in the product stream depends very much on the temperature using both Pt and Pt-Sn catalysts.

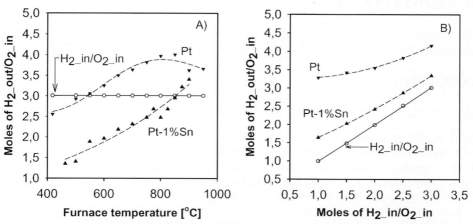

Fig. 3. Comparison between amount of moles of H_2 in the feed and in the product stream related to the amount of O_2 in the feed. Total flow rate: 2000 Nml/min. A) Effect of the temperature. Feed [Nml/min]: C_2H_6 (308), air (733), H_2 (462) and Ar (497). B) Effect of the H_2_in/O_2_in. Feed [Nml/min]: C_2H_6 (308), air (733), H_2 (154–462) and Ar (805-497).

The H_2_out/O_2_in ratio increases with increasing temperature. In the case of the Pt catalyst, a temperature of 550 °C is high enough to obtain H_2_out/O_2_in ratio equal to H_2_in/O_2_in, while a temperature of 850 °C is required to reach the same point using the Pt-1wt.%Sn catalyst. The experiments with various H_2_in/O_2_in ratios in the feed were performed at 850 °C. Fig. 3B shows that higher amount of H_2 moles is coming out compared to what was fed for both the Pt and the Pt-Sn catalyst. To obtain an even higher amount of hydrogen in the product gas, the H_2_in/O_2_in ratio can be reduced. This is clearly shown by the increasing difference between H_2_in/O_2_in and H_2_out/O_2_in ratios with decreasing H_2_in/O_2_in ratios (Fig. 3B).

4 CONCLUSIONS

The addition of hydrogen contributed to a lower production of carbon oxides and to a higher production of ethene. The ethene selectivity was substantially increased also by addition of Sn to the Pt monolithic catalyst. The Pt-Sn catalyst also showed a significant effect on the formation of carbon monoxide and on the selectivity of carbon dioxide, which were lowered remarkably. Further addition of Sn above 1wt.% to the Pt monolithic catalysts did not affect the selectivities of products to any large degree. The selectivity of methane was enhanced by the addition of hydrogen. It has been shown that the amount of hydrogen in the product stream depends strongly on the temperature and on the H_2/O_2 ratio in the feed.

ACKNOWLEDGEMENT

The financial support from the Norwegian Research Council and the Norwegian University of Science and Technology in Trondheim is greatly acknowledged.

REFERENCES

[1] R. Lødeng, O. Lindvåg, S. Kvisle, H. Reier-Nielsen, A. Holmen, Appl. Catal., 187 (1999) 25.
[2] M. Huff, L.D. Schmidt, J. Phys. Chem., 97 (1993) 11815.
[3] A. Beretta, E. Ranzi, P. Forzatti, Catal. Today, 64 (2001) 103.
[4] D.A. Henning, L.D. Schmidt, Chem. Eng. Sci., 57 (2002) 2615.
[5] O.V. Buyevskaya, D. Wolf, M. Baerns, Catal. Today, 62 (2000) 91.
[6] T. Gjervan, R. Prestvik, A. Holmen, Catalytic Reforming. In: Basic Principles in Applied Research (M. Baerns (Ed.)). Springer Series in Chemical Physics, 75 (2004) 125.
[7] L. Bednarova, C.E. Lyman, E. Rytter, A. Holmen, J. Catal., 211 (2002) 335.
[8] R.D. Cortright, J.A. Dumesic: J. Catal., 148 (1994) 42.
[9] G.C. Bond, R.H. Cunningham: J. Catal., 103 (1996) 328.
[10] J.H. Sinfelt, J.L. Carter, D.J.C. Yates: J. Catal., 24 (1972) 283.
[11] B. Silberova, R. Burch, A. Goguet, C. Hardacre, A. Holmen: J. Catal., 219(1) (2003) 206.
[12] A. Sachdev, J. Schwank: Proc. 9th Int. Cong. Catal., 3 (1988) 1275.

Studies in Surface Science and Catalysis, volume 147
X. Bao and Y. Xu (Editors)
691

Cr_2O_3/ZrO_2 catalysts for CO_2 dehydrogenation of ethane to ethylene

Shaobin Wang[a] and Kazuhisa Murata[b]

[a]Department of Chemical Engineering, Curtin University of Technology, GPO Box U1987, Perth, WA 6845, Australia. Email: wangshao@vesta.curtin.edu.au

[b]Research Institute for Green Technology, AIST, Tsukuba, Ibaraki 305-8565, Japan

ABSTRACT

Several Cr_2O_3 catalysts supported on zirconia and its modified forms were prepared, characterized and tested for oxidative dehydrogenation of ethane into ethylene by CO_2. The effects of sulfation and tungstation of zirconia on the catalytic activity were investigated. Cr_2O_3/ZrO_2 exhibited high activity in this reaction with 60% ethylene selectivity at 57% ethane conversion at 650 °C. Addition of WO_3 reduced the surface area, redox potential and acidity, resulting in decreases in activity and selectivity, whereas SO_4^{2-} additive could enhance the ethylene selectivity and yield due to the change of property of acid sites.

1. INTRODUCTION

Ethylene is an important raw material for petrochemical industry and plastics production. Oxidative dehydrogenation of ethane by oxygen into ethylene has been proposed as a good alternative to the process of thermal cracking of ethane because it provides several advantages, such as lower operation temperature and less coking [1, 2]. However, over-oxidation of ethylene to carbon oxides is a serious concern. In the past several years, some researchers have attempted to use weak oxidants like CO_2 [3-8] or N_2O [9-11] for the oxidative dehydrogenation of hydrocarbons. We have found that zirconia-supported Cr_2O_3 exhibits good activity for dehydrogenation of ethane and CO_2 can promote the dehydrogenation [12]. Sulfation and tungstation of zirconia will change the surface properties of the catalysts thus promoting the activity in oxidative dehydrogenation of ethane by oxygen [13, 14]. Here, we report our further investigation on the effect of modification of zirconia by sulfation and tungstation on their catalytic behaviors in the dehydrogenation of ethane by carbon dioxide.

2. EXPERIMENTAL

ZrO_2 samples were prepared by heat-treatment of the amorphous ZrO_2 obtained by precipitation of $ZrO(NO_3)_2$. The sulfated and tungstated zirconia samples were then prepared by wetness impregnation of ammonium sulfate and ammonium tungstate salts on the amorphous ZrO_2 with 6 wt% loading of sulfate or WO_3, calcination at 700 °C for 3 h. All chromium-based catalysts were prepared by impregnation of $Cr(NO_3)_3$ at a Cr_2O_3 loading of 5 wt% on the supports, calcination at 700 °C for 3 h.

BET surface area of the catalysts was obtained by nitrogen adsorption at − 196 °C. XRD measurements were conducted on a Philips PW 1800 X-ray diffractometer at 40 kV and 40 mA. XPS measurements were carried out on a PHI 5500 ESCA system (Perkin-Elmer) with Mg Kα as radiation source. Temperature-programmed reduction (TPR) experiments were conducted in a fixed-bed reactor loaded with 0.5 g samples. The samples were reduced in 10% H_2/Ar flow at a rate of 30 ml/min from ambient temperature to 700 °C at a heating rate of 3 °C/min. The TPD profiles were obtained on a special NH_3-TPD apparatus under vacuum conditions with the temperature varying from 100-800 °C at a heating rate of 10 °C/min.

The oxidative dehydrogenation of ethane by CO_2 was performed at atmospheric pressure on a fixed-bed reactor packed with 1 g catalysts and 2 g quartz sands. The reactant stream C_2H_6: CO_2: N_2= 10%: 50%: 40% was introduced into the reactor at a flow rate of 60 ml/min while no back-mixing effect was found. The reaction temperature ranged between 500-650°C. The products were analyzed by two gas chromatographs (Shimadzu, GC-8A).

3. RESULTS AND DISCUSSION

3.1. Catalyst characterization

Table 1 presents the surface areas of the Cr-based catalysts prepared. It is seen that ZrO_2 has the lowest surface area and the surface area of the sulfate modified Cr-based catalyst is about the same as that of the unpromoted Cr/ZrO_2 catalyst. However, tungstated Cr/ZrO_2 exhibits much lower surface area. These results indicate that sulfation has little effect on the surface area while tungstation induces a remarkable reduction in the surface area.

The XRD patterns of ZrO_2 and Cr-based catalysts are shown in Fig.1. The calcined support, ZrO_2, presents crystalline phase of monoclinic ZrO_2. For Cr/ZrO_2, tetragonal ZrO_2 is the major phase with minor monoclinic ZrO_2 while two ZrO_2 phases, monoclinic and tetragonal ZrO_2, coexist in those modified Cr/ZrO_2 catalysts. However, the two phases show a different ratio. Monoclinic ZrO_2 is the major phase for $Cr/S-ZrO_2$ whereas the tetragonal ZrO_2 dominates

Table 1
Physico-chemical properties of chromium oxide based catalysts.

Catalyst	S_{BET} (m^2/g)	NH_3 $(\mu mol/g)$	Cr Concentration (%)	Cr/Zr	Cr^{3+} BE (ev)	Cr^{6+} BE (ev)
ZrO_2	22	-	-	-	-	-
Cr/ZrO_2	62	20.7	4.5	0.197	576.7	579.1
$Cr/W\text{-}ZrO_2$	36	8.4	3.2	0.187	577.0	579.0
$Cr/S\text{-}ZrO_2$	60	14.0	2.8	0.134	575.8	577.4

the ZrO_2 phase in $Cr/W\text{-}ZrO_2$. These results suggest that the additives on ZrO_2 can influence the transformation of ZrO_2 phases, favoring the stability of tetragonal phase. For all zirconia-supported catalysts, no significant Cr_2O_3 diffraction peaks can be observed, probably due to low amount and well-dispersed Cr_2O_3 particles on ZrO_2 surface.

The surface Cr species on the various Cr/ZrO_2 catalysts were determined by XPS and their concentrations and Cr_{2p} spectra are shown in Table 1 and Fig. 2, respectively. Cr/ZrO_2 has the largest Cr surface concentration while $Cr/S\text{-}ZrO_2$ shows the lowest surface Cr value. The ratio of Cr/Zr also indicates the greatest dispersion of Cr on Cr/ZrO_2. The Cr_{2p} spectra demonstrate that three catalysts display somewhat different XPS spectral patterns. Two Cr species can be found by curve fitting and they are assigned to Cr^{3+} and Cr^{6+}, whose binding energies (BE) are given in Table 1. The peaks in the $Cr/S\text{-}ZrO_2$ spectrum show a shift to lower binding energy while others shows a similar position in BE values. The spectra also show that the concentrations of Cr^{3+} and Cr^{6+} are different for three Cr-based catalysts.

The ammonia adsorption and TPD profiles over the Cr-based catalysts are presented in Table 1 and Fig. 3, respectively. As seen, sulfation and tungstation

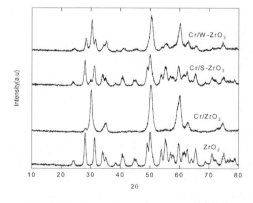

Fig. 1. XRD patterns of ZrO_2 and Cr-based catalysts.

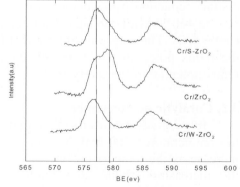

Fig. 2. XPS Cr_{2p} profiles of Cr-based catalysts.

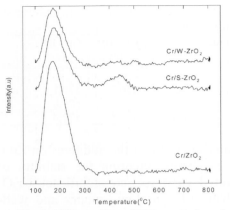

Fig. 3. TPD profiles of NH$_3$ on Cr-based
catalysts.

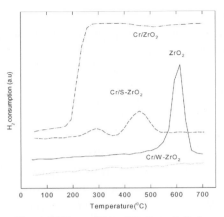

Fig. 4. TPR profiles of ZrO$_2$ and Cr-based
catalysts.

also reduce the ammonia adsorption and the addition of WO$_3$ results in the
greatest reduction in acidity. From NH$_3$-TPD profiles, one can see that Cr/ZrO$_2$
exhibits only one strong desorption peak occurring at 100 - 300 °C. Two broad
peaks can be observed in the TPD profile of Cr/S-ZrO$_2$. The stronger one
appears at 100 - 300 °C, the same position as that on Cr/ZrO$_2$, and the weak one
occurs at 360 - 500 °C. For Cr/W-ZrO$_2$, three peaks are present in the TPD
profile. The strongest one appears at 100 - 300 °C while the other two weak
peaks appear at 380 - 450 °C and 480 - 520 °C, respectively.

The TPR profiles of the calcined ZrO$_2$ and Cr-based catalysts are illustrated
in Fig. 4. The TPR profiles of ZrO$_2$ and the Cr-based catalysts display
significantly different behavior. ZrO$_2$ shows a strong reduction peak only at 550
- 650 °C. Cr/ZrO$_2$ exhibits strong and broad reduction peaks after 200 °C. Two
distinct reduction peaks occur on the TPR profile of Cr/S-ZrO$_2$, centred at 290
°C and 460 °C while no significant reduction peak appears on the TPR profile of
Cr/W-ZrO$_2$. Those results suggest that impregnation of Cr$_2$O$_3$ on ZrO$_2$ enhances
the redox potential while modification of Cr/ZrO$_2$ significantly decreases the
redox potential of the catalysts.

3.2. Catalytic performance

Table 2 gives the results of catalytic activity and selectivity of all catalysts at
the temperatures 500-650 °C. One can see that ethane and CO$_2$ conversions
increase as the temperature increases while ethylene selectivity decreases with
the increasing ethane conversion. The consumption of CO$_2$ suggests the
involvement of CO$_2$ in the reaction. The calcined ZrO$_2$ exhibits little activity at
low temperatures but can induce reaction at high temperature of 650 °C, giving
93 % ethylene selectivity at 10 % ethane conversion and also producing some

oxygenates. Cr/ZrO_2 shows a high activity, but low selectivity to ethylene, especially at higher temperatures. At 650 °C, ethane conversion is 57 % while ethylene selectivity is only 60%, giving ethylene yield of 35%. Addition of sulfate or tungstate oxide on Cr/ZrO_2 shows a different effect on catalytic behavior. $Cr/W-ZrO_2$ exhibits lower ethane conversion and ethylene selectivity. Although ethane conversion is reduced on $Cr/S-ZrO_2$ while ethylene selectivity is improved. Due to the higher selectivity to ethylene, $Cr/S-ZrO_2$ produces even higher ethylene yield at high temperatures (600-650 °C). Therefore, the catalytic activity of ZrO_2 supported catalysts presents an order as $Cr/ZrO_2 > Cr/S-ZrO_2 > Cr/W-ZrO_2 > ZrO_2$ in terms of ethane conversion.

It has been proposed that the dehydrogenation and oxidative dehydrogenation of ethane due to the introduction of CO_2 and its decomposition to produce surface oxygen species will be the parallel reaction paths [12]. It is well known that selective oxidation reactions on oxides proceed through a redox mechanism. The acid-base and redox properties of the oxide surface play an important role in catalytic activity. Therefore, the different activity of Cr-based catalysts can be attributed to the varying chromium structure and redox properties of the catalysts prepared. From Table 1, it is seen that tungstation of Cr/ZrO_2 significantly reduces the surface area while sulfation shows little change on the surface area of Cr/ZrO_2. The order of surface area is the same as the ethane conversion of the catalysts. The amounts of ammonium adsorption on three Cr_2O_3-based catalysts show that $Cr/W-ZrO_2$ and $Cr/S-ZrO_2$ have lower

Table 2

Catalytic activity and product selectivity over chromium oxide based catalysts at 500-650 °C.

Catalyst	Temp (°C)	Conv (%)		Sel (%)					Yield (%)
		C_2H_6	CO_2	C_2H_4	CH_4	C_3H_8	EtOH		C_2H_4
ZrO_2	500	0.1	1.0	81.8	18.2	0	0		0.1
	550	0.9	1.9	97.9	2.1	0	0		0.9
	600	1.8	2.4	84.0	16.0	0	0		1.5
	650	10.6	4.0	93.4	6.1	0.05	0.5		9.9
Cr/ZrO_2	500	10.3	4.6	95.3	4.6	0.03	0		9.8
	550	22.9	10.2	87.2	12.7	0.05	0		20.0
	600	37.9	19.2	74.6	25.3	0.09	0		28.3
	650	57.3	29.1	60.4	39.4	0.10	0.08		34.6
$Cr/W-ZrO_2$	500	7.4	4.3	94.0	5.7	0.31	0		6.9
	550	18.5	7.2	84.9	14.0	0.72	0.29		16.0
	600	33.6	15.0	67.0	31.6	0.85	0.51		22.6
	650	49.0	23.0	54.6	44.7	0.47	0.27		26.7
$Cr/S-ZrO_2$	500	7.0	2.2	94.3	5.7	0	0		6.6
	550	17.5	8.9	91.3	8.7	0.02	0		16.0
	600	35.0	17.0	84.2	15.7	0.06	0		29.4
	650	50.6	21.9	76.0	23.9	0.10	0		38.4

acidities than that of Cr/ZrO$_2$, similar to the order of ethane conversion. While the profile of TPD on Cr/S-ZrO$_2$ shows a different pattern from that of the other two catalysts, which can contribute to the difference in ethylene selectivity. The temperature-programmed reduction by hydrogen also shows that ZrO$_2$ and its supported Cr$_2$O$_3$ catalysts demonstrate varying redox potentials. ZrO$_2$ only exhibits one reduction peak centered at 650 °C while others show two broad peaks both at lower temperatures before 500 °C, which indicates a higher redox potential. The XPS results also show a different pattern of Cr^{6+}/Cr^{3+} on three Cr-based catalysts with Cr/ZrO$_2$ showing the highest value, which favors the redox cycle of Cr^{6+}/Cr^{3+}.

4. CONCLUSION

Cr$_2$O$_3$/ZrO$_2$ based catalysts are effective for oxidative dehydrogenation of ethane with CO$_2$. Sulfate shows a better effect on catalytic activity by enhancing the ethylene selectivity while tungstate oxide suppresses the activity and ethylene selectivity. Addition of sulfate changes the property of acid sites responsible for the ethylene selectivity while addition of tungstate oxide reduces the surface area, acidity, and redox potential resulting in decreases in activity and selectivity to ethylene.

REFERENCES

[1] E.A. Mamedov and V. Cortes-Corberan, Appl. Catal. A., 127 (1997) 1.
[2] T. Blasco and J.M. Lopez-Nieto, Appl. Catal. A., 157 (1997) 117.
[3] S. Wang, K. Murata, T. Hayakawa, S. Hamakawa and K. Suzuki, Catal. Lett., 63 (1999) 59.
[4] K. Nakagawa, M. Okamura, N. Ikenaga, T. Suzuki and T. Kobayashi, Chem. Commun., 1998, 1025.
[5] I. Takahara and M. Saito, Chem. Lett., 1996• 973.
[6] F. Solymosi and R. Nemeth, Catal. Lett., 62 (1999) 197.
[7] S. Wang, K. Murata, T. Hayakawa, S. Hamakawa and K. Suzuki, Chem. Lett., 1999, 569.
[8] L. Xu, J. Liu, Y. Xu, H. Yang, Q. Wang and L. Lin, Appl. Catal. A., 193 (2000) 95.
[9] E.M. Thorsteinson, T.P. Wilson, F.G. Young and P.H.J. Kasai, J. Catal., 52 (1978) 116.
[10] L. Mendelovici and J.H. Lunsford, J. Catal., 94 (1995) 37.
[11] A. Erdohelyi and F. Solymosi, J. Catal., 135 (1992) 563.
[12] S. Wang, K. Murata, T. Hayakawa, S. Hamakawa and K. Suzuki, Appl. Catal. A., 196 (2000) 1.
[13] S. Wang, K. Murata, T. Hayakawa, S. Hamakawa and K. Suzuki, Chem. Commun., 1999, 103.
[14] S. Wang, K. Murata, T. Hayakawa, S. Hamakawa and K. Suzuki, Catal. Lett., 59 (1999) 187.

Studies in Surface Science and Catalysis, volume 147
X. Bao and Y. Xu (Editors)

Effect of support on chromium based catalysts for the oxidative dehydrogenation of C_2H_6 with CO_2 to C_2H_4

Hong Yang[*], Shenglin Liu, Longya Xu, Sujuan Xie, Qingxia Wang, Liwu Lin

Dalian Institute of Chemical Physics, the Chinese Academy of Sciences, Dalian 116023, China

ABSTRACT

The oxidative dehydrogenation of ethane (ODE) with CO_2 to C_2H_4 has been studied over a series of Cr-based catalysts using SiO_2, Al_2O_3, (MCM-41 zeolite) MCM-41, MgO and Silicate-2 (Si-2) as the supports. TPR, NH_3-TPD, and EPR characterizations of catalysts were carried out to investigate the reduction property of Cr species on different supports, the acidities of catalysts and Cr species of $6Cr/SiO_2$ catalysts, respectively.

1. INTRODUCTION

Up to now, the production of C_2H_4 from C_2H_6 has been carried out by the conventional steam cracking process, in which a very high temperature (about 1123 K) and large amount of steam were needed. In the last decade, some researchers have become interested in the oxidative dehydrogenation of C_2H_6 in the presence of oxygen as an alternative to the conventional process [1]. Recently, a new process for the production of C_2H_4 from C_2H_6 has been reported [2, 3], which is realized by the oxidative dehydrogenation of ethane (ODE) with CO_2. In this process, CO_2, one of the major green house gases, has been considered as a mild oxidant replacing oxygen. In our laboratory, Xu et al.[4] have studied the performance of a K-Fe-Mn/Silicate-2 catalyst for the ODE with CO_2 to C_2H_4. Wang et al.[5,6] investigated the reaction over a 8 wt.% Cr_2O_3/SiO_2 catalyst at 923 K when the molar ratio of CO_2 to C_2H_6 in the feed was 5. They reported that the acidity/basicity and redox property of catalysts determines the catalytic activity in the dehydrogenation of ethane by CO_2. In addition, Solymosi et al.[7] investigated the oxidative dehydrogenation of ethane by carbon dioxide over a Mo_2C/SiO_2 catalyst at 823-923K.

In the present paper, we investigated the reaction performance of the ODE with CO_2 to C_2H_4 reaction on several supported chromium catalysts on SiO_2,

Al$_2$O$_3$, MgO, (MCM-41 zeolite) MCM-41 and Silicate-2 (Si-2) at low temperature of 923 K. Meanwhile, the supported chromium catalysts were characterized by TPR, NH$_3$-TPD, and EPR.

2. EXPERIMENTAL

2.1 Catalyst preparation and activity measurement

All supported chromium catalysts were prepared by an impregnation method using an appropriate amount of chromium nitrate aqueous solution. The catalysts were then dried in air at 393 K for 1 hour and calcined in argon at 923 K for 2 h. Cr/support catalyst is denoted as x Cr/support, where x is the Cr content in weight percent.

Under atmospheric pressure, the catalysts were tested in a fixed-bed flow micro-reactor. The reactant stream, consisting of 47.5% ethane, 47.5% carbon dioxide and 5% argon (or C$_2$H$_6$/Ar = 50/50) was introduced into the reactor at a flow rate of 20 ml/min. Gas products were sampled after 20 min time on stream (Tos) reaction and analyzed by a gas chromatograph equipped with a thermal conductivity detector (TCD). An internal standard method with argon as the internal standard gas was used to calculate the conversion of C$_2$H$_6$, C$_2$H$_4$ yield and coke selectivity.

2.2 Catalyst characterization

For acid measurement of the catalyst, ammonia temperature-programmed desorption (TPD) was carried out in a conventional flow system equipped with a TCD. TPR of the sample was conducted as following. The sample was contacted with a 30ml/min flow of 10%H$_2$/Ar and heated, at a rate of 13.8 K/min, to a final temperature of 1023 K. The hydrogen consumption was monitored by a TCD after the removal of the water formed.

EPR studies were performed on a JEOL ESEO3X X-band spectrometer at room temperature. A microwave frequency ν of 9.42GHz with a power of 1 mW was used. The g factor was calculated by taking the signal of manganese as an internal standard. A specially designed EPR cell, which can be used at the same time as a fixed-bed flow reactor, was employed. After the reaction or the pretreatment, the reactor was transferred into the resonance cavity of the EPR spectrometer without contacting air. Therefore, it is possible to detect and measure the EPR signals of the sample under the conditions similar to the ODE with CO$_2$ to produce C$_2$H$_4$.

3. RESULTS AND DISCUSSION

3.1 Influence of the support of the chromium based catalysts on the reaction activity of ODE with CO$_2$

Table 1
Results of ODE with CO_2 over chromium catalysts with different supports[a].

Catalyst	Conv. C_2H_6 (%)	Conv. CO_2 (%)	C_2H_4 yield (%)	Sele. (%)			
				C_2H_4	CH_4	CO	Coke
6Cr/SiO$_2$	32.1	23.0	24.8	77.3	10.2	10.4	2.1
6Cr/MCM-41	21.1	12.6	15.9	75.4	9.2	10.6	4.8
6Cr/Al$_2$O$_3$	24.6	15.1	14.5	58.9	13.5	22.5	5.1
6Cr/MgO	4.8	4.5	3.2	66.7	5.0	27.5	0.8
6Cr/Si-2	2.9	1.9	2.8	95.8	1.8	2.1	0.7

[a]$T = 923$ K, $P = 1$ atm, GHSV = 1200 h^{-1}, $CO_2/C_2H_6 = 1/1$, Tos = 20 min

Cr-based supported catalysts have been extensively utilized for the dehydrogenation of light alkanes such as butane dehydrogenation [8]. The reaction performances of the ODE with CO_2 to produce C_2H_4 over a series of Cr-supported catalysts using SiO_2, Al_2O_3, MCM-41, MgO and Si-2 as supports were investigated. The results are shown in Table 1. The main products are C_2H_4, CH_4 and CO. High C_2H_6 conversion and C_2H_4 yield are obtained on catalysts of 6Cr/SiO$_2$, 6Cr/MCM-41 and 6Cr/Al$_2$O$_3$, which give > 20% C_2H_6 conversion and > 14% yield of C_2H_4. And a low C_2H_6 conversion (<5%) appears over the 6Cr/MgO and 6Cr/Si-2 catalysts. The catalyst activities such as C_2H_6 conversion decreased in the order 6Cr/SiO$_2$ >> 6Cr/Al$_2$O$_3$ > 6Cr/MCM-41 >> 6Cr/MgO > 6Cr/Si-2. Wang et al.[5] compared the catalytic activity over Cr_2O_3 catalysts on metal oxides for the ODE with CO_2 process. They reported that the order of activity follows Cr_2O_3/SiO$_2$ > Cr_2O_3/ZrO$_2$ > Cr_2O_3/Al$_2$O$_3$ > Cr_2O_3/TiO$_2$. And the nature of the support exerts an influence upon chromium species and redox properties, which in return determine the catalytic behaviour of the supported chromium catalysts.

3.2 TPR characterization of chromium supported catalysts

The TPR characterization for the 6Cr/SiO$_2$, 6Cr/Al$_2$O$_3$, 6Cr/MCM-41, 6Cr/MgO and 6Cr/Si-2 catalysts was carried out to investigate the reduction properties of the chromium supported catalysts. The result is shown in Fig. 1. It is shown that different chromium-based catalyst presented different TPR patterns. As shown in Fig.1, the reduction peaks of all catalysts may be classified into two types. One is denoted as the low temperature reduction peak (peak I) between 473 and 653 K and another as high temperature peak (peak II) between 653 to 923 K. The areas of peak I and II for each sample can be calculated based on the baseline of each profile (not shown). The total peak areas of the catalysts decrease in the order 6Cr/SiO$_2$ > 6Cr/Al$_2$O$_3$ > 6Cr/MCM-41 > 6Cr/MgO > 6Cr/Si-2. When the reaction performance (Table 1) is correlated with the results of TPR (Fig. 1), it is of interest to note that the decrease sequences of C_2H_6 conversion of the ODE with CO_2 reaction over the catalysts is identical to that of total peak areas of the catalysts. Furthermore,

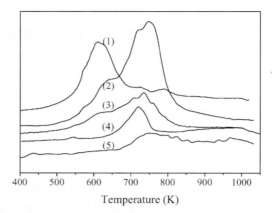

Fig. 1 TPR profiles of the Cr_2O_3 catalysts supported on different supports
(1) 6Cr/Al$_2$O$_3$, (2) 6Cr/SiO$_2$, (3) 6Cr/MCM-41, (4) 6Cr/MgO, (5) 6Cr/Si-2

6Cr/SiO$_2$ catalyst has the strongest peak II among the catalysts investigated, which shows highest ethane conversion and ethylene selectivity. This result indicates that reduction peak II has a close relationship with ODE reaction and suggests that the species giving rise to this peak facilitates the ethylene production.

3.3 TPD characterization of Cr-supported catalysts

It has been reported [5] that the catalytic activities of supported chromium catalysts are strongly affected by the acidity of the oxide supports. Here, the acid of the above typical catalysts was determined by NH$_3$-TPD, which was shown in Fig. 2. The acid intensities of all the catalysts follow the order of 6Cr/MCM-41 > 6Cr/Al$_2$O$_3$ > 6Cr/SiO$_2$ > 6Cr/MgO ~ 6Cr/Si-2, while the catalyst activities such

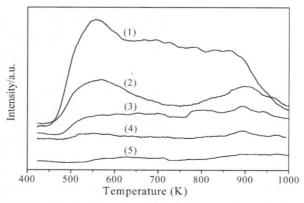

Fig. 2 TPD profiles of the Cr_2O_3 catalysts supported on different supports
(1) 6Cr/MCM-41, (2) 6Cr/Al$_2$O$_3$, (3) 6Cr/SiO$_2$, (4) 6Cr/Si-2, (5) 6Cr/MgO

as C_2H_6 conversion for the ODE with CO_2, decrease as $6Cr/SiO_2$ > $6Cr/Al_2O_3$ > 6Cr/MCM-41 >> 6Cr/MgO> 6Cr/Si-2. According to the above results suggest that moderate acidity of the catalyst is beneficial to convert C_2H_6 and produce C_2H_4.

3.4 EPR characterization of Cr-supported catalysts

At first, EPR measurement was performed for the catalyst after O_2 (1200 h^{-1}) pretreatment at 923 K for 40 min. A weak δ signal and a sharp γ signal could be detected (Fig. 3). The δ signal is at g = 3.6 and it is believed to be connected with isolated Cr^{3+} ions in a distorted octahedron [9]. The γ signal is thought to be made up of two signals, $γ_1$, $γ_2$. The $γ_2$ signal could be due to Cr^{5+} in a surrounding with a lower symmetry than that corresponding to $γ_1$ signal, but it could also be due to Cr^{4+} in a tetrahedral coordination [9].

When the $6Cr/SiO_2$ catalyst was pretreated with O_2 at 923 K for 40 min and subsequently treated with C_2H_6/CO_2 (C_2H_6/CO_2 =1/1 mol, 1200 h^{-1}, purity: 99.99%) at 423 K for 1h, the β signal located at g=1.98, which is broad and attributed to clusters of Cr^{3+} ions which are antiferromagnetically [9], begins to appear. With increase of the temperature to 523 K, the intensity of β signal reaches a maximum. As the temperature is further increased to 923 K, the intensity of β signal does not change obviously. The δ signal is rather weak in EPR spectra of the $6Cr/SiO_2$ catalyst under C_2H_6/CO_2, and it often overlapped with the broad β signal. That is, the isolated Cr^{3+} in a distorted octahedron is very small in number in the studies.

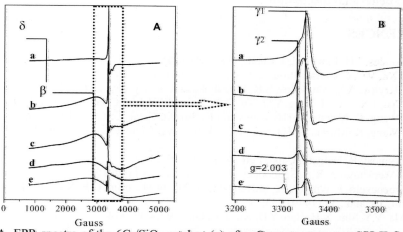

Fig. 3 A. EPR spectra of the $6Cr/SiO_2$ catalyst (a) after O_2 pretreatment at 923 K for 40 min; and subsequently treated with C_2H_6/CO_2 at (b) 423 K for 1 h; (c) 523 K for 1 h; (d) 723 K for 1 h; (e) 923 K for 1 h; B (detail spectra of A)

A new signal of g = 2.003 was clearly detected when the temperature reached 923 K. This comes from the free electrons of carbonaceous deposits formed on the catalyst from deep dehydrogenation of ethane and is similar to the signal corresponding to carbonaceous deposits during methane dehydro-aromatization under non-oxidative conditions [10]. The EPR spectra thus revealed that Cr^{3+}, Cr^{4+} and Cr^{5+} oxides were the major chromium species on the $6Cr/SiO_2$ catalyst under reaction conditions, which maybe responsible for ODE with CO_2 to produce C_2H_4, and form carbonaceous deposits on the catalyst during reactions at 923 K.

4. CONCLUSIONS

6wt% Cr/ SiO_2 catalyst showed the highest activity in the ODE with CO_2 reaction at 923 K among the catalysts using different supports. The decrease sequence of catalytic activity of the catalysts is identical to that of total peak areas of the catalysts in TPR, $6Cr/SiO_2$ having biggest peak II area which showed highest C_2H_4 yield. NH_3-TPD results indicate that moderate acidity of the catalyst is beneficial to the ODE reactions. EPR characterizations show that Cr^{3+}, Cr^{4+} and Cr^{5+} chromium oxides species maybe responsible for ODE with CO_2 to produce C_2H_4.

ACKNOWLEDGMENTS

The authors show great appreciation to Prof. X. H. Bao and Dr. Z. P. Qu for their help and beneficial discussions.

REFERENCES

[1] E. Morales, J.H. Lunsford, J. Catal., 118 (1989) 255.
[2] Y.D. Xu, V.C. Corberan. Prog. Nat. Sci., 10 (2000) 22.
[3] O.V. Krylov, A.K. Mamedov, S.R. Mirzabekova, Catal. Today, 24 (1995) 371.
[4] L.Y. Xu, J.X. Liu, H. Yang, Y.D. Xu, Q.X. Wang, L.W. Lin, Catal. Lett., 62(1999)185.
[5] S.B. Wang, K. Murata, T. Hayakawa and K. Suzuki. Catal. Lett., 63 (1999) 59.
[6] S.B. Wang, K. Murata, T. Hayakawa, S. Hamakawa, K. Suzuki, Appl. Catal. A, 196 (2000) 1.
[7] F. Solymosi, R. Nemeth, Catal. Lett., 62 (1999) 197.
[8] B.M. Weckhuysen, R.A. Schoonheydt, Catal. Today, 51 (1999) 223.
[9] C. Groeneveld, P. Wittgen, A. M.Kersbergen, P.L. M.Mestrom, C.E. Nuijten, G.C.A. Schuit, J. Catal., 59 (1979) 153.
[10] D. Ma, Y.Y. Shu, X.H. Bao, Y.D. Xu, J. Catal. 189 (2000) 314.

Studies in Surface Science and Catalysis, volume 147
X. Bao and Y. Xu (Editors)

Partial oxidation of ethane-containing gases

V.S. Arutyunov[a], E.V. Sheverdenkin[a], V.M. Rudakov[b], V.I. Savchenko[b]

[a]Semenov Institute of Chemical Physics, Russian Academy of Sciences, Kosygina, 4, Moscow 119991, Russia.

[b]Institute of Problems of Chemical Physics, Russian Academy of Sciences, Chernogolovka, Moscow Region., 142432, Russia.

ABSTRACT

Experimental data on high pressure partial oxidation of ethane and methane-ethane mixtures are presented. It was shown that even at moderate pressures no higher than 30 atm it is possible to obtain relatevely high yields of metanol and formaldehyde. The possibility to have significantly lower operating pressure makes any future technology much attractive.

1. INTRODUCTION

The direct oxidative conversion of hydrocarbon gases into liquid oxygenates (methanol, formaldehyde, ethanol, etc.) is a promising route to involve remote natural gas resourses and those of low resourse gas fields in chemical processing. It gives the possibility to use low scale highly automated installations for converting gas phase hydrocarbons into much easily transportable liquid products. The kinetics of direct methane oxidation to methanol (DMTM) at high pressures has been intensively studied in recent years [1-3], but there are a few studies dealing with heavier alkanes.

Meanwhile, there are essential resourses of such gases. Transportation or processing of them with existing technologies of oil and gas production and treatment in many remote areas is unprofitable. More than 40 billions m^3 of such gases are vented or flared annually throughout the world, causing losses of valuable feedstock and serious ecological problems. The main products of direct oxidation of hydrocarbon gases with high content of heavier alkanes, alongside with methanol and formaldehyde, are C_2-C_4 alcohols, aldehydes, carbon acids and other oxygenates.

The aim of the present study is to obtain necessary kinetic data on high pressure oxidation of ethane and methane-ethane mixtures for more reliable technico-economical assessments of possibilities to design profitable process of direct oxidation of ethane containing gases, including associated gases.

2. EXPERIMENTAL

The study of direct oxidation of ethane was conducted under flow conditions. More detailed description of apparatus and our recent data on direct oxidation of methane are given elsewhere [4]. The stainless steel reactor (i.d. 10 mm) with three independently heated sections (200 mm each) and internal quartz insertions with i.d. 7 mm lets us to operate at pressures up to 100 atm and gas flows up to 1 m^3/h. Typical experiments with ethane were accomplished at pressures 20-36 atm, maximal reaction temperature 450-550°C and gas flow Q = 400 – 450 l/h. Ethane-methane and ethane-nitrogen mixtures were investigated in a wider pressure range up to 70 atm. Ethane from cylinders was used without purification and had the following composition: ethane – 91.2%, methane – 8.3%, propane-butane fraction – 0.5%. Both oxygen and air were used as an oxidant with total oxygen concentration varied from 1.3 to 13.3%. The reaction was accomplished in a middle section of reactor with the first section used for preliminary gas heating (<200°C) and mixing. In all experiments we made provision for full oxygen conversion. The temperature control was ensured by a row of successive thermocouples in each section. Gas phase was analyzed by GC probe sampling at both ends and at the middle of reaction section. The liquid phase was separated downstream the reactor and analyzed by GC.

3. RESULTS AND DISCUSSION

The dependences of product selectivity and yield at high pressure homogeneous oxidation of ethane on temperature, pressure, initial concentration of oxygen and discharge of reaction mixture through the reactor were investigated. The

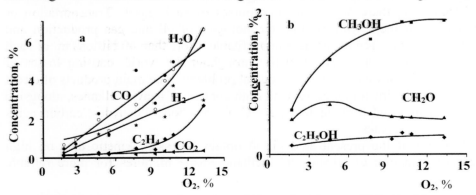

Fig. 1. Dependence of gas phase concentrations of products on [O$_2$]$_0$. P = 26 atm, T_{max} ≈ 500°C, Q = 400 l/h.

significant decrease in the temperature of beginning of the reaction was observed comparatively to methane oxidation. The principal products were the same as at methane oxidation. The main gas phase products were CO, H_2, C_2H_4, CO_2 and main liquid products were CH_3OH, CH_2O, C_2H_5OH and H_2O (Fig. 1.).

It is worth to note prominent increase in ethylene formation with oxygen. The ratio CO/CO_2 was approximately three times higher relatively to that at methane oxidation. The concentration of methanol in liquid products decreases insufficiently in comparison with methane oxidation; however, concentrations of formaldehyde and ethanol are several times higher. At optimal conditions the concentration of organic products in liquid phase was up to 75% which is significantly higher than at oxidation of methane (<50%). The main part of this increase is due to increase in formaldehyde formation up to 30%, which is several times higher than at oxidation of methane (~5-8%). However, the concentration of formaldehyde drops significantly with increase of initial oxygen (Fig. 2).

In comparison with methane oxidation the concentration of ethane increases by a factor of ten, and equals 10%. Besides, minor quantities of propanol on a level 0.2 - 0.5% were registered. In contrast to methanol, the percentage of C_2-C_3 alcohols in liquid was approximately stable in the whole range of oxygen change from 1.3 to 13.3%. The formation of formic acid at a level 0.5 − 1.5% was also detected.

The total yield of liquid organic products rises with oxygen up to $[O_2] \approx 10\%$. The yield of methanol gains its maximum at $[O_2] \approx 5\text{-}6\%$ and equals

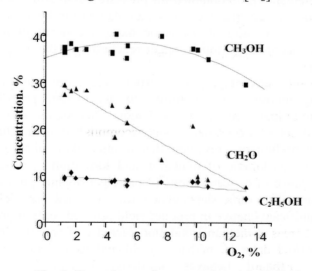

Fig. 2. Dependence of concentrations of principal organic products in liquid phase on $[O_2]_0$. $P = 26$ atm, $T_{max} \approx 500°C$, $Q = 400$ l/h.

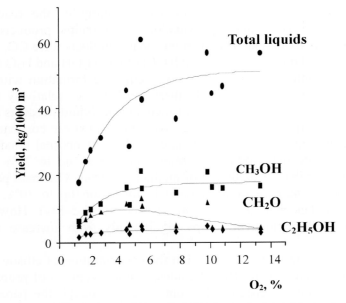

Fig. 3. Dependence of yield of principal organic products
on $[O_2]_0$. $P = 26$ atm, $T_{max} \approx 500°C$, $Q = 400$ l/h.

approximately the same value 20 kg/1000 m³ of gas flow as at methane oxidation. The yield of formaldehyde has a maximum at $[O_2] \approx 4$-5% (Fig. 3.).

The ethanol/methanol ratio obtained for pressure 30-36 atm is approximately 0.2-0.3. This value is in accordance with data obtained for the same pressures in [5,6]. In [6] at $P = 34$ atm at the same oxygen concentrations but lower temperatures and, accordingly, much longer reaction time the maximal value of this ratio was 0.47.

The significant advantage of partial oxidation of ethane and ethane-containing mixtures is possibility to accomplish process at lower pressures. To investigate this possibility more thoroughly and in a wider pressure range, a set of experiments was accomplished with ethane-nitrogen mixtures. It was preliminary revealed that dilution with nitrogen causes no principal influence on kinetic of oxidation and formation of products. Fig. 4 shows the dependence of concentrations of principal organic products in liquid phase on pressure. The same dependence was obtained for yields of these products. The significant change in product yield proceeds only if pressure drops below 30 atm, whereas at methane oxidation decrease of pressure from 70 to 30 atm leads to twofold drop in methanol yield and its concentration in liquid products [1-3]. Ethanol behaves similarly. Formaldehyde yield and concentration remain practically constant up to 30 atm and at further pressure drop even increase. So at $P < 15$ atm it becomes the main reaction product.

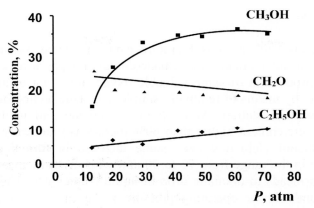

Fig. 4. Dependence of concentrations of principal organic products in liquid phase on pressure at oxidation of ethane-nitrogen mixtures. $T_{max.} = 460^{\circ}C$. $[C_2H_6]_0 = 28.5\%$, $[O_2]_{0.} = 11.5\%$.

Experiments show practically no change in alcohols formation at more than threefold decrease of gas flow from 650 to 180 l/h. But concentration of formaldehyde has decreased in two times. It is probably due to increase of its heterogeneous decay with increase in residence time in reactor.

The oxidation of methane-ethane mixtures shows complex mutual interaction of these hydrocarbons: at constant pressure yields of formaldehyde and ethanol steadily decrease with the increase in methane share in mixture, and the yield of methanol has a maximum at methane concentration 40-50% (Fig. 5).

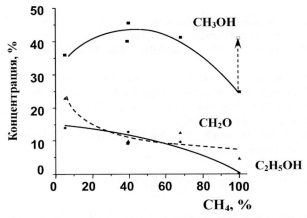

Fig. 5. Dependence of concentration of organic products in liquid phase on methane content in methane-ethane mixture. $P = 35 - 45$ atm. $T_{max} = 450–500^{\circ}C$. $[O_2]_0 = 3,6-4\%$. Arrow – the increase of CH_3OH at $P = 70$ atm.

4. CONCLUSION

The obtained results and previous literature data on ethane partial oxidation [5-10] let us to consider ethane and ethane-containing gases as a promising resource for the production of some valuable oxygenates, first of all methanol and formaldehyde, by means of relatively simple one stage technology of direct homogeneous partial oxidation. At oxidation of ethane and methane-ethane mixtures with ethane content higher than 60% there was practically no drop in methanol and ethanol yield and even some increase in formaldehyde with pressure decrease in the range from 70 to 30 atm whereas at methane oxidation such pressure drop leads to dramatic drop in liquid oxygenates. The possibility to have significantly lower operating pressure makes any future technology much attractive. The technical-economical assessment of such conceptual low scale process was made.

ACKNOWLEDGEMENTS

The financial support from Russian Foundation for Basic Research (Grant N 03-03-32105) is gratefully acknowledged.

REFERENCES

[1] V.S. Arutyunov, V.Ya. Basevich, V.I. Vedeneev, Ind. Eng. Chem. Res., 34 (1995) 4238.
[2] V.S. Arutyunov, V.Ya. Basevich, V.I. Vedeneev, Russian Chemical Reviews, 65 (1996) 197.
[3] V.S. Arutyunov, O.V. Krylov, Oxidative Conversion of Methane, Nauka, Moscow, 1998 (in Russian).
[4] V.S. Arutyunov, V.M. Rudakov, V.I. Savchenko, E.V. Sheverdenkin, O.G. Sheverdenkina, A.Yu. Zheltyakov, Theoretical Foundations of Chemical Engineering, 36 (2002) 472.
[5] H.D. Gesser, N.R. Hunter, L.A. Morton, and P.S. Yarlagadda, Energy and Fuels, 5 (1991) 423.
[6] A.N. Shigapov, N.R. Hunter, and H.D. Gesser, Catal. Today, 42 (1998) 311.
[7] D.M. Newitt, A.M. Bloch, Proc. Roy. Soc., London, A140 (1933) 426.
[8] D.M. Newitt, P. Szego, Proc. Roy. Soc., London. A147 (1934) 555.
[9] D.M. Newitt, Chem. Rev., 21 (1937) 299.
[10] V.S. Arutyunov, V.Ya. Basevich, V.I. Vedeneev, Yu.V. Parfenov, O.V. Sokolov, V.E. Leonov, Catal. Today, 42 (1998) 241.

Studies in Surface Science and Catalysis, volume 147
X. Bao and Y. Xu (Editors)

Dehydro-aromatization of methane under non-oxidative conditions over MoMCM-22: effect of surface aluminum removal

A. C. C. Rodrigues and J. L. F. Monteiro

NUCAT/PEQ/COPPE/UFRJ, P.O. Box 68502 – 21945-970 – Rio de Janeiro – Brazil – e-mail: monteiro@peq.coppe.ufrj.br

ABSTRACT

The dehydro-aromatization of methane was studied over molybdenum carbide supported on HMCM-22 zeolite, with high benzene selectivity at 973 K. Extensive aluminum removal promoted an increase in benzene selectivity (to more than 80%) and in stability. That increase in stability was attributed to the decrease of coke production on the surface, thus reducing pore blockage.

1. INTRODUCTION

The non-oxidative conversion of methane into benzene has recently attracted considerable attention [1-7]. In this reaction hydrogen is also produced. Over MoHZSM-5, methane conversion and selectivities approach thermodynamic equilibrium for the conditions usually reported in the literature for temperatures in the range 973-1073 K. It is normally accepted that Bronsted acid sites, molybdenum species and their location, and the zeolite channel structure are important to catalytic performance.

The molybdenum carbide species have been identified as the active center for methane activation, whereas surface acid sites are responsible for aromatization of the intermediates [2]. Aromatization also leads to coke formation, causing catalyst deactivation.

Ding et al [8] have shown that silanation of external acid sites of ZSM-5 increases the activity, the selectivity and the stability of Mo/H-ZSM-5 catalysts.

In this work, a HMCM-22 zeolite was employed as support for the Mo-species. This zeolite has pores smaller than those of ZSM-5 and expectedly coke formation and pore blockage (observed on Mo-HZSM-5) could be reduced, as previously shown [9,10]. Chemical surface aluminum removal was also carried out, aiming at reducing the activity of the external surface. This removal was performed by treating the sample with ammonium hexafluorosilicate thus substituting silicon for aluminum on the external surface in contrast to the acid treatment employed previously by other authors [10].

2. EXPERIMENTAL

2.1. Catalyst Preparation

The HMCM-22 zeolite was synthesized according to [11]. The sample was dried and calcined at 823K for 10h. Dealumination was carried out with a 0.35M $(NH_4)_2SiF_6$ solution at 373K for 3h (HMCM-22D). Both samples were impregnated (incipient wetness) with a solution of ammonium heptamolybdate to obtain samples with 2% Mo (w/w). The resulting materials were calcined at 773K for 6 h and will be refered to as MoHMCM-22 and MoHMCM-22D, respectively.

2.2. Characterization

The chemical composition was determined by X-ray fluorescence spectrometry on a Rigaku spectrometer. BET specific surface areas, micro- (t-plot) and mesopore (BJH) volumes were determined by N_2 adsorption at 77 K on a Micromeritics ASAP 2000. XRD data were taken in a Rigaku Dmax Ultima+ diffractometer with CuKα radiation, 40 kV and 40 mA. The diffractograms were recorded from 5° to 40° with 0.05° steps. NH_3 temperature-programmed desorption (NH_3-TPD) was carried out in a microflow reactor operating at atmospheric pressure, in the range 393K-973K, at 20Kmin^{-1} and keeping at 973K for 0.5h. The outflowing gases were analysed by on-line mass spectrometry using a Balzers quadrupole spectrometer (model PRISMA-QMS 200). The samples were also analyzed by infrared spectroscopy with pyridine adsorption in a FTIR Perkin Elmer 2000 spectrometer with a resolution of 4 cm^{-1}. The sample was evacuated at 673 K for 3h and pyridine was adsorbed at 393 K. Desorption was carried out at 523 and 623 K. After the catalytic tests, the samples were oxidized in a flow of 5%O_2/He (TPO), in the range 298K-1073K, using a heating rate of 5Kmin^{-1} and keeping at 1073K for 0.5h. The outflowing gases were also accompanied by on-line mass spectrometry.

2.3. Reaction conditions

The catalytic tests were performed in a fixed-bed microreactor with pure methane at atmospheric pressure and 973K, at a space velocity of 1.7h^{-1} for up to 50h time-on-stream. Before the catalytic tests, the samples were activated in a H_2/CH_4 mixture (5:1) from ambient temperature to 923K (2h). The activation conditions were optimized by the use of mass spectrometry (Prisma, Balzers). Hydrocarbons were analyzed by on-line FID gas chromatography, while the hydrogen formed and Ar, used as an internal reference [12], were analyzed by on-line TCD gas chromatography.

3. RESULTS

The calcined HMCM-22 zeolite had a Si/Al ratio (SAR) of 27, a specific surface area of about 500 m^2g^{-1} and a microporous volume of 0.22 cm^3g^{-1}. Aluminum removal raised the global SAR to 78 and caused some changes in the specific surface area and in the microporous volume (to 420 m^2g^{-1} and 0.17 cm^3g^{-1}, respectively), indicating that some dealumination within the porous system also took place. Al removal had no significant effect on cristalinity. The introduction of molybdenum species promoted a small decrease of both the specific area and the micropore volume (to 460 m^2g^{-1} and 0.20 cm^3g^{-1}, and to 390 m^2g^{-1} and 0.16 cm^3g^{-1} for samples MoHMCM-22 and MoHMCM-22D, respectively). Molybdenum content was close to the desired one (2.4 and 2.1 wt%, for samples MoHMCM-22 and MoMCM-22D, respectively), as confirmed by X-ray fluorescence spectrometry.

The X-ray diffraction patterns showed a good dispersion of molybdenum species since no other phase (such as molybdenum oxide) besides HMCM-22 could be observed.

Fig. 1 shows the NH₃-TPD profiles for samples HMCM-22 and HMCM-22D (parent and dealuminated zeolites). They were deconvoluted into three peaks (Table 1) at 540-560K, 680-690K and 800-850K. The first peak was attributed to weakly adsorbed NH_3 species [10,13], possibly on the outer surface, and the second one to NH_3 desorption from Bronsted sites. Since zeolites treated with ammonium hexafluorosilicate usually do not exhibit extraframework species, the third peak cannot be associated to Lewis sites as done by other authors [10,13]. The infrared results (Fig. 2) confirm the rather small concentration of Lewis sites and the presence of very strong Bronsted sites. So, the third peak in the NH₃-TPD profiles was attributed to a second type of Bronsted acid site, tentatively associated to those sites inside the small pores of MCM-22.

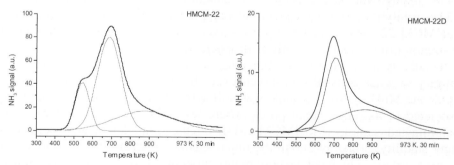

Fig. 1 – NH₃-TPD profiles for samples HMCM-22 and HMCM-22D

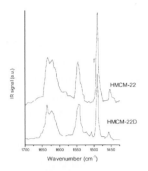

Fig. 2 – Infrared spectra after pyridine desorption at 623K

Table 1
Amount of acid sites of HMCM-22 and HMCM-22D by NH_3-TPD

Sample	NH_3 desorption ($\mu mol.g^{-1}$)			
	Peak 1	Peak 2	Peak 3	Total
HMCM-22	302	1000	539	1841
HMCM-22D	4	131	268	403

It can be seen that dealumination significantly reduced all three types of NH_3 adsorption sites. The amount of NH_3 desorbed from sample HMCM-22D is in accordance with the theoretical value for Bronsted sites (418 $\mu moles.g^{-1}$), supporting the assignment done to the third peak. For sample HMCM-22, the sum of the second and the third peaks gave a value higher than the theoretical one (1200 $\mu moles.g^{-1}$), indicating that some physisorbed NH_3 also contributed to the second peak, probably due to the occurence of a desorption/readsorption process.

The reaction products observed were hydrogen, C2 species (ethane and ethylene), benzene, toluene, naphthalene and polycondensed hydrocarbons (C11+: methylnaphthalene, fluorene, phenanthrene and anthracene).

The catalytic performance of the samples is shown in Fig. 3. Sample MoHMCM-22 showed an activity that stayed relatively constant for the first 10h on stream. From then on its activity dropped steadily and the inflexion point can be tentatively associated to the blocking of the mouth of the pores. However, contrary to what is usually observed over HZSM-5, after 20h time on stream (TOS) the MoHMCM-22 sample was still active (Fig. 3). This behavior can be associated to its rather small pores, where less coke is expected to form. The dealuminated sample (MoHMCM-22D), on the other hand, showed a much larger stability, probably due to the much smaller concentration of acid sites on its outer surface, which delays pore mouth blocking.

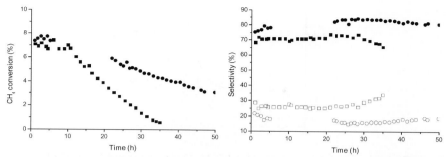

Fig. 3 – Catalytic performance – Methane conversion (square – MoHMCM-22, circle – MoHMCM-22D) and selectivity to products (solid – benzene and open – naphthalene)

Fitting of a first-order deactivation kinetics gives a deactivation constant of $0.02 \ h^{-1}$ for the latter sample, a value similar to that of [8]. In spite of the difference between the acid sites density of the two samples, the original activity was roughly the same thus suggesting that the concentration of acid sites was not the limiting factor.

The MoHMCM-22 sample presented a high (around 70%) and constant selectivity to benzene (Fig. 3). The sample submitted to dealumination showed an increase in benzene selectivity (to 80%), with a decrease in the selectivity to naphthalene. Only traces of polycondensed hydrocarbons were observed over the two samples. As to toluene, it was no longer observed after about 20h on stream over both samples.

As mentioned before, the higher stability shown by sample MoHMCM-22D was attributed to its lower coke formation. Fig. 4 shows the TPO results for the samples used in the present work, which confirmed this assertion. Recently, Ma et al. [14] using TPO and TPH claimed the existence of at least three different carbonaceous deposits on MCM-22: carbidic carbons (molybdenum carbide), molybdenum-associated coke, and aromatic-type coke on acid sites (possibly of two kinds). This result agrees with a similar assignment done for MoHZSM-5 zeolite [15]. TPH studies [14] also showed that the first two coke species related to molybdenum atoms are oxidized together, with only one peak on the TPO profiles. Deconvolution of the curves in Fig. 4 (not shown) indicates that the first peak (coke associated to Mo) is nearly the same for the two samples while the high temperature peak is strongly reduced by dealumination. The latter has been shown to be associated to acid sites and to be responsible for deactivation [16].

Thermogravimetric analyses showed coke contents of 15 and 10 wt% for samples MoHMCM-22 and MoHMCM-22D, respectively, in complete agreement with TPO results.

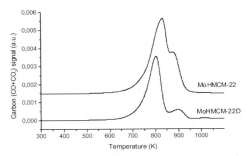

Fig. 4 – TPO of coke formed on samples MoHMCM-22 (+0.0015) and MoHMCM-22D

4. CONCLUSION

Mo containing HMCM-22 zeolite with reduced surface aluminum concentration seems to be a promising catalyst for methane dehydro-aromatization. The removal of aluminum atoms (acid sites) from the surface of HMCM-22 zeolite by ammonium hexafluorosilicate increased its selectivity to benzene and its stability by reducing coke formation, and, as a result, pore blockage. A decrease in naphthalene production was also observed. All types of coke species decreased after aluminum removal, especially the high temperature aromatic coke.

REFERENCES

[1] L. Wang, J. Huang, L. Tao, Y. Xu, M. Xie, G. Xu, Catal. Lett., 21 (1993) 35
[2] S. Liu, L. Wang, R. Ohnishi, M. Ichikawa, J. Catal., 181 (1999) 175
[3] O. Rival, B. P. A Grandjean, C. Guy, A. Sayari, F. Larachi, Ind.Eng.Chem. Res., 40 (2001) 2212
[4] C. Bouchy, I. Schmidt, J.R. Anderson, C.J.H. Jacobsen, E.G. Derouane, S.B. Derouane-Abd Hamid, J. Mol. Catal. A. Chem., 163 (2000) 283
[5] B.M. Weckhuysen, D. Wang, M.P. Rosynek, J.H Lunsford, Journal of Catalysis, 175 (1998) 338
[6] R. Borry III, E. C. Lu, Y. Kim, E. Iglesia, Stud. Surf. Sci. Catal., 119 (1998) 403
[7] W. Li, G.D. Meitzner, R. Borry III, E. Iglesia, J. Catal., 191 (2000) 373
[8] W. Ding, G.D. Meitzner, E. Iglesia, J. Catal., 206 (2002) 14
[9] Y. Shu, D. Ma, L. Xu, Y. Xu, X. Bao, Catal. Lett., 70 (2000) 67
[10] Y. Shu, R. Ohnishi, M. Ichikawa, Catal. Lett., 81 (2002) 9
[11] A. Corma, C. Corell, J. Pérez-Pariente, Zeolites, 2 (1995) 15
[12] S. Liu, L. Wang, Q. Dong, R. Ohnishi, M. Ichikawa, Chem. Commun., (1998) 1217
[13] S. Unverricht, M. Hunger, S. Ernst, H.G. Karge, J. Weitkamp, Stud. Surf. Sci. Catal., 84 (1994) 37
[14] D. Ma, D. Wang, L. Su, Y. Shu, Y. Xu, X. Bao, J. Catal., 208 (2002) 260
[15] W. Ding, S. Li, G.D. Meitzner, E. Iglesia, J. Phys. Chem. B, 105 (2001) 506
[16] D. Ma, Y. Shu, X. Han, X. Liu, Y. Xu, X. Bao, J. Phys. Chem. B, 105 (2001) 1786

Studies in Surface Science and Catalysis, volume 147
X. Bao and Y. Xu (Editors)
©2004 Elsevier B.V. All rights reserved.

The role of coke in the deactivation of Mo/MCM-22 catalyst for methane dehydroaromatization with CO_2

Jie Bai, Shenglin Liu, Sujuan Xie, Longya Xu[*]**, and Liwu Lin**

State Key Laboratory of Catalysis, Dalian Institute of Chemical Physics,
Chinese Academy of Sciences, P.O. Box 110, Dalian 116023, China

ABSTRACT

The effect of space velocity on reaction performance and coke deposition over 6Mo/MCM-22 catalyst in methane dehydro-aromatization (MDA) with CO_2 were studied. The characterization of catalysts reacted at different space velocity after the same amount of methane feed by TG, TPO and Benzene/NH_3-TPD techniques suggested that the inert coke maybe responsible for the deactivation of catalyst because of its blockage effect for pore system.

1. INTRODUCTION

Currently, a great challenge and intriguing problem in heterogeneous catalysis is the catalytic conversion of methane into higher hydrocarbons and/or aromatic compounds as alternative feed stocks for the petrochemical industry. Since the non-oxidative conversion of methane to aromatics on Mo/ZSM-5 was reported in 1993 [1], many catalysts have been studied in conjunction with their reactivities, reaction mechanisms and catalyst characterizations [2-7]. Another interesting catalyst, Mo/MCM-22 reported by Bao et al., exhibits a higher benzene selectivity and better stability in comparison with the Mo/ZSM-5 [8].

Nevertheless, it was found that methane conversion and rates of benzene formation drastically decreased after a few hours owing to significant coke formation. As for the role of coke in the deactivation of a specific catalyst had not been studied in detail up to now the 6Mo/MCM-22 catalysts reacted at different space velocities after the same amount of methane feed were chosen in this study. To determine the character of coke deposited on reacted catalysts, TG and TPO methods were adopted. Furthermore, the influence of coke deposition on the aromatization ability of catalyst was studied by Benzene/NH_3-TPD techniques.

[*]Corresponding author: e-mail: lyxu@dicp.ac.cn

2. EXPERIMENTAL

2.1 Catalyst preparation and evaluation

The 6Mo/MCM-22 catalyst was prepared by the conventional impregnation (incipient wetness) procedure with HMCM-22 ($SiO_2/Al_2O_3=30$, synthesized according to reference [9]) zeolite powder in $(NH_4)_6(Mo_7O_{24})4H_2O$ aqueous solution. After the impregnation, the sample was dried at $110°C$, and calcined at $500°C$ in air for 6h. The catalyst is denoted as 6Mo/MCM-22, and the first number is the weight percent of Mo loading.

The catalytic tests were carried out under atmospheric pressure of CH_4 with CO_2 in a continuous flow micro-reactor system equipped with a quartz tube (i.d. 8 mm) packed with 0.4g of catalyst pellets of 16~32 mesh. After flushing with He at $720°C$ for 30 min, a feed gas mixture of 90% CH_4, 1.8% CO_2 and 8.2% Ar, used as an internal standard for analysis, was introduced into the fixed-bed reactor under different flow rates (space velocity= 750, 1500, 2250 and 3000 ml (g cat. h)$^{-1}$, et al.) through a mass flow controller. Hydrocarbons including alkanes (and/or alkenes) such as CH_4, C_2H_4, C_2H_6 and aromatics such as benzene, toluene, xylene and naphthalene in the tail-gas were analyzed by an on-line gas chromatograph (Varian CP 3800) equipped with a column of SE-30 and a flame ionization detector (FID). A Hayesepe-D column, a 13X column and a thermal conducting detector (TCD) were also placed on the same chromatograph for the on-line analysis of H_2, Ar, CO, CH_4, CO_2, C_2H_4, C_2H_6, etc. Conversion of methane, selectivity of hydrocarbons, CO and coke formation on the catalyst were calculated according to the mass balance of hydrocarbons [10], and the CO originating from CO_2 was excluded because CO_2 was completely consumed on the catalyst.

2.3. TPO and TG measurement

The characterization of the coke deposited on reacted catalyst was carried out by the temperature-programmed oxidation (TPO) technique. Each sample (100mg) was loaded to a U-type quartz micro-reactor. After purging in a helium flow (30ml/min) at 500^0C for 1h, the temperature of the sample was lowered to 50^0C and the helium flow was switched to a flow composed of 2% O_2 and 98% He at a flow rate of 20ml/min. The total oxidation products were detected by an on-line gas chromatograph equipped with a TCD and the temperature increased at 4^0C /min.

The thermo-gravity (TG) measurements were carried out on a Perkin-Elmer TGA instrument by heating the sample from room temperature to 800^0C at a heating rate of 4^0C /min under O_2 stream with a flow rate of 40ml/min.

2.3. TPD measurement

Benzene-TPD (temperature programmed desorption) measurements were performed on a conventional setup equipped with a thermal conductivity detector (TCD). The sample (100mg) was first flushed with a helium flow (20ml/min) at 500^0C for 30 min, and then cooled to 100^0C. After that, it was saturated with benzene until equilibrium. When the baseline of the integrator was stable, Benzene-TPD was started at a heating rate of 10^0C /min, and the amount of desorbed benzene from the sample were detected on-line by TCD. NH_3-TPD measurements were almost the same to that of Benzene-TPD except for different temperature ramp (from 150^0C to 600^0C) and probe (NH_3).

3. RESULTS AND DISCUSSION

3.1. Methane dehydro-aromatization over 6Mo/MCM-22 catalyst at different space velocities

Fig. 1 shows the dependence of methane conversion and hydrocarbon selectivity in the methane conversion on 6Mo/MCM-22 at 720℃ by varying the space velocity of methane from 750 to 3000 ml/gcat.h. By increasing the flow rates of methane mentioned above, the initial methane conversion decreased from 13.2% to 11.0%, similar to the phenomenon reported by Xu et al [11]. In addition, the methane conversion maintained as high as 11.0% even at 50h time on stream when the space velocity was 750ml/gcat.h. While this duration was only 15h and 5h when the space velocity was doubled and tripled, respectively. It is also interesting to notice the variance of the hydrocarbon selectivity with the methane space velocity. When the MDA process was carried out at 750ml/gcat.h or 1500ml/gcat.h, the selectivity of hydrocarbon increased after passing through the initial stage and then maintained at about 75% steadily, despite the different

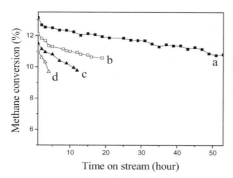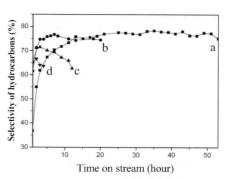

Fig. 1. The comparison in methane conversion (Left) and hydrocarbon selectivity (Right) over 6Mo/MCM-22 at different space velocity
(a: 750ml/gcat.h; b: 1500 ml/gcat.h; c: 2250ml/gcat.h; d: 3000ml/gcat.h)

time to attain the stable value of these two space velocities. However, as methane space velocity increased to 2250ml/gcat.h and 3000ml/gcat.h, there was no stable value in the selectivity of hydrocarbons and the maximal value decreased to 72% and 68% respectively. All these results indicate that the space velocity can remarkably affect the performance of the 6Mo/MCM-22.

3.2. Characterization of coke deposited on reacted 6Mo/MCM-22 catalysts by TG and TPO techniques

To study the role of coke in deactivation, a series of reacted 6Mo/MCM-22 catalysts were prepared under reaction conditions mentioned in Fig.1 after 6000 ml of methane introduction. It is interesting to find that, although the space velocity was different, the total amount of coke deposited on each sample (detected by TG technique) was similar, i.e. about 8-9 wt%.

The TPO profiles of the identical samples were shown in Fig. 2. A sharp peak at low-temperature area and a broad shoulder at high-temperature area appeared in each TPO profile, which was similar to that of reacted Mo/ZSM-5 catalysts reported by Ichikawa et al [12]. Based on their suggestion, we assigned the low-temperature peak to the oxidation of active coke associated with molybdenum, and the higher one to carbonaceous deposits related to acid sites.

It is obviously that, the position and area of these two peaks in each curve were not the same. The high and low peaks' temperatures in each curve (deconvoluted by the Gauss curve fitting method, not shown) are 563℃ and 503℃, 552℃ and 484℃, 544℃ and 473℃, 539℃ and 467℃, respectively, i.e. the oxidation temperature of each kind of coke deposited on the catalyst shifts to lower values with increasing methane space velocity. Furthermore, the ratios of the area of high-temperature peak to the lower one for a-d curves in figure 2 are1:3.25, 1:2.25, 1:1.95, 1:1.56, respectively. Because the total amount of coke

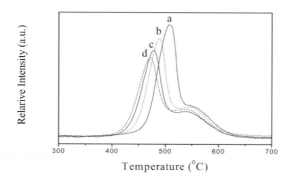

Fig. 2. The TPO profiles of 6Mo/MCM-22 reacted at 750 ml/gcat.h (a), 1500 ml/gcat.h (b), 2250 ml/gcat.h (c) and 3000 ml/gcat.h (d)

deposited on these samples were almost the same, it can be easily inferred that the weight of acid sites related coke increased when the space velocity rose.

3.3. Characterization of reacted 6Mo/MCM-22 catalysts by TPD method

Fig. 3 showd the NH_3-TPD profiles of fresh 6Mo/MCM-22 and the four samples mentioned in 3.2. Compared to that of fresh catalyst, the remained amount of acid sites in reacted catalyst was much less, indicating the coverage effect of coke deposition on acid sites. In addition, the almost similar decrease of acid sites in catalysts reacted under different methane space velocities indicated that the coverage of acid sites was dependent on the weight of total coke deposition.

Being different from NH_3 absorption, the absorption of benzene needs not only acid sites, but also a pore system with the aperture at least equal to the size of benzene, so the amount of benzene that a MDA catalyst can absorb reflects its ability in the formation and transportation of benzene. To eliminate the cohesion of benzene in the samples, the temperatures of adsorption and starting of desorption were set at 100°C. The benzene-TPD profiles of the 6Mo/MCM-22 before and after the introduction of 6000ml methane at different space velocities were shown in Fig. 4. It is evident that the amount of benzene absorbed by a reacted catalyst was much lower than that absorbed by the fresh catalyst. Furthermore, the decreasing degree of each sample was not the same. For the catalysts reacted at the space velocity of 2250 or 3000ml/gcat.h, the ratio in the amount of benzene adsorption to that of the fresh catalyst is about 1/6, whereas this value was doubled (about 1/3) for the catalyst reacted at the space velocity of 750 or 1500ml/gcat.h. It may be proposed that, when methane space velocity

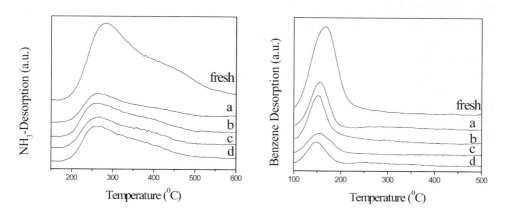

Fig. 3. (Left) The NH_3-TPD of fresh and the catalyst reacted at different space velocity
Fig. 4. (Right) The Benzene-TPD of fresh and the catalyst reacted at different space velocity
(in Figs.3 and 4, a :750 ml/gcat.h, b :1500 ml/gcat.h, c :2250 ml/gcat.h , d :3000ml/gcat.h)

was equal or greater than 2250ml/gcat.h, the rate-determining-step of MDA had changed from the dissociation of methane to the transference of active hydrocarbon species (CHx) into pore. As parts of CHx could not enter into the pores to be transformed to aromatics, they had to deposit at the outer space of the pores, which led to the blockage of the pores and the lost of aromatization ability.

Because the weight percent of coke deposited on catalysts reacted at different space velocity was all in the range of 8~9(wt)%, it is obvious that the remarkable difference in the catalyst deactivation could be attributed to the proportion of different kind of coke. Although molybdenum related coke was dominating in the coke deposition, its effect in catalyst deactivation was relatively small because methane could be continuously activated into CHx even when multiple layers of coke covered active molybdenum species [13]. Based on the fact that the lost of aromatization ability of 6Mo/MCM-22 was resulted from the blockage of pore mainly originated from the deposition of acid sites related coke, it is reasonably inferred that the inert coke related to acid sites played a key role in the catalyst deactivation. To prolong the stability of a MDA catalyst, endeavor should be taken to restrict the formation of coke related to acid sites.

REFFRENCES

[1] L.S. Wang, L.X. Tao, M.S. Xie, G. Xu, J.Huang, Y.D. Xu, Catal. Lett. 21 (1993) 35
[2] R.W. Borry III, Y.-H. Kim, A. Hunffdmith, J.A. Renmer, and E. Iglesia , J. Phys. Chem. B, 103 (1999) 5787
[3] W. Li, G.D. Meitzner, R.W. Borry III, and E. Iglesia, J.Catal. 191 (2000) 373
[4] Y.D. Xu, S.T. Liu, L.S. Wang, M.S. Xie, X.X. Guo, Catal. Lett. 30 (1995) 135
[5] F. Solymosi, A. Szoke, J. Cserenyi, Catal. Lett. 39 (1996) 157
[6] S.T. Liu, L.S. Wang, R. Ohnishi, and M. Ichikawa, J. Catal. 181(1999) 175
[7] Y.D. Xu, Y.Y. Shu, S.T. Liu, J.Huang, X.X. Guo, Catal. Lett. 35 (1995) 233
[8] Y.Y. Shu, D. Ma, L.Y. Xu, Y.D. Xu, X.H. Bao, Catal. Lett. 70 (2000) 67
[9] S.J. Xie, Q.X. Wang, L.Y Xu, W. Chen, J. Bai, L.W. Lin, Chin. J. Catal. 6 (1999) 583
[10] Y. Lu, Z.S. Xu, L.W. Lin, L.F. Zang, Y.M. Gao, Z.J. Tian, T. Zhang, Chin. J. Catal. 3 (1999) 273
[11] B.L. Tian, H. Liu, L.L. Su, Y.Y. Shu, Y.D. Xu, Chin. J. Catal. 22 (2001) 124
[12] R. Ohnishi, S.T. Liu, D. Dong, L.S. Wang, M. Iichikawa, J. Catal. 182 (1999) 92
[13] D. Ma, D.J. Wang, L.L. Su, Y.D. Shu, Y. D. Xu, X.H. Bao, J.Catal. 208 (2002) 260

Studies in Surface Science and Catalysis, volume 147
X. Bao and Y. Xu (Editors)
©2004 Elsevier B.V. All rights reserved.

Production of isobutene from n-butane over Pd modified SAPOs and MeAPOs

Yingxu Wei, Gongwei Wang, Zhongmin Liu[*]**, Lei Xu and Peng Xie**

Natural Gas Utilization & Applied Catalysis Laboratory, Dalian Institute of Chemical Physics, Chinese Academy of Sciences, P.O.Box 110, Dalian 116023, China

ABSTRACT

Pd modified AlPO-11 and SAPO-5, 11, 34 were used in the direct transformation of n-butane to isobutene. The effect of acidity and porous structure on the supported Pd and the catalytic performance were discussed. For higher isobutene selectivity, some metals, such as Ti, Fe, Mg, Co and Mn, was incorporated into AlPO-11 framework and isobutene selectivity of 34.86% can be obtained over Pd/MnAPO-11. A combined catalyst system was used for a further improvement of the isobutene selectivity.

1. INTRODUCTION

Some SAPOs molecular sieves, such as SAPO-5, 11, 34, which possesses AFI, AEL and CHA topology respectively, have been proved to be very active and selective catalysts for many reactions [1,2]. SAPO-11 and metal substituted SAPO-11were reported as selective catalyst for isomerization of n-butene [3, 4].

Different processes for isobutene production including skeletal isomerization of n-butene received much attention recently with the increasing demand of isobutene in industry [5-7]. Considering the abundant supply of n-butane from natural gas and refinery streams, n-butane is the preferred raw material for isobutene production.

In the present work, the catalysts of Pd modified AlPO and SAPOs were prepared and used in dehydroisomerization of n-butane to isobutene. The effects of acidity and porous structure of the support on the catalytic performance were investigated in detail. To improve the isobutene selectivity, Ti, Mg, Fe, Co and Mn were incorpated into AlPO-11 and a combined catalyst was used for this reaction.

2. EXPERIMENTAL

2.1 Synthesis of molecular sieve

AlPO-11, SAPO-5, 11, 34, and MeAPO-11s were synthesized by hydrothermal method, following the procedure reported in the literature [8]. The products were filtrated, washed, and dried at 373 K for 3 h, then calcined at 823 K for 6 h to completely remove the template.

2.2 Catalyst preparation

Pd modified catalysts were prepared by impregnating the calcined molecular sieves with a solution of $Pd(NH_3)_4Cl_2$. The samples were then dried at 373 K for 10 h and calcined at 773 K in air for 4 h.

2.3 Characterization

Crystallinity and phase purity of the as-synthesized samples were characterized by powder X-ray diffraction. The chemical composition of the samples was determined with X-ray fluorescence technique. Surface area and porosity measurements were carried out by physical adsorption. Acidity of the samples was characterized using temperature programmed desorption (TPD) spectra of ammonia.

2.4 Catalytic testing

The catalytic tests were performed using a fixed bed reactor system at atmosphere pressure. Before reaction, the catalysts were reduced in situ with H_2 (60 cm^3/min) at 773 K for 1 h, then the atmosphere was switched from H_2 to the feed, a mixture of H_2 and n-butane (the molar ratio of H_2 /n-butane was 2). The weight hourly space velocity (WHSV) was 1.98 h^{-1} for n-butane. The reaction products were analyzed on-line by a Varian GC3800 gas chromatograph equipped with a FID detector and an Al_2O_3 capillary column.

3. RESULTS AND DISCUSSION

For the samples of SAPO-11, SAPO-34 and SAPO-5, the position and the intensity of the diffraction peaks of their XRD patterns were identical to those reported in the literature [8]. XRD patterns of AlPO-11 and metal-containing AlPO-11 showed in Fig.1 indicate that they are highly crystalline with no obvious impurity phase.

The chemical composition, BET and microporous volume of the synthesized AlPO and SAPO samples are listed in Table 1 for comparing. With modulating the composition of the starting gel, three SAPOs with nearly the same chemical composition of the crystalline products were obtained. Pore geometry difference exists between SAPO-5, 11 and 34 molecular sieves. Among the four samples,

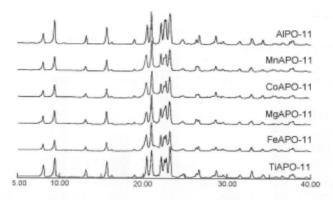

Fig.1. The XRD patterns of the synthesized AlPO-11 and MeAPO-11s

SAPO-34 possessed the highest specific area and microporous volume.

The reaction results of Pd modified SAPO-5, SAPO-11 and SAPO-34 in Fig. 2 showed the effect of pore geometry on the catalytic properties. For Pd/SAPO-11, n-butane was mainly transferred to isobutene and other butenes. A small quantity (<20%) of cracking and hydrogenolysis products was also obtained. But for Pd/SAPO-34 and Pd/SAPO-5, an abrupt change in product distribution and catalytic activity was observed. Though the dehydrogenation product was also obtained, the simultaneous skeletal isomerization is virtually suppressed over the two samples. The pores of the catalyst are thought to accommodate the intermediate of isomerization and the metal particle with dehydrogenation function, so pore geometry of the molecular sieve may play an important role for the catalytic performance. Among the catalysts, SAPO-34 has the largest micropore volume and area, but its 8-member ring pore opening may sterically inhibit the isomerization of n-butane or n-butene; for Pd/SAPO-5, with the 12-ring pore-opening molecular sieve, no sterical inhibition for isomerization exists in this catalyst, the low isomerization selectivity may come from its weak acidity.

Table 1

Molar compositions, specific area and microporous volume of the crystalline products

Product	Mole composition	Specific area (m^2/g)	Microporous volume (cm^3/g)
SAPO-34	$Al_{0.50}P_{0.44}Si_{0.06}O_2$	554	0.273
AlPO-11	$Al_{0.50}P_{0.50}O_2$	120	0.063
SAPO-11	$Al_{0.50}P_{0.44}Si_{0.06}O_2$	137	0.067
SAPO-5	$Al_{0.50}P_{0.44}Si_{0.06}O_2$	158	0.063

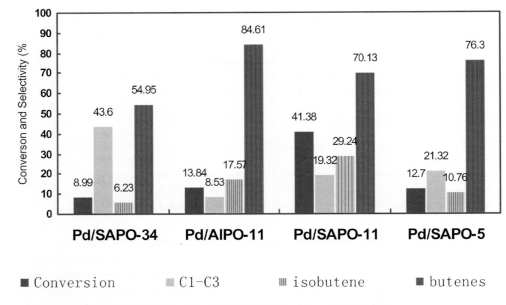

Fig. 2 The catalytic performance of Pd/AlPO-11 and Pd/SAPOs.

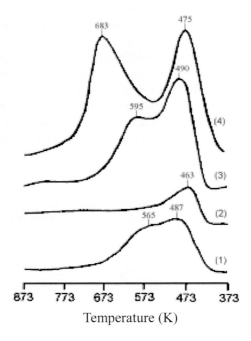

Fig. 3 NH₃-TPD patterns of AlPO-11 and SAPOs
(1) SAPO-5 (2) AlPO-11 (3) SAPO-11 (4) SAPO-34

For Pd/AlPO-11 and Pd/SAPO-11, the molecular sieve as the support possessing the same AEL topology, the difference of the catalytic performance should be attributed to the acidity. AlPO-11 showed the weakest acidity (see Fig. 3), which results from the absence of acidic sites over the neutral framework, so over the catalyst of Pd/AlPO-11, the dehydrogenation catalyzed by noble metal is prominent; while over the Pd/SAPO-11, with the medium-acidic support, the dehydrogenation and isomerization can happen simultaneously.

Comparing the NH_3-desorption peaks in the Fig 3, the acid strength roughly followed the order that: SAPO-34>SAPO-11>SAPO-5; while we know the order of pore opening is: SAPO-34<SAPO-11, AlPO-11<SAPO-5. From the characterization results and catalytic performance, poor activity and selectivity for isobutene were obtained over the catalysts of Pd/SAPO-5 with weak acidity and over Pd/SAPO-34 with narrow pore opening. Pd/SAPO-11 with 10-member ring pore opening and medium acidity showed the highest activity and selectivity for isobutene. The acidity of support also has an effect on the metallic properties of the bifunctional catalysts. The dehydrogenation selectivity followed the order of Pd/AlPO-11>Pd/SAPO-5>Pd/SAPO-11>Pd/SAPO-34. The butenes selectivity order is opposite to the order of acidity.

Metal incorped AlPO-11 have been proved to be selective catalysts for n-butene isomerization [3,4]. In this study, MeAPO-11 also showed special catalytic properties in n-butane's dehydroisomerization. More isobutene selectivity can be obtained over the sample of Pd/MeAPO-11, especially for Pd/MnAPO-11, the isobutene selectivity attained to 34.86% (see table 2).

To improve the isobutene selectivity, two consecutive catalysts, a dehydroisomerization catalyst of Pd/SAPO-11 in the upper part and a complementary skeletal isomerization catalyst of SAPO-11or MnAPO-11 in the lower part were used in one fixed catalyst bed. The catalytic performance of n-butane transformation at 823 K over these dual-layered catalysts are showed is shown in Table 3. Compared with the catalytic performance of Pd/SAPO-11, a relatively high selectivity towards isobutene could be obtained over the combined catalysts bed.

Table 2
The catalystic properties of Pd/AlPO-11 和 Pd/MeAPO-11 at 773 K

Catalyst	Conversion (%)	Selectivity (%)				
		$iC_4^=$	Total $C_4^=$ a	iC_4^0	C_1-C_3 [b]	C_5^{+c}
Pd/TiAPO-11	15.87	30.38	83.11	7.49	8.56	0.84
Pd/FeAPO-11	15.33	30.70	83.20	7.69	8.69	0.42
Pd/MgAPO-11	16.27	32.64	82.91	7.19	9.59	0.31
Pd/CoAPO-11	18.89	33.61	81.54	10.65	7.81	0
Pd/MnAPO-11	20.14	34.86	82.52	9.63	7.85	0

Table 3
Catalytic performances of catalysts in a two-layered bed at 823 K

Catalyst	Conversion (%)	Isobutene yield (%)	Selectivity(%)			
			$iC_4^=$	Total $C_4^=$	iC_4^0	Other [a]
Pd/SAPO-11(upper) +SAPO-11(lower)	28.62	9.27	32.40	76.62	6.73	16.65
Pd/SAPO-11(upper) +MnAPO-11(lower)	27.10	9.65	35.63	77.14	7.32	15.54

[a]Other products: C_1-C_3(CH_4+C_2H_6+C_2H_4+C_3H_8+C_3H_6), C_5 and products higher than C_5.

3. CONCLUSION

SAPO-5, 11, 34 and metal-containing AlPO-11 were synthesized. Characterization results showed that the acid strength order is: SAPO-34>SAPO-11>SAPO-5>AlPO-11. The Pd modified catalysts based on these molecular sieves were used for the dehydroisomerization of n-butane to isobutene. Poor activity and selectivity to isobutene were obtained over the catalyst based on the SAPO-5 or AlPO-11 with weak acidity and SAPO-34 with narrow pore opening. The Pd modified SAPO-11 with 10-member ring pore opening showed highest activity and selectivity to isobutene among the four catalysts. A medium strong acidity and suitable pore geometry may be needed in this transformation of n-butane to isobutene in one step. Metal incorporation into AlPO-11 generated good isomerization properties, and 34.86% isobutene selectivity was obtained over Pd/MnAPO-11. With combined catalysts Pd/SAPO-11 and MnAPO-11, the isobutene selectivity can be further improved.

REFERENCES

[1] I.M. Dahl and S. Kolboe, J. Catal., 161 (1996) 304.
[2] J.M. Campelo, F. Lafont and J. M. Marinas, Appl. Catal., 152 (1997) 5.
[3] A. Vieira, M.A. Tovar, C. Pfaff, B. Mendez, C.M. Lopez, F.J. Machado, J. Goldwasser and M. M. Ramirez de Agudelo, J. Catal., 177 (1998) 60.
[4] S-M Yang, J-Y Lin, D-H- Guo and S-G Liaw, Appl. Catal. A: General 181 (1999) 113.
[5] R. Byggningsbacka, N. Kumar and L.-E. Lindfors, Catal. Lett. 55 (1998) 173.
[6] B. Didillion, C. Travers and J-P Burzynski, US Patent No. 5 866 746 (1999).
[7] G.D. Pirngruber, K. Seshan and J.A. Lercher, J. Catal., 186 (1999) 188.
[8] B.M. Lok, C.A. Messina, R.L. Patton, R.T. Gajek, T.R. Cannan and E.M. Flanigen, US Patent No. 4 440 871 (1984).

Author Index

STUDIES IN SURFACE SCIENCE AND CATALYSIS

Advisory Editors:
B. Delmon, Université Catholique de Louvain, Louvain-la-Neuve, Belgium
J.T. Yates, University of Pittsburgh, Pittsburgh, PA, U.S.A.

732

734

738

739